Pollution Control Handbook 2009

environmental protection uk

Pollution Control Handbook

First published 1978
Revised annually since 1985

Published in Great Britain by
Environmental Protection UK, 44 Grand Parade, Brighton BN2 9QA
Tel: 01273 878770
Email: admin@environmental-protection.org.uk
http://www.environmental-protection.org.uk

ISBN 978-0-903474-63-4

Cover images:
Main photo: Lancaster by David Pickavant
additional images by (l to r): Paolo Crosetto, Andrew Lewis, Martin Edmondson, Russell James Smith, David Pickavant

This edition of the Pollution Control Handbook was finalised between September and December 2008, with most legislation, (both in force and pending at the time of writing) included.

The Pollution Control Handbook is widely used by central and local government, industry, pollution control specialists, consultants, researchers and students. Its aim, therefore, is to provide an overview of environmental and pollution control legislation and related issues that is accessible to readers with widely varying knowledge and expertise. It does not, however, set out to provide a detailed statement of the law and specialist professional advice should always be sought in dealing with specific problems, or before taking legal action.

Environmental Protection UK can accept no legal responsibility for any inaccuracies which may have arisen.

Pollution Control Handbook 2009

I am delighted to write the foreword to the Pollution Control Handbook 2009. For over 30 years the handbooks, and the reference books which preceded them, have been an invaluable source of information for environmental professionals. The Environment Agency, together with our colleagues in the Scottish Environment Protection Agency, is happy to sponsor the latest edition.

Environmental Protection UK's vision of 'a cleaner, quieter, healthier world' complements the Environment Agency's vision of 'a healthy, rich and diverse environment in England and Wales, for present and future generations'. These are not empty buzzwords – both organisations are pursuing these goals assiduously. Great improvements have been made in environmental quality over the past 30 years. Rivers are much cleaner and fish have returned to waters which previously supported little more than bloodworms and anaerobic bacteria. Industrial emissions of sulphur have been reduced by over 90% and lead emissions from motor vehicles are a thing of the past. Wastes, too, are more carefully managed while landfills receiving hazardous wastes are much fewer in number and strictly regulated.

Despite these improvements we still face formidable environmental problems. Foremost amongst these is climate change, which will affect every living thing on the planet. We will need all the technical ingenuity and political will we can muster to minimise its impact, including measures to capture and store carbon emissions from new coal-fired power stations. We must overcome technical and political barriers to minimising and recovering wastes, generating energy and achieving serious reductions in energy demand.

Levels of nitrogen oxides and particulates are already too high in many places and it will be increasingly difficult for the UK to meet environmental quality standards as traffic levels grow and imprudent infrastructure developments are permitted. Some measures to reduce climate change may have adverse effects on air quality so we need to strike a balance. Background levels of ozone are also rising, with a potential threat to agricultural productivity looming as well as implications for human health.

If we are to tackle these problems effectively, we must work in partnership - government, industry and communities. We need to champion the environment in the face of deteriorating economic conditions; convince industry that it is in everyone's interest to maximise resource and energy efficiency, and allay the fears which some communities have over developments such as wind turbines and waste to energy plants. The planning process is crucial since how we structure our communities determines our demand for resources, land, and energy as well as the amounts of pollution which we generate.

I look forward to continuing co-operation between our organisations in addressing these crucial problems and welcome the publication of this Handbook.

Rt Hon Lord Smith of Finsbury
Chairman, Environment Agency
3 December 2008

Pollution Control Handbook 2009 – Update

This edition of the Pollution Control Handbook was finalised between September and December 2008; summarised below are legislative and other developments which were too late for inclusion in the appropriate chapter.

CHAPTER 1

Planning, England & Wales (1.16, page 17);

Planning Act 2008 – (summarised pages 17-18) creates an integrated planning system for major infrastructure; National Policy Statements (NPSs) will detail national infrastructure priorities in areas such as energy, aviation, road and rail transport, water and waste, integrating social, economic and environmental policies. Developers will be required to consult local communities and other key stakeholders prior to submitting applications and to conduct environmental assessments as required by the EIA Regulations. Decisions on nationally significant applications will be taken by a new Infrastructure Planning Commission, based on NPSs. Eleven NPSs are planned: overarching energy (providing an overview of the other 5 energy NPSs); fossil fuels; renewable energy; electricity networks; gas and downstream oil infrastructure; nuclear power; ports; national networks (strategic highway and rail networks); aviation; water supply and waste water treatment; hazardous waste (excluding nuclear waste).

Planning and Energy Act 2008 – enables local authorities in England and Wales to include in their local development plans policies imposing reasonable requirements for a proportion of energy used in development in their area to be from renewable sources, to be low carbon energy, and to comply with energy efficiency standards in excess of building regulation requirements.

Planning, Scotland (1.17, page 20)

The Scottish Government is proposing (news release, 28.10.08) a number of reforms to the planning system aimed at improving efficiency and ensuring planning is geared towards supporting increased sustainable economic growth. Scottish Planning Policy documents are to be streamlined into a single document – the first two parts on core principles and the key elements of the planning system were also published on 28.10.08; the third part – a statement of policy on a number of development themes – is to be published in the Spring 2009.

Wind Farm Planning Advice Note (No. 45) – was published by the Scottish Government in November 2008. It will help councils and national park authorities prepare their own guidance for considering wind farm applications. As such it will help authorities identify areas where proposals are likely to be supported, areas of significant protection (e.g. SSSIs), and areas where other constraints will have to be satisfactorily addressed.

Planning, Hazardous Substances (1.19, page 22)

Guidance on appeals relating to hazardous substances consents has been published by the Planning Appeals Inspectorate and is available on www.planning-inspectorate.gov.uk

Statutory Nuisance, Scotland (1.29, page 36)

The Scottish Government has published draft guidance on the application of the two new statutory nuisances (artificial light and insect infestation) introduced into the Environmental Protection Act 1990 by the Public Health etc (Scotland) Act 2008. It also provides procedural guidance on existing statutory nuisances and on other matters which raise similar issues to statutory nuisances but are dealt with by other existing legislation.

CHAPTER 2

REACH (2.15.2, page 75)

The REACH Enforcement Regulations 2008 (SI 2008/2852) came into force on 1 December 2008. They apply throughout the UK and provide for the enforcement of EU Regulation 1907/2006 concerning the Registration, Evaluation, Authorisation and Restriction of Chemicals regulation.

CHAPTER 3

Climate Change Act 2008 (3.7.3, page 87)

This Act received Royal Assent on 26 November 2008. Key provisions include the setting of a legally binding target of a reduction in emissions of greenhouse gases of at least 80% by 2050, and a reduction in carbon dioxide emissions of at least 26% by 2020, against a 1990 baseline. The Act also introduces a carbon budgeting system which caps emissions over five year periods, with three budgets set at a time. The first three, which will run from 2008-12, 2013-17 and 2018-22, must be set by 1 June 2009.

Climate Change Bill, Scotland (3.7.3, page 88)

The Scottish Government published its draft Climate Change Bill in December 2008. This also proposes a cut in greenhouse gas emissions of 80% by 2050 compared to 1990, with an interim target reduction of 50% by 2030. Annual emission reduction targets are to be set for each calendar year from 2010 to 2050, and from 2020 the annual reduction target will be at least 3%.

Prevention of Air Pollution from Ships (3.29, page 137)

The Merchant Shipping (Prevention of Air Pollution from Ships) Regulations 2008 (SI 2008/2924), which came into force on 8 December 2008, implement Annex VI to the MARPOL Convention controlling emissions of ozone depleting substances, nitrogen oxides, sulphur oxides, VOCs, as well as shipboard incineration and fuel oil quality.

CHAPTER 4

Environmental Noise Regulations (4.6, page 148)

The following documents have been published since this section of the Pollution Control Handbook was finalised:

England: Draft Action Planning (Airports) – guidance for airport operators – Defra (September 2008).

Wales: Noise Action Planning Guidance (Agglomerations, Roads and Railways), together with draft Noise Action Plans – Welsh Assembly Government (November 2008).

Scotland: Summary of the Environmental Report prepared as part of the Strategic Environmental Assessment of draft Noise Action Plans in Scotland – Scottish Government (October 2008).

CHAPTER 5

Waste Electrical and Electronic Equipment (5.12.1, page 186)

The EU Directives on the collection and recycling of waste electrical and electronic equipment (WEEE) and on the restriction of hazardous substances in EEE (ROHS) are being recast. The proposed revised WEEE Directive will set a new binding target for the collection of WEEE of 65% of the average weight of EEE placed on the market over the previous two years in each Member State; registration and reporting requirements for producers are to be harmonised and national registers made inter-operable. Revisions to the ROHS Directive will extend it to cover medical devices and monitoring and control instruments. A list of "priority" substances posing particular concerns when used in EEE will be assessed in line with REACH and a possible ban in the future.

Hazardous Waste Regulations, England and Wales (5.23.2, page 203)

Defra published proposals for amending the HWR for England in November 2008 (with the Welsh Assembly Government expected to issue similar proposals). These will clarify that where asbestos waste is produced at domestic premises, the contractor engaged to deal with the waste is subject to the HWR, but the occupier of the premises is not. Where householders put domestic hazardous waste out separately for collection or deliver it to a civic amenity site or recycling centre, it must be kept separate and is subject to the Regulations. All low risk premises producing less than 500 kg of HW in any 12 month period will be exempt from notification. A person collecting HW from multiple premises on a single journey will be required to use the multiple collection procedure and paperwork. Common provisions on notifications are to be amended so that information required to be released by the Environment Agency relates only to the premises producing HW, and does not include information about the actual notifier.

Transfrontier Shipment of Radioactive Waste (5.39.5, page 228)

The Transfrontier Shipment of Radioactive Waste and Spent Fuel Regulations 2008 (SI 2008/3087) came into force on 25 December 2008, revoking the 1993 Regulations. The Regulations are summarised on page 229 of the Pollution Control Handbook.

CHAPTER 6

Flood Risk Management, Scotland (6.8.5, page 252)

The Flood Risk Management (Scotland) Bill was introduced to Parliament on 29 September 2008. As well as implementing the EU's Floods Directive, the Bill seeks to place a duty on Ministers and SEPA and responsible authorities to exercise their flood risk management function with a view to reducing overall flood risk. SEPA, together with the responsible authorities, will be required to prepare flood risk assessments, flood risk and flood hazard maps, and flood risk management plans.

Drinking Water, Northern Ireland (6.9.3, page 253 & 6.17.1, page 274)

In 2007, 99.85% (2006: 99.76%) complied with microbiological standards. There was however a decrease in compliance achieved at consumers' taps – 99.30% compared to 99.34% in 2006. (12th Annual Report of the NI Drinking Water Inspectorate, November 2008).

The Department for Regional Development issued proposals for amending the Water Supply (Water Quality) (NI) Regulations 2007 (SR 2007/147) in November 2008. The amendments will require monitoring of raw water quality, supplement the NI Environment Agency's raw water monitoring programme and enable action to be taken against NI Water in the event of untreated water entering the public supply.

Nitrate Pollution, Scotland (6.12.1(b), page 269)

The Action Programme for Nitrate Vulnerable Zones (Scotland) (Amendment) Regulations 2008 (SSI 2008/394) make a number of minor amendments to SSI 2008/29; these include amending reg. 9 so that it applies to the storage of livestock manure other than slurry (rather than "solid" manure).

Marine and Coastal Access Bill (6.24.1, page 283)

The above Bill was introduced in the House of Lords on 4 December 2008. The Bill introduces a new system of marine management, covering a new planning system for setting out the Government's strategic objectives for the marine environment and developing marine plans to set out in more detail what should happen where in each area of the sea. The Bill makes changes to the way licensing of activities in the marine environment is undertaken and provides for the designation of conservation zones. It makes changes to the way fisheries are managed, allows for designation of an Exclusive Economic Zone for the UK and for the creation of a Welsh Zone in the sea adjacent to Wales.

■ Contents

CHAPTER 1

CHAPTER 2 - INDUSTRIAL REGULATION

Radioactive Waste

1 GENERAL

Regulatory agencies

Sustainable development

Biological diversity

Environmental liability

Environmental crime

Land use planning

Environmental appraisal and assessment

Environmental risk assessment

Access to environmental information

Environmental management and audit systems

Local environmental quality

1.1 INTRODUCTION

The principles underlying the UK approach to pollution control evolved over many years, with the term "Best Practicable Means" (BPM) providing the fundamental basis for almost 150 years. With integrated pollution control (IPC), came the requirement to select the "Best Practicable Environmental Option" and "Best Available Techniques Not Entailing Excessive Cost". The need to apply a holistic approach in controlling pollution has now been taken a step further with the implementation of the EU Directive on Integrated Pollution Prevention and Control (IPPC – see chapter 2) which, as well as requiring the use of "Best Available Techniques" (BAT), requires a much wider range of environmental impacts to be taken into account, such as noise, energy use, waste management and prevention of accidents.

This chapter describes some of the general principles and other issues which can impact upon pollution control legislation, such as sustainable development, environmental risk assessment and land-use planning; it also covers those areas of legislation which cut across individual pollution media, including statutory nuisance legislation, land-use planning and environmental assessment, and access to environmental information.

1.2 EARLY CONTROLS

1.2.1 Alkali Acts & Best Practicable Means

Most of the UK's early pollution control measures were enacted as a result of the Industrial Revolution of the late 18th and 19th centuries, although both air and water pollution had been identified as problems many hundreds of years before (see chapters 3 and 6 respectively). The concept of "Best Practicable Means" (BPM) – which for many years was a fundamental principle of pollution control legislation – was often used to describe the whole approach of UK anti-pollution legislation towards industrial emissions, and can still be used as a defence in respect of nuisance from industrial, trade or business premises (see later this chapter, 1.29 and chapter 4, 4.9.2).

BPM was first used in a law applying in Leeds in 1842 under which a fine could be imposed on offenders who had not used the Best Practicable Means to prevent or abate smoke nuisance. In 1874 an amendment to the *Alkali Act 1863* required the use of BPM to prevent the discharge of all noxious or offensive gases arising from alkali works.

A schedule to the *Health and Safety (Emissions into the Atmosphere) Regulations 1983* (revoked in December 1996) listed all those processes ("works") which had to be registered with the pollution inspectorate; as an essential prior condition to registration, operators of scheduled works had to satisfy the inspectorate that the works were provided with the BPM for preventing the escape of noxious or offensive substances, and for rendering all such emissions harmless and inoffensive. BPM took into account the cost of pollution abatement and its effect on the viability of the industry concerned.

The essential elements of BPM so far as emissions to air were concerned could be defined as follows:

- no emission could be tolerated which constituted a recognised health hazard, either short or long term;
- emissions in terms of both concentration and mass, had to be reduced to the lowest practicable amount taking into account local conditions and circumstances, current state of knowledge on control technology and effects of substance emitted, financial considerations, and the means to be employed;
- having secured the minimum practicable emissions, the height of discharge should be arranged so that the residual emission was rendered harmless and inoffensive by dilution and dispersion.

The Control of Pollution Act 1974 extended the principle to noise control. Under s.72 it is a defence to prove that the BPM has been used to mitigate the effects of excessive noise from, for example, a factory. Section 80(7) of the *Environmental Protection Act* also allows a defence of BPM in respect of an action for nuisance from industrial, trade or business premises.

When working out the control details for any particular type of process, the pollution inspectorate would discuss ways of reducing emissions with representatives of the industry concerned, using their knowledge of control techniques and their own technical expertise. The conclusions were published in *Notes on Best Practicable Means*, which described treatment plant to be used and its maintenance, methods of operations and possibly a "presumptive limit" for emissions. Presumptive emission limits presumed that if an industry was discharging to the prescribed limit, then it would be meeting the BPM requirement. However, the intention was not that industry should pollute up to the limit, but rather that they should do whatever was practicable to reduce emissions below the limit.

BPM notes issued early in 1988 reflected for the first time European Community legislation on industrial emissions, in particular the 1984 framework Directive on air pollution from industrial plant (84/360/EEC); BPM notes were not enshrined in legislation so the limits could be tightened by the inspectorate as required to keep up with advances in abatement technology, or in scientific understanding of the effects of pollutants.

1.2.2 Pollution Inspectorates

(a) England and Wales

Until 1 April 1987, the Alkali Inspectorate (renamed HM Industrial Air Pollution Inspectorate in 1983) was responsible for regulating processes scheduled under the *Alkali Acts*. From 1984-1987 the Inspectorate, together with the Factory Inspectorate, came under the Health and Safety Executive. Before that and from 1 April 1987 to 1 April 1996, the Inspectorate came under the direct control of the former Department of the Environment.

The Royal Commission for Environmental Pollution's fifth report *Air Pollution Control: An Integrated Approach* (1976) had highlighted the potential for pollution to affect more than one media, and that reduction of a release to one environmental medium could have implications for another. The case was made for a unified inspectorate "to ensure an integrated approach to difficult industrial problems at source, whether these affect air, water or land". This finally resulted in the establishment on 1 April 1987 of HM Inspectorate of Pollution (HMIP), and brought together the inspectorates for air pollution, radiochemicals and hazardous waste; it was also responsible for the control of water pollution from particularly polluting processes. An important area of HMIP's work was to develop a method for determining the Best Practicable Environmental Option (BPEO) for dealing with industrial processes and the wastes which they produce. On 1 April 1996, HMIP's responsibilities were transferred to the Environment Agency (see 1.3.1 below).

(b) Scotland

Until 1 April 1996 HM Industrial Pollution Inspectorate (HMIPI) was responsible for enforcing legislation controlling emissions from industrial premises, for the *Radioactive Substances Act 1993* and for the implementation of IPC (together with the River Purification Authorities). The Inspectorate also provided scientific and technical advice on industrial pollution generally, radioactive waste management and environmental radioactivity. HMIPI contributed to the work of the Hazardous Waste Inspectorate (Scotland), which was responsible for the oversight of waste management in Scotland. Both HMIPI and the HWI worked within the Scottish Office. On 1 April 1996 HMIPI's responsibilities were transferred to the Scottish Environment Protection Agency (see 1.3.2 below).

(c) Northern Ireland

Enforcement of the *Alkali Acts* and subsequent pollution control legislation was the responsibility of the Alkali Inspectorate in Northern Ireland. The regulation and enforcement of pollution control legislation is largely the responsibility of the Department of the Environment, Northern Ireland, through the Northern Ireland Environment Agency, which replaced the Environment and Heritage Service, on 1 July 2008 (see 1.3.3 below).

REGULATORY AGENCIES

The main bodies involved in implementing and enforcing pollution control legislation are given below, together with an outline of their responsibilities. More specific detail is given in the individual chapters.

1.3 ENVIRONMENT AGENCIES

Proposals for establishing independent environment agencies for England and Wales and for Scotland were published for consultation in 1991 and confirmed in July 1992. Such Agencies would, it was suggested, be better able to develop a consistent and cohesive approach to environmental protection across all media as envisaged by the implementation of Integrated Pollution Control and taking account of the principles of sustainable development. The Agencies, which formally took over their pollution control and other functions on 1 April 1996, were established by the *Environment Act 1995*, Part I. The Act received Royal Assent on 19 July 1995. In exercising their powers, s.39 of the Act requires both Agencies to take into account the likely costs and benefits of exercising (or not exercising) a power.

The Northern Ireland Environment Agency was established on 1 July 2008 taking over the powers and responsibilities of the Environment and Heritage Service.

1.3.1 Environment Agency, England & Wales

The Environment Agency (in Welsh - Asiantaeth yr Amgylchedd) took over the responsibilities of HM Inspectorate of Pollution, the National Rivers Authority and the waste regulatory functions of local authorities on 1 April 1996. It is a non-departmental public body; members of the board are appointed by the Secretary of State for Environment, Food and Rural Affairs and the Secretary of State for Wales. The Agency's head office in Bristol is responsible for policies, standards, ensuring a consistent approach to environmental protection and overall financial control. There are eight regional offices arranged on the basis of water catchment areas for water management and political boundaries for pollution control and prevention and waste regulation; these latter have been drawn as close as possible to water catchment boundaries, using district rather than county boundaries where these provide a better match. In Wales, however, the boundary follows that of the Principality. The pollution control boundary is also the local "public face" of the Agency. Area offices within each region are largely responsible for the operational side of the Agency's work.

More about the Environment Agency on www.environment-agency.gov.uk

(a) Aims and Objectives

The principal aim of the Agency in discharging its functions to protect and enhance the environment is to make a contribution towards "attaining the objective of achieving sustainable development" (s.4(1) of the Act); s.4(2) requires Ministers to give the Agency statutory guidance on its objectives (including guidance on its contribution to sustainable development) to which the Agency must have regard. Statutory Guidance issued in December 2002 (replacing that issued in November 1996) details the Government's objectives for the Agency, the delivery of which will also contribute to the achievement of sustainable development (see box). The Agency is required to give priority to work which is directly related to, or enhances, delivery of its statutory objectives, its statutory duties (regulatory role) and to those areas or activities in which it has particular expertise or skills. The Agency has also been given objectives specific to each of its main areas of responsibility including process industry regulation, water quality and water resources, waste management, land contamination and radioactive substances.

(b) Functions

In considering or formulating proposals relating to its functions (except pollution control – see next para) the Agency or Ministers should ensure that these will *further* the conservation and enhancement of natural beauty, flora and fauna etc (s.7(1)(a) of the Act).

In carrying out its pollution control responsibilities, the main objective is to prevent or minimise, remedy or mitigate the effects of pollution of the environment (s.5(1)); as well as ensuring it has sufficient information regarding the level, or potential level, of pollution of the environment, the Agency must also follow developments in technology and techniques for abating and preventing pollution or its effects. In formulating or considering proposals for pollution control the Agency or Ministers should have regard to the *desirability* of conserving and enhancing natural beauty, flora and fauna etc (s.7(1)(b)). Section 81 of the Act also requires the Environment Agency to have regard to the National Air Quality Strategy in discharging its pollution control functions.

Government Objectives for the Agency

- Protect or enhance the environment, taken as a whole, in a way which takes account (so far as is consistent with the Agency's legal obligations) of economic and social considerations, so as to make the contribution towards achieving sustainable development which the Secretary of State considers appropriate, as set out in [the] guidance.
- Adopt an integrated approach to environmental protection and enhancement, which considers impacts of substances and activities on all environmental media, on natural resources, and where appropriate on human health.
- Discharge the Agency's functions in an economical, efficient and effective manner and to organise its activities in ways which reflect good management practice and provide value for money.
- Meet high standards of professionalism (based on sound science, information and analysis of the environment and on processes which affect it), transparency, consistency and environmental performance.
- Conduct its affairs in an open and transparent manner in full compliance with the requirements of all relevant statutory provisions and codes of practice relating to the freedom of, and public access to, environmental and other information and to make such information broadly available subject to legislative constraints.
- Ensure that regulated individuals and organisations comply with relevant legislation.
- Develop in conjunction with Government a risk-based, proportionate, efficient and cost-effective approach to the regulatory process; follow better regulation principles; and evaluate and where necessary improve the operation of regulation.
- Provide timely and high quality advice to Government, grounded in the Agency's technical expertise and operational knowledge, including where appropriate in relation to the development and implementation of Government policy and strategy, the implementation of international, European and domestic legislation and in European Union negotiations.
- Reflecting on and building upon the principles of public accountability, develop a close and responsive partnership with the public, local authorities and other representatives of local communities, regional chambers and other regional bodies, other public bodies and regulated organisations, and adopt effective procedures to manage these relationships.
- Collect data of appropriate quality and prepare and disseminate information in a timely fashion for monitoring and reporting on all areas of Agency responsibility.
- Monitor and produce periodic reports on the state of the environment, in collaboration with others as appropriate.
- Undertake research necessary to support the Agency's functions and the delivery of its objectives, in a manner which is consistent with and complementary to the Government's research programme and takes account of research undertaken by others.

With regard to water, the Agency also has a duty "to promote the conservation and enhancement of the natural beauty and amenity of inland and coastal waters; the conservation of flora and fauna ... dependent on an aquatic environment; and the use of such waters and land for recreational purposes."

(c) Pollution Control Responsibilities

- Authorisations, licences and consents for emissions, discharges and disposals to air, water and land; monitoring compliance and enforcement, including prosecutions under water legislation.
- Permitting of installations and enforcement of the *Environmental Permitting Regulations 2007* (which came into force on 6 April 2008, implementing a number of industrial pollution Directives, including the IPPC Directive, and those on large combustion plant, landfill, waste incineration, waste electrical and electronic equipment and end of life vehicles (see chapter 2, 2.4).
- Registration of waste carriers; regulating the import and export of waste and control over the movement of waste, and the regulation of special waste; regulation of radioactive waste accumulation and disposal; assessing waste disposal needs and priorities; production of technical guidance on waste management (see chapter 5).
- Regulation of contaminated land designated as "special sites", and of those contaminated by radioactive materials; report on the state of contaminated land and, as necessary, produce site-specific guidance for local authorities (see chapter 5).
- Monitoring environmental conditions and publishing relevant statistics; research; advice to government in setting environmental quality and other standards and proposals for pollution prevention measures.
- Advice and guidance to industry and others on best environmental practice.

The Agency is also responsible for administering registration and exemption schemes for producer responsibility in accordance with regulations, and monitoring and enforcement of associated obligations (*Environment Act*, ss.93-95 – see chapter 5, 5.33).

(d) Enforcement and Prosecution

The Agency's Enforcement and Prosecution Policy (first issued in 1998) outlines the circumstances under which the Agency will normally prosecute, and the general principles under which it operates in relation to enforcement and prosecution; these are:

- Proportionality: enforcement action should be proportionate to the risks posed to the environment and to the seriousness of any breach of the law;
- Consistency: while this is important, variables such as environmental impact and offending history can also be taken into account;
- Transparency – as to why enforcement action has been taken;
- Targeting – of regulatory effort at those areas presenting the greatest risk to the environment.

(Policy EAS/80001/1/1, issued July 2008, available on the Agency's website)

The Legislative and Regulatory Reform Act 2006, which received Royal Assent on 8 November 2006, establishes statutory principles of good regulation to which regulators must have regard. These principles, which are based on the recommendations of Philip Hampton's report, *Reducing Administrative Burdens: effective inspection and enforcement* (produced at the request of the Chancellor of the Exchequer, and published in 2005) are that

- Regulatory activities should be carried out in a way which is transparent, accountable, proportionate and consistent; and
- Regulatory activities should be targeted only at cases in which action is needed.

The Regulators' Compliance Code, published by the Department for Business, Enterprise & Regulatory Reform (BERR) in December 2007, and which came into force on 6 April 2008, is a statutory code issued under s.21(1) of the Act, and provides guidance on the Hampton Principles. Regulators across the UK have a duty to have regard to the Code in developing their policies or principles or in setting standards or giving guidance. The Code is available at www.berr.gov.uk/files/file45019.pdf

(e) Consultation

Under s.8 of the Act, the Nature Conservancy Council for England, the Countryside Council for Wales or National Park (or Broads) Authority should notify the Environment Agency of any sites of special interest which might be affected by any of the Agency's activities or an authorisation. Where such notification has been received, the Agency must consult the relevant body before carrying out or authorising the activity. The Agency is also required to consult certain bodies before issuing a permit, authorisation, or discharge consent.

The Environment Agency has initiated a procedure for extending public consultation on applications for environmental licences which raise matters of significant local concern; in such cases, the Agency will consult widely with local public and relevant representative groups both on the application and on its proposed decision; its final decision is then published together with the reasons for that decision. In some cases the Agency will hold public meetings where the applicant would have to make a presentation on its application. Applications subject to enhanced consultation are likely to include: large industrial plant (e.g. power stations, chemical plants, steel works); hazardous waste incinerators; burning waste as a fuel; operating new nuclear facilities; increasing discharges of radioactive substances; landfill sites; processes leading to significant discharges to bathing waters, and those likely to be of major public concern or which have, in the past, led to complaints.

The Agency is required to set up and consult Regional Environment Protection Advisory Committees (REPACs), Regional and Local Fisheries Advisory Committees and Regional Flood Defence Committees in England and Wales. Members of the REPAC will normally include a representative from the fisheries and flood defence committees, industry, local authorities, environmental organisations and persons with a particular interest in the air and land aspects of the Agency's work.

An Advisory Committee for Wales advises the Secretary of State on matters relating to the Agency's functions in Wales.

(f) Pollution Inventory

The Agency maintains an Inventory of over 200 pollutants released to air, water and land from the (approx.) 2,000 industrial sites it regulates in England and Wales, enabling the public and others interested to find out about pollution from industrial and other sources in their own locality and nationally. Operators of all IPC processes and IPPC installations, sites authorised to dispose of radioactive substances, larger sewage treatment works and waste management sites moving to PPC are required to submit annual reports on releases of individual substances to the Agency.

The Pollution Inventory can be accessed on www.environment-agency.gov.uk/pi,

The Agency also collates data on emissions of 50 substances from installations covered by IPPC for inclusion in the European Pollutant Emission Register (EPER), which has been established under the IPPC Directive (see 2.1). Local authorities will need to submit data from those IPPC sites which they regulate.

The Environment Agency, together with SEPA and the Northern Ireland Environment Agency, is developing a national Pollutant Release and Transfer Register as required under the Aarhus Convention (see below 1.24); this will enable comparisons of substance releases across the UK. See also chapter 2, 2.1.3.

1.3.2 Scottish Environment Protection Agency

SEPA took over the responsibilities of HM Industrial Pollution Inspectorate, the River Purification Authorities and the Hazardous Waste Inspectorate on 1 April 1996; it also took over responsibility from local authorities for waste regulation and for the control of industrial air pollution under Part I of the EPA; it is also responsible for the control of all installations regulated for Integrated Pollution Prevention and Control. Regional boards (covering the East, West and North of Scotland) provide a link with local interest groups. SEPA's headquarters are in Stirling.

For more about SEPA visit www.sepa.org.uk

(a) Aims and Objectives

SEPA's overall aim is "to provide an effective and integrated environmental protection system for Scotland which will both improve the environment and contribute to Scottish Ministers' goal of sustainable development". Section 31 of *the Environment Act* requires the Secretary of State to give SEPA Guidance, to which it must have regard, on the objectives to be pursued in carrying out its functions; this includes Guidance on SEPA's contribution to attaining the objective of sustainable development (2004, replacing Guidance published 1996).

In considering or formulating proposals relating to its functions, SEPA and the Secretary of State have a duty to have regard to the desirability of conserving and enhancing the natural heritage of Scotland, and of protecting and enhancing buildings, sites and objects of interest (e.g. historic, architectural) and the effect which any proposals might have on the natural heritage etc (s.32).

In carrying out its pollution control responsibilities, the main objective is to prevent or minimise, remedy or mitigate the effects of pollution of the environment (s.33(1)); as well as ensuring it has sufficient information regarding the level, or potential level, of pollution of the environment, SEPA must also follow developments in technology and techniques for abating and preventing pollution or its effects. Section 81 of the Act requires SEPA to have regard to the National Air Quality Strategy in discharging its pollution control functions.

With regard to water, SEPA has a duty (s.34) to promote the cleanliness of rivers, inland, ground and tidal waters in Scotland and to promote the conservation and enhancement of the natural beauty and amenity of inland and coastal waters and the conservation of flora and fauna dependent on an aquatic environment.

Where land has been designated as a Natural Heritage Area or Scottish Natural Heritage considers an area of land to be of special interest, and which could be affected by an activity or authorisation of SEPA, it should notify SEPA of the fact. Where SEPA has received such notification, it must consult Scottish Natural Heritage prior to carrying out or authorising any activity on the land in question (s.35).

(b) Pollution Control Responsibilities

SEPA is responsible for regulating all installations under the *Pollution Prevention and Control (Scotland) Regulations 2000*. SEPA is also responsible for

- Controlling and regulating waste management activities, including landfill sites.
- Regulating the use and disposal of radioactive materials and waste.
- Dealing with land designated as a special site owing to the nature of contamination; inspection and remediation of sites thought to be contaminated with radioactive materials.
- Regulating polluting discharges to water, and for protecting and improving the water environment.

(c) Enforcement and Prosecution

The aims of SEPA's enforcement policy (a revised version of which was launched in March 2008, and is available on its website) are to ensure

- The protection and improvement of the environment and prevent harm to human health;
- Compliance with environmental legislation;
- That enforcement action is proportionate;
- That a consistent approach to enforcement is adopted throughout SEPA.

SEPA's approach to enforcement enables it to take account of the seriousness of the offence, and the attitude and history of the offender; thus it aims to ensure conformity with the law by means of a *compliance strategy* (e.g. development of relationship with non-compliant operator, informal and formal warnings, duty of care and awareness raising) (see also chapter 2, 2.5.9(b); and a *deterrence strategy* (i.e. detecting breaches of the law, identifying the responsible person and punishing offenders, e.g. suspension of licence, prosecution, to act as a deterrent to others).

SEPA's presumption will, however, be to use enforcement action where there is observed harm to – or potential to harm – the environment or human health, a breach of licence conditions or where activities are being carried on without a licence. Action will also be taken to deal with knowingly criminal or negligent acts which damage or threaten the environment or which break the law, and persistent breaches of conditions will not be tolerated.

Under the Scottish legal system, SEPA (unlike the Environment Agency) is unable to take offenders to court itself, and instead refers cases to the Procurators Fiscal for prosecution. In 2004, however, the Scottish Government announced that a specialist environmental prosecutor is to be appointed to each of the regional offices of the Crown Office and Procurators Fiscal Service (COPFS) and that they will, wherever possible be responsible for bringing environmental cases to court. It is hoped that this initiative will result in more effective enforcement of environmental law. SEPA and the COPFS have agreed a Protocol (June 2006) covering submission, processing and monitoring of prosecution reports; this sets outs their respective responsibilities, liaison arrangements and training, as well as disclosure of information and publicity.

The Legislative and Regulatory Reform Act 2006, which provides for a statutory code of practice on enforcement principles, also applies in Scotland (see 1.3.1d above).

(d) Pollution Inventory

The Scottish Pollutant Release Inventory (SPRI) was established in 2002 and developed into an internet based system providing publicly accessible information on chemical and pollutant releases into the environment from all activities regulated for IPPC; this includes data on emissions of the 50 substances required by the Decision establishing the European Pollutant Emission Register (see chapter 2, 2.1.2). SEPA is responsible for compiling the inventory which can be accessed on http://www.sepa.org.uk/spri/index.htm.

SEPA passes information from the SPRI to the Environment Agency which is responsible for compiling the UK return to the Commission.

1.3.3 Northern Ireland Environment Agency

The Department of Environment (Northern Ireland) has overall responsibility for environmental protection and pollution control, conservation, and historic buildings and monuments. The Northern Ireland Environment Agency (NIEA) was established on 1 July 2008, replacing the Environment and Heritage Service. It is an Executive Agency of the DOE, and takes the lead in implementing and regulating pollution control; its overall aim is "to protect, conserve and promote the natural and built environment for the benefit of present and future generations".

On 26 November 1997 *The Industrial Pollution Control (Northern Ireland) Order 1997* (SI 2777, NI 18) was made, introducing integrated pollution control and local air pollution control regimes similar to those introduced through the EPA 1990 in the rest of the UK. The Regulations implementing the legislation came into force on 2 March 1998; new processes were required to apply for an authorisation immediately, with transfer of existing processes completed by 31 December 2002. There are three levels of control – Part A processes with high pollution potential are regulated for IPC "integrated central control" by the NIEA (Industrial Pollution and Radiochemicals Inspectorate, IPRI); IPRI also regulates Part B processes (i.e. those with the potential to cause serious air pollution – "restricted central control"), with district councils regulating Part C processes (those with significant but less potential for air pollution). The NIEA is also responsible for control, disposal and transport of radioactive substances;

The IPCO will be repealed following full implementation of PPC in Northern Ireland.

The Environment (Northern Ireland) Order 2002 (SI 3153, NI 7), made on 17 December 2002, implements the 1996 Directives on IPPC and on Air Quality Management and Assessment. *The Pollution Prevention and Control Regulations (Northern Ireland) 2003* (SR 2003/46), brought into force on 31 March 2003 contain the detailed requirements of the Directive. Part A and B installations are regulated by the Chief Inspector [Industrial Pollution and Radiochemicals Inspectorate], and Part C by district councils.

The Northern Ireland Pollution Inventory contains data on industrial pollutant emissions in Northern Ireland. Arrangements will need to be made in Northern Ireland for collecting emissions data as required by the Decision establishing the European Pollutant Emission Register (see chapter 2, 2.1.2).

As part of its Better Regulation Programme, the NIEA has set five guiding principles which will inform both the enforcement of existing regulations and the development of new legislation:

- Proportionate or risk based – regulations and actions should be appropriate to the risk they aim to address;
- Accountable – decisions should be justifiable and clear standards and criteria on which actions are based should be published;
- Consistent – rules and standards should be constant and fairly applied;
- Transparent – policies and processes should be clearly explained and understood by all stakeholders
- Targeted – regulatory activity should focus on the problem and on delivering the desired environmental outcomes.

(*Better Regulation for a Better Environment*, Environment and Heritage Service, March 2008, published September 2008)

The Legislative and Regulatory Reform Act 2006, which provides for a statutory code of practice on enforcement principles, also applies in N. Ireland (see 1.3.1d above). The Department for Business, Enterprise and Regulatory Reform (BERR) has published (2008) Guidance on the Act; this covers the establishment of the Local Better Regulation Office and the coordination of regulatory enforcement, regulatory sanctions, and the requirement to remove unnecessary regulatory burdens.

More about pollution control responsibilities at www.ni-environment.gov.uk

1.4 LOCAL AUTHORITIES

Local authorities – usually environmental health departments – have a wide range of responsibilities covering the whole spectrum of pollution control and environmental protection; these are described in more detail in the individual chapters of this book, but include:

- Local air quality management;
- Inspecting their areas to identify contaminated land, and subsequent regulatory action leading to remediation;
- (E & W) – permitting of Part A2 installations for integrated pollution prevention and control (LA-IPPC) and of Part B for air pollution control (LA-PPC) under *the Environmental Permitting Regulations 2007*; district councils in Northern Ireland are also responsible for local air pollution control (Part C processes and installations);
- Industrial and domestic smoke control under *the Clean Air Act* (and similar in NI);
- Local environmental quality issues – including, statutory nuisance, control of noise pollution, and litter;
- Collection and disposal of household waste;
- Planning.

The Public Health (Control of Disease) Act 1984 gives Port Health Authorities which have been constituted as a Port Health District similar responsibilities and powers to local authorities for legislation relating to public health and the control of pollution (including Part B installations under the *Environmental Permitting (England and Wales) Regulations 2007.*

1.4.1 Local Better Regulation Office

Part 1 of *the Regulatory Enforcement and Sanctions Act 2008* (RESA) establishes the Local Better Regulation Office (LBRO), a statutory body whose role is to support local authorities to make regulation simpler and more effective and to ensure more consistency in local authority regulation and enforcement, particularly for those organisations which operate across more than one local authority boundary. Part 2 of RESA establishes the statutory Primary Authority Scheme, with s.26 enabling the LBRO to register a local authority as the Primary Authority with responsibility for advising organisations that operate across more than one local authority on issues relating to trading standards, environmental health and some fire safety functions. The Primary Authority Scheme also applies in Scotland and Northern Ireland to those parts of trading standards, environmental health and fire safety legislation which has not been devolved to either the Scottish Government or Northern Ireland Assembly.

The Primary Authority Scheme (Definition of Enforcement Action) Order 2009, which is due to come into force on 6 April 2009, lists the legislation covered by the scheme, and the circumstances when the Act need not be applied (e.g. to avoid a significant risk of serious harm to human health or the environment) (draft published by BERR, September 2008).

Where the LBRO has registered a local authority as an organisation's Primary Authority, any other local authority that proposes to take enforcement action against that organisation must contact the Primary Authority first. The Primary Authority may block the enforcement action if it believes it to be inconsistent with advice or guidance previously given to the organisation; where a local authority and Primary Authority cannot agree on action, the matter may be referred to the LBRO.

The Primary Authority Scheme (Procedure for References to LBRO) 2009, which is also due to come into force on 6 April 2009, outlines the procedure to be followed (draft published by BERR, September 2008).

Local authorities will also need to have regard to the Regulators Compliance Code which embodies the Hampton Principles of enforcement, and which came into force in April 2008 (see above 1.3.1d).

1.5 HEALTH AND SAFETY EXECUTIVE

On 1 April 2008 the Health and Safety Commission and its executive arm the Health and Safety Executive were abolished and a new Health and Safety Executive established. *The Legislative Reform (Health and Safety Executive) Order 2008* amends *the Health and Safety at Work etc Act 1974* to set out the functions and powers of the HSE and amend the regulations for which the HSE is the enforcing authority.

The HSE's mission is "to protect people's health and safety by ensuring risks in the workplace are properly controlled." As such HSE enforces health and safety law for the majority of industrial premises for the protection of both the workforce and the general public. Among the regulations for which it is the either the Competent Authority or the enforcing authority, are

- The Control of Major Accident Hazards Regulations (chapter 2, 2.10)

- The Control of Substances Hazardous to Health Regulations (chapter 2, 2.11)
- REACH (chapter 2, 2.15)
- The Control of Noise at Work Regulations 2005 (see chapter 4)
- Licensing authority for nuclear installations (see chapter 5).

http://www.hse.gov.uk

1.6 CONSERVATION BODIES

1.6.1 Natural England

Established by *the Natural Environment and Rural Communities Act 2006*, Natural England is responsible for "conserving, enhancing and managing England's natural environment for the benefit of current and future generations. As from 1 October 2006, Natural England took over the functions and responsibilities of English Nature, the landscape, access and recreation elements of the Countryside Agency and the environmental land management functions of the Rural Development Service.

The Act also makes provision in respect of biodiversity, pesticides harmful to wildlife and the protection of birds; it amends the functions and constitution of National Park Authorities, the functions of the Broads Authority and the law on rights of way. A Commission for Rural Communities is established to provide information and advice, to monitor and to report to government and others on issues and policies affecting rural issues.

Part 2 of the Act, which also covers Wales, Scotland and Northern Ireland, reconstitutes the Joint Nature Conservation Committee, which comprises Natural England, the Countryside Council for Wales, Scottish Natural Heritage and, in Northern Ireland, the Council for Nature Conservation and the Countryside. In discharging their functions regarding conservation and fostering the understanding of nature conservation, they must have regard to actual or possible ecological changes, and the desirability of contributing to sustainable development.

http://www.naturalengland.org.uk

1.6.2 Countryside Council, Wales

The Countryside Council for Wales is the Government's statutory adviser on sustaining natural beauty, wildlife and the opportunity for outdoor enjoyment in Wales and its inshore waters. It is a statutory consultee for applications for environmental permits, licences and authorisations etc.

http://www.ccw.gov.uk

1.6.3 Scottish Natural Heritage

SNH is a Government body, responsible to Scottish Ministers. It has duties to conserve and enhance Scotland's natural heritage, help people to understand it, and to encourage others to use it sustainably. The SNH Board determines the objectives, strategies and policies of SNH in the light of statutory obligations and guidance from the Scottish Government. It is a statutory consultee for environmental permits and licences, authorisations etc.

http://www.snh.gov.uk

1.6.4 Northern Ireland Environment Agency

The NIEA (established on 1 July 2008, replacing the Environment and Heritage Service) is an Executive Agency of the NI Department of Environment. It takes the lead in advising on, and in implementing the Government's environmental policy and strategy in NI. It has a wide range of responsibilities aimed at promoting the key themes of sustainable development, biodiversity and climate change.

The Council for Nature Conservation and the Countryside is a statutory advisory council which advises the Department of the Environment on countryside and nature conservation issues.

http://www.ni-environment.gov.uk

SUSTAINABLE DEVELOPMENT

The term "sustainable development" has now become the starting point – or key principle – in formulating environmental policy and legislation. It was first introduced in the 1987 report of the World Commission on Environment and Development, *Our Common Future* (the Brundtland Report); this defined sustainable development as "meeting the needs of the current generation without compromising the ability of future generations to meet their own needs".

1.7 EU STRATEGY

The EU's first sustainable development strategy (2001) – *A sustainable Europe for a better world: A European strategy for sustainable development* – proposed measures aimed at promoting sustainable development in a number of important areas, including the following relating to the environment: climate change, management of natural resources and mobility and transport. In June 2006, EU Member States agreed a renewed strategy based on the Commission Review which had been published in December 2005. Its overall aim is to "identify and develop actions to enable the EU to achieve continuous improvement of quality of life both for current and for future generations, through the creation of sustainable communities able to manage and use resources efficiently and to tap the ecological and social innovation potential of the economy, ensuring prosperity, environmental protection and social cohesion." As well as outlining guiding policy principles, the EU SDS identifies seven key challenges (with targets, objectives and actions outlined for each):

- ***Climate change and clean energy:*** to limit climate change and its negative effects to society and the environment.
- ***Sustainable transport:*** to ensure that transport systems meet society's economic, social and environmental needs whilst minimising their undesirable impacts on the economy, society and the environment, with particular emphasis on reducing greenhouse gas emissions from transport.
- ***Sustainable consumption and production:*** to promote sustainable consumption and production patterns – an action plan is to be drawn up in 2007 with the aim of raising awareness of unsustainable production and consumption habits and promoting change.

- ***Conservation and management of natural resources:*** to improve management and avoid over-exploitation of natural resources, recognising the value of ecosystem services.
- ***Public health:*** to promote good public health on equal conditions and improve protection against health threats – this includes a commitment by 2020 to ensure that pesticides and chemicals are used and handled in ways that do not pose significant threats to human beings. The Commission is also to draw up a strategy for improving indoor air quality.
- ***Social inclusion, demography and migration:*** to create a socially inclusive society by taking into account solidarity between and within generations and to secure and increase the quality of life of citizens as a precondition for lasting individual well-being.
- ***Global poverty and sustainable development challenges:*** to actively promote sustainable development worldwide and ensure that the EU's internal and external policies are consistent with global sustainable development and its international commitments.

1.8 UK STRATEGY

1.8.1 Securing the Future, the UK Strategy

Securing the Future published in March 2005, sets out the Government's framework for sustainable development for the whole UK; it updates and builds on an earlier Strategy published in 1999 and aims to show people how they can become involved in making sustainable choices. The Strategy is supported by all the Devolved Administrations, who have formulated specific principles and action plans reflecting the shared priorities of the Strategy, which are:

- ***Sustainable production and consumption***, through for example encouraging: improved product design; resource efficiency; advice for consumers on sustainable consumption; commitment on sustainable procurement in the public sector; support for innovation – new products, materials and services; building partnerships with key business sectors; review of waste strategy with more emphasis on waste as a resource.
- ***Climate change*** – development of measures to meet Government target of reducing carbon dioxide emissions 20% below 1990 levels (see also chapter 3, 3.7); for example, making climate change a priority at international meetings; launch of Climate Change Communications initiative to increase public understanding of climate change issues; development of code for sustainable buildings; launch of scheme enabling government departments to offset carbon impacts of their air travel through investment in renewable energy and energy efficiency projects.
- ***Natural resource protection***, through, for example: working with international partners to reduce rate of biodiversity loss worldwide; launch of environmental stewardship scheme to encourage farmers to deliver environmental benefits; creation of new agencies to manage the marine and terrestrial environments; produce integrated policy approach for protecting and enhancing natural resources.
- ***Sustainable communities***, through, for example: placing sustainable development at the heart of the land-use planning system; joining up effectively at local level around the vision of sustainable communities with Sustainable Community Strategies and Local Area Agreements; enabling people to have a say in how their neighbourhoods are run; new powers for local authorities under *the Clean Neighbourhoods and Environment Act.*

The Sustainable Development Commission, established jointly by the Government and the Devolved Administrations, is charged with reviewing progress towards achieving sustainable development, building consensus on the actions needed for further progress, and identifying those factors which may be undermining progress. The SDC is also charged with monitoring the performance of government in delivering sustainable development. For further information, see www.sustainable-development.gov.uk.

Both the Environment Agency and SEPA are charged with contributing towards attaining the objectives of sustainable development in carrying out their functions, with statutory guidance published in 2002 (Sc, 2004) outlining how this should be done.

The Local Government Act 2000, which received Royal Assent in July 2000, requires all local authorities in England and Wales to prepare a community strategy for promoting or improving the economic, social and environmental well-being of their community (s.4); this was brought into force on 18 October 2000 and also requires that the strategy contributes to the achievement of sustainable development in the UK. Local authorities are also given a new power to do anything they consider likely to promote or improve the economic, social or environmental well-being of their area; this is largely intended for use in conjunction with community strategies and in deciding how to use this power local authorities in England must have regard to the *Guidance on the Power to Promote or Improve Economic, Social or Environmental Well-Being* (May 2001). In preparing its strategy, the local authority should consult and involve appropriate people and those in England must have regard to the *Guidance on Preparing Community Strategies* from the Secretary of State (December 2000). Separate guidance is expected for Welsh local authorities.

The Greater London Authority Act 1999 requires the Mayor to draw up a number of strategies for London (including on air quality, noise, municipal waste management) and to have regard to the effect of those strategies on the achievement of sustainable development in the UK.

Sustainable Development Indicators

Progress towards achieving sustainable development is measured against a set of indicators (currently 68) which were outlined in the Strategy, where the impacts of production and consumption can have significant environmental impacts. For the environment, these include greenhouse gas emissions, sulphur dioxide and nitrogen oxides emissions; rivers of poor or bad quality; commercial and industrial waste arisings and household waste not recycled; electricity produced and fossil fuel use by electricity generators and associated emissions; agricultural output and use of fertilisers and methane emissions. Other areas for which indicators (currently 68) have been developed include housing, education and health. *Sustainable Development Indicators in your Pocket 2008* (Defra, 2008) provides an updated assessment on all the indicators; it is also available online at www.sustainable-development.gov.uk.

1.8.2 Learning to live Differently, Wales

Under the *Government of Wales Act 1998* (s.121), the Welsh Assembly is required to make a scheme on how it proposes to promote sustainable development principles in carrying out its work and to publish annual reports on progress. In November 2000 the Assembly adopted a statutory sustainability scheme – *Learning to Live Differently* – outlining how the Assembly will implement this duty; as well as integrating the principles of sustainable development into its own work, the Assembly will seek to influence others to do the same; from Spring 2002, all papers (policy proposals, grant requests etc) submitted to the Assembly have been required to take account of the principles of sustainable development and how it is to be implemented in practice. Also by Spring 2002, all public bodies sponsored by the Assembly had to show how they intend integrating sustainable development in their work.

An Action Plan covering the period 2004-2007 includes a number of commitments aimed at increasing the use of renewable energies in Wales; among these are that

- by 2010 100% of electricity used in Assembly buildings to be supplied from renewable sources or good quality embedded generation; the Assembly, its agencies and the NHS in Wales are to report annually on energy usage;
- from January 2005, all contracts for new or refurbished public buildings to specify that the design must achieve as a minimum the BREEAM 'very good' standard and wherever possible its 'excellent' rating;
- by 2006, pilot projects to be established exploring the potential of using renewable energy solutions in policies and programmes aimed at tackling fuel poverty.

Other parts of the Action Plan include commitments to research on barriers and benefits to promoting the uptake of alternative fuels; introduction of agri-environment schemes, and development of specifications for school meals contracts which encourage use of seasonal foods.

Progress in achieving sustainable development is measured using a number of sustainability targets and indicators – *Sustainable Development Indicators for Wales 2008* (Welsh Assembly Government) provides information on progress towards meeting the indicators. Among the headline environmental indicators are: greenhouse gas emissions; air and river quality; waste generation and management; renewable energy; and an indicator to track the global impact of resource consumption in Wales (measuring Wales's "ecological footprint").

A draft report reviewing the Welsh Sustainability Scheme and Action Plan was recently submitted to the Welsh Assembly Government; this also suggest measures and actions for the next four years to help achieve sustainable development (www.sustainwales.com).

The Welsh Assembly Government has also published (May 2006) an *Environment Strategy for Wales* which sets the strategic direction for the next 20 years; its aim is to see the "distinctive Welsh environment thriving and contributing to the economic and social well-being and health of all the people of Wales". The strategy covers climate change, sustainable resource use, distinctive biodiversity, land- and seascapes; local environment and environmental hazards.

1.8.3 Sustainable Scotland

Sustainable Scotland outlines the Scottish Government's approach to achieving sustainable development which is seen as an integral part of all its policies and programmes; as a first step sustainable development indicators are being developed for waste, energy and travel. The Scottish Government has also published (September 2004) an Advisory Note for local authorities relating to the duty of Best Value in *The Local Government in Scotland Act 2003* requiring that local authorities "discharge their duties under this section in a way which contributes to the achievement of sustainable development". A revised Strategy reflecting the shared priorities of the UK Strategy was published in December 2005: *Choosing our Future: Scotland's Sustainable Development Strategy* sets out a programme of action for improving quality of life, protecting natural resources and reducing Scotland's impact on the global environment. It aims to put in place a clear methodology for reducing Scotland's global footprint, suggesting that this will be achieved alongside economic growth, but that Scotland's future will involve a more sustainable economic growth in a low-carbon economy. Key elements of the strategy include:

- Reducing Scotland's Global Impact – as well as supporting the UK government's goal of reducing carbon dioxide emissions, ambitious targets have been set to ensure 18% of electricity generated in Scotland will come from renewable sources by 2010, rising to 40% by 2020.
- Transport – a new National Transport Strategy was to be published in 2006; sustainable Travel Plans are to be promoted, with substantial investments to be made in public transport infrastructure and services.
- Biodiversity loss is to be halted, and natural resources managed sustainably; attention to be given to ensuring the coastal, sea and water environments continue to meet the social and economic needs of communities whilst protecting, conserving and enhancing species and habitats.
- Waste – continued investment in recycling, waste treatment and prevention will aim to ensure Scotland meets its targets for recycling and composting of municipal waste, currently set at 25% by 2006, 30% by 2008 and 55% by 2020; promote consumer awareness of waste issues and promote sustainable consumption.
- Thriving communities – good quality open spaces and streetscapes will be supported as they can help to promote sustainable lifestyle choices such as increasing exercise through walking, and cycling. Environmental regeneration of the most deprived neighbourhoods will be a priority, and access to funding streams (which cross environmental, developmental and regeneration budgets) has been simplified. Scotland will continue to pioneer an assets-based approach to community regeneration in the Highlands and Islands, enabling communities to build assets and provide services resulting in local renewable energy programmes, regeneration programmes and other local approaches to local needs.

1.8.4 Northern Ireland

Northern Ireland's first Sustainable Development Strategy, *First Steps Towards Sustainability*, was published in May 2006. The Strategy, which aims to encourage all sectors to live in an environmentally sustainable way, is underpinned by six guiding principles:

- Living within environmental limits – respecting the limits of the planet's environment, resources and biodiversity.

- Ensuring a strong, healthy and just society – meeting the diverse needs of all people in existing and future communities.
- Achieving a sustainable economy – that provides prosperity and opportunities for all, and which embodies the principle of the polluter pays, but incentivises efficient resource use.
- Promoting good governance – which engages people's creativity, energy and diversity.
- Uses sound science responsibly – ensuring policy is developed and implemented on the basic of strong scientific evidence while taking into account scientific uncertainty through the precautionary principle.
- Promoting opportunity and innovation.

The Strategy adopts the priority areas for action identified in the UK Strategy (see above 1.8.1), and adds a further priority area: to create sustainable communities that embody the principles of sustainable development at the local level; and governance for sustainable development.

BIOLOGICAL DIVERSITY

The United Nations' Convention on Biological Diversity reflects worldwide concern that human activities are changing and destroying habitats and natural ecosystems on an increasing scale, with unprecedented loss of species. Biological diversity is defined as "the variability among living organisms from all sources including, inter alia, terrestrial, marine and other aquatic ecosystems and the ecological complexes of which they are part; this includes diversity within species, between species and of ecosystems". Biodiversity is not restricted to rare or threatened species, but includes the whole of the natural world.

A report presented at the ninth meeting of the parties to the Convention in May 2008 suggests that 11% of the world's terrestrial biodiversity could be lost by 2050. Between 1970 and 2005, there has been "unprecedented loss" of biodiversity, with an overall fall in the population trends measured by the Living Planet Index of 27% (*Economics of Ecosystems and Biodiversity Study*).

More: www.cbd.int/cop9

1.9 CONVENTION ON BIOLOGICAL DIVERSITY

As well as the Convention on Climate Change (see chapter 3, 3.5), the United Nations Conference on Environment and Development held in Brazil in 1992 (The Earth Summit) also agreed the Convention on Biological Diversity. This aims to protect and preserve endangered plants and species on land and in the oceans and has three main goals: the conservation of biodiversity; the sustainable use of the components of biodiversity; and the fair and equitable sharing of the benefits arising from the use of genetic resources.

In 2002 Parties to the Convention committed themselves "to achieve by 2010 a significant reduction of the current rate of biodiversity loss at the global, regional and national level as a contribution to poverty alleviation and to the benefit of all life on earth".

The Convention came into force on 29 December 1993 and has now been ratified by 191 parties, including the EC and the UK. The Convention commits governments to conserve and to use biodiversity in a sustainable way, and to develop national biodiversity strategies and action plans. Governments are required to draw up a list of "protected areas"; use of resources in protected areas – e.g. exploitation of plants for medicines – would need to be paid for with financial assistance additional to current levels of development assistance.

At the ninth meeting of the Parties to the Convention held in May 2008 agreement was reached on

- Putting in place the mechanisms for developing an international sustainability standard for biofuels.
- An indefinite moratorium on dumping nutrients, such as iron and nitrogen, into the oceans to promote the growth of phytoplankton and absorb excess carbon dioxide.
- A detailed road map leading to an international regime by 2010 to ensure that people in developing countries benefit from discoveries using their rich biodiversity and genetic resources.

More on http://www.cbd.int

1.9.1 Cartagena Protocol

A Protocol to the Convention – the Cartagena Protocol on Biosafety – came into force on 11 September 2003 and has now been ratified by 147 Parties (including the UK).

The Protocol addresses the potential risks of transboundary movements (through trade or by accident) of living genetically modified organisms (LMOs). Exporters will be required to obtain the prior informed consent – or, the term used in the Protocol, Advanced Informed Agreement – from the importing state. Detailed information about the shipment, including an environmental and health risk assessment must be provided and authorisation obtained from the importer prior to its arrival. Importing countries have 270 days to reach a decision (though they can ask for longer) and must inform the Biosafety Clearing House of their decision. The Clearing House will maintain details of shipments and of risk assessments. The Protocol includes labelling requirements and enables importing countries to adopt a "precautionary" approach in deciding whether or not to authorise a shipment – thus if there is any scientific uncertainty, they can refuse to proceed with the shipment.

1.10 EU BIODIVERSITY STRATEGY

The EU's first Biodiversity Strategy was published in 1998 as a Communication to the Council and the Parliament and was endorsed by both bodies the same year. It focussed specifically on the integration of biodiversity themes into relevant sectoral policies and was built around four themes:

- Conservation and sustainable use of biological diversity.
- Sharing of benefits arising out of the utilisation of genetic resources.
- Research, identification, monitoring and exchange of information.
- Education, training and awareness.

A further Communication published in 2001 (COM(2001) 162) outlines four sectoral biodiversity action plans – conservation and natural resources, agriculture, fisheries, and economic and development cooperation – aimed at halting the loss of biodiversity in the EU by 2010.

In May 2006, the Commission adopted a further Communication on Biodiversity (COM(2006) 216) which acknowledged that action to date had been insufficient and set as a long term aim "the conservation and sustainable use of biodiversity, and thereby the maintenance of ecosystem services for human well-being. An Action Plan setting out the responsibilities of both EU institutions and Member States in halting the loss of biodiversity, specifying indicators to monitor progress, and a timetable for evaluation, accompanied the Communication. Short to medium term aims are to:

- reinforce EU actions to halt the loss of biodiversity in the EU by 2010;
- optimise the EU contribution towards significantly reducing the rate of biodiversity loss worldwide by 2010; and
- accelerate progress towards the recovery of habitats and natural ecosystems in the EU over the period to 2013.

1.11 UK BIODIVERSITY ACTION PLANS

The UK's Biodiversity Action Plan (published in 1994) outlines the measures the UK will take to conserve species and habitats; to develop public awareness and understanding of biodiversity; and to contribute to biodiversity work in the European and global context.

More recently, *Conserving Biodiversity – the UK Approach*, published by the UK Government and the Devolved Administrations in October 2007, calls for joined-up thinking across the public, voluntary and business sectors. Shared priorities for action include developing data, taking targeted action for priority species and habitats, engaging people and encouraging behaviour change.

More on www.defra.gov.uk/wildlife-countryside/

1.11.1 England

Under s.74 of *the Countryside and Rights of Way Act 2000*, both the Secretary of State and National Assembly for Wales had a duty to have regard to the conservation of biodiversity when carrying out their functions, and were required to publish lists of the species and habitats of principal importance for the conservation of biodiversity. This has now been replaced by ss.40 and 41 of *the Natural Environment and Rural Communities Act 2006*.

Section 40 places a duty on public authorities in England and Wales to exercise their functions in a way which is consistent with conserving biodiversity; public authorities include Ministers, National Assembly for Wales, public bodies (government departments, local authorities, planning authorities), and statutory undertakers. Guidance on the duty was published by Defra in May 2007

(www.defra.gov.uk/wildlife-countryside/biodiversity/index.htm)

Section 41 of the Act requires the Secretary of State (as regards England) to publish a "list of the living organisms and types of habitat which ... are of principal importance for the purposes of conserving biodiversity". In drawing up the list Natural England must be consulted. The lists must be published and reviewed and revised as necessary.

*Working with the grain of nature – A biodiversity strategy for **England*** (Defra, 2002) sets out sustainability policies and objectives and a five year action plan to make biodiversity a fundamental consideration in agriculture, water and wetlands, woodland and forestry, marine and coastal management, and in urban areas. It also looks at ways of engaging society as a whole in understanding the needs of biodiversity and what can be done to conserve and enhance it.

In July 2007, Defra published a number of proposed environmental performance indicators for local authorities in England. For biodiversity, this is to be shown as the percentage of sites where a positive biodiversity outcome has been delivered, measured against specific targeted actions within the authority's Local Biodiversity Action Plan.

Defra, together with Natural England, the Environment Agency and other stakeholders have also developed *The Ecosystems Approach Action Plan*. Published in December 2007, the Action Plan highlights the link between environmental, social and economic well being, with priority actions including

- promoting joined up working within Defra and the Defra network to deliver environmental outcomes more effectively;
- identifying opportunities for mainstreaming an ecosytems approach;
- through case studies, demonstrating the benefits of an ecoystems approach;
- developing ways of valuing ecosystem services.

More on www.defra.gov.uk/wildlife-countryside

The Mayor's Biodiversity Strategy – Connecting with London's Nature (published July 2002) is one of seven that he is required to publish. It outlines his vision for protecting and conserving London's natural open spaces; it includes a number of policy statements aimed at ensuring that there is no overall loss of wildlife habitat in London and that more open spaces are created and made accessible.

More on http://www.london.gov.uk/mayor/strategies/biodiversity

1.11.2 Wales

As noted above s.40 of *the Natural Environment and Rural Communities Act 2006* places a duty on public authorities in England and Wales, including the National Assembly for Wales to exercise their functions in a way which is consistent with conserving biodiversity; s.42 requires the National Assembly for Wales to publish a "list of the living organisms and types of habitat which ... are of principal importance for the purposes of conserving biodiversity". In drawing up the list, which must be published and reviewed and revised as necessary, the Countryside Council for Wales must be consulted. A List of Species of Principal Importance for the Conservation of Biological Diversity was published by the Assembly in November 2002 under s.74 of *the Countryside and Rights of Way Act 2000*.

The Environment Strategy for Wales *Our Environment, Our Future, Your Views*, published together with an Action Plan in

May 2006 by the Welsh Assembly Government sets out to achieve "an environment which is clean, healthy, biologically diverse and valued by the people of Wales".

More: http://new.wales.gov.uk/topics/environmentcountryside/

1.11.3 Scotland

Scotland's Biodiversity: It's in your Hands (May 2004) outlines a vision for the future health of Scotland's biodiversity, setting a 25 year framework for action with a key theme being the need to raise public consciousness and reinforce the link between people and biodiversity. Indicators have been developed to measure progress towards achieving the objectives of the Strategy by 2030, which are:

- ***Species and habitats:*** to halt the loss of biodiversity and continue to reverse previous losses through targeted action for species and habitats;
- ***People:*** to increase awareness, understanding and enjoyment of biodiversity, and engage many more people in conservation and enhancement;
- ***Landscapes and ecoystems:*** to restore and enhance biodiversity in all its urban, rural and marine environments through better planning, design and practice;
- ***Integration and coordination:*** to develop an effective management framework that ensures biodiversity is taken into account in all decision making;
- ***Knowledge:*** to ensure that the best new and existing knowledge on biodiversity is available to all policy makers and practitioners.

More: http://scotland.gov.uk

1.11.4 Northern Ireland

Northern Ireland's Biodiversity Strategy was published in August 2002 and sets as a broad aim to arrest and reverse the trend of declining biodiversity. The Strategy was informed by the work of the Northern Ireland Biodiversity Group which identified a number of areas having a major impact on biodiversity, making recommendations for action in each area up to 2016. These areas are: agricultural systems and support; forestry and woodland management; coastal and marine management; freshwater use and management; construction and development; tourism and recreation; peatland management; and introduced species and genetic material. The Strategy stresses the need for policy in other areas to take account of the need to conserve biological diversity, and for all sectors of society to be involved. Progress towards halting and reversing the decline in biodiversity is to be measured through regular Countryside Surveys.

More: http://www.ni-environment.gov.uk/biodiversity/issues.htm

ENVIRONMENTAL LIABILITY

In March 1993 the European Commission published for discussion a Green Paper (COM(93) 47) on Remedying Environmental Damage. The rationale for the document was the principle that the polluter should pay for any damage caused to the environment; the paper proposed strict (or no fault) liability where the polluter is known and the establishment of a compensation fund to cover damage where the polluter is not known or where damage is the result of pollution and cannot be attributed to a single party.

The Green Paper was succeeded by a White Paper – COM(2000) 66 – published in February 2000, which confirmed the principle that the "polluter should pay" and suggested that a liability regime would also enhance two other important principles of environmental protection – prevention and the precautionary principle, and thus result in more responsible behaviour on the part of business. The White Paper concluded that a framework Directive would be the most appropriate way forward and the Commission's proposal for a Directive was published in 2002, and finally adopted on 30 April 2004.

1.12 EU DIRECTIVE

Directive 2004/35/CE *on environmental liability with regard to the prevention and remedying of environmental damage*, should have been transposed into national law by 30 April 2007; the Directive covers all professional and commercial activities (but excludes nuclear installations) which cause "significant environmental damage" to

- Species and natural habitats protected under the Birds and Habitats Directives, where this has significant adverse effects on their conservation. Member states may extend the definition to include wildlife sites protected under national legislation, e.g. sites of special scientific interest.
- Land – as a result of the introduction of chemicals, organisms and micro-organisms which pose significant risk to human health.
- Water bodies – significantly affecting their status or ecological potential as defined under the *Water Framework Directive.*

The Directive includes the following provisions:

- Only damage caused by an emission, event or incident after 30 April 2007 is covered; however it does not apply to damage caused by an emission, event or incident after 30 April 2007 which is derived from an activity that took place and finished before that date; nor does it apply to damage caused by an emission, incident or event which happened more than 30 years before.
- It does not apply to diffuse sources of pollution (such as air pollution), liability for which cannot easily be apportioned.
- "Strict liability"* is introduced for environmental damage to land, water and biodiversity caused by occupational activities subject to EU legislation listed in annex III to the Directive; these include those covered by legislation dealing with emission limits for hazardous substances; dangerous substances and preparations; prevention and control of risks from accidents; transfrontier shipment and disposal of hazardous and other waste, landfill and incineration.
- "Fault based liability"* is introduced for damage to protected species and natural habitats from all other occupational activities.

(* "Strict liability" – sufficient to establish a causal link between the occupational activity and the environmental damage. "Fault based liability" – the damage has been caused through deliberate action or negligence.)

- Member States may choose to allow a defence that the environmental damage had been caused by an emission or event permitted under national pollution laws; a further defence which Member States may choose whether to allow is that the activity was "not considered likely to cause environmental damage according to the state of scientific and technical knowledge at the time".
- The operator of the activity causing the damage is liable; if the polluter cannot be identified, or is not required to pay for remedial action or compensation under the terms of the Directive, the State may pay, but it is not obliged to do so.
- Member States are to encourage the development of financial guarantees; the Commission will review this matter in a report six years after the Directive enters into force, and will consider a gradual approach towards the development of a mandatory system, a ceiling for liability and the exclusion of low risk activities from mandatory insurance.
- Authorities in Member States would be responsible for ensuring restoration to "baseline condition" of damage to biodiversity and water, including the removal of any harm to human health. Where the authority has to take preventative or restorative action itself, it should recover the costs from the operator. The competent authority may require an operator to act to deal with an imminent threat of environmental damage, or do so itself and recover the costs.
- Persons who are adversely affected, or likely to be so affected, by environmental damage may request that the competent authority take appropriate action; "qualified entities" (e.g. public interest groups meeting certain criteria) would have the right to seek a judicial or other official review in those cases where it is felt that the competent authority had not acted at all or not acted "properly".

1.13 UK Regulations

1.13.1 England and Wales

The [draft] Environmental Damage (Prevention and Remediation) Regulations 2008, of which there are separate, but similar, Regulations for England and Wales were issued for consultation, together with draft guidance, in March. Defra and the Welsh Assembly Government were hoping that the final Regulations would come into force in December 2008.

A key purpose of the Regulations is to make the operator of an activity solely responsible for putting right any environmental damage, or threat of environmental damage (and the costs of so doing), caused by his activities – i.e. imposing the polluter pays principle. The operator will also be liable for any remedial costs incurred by the enforcing authority unless he can demonstrate that the environmental damage (or threat of environmental damage) was caused by a third party.

The Regulations will apply to environmental damage, which is damage to

- Protected species and habitats listed in the Birds Directive (79/409/EEC) and the Habitats Directive (92/43/EEC) – mainly those on Natura 2000 sites – with the Regulations also protecting those on Sites of Special Scientific Interest notified under *the Wildlife and Countryside Act 1981*. Schedule 1 to the Regulations details the threshold of damage for species and habitats on and off SSSIs.
- Surface water or ground water, including any adverse effect on the environment of surface or groundwaters resulting in the deterioration of status as defined in the Water Framework Directive (see chapter 6, 6.2), whether ecological, chemical or quantitative.
- Land – relates to contamination which brings about a significant risk to human health from an activity listed in Sch.3 to the Regulations (including those covered by environmental permits and authorisations); or contamination through the introduction of organisms or micro-organisms.

The Regulations are to be enforced by various agencies, largely in accordance with current responsibilities:

- Environment Agency – activities licensed under *the Environmental Permitting Regulations 2007*; also water damage and biodiversity damage in waters other than the sea.
- Natural England, the Countryside Council for Wales – damage to biodiversity.
- Local authorities – land damage.
- Secretary of State in England and Welsh Ministers in Wales – damage at sea.

The Regulations reproduce the Directive in imposing strict liability on those engaged in Annex III activities (see above; Sch.2 to the Regulations). The Regulations provide that other enactments regarding environmental damage continue to apply – for example if land has been contaminated but not to the extent that involves a risk to human health (and thus the ELD Regulations do not apply), then remediation could be required under Part IIA of *the Environmental Protection Act 1990*; this would also be the case if the damage could be proved to be have been caused by an emission which predates the ELD Regulations. Similarly if an operator acting in accordance with his permit has caused the damage, the Environment Agency could issue a Works' Notice (under s.161 *of the Water Resources Act 1991* – see chapter 6, 6.14).

The ELD Regulations do not apply to environmental damage caused by an emission or incident which took place and finished before the Regulations came into force (even if the damage continues to exist); nor do they apply to damage caused after the Regulations come into force, even if the emission post-dates the coming into force of the Regulations if this has resulted from an operation which ceased before they came into force. Liability expires after 30 years.

The Regulations do not cover oil pollution; the 1992 Protocol to the Convention on Civil Liability for Oil Pollution Damage already imposes strict liability and compulsory insurance. Nor do they cover "natural disasters" such as hurricanes or exceptional flooding (but regular seasonal flooding is included). Commercial sea fishing carried out in compliance with the Common Fisheries Policy is also excluded.

Operators are under a duty to take all practicable steps to deal with imminent threats of environmental damage and unless he has successfully dealt with the threat, must notify the appropriate enforcing authority with details of the incident. A notice may be served on the operator telling him what to do to deal with the threat. Defra guidance suggests that an "imminent threat" is one where "there is a sufficient likelihood that environmental damage will occur in the near future" if action is not taken to remedy the situation.

Where damage has been caused and there are reasonable grounds to believe that this is or will become environmental damage, operators must take immediate steps to control and contain whatever is causing the damage so as to prevent it getting worse – detailed analysis of the precise nature of the damage should not delay the taking of measures to control the damage. Where the operator fails to take the right measures, the enforcing authority can serve a notice on the operator telling him what to do. If it becomes clear that emergency action is needed to contain the damage, then the enforcing authority can require the operator to take this (reg. 12).

Where it is reasonable to believe that environmental damage has been caused, the enforcing authority should carry out an assessment to assess the extent of the damage and whether it is indeed environmental damage. If environmental damage is identified, the costs of carrying out the assessment are recoverable from the operator. Having identified the environmental damage the enforcing authority should identify the responsible operator and notify him that a remediation notice is to be served.

The operator may appeal against the notification on the following grounds:

- that he did not cause the damage
- that it does not exceed the thresholds in the Regulations
- or that it is an exempt activity – see above.

The operator may also appeal on the grounds that

a) the damage was caused as a result of compliance with an instruction from a public authority.

b) he was not at fault or negligent – and had a relevant permit and was operating in accordance with it, or that the emission was not considered likely to cause environmental damage according to the state of scientific knowledge at the time.

(the latter defence does not apply in Wales in relation to damaged caused by the release of GMOs; the Welsh Assembly Government is also proposing to extend liability to the holder of the permit, consent or authorisation for environmental damage caused by deliberate release of any GMOs in Wales.)

c) while he caused the damage, the actions of another person contributed to it, despite the fact that appropriate safety measures were in place.

These defences only apply to the requirement to remediate – not to the requirement to carry out emergency measures to prevent further damage (under reg. 12). Where these defences do apply, there is no obligation on the operator to carry out any remediation.

Where no appeal is made against a notification, or it is unsuccessful, the enforcing authority will serve a remediation notice specifying the steps to be taken (taking account of Sch. 4 to the Regulations); before doing so the operator will be invited to submit his own remediation plan, which it is open for the enforcing authority to endorse thus saving time and money for both sides.

An operator may appeal against a remediation notice to the Secretary of State (or Welsh Ministers, and in most cases the notice will be suspended pending determination of the appeal.

Regulation 18 (Wales reg.20) implements Article 12 of the Directive whereby persons with a "sufficient interest" can request action from the enforcing authority if they perceive a threat of environmental damage to something of concern to them; the draft guidance suggests that persons of sufficient interest may include established NGOs, residents' associations and neighbourhood groups, anglers, ramblers or birdwatchers, and registered charities whose objectives include conservation of the environment. Those persons should give details of their interest to the enforcing authority, together with details of the threat or damage; the enforcing authority must consider the complaint, and inform the complainants of the steps they will take; the authority should also inform the responsible operator, inviting their comments.

It is an offence to contravene the terms of a remediation notice; conviction in the Magistrates Court may result in a fine not exceeding the statutory maximum and/or up to three months' imprisonment. Conviction on indictment may result in an unlimited fine and/or up to two years' imprisonment.

1.13.2 Scotland

The Scottish Government have also published draft *Environmental Liability (Prevention and Remediation) (Scotland) Regulations*, together with draft guidance. These are similar to the draft Regulations and guidance published for England and Wales, but the following should also be noted:

- Species and sites protected for damage to biodiversity include Special Areas of Conservation and Special Protected Areas.
- Competent Authorities – Scottish Natural Heritage in relation to environmental damage (or an imminent threat of environmental damage) to protected species or natural habitats; SEPA for environmental damage (or imminent threat of such damage) to water (up to 3 nautical miles) or land; Scottish Ministers in respect of damage (or the imminent threat of such damage) to protected species or natural habitats within the marine environment.
- Environmental damage caused by the release of GMOs – it is not a defence to cite the existence of a relevant permit and operating in accordance with it, or that the emission was not considered likely to cause environmental damage according to the state of scientific knowledge at the time. Also the threshold for liability for past damaging events in respect of GMOs is extended to 75 years.

1.13.3 Northern Ireland

At the time of writing (October 2008), consultation on draft Regulations was awaited.

1.14 THE LUGANO CONVENTION

The Council of Europe has already agreed (March 1993) a *Convention on Civil Liability for Damage Resulting from Activities*

Dangerous to the Environment (the Lugano Convention). Although some countries have signed the Convention, it requires ratification by three before it comes into force. (In *Paying for our Past*, DOE, 1994, it was stated that there were no plans for the UK to sign the Convention.) The Convention applies to a wide range of industries, including research laboratories dealing with genetically modified organisms, who will be liable for any "intolerable" damage caused even though pollution legislation has been complied with. It provides for compensation to persons and property and for economic loss as a result of environmental damage; operators are required to have insurance cover and the Convention provides for environmental organisations to take legal action in pursuit of the Convention's requirements.

ENVIRONMENTAL CRIME

1.15 EU Proposal

COM(2007)51, published in February 2007, is a *proposal for a Directive on the protection of the environment through criminal law*; in March 2008 it was transmitted to the European Parliament for an Opinion under the Co-decision procedure. One of the aims of the proposal is to provide consistency throughout the EU both in the definition of environmental crime and the level of sanction. Member States will be required to ensure that a range of activities (see below) already prohibited by EU or national legislation are considered criminal offences when committed intentionally or with serious negligence. Particularly serious environmental crimes would be punishable by up to five years imprisonment and for companies a fine of at least Euros 750,000. "Serious" crimes will include those that have resulted in death or serious injury to a person or substantial damage to air, soil, water, animals or plants, as well as those crimes committed by criminal organisations. Member States would have the option of imposing supplementary or alternative sanctions, such as obliging the offender to clean up or reinstate the environment, or closing down the business.

Among the environmental crimes to be covered by the Directive are the following:

- the unlawful treatment, transport, export or import of waste, including hazardous waste;
- the unlawful emission of hazardous substances in to the air, soil or water;
- the unlawful trade in endangered species;
- the unlawful trade in or use of ozone depleting substances;
- the unlawful operation of a plant in which a dangerous activity is carried out or in which dangerous substances or preparations are stored or used.

While oil spills are not explicitly excluded from this proposal, they are expected to be covered by a separate proposal to amend Directive 2005/35 on ship source pollution.

LAND USE PLANNING – AN OVERVIEW

The Communities and Local Government Department (CLG) is responsible for planning policy, with its remit also including local government, housing and urban regeneration.

Planning policy for Scotland, Wales and Northern Ireland is the responsibility of the relevant Devolved Administration.

For information on all aspects of the planning system visit the Town and Country Planning Association's website at www.tcpa.org.uk, and the Planning Portal website at www.planningportal.gov.uk

The ***European Union*** has no Directive on land use planning. However, it does have a *European Spatial Development Framework Perspective* which influences the administration of structural funds and programmes (e.g. the *6th Environmental Action Programme*), and which Member States should have regard to when making planning law. It also has indirect influence on planning by way of legislation relating to other areas (e.g. water, noise, human rights).

1.16 ENGLAND AND WALES

The Planning and Compulsory Purchase Act 2004, received Royal Assent on 13 May 2004, and has been brought into force through Commencement Orders during 2004-2006. The Act covers England and Wales, with the following exceptions: Parts 1 and 2 (regional functions and local development) apply in England only; Part 6 reforms the development system in Wales; chapter 2 of Part 7 (Crown immunity) relates to Scotland and Scottish Ministers are responsible for its commencement. The Act repeals Part 2 of *the Town and Country Planning Act 1990* which covered development plans and makes a number of other amendments and additions to that Act.

All those with plan-making functions under Parts 1, 2 and 6 of the Act have a duty (Part 3 of the Act) to exercise their functions with the objective of contributing to sustainable development, and must have regard to any guidance issued by the Secretary of State. In this respect, Planning Policy Statement (PPS) 1, *Delivering Sustainable Development*, published February 2005, provides guidance on how the Government's four aims of sustainable development – economic development, social inclusion, environmental protection, and the prudent use of natural resources – can be achieved through the planning system.

Planning Bill - England

This Bill, published by the Government in November 2007 follows its White Paper *Planning for a Sustainable Future* (Communities & Local Government, May 2007) which set out proposals for reform of the planning system in England. The Government was hoping the Bill would receive Royal Assent in autumn 2008.The Bill's main provisions are aimed at

- **Streamlining decision-making on major infrastructure projects:** national policy statements will set out key infrastructure needs (e.g. waste, energy, transport, nuclear power plant) for the next 10-25 years; the NPSs would be subject to extensive public consultation and sustainability appraisal. An independent Infrastructure Planning Commission (IPC) consisting of experts from different sectors will lead inquiries and take decisions on major projects on the basis of their consistency with the aims of the national policy statements. An amendment to the Bill would confer statutory authority for carrying out development for which consent is granted.

- **Simplifying the local planning system** for householders to make it easier to make home improvements (such as extensions, conservatories, microgeneration devices*) where there is little or no impact on neighbours; simpler information requirements for all applications, with minor amendments (e.g. repositioning of a door) to be made without need for full planning application.
- **Climate change** – local development plans will need to include policies on mitigating and adapting to climate change.
- **Ensuring the planning system continues to support "vibrant" town centres** with new planning rules which enable councils to prioritise town centres over out-of-town shopping centres.

**The Town and Country Planning (General Development) (Amendment) (England) Order 2008* (SI 2008/675), which came into force on 6 April 2008, removes the need for household micro-renewables to have planning permission, allowing their installation through permitted development rights subject to various restrictions (e.g. height, size location). The Order covers specified types of microgeneration equipment on or within the curtilage of detached dwelling houses; these include solar PV or solar thermal equipment (wind turbines), ground source heat pumps and water source heat pumps and biomass heating systems. Properties near air source heat pumps and wind turbines should not be exposed to noise above 45 decibels (Microgeneration Certification Scheme). Planning permission will still be required for wind turbines on semi-detached or terraced housing.

more: www.communities.gov.uk

1.16.1 The Planning & Compulsory Purchase Act 2004

Note: the Department for Communities and Local Government took over responsibility for planning in May 2006 from the former Office of the Deputy Prime Minister.

(a) Regional Spatial Strategy

Part 1 of the Act (brought into force August/September 2004) applies In England only and requires a *regional spatial strategy* (RSS) (except in London) to be prepared for each region by the regional planning body (RPB); RSSs, which take the place of regional planning guidance, must relate to the Secretary of State's policies for development and land use. The RPB – of which at least 60% should comprise members of county, district, or metropolitan district councils, national park or Broads authorities – is responsible for reviewing and monitoring implementation of the RSS, preparing revisions as necessary, holding local examinations in public on the revisions and submitting it to the Secretary of State. At all stages the RPB must seek advice from relevant county councils and other authorities with strategic planning expertise.

The Town and Country Planning (Regional Planning) (England) Regulations 2004 (SI 2004/2203), which came into force on 28 September 2004, outline the criteria for recognition of RPBs, the form and content of the draft revision of RSSs and procedure for the preparation of the revision of the RSS. PPS 11 (September 2004) sets out the policies applying in England (except London) which RPBs will need to take account of in the preparation of revisions to RSSs.

In London, the Mayor is responsible for the preparation of a *spatial development strategy*. He has a duty to consult the Assembly and the functional bodies about his proposals and must have regard to any comments submitted to him. He is also required (s.29 of *the Greater London Authority Act 2007*) to prepare a statement for the Chair of the Assembly showing how any comments received have been taken into account or setting out his reasons for not accepting comments submitted.

All land use and spatial plans must be accompanied by a sustainability appraisal which fully complies with the SEA Directive (see below, 1.20) – interim advice was published by the ODPM in March 2005 (*Sustainability Appraisal of Regional Spatial Strategies and Local Development Frameworks*).

Where it is thought that a planning application may impact on the implementation of the RSS the planning authority should consult the RPB.

The National Assembly for Wales in its Spatial Plan (see d below) defines spatial planning as "the consideration of what can and should happen where. It investigates the interaction of different policies and practice across regional space, and sets the role of places in a wider context. It goes beyond 'traditional' land-use planning and sets out a strategic framework to guide future development and policy interventions, whether or not these relate to formal land-use planning control."

(b) Local Development

Part 2 of the Act, which also applies in England only (brought into force August/September 2004), requires each local planning authority (LPA) to prepare and maintain *a local development scheme*; this will set out what *local development documents* (LDD) the authority should prepare, timetable for preparation and whether or not the LDD is to be prepared jointly with one or more authorities. A statement of community involvement (NI Art.3, 2006 Order) should also be prepared outlining the local planning authority's policies for involving interested parties in development matters in their area. LDDs, which replace local plans, unitary development plans and structure plans, must be prepared in conformity with the RSS. A sustainability appraisal of the proposals in each LDD must be carried out.

The local development plans produced by London Boroughs should be prepared in conformity with the London spatial development strategy; section 30 of *the Greater London Authority Act 2007* will permit the Mayor of London to direct changes to boroughs' LDSs, subject to override by the Secretary of State.

County councils are responsible for preparing and maintaining a minerals and waste development scheme in respect to any part of their area for which there is a district council. LDDs prepared under such schemes must also set out the authority's policies relating to minerals and waste development.

The Town and Country Planning (Local Development) (England) Regulations 2004 (SI 2004/2204), which came into force on 28 September 2004, prescribe the form and content of the local development scheme, the procedure to be followed in bringing it to effect, the documents which must be development plan documents, as well as the form and content of LDDs. The Regulations also cover the preparation of joint LDDs and the operation of joint committees. The Regulations apply to county councils for the purposes of minerals and waste development planning.

PPS 12, *Local Development Frameworks* (September 2004) sets out the policies which apply throughout England, focusing on procedural policy and the process of preparing LDDs.

Guidance on Assessing the Soundness of Statements of Community Involvement, and on Assessing the Soundness of Development Plan Documents was published in 2005 by the then ODPM and can be found on the Planning Portal (www.planningportal.gov.uk). They clarify the role of inspectors, local planning authorities, government offices, regional planning bodies and the wider community. They set out the evidence requirements of the examination process and provide model representation forms and guidance notes which local authorities can give to anyone wishing to make representations.

(c) Development Control

Part 4 of the Act (brought into force 2004/2005) inserts additional sections into *the Town and Country Planning Act 1990* enabling local planning authorities in both England and Wales to introduce local permitted development rights through local development orders. The Secretary of State may make development orders and regulations prescribing the procedure for making applications for planning permission and certain consents, set a timetable for applications subject to appeal or which have been "called in" for decision, and prescribe fees and charges.

The Secretary of State may direct that a planning authority refer a planning application for a major infrastructure project to her for decision if the development is considered to be of national or regional importance. The different elements of inquiries into such projects can be heard concurrently. (*The Town and Country Planning (Major Infrastructure Project Inquiries Procedure) (England) Rules 2005*, SI 2005/2115, brought into force 24 August 2005.)

Other amendments made to the 1990 Act through Part 4 of the PCPA 2004, brought into force by Commencement Order on 24 August 2005 include the following:

- A local authority may decline to determine repeat applications if it is of the opinion that nothing has significantly changed since its earlier consideration and refusal (s.43) (NI Art.9, 2006 Order).
- Most planning permissions will be granted for three years (instead of 5), except where the local authority agrees to a longer period (s.51).
- Statutory consultees have a duty to respond to consultation within 21 days to help speed-up planning decisions (s.54).

Local planning authorities now have 13 weeks in which to determine major planning applications before an applicant can appeal on the basis of non-determination. (Commencement Order no. 8, SI 2006/1061).

Sections 31-36 of *the Greater London Authority Act 2007*, which were commenced on 6 April 2008 (SI 2008/582) give the Mayor of London powers to direct that he should be the planning authority for the purposes of determining planning applications of major strategic importance to London, and to lead on planning obligations and planning contributions for those applications he determines. Schedule 1 to *the Town and Country Planning (Mayor of London) Order 2008* (SI 2008/580) identifies the categories of projects which may be referred; these include new housing, waste facilities and transport networks.

(d) Development Plans – Wales

Part 6 of the PCPA applies in Wales. Each local planning authority is required to prepare a local development plan, focusing on the authority's objectives for the use and development of land in the area and general policies for implementation. Public participation in the formulation and expedition of such plans is to be enhanced through community involvement schemes and timetables agreed between the local planning authority and the National Assembly for Wales.

In preparing their local plans, local planning authorities must have regard to *People, Places, Futures – The Wales Spatial Plan*, finalised by the National Assembly for Wales in November 2004; this is available on the NAW website – www.wales.gov.uk.

(e) Crown Immunity

Part 7 of the Act, some of which was brought into force on 6 August 2004, ends Crown Immunity in the planning system in England, Wales and Scotland, and makes provision in relation to certain planning applications by or on behalf of the Crown, and in respect of the enforcement of planning controls in England and Wales. *The Town and Country Planning (Application of Subordinate Legislation to the Crown) Order 2006*, a draft of which was published in September 2005, will give effect to the provisions of Part 7. Guidance for local authorities and Crown bodies on how applications from the Crown should be handled is to be produced.

Similar provision is made in Northern Ireland in the *Planning Reform (Northern Ireland) Order 2006*.

(f) Compulsory Purchase

Under Part 8 of the Act, local authorities, joint planning boards and National Park authorities may acquire land by compulsory purchase to facilitate development if such acquisition will be of economic, social or environmental benefit to their area – these powers were commenced on 31 October 2004. The Act outlines the categories of person entitled to have their objections heard in relation to a proposed compulsory purchase, the procedure for such representations, and provisions relating to compensation.

The Compulsory Purchase of Land (Prescribed Forms) (Ministers) Regulations 2004 (SI 2004/2595) prescribe forms for use in the preparation of compulsory purchase orders; further regulations will set out the written representations procedure for considering objections where remaining objectors agree to such a procedure as an alternative to holding a public local inquiry. Circular 06/2004, published by the ODPM in November 2004 outlines how local authorities should use their new powers of compulsory purchase.

1.16.2 Planning Guidance

England

All those with a responsibility for preparing development and planning documents, spatial strategies etc must have regard to any relevant statutory guidance or planning policy statements as directed by the Secretary of State. Planning Policy Statements and Guidance Notes are published on the website of the Department for Communities and Local Government www.communities.gov.uk. As well as PPSs, the site includes minerals policy guidance notes (being redrafted as minerals policy statements), regional policy guidance notes, a list of research reports and best practice advice.

The ODPM also published (2005) T*he Planning System: General Principles*; this provides advice and information about the operation of the planning system, updated to reflect the PCPA, which had previously been included in PPG1.

Both PPGs (which are now being revised into PPSs) and PPSs are listed in the Further Reading section of the Pollution Control Handbook, with the following also being of particular relevance to sustainable development and to pollution control:

- PPS 1 *Delivering Sustainable Development* (2005) – sets out an integrated approach to planning for sustainable development that takes full account of the need to achieve social inclusion, protection and enhancement of the environment, the prudent use of natural resources and sustainable economic development.
- PPS: Planning and Climate Change, published as a supplement to PPS 1 in December 2007, sets out how the Government expects participants in the planning process, including both local authorities and developers to work towards the reduction of carbon emissions in the location, siting and design of new development. Thus, it outlines how regional and local planning can best support achievement of the government's zero-carbon targets [for new housing by 2016] while meeting community needs for economic and housing development.
- PPS 10 *Sustainable Waste Management* (2005) – puts the waste hierarchy at the centre of policy, with greater emphasis on waste as a resource; the need to deliver a better match between the waste communities generate, and the [local] facilities needed to manage that waste is emphasised. Thus regional planning bodies and local authorities (through their planning strategies) will need to plan for managing their waste, and ensuring that it is disposed of as near as possible to its place of production (see also chapter 5, 5.4). A "Companion Guide" to PPS10, published June 2006, provides further advice for planning authorities, developers and communities on the delivery of appropriate waste management facilities.
- PPS 12 *Local Spatial Planning* (2008) – explains what local spatial planning is, the key ingredients of local spatial plans and how they should be prepared; it should be taken into account by local planning authorities in preparing development plan documents and other local development documents.
- PPS 23 *Planning for Pollution Control* (2004) – provides guidance on how major new systems for pollution control and the management of contaminated land should be taken into account when considering proposals for development. Both local planning authorities and regional planning bodies will need to have regard to the policies in PPS23 in the preparation of local development documents and regional spatial strategies, respectively. Annexes cover pollution control, air and water quality, and development on land affected by contaminated land.
- PPG24 *Planning and Noise* (1994) – outlines how noise should be taken into account in the planning process (see chapter 4, 4.8.1).

Wales

Planning Policy Wales (PPW), published in 2002, sets out the land use policies of the Welsh Assembly Government which should be taken into account by local planning authorities in the preparation of Unitary Development Plans. PPW is supplemented by a series of Technical Advice Notes (TAN) – see Further Reading at the back of the *Pollution Control Handbook*.

Ministerial Interim Planning Policy Statement 01/2008 updates the PPW to promote sustainability through good design, requiring the design process to ensure that development contributes to tackling the causes of climate change and to effective adaptation to climate change impacts. It should also promote the efficient use of resources, maximising energy efficiency, minimising the use of non-renewable resources, as well as the generation of waste and pollution.

More at www.planningportal.gov.uk.

1.17 SCOTLAND

Following its White Paper published in June 2005 outlining proposals for modernising the planning system in Scotland, the Scottish Government published a planning Bill which received Royal Assent on 20 December 2006. *The Planning etc (Scotland) Act 2006* places a formal obligation on Scottish Ministers and planning authorities to exercise their functions with the objective of contributing to sustainable development. The public is given an increased and effective role in planning decisions at an early stage, with new measures encouraging public participation in planning, including on major applications, and ensuring greater transparency in the handling of local authority interest cases.

1.17.1 Planning etc (Scotland) Act 2006

Much of the 2006 Act amends, replaces various sections of, or introduces new provisions into, *the Town and Country Planning (Scotland) Act 1997*. It is being brought into force through Commencement Orders. As well as those parts covered below, the Act also makes changes to the way planning applications are handled, introduces provisions to allow local businesses to invest collectively in improvements to the area in which they operate "business improvement districts" (brought into force 1 April 2007, and powers for Ministers to assess the performance of planning authorities.

(a) National Planning Framework

Section 1 of the Act, brought into force on 1 April 2007, inserts a new Part 1A into the 1997 Act and requires Scottish Ministers to prepare and publish a National Planning Framework setting out in broad terms how Scottish Ministers consider that the development and use of land could and should occur; it should contain a strategy for Scotland's spatial development, and a statement of what Scottish Ministers consider to be priorities for that development; it may identify specific developments or classes of development which are to be defined as 'national development'. The Act places a duty on Scottish Ministers to prepare, publish, review and revise the National Planning Framework (NPF) at least every five years; if it is decided not to revise the NPF, reasons for this should be made public. A "public participation" statement must be published setting out when and with whom consultation on its preparation or review is likely to take place, and the steps to be taken to involve the general public. The Act allows 60 days for Parliamentary consideration of the proposed NPF and requires Ministers to lay a copy of the published version before Parliament.

Scotland's first (non-statutory) National Planning Framework was published in 2004, and set out a development strategy to 2025. As required under the 2006 Act, Scottish Ministers have now published for discussion a second National Planning Framework (NPF2) which aims to guide Scotland's development to 2030. NPF2 takes forward the spatial aspects of the Scottish Government's policy commitments on sustainable economic growth and climate change, and identifies a number of major national transport, energy and environmental infrastructure projects for further environmental assessment and consultation. The final version of NPF2 was expected before the end of 2008.

(b) Development Plans

Part 2 of the 2006 Act relating to strategic development planning and local development planning replaces Part 2 of the 1997 Act, and was expected to be brought into force in its entirety in January 2009. Section 4(1) of the Act enables ministers to establish "strategic development planning authorities", each of which is to draw up a strategic development plan and to keep the plan under review.

This section of the Act was brought into force by Commencement Order (No. 3, SSI 2008/164), with *Strategic Development Plan Authority Designation Orders*, which came info force on 25 June 2008, designating the following SDPAs:

- SSI 2008/195 (No. 1 Order) – the Councils of East Dunbartonshire, East Renfrewshire, Glasgow, City, Inverclyde, North Lanarkshire, Renfrewshire, South Lanarkshire and West Dunbartonshire.
- SSI 2008/196 (No. 2 Order) – Aberdeen City and Aberdeenshire Councils.
- SSI 2008/197 (No. 3 Order) – the Councils of Angus, Dundee City, Fife, and Perth & Kinross.
- SSI 2008/198 (No. 4 Order) – the Councils of City of Edinburgh, East Lothian, Fife, Midlothian, Scottish Borders, and West Lothian.

The SDPAs had until 25 September 2008 to submit plans to Scottish Ministers proposing the boundary for their SDP areas (with individual planning authorities having the right to submit alternative plans if they do not agree with the proposed boundary). Statutory Guidance on Strategic Development Planning Authorities (Planning Circular 2/08) was issued (under s.4(7) of the Act) in April 2008, and is available at www.scotland.gov.uk/planning

SDPs, to be drawn up in conformity with the NPF, will include

- a vision statement setting out in broad terms the authority's views as to how the development of the SDP area could and should occur, and matters which could affect that development (i.e. the principal physical, economic, social and environmental characteristics of the area; purposes for which the land is used; population, infrastructure etc);
- a spatial strategy setting out in broad terms proposals as to the development of the land with the SDP area;
- an analysis of the relationship of the vision statement and spatial strategy to general proposals for development and other use of land in districts adjoining its area;
- other matters as prescribed, or as the SDPA considers appropriate.

The Act details procedures for the preparation and production of plans, consultation and publication. Scottish Ministers may approve, reject or amend plans.

Local Development Plans are to be prepared by planning authorities as soon as soon as practicable after the relevant provision of the Act is brought into force, and thereafter at intervals of no more than five years. They should take account of the NPF and must include a spatial strategy detailing the authority's policies and proposals for the development of the land, such other matters as may be prescribed or which the authority considers it appropriate to include. Where the land is not in a strategic development plan area, the authority should set out in a vision statement its views of the how the land could and should develop and matters which might affect the development. Again the Act sets out the procedures to be followed in the preparation and adoption of the plan, including consultation and publication arrangements.

Action programmes must accompany each SDP and LDP showing how each plan is to be implemented – these must be published within three months of the approval or adoption of the plan and should be updated at least every two years.. The form and content and procedures for their preparation is to be set out in Regulations.

(c) Development Management

Section 5 inserts s.26A into the 1997 Act to introduce a new hierarchy for proposed developments which will determine the decision-making process, with Ministers being given powers to make Regulations to describe major and local developments.

- ***National developments*** (identified within the National Planning Framework, considered to be of national strategic importance) decision to be taken by Scottish Ministers, with the full involvement of the Parliament;
- ***Major developments*** (large-scale, though not of national importance, identified in development plans) – decision by local authorities;
- ***Local developments*** (e.g. smaller housing, commercial and large householder developments) – decisions on non-controversial applications to be taken by planning officers; decisions on proposals which conflict with the development plan or require an EIA would continue to be dealt with by elected officials.

The Scottish Government is proposing (March 2008) to allow energy generating equipment, including solar panels, heat pumps and wind turbines to be installed without the need for planning permission (Permitted Development Rights), so long as they meet certain criteria. The proposals would allow ground source heat pumps and water source heat pumps anywhere.

Wind turbines on domestic buildings will not be allowed in conservation areas or world heritage sites; elsewhere they would not be permitted under PD if more than 3m above the top of the roof, with a blade length of 1.2 metres or (it is proposed) within 100 metres of another property. PD for freestanding turbines would be limited to those with an overall height limit of 11.1 metres and at least 12 metres from the boundary.

1.17.2 Planning Guidance

Scottish Planning Policies (SPPs) provide statements of Scottish Government Policy on nationally important land-use and planning matters, supported by circulars containing guidance on policy implementation. Planning Advice Notes provide advice on good practice and other relevant information. SPPs are replacing National Planning Policy Guidelines. The National Planning Framework sets out the strategy for Scotland's long-term spatial development and has the same status as the SPPs.

More: www.scotland.gov.uk/planning

1.18 NORTHERN IRELAND

The Planning (Northern Ireland) Order 1991 provides the main legal basis for planning in Northern Ireland. It has been amended by *the Planning Reform (Northern Ireland) Order 2006*, made on 9 May 2006 (SI 2006/1252, NI 7), which reflects many of the measures introduced by *the Planning and Compulsory Purchase Act 2004* – see above. In November 2007, the Environment Minister announced an intention to bring forward proposals for a wide-ranging reform of the planning system in NI; a formal consultation paper was expected during 2008.

Part I of the Order contains introductory provisions; Part II deals with reforms aimed at improving the management of the planning process, promoting sustainable development, enhancing transparency and community involvement and strengthening enforcement powers. Part III deals with correction of errors in planning decisions, with Part IV amending the 1991 Order to remove Crown immunity from planning decisions. Part V covers mineral planning permissions and Part VI amendment and repeal provisions.

The Planning Service, an agency of the NI Department of the Environment is responsible for overall administration of the NI planning system, including the development and implementation of policies and development plans in NI.

Further information available on www.planningni.gov.uk. Guidance covering Northern Ireland can be found at www.planningni.gov.uk.

1.19 PLANNING – HAZARDOUS SUBSTANCES

1.19.1 Hazardous Planning Consent

The Planning (Hazardous Substances) Act 1990 consolidates and amends certain provisions of the *Town and Country Planning Act 1971* and the *Housing and Planning Act 1986*. Under the Act, the presence of a hazardous substance on, over or under land above a specified quantity (the controlled quantity) will require a consent from the hazardous substances authority (HSA). This Act provides for local authorities to act as the HSA and to take into account the current and contemplated use of the land and its surroundings when determining applications. The Act provides for public registers to be maintained containing full details of applications, consents, conditions, etc. All establishments which fall within the COMAH regime require a hazardous substances consent (see chapter 2, 2.10).

The Planning (Hazardous Substances) Regulations 1992 (S1 656), which came into force on 1 June 1992, implement the 1990 Act in England and Wales. Similar regulations apply in Scotland and Northern Ireland: the *Town and Country Planning (Hazardous Substance) (Scotland) Regulations 1993* (S1 323), effective 1 May 1993, and the *Planning (Hazardous Substances) Regulations (Northern Ireland) 1993* (SR 2003/275, amended SR 2005/320).

Storage of certain hazardous substances above the controlled quantity, separately or in aggregate (if of the same generic EC classification as specified in the Regulations) requires a consent from the HSA; the procedure for obtaining a consent is set out in the Regulations. This is similar to the procedure for applying for permits under other legislation: a prescribed form must be used for the application which must be accompanied by the appropriate fee; the application must be publicised locally and statutory consultees asked for their views. In the case of refusal of an application, there is a right of appeal to the Secretary of State – in England these are dealt with by the Planning Inspectorate. (CLG consultation, August 2007. Applications, consents and their conditions, modifications, revocations, etc will all be put on a public register. Guidance from the Planning Inspectorate outlines the grounds for appeal and the procedure.

The Regulations do not apply to controlled wastes or radioactive waste. These are subject to control under the *Environmental Protection Act 1990* and *Radioactive Substances Act 1993*. Nor do they cover the storage of explosives where the Health and Safety Executive administer the licensing system. Contravention of the Regulations is subject to a maximum fine of £20,000 on summary conviction or an unlimited fine on conviction on indictment.

1.19.2 Planning for Hazardous Substances

The 1990 Act requires planning authorities to take the prevention of major accidents (and the limiting of the consequences of such accidents) into account when preparing plans. The *Planning (Control of Major-Accident Hazards) Regulations 1999* (SI 981), which came into effect in England and Wales on 20 April 1999, amend both *The Planning (Hazardous Substances) Act* and the Regulations which implement those requirements of the 1996 EU Directive on the Control of Major Accident Hazards (see chapter 2, 2.10.2) relating to the use of land-use controls in the siting of new hazardous installations, including the consent requirements. Planning authorities must take land-use controls into account when considering modifications to sites or other new developments in the vicinity of such sites; they must also comply with the Directive's objectives that development decisions take account of the need to ensure danger to human health or the environment from accidents at major hazard sites is minimised. In particular, planning authorities are required to consider hazards of major accidents when determining land allocations and to maintain appropriate distances between establishments and residential areas, areas of public use and areas of particular sensitivity and/or interest in the long term; in the case of existing establishments, to impose additional measures in accordance with Article 5 of the Directive so as not to increase risks to people.

Similar Regulations for Northern Ireland (2000 - SR 101) came into force on 24 April 2000, and in Scotland on 6 July 2000 (2000 - SSI 179).

ENVIRONMENTAL APPRAISAL & ASSESSMENT

Environmental Impact Assessment (EIA) is a generic term for describing a tool for systematically examining and assessing the impact and effects of development on the environment. In Europe, the term relates specifically to the assessment of discrete development *projects*, whilst the term Strategic Environmental Assessment (SEA) relates to that of *plans, policies and programmes*.

1.20 STRATEGIC ENVIRONMENTAL ASSESSMENT

1.20.1 SEA Directive

In June 2001 the European Council adopted a *Directive on the assessment of certain plans and programmes on the environment* (2001/42/EC), with the aim of ensuring a uniform and high standard of assessment throughout all Member States. The Directive, which had to be implemented by Member States by 21 July 2004, applies to plans and programmes whose formal preparation began after 21 July 2004, and to those which were already in preparation by that date but had not been adopted or submitted to legislative procedure by 21 July 2006.

Competent authorities in Member States are required to undertake a systematic SEA of all plans and programmes to assess their environmental implications prior to adoption of the plan or programme concerned; this will comprise both the preparation of an environmental report (ER) and consideration of responses arising from formal consultation.

SEA is mandatory for plans and programmes for agriculture, forestry, fisheries, energy, industry, transport, waste management, water management, telecommunications, tourism, town and country planning, or land use and that "set the framework for future development consent" of projects listed in the Directive on EIA (85/337/EEC, as amended). The Directive also provides for the mandatory SEA of plans and programmes deemed to need appropriate assessment under the *1992 Habitats Directive* (92/43/EEC). The need for an SEA is discretionary for both "minor modifications" to the above-mentioned types of plans and programmes and to those determining "the use of small areas at local level". Member States may determine whether any other types of plans/programmes require SEA, and also whether these are likely to have significant effects on the environment (either through a case-by-case examination and/or by specifying types as such), taking into account the views of relevant authorities and the specific criteria set out in Annex II of the Directive (see below). Financial, budgetary, national civil defence or civil emergency plans and programmes are not subject to SEA.

Information Comprising an Environmental Report for the Purposes of SEA (Annex I to Directive 2001/42/EC)

(a) an outline of the contents, main objectives of the plan or programme and relationship with other relevant plans and programmes;
(b) the relevant aspects of the current state of the environment and the likely evolution thereof without implementation of the plan or programme;
(c) the environmental characteristics of areas likely to be significantly affected;
(d) any existing environmental problems which are relevant to the plan or programme including, in particular, those relating to any areas of a particular environmental importance, such as areas designated pursuant to Directives 79/409/EEC and 92/43/EEC;
(e) the environmental protection objectives, established at international, Community or Member State level, which are relevant to the plan or programme and the way those objectives and any environmental considerations have been taken into account during its preparation;
(f) the likely significant effects* on the environment, including on issues such as biodiversity, population, human health, fauna, flora, soil, water, air, climatic factors, material assets, cultural heritage including architectural and archaeological heritage, landscape and the interrelationship between the above factors;
(g) the measures envisaged to prevent, reduce and as fully as possible offset any significant adverse effects on the environment of implementing the plan or programme;
(h) an outline of the reasons for selecting the alternatives dealt with, and a description of how the assessment was undertaken including any difficulties (such as technical deficiencies or lack of know-how) encountered in compiling the required information;
(i) a description of the measures envisaged concerning monitoring in accordance with Article 10;
(j) a non-technical summary of the information provided under the above headings.

* These effects should include secondary, cumulative, synergistic, short, medium and long-term permanent and temporary, positive and negative effects.

Criteria for Determining the Likely Significance of Effects for the Purposes of SEA (Annex II to Directive 2001/42/EC)

1. The characteristics of plans and programmes, having regard, in particular, to
- the degree to which the plan or programme sets a framework for projects and other activities, either with regard to the location, nature, size and operating conditions or by allocating resources,
- the degree to which the plan or programme influences other plans and programmes including those in a hierarchy,
- the relevance of the plan or programme for the integration of environmental considerations in particular with a view to promoting sustainable development,
- environmental problems relevant to the plan or programme,
- the relevance of the plan or programme for the implementation of Community legislation on the environment (e.g. plans and programmes linked to waste management or water protection).

2. Characteristics of the effects and of the area likely to be affected, having regard, in particular, to
- the probability, duration, frequency and reversibility of the effects,
- the cumulative nature of the effects,
- the transboundary nature of the effects,
- the risks to human health or the environment (e.g. due to accidents),
- the value and vulnerability of the area likely to be affected due to:
  - special natural characteristics or cultural heritage,
  - exceeded environmental quality standards or limit values,
  - intensive land-use,
  - the effects on areas or landscapes which have a recognised national, Community or international protection status.

An environmental report which identifies, describes and evaluates the likely significant effects on the environment of implementing the plan or programme together with reasonable alternatives, is also required; this should take into account the objectives and geographical scope of the plan or programme. Annex I of the Directive refers to the information to be given for this purpose (see page 23).

Following adoption, implementation of the relevant plans and programmes must be monitored to identify unforeseen effects and facilitate remedial action. The Directive also includes requirements for consultation to take place with authorities, the public and other Member States within an appropriate timescale, and for decision-makers to take these and the ER into account during determination.

ODPM (now DCLG) and the Devolved Administrations have produced *A Practical Guide to the SEA Directive* (September 2005); other UK guidance, including that for specific types of plans and programmes is listed in the "Further Reading" section of the *Pollution Control Handbook*.

1.20.2 Regulations (E, W, NI)

The Environmental Assessment of Plans and Programmes Regulations 2004 (SI 2004/1633), which came into force on 20 July 2004 implement the Directive in ***England***, and also the ***UK*** where plans relate to England with any other part of the UK. The relevant Regulations for ***Wales*** (SI 2004/1656, W.170) took effect on 12 July and those for ***Northern Ireland*** (SR 2004/280) on 22 July 2004 – all these Regulations cover plans and programmes relating to that country only. Similar Regulations for ***Scotland*** (SSI 2004/258) were repealed and their provisions included in *the Environmental Assessment (Scotland) Act 2005* (see 1.20.3 below).

As required by the Directive responsible authorities are required to prepare an SEA for all plans and programmes falling within the scope of the Directive (see above) for which the first formal preparatory act occurred after 21 July 2004. Relevant plans and programmes begun on or before 21 July 2004 but not adopted or submitted to legislative procedure by 22 July 2006 need not be subject to an SEA if the responsible authority decides it would not be feasible and informs the public of its decision. For those plans or programmes for which an SEA is discretionary (see above), the responsible authority having consulted statutory consultees may take the decision as to whether an SEA is necessary; the Regulations (not NI) also provide for the Secretary of State to call in such plans and programmes for decision.

In England the Countryside Agency, English Heritage, Natural England and the Environment Agency (statutory consultees) should be consulted about all plans and programmes subject to assessment; they should also be consulted as part of the screening procedure when the scope and detail of the environmental report is being decided, and on the actual ER, and on any transboundary cases. Relevant bodies in Scotland, Wales and N. Ireland should be consulted on those plans and programmes which relate to England and part of their territory. The public should also be given the opportunity to comment on draft plans and programmes and ERs – these documents should also be made available at the main office of the responsible authority. (Statutory consultees in Scotland: Scottish Ministers, SNH & SEPA; Wales: Welsh Assembly, Environment Agency & Countryside Council; NI Environment Agency.)

Plans or programmes falling within the scope of the SEA Directive may not be adopted or submitted to legislative procedure until the requirements concerning environmental reports, including consultation and taking account of consultation, have been completed.

The Regulations also implement the Directive's requirements for informing statutory consultees etc when plans and programmes have been adopted, for making available details of how environmental considerations have been integrated into the plan or programme, and the reasons for choosing the particular plan or programme, and the monitoring measures to be implemented. The responsible authority should also take appropriate steps to bring the adopted plan or programme and other relevant information to the attention of the public.

1.20.3 EA (Scotland) Act 2005

Scottish Regulations implementing the SEA Directive were revoked and their provisions included in *the Environmental Assessment (Scotland) Act 2005*, which was brought into force by Order on 20 February 2006 (SSI 2006/19). This requires SEA for all new strategies, plans and programmes relating to Scotland developed by the public sector (defined as "any person, body or office-holder exercising functions of a public character"). Plans and programmes which relate partly to Scotland and another part of the UK are subject to the 2004 UK Regulations (see above).

Part 1 of the Act requires "Responsible Authorities" (i.e. the authority in charge of the qualifying plan or programme) to ensure an environmental assessment is carried out during the preparation of a "qualifying" plan or programme – these are those relating to agriculture, forestry, fisheries, energy, industry, transport, waste and water management, telecommunications, tourism, town and country planning and land use; also those which set the framework for future development consent of various types of project listed in Sch.2 to the Act. (References to plans and programmes includes strategies.) The Act exempts from the need for SEA plans and programmes relating to individual schools; Ministers may, by order, add further exemptions covering other types of plan or programme which experience over time proves do not need EA, and which, in the opinion of Scottish Ministers, are likely to have no effect or minimal effect in relation to the environment.

Plans and programmes outside the scope of the SEA Directive should be pre-screened by the Responsible Authority (RA) to determine whether, in its opinion, it will have no or minimal environmental effect and need not therefore be subject to SEA; schedule 2 to the Act sets out the criteria for determining environmental significance, and guidance is also to be produced. Where it is decided that an SEA is not necessary, the RA should notify statutory consultees (Scottish Ministers, SEPA and Scottish Natural Heritage), giving details of the plan and programme. A register of such notifications is to be kept.

Certain other plans and programmes – mainly those which determine the use of small areas at local levels or minor modifications of a plan or programme – must be formally screened to establish whether or not they need to be subject to full environmental assessment – again the criteria in Sch.2

should be used in reaching a decision and the statutory consultees consulted before any final determination is made. For all those plans and programmes which are required to be formally screened, the RA should submit a summary of its views as to whether a particular plan or programme is likely to have significant environmental effects (taking account of the criteria in Sch.2) to statutory consultees.

Statutory consultees have 28 days to respond to RAs. Where they are in agreement, the RA will make the relevant determination; where they cannot reach agreement the matter will be referred to Scottish Ministers. A copy of the final determination, with reasons for the determination, should be sent to statutory consultees within 28 days; it should also be publicised on the RA's website and steps taken within 14 days to bring it to the attention of the public (e.g. via a local newspaper).

For those plans and programmes which require full SEA, the RA will need to prepare an environmental report which identifies, describes and evaluates the likely significant effects on the environment of implementing the plan and programme, and outlines reasonable alternatives; the objectives of the plan or programme and its geographical scope should be taken into account. Schedule 3 to the Act outlines the information to be included in the environmental report. Before deciding on the scope and level of detail to be included in the environmental report and the consultation period, the RA must send consultation authorities a summary of the draft plan or programme to enable them to take a view – they have five weeks in which to respond back to the RA.

Within 14 days of preparing the environmental report, the RA should send it to consultation authorities for comment and publicise it locally, and facilitating early and effective opportunity to participate in the process.

In preparing plans or programmes, the RA must take account of the environmental report and of any views expressed. Following adoption, both it and the environmental report should be made publicly accessible, it should be publicised in a local newspaper and on the RA's website. The RA is also responsible for monitoring the significant environmental effects of the implementation of the plan or programme; if any unforeseen adverse effects are identified, appropriate remedial action should be undertaken.

1.21 ENVIRONMENTAL IMPACT ASSESSMENT

1.21.1 EIA Directive

The 1985 *Directive on the assessment of the effects of certain public and private projects on the environment* (85/337/EEC) ("the 1985 Directive") requires competent authorities within member states to request an EIA for development projects likely to have a significant effect on the environment; it also requires them to prescribe the form of the subsequent assessment (so as to include a full description of the project and mitigation measures to be taken to avoid, reduce or remedy adverse effects). Annex I of the Directive provides a list of projects (e.g. large power stations, airports, oil refineries) for which EIA is mandatory, and Annex II provides a list of projects (e.g. pig farms, glass-making plants, paper mills) for which EIA is necessary if member states consider them to be so (i.e. discretionary). The 1985 Directive also requires public consultation about such development proposals to take place, and that decision-takers take subsequent representations into account when determining applications for development.

The 1985 Directive has been amended by Directive 97/11/EC, which was formally adopted in March 1997 and had to be implemented nationally by 14 March 1999. This amended the 1985 Directive as follows:

- additions, transfers and modifications of project types within both annexes;
- competent authorities to provide screening determinations for Annex II projects;
- member states may introduce screening thresholds for Annex II projects;
- developers are entitled to scoping opinions as appropriate;
- wider consultation process for projects with potential transfrontier effects;
- provision of additional annex containing screening criteria for Annex II projects (criteria to include (a), is the project likely to have "significant" environmental effects, on the basis of pre-set thresholds; and (b), is it likely to have a "significant effect" on an area designated as a special protection zone under other EU legislation;
- the ES should include an outline of alternatives considered and reasons for choice;
- competent authorities are required to make public a statement of the main reasons for decisions, together with descriptions of main mitigation measures;
- provision of more guidance on the information to be provided.

1.21.2 EIA for Transboundary Projects

The 1997 Directive also implements the UNECE Convention on EIA in a Transboundary Context (the Espoo Convention), which sought to facilitate a wider, more transparent and comprehensive consultation process for projects with cross-boundary effects (e.g. off-shore oil production, new reservoirs or dams). It gives those likely to be affected by a project the right to be notified about the project, to participate in the EIA and then to be consulted about the transboundary impacts and measures to mitigate them. The Convention came into force in September 1997 and has now been ratified by 42 countries (including the European Community on 24 June 1997 and the UK on 10 October 1997). A Protocol under the Convention on *Strategic Environmental Assessment*, agreed on 21 May 2003, will come into force following ratification by 16 countries (there are currently nine ratifications, including the European Community, and 38 signatories).

1.21.3 EIA of Planning Projects in the UK

For development that requires planning permission, the *Town and Country Planning (Environmental Impact Assessment) (England and Wales) Regulations 1999* (SI 1999/293), which came into force on 14 March 1999, implement the 1985 Directive (as amended) and revoke the earlier *Town and Country Planning (Assessment of Environmental Effects) Regulations 1988*. The *Environmental Impact Assessment (Scotland) Regulations 1999* (SSI 1) as amended, and *the Planning (Environmental Impact Assessment) Regulations (Northern Ireland) 1999* (SR 73) do likewise for Scotland and Northern Ireland respectively.

Projects for which EA is Mandatory (Annex I to Directive 97/11/EC)

- Power stations with a heat output of 300 MW or more, nuclear power stations and nuclear reactors (including decommissioning and dismantling);
- Crude oil refineries; facilities for gasification and liquefaction of coal or bituminous shale;
- Asbestos plants;
- Construction of motorways and express roads and widening of existing dual and single carriageways where new or widened section is more than 10 km in length;
- Long distance railway lines and airports with a planned runway length of 2100 metres or more;
- Inland waterways and inland waterway ports which allow the passage of vessels over 1,350 tonnes;
- Developments for the disposal of special waste, including incinerators, chemical waste treatment and landfill facilities, radioactive waste disposal sites and toxic waste facilities; non hazardous waste facilities with capacity of more than 100 tonnes per day;
- Projects for the reprocessing of irradiated nuclear fuels, for the production or enrichment of fuel, and the processing or storage of nuclear waste;
- Integrated chemical facilities (producing inorganic or organic chemical products, fertilisers, biocides, pharmaceuticals or explosives;
- Groundwater abstraction or artificial groundwater recharge with annual volume of at least 10 million cubic metres;
- Plant for the transfer of water between river basins;
- Waste water treatment plant with capacity exceeding 150,000 population equivalent;
- Extraction of oil (> 500 tonnes per day) and natural gas (> 500,000 m^3 per day);
- Gas, oil and chemical pipelines > 800 mm diameter and > 40 km in length;
- Dams storing in excess of 10 million cubic metres;
- Intensive pig and poultry rearing units;
- Industrial plant for production of pulp from timber or similar fibrous materials;
- Quarries and open cast mines > 25 hectares and peat extraction sites exceeding 150 hectares;
- Construction of overhead power lines with voltage of 220 kV or more and length > 15 km;
- Oil, petrochemical or chemical products storage facilities with capacity of > 200,000 tonnes.

The 1999 Regulations transpose Annex I and II projects into Schedules 1 and 2 respectively, and require the submission of an ES with applications for planning permission for certain types of development (i.e. "EIA development"). In terms of *content*, the 1999 Regulations centre on the production of an ES, a document or series of documents assessing the likely environmental impact of a development proposal and the significance of the effects arising from its impact. In terms of *process*, the 1999 Regulations provide for the respective stages of screening, scoping, preparing an ES, evaluating the ES and determining the application concerned. The Regulations also provide for copies of environmental statements to be placed on the planning register and to be made publicly available, for appeals, and for consultation with EU member states about projects with significant transboundary effects.

The 1999 Regulations (SI 1999/293) have been amended in England by SI 2006/3295 to give effect to Art.3 of the 2003 Public Participation Directive (see 1.25.2) insofar as it affects public participation in the decision making process for applications and appeals relating to development for which EIA is required. They also apply the requirements of the 1985 EIA Directive, which requires the assessment of the likely major environmental effects of major new developments, to local development orders. SSI 2006/614 amends the 1999 Scottish Regulations to implement Art.3 of the Public Participation Directive in Scotland, with SR 2008/17 applying in Northern Ireland. *The Town and Country Planning Regulations 2008* (SI 2008/2093) make further amendments to require environmental statements at both the outline planning permission stage and for applications for subsequent approval of "reserved matters", with similar Regulations in Wales (SI 2008/2335, W.198), Scotland (SSI 2007/484), and NI (SR 2008/372).

DETR Circular 2/99 on *Environmental Impact Assessment* provides guidance on how LPAs should interpret and implement the 1999 Regulations. In Scotland, Planning Circular 8/07 replaces Circular 15/99, to take account of recent amendments to the Regulations, including the incorporation of public participation requirements into EIA legislation, and to provide updated guidance in the light of emerging case law. In June 2006, the DCLG published proposals for amending and updating Circular 2/99 to reflect the incorporation of public participation requirements into EIA legislation and to update screening criteria and advice on using some form of checklist to ensure that proper record is kept of the considerations taken into account when screening.

The 1999 Regulations and Guidance do not cover projects that lie outside the scope of the statutory planning system (i.e. projects subject to alternative development consent procedures – such as new railway lines, harbours, pipelines. Separate Regulations have been made covering EIA for such projects, and include Regulations on: land drainage improvement works; offshore petroleum production and pipelines; electricity works; Channel Tunnel rail link; fish farming in marine waters; harbour works; roads; certain water management projects in the agriculture sector; minerals sites; agriculture and afforestation projects. It should be noted that, confusingly, the EIA for some of these projects is termed "appraisal". Separate guidance also applies (in some cases) to these projects (e.g. the Department for Transport's *Design Manual for Roads and Bridges*).

Environmental Risk Assessment

Environmental risk assessment is a management tool for organising and analysing the available information on an environmental problem. It has some aspects in common with other decision-making tools, such as strategic environmental assessment (SEA) and environmental impact assessment (EIA) – see 1.20 & 1.21 above – though its explicit treatment of probability and uncertainty makes it ideally suited to distinguishing between adverse environmental impacts (or consequences) that could occur, and the likelihood (probability) of the impacts actually occurring. This is a function not explicitly performed by EIA and is an important distinction for those charged with managing risk because separate strategies exist for managing the probability and consequences of environmental impacts.

1.22 OVERVIEW

Operators and developers of a wide range of programmes, plans and projects are required to submit environmental risk assessments to local authorities and the environment agencies in support of their activities. Risk assessments help regulators assess the significance of the risks associated with an activity and then identify the risk management measures that are required to reduce risks to, and from, the environment and to human health. Regulators stipulate these measures as "conditions" in their permits to develop or operate. Preparation of the environmental risk assessment is usually the applicant's responsibility, in line with the polluter pays principle. It is a living document requiring updating as and when modifications to an operation are proposed, or when new or additional information, relevant to the operation, becomes available.

Not all activities necessarily require complex risk assessment though – for example, where the risk is undisputedly negligible or the impacts well understood, or where there exist accepted mechanisms for control, a simple risk screening may be all that is necessary. As uncertainty increases and the likelihood of severe consequences become less clear, however, a formalised process of risk assessment assists in understanding the severity of the risk and how to manage it. Advantages to risk assessment include:

- the distinction made between consequences (impacts) and the likelihood of occurrence;
- a structured approach to assessing risk and thus establishing a logical basis for managing risk;
- providing a basis for the prioritisation of risk management actions and the targeting of regulatory effort;
- the recording of decisions for future use;
- ensuring decision processes, and their underlying logic, are transparent for others to appraise.

Guidance from the Health and Safety Executive (*Five Steps to Risk Assessment*, revised 2006) provides simplified advice, together with examples of what a risk assessment might look like, on the five key elements to an effective risk assessment:

- identifying the hazards;
- deciding who might be harmed and how;
- evaluating the risks and deciding on precautions;
- recording findings and implementing them;
- ensuring they are reviewed at regular intervals.

1.22.1 Policy Context

In the UK, the application of environmental risk assessment within pollution control features increasingly as an explicit requirement of environmental planning and permitting and within the requirements of European and domestic legislation. Overarching guidance to practitioners within and beyond Government on risk assessment was first published in 1995, in the then Department of Environment's "*Guide to Risk Assessment and Risk Management for Environmental Protection*". Revised Departmental "*Guidelines on Environmental Risk Assessment and Management*" for England and Wales were issued in 2000, with the Department for Environment, Food and Rural Affairs (Defra) re-endorsing their use in 2002.

Good problem definition, use of a staged and tiered approach, the principles of proportionality and consistency of use, the explicit treatment of uncertainty and the need for presentational transparency are common themes in the revised Guidelines, referred to as "Green Leaves II". The guidance provides an over-arching risk management framework to which specific risk assessment guidance, such as that for waste management regulation (chapter 5, 5.18.5), genetically modified organisms (chapter 3, 3.27), historically contaminated land (chapter 5, 5.41.4), groundwater protection (6.11.3) and major accident hazards (chapter 2, 2.10.2), for example, can refer. The guidance fulfils, in part, Defra's and the Environment Agency's commitments under the modernising government agenda to make public their risk management frameworks.

1.22.2 General Powers

Environmental risk assessments are performed by operators, developers and permit applicants (or their professional advisors) in support of their activities. The regulator has the role of technically reviewing the risk assessment in the context of the statute to inform its decision on authorising the specific activity. The regulator usually retains a right to undertake its own independent environmental risk assessment, should it so wish. Permit applicants use risk assessments to evaluate where pollution control measures are required and in turn identify the type and level of financial investment (cost) required for risk reduction (benefit). Risk assessments are used by the regulator to:

- assess the magnitude and significance of the risk posed by the activity;
- identify the key drivers of the risk;
- prioritise risk reduction measures; and
- assess the level of residual risk following application of these measures.

Risk reduction is then secured by applying conditions on the environmental permit. Measures may range from technological interventions (e.g. the application of best available technology) to a requirement for environmental management systems. Adherence to these measures is monitored and enforced through regulatory inspection and related regulatory mechanisms.

Regulators expect operators to adhere to the general and specific guidance issued on environmental risk assessment and management. In reviewing an applicant's submission, the regulator evaluates it from a regulatory and technical perspective, assessing the quality and suitability of the submission, by reference to statutory requirements, the supporting science and relevant published guidance. On receipt of the operator's risk assessment, and contingent on the type of facility and quality of submission, the regulator will normally:

- conduct a technical and regulatory review of the submission, consulting internally and externally as required, and probing the submission for its technical soundness and completeness by reference to the legislation, available guidance and current state of the science base;
- request further work of the operator, if deemed necessary;
- consider the need for an independent review of the operator's risk assessment where there are reasonable differences of opinion or where a second opinion might prove valuable;
- consider the need for an independent risk assessment for any operation or proposal; and

- determine the risk management measures required, or which should be considered, to mitigate the risks to an acceptable level of residual risk.

Whilst individual legislation specifies the use of environmental risk assessment for specific regulatory processes, the environment agencies have general duties and powers (for the Environment Agency, s.4 & 5 of the *Environment Act 1995*) they can use to request risk assessments of operators, permit holders and applicants. Furthermore, the adoption of various principles, tools and techniques for environmental decision-making feature in the Ministerial Guidance provided to the Environment Agency on its contribution to sustainable development. In short, the agencies may generally request a risk assessment for any activity that they regulate where pollution of the environment is suspected. The general expectations of environmental risk assessment work submitted for regulatory review are set out in the revised *Guidelines on Environmental Risk Assessment and Management (2000).*

1.23 ERA FOR POLLUTION CONTROL

Individual statutory instruments set out the context and objectives of the legislation and usually state whether risk assessment is to be applied and for what purpose. Supplementary guidance, whether statutory or non-statutory, may describe the detail on how risk assessment should be used and what the regulatory expectations are. Further technical guidance is sometimes developed in support of this. So, for example, planning legislation is concerned with the appropriateness of development and the risks it may pose in a spatial and local context; environmental permitting for PPC with the application of environmental technology and management to minimise risks from process facilities; and waste legislation with minimising risks to human health and the environment from waste management facilities. Safety cases for new or existing plants are primarily concerned with contingency planning and the avoidance of major accidents.

Some of the more common regulatory applications of risk assessment, the specific context and the guidance that supports their application are discussed below.

1.23.1 Environmental Planning

For certain development projects such as development of waste management facilities (incinerators, landfill, hazardous waste facilities), it has become commonplace for applicants to submit a project-level environmental risk assessment as part of the environmental impact assessment required under planning legislation (see 1.16). Risk assessment offers greater resolution over EIA with respect to the relative likelihood of effects that the development may pose. It allows planners and their consultees the opportunity to scrutinise in detail the relative significance of potential impacts (see "*Environmental Impact Assessment (EIA) – A Handbook for Scoping Projects, 2002*). With respect to the potential impacts of development projects on human health, the increasing application of health impact assessment (HIA) is seeing the outputs of project level EIA, risk assessment and HIA converge.

Risk assessments submitted during planning may be refined, adapted or expanded upon further at the environmental permitting stage contingent on the requirements of the permitting regime. At the permitting stage, the environment agencies have powers to require permit applicants to furnish additional information, within a specified period, for the purposes of determining the application. Under PPC, for example, a Schedule 4 notice may be used to set out the Environment Agency's expectations for further information on, among other matters, the risk assessment related to a permit application.

1.23.2 IPPC and COMAH

Risks from the process sector have historically been addressed through the Integrated Pollution Control (IPC) regime. The IPC approach was largely effects-based and driven by the need to render releases from regulated processes to the environment harmless. In risk terms, the approach used environmental criteria, and their exceedance, as a surrogate for assessing the consequences and probability, respectively, of environmental harm.

Under IPPC (see 2.4 and 2.5) and COMAH legislation (see 2.10), risk assessment assumes a more formalised role. For COMAH, there is a fundamental requirement for operators to undertake an environmental risk assessment in a systematic way and to clearly demonstrate that risks have been identified, and all necessary measures put in place, to prevent major accidents and limit their consequences if they do occur. Each site is different and a systematic approach allows for the identification of the most important high-risk accident scenarios and the prioritisation of resources, resulting in a transparent, proportionate approach to the management of major hazards from dangerous substances. As a general principle, IPPC requires that all industrial operators applying for a permit are required to assess the risk of accidents and their consequences and that necessary measures be taken to prevent accidents and limit their environmental consequences.

Guidance for risk practitioners on the scope of environmental risk assessments for these regimes is provided in "*Preparing Safety Reports: Control of Major Accident Hazards Regulations, HSG190*" and "*Guidance on the Environmental Risk Assessment Aspects of COMAH Safety Reports*". Under COMAH, a formal risk assessment is the principal component of the safety report produced by operators of major process facilities. The operator is concerned with the probability of an initiating event or combination of events that could lead to a major accident with consequences for the safety of on-site staff, the public and the wider environment. The general duty is to 'take all measures necessary to prevent major accidents and limit their consequences to persons and the environment'. It is recognised that risks cannot be completely eliminated, but that demonstrable measures are required for prevention and mitigation. The approach is based on the concept of risk tolerability – requiring that measures be taken to reduce the likelihood of hazards and to limit their consequences until further reduction of risks cannot be justified; that is, until the risks are 'as low as reasonably practicable' (ALARP). This implies a trade-off between the costs of risk reduction and the benefits obtained.

There is some element of overlap between IPPC and COMAH for certain sites and there is recognition that certain information required for both regimes may be interchangeable. However, the accident provisions for IPPC may fall beneath the threshold for COMAH classification. Whilst IPPC risk assessments may require consideration to be given to smaller incidents, these may still have significant impacts on the environment, individually or cumulatively.

1.23.3 Waste Management

Environmental risk assessment is fundamental to all phases of development for waste management facilities, from the strategic planning level through to the regulation of an individual facility. At the strategic level, risk assessment informs decisions about land use, and underpins assessment of the environmental impact associated with the site location considered through the development planning process. In the context of environmental regulation, risk assessment is used to enable the operator and the environment agencies to identify whether, and what risk management options, or mitigation measures, are required to adequately prevent, control, minimise and/or mitigate the identified risks to the environment from the facility. These measures are stipulated for waste management licences as licence conditions or in the working plan, and for PPC in permits or as conditions.

Quantitative risk assessments, using the "LandSim" probabilistic risk assessment tool, for example, have represented good practice now for many years for the assessment of hydrogeological risks from landfill facilities. The risk management measures subsequently addressed by specific licence conditions and working plan specifications are provided in the Environment Agency's library of licence conditions, working plan specifications and in the "shell licensing kits" that support them. The complexity of the measures required depend upon the type and magnitude of risks that the operations present to the environment. Risk management measures may be relatively simple, such as operational procedures requiring simple action and documentation, or more complex, such as engineered systems with fully documented and quality assured stages of design, construction, testing and validation, operation and maintenance.

Some waste management facilities, including landfill facilities, now fall within *the Environmental Permitting Regulations* (EPR) (E & W) or the *PPC Regulations* (Sc; NI) as prescribed by Schedule 1, section 5. Under the EU Landfill Directive, all landfills are now required to adhere to requirements regarding their design and operation, with waste management licences being replaced with new PPC permits that comply with the Directive's requirements. Schedule 10 to the EPR (E & W) and *2003 Landfill Regulations* (Sc; NI) (see 5.29.1), require operators to comply with the Landfill Directive and to demonstrate that necessary measures are taken to protect the environment and human health and to prevent accidents.

1.23.4 Contaminated Land

Risks from land contamination have historically been addressed on a suitable for use basis with most sites being assessed for their future use under the planning regime. With the introduction of Part IIA of the *Environmental Protection Act 1990* ("the contaminated land regime") in 2000, an increased awareness by regulators and industry of the risks posed by land based on its current use has developed.

Part IIA has made the role of risk assessment in contaminated land decision-making more explicit. It firmly establishes the role of the conceptual model and the source-pathway-receptor relationship. A tiered approach to risk assessment is used with defined stages and roles for risk screening, generic and detailed quantitative risk assessment as well as formalised options appraisal for risk management (in accordance with Green Leaves II).

Technical guidance is available to assist regulators and industry to assess risks to human health (Defra and Environment Agency, 2002), controlled waters (Environment Agency, 1999), and buildings (Environment Agency, 2001). In addition, more general guidance covers the development of soil and groundwater sampling strategies (Environment Agency, 2001) and the communication of risk (SNIFFER, 1999) – see Further Reading at the end of the *Pollution Control Handbook*.

Planning and redevelopment of sites affected by land contamination still represents the most cost-effective and beneficial way of dealing with such risks in the longer term. One of the likely benefits of Part IIA is that its explicit approach to the assessment of risks will raise the standards of such assessments under the planning regime, as its principles become more familiar among practitioners.

1.23.5 Radioactive Waste Performance Assessment

The Radioactive Substances Act 1993 (RSA 93) provides the framework for controlling the creation and disposal of radioactive wastes so as to protect the public from hazards that may arise from their disposal to the environment. Regulatory guidance for disposal facilities has been published under RSA93 – *Disposal Facilities on Land for Low and Intermediate Level Radioactive Wastes: Guidance on Requirements for Authorisation (1997)*. The guidance sets down two criteria for assessment of the radiological safety of radioactive waste disposal facilities:

- a dose limit for a facility's operational phase; and
- a post-closure radiological risk target.

Post-closure, a radiological risk target is considered an appropriate protection standard because of the uncertainties inherent in assessment of future performance of a disposal system. The assessed radiological risk from a facility to a representative member of the potentially exposed group at greatest risk should be consistent with a risk target of 10^{-6} per year. Radiological risk is the product of the probability that a given dose will be received and the probability that the dose will result in a serious health effect, summed over all situations that could give rise to exposure to the group.

If, for a chosen facility design, the assessed risk exceeds the risk target, the developer should show that the design is "optimised" and that the radiological risk has been reduced (benefit) to a level that represents a balance between radiological and other factors, including social and economic factors (cost). This is consistent with the "as low as reasonably achievable (ALARA)" approach. Where the risk is below the risk target and the regulator is satisfied that the safety case has a sound scientific and technical basis and that good engineering principles and practice are being applied, then no further reductions in risk need be sought.

A radiological risk assessment provided by the operator or developer is likely to be an important part of the post-closure safety case, although the relative importance of quantitative and qualitative arguments will change as uncertainties increase with the evolution of the disposal system over time. The regulatory guidance does not prescribe the form of risk assessment to be applied but notes that it may be based on probabilistic techniques. The regulatory guidance states that a risk assessment or any other technical assessment is unlikely to be sufficient on its own to provide a satisfactory demonstration of safety. Sufficient assurance of safety over the very long timescales that may need to be considered, possibly 100s of 1000s years, is likely to be achieved only through multiple and complementary lines of scientific reasoning. All the separate lines of reasoning that a developer might present in support of safety will, in different ways and to different degrees, inform the regulatory decision on the authorisation of the disposal facility.

1.23.6 Groundwater Regulations

Activities likely to lead to the direct or indirect discharge of List I or List II substances, as defined by the Groundwater Directive, require prior authorisation (see 6.11.3). Many activities authorised under the *Groundwater Regulations* are intermittent agricultural discharges or disposals, and due to the large numbers (over 11,000 as of 2002) of authorisations, a tiered system of risk assessment consistent with the "*Guidelines on Environmental Risk Assessment and Management*" above has been developed. This approach allows the Environment Agency to match the scale of the operation with the complexity of the assessment. A simple risk screening system is applied to the bulk of the applications. This risk screening system (tier 1) uses several indicators that are readily ascertained from the application forms and readily accessible national data sets to score the application and determine whether the application for discharge can be:

- approved subject to standard conditions;
- refused; or
- needs to be supported by further data or a more detailed analysis of the risk.

A second level risk assessment tool for land spreading (the majority of activities) has now been developed that relies on soil property data and a soil leaching algorithm. Other tools such as the Environment Agency's ConSim software package that takes account of processes within the unsaturated zone are available for more detailed assessment of point source disposals.

The close relationship between Codes of Good Practice and risk assessment is stressed. Many of the risk assessments are predicated on adherence to good practice, as noted in recognised codes, a number of which are now statutory. The assumption of good practice is reflected in standard conditions on authorisations and a system of site inspection to check for compliance to the terms of authorisations.

ACCESS TO ENVIRONMENTAL INFORMATION

The establishment of public registers under, among others, the *Environmental Protection Act 1990*, has made a wide range of environmental information held by enforcement authorities publicly available.

Information on the registers includes applications for (and actual) permits and consents, enforcement notices, monitoring results. Only matters of genuine commercial confidentiality and national security may remain secret. Among the regulatory areas covered by registers are Integrated Pollution Prevention and Control and local air pollution control, waste management, water pollution and radioactive substances. In addition, the *Environment and Safety Information Act 1988* provides for the maintenance of registers of notices relating to environmental and public safety matters issued under, for example, the *Health and Safety at Work Act 1974* and the *Food and Environment Protection Act 1985.*

1.24 UNECE CONVENTION

This Convention, the full name of which is the *Convention on Access to Information, Public Participation in Decision Making and Access to Justice in Environmental Matters*, was adopted in June 1998 in Aarhus, Denmark; it aims to strengthen the role of members of the public and environmental organisations in environmental protection issues, and provides for greater access to information held by public authorities. It came into force on 30 October 2001 and has now been ratified by 41 countries (including the UK in February 2005).

The Convention covers:

- **Access to information** held by public authorities and companies providing public services – the definition of environmental matters is rather wider than that in either the 1990 EU Directive or the UK Regulations, and includes issues which may form part of the decision making process connected with the environmental issue, such as land use planning, health and safety and economic issues; information on biological diversity and genetically modified organisms are also covered by the Convention. It covers environmental data held by public authorities at central, regional and local level. Criteria for withholding information are very tightly defined, but include information which could adversely affect international relations, security or defence; information should normally be released within one month of a request (two in complex cases) and there is a right of appeal in the event of refusal. Citizens can apply for information held by the authorities in their own or another country which is a party to the Convention. A "reasonable" charge may be made for supplying information; a schedule of charges should be published which also outlines when a charge will be made and those instances when payment in advance will be required.
- **Public participation** in decision making – the Convention lists projects where this would be obligatory; they include developments relating to waste management, energy and chemical industries and intensive farming. The public should be given the opportunity to comment either directly or through a representative body, and the results of such consultation taken into account as much as possible. A 2005 amendment to the Convention (which has yet to come into force) adds the deliberate release into the environment and placing on the market of GMOs to those areas on which the public has a right to participation in the decision-making procedure.
- **Access to justice** – the Convention aims to make it easier for people to take action through the courts or other public authorities if they feel that national environmental law is being contravened. "Adequate and effective remedies" must be available if the case is proved.

In May 2003, the Kiev Protocol on Pollutant Release and Transfer Registers was adopted and signed by 36 states (including the UK) and the European Community. Its objective is "to enhance public access to information through the establishment of coherent, nationwide pollutant release and transfer registers". The Protocol will enter into force 90 days after the 16th ratification; as at 1 July 2008, there were nine ratifications and 38 signatories. See also chapter 2, 2.1.3.

1.25 EU DIRECTIVES

To enable its ratification of the Aarhus Convention, the European Community has adopted two relevant Directives: 2003/4/EC which is a new Directive on public access to environmental information, and 2003/35/EC which provides for public participation in the preparation of environmental plans, programmes and projects with significant environmental impacts. A further proposal for a Directive on access to justice in environmental matters (COM(2003) 624), published in October 2003, will implement the remaining aspects of the Convention. This will enable members of the public or "qualified entities" (e.g. certain environmental protection organisations) to challenge in the courts perceived "administrative acts or omissions" of EU laws or national legislation implementing such laws.

Regulation 1367/2006 of September 2006 applies the provisions of the Aarhus Convention to EU institutions and other bodies. Decision 2005/370/EC of 17 February 2005 ratifies the Convention on behalf of the EU.

1.25.1 Access to Information

Directive 2003/4/EC *on public access to environmental information* came into force on 28 January 2003 and had to be implemented by 14 February 2005, the date on which the original Directive (90/313/EEC) was repealed. The 2003 Directive updates the 1990 Directive to take account of developments in information technology and to specify more clearly information access procedures and enabled the Community to ratify the UNECE Convention on Access to Environmental Information.

Regulation 1367/2006 of 6 September 2006 places an obligation on EU institutions and bodies to comply with the rules of the Aarhus Convention, including those regarding public participation in decision-making; in this latter regard, a time limit of at least eight weeks must be set for receiving comments from persons or organisations with an interest in EU environmental projects; at least four weeks' notice of hearings or meetings must be given. The Regulation entered into force on 28 September 2006 and applies from 28 June 2007.

The definition of "environmental information" includes information in any form on the state of the environment; on factors, measures or activities affecting or likely to affect the environment and on those designed to protect it, on emissions, discharges and other releases into the environment including waste and radioactive waste, on the cost benefit and economic analyses used within the framework of such measures or activities; and on the state of human health and safety. As well as national, local and regional government and public administrations, the definition of public authorities, to whom the Directive applies, includes bodies carrying out functions or providing services directly or indirectly related to the environment; thus the Directive may well cover, for example, private companies providing water supply or refuse collection services. The Directive also covers environmental information held on behalf of public authorities; information should be made available as soon as possible or at the latest within one month of a request being made, two months for particularly complex requests.

Information should only be withheld in those circumstances where the benefits of non-disclosure, or protection of the public interest, clearly outweigh the public interest in the information. Information held by public authorities on emissions, discharges and other releases to the environment which relate to Community legislation may not be withheld for reasons of commercial or industrial confidentiality – this restriction only applies to situations covered by national legislation; personal data is also exempt. Where a request for information has been directed to the wrong authority, the authority should wherever possible refer it to the correct body. Authorities will be allowed to charge a "reasonable" amount for supplying information but may not make payment in advance a condition for releasing information; a schedule of charges should be published. Where a request for information is refused, the authority must give its reasons in writing; where access is refused, there should be a right of appeal either through the courts or a body set up to deal with appeals.

1.25.2 Public Participation

Directive 2003/35/EC, formally adopted on 26 May 2003 provides for the public to be given early and effective opportunities to participate in the preparation and modification or review of environmental plans, programmes and projects with significant environmental impacts, thus enabling the ratification of the Aarhus Convention by the Community. The Directive has been implemented in the UK through the amendment of relevant Regulations to ensure the public are given early and effective opportunity to participate.

1.26 UK FREEDOM OF INFORMATION ACTS

The Freedom of Information Act 2000, received Royal Assent on 30 November 2000 and has been progressively brought into force, with full implementation from 1 January 2005. It extends the right to official information on environmental (and other) matters to that held by a wide range of public authorities at both central and local government levels; it applies in England and Wales and in Northern Ireland. In Scotland this Act applies only to those public authorities not carrying out devolved functions as defined by the *Scotland Act 1998*. The Act supersedes the 1994 Code of Practice on Access to Government Information under which members of the public could apply for access to information held by public bodies answerable to the Parliamentary Ombudsman.

The Freedom of Information (Scotland) Act 2002, which was passed on 28 May 2002, provides a general right to information held by Scottish public authorities, which are listed in a schedule to the Act. The Scottish Information Commissioner is responsible for supervising the environmental information regime, and for determining appeals relating to the failure or refusal to provide information. Regulations enabling Scotland to comply with the Aarhus Convention have been made under s.62 of the Act.

Schedule 1 to the 2000 Act lists the main public authorities to which the Act applies, and includes both the House of Commons and the House of Lords, the National Assembly of

Wales, courts and tribunals, regulatory bodies, the police, educational and health bodies, conservation bodies etc. Schedule 1 has been added to by *the Freedom of Information (Additional Public Authorities) Order* – 2005 (SI 2005/3593) and 2008 (SI 2008/1271); *the Freedom of Information (Removal of References to Public Authorities) Order 2005* (SI 2005/3594) removes a number of bodies from the Schedule.

An application for information should normally be dealt with "promptly" and in any case within 20 working days from its receipt. Authorities may charge a fee (to be set in Regulations, but possibly up to 10% of the marginal costs of dealing with the request); applicants will have up to three months to pay the requested fee and authorities need not supply any information until the fee has been paid.

The Act requires public authorities to set out in a "Publication Scheme" the types of information that they will make routinely available, how this information can be accessed, and whether a charge is likely to be made. The Publication Schemes, which have to be submitted to the Information Commissioner for approval, were brought into effect on a rolling programme between 2002-2004.

While there is a presumption towards granting access, in deciding whether to accede to a request, the authority will consider whether the public interest in disclosure outweighs that of withholding; exemptions include such issues as national security and defence, international relations, commercial confidentiality, investigations into possible criminal offences, accident investigations, or documents relating to the "formulation or development of government policy".

The Information Commissioner appointed under the Act has powers to determine appeals where access to information has been refused and to order the release of information; decisions relating to what is or is not in the public interest will, however, continue to be taken by Ministers.

1.26.1 Environmental Information Regulations (E, W, NI)

The Environmental Information Regulations 2004 (SI 2004/3391), which cover England, Wales and Northern Ireland, came into force on 1 January 2005. They are in accord with the FOI Act 2000, replacing the *Environmental Information Regulations 1992* (SI 1992/3240; SR 93/45 & amendments). Separate Regulations have been made in Scotland – see below. The Regulations have enabled the UK to ratify the 1998 UNECE Convention and implement the 2003 EU Directive on public access to environmental information (1.24 & 25 above).

The Regulations for E, W & NI include the following measures:

- 'Public authority' includes any minister of the crown, government department or local authority, any person that is from time to time a 'public authority' within the meaning of the FOI Act, is carrying out functions of public administration in relation to the environment, and any person with public responsibilities or functions of public administration in relation to the environment. An organisation or person contracted to carry out a particular piece of research/review on behalf of a public authority would also be covered by the EIR Regulations in respect of functions relating to the contracted work (Decision Notice from Information Commissioner, 2006).
- Public authorities have a duty to make available environmental information on request and to provide advice and assistance; if a request for information is unclear, the authority should ask the applicant to clarify the request and should assist the applicant in doing so.
- The time-limit for dealing with most requests for information will be 20 working days (40 working days for particularly complex requests or requests involving large amounts of material); *the Freedom of Information (Time for Compliance with Request) Regulations 2004* (SI 2004/3364), enable the period for compliance to be extended to no more than 60 working days in specified circumstances.
- A request may be refused if it is thought that disclosure would adversely affect international relations, defence, national security or public safety; justice or disciplinary reasons; confidentiality of commercial or industrial information; the interests of the person who provided the information; the protection of the environment to which the information relates. However even in these cases information can only be withheld if the public interest in doing so clearly outweighs the public interest in disclosure.
- Where a request for access to information is refused, the applicant should be informed in writing, together with the reasons for refusal, no later than 20 days after the date of receipt of the request.
- A charge may be made in respect of the costs of supplying the information, with supply of the information made conditional on payment of the charge; *the Freedom of Information and Data Protection (Appropriate Limit and Fees) Regulations 2004* (SI 2004/3244) prescribe certain fees and costs to be taken into account.

The definition of ***environmental information*** follows that in the Aarhus Convention, and includes:

- The state of the elements of the environment, such as: air and atmosphere; water; soil, land, landscape, flora and fauna (including human beings); natural sites, biological diversity and its components, GMOs;
- Factors such as substances, energy, noise, radiation or waste, likely to affect the above;
- Emissions, discharges and other releases into the environment;
- Measures, such as policies, legislation, plans and programmes, affecting or likely to affect, or intended to protect the environment; cost-benefit and other economic analyses and assumptions used in environmental decision-making;
- The state of human health and safety, conditions of human life, cultural sites and built structures, inasmuch as they are or may be affected by or through the state of the elements of the environment.

Section 48 of the FOI Act outlines the powers and functions of the Information Commissioner who is also charged with promoting good practice in the way the Regulations are complied with. If the Commissioner is of the opinion that a public authority is not following good practice, he may issue a "practice recommendation" (s.48 of the Act) specifying the steps to be taken. The Commissioner also deals with complaints relating to refusal to supply information or that a request has not been dealt with in accordance with the Regulations and may issue an "information notice" requesting further details of the matter under complaint. The Commissioner will notify both the complainant and the public authority of his decision through a "decision notice" – where a complaint is upheld the notice will outline the steps to be taken to comply with the notice; the notice will also outline the right of appeal to the Information Tribunal. Where a public authority has failed to comply with the Regulations, the Information Commissioner may issue an enforcement notice detailing the steps to be taken to remedy the matter, and noting the right of appeal.

Both complainants and public authorities have a right of appeal to the Information Tribunal in respect of decision notices; public authorities also have a right of appeal with regard to both information and enforcement notices. There is a further right of appeal with regard to the Tribunal's decision to the High Court (Scotland: Court of Session).

Defra has published guidance on the Regulations, and on who is covered by the Regulations; a Code of Practice provides guidance to public authorities as to the practice that it would be desirable for them to follow in discharging their functions under the Regulations.

1.26.2 Environmental Information Regulations (Scotland)

The Environmental Information (Scotland) Regulations 2004 (SSI 2004/520) – made under s.62 of the Scottish Act also came into force on 1 January 2005 and are similar to those applying in the rest of the UK. Information may only be withheld if it would "substantially prejudice" the interests of the body concerned. The Scottish Information Commissioner is also responsible for determining appeals for the failure/refusal to provide environmental information on request.

The Scottish Government has published guidance to accompany the Regulations and a code of practice for public authorities to follow in connection with the discharge of their functions under the legislation.

ENVIRONMENTAL MANAGEMENT & AUDIT SYSTEMS

A responsible attitude towards the environment is now recognised by many businesses – industrial, manufacturing or service – to be not only good for the environment but also good for their business in terms of how they are viewed by their customers.

The Government's Position Statement on environmental management systems (Defra, 2005) says that a formal EMS can help businesses – in both the public and private sector – to "both understand and describe their organisation's impacts on the environment, to manage these in a credible way and to evaluate and improve their performance in a verifiable way. Properly implemented EMSs will help with managing risks, liabilities and legal compliance".

1.27 ENVIRONMENTAL MANAGEMENT

ISO 14001 and supporting documents, developed by the International Standards Organisation, provide guidance on implementing an environmental management system. They provide general guidelines on principles, systems and supporting techniques and outline the requirements for companies wishing to certify to the standard. A number of other documents in the series outline standards for environmental auditing. These standards have been formally recognised by the Commission with the amendment to the 1993 EU Regulation setting requirements compatible with ISO 14001 (see below). ISO 14001 is itself under revision; issues under discussion include a mandatory requirement for environmental reports to be made publicly available, as is the case under EMAS, clarification of what is meant by continuous environmental improvement and a more explicit requirement to comply with environmental legislation.

Companies or organisations wishing to be certified under ISO must ensure that their environmental management system takes account of the following:

- the development of an environmental policy;
- a commitment to comply with relevant environmental legislation and regulations;
- products or services are produced, delivered and disposed of in an environmentally friendly manner – and thus any adverse effects on the environment are minimised;
- expenditure on environmental protection is timely and effective, and that planning for future investment and growth reflects market needs on the environment – that best available technology is used where appropriate and economically viable;
- the principles of sustainable development;
- objectives and targets aimed at continuous environmental improvement; evaluation procedures; staff training and awareness.

Where the company has identified a "significant environmental effect" this should be entered on a register. All participating companies should carry out an environmental audit at least every three years or annually if there is "particular potential to cause environmental harm".

Certification for compliance with the standard is voluntary and may cover all or part of a company's sites or operations. In the UK certification is carried out by external assessors ("certifiers") who are formally accredited under a scheme run by the UK Accreditation Service. UKAS, which is also responsible for accrediting verifiers for the environmental statements required under the Eco-Management and Audit Regulation (see below), is the national accreditation body for the UK. It was formed as a result of amalgamation of the National Certification Accreditation Board (NACCB) and the National Measurement Accreditation Service (NAMAS).

British Standard BS 8555 (2004) – Steps to Environmental Management Systems – has been designed for small to medium sized enterprises enabling them to implement an EMS by means of a phased approach. There are six key phases:

- Commitment and establishing the baseline;
- Identifying and ensuring compliance with legal and other requirements;
- Developing objectives, targets and programmes;

- Implementation and operation of the EMS;
- Checking, audit and management review;
- ISO 14001 or EMAS registration.

The Institute of Environmental Management and Assessment "Acorn" inspection scheme enables organisations to gain accredited inspection and recognition for their achievements at each stage, with the final stage being formal registration to ISO 14001 or EMAS. More details on www.iema.net/acorn

1.28 ENVIRONMENTAL AUDITING

The European Union's first Council Regulation (1836/93) allowing Voluntary Participation by Companies in the Industry Sector in a Community Eco-Management and Audit Scheme (EMAS) came into force in May 1995; its aim was to encourage industrial companies to continually evaluate and improve their environmental performance. An amendment Regulation (761/2001) adopted in March 2001 (and which came into force on 27 April 2001), updates and replaces the 1993 Regulation to, among other things, make it compatible with ISO 14001 (the text of ISO 14001 is included in the Regulation); certification to ISO 14001 will go some way towards satisfying the requirements of EMAS. The scheme has been extended to cover non-industrial companies and organisations, including SMEs (e.g. agricultural and financial sectors) whose activities can have a direct or indirect effect on the environment. Organisations may still register one or all their sites; where more than one is registered, each will have to comply with the Regulation; organisations operating in a number of Member States will have to register in each country. The revised Regulation also strengthens requirements on those carrying out environmental audits. A Recommendation detailing guidelines for implementing the Regulation was adopted by the European Commission in July 2003. A proposal for a new Regulation was published by the European Commission in mid-2008 – see below.

Companies satisfying the requirements of the Regulation and registering under the Scheme can use a "statement of participation". This cannot be used on actual products or packaging, or to advertise the product. The Regulation has, however, introduced a logo as a means of promoting EMAS more widely and organisations can use this in advertising their products, services and activities and on their letterhead.

The Institute of Environmental Management and Assessment is the UK Competent Authority; companies may choose whether or not to participate in the scheme and indeed whether to register all or only some of their sites. Companies registering under EMAS are required to carry out an assessment of operations at each site and the impact on the environment; this forms the basis of an environmental protection system (including an environmental policy, environmental management system and audit programme). Issues to be covered include energy policy, waste and water management, product planning, safety, staff training and involvement in environmental issues, information to be made public and complaints handling; objectives and targets should be set with a view to "reasonable continuous improvement of environmental performance" and reducing environmental impacts. Continual improvement is defined as "the process of enhancing, year by year, the measurable results of the EMS related to an organisation's management of its significant environmental impacts, based on its environmental policy, objectives and targets. . ." An environmental audit must be carried out not less than every three years and the resulting environmental statement verified; in the intervening years an environmental statement should be published. The Regulation requires verification of the annual statements if it reports any changes from the 3-year statement.

Accredited verifiers ensure compliance with all aspects of the Regulation and validate the environmental statement if satisfied with the way the company has complied with the Regulation. The verifier and the auditor should be independent of each other.

The UK Accreditation Service is responsible for accrediting verifiers for the environmental statements.

Proposed EU Regulation

COM(2008) 402, published by the European Commission in mid-2008, is a Proposal for a Regulation on the voluntary participation by organisations in a Community eco-management and audit scheme. Following adoption the 1993 Regulation and subsequent amendments will all be repealed. The objective of the new proposal is to strengthen EMAS so that it is recognised as a benchmark for environmental management systems, thus encouraging more organisations to apply for registration. Thus the main modifications proposed are:

- Organisations would need to comply with environmental legislation prior to initial EMAS registration.
- Mandatory environmental reporting using core performance indicators to be specified in the Regulation.
- The Commission is to develop sectoral guidance on best practice in environmental management; EMAS verifiers would be required to have regard to the guidance, and organisations seeking EMAS registration would be encouraged to use the guidance for setting up their environmental management system and defining their environmental targets.
- Rules and procedures for accreditation and verification are to be harmonised throughout the EU.
- The scheme is to be opened up to organisations outside the EU who would be able to register in a member state.

LOCAL ENVIRONMENTAL QUALITY

The quality of people's local environment and the protection of health has been recognised through statute for many centuries – indeed King Edward I firmly believed that smoke affected his health and issued proclamations forbidding the use of coal while Parliament was sitting, and it is related that a man was actually hanged in the 14th century for causing a smoke nuisance!

There are three types of nuisance: statutory nuisance where a particular nuisance has been made so by statute; and public and private nuisance which are within Common Law.

Current legislation for statutory nuisance in England, Wales and Scotland, is contained in Part III of *the Environmental Protection Act 1990*, as amended, and enforced by local authorities, with controls on various types of noise nuisance regulated under *the Control of Pollution Act 1974* (see chapter 4). In England and Wales, *the Anti-Social Behaviour Act 2003* provides local authorities and other enforcement authorities with additional powers to deal with various aspects of anti-social – or nuisance – behaviour, including fly-tipping, abandoned vehicles, spray painting, noisy premises, graffiti and fly-posting. There is similar (2004) legislation in Scotland; this also increases the fine for certain noise and nuisance offences to £40,000 – see 1.33 below.

The Noise Act 1996 (as amended) enables action to be taken in respect of noise nuisance from dwellings between 11.00 pm and 7.00 am in England and Wales, with the Scottish *Anti-Social Behaviour Act 2004* including a similar provisions (see chapter 4, 4.9.4 and 4.10).

The Clean Neighbourhoods and Environment Act 2005 (which also covers E & W and received Royal Assent in April 2005) includes measures to tackle a range of environmental crimes and anti-social behaviour – see below, 1.34 for a summary of those parts of the Act which affect local environmental quality. The Act also amends the EPA 1990 to make nuisance from insects and from artificial light statutory nuisances. Similar provision for Scotland is made by sections 109 and 118 of *the Public Health etc. (Scotland) Act 2008* which are expected to be brought into force in early 2009.

In Northern Ireland, statutory nuisances are regulated under *the Public Health (Ireland) Act 1878*, as amended (which also includes powers for local authorities to serve abatement notices; noise nuisance is regulated under *the Pollution Control and Local Government (Northern Ireland) Order 1978* – see below 1.30.

1.29 STATUTORY NUISANCE (E, W, Sc)

Part III of *the Environmental Protection Act 1990* (EPA) was brought into force in England and Wales from 1 January 1991, and in Scotland from 1 April 1996 as a result of implementation of s.107 and Schedule 17 of the *Environment Act 1995* which extended the relevant provisions of the EPA to Scotland.

Section 79(1) of the EPA, as amended, defines the following statutory nuisances:

a) any premises in such a state as to be prejudicial to health or a nuisance;
b) smoke emitted from premises so as to be prejudicial to health or a nuisance;
c) fumes or gases emitted from premises so as to be prejudicial to health or a nuisance (from private dwellings only);
d) any dust, steam, smell or other effluvia arising on industrial, trade or business premises and being prejudicial to health or a nuisance;
e) any accumulation or deposit which is prejudicial to health or a nuisance;
ea) any water covering land or land covered with water which is in such a state as to be prejudicial to health or a nuisance; (Sc – see below)
f) any animal kept in such a place or manner as to be prejudicial to health or a nuisance;
fa) any insects emanating from relevant industrial, trade or business premises and being prejudicial to health or a nuisance; (E & W – see below)
faa) any insects emanating from premises and being prejudicial to health or a nuisance; (Sc – see below))
fb) artificial light emitted from premises so as to be prejudicial to health or a nuisance; (E & W – see below)
fba) artificial light emitted from (i) premises, (ii) any stationary object [including a stationary object located on premises], so as to be prejudicial to health or a nuisance; (Sc – see below)
g) noise emitted from premises so as to be prejudicial to health or a nuisance;
ga) noise that is prejudicial to health or a nuisance and is emitted from or caused by a vehicle, machinery or equipment in a street or in Scotland, road;
h) any other matter declared by any enactment to be a statutory nuisance. (See, for example, s.259, *Public Health Act 1936* regarding ponds and watercourses – *Pollution Control Handbook*, chapter 6, 6.7.1)

Where an installation is regulated under the EPA, a local authority may only take action for nuisance defined in (b), (d) or (e) with the consent of the Secretary of State if proceedings can be brought under the EPA Part I (s.79(10); *the Pollution Prevention and Control Act 1999*, (Sch.2, para 6), extends this to Regulations made under the PPC Act, with the PPC Regulations (sch.10, para 7) adding paras (g) and (ga) above, except for Part B. Nuisance action may however be taken for activities not covered by the Regulations even though they are on the site of a regulated installation.

Action for smoke emissions can be taken under the *Clean Air Act 1993* (CAA) in certain instances and cannot therefore be dealt with as a statutory nuisance under the EPA. These instances are:

- smoke emitted from a chimney of a private dwelling in a smoke control area (s.20 of CAA);
- dark smoke from a chimney of a building or a chimney serving the furnace of a boiler or industrial plant attached to a building or installed on any land (s.1 of CAA);
- smoke from a railway locomotive steam engine (s.43 of CAA);
- dark smoke caused by industrial or trade burning (s.2 of CAA).

Also excluded from statutory nuisance action are dust, steam, smell etc (d above) from a railway locomotive steam engine and noise (g above) from aircraft other than model aircraft. Crown defence premises and those occupied by visiting forces are excluded from action under (b)-(g) above. Statutory nuisance legislation cannot be used to take action against land which is in a contaminated state (EPA, s.79(1A), *Env. Act 1995*, Sch.22, para 89, effective 1 April 2000).

It should be noted that the EPA interprets "premises" as including "land". Equipment (ga above) includes musical instruments. Guidance from the Department of the Environment, Transport and the Regions (Env. Circular 9/97, WO 42/97) suggests that loudspeaker tannoys, loudhailers, radios and "ghetto-blasters" should also be regarded as equipment.

There is a defence (ss.80(8) & 82(10) of the EPA) of Best Practicable Means in respect of nuisance from industrial, trade or business premises. This is interpreted as practicable given local conditions, technical knowledge and financial implications; means includes design, installation and maintenance of plant, buildings and structure and periods of operation.

England and Wales

Sub-sections fa and fb were added by ss.101-103 of *the Clean Neighbourhoods and Environment Act 2005*, which covers England and Wales, and in England were brought into force on 6 April 2006 and in Wales on 31 January 2007. Land used for agricultural, arable, market gardening etc is excluded from the definition of premises so far as nuisance from insects (s.79(1)(fa)) is concerned, as is land included in an SSSI; *the Statutory Nuisance (Insects) Regulations 2006* (SI 2006/770) – and in Wales *the Statutory Nuisances (Miscellaneous Provisions) (Wales) Regulations 2007* (SI 2007/117, W.8) – prescribe that land in respect of which payments are made under land management schemes listed in a schedule to the Regulations is also excluded. Nuisance lighting (s.79(1)(fb)) excludes that from airports, harbours, railway and tramway premises, bus stations, public service vehicle and good vehicle operating centres, lighthouses, prisons and premises occupied for defence purposes.

Guidance on the two new statutory nuisances has been included in Defra's Guidance on *the Clean Neighbourhoods and Environment Act 2005*, and can be downloaded from Defra's website; similar guidance is available on the Welsh Assembly Government's website. Further sources of information on both the legislation on nuisance from artificial light, and ways of avoiding a nuisance are listed in the "Further Reading" section of the *Pollution Control Handbook*.

Section 103 of the CN&EA extends the defence of use of "best practicable means" to cover statutory nuisance from insects and from artificial light from industrial, trade or business premises and in the case of light from relevant outdoor sports facilities (which are not also businesses, e.g. local authority sports grounds); *the Statutory Nuisances (Artificial Lighting) (Designation of Relevant Sports) (England) Order 2006* (SI 2006/781) – in, Wales *the Statutory Nuisances (Miscellaneous Provisions) (Wales) Order 2007* (SI 2007/120, W.10) – list the sports, the facilities for which will be able to use the defence of BPM in any action for statutory nuisance.

Steam railway locomotives and recreational steam vessels are exempt from nuisance action in respect of steam and smoke; steam road vehicles (such as traction engines and steamrollers) used on a public highway are also exempt.

So far as action to deal with bonfire smoke is concerned, in certain instances it would be possible to use the *Highways (Amendment) Act 1986*: this makes it an offence to light a fire, permit or direct one to be lit, the smoke from which injures, interrupts or endangers anyone using a highway.

Scotland

Part 9 of *the Public Health etc. (Scotland) Act 2008* – to be brought into force by Order (expected early 2009) – amends s.79(1) of the EPA to add insect nuisance (faa), artificial light nuisance (fba) and any water covering land or land covered with water (ea) to the list of statutory nuisances; the Act further amends s.79(1) to enable Scottish Ministers to add to the list of statutory nuisances, or vary their description by Regulations; before making such Regulations Scottish Ministers should endeavour to consult relevant associations of local authorities and such other persons as they consider appropriate. Section 80 of the EPA is amended to allow local authorities to issue a fixed penalty notice offering the person who they believe has committed an offence to discharge their liability to conviction through payment of a fixed penalty.

The amendments do not apply to insect nuisance caused by "insects that are wild animals" listed in Sch. 5 to *the Wildlife and Countryside Act 1981*, and premises do not include SSSIs or other places that Scottish Ministers may prescribe. Artificial light emitted from lighthouses is excluded from statutory nuisance. So far as the statutory nuisance of any water covering land or land covered by water is concerned, land includes structures (other than buildings) in, on or over land, but not mains or pipes for carrying a water supply; any part of the public sewerage system; any other sewers, drains or other pipes, for carrying sewage; the foreshore (i.e. the land between high and low water marks); or the seabed.

1.29.1 Action by Local Authorities

Local authorities have a duty to inspect their areas from time to time to detect whether a nuisance exists or is likely to occur or recur (s.79(1) of EPA). (Port Health Authorities have the same powers to take action for nuisance, except for noise.) An authority must also take such steps as are reasonably practicable to investigate any complaint of statutory nuisance from a person living in its area. Section 81(2) enables a local authority to take action against a statutory nuisance outside its area if it appears to be affecting any part of its area.

Where the local authority is satisfied that a statutory nuisance exists, or is likely to occur or recur, it must serve an abatement notice on the person responsible for the nuisance or if that person cannot be found, on the owner or occupier of the premises (s.80(1)). If the nuisance is as a result of a structural problem, then the notice should be served on the owner of the premises. The *Noise and Statutory Nuisance Act 1993* amends the procedure for serving an abatement notice in respect of a nuisance from or caused by an unattended vehicle, machinery or equipment (VME) on the street to enable the notice to be affixed directly to it (see chapter 4, 4.11.1). Section 86 of *the Clean Neighbourhoods & Environment Act 2005* (brought into force 6 April 2006 in England), amends s.80 to enable a local authority in E & W to defer the issuing of an abatement notice in respect of noise for seven days while efforts are made to persuade the person to abate the nuisance.

The notice should impose all or any of the following requirements (s.80(1)):

a) the abatement of the nuisance or prohibiting or restricting its occurrence or recurrence;
b) the carrying out of such works and other steps necessary to abate the nuisance.

The notice should specify the time or times within which the requirements of the notice must be complied with; it should also state that there is a right of appeal and that any appeal against a notice should be made within 21 days to the Magistrates' Court (Sheriff in Scotland) – see below.

Failure to comply with the terms of an abatement notice without reasonable excuse may result in prosecution in the Magistrates' Court (Sheriff in Scotland) (s.80(4)). Conviction may result in a maximum fine of £5,000, plus a daily fine of £500 for each day on which the offence continues after conviction. Where the conviction is for noise from industrial, trade or business premises, the maximum fine is £20,000 (Sc: £40,000, *ASB Act 2004*).

If an abatement notice is not complied with, the local authority may take the necessary steps to abate the nuisance itself (s.81(3)). In the case of action to abate a noise nuisance (EPA, s.79(1)(g)), this can include seizure of the noise-making equipment – s.10(7) and Sch.1 of the *Noise Act 1996*; the procedure is set out in chapter 4, 4.12.2. (Note – the *Noise Act* does not apply in Scotland, but similar provisions are introduced through the Scottish ASB Act – see 4.10.)

Any expenses reasonably incurred by a local authority in England and Wales in abating or preventing the recurrence of a statutory nuisance may be recovered by them from the person responsible for the nuisance (s.81(4)). The *Noise and Statutory Nuisance Act 1993* added s.81A to the EPA enabling local authorities in England and Wales to charge interest from the date of serving the payment notice until full payment is received. If no payment has been received after 21 days, then the expenses plus accrued interest become a legal charge on the property. The charge on the property remains until the debt has been paid in full; if the property is sold, then the charge passes to the new owners. The person on whom the notice is served may appeal to the county court within 21 days of its service. Section 81B of the EPA enables local authorities in England and Wales to require the debt to be paid in instalments over a period of up to 30 years.

Under *The **London** Local Authorities Act 2004*, authorised officers may issue a fixed penalty notice for contravention or failure to comply with an abatement notice. Offenders have 14 days in which to pay the fixed penalty fine, during which time no other prosecution action may be taken for that offence.

Future Development

The Primary Authority Scheme (Definition of Enforcement Action) Order, a draft of which was issued for consultation by the Department of Business, Enterprise and Regulatory Reform in September 2008, will require a local authority to notify the "Primary Authority" in certain cases before serving an abatement notice under s.80 of the EPA. This would apply when a local authority, which has not been designated a Primary Authority is planning to serve an abatement notice on a business or other organisation that operates both in its area and in the area of other local authorities – for more details of the Primary Authority Scheme, see earlier this chapter, 1.4.1. The consultation paper does, however, suggest that a local authority would not need to notify the Primary Authority where "urgent action needs to be taken, in relation to relatively minor harm" – this it suggests would include responding to cases of nuisance, such as silencing an intruder alarm.

1.29.2 Action by Individuals

Action to abate a statutory nuisance may also be taken by an individual through the Magistrates' Court (Sheriff in Scotland) (EPA, s.82). The nuisance must exist or be likely to recur. Prior to taking action, the complainant must notify the alleged nuisance-maker that it is intended to take court action.

If proved, the court will serve an order which

a) requires abatement of the nuisance within a specified time, and the carrying out of any necessary works; and/or
b) prohibits a recurrence of the nuisance and requires the carrying out of any necessary works within a specified time to prevent a recurrence.

Again, failure to comply with an order without reasonable excuse is an offence and results in a fine, plus a daily fine for each day on which the offence continues after conviction.

1.29.3 Statutory Nuisance Appeals

An appeal against an abatement notice served under ss.80 and 80A of the *Environmental Protection Act 1990* (as amended by the *Noise and Statutory Nuisance Act 1993*) should be lodged with the Magistrates' Court within 21 days of service of the notice.

The Statutory Nuisance (Appeals) Regulations 1995 (SI 2644), which apply in England and Wales, came into force on 8 November 1995, replacing 1990 Regulations; the 1995 Regulations are very similar to the 1990 Regulations, their main effect being to add further grounds for appeal against an abatement notice. The Regulations also outline the procedure to be followed when the appellant claims the notice should have been served on another person, and the action which the court may take to give effect to its decisions. Similar Regulations – the *Statutory Nuisance (Appeals) (Scotland) Regulations 1996* (SI 1076, S.166) – came into force on 2 May 1996 in Scotland, with 1983 Regulations covering appeals in respect of noise.

Grounds for appeal include:

- unreasonable amount of time given for complying with a notice, or unreasonable requirements;
- an error or other irregularity in the notice;
- that it could or should have been served on someone else; in this instance the appellant should give a copy of the appeals notice to any other person referred to in it;
- in the case of a notice served on a trade or business premises that the best practicable means was used to prevent or counteract the effects of the nuisance. This can be used in defence of the following nuisances: premises in such a state as to be prejudicial to health etc; smoke emitted from a chimney; dust, steam, smell or other effluvia; the manner or place in which animals are being kept; noise; and noise from or caused by vehicles, machinery or equipment.

An appeal may also be lodged if it is felt that the requirements of an abatement notice in respect of noise from premises or VMEs are more onerous than the terms of a valid consent (e.g. for construction work or for the use of loudspeakers).

2006 Amendment Regulations covering England (SI 2006/771), which came into force on 6 April 2006 enable a defence of best practicable means to be used in an appeal against an abatement notice in respect of nuisance from insects emanating from relevant industrial, trade or business premises or artificial light from industrial, trade or business premises or that used for illuminating an outdoor relevant sports facility. Insofar as insects are concerned (CN&EA, s.101(5)) relevant land includes land used as arable, grazing, meadow or pasture land, that used as osier land, reed beds or woodland, for market gardens, nursery grounds or orchards, land included in an SSI, and covered by, and the waters of, rivers and watercourses, and land forming part of an agricultural unit; *the Statutory Nuisance (Insects) Regulations 2006* (SI 2006/770) prescribe those types of land which can be considered to form part of an agricultural unit. "Relevant" sports facilities are listed in *the Statutory Nuisances (Artificial Lighting) (Designation of Relevant Sports) (England) Order 2006* (SI 2006/781). Similar provisions cover Wales – SI 2007/117. W.8 and SI 2007/120, W.10.

Where an appeal has been lodged, the notice will be suspended pending determination of the appeal if compliance would result in expenditure or the noise was caused due to carrying out a task imposed by law, except

- where there is a risk of injury to health;
- the cost of carrying out the work required by the notice is not considered disproportionate to the expected benefit;
- there is no point in suspending the notice because of its limited duration.

The Court also has powers to dismiss an appeal, or quash or vary the terms of an abatement notice.

1.30 NORTHERN IRELAND

The Public Health (Ireland) Act 1878 gives local councils responsibility for monitoring their areas for odour nuisance, and to investigate complaints from the public, and may serve an abatement notice. Subsequent Acts have added to the range of statutory nuisances, including *the Pollution Control and Local Government (Northern Ireland) Order 1978*. This latter also details local authority powers for controlling noise nuisance which are similar to procedures in England, Scotland and Wales under Part III of the EPA and COPA 1974.

There is no legislation on nuisance from lighting or insects in Northern Ireland.

1.31 PUBLIC NUISANCE

A public nuisance is both a tort (civil wrong) and a crime punishable by law, whereas private nuisance is a tort, a wrong for which there is remedy by compensation or damages. Where nuisance is alleged, the court will consider a variety of elements including the nature of the facts complained of, the location, and whether or not the complainant has lost the right to complain through acquiescence. The harm, inconvenience or discomfort suffered by the complainant must be material, and continuing or recurrent; the courts will also take into account the nature of the locality: what is acceptable in an area of substantial industrial activity may not be acceptable in the heart of a residential area; and indeed what is acceptable in a rural area may not be acceptable in an urban situation. Public nuisance takes into account primarily the number of persons affected. The nuisance should affect "all persons who come within the sphere of its operation".

In cases of public nuisance, the Attorney General or other enforcing bodies including local authorities' environmental health departments can make criminal prosecution; the Attorney General can institute civil proceedings alone or at the request of a local authority or individual. Section 222 of the *Local Government Act 1972*, provides that a local authority "in the case of civil proceedings may institute them in their own name". Local authorities may also use section 222 to apply for an injunction to stop anti-social behaviour that can be shown to be a public nuisance.

1.32 PRIVATE NUISANCE

Private nuisance actions are more common than public nuisance cases and rely on damage or unreasonable interference with an individual's right to use and enjoy their property. Action may be brought by the aggrieved person in the civil court (either the County or the High Court). Only the occupier of the land or person having a requisite interest is entitled to sue. This may include owner-occupiers and tenants but not relatives or visitors even if living with the occupier.

In any proceedings, the person causing the nuisance will be responsible whether or not the nuisance arises from the property he/she occupies, provided that it affects the plaintiff's property. Ownership of land does not necessarily mean that the owner will be held responsible for the actions of others who may cause nuisance. However, the occupier of property may be liable for nuisance even if he/she did not cause it, if he/she allows it to continue during his/her occupation. Remedial action is principally an injunction (a court order restraining the convicted person from committing or continuing the act or omission complained about) and damages.

Eight important principles have been established in court in relation to private nuisance:

a) There must be material interference with property or personal comfort;
b) It is no defence for the defendant to show that all reasonable steps and care have been taken to prevent the nuisance;
c) The nuisance need not be injurious to health;
d) Temporary or transient (noise) nuisance will not generally be accepted as a nuisance;
e) The courts do not seek to apply a fixed standard of comfort;
f) It is no defence to show that the plaintiff came to the nuisance;
g) The courts will not interfere with building operations conducted in a reasonable manner;
h) Contrary to the general rule in the law of tort, malice may be a significant factor.

Generally speaking, nuisances caused by a person or organisation carrying out a duty imposed by statute is exempt from nuisance action.

1.33 ANTI-SOCIAL BEHAVIOUR (E, W, Sc)

The Anti-Social Behaviour Act 2003, which covers England and Wales, received Royal Assent on 20 November 2003, with a similar 2004 Act for Scotland receiving Royal Assent on 26 July 2004. Both Acts cover a wide range of issues from parental responsibilities, public order and trespass, and anti-social behaviour from tenants, graffiti and flyposting etc. So far as the environment is concerned the Acts cover the following:

- Noise nuisance (see also chapter 4): in E & W the provisions of *the Noise Act 1996* are amended to enable all local authorities to investigate and deal with night-time noise; Part 5 of the Scottish Act enables local authorities in Scotland to specify "noise control periods" when the provisions will apply; local authorities and police constables will be given powers to issue warning and fixed penalty notices, and to confiscate noise-making equipment.

- Closure of noisy premises (see also chapter 4): In E & W, local authorities may order the closure for 24 hours of licensed premises or premises with a temporary events licence to prevent noise nuisance.

- Litter & flytipping (see below, 1.35). In Scotland, the police (as well as local authorities) may also issue FPNs and it is no longer necessary to catch an offender "in the act" of committing an offence. In E & W, the ASB Act amends the *Control of Pollution (Amendment) Act 1989* with regard to the powers of waste collection authorities to investigate and deal with flytipping.

- In Scotland, the maximum penalty available in summary proceedings for a range of environmental offences has been increased to £40,000; these include certain offences under the PPC and Landfill Regulations, the EPA, COPA and *the Water Environment and Water Services (Scotland) Act 2003.*

(*The Anti-Social Behaviour (Northern Ireland) Order 2004*, SI 2004/1988, NI 12, provides for the implementation and operation of anti-social behaviour orders.)

1.34 CLEAN NEIGHBOURHOODS & ENVIRONMENT ACT 2005 (E, W)

This Act (CN&EA) received the Royal Assent on 8 April 2005; it covers England and Wales and has largely been brought into force – see relevant chapters of the *Pollution Control Handbook*. It contains a range of measures aimed at improving the quality of the local environment by giving local authorities and the Environment Agency additional powers in particular to deal with waste and litter and with noise and other nuisances, including:

- Sections 18-21 of the Act extend and amend EPA litter controls, making it an offence to drop litter anywhere, including bodies of water such as rivers and lakes, regardless of ownership. Other amendments relate to fixed penalties and litter control notices (see below).
- Section 35 of the CN&EA amends s.1 of *the Control of Pollution (Amendment) Act 1989* to remove the defence that waste being transported illegally was done so under employer's instructions (brought into force 7 June 2005). Further amendments relate to waste carriers' registration, with new provisions which enable a constable or authorised officer of the regulatory authority to stop, search & seize a vehicle that it is reasonably believed is being used unlawfully for the transportation of waste (chapter 5, 5.34.1).
- Section 33 of the EPA is amended by s.40 of the CN&EA to remove the defence that waste deposited or disposed of illegally was done so on employer's instructions and increase the penalties for conviction under s.33. Other amendments and additions relating to s.33 of the EPA include powers for the court to require offenders to pay costs incurred for both investigation and removal of the waste and the forfeiture of the vehicle used in the offence (see chapter 5, 5.34.1). Other amendments of relevance to waste regulation include: s.47 of the CN&EA repeals s.32 of the EPA requiring local authorities to contract out waste disposal functions; s.49 (England only) of the CN&EA amends s. 52 of the EPA regarding payments for recycling and disposal. Section 54 of the CN&EA provides for regulations to be made requiring developers and contractors of demolition projects to draw up site waste management plans for managing and disposing of the waste created.
- Sections 69-76 of the CN&EA introduce new powers enabling local authorities to deal with audible intruder alarms; see chapter 4.
- Section 84 & Sch.1 of the CN&EA amend *the Noise Act 1996*, extending the powers of local authorities to take action at night from licensed premises, including those with a temporary event notice under *the Licensing Act 2003* – see chapter 4.
- Sections 101-103 amend s.79(1) of the EPA to make nuisance from artificial light and from insects statutory nuisances; s86 amends s.80 to allow a local authority to defer the issuing of an abatement notice in respect of noise nuisance; see above 1.29.1.

1.35 LITTER

Part IV of *the Environmental Protection Act 1990* introduced strict controls in England, Scotland and Wales to deal with litter; these have now been strengthened in England and Wales by Part 3 of *the Clean Neighbourhoods and Environment Act 2005* (CN&EA) which makes various amendments to Part IV. Prior to the EPA, litter was mainly regulated under the *Control of Pollution Act 1974* and in England and Wales under the *Litter Act 1983* and in Scotland under provisions contained in the *Local Government and Planning (Scotland) Act 1982*; all remain of relevance though certain sections are repealed as a result of the new legislation.

In Northern Ireland the comparable legislation is the *Litter (Northern Ireland) Order 1994* which came into effect on 1 October 1994, and *the Local Government (Miscellaneous Provisions) (NI) Order 1985*. These enable councils to take action in relating to littering offences, fly-tipping, fly-posting and graffiti.

Guidance (England only, published by Defra, 2006) on Part IV of the EPA, as amended by the CN&EA, has been included as part of the document providing guidance on the CN&EA. This also includes guidance on the use of fixed penalty notices issued for various offences under the EPA, as amended by the CN&EA.

Section 89 of the EPA (Art. 7 of 1994 NI Order) imposes a duty on those with responsibility for "relevant land" and "relevant highways" to ensure that these are as far as practicable kept clear of litter and refuse. These include local authorities as respects any highway for which it is responsible, Secretary of State as regards trunk roads or other special roads, principal litter authorities, Crown authorities, statutory undertakers in respect to their relevant land, and governing bodies for educational establishments. All those having a duty are required to have regard to the code of practice (see below 1.35.8).

What actually constitutes litter is not strictly defined – section 87 of the EPA says: "If any person throws down, drops or otherwise deposits ... and leaves any thing whatsoever in such circumstances as to cause, or contribute, or tend to lead to, the defacement by litter ..." Section 86(14) enables the Secretary of State to apply Part IV to animal droppings, and this has been done by *the Litter (Animal Droppings) Order 1991* (SI 1991/961). Section 27 of *the Clean Neighbourhoods and Environment Act 2005* (CN&EA) in England and Wales clarifies that for the purposes of Part IV of the EPA, litter specifically includes cigarettes, cigars and like products and discarded chewing gum (including bubble gum).

1.35.1 Offence of Littering

Under s.87 of the EPA (Art.3 of 1994 NI Order) it is an offence to drop or leave litter in the open air anywhere to which the public are entitled to have free access; this includes covered places which are open to the air on at least one side to which the public have free access and relevant land owned by local authorities, statutory undertakers, designated educational establishments and Crown land. Section 18 of the CN&EA (E & W, brought into force 7 June 2005) extends s.87 to cover all land, regardless of ownership, including rivers and lakes and anywhere above the low-water mark on beaches; it does not include covered places open to the air on at least one side to which the public do not have access.

Responsibility for enforcement of the legislation lies with local authorities – principal litter authorities. Section 19(4) of the CN&EA (E & W) extends the definition of litter authority to all parish and community councils.

An offence under s.87 (NI, Art.3) may lead to a maximum fine of £2,500; for businesses, the duty of care applies (see 5.17) and if found guilty of littering the business may be liable (depending on the offence) to a fine of up to £20,000. It is a defence to prove that the littering was authorised by law, or done by or with the consent of the owner, occupier or other authority or person having control of the area; in the case of a watercourse, lake or pond, this includes ownership of all the surrounding land.

1.35.2 Fixed Penalty Notices

Section 88 of the EPA (Art.6 of 1994 NI Order) enables the enforcement authority – principal litter authority – to issue a fixed penalty notice in respect of an offence of littering under s.87. Section 19 of the CN&EA (brought into force E, 14 March/6 April 2006; W 16 March 2006 & 15 March 2007) amends s.88 to enable a principal litter authority to set the amount of fixed penalty in its area; *the Environmental Offences (Fixed Penalties (Miscellaneous Provisions) Regulations 2006* (SI 2006/783, E, effective 6 April 2006: W: SI 2007/739, W.67, 15 March 2007) specify that this should be not less than £50 and not more than £80; where no amount has been so specified the CN&EA sets it at £75. Under *the Litter and Dog Fouling (Fixed Penalty) (Wales) Order 2004* (SI 2004/909, W.92) the fine is currently £75.

In Scotland s.88 is implemented through *the Litter (Fixed Penalty Notices) (Scotland) Order 2004* (SSI 2004/427, which also revoked an earlier 1991 Order), with SSI 2003/268 fixing the amount of fine at £50. If it is not paid within 14 days, the offender is liable to prosecution, for which the maximum fine on conviction is £2,500. In Northern Ireland SR 2004/73 has similar effect.

The Anti-Social Behaviour Act 2003 (E & W) Act also includes provisions enabling local authorities to issue fixed penalty notices (£50) to persons defacing property/walls with graffiti or putting up fly-posters. Scotland's 2004 ASB Act enables Scottish Ministers to vary the amount of FPNs for litter or fly-tipping to £200 and extends the power to issue FPNs to the police.

1.35.3 Litter Control Areas

Section 90 & Sch. 5 of the EPA which enabled principal litter authorities to designate as Litter Control Areas the main categories of land, prescribed by the Secretary of State, to which the public is entitled or permitted to have access, has been repealed in E & W by s.20 of the CN&EA and replaced by new ss.92A-C – see above 1.34.

Section 90 does however still apply in Scotland (*The Litter Control Areas Order 1991* (SI 1325), as amended in 1997 by SI 633) – N. Ireland, Art.10 of the 1994 Order and *The Litter Control Areas Order (Northern Ireland) 1995.* Land in such areas must be kept free of litter and refuse to the standards required by the designated zone into which the land falls (see below). Where standards are not met, members of the public may apply to the Magistrates' Court (Sheriff Court in Scotland) for a litter abatement order (s.91; Art.11 1994 NI Order), having first given the offender five days notice of intention to do so. Failure to comply with a litter abatement order without adequate excuse may lead to further prosecution and a maximum fine of £2,500, plus a maximum daily fine of one-twentieth for each day the offence continues after conviction.

1.35.4 Litter Abatement Notices

Section 92 of the EPA (Art.12 of 1994 NI Order) enables principal litter authorities to issue a litter abatement notice where they are of the view that any of those with a duty to keep their relevant land free of litter and refuse are failing to do so. This may impose a requirement that litter and refuse be cleared within a certain time and/or place a prohibition on permitting the land to become defaced with litter and refuse again. Local authorities should take into account guidance given in the Code of Practice on Litter and Refuse (see below) for the particular type of land. It is an offence not to comply with an abatement notice, subject to a maximum fine of £2,500, increasing by one-twentieth each day that the offence continues after conviction. In any proceedings it is a defence for the defendant to prove that he has complied with his statutory duty; the Code of Practice is admissible as evidence and where relevant must be taken into account.

If a notice is not complied with, the local authority may clear up the land itself, recovering the costs from the offender. Section 56 of *the Anti-Social Behaviour Act 2003* (brought into force 31 March 2004) amends s.92(10) of the EPA to enable local authorities in E & W to clear litter from Crown land or that owned by statutory undertakers, recovering the costs through the courts.

1.35.5 Litter Clearing Notices

New sections 92A-C of the EPA, inserted by s.20 of the CN&EA, and brought into force in England on 6 April 2006 (W, 15 March 2007), enable a principal litter authority in E & W to serve a litter clearing notice on an occupier of any land – public or private – in the open air requiring the land to be cleared of litter and refuse, specifying the standard to which the land must be cleared, and to take steps to prevent further littering; if unoccupied the notice may be served on the owner of the land. Section 92B gives a right of appeal to the Magistrates Court against a notice. Section 92C makes it an offence not to comply with a notice and enables a principal litter authority to take steps to clean up the land itself where a notice has not been complied with – at least 28 days should be allowed for compliance –

recovering costs from the person on whom the notice was served. Offences under s.92C may be dealt with by the issuing of a fixed penalty notice for £100 (or amount specified by enforcing authority).

1.35.6 Street Litter Control Notices

Sections 93 and 94 of the EPA (Art.13 of 1994 NI Order), brought into force on 1 July 1991 through the *Street Litter Control Notices Order 1991* (SI 1991/1324, as amended by SI 1997/632) empower local authorities to issue Street Litter Control Notices to various premises with a street frontage which may cause a litter problem. These include take-away food premises, service stations, cinemas, theatres and other recreational premises; banks and building societies with cash machines on an outside wall are also covered. Section 21 of the CN&EA (brought into force: E, 6 April 2006, W, 15 March 2007) extends these sections in E & W to cover vehicles, stalls and other moveable street vending structures. *The Street Litter Control Notices (England) (Amendment) Order 2007* (SI 2007/1524), which came into force on 1 July 2007, amends the description of commercial and retail premises in the 1991 Order to "premises used wholly or partly for the sale of food or drink for consumption on the premises".

A Street Litter Control Notice should set out "reasonable requirements" for dealing with the problem, such as provision of litter bins or clearance of litter at specified times or intervals. The notice may relate to any kind of litter covered by the legislation, including smoking-related litter. Twenty-one days' notice of intention to serve a Notice must be given and any representations made taken into account. Any appeal against the terms of a Notice should be made to the Magistrates'/Sheriff Court.

The CN&EA substitutes s.94(8) & (9) to make non-compliance with a Street Litter Control Notice an immediate offence, subject to a maximum fine on conviction of £2,500; alternatively a fixed penalty notice may be issued for £100 (or amount specified by the local authority. (In Scotland the enforcing authority should apply to the Sheriff Court for an order requiring compliance with the Notice within a specified period; further non-compliance may then lead to prosecution and fine.)

1.35.7 Distribution of Free Literature

New section 94B & Sch. 3A of the EPA inserted by s.23 of the CN&EA, brought into force on 6 April 2006 in England (W, 15 March 2007), gives principal litter authorities powers to control the distribution of free literature to prevent it from becoming litter. Principal litter authorities may designate the areas in which such controls apply and establish a consent system, thus making it an offence, subject to a fine if convicted of up to £2,500, to distribute free literature without a consent. The authority may instead issue a fixed penalty notice in the amount of £75 (or other amount set by the authority) in relation to an offence under this section. Material distributed for charitable, religious or political purposes is exempt from the offence, as is putting the literature inside buildings or through letterboxes or distributed entirely within a public service vehicle. If an offence is committed, the authority may seize the literature (which can be reclaimed by the owner on application to the Magistrates Court).

1.35.8 Codes of Practice on Litter & Refuse

In England a new Code of Practice was published in April 2006 (replacing the original 1991 code, amended 1999) which takes account of amendments to the EPA made by *the Clean Neighbourhoods and Environment Act 2005*; Part 1 (the code of practice) aims to encourage those with a duty to manage their land within acceptable cleanliness standards, with an emphasis on appropriate management, rather than how often it is cleaned; the code is admissible in evidence in any proceedings. Part 2 contains advisory standards for graffiti and fly-posting, and Part 3 the legislative framework.

The standards of cleanliness are

- *Grade A:* No litter or refuse;
- *Grade B:* Predominantly free of litter & refuse, apart from some small items;
- *Grade C:* Widespread distribution of litter and/or refuse, with minor accumulations;
- *Grade D:* Heavily affected by litter and/or refuse, with significant accumulations.

Standards for removing detritus from metalled highways and hard surfaces are similar.

Maximum response time for restoring different types of area to grade A if it falls below grade B, are as follows:

- High intensity of use (busy public areas): half a day.
- Medium intensity of use (housing/'everyday areas' occupied by people most of the time): 1 day.
- Low intensity of use (lightly trafficked areas that do not impact on people's lives most of the time): 14 days.
- Areas with special circumstances (where issues of health and safety and reasonableness and practicality are dominant considerations): 28 days or as soon as reasonably practicable.

Duty bodies in England – those with a duty under s.89 of the EPA – were expected to have completed re-zoning their area within a year of the Code taking effect (i.e. by 6 April 2007).

The Scottish Government issued a revised code of practice for Scotland in October 2006. In Wales consultation on a revised code of practice incorporating changes to litter legislation introduced by the CN&EA closed on 2 March 2007, and a new code is awaited. This will provide statutory guidance for those with a duty to clear litter and refuse from their land, and help other land managers to maintain a high environmental quality. Both codes are similar to that for England.

A voluntary code of best environmental practice for the fast food industry (developed by ENCAMS for Defra, November 2004) provides guidance on the practical steps outlets can take to ensure that litter is disposed of properly and to encourage their customers to dispose of litter responsibly.

In Northern Ireland Art.9 of *the Litter (Northern Ireland) Order 1994* Order requires a code to be drawn up.

1.36 CONTROL OF PROCESS ODOURS

Most complaints about odour relate to annoyance about smells caused by emissions from a factory or farm reducing enjoyment of homes and gardens or making working conditions unpleasant. It is now appreciated that annoyance is the combined result of a number of odour properties.

1.36.1 Process Odour Composition

It is usual to describe the odour by the nature of the process causing it but process odours, especially where animal or vegetable matter is heated, are often complex mixtures in air of many substances present at very low concentrations. These substances may be individually malodorous and will occur with large numbers of other compounds including hydrocarbons which cause little or no odour response. Recent years have seen the development of techniques for the sampling and analysis of such mixtures containing perhaps only parts per million (ppm) or even parts per billion (ppb) quantities in air.

Analysis of the process odour is likely to involve pre-concentration on an adsorbent and then separation on a gas chromatographic column, followed by identification of the separated components. The most positive and comprehensive method of identification is a combined gas chromatograph-mass spectrometer linked to a computer-based data system. If part of the flow of separated components is delivered to an "odour port", then the odour, if any, of each component can be described.

1.36.2 Odour Properties & Effects

Odour intensity is the strength of the perceived odour sensation and depends in a complex way on the concentration of the odorous substances present. It can be measured by comparison with reference samples but often an alternative measure of odour strength, the dilution to threshold value, is determined.

Odour character is the property that enables one to distinguish between different odours and it is said that an odour with a distinct recognisable character tends to be more annoying.

The acceptability – or "hedonic tone" – of an odour is an important factor in judging whether or not it is offensive or a nuisance. Not only the nature of the odour is taken into account (whether it is pleasant or unpleasant), but also its intensity, time and duration of release. It should also be borne in mind that a normally "pleasant" odour becomes less acceptable when exposed to it frequently; for example, a "pleasant" odour from a fragrance factory may be unacceptable if persistent in a residential area.

1.37 REGULATORY CONTROLS

1.37.1 Odour as a Statutory Nuisance

The Environmental Protection Act 1990 replaced and repealed those sections of the *Public Health Act 1936* dealing with statutory nuisance, together with the whole of the *Public Health (Recurring Nuisances) Act 1969*. The "offensive trades" provisions of the PHA 1936 were repealed in 1995. Statutory nuisance – which includes "smell or other effluvia" – is now covered by Part III of the EPA (see above).

A High Court ruling in summer 2003, that sewage works can be considered as "premises", means that odour from sewage works can be dealt with under statutory nuisance legislation.

In England & Wales Defra and the Welsh Assembly Government published a voluntary code of practice in April 2006. This provides advice for sewage treatment works operators and regulators on understanding the odour, factors to be taken into account in assessing how much of a nuisance it is to the surrounding area, managing complaints, and ways of abating or limiting the odour. Action against odour from sewage treatment works can be taken under the statutory nuisance legislation – see above.

1.37.2 Integrated Pollution Prevention & Control

Installations regulated under *the Environmental Permitting Regulations (England and Wales) 2007*, or *the Pollution Prevention and Control Regulations* (Sc. 2000, NI 2003) require a permit from the appropriate regulatory authority. Some – such as the food and drink industry and intensive livestock units – are now regulated for all their emissions – including odour – for the first time. All installations are required to ensure that the "Best Available Techniques" are used for preventing or minimising odour emissions, with permits including a condition requiring operators to ensure there are no offensive odour emissions beyond their site boundary – see chapter 2, 2.4 and 2.5. Thus operators will need to assess the potential for odour and put in place the appropriate controls rather than simply responding to complaints. The IPPC Directive does not exempt "trivial emissions" from regulation.

UK-wide guidance on odour assessment and control with particular relevance to IPPC applications was issued in 2002: H4 draft horizontal guidance for odour, parts 1 and 2 applies to all IPPC sectors. Part 1 outlines the main considerations to be taken into account in the regulation and permitting of relevant activities, and part 2 details assessment and control techniques. Technical Guidance Note SRG6.02 (farming) provides guidance on odour management at intensive livestock installations.

NETCEN has also produced guidance on the control of odour and noise from commercial kitchen exhaust systems, providing information on best practice techniques for minimising problems.

1.37.3 Water Services etc (Scotland) Act 2005

Provisions included in this Act enable Scottish Ministers to issue statutory codes of practice to deal with nuisance from sewage works, which include a duty of compliance on the operator and enforcement obligations on the regulator.

The Sewerage Nuisance (Code of Practice) (Scotland) Order 2006 (SSI 2006/155) which came into force on 22 April 2006 includes a statutory code of practice on the assessment and control of odour from waste water treatment works. It applies to existing WWTW in operation on 22 April 2006 and to new ones coming into operation after that date and requires the best practicable means to be applied to control odour emissions from contained and fugitive sources and to ensure that emissions do not create an odour nuisance beyond the boundary; see also chapter 6, 6.20.2.

Section 115 of *the Public Health etc (Scotland) Act 2008* (which is expected to be brought into force in early 2009), inserts a new sub-section in the 2005 Act. A local authority enforcing the sewerage code will have powers of entry to premises for the purposes of establishing whether a statutory nuisance exists or for carrying out of remedial action under Part III of the EPA.

2 CHAPTER

Integrated pollution prevention and control

Environmental permitting

Major accident hazards

Control of substances hazardous to health

Control of asbestos

Control of lead at work

REACH

INTEGRATED POLLUTION PREVENTION AND CONTROL

On 24 September 1996, EU Environment Ministers formally adopted a *Directive on Integrated Pollution Prevention and Control* (96/61/EC). It came into force on 30 October 1996 and had to be implemented by 30 October 1999 when it applied to all new installations and those undergoing a substantial change. This Directive has now been repealed and its provisions together with those in subsequent amendments, codified into a single Directive, 2008/1/EC of 15 January 2008.

Following a review of the Directive, the Commission has now published (December 2007) a *proposal for a Directive on industrial emissions (integrated pollution prevention and control)*; as well as revising the IPPC Directive, the proposal incorporates various other Directives on industrial emissions, including those on VOCs (1999), waste incineration (2000), large combustion plant (2001), and Directives regulating the titanium dioxide industry – see below 2.1.4.

In England and Wales the IPPC Directive is now implemented through *the Environmental Permitting (England and Wales) Regulations 2007*, which came into force on 6 April 2008, replacing and revoking *the Pollution Prevention and Control (E&W) Regulations 2000*, as amended (see 2.4 below). In Scotland, the IPPC Directive has been implemented through *the Pollution Prevention and Control (Scotland) Regulations 2000*, with similar 2003 Regulations applying in Northern Ireland (see 2.5 below).

2.1 IPPC DIRECTIVE – 2008/1/EC

Directive 2008/1/EC amalgamates the original 1996 IPPC Directive and its subsequent amendments into a single instrument. Codified instruments do not make substantive changes to the original text, but may make corrections and amendments for the purposes of clarification, linguistic problems etc.

Existing installations (i.e. operating before 31 October 1999 or for which an authorisation existed or had been requested and the installation was brought into operation by 31 October 2000) had to comply with the Directive by 31 October 2007. Existing installations wishing to carry out a substantial negative change (i.e. one with significant negative effects on human beings or the environment) between implementation of the Directive and 31 October 2007 had to apply for a permit covering that part of the installation to be substantially changed.

The main purpose of the IPPC Directive is "to achieve integrated prevention and control of pollution" from a wide range of industrial and agricultural activities. This is to be done by preventing, or where that is not practicable, reducing emissions to the air, water and land by potentially polluting industrial and other installations "so as to achieve a high level of protection of the environment taken as a whole".

Annex I to the Directive (see Appendix 2.1) lists the activities to which the Directive applies; these are: energy industries; production and processing of metals; mineral industry; chemical industry; waste management; and other activities. This latter category includes pulp production, paper and board production, pre-treatment or dyeing of fibres, tanning, slaughterhouses and disposal of animal carcases, various food processes and intensive poultry and pig rearing installations.

An application for a permit must be made to the competent authority (i.e. enforcing authority); where there is more than one "competent authority" (e.g. Environment Agency and local authority), steps must be taken to ensure an integrated approach to the procedure for granting a permit. The permit should take account of the whole environmental performance of the plant and thus conditions attached to it will relate to the operation of the plant, including

- emission limit values for certain substances and preparations listed in Annex III to the Directive based on best available techniques (BAT) (see 2.1.3 below) – all installations had to be operating to a BAT-based permit by November 2007;
- local environmental conditions, geographical location and other technical characteristics;
- consumption of raw materials, energy efficiency, heat, noise, light and vibration, and accident prevention;
- waste management practices to avoid pollution following closure of the site.

Permit conditions should ensure that there will be no breach of EU environmental quality standards or other EU legislation and ensure "a high level of protection for the environment taken as a whole". The permit must also contain monitoring requirements, specifying measurement methodology and frequency. Permits should be reviewed periodically and where necessary updated in line with BAT or to tighten permit conditions for various reasons. The Directive allows for Member States to use "general binding rules" as an alternative to individually tailored conditions.

Technical Guidance – or "Best Available Techniques Reference Documents" – for each of the sectors covered by the Directive have been drafted on behalf of the Commission and, once adopted, must be taken into account by competent authorities – a full list of BREFs, including their status, is at Appendix 2.3; the list as well as the complete documents can also be accessed on the website of the European IPPC Bureau – http://eippcb.jrc.es. The Bureau has been established to ensure full exchange of information between Member States on BAT. The Bureau is to review all BREFs over the next few years to ensure emission levels remain consistent with BAT.

The European Commission published *Guidance on the Interpretation and Implementation* of the IPPC Directive in April 2007. This and further information on the IPPC Directive and associated issues can be found on the IPPC web page of the Commission's Environment DG – http://ec.europa.eu.int.

2.1.1 Proposal for Industrial Emissions Directive

Published by the Commission in December 2007, COM(2007) 844 will revise the IPPC Directive to take account of the results of the Commission's review of its implementation and will incorporate a number of other Directives, including those on VOCs (1999), waste incineration (2000), large combustion plant (2001), and various Directives regulating the titanium dioxide industry. These Directives will all be repealed following adoption of the revised legislation. Areas of concern identified by the review and which are addressed in the proposed Directive include:

- Clarification and strengthening of the concept of BAT; BAT reference documents (BREFs) must be used to set permit conditions and emission limit values may not exceed emission levels associated with the BAT; derogations based on strictly defined criteria will be permitted, but must be justified through a publicly available assessment of the costs and environmental effects.
- Revision of minimum emission limit values for large combustion plant and installations producing titanium dioxide to align them with BAT standards.
- New provisions on inspection and environmental improvements. Operators will be required to regularly report on compliance with permit conditions; conditions will have to be updated following adoption of a new or updated BREF to take account of developments in BAT. Member States will also have to provide for a system of environmental inspections.
- Simplifying and clarifying certain provisions on permitting, monitoring and reporting to cut administrative burdens. There is to be a new provision requiring periodic monitoring of soil and groundwater on the sites of all installations to protect them from contamination. Operators will be required to provide a baseline report on the state of soil and groundwater and upon definitive cessation of activities will be required to remediate the site and return it to the baseline state.
- Extending and clarifying the scope and provisions of the legislation to better contribute to the objectives of the Thematic Strategies.

The new Directive will basically cover the same installations and activities as the current Directives (extending the scope of some activities, e.g. to LCP between 20-50 MW), with chapter I laying down general provisions applying to all industrial activities covered, chapters III-VI contains minimum technical requirements for LCP, waste incineration, solvents and titanium dioxide installations; chapter VII contains provisions on competent authorities, reporting requirements etc.

In May 2008 Defra and the Devolved Administrations published a number of consultation documents seeking comments on the Commission's proposals to help inform the UK Government's formal position on the proposals.

2.1.2 European Pollutant Emission Register

A Decision (2000/479/EC) establishing the Register, which is required under the IPPC Directive, was approved in July 2000. Member States are required to collect data from IPPC facilities (defined as an "industrial complex with one or more installations on the same site where one operator carries out one or more [IPPC] activities") on emissions to air and water of 50 substances present in quantities above specified thresholds from 56 industrial activities. The EPER was formally launched in February 2004, by the Commission and the European Environment Agency with data for 2001 (or 2002 or 2003 if unavailable for 2001) from around 10,000 industrial facilities covering the (as at that date) 15 EU Member States, Norway and Hungary. The Register, which is managed by the EEA, is updated every three years; emissions data for 2004 (required in June 2006), including from those Member States which joined the EU in May 2004, covering some 12,000 sites was published on the website at the end of 2006.

The EPER can be accessed on www.eper.ec.europa.eu

The EPER is to be replaced by the E-PRTR once this latter goes online in 2009.

2.1.3 Pollutant Release and Transfer Register

EU Regulation 166/2006 (adopted 18 January 2006, and which came into force on 24 February 2006) establishes an integrated "Pollutant Release and Transfer Register" (E-PRTR) in line with the UNECE Protocol of the same name which was made under the Aarhus Convention (see chapter 1, 1.24). The new Register, which is to be in the form of a publicly accessible electronic database, will cover more than 90 substances released to air, water and land from around 65 industrial activities covered by the IPPC and hazardous waste Directives; it will also include emissions from road traffic, domestic heating and agriculture, and transfers of waste and waste water from industrial facilities to other locations. The first reporting year is 2007, with Member States required to submit (by electronic transfer) the required data to the Commission within 18 months of the end of the year (i.e. by mid-2009); the Commission, assisted by the European Environment Agency, is required to incorporate the data into the E-PRTR within a further 3 months – thus the first reports should be available online by September 2009. Thereafter, Members States will be required to submit their data to the Commission within 15 months of the end of the reporting year, with the Commission having a further month to incorporate it into the PRTR.

In accordance with the Aarhus Convention, the public have a right of access to the European PRTR, and should be given early and effective opportunities to participate in its development. The Convention also requires the establishment of a publicly accessible national (i.e. UK) PRTR.

More at: http://europa.eu/scadplus/leg/en/lvb/l28149.htm

UK Implementation

Defra, in consultation with the Devolved Administrations, has published proposals (September 2007) for implementing the E-PRTR in England, Wales & Northern Ireland and for establishing the national PRTR (required by the Convention); the Scottish Government has issued similar proposals for implementing the E-PRTR in Scotland, as well as seeking comments on the national PRTR.

The PRTR extends reporting requirements to a number of new sectors, including Part A2 facilities and certain others not regulated as Part A1 (E&W) or Part A (NI & Sc). For England, Wales & NI, it is proposed that the relevant regulatory authority (local authorities or Environment Agency, NIEA or district councils) be responsible for enforcement of the E-PRTR, ensuring that operators fulfil their obligations for reporting emissions data. In Scotland SEPA will be the competent authority.

For the first reporting year (2007) the new sectors were required to submit data to the relevant competent authority by 28 February 2008, and by the same date every year thereafter.

(For the first year only operators of local authority regulated facilities in E&W had until 28 April 2008 to submit data.) In a letter to all Chief Executives (13 March 2008), Defra advises that a condition be inserted in all relevant permits requiring operators to comply with any Information Notice served on them for the purposes of complying with their obligation to report pollutant releases and off-site waste transfers in compliance with the EU Regulation, and that non-compliance would constitute a breach of their permit. (*The Environmental Permitting (Transitional – European Pollutant Release and Transfer Register) (England) Direction 2008 refers.*)

Annually, the competent authorities (including SEPA), and Defra for local authorities in E&W, will be required to compile E-PRTR data in a format compatible with that required for submission to the European Commission; this will be submitted to the Environment Agency who will act as the coordinating authority.

In compliance with the Aarhus Convention, it is proposed to establish a web-based UK-PRTR. Members of the public will be able to search and identify releases and transfers by activity, geographical location, operator, pollutant & waste; they will also be able to search by environmental media into which the pollutant is released; by destination of waste transfer and, where appropriate the disposal and recovery operation for waste.

2.1.4 Best Available Techniques (BAT)

The IPPC Directive, requires competent authorities to ensure that "installations are operated in such a way that all the appropriate preventive measures are taken against pollution, in particular through the application of best available techniques ...". In seeking through the application of BAT to balance costs to the operator against benefits to the environment, the Directive defines BAT as

> "the most effective and advanced stage in the development of activities and their methods of operation which indicates the practical suitability of particular techniques for providing in principle the basis for emission limit values designed to prevent and, where that is not practicable, generally to reduce emissions and the impact on the environment as a whole.
>
> - "available techniques" means those techniques which have been developed on a scale which allows implementation in the relevant industrial sector, under economically and technically viable conditions, taking into consideration the costs and advantages, whether or not the techniques are used or produced inside the Member State in question, as long as they are reasonably accessible to the operator;
>
> - "best" means, in relation to techniques, the most effective in achieving a high general level of protection of the environment as a whole;
>
> - "techniques" includes both the technology used and the way in which the installation is designed, built, maintained, operated and decommissioned."

Both *the Environmental Permitting Regulations* (E & W) *and the Pollution Prevention and Control Regulations* (Sc and NI) require Regulators to include conditions based on BAT in permits.

The European Commission has produced a series of BAT Reference Documents (BREF Notes) defining BAT for the individual sectors covered by the IPPC Directive – see Appendix 2.3 which lists final and draft BREF notes, as well as Environment Agency guidance. BREF notes, which can also be seen on http://eippcb.jrc.es, form the basis for Environment Agency guidance notes on IPPC activities (see also Appendix 2.3). BREF notes are now being reviewed to ensure that they reflect technological developments since their formal adoption.

2.1.5 Environmental Inspections

EU Recommendation (2001/31/EC) adopted on 4 April 2001 sets minimum criteria for environmental inspections for all industrial installations and other facilities subject to environmental licensing or permitting; the aim is to ensure more consistent application of EU legislation across Member States. It recommends that Member States should draw up plans specifying what environmental inspections should cover and that these should be publicly available. Site inspection reports should be made publicly available within two months of the inspection. The Commission is currently reviewing implementation of the Recommendation and is considering the inclusion of legally binding requirements for inspection in various Directives, such as the IPPC Directive.

2.2 POLLUTION PREVENTION AND CONTROL ACT 1999

This Act, which received Royal Assent on 27 July 1999 enables Regulations to be made in England, Wales and Scotland implementing the 1996 EU Directive on Integrated Pollution Prevention and Control (IPPC). In so doing, it provides for the repeal of Part I of the *Environmental Protection Act 1990* under which processes were regulated for either Integrated Pollution Control or Local Air Pollution Control.

The PPC Act also makes provision for improved environmental regulation of off-shore gas and oil installations (see chapters 3 and 6).

Section 4 of the Act which enabled waste licences issued under the *Control of Pollution Act 1974* and which had expired to be treated as though they were still in force has been repealed by *the Environmental Permitting (England and Wales) Regulations 2007.*

2.2.1 Designation of Directives for Purposes of PPC

Section 2 of the PPC Act empowers the Secretary of State (Scottish Ministers) to make Regulations under the Act for the purposes of transposing, for example, EU Directives, and in this respect the following Orders have been made:

England & Wales

SI 2007/2247, which came into force on 18 August 2007 (and revoking SI 2001/3585, 2002/2528 and 2003/948), designates the following Directives for the purposes of the Act: 87/217/EC (prevention and reduction of environmental pollution by asbestos; 92/112 (waste from the titanium dioxide industry); 94/63/EC (control of VOCs resulting from storage of petrol & distribution to service stations); 99/13/EC (limitation of VOC emissions from use of solvents); 99/31/EC (landfill); 2000/53/EC (end-of-life vehicles); 2000/76/EC (waste incineration); 2001/80/EC (large combustion plant); 2001/81/EC (national emissions ceilings); 2002/96/EC (WEEE); 2006/12/EC (waste); 2006/66/EC (batteries & accumulators). SI 2008/2549, which came into force on 6 October 2008, designates Directive 2006/21/EC (mines and quarries waste), and 2008/1/EC (consolidated IPPC Directive).

Scotland

The following Orders have been made designating Directives for the purposes of the PPC Act:

- SSI 2002/488, effective 04.11.02, designates Directives 2001/80/EC (large combustion plant) and 2001/81/EC (national emissions ceilings).
- SSI 2003/185, effective 01.04.03, designates Directive 99/31/EC (landfill).
- SSI 2003/204, effective 01.04.03 designates the Incineration Directive.
- SSI 2003/600, effective 15.01.04, designates Directive 99/13/EC (solvent emissions).
- SSI 2005/461, effective 05.10.05, designates Directive 2003/35/EC (public participation).
- SSI 2008/86, effective 17.03.08, designates Directive 2006/66/EC (batteries and accumulators).

2.2.2 Penalties – Emissions Trading

Part 2 of *the Waste and Emissions Trading Act 2003*, which received Royal Assent on 13 November 2003, inserts an additional paragraph into Schedule 2 of the PPC Act. This has the effect of allowing penalties to be levied on participants in the UK Greenhouse Gas Emissions Trading Scheme (see chapter 3, 3.6.2) – or any future emissions schemes – who fail to comply with reduction targets.

Better Regulation Review

As part of the Government's "Better Regulation Review", Defra, the Scottish Government & the Welsh Assembly Government have been looking at the regulation of installations under the PPC Regulations whose activities are not required to be regulated under the IPPC Directive; they are looking at whether other regulatory regimes (e.g. nuisance, or the Clean Air Act) would deliver satisfactory regulation while still meeting environmental objectives, EU and international obligations. Following consultation in early 2006, Defra, the SG and WAG have now refined the criteria to be applied in assessing the scope for simplifying or using alternative approaches. For the second stage of the review the Local Authority Unit and Environment Agency are assembling data on the pollution impact of each industry sector under review; there will be a separate analysis of whether the sectors are amenable to alternative regulatory models and the practical, environmental and health impacts of doing so. Further consultation is expected. Details of the review can be found on Defra's website on the PPC (local authorities) pages.

2.3 THE ENVIRONMENT (NORTHERN IRELAND) ORDER 2002

The Environment (Northern Ireland) Order 2002 (SI 2002/3153), made on 17 December 2002, implements the IPPC Directive in Northern Ireland, together with *the Pollution Prevention and Control Regulations (Northern Ireland) 2003* (SR 2003/46), as amended, which came into force on 31 March 2003; these are similar to the Regulations applying in the rest of the UK. These Regulations are referred to in the text at 2.5; the Industrial Pollution Inspectorate is responsible for the regulation of Part A installations for IPPC and Part B installations for air pollution; Part C installations will continue to be regulated by local authorities for air pollution control.

The Environment (Designation of Relevant Directives) Order (Northern Ireland) 2003 (SR 2003/209) designates the following Directives for the purposes of the 2002 Order: solvents (1999); end of life vehicles and waste incineration (2000); and large combustion plants (2001).

Part III of the 2002 Order makes provision for local authorities to review and assess local air quality – see chapter 3, 3.11.

2.4 ENVIRONMENTAL PERMITTING REGULATIONS (EPR)

During 2006 and 2007 Defra, the Welsh Assembly and the Environment Agency consulted on proposals to combine the IPPC, LA-IPPC and waste management licensing regimes, with a view to simplifying and standardising the mechanics of regulation through the development of a single permitting and compliance system that can also target resources where they are needed most – i.e. adopts a risk-based approach.

The Environmental Permitting (England and Wales) Regulations 2007 (SI 2007/3538) came into force on 6 April 2008, replacing and revoking the PPC Regulations, as well as various other Regulations covering industrial emissions and waste management; *the 1994 Waste Management Regulations* have been amended to reflect the fact that waste operations – including exemptions and licensing – are now regulated under the new Regulations.

As well as "bespoke" permits for higher risk activities, operators of facilities for which standard rules – or general binding rules – have been prepared will have the option of applying for a "standard rules permit", for which a lower fee will be charged.

Charging schemes will continue to reflect cost recovery (polluter pays principle), be proportionate to the level of regulatory effort required and risk-based; charges for standard rules permits are lower. All associated regulatory and other guidance, forms and documentation has been revised to reflect the EP Regulations.

A second phase of the environmental permitting programme (EPP2) is aiming to integrate and streamline further permitting regimes, including those for water discharge consents, groundwater authorisation, water abstraction, radioactive substances and some aspects of the licensing of waste carriers and brokers. A consultation document was expected before the end of 2008.

Defra and the Department for Communities and Local Government (DCLG) have also issued a joint consultation (September 2006) which looks at the interface between planning and pollution control in England and considers options for streamlining processes – see chapter 1.

There are currently no similar proposals for Scotland or Northern Ireland.

2.4.1 Overview

The Environmental Permitting (England and Wales) Regulations 2007 (EPR) (SI 2007/3538), which came into force on 6 April 2008, set out a single set of procedural rules covering IPPC and waste management permitting. An environmental permit is required for all those installations and mobile plant carrying out an activity listed in Annex 1 to the IPPC Directive (see Appendix 2.1). Schedule 1 to the Regulations details those to be controlled for all emissions to air, land and water (Part A1 regulated for IPPC and Part A2 (LA-IPPC), and Part B installations to be regulated for air pollution (LAPPC). A permit is also required for waste operations & mobile plant (involving the recovery or disposal of waste), unless specifically exempted (Schedules 2 & 3 of the EPR).

The Environment Agency is the regulator for Part A1 installations and mobile plant and most waste operations. Local authorities regulate Part A2 installations and mobile plant, and local authorities (or the Port Health Authority) Part B installations. Local authorities also regulate waste operations which are carried out as part of a Part A2 or B installation or mobile plant.

Regulations made under *the Environmental Protection Act 1990*, which implemented Integrated Pollution Control, Local Air Pollution Control and waste management licensing have been revoked, including:

- *Environmental Protection (Prescribed Processes & Substances Regulations 1991)*, SI 1991/472 and all amendments
- *Environmental Protection (Applications, Appeals & Registers) Regulations 1991*, SI 1991/507, and all amendments
- *Waste Management Licensing (Amendment) Regulations*: SI 1995/1950, 1996/1279, 1997/2203, 2002/674, 2003/595, 2005/1728; WSI 2002/1087, 2003/780, 2004/70
- *Waste Management Regulations 1996* (SI 1996/634)

Schedules attached to the Regulations specify the additional requirements of a number of EU Directives which must be complied with; as well as the IPPC Directive, these include the Directives on large combustion plant, waste, incineration, landfill, solvents, WEEE, petrol vapour recovery and asbestos; the following Regulations have been revoked:

- *Pollution Prevention & Control (E & W) Regulations 2000*, SI 2000/1973, and all amendments
- *Landfill (E & W) Regulations 2002*, SI 2002/1559, and all amendments, and *the Pet Cemeteries (E & W) Regulations 2007*, SI 2007/2596)
- *Large Combustion Plants (E & W) Regulations 2002*, SI 2002/2688
- *Waste Incineration (E & W) Regulations 2002*, SI 2002/2980
- *Environmental Protection (Prescribed Processes & Substances) (Amendment) (Petrol Vapour Recovery) Regulations 1996*, SI 1996/2678

The Regulations are to be amended to bring quarries and mines within the permitting system, implementing EU Directive 2006/21/EC (see chapter 5, 5.13). This requires operators of mines and quarries to store extractive wastes without endangering human health and the environment, to have a site waste management plan and a major accident prevention policy.

The EPR provide for existing permits and waste management licences to automatically become environmental permits under the new Regulations without the need for a new application. Applications for permits or waste management licences made prior to the new Regulations taking effect will be determined under the "old" system, and once determined will then be treated as an environmental permit under the "new" system. The transitional arrangements also provide for a holder of an existing waste management licence to be considered as the operator under the EPR (reg.69(2)) – on the occasion of a licence (i.e. permit) transfer, it will need to be transferred to the person who has control of the facility (the operator).

Operators of new facilities will need to obtain a permit, or register a waste exemption under the EPR before they can begin operations. Current waste exemptions registered under *the Waste Management Licensing Regulations* will automatically become registered under the EPR when these come into force.

Note the PPC Regulations and those under the EPA remain in force for deciding applications made prior to the EPR coming into force, for notices served and for determining appeals made prior to 6 April 2008.

It is an offence to operate a regulated facility without the correct permit, or in contravention of its conditions (reg.12).

2.4.2 Guidance

Environmental Permitting Core Guidance, published by Defra in early 2008, describes the permitting and compliance requirements for all [Environment Agency] regulated facilities. Although this is not statutory guidance, it sets out the views of the Secretary of State and Welsh Ministers as to how the Regulations should be applied. The Environment Agency has also set out how it will implement the Regulations. Defra has published separate Guidance on each of the Directives which have been transposed through the EP Regulations and how the requirements of each Directive should be applied. For a list of available Guidance, which can be downloaded via the Environmental Permitting pages of Defra's website see Appendix 2.3 of the *Pollution Control Handbook*.

The Environment Agency's *Getting the Basics Right: How to comply with your environmental permit* (April 2008) describes the standards and measures businesses should take to control the risk of pollution from "the most frequent situations" in waste management and process industry. It includes advice on

- Managing activities – operations, maintenance, competent personnel, technical competence schemes; accident management plan.
- Operations – permitted activities, waste acceptance.
- Emissions and monitoring – emissions to water, air and land; fugitive or diffuse emissions; odour; noise and vibration; pest control.
- Information – records and reporting.

- Annexes – standards for particular activities; activities for which odour is a key issue; requirements for IPPC Directive activities.

The *Secretary of State's Guidance: General Guidance Manual on Policy and Procedures for A2 and B Installations* has been revised to reflect the EPR, and to incorporate AQ guidance notes. Published in early 2008 (and amended July 2008), it is available both in hard copy and via Defra's website – note, updates will only be incorporated into the website version of the manual. The *General Guidance Manual*, which constitutes statutory guidance to local authority regulators under EP reg.64, is in four parts

- Part A – guidance on policy and procedure for permitting A2 & B installations, including connections with other legislation and EU Directives (e.g. solvents, paints, petrol vapour recovery, asbestos) and the European Pollutant Release and Transfer Register;
- Part B – contains annexes going into more detail on issues dealt with in Part A of the guidance (e.g content of applications, conditions of permits etc);
- Part C – specimen application, surrender, transfer and variation forms for both A2 and B installations;
- Part D – specimen notices, including variation, suspension and revocation notices etc.

For further guidance, details of EU BREF notes etc, see Appendix 2.1-2.4 of the *Pollution Control Handbook*.

2.4.3 Waste Operations (regs. 4 & 5)

Certain waste operations are excluded from the need for a permit because they are already regulated under other regimes; these include

- the disposal or recovery of waste where that operation is covered by licence under Part II of *the Food and Environmental Protection Act 1985*;
- the disposal of liquid waste under a consent under Part III of *the Water Resources Act 1991*;
- disposal of agriculture in or on land under an authorisation under reg. 18 of *the Groundwater Regulations 1998*;
- disposal or recovery of waste which is not to be treated as industrial or commercial waste by virtue of *the Controlled Waste Regulations 1992*.

In addition various waste activities are exempt from the need for a permit (Sch.3), but will nevertheless require to be registered with the appropriate Exemption Registration Authority – depending on the activity (Sch.2, para 2) this will either be the Regulator or the authority responsible for authorisation under the *Animal By-Products Regulations*. To qualify for exemption the waste operation is usually of low risk and must be carried out in accordance with Article 4 of the Waste Framework directive, i.e. without endangering human health, and without using processes or methods likely to harm the environment; it must not cause noise or odour nuisance or adversely affect the countryside or places of special interest. Exemptions do not usually apply to hazardous waste and the person carrying on the exempt activity must have the permission of the occupier of the land or be entitled to use the land for the activity.

Exemption Registration Authorities are required to establish and maintain a register of all exempt activities for which they are the registration authority (Sch.2, para 4); this should include name and address of establishment or undertaking carrying on the exempt activity, the waste operation which constitutes the exempt activity and the place at which it is carried on, with additional details required in respect of the recovery of scrap metal or the dismantling of waste motor vehicles (Sch.2, para 6). Registers are open for the public to inspect free of charge at reasonable hours, with copies to be made available at a reasonable charge. Exemption Registration Authorities are also responsible for removing entries from the register when they become aware that an establishment or undertaking is no longer carrying on the activity, it ceases to be an exempt activity or is unable to verify particulars in accordance with Art. 6(2) of the ELV or WEEE Directives (as appropriate).

Certain exempt activities – termed "notifiable exempt activities" – are considered to be of higher environmental risk and require establishments or undertakings carrying out the activity to provide the Exemption Registration Authority with a notice, together with the appropriate fee, giving particulars of the establishment or undertaking, and quantities of waste to be recovered or disposed. These are:

- sludge on land*
- waste for the benefit of land*
- land reclamation improvement*
- sewage and water treatment works
- waste for construction
- repair or refurbishment of WEEE
- burning at a dock of plant tissue waste and wood

*to qualify for exemption, the establishment or undertaking must certify that the activity will result in benefit to agriculture or ecological improvement and be consistent with article 4 of the Waste Framework Directive (Sch.2, para 8(2)).

Guidance on registering exempt activities is available on the Environment Agency's website, and see Appendix 2.3.

Every 12 months, the establishment or undertaking must by notice in writing confirm that they wish to continue to carry on the exempt activity, providing details of the quantity of waste to be disposed of or recovered, as well as any other information required by the Registration Authority (Sch.2, para 10). Notices relating to notifiable exempt activities should be entered on the register, or if the Authority declines to do so it should notify the establishment or undertaking within 25 working days stating its reasons for refusal. However prior to entering notices in respect of the repair or refurbishment of WEEE on the register, the Authority should carry out an inspection of the establishment or undertaking and should satisfy itself that the best available treatment, recovery and recycling techniques are being used. Entries should be removed from the register if no renewal notice is received or the Authority decides within 25 working days (WEEE, 2 months) of receiving a renewal notice to refuse it (Sch.2, para 11).

Establishments and undertakings carrying out notifiable activities are required to keep records detailing the quantity, nature, origin and destination and, where relevant, treatment method of all waste disposed or recovered. Records should be kept for two years (Sch.2, para 12).

Exemption Registration Authorities are required to carry out periodic inspections of establishments and undertakings, and every 12 months in respect of certain scrap metal operations involving waste motor vehicles and in respect of the treatment of WEEE.

Review of Exemptions

In July 2008, Defra, the Welsh Assembly and the Environment Agency published proposals for exemptions from environmental permitting. This follows an initial consultation in March 2007 and aims to provide a more risk based, balanced approach to regulating low risk waste recovery and disposal operations. The proposals affect the following areas:

- Construction: some activities (e.g production or use of more than 500 tonnes of aggregates from inert waste) will require a standard permit.
- Agriculture: some exemptions (e.g. burning up to 10 tonnes per day of plant matter) will still operate under an exemption but will need to re-register every three years.
- Waste management: varying limits on the quantities of waste stored in order to register a simple exemption, depending on whether the waste is stored and processed inside or out.
- Some waste activities that have operated without a permit or exemption will need to register an exemption; these include clinical waste, clothing, construction and demolition waste, electrical equipment waste, food waste from retailers and producers, furniture and household items, hazardous wastes, oil, production of small amounts of biodiesel from used cooking oil.

The consultation was open for comment until 23 October. Amending regulations are due to come into force on 1 October 2009.

2.4.4 Grant of Environmental Permits

(regs. 13-19 & Sch.5)

There are two types of environmental permit: "bespoke permits" which are individually tailored to an operator containing conditions specific to the installation, and Standard Permits for less complex operations which include a single condition referring to the standard rules with which the operator must comply. To date the Environment Agency have developed standard permits for a range of low and medium risk waste operations – see below 2.4.7.

Applying for a Permit

An application to operate a regulated facility should be made on a form provided by the appropriate regulator, and can be made electronically. Schedule 1 (Part 2) to the EPR lists Part A1, A2 and B activities, installations and mobile plant requiring a permit. The application should be made by the operator – defined (reg. 7) as the person who has, or will have, control over the operation of the regulated facility. Where different operators run different parts of a single regulated facility, each will need to apply for a permit covering their part of the operation.

Defra/WAG's core guidance suggests that an application for a permit, or to make a substantial change, be submitted prior to the start of construction of a facility, thus ensuring that operational and management techniques meet the requirements of the Regulations. It is also suggested that where planning permission is required, this application should be made at the same time to enable both applications to proceed in a coordinated manner. For novel or very complex installations, it is suggested that a "staged" application procedure be used; the use of this procedure must be agreed with the Regulator beforehand, with applications submitted at the design stage of each phase of the construction – Part A staged applications will be open to consultation in the usual way and all will be put on the Public Register. The applicant will still need to submit a consolidated version of all the applications for determination in the usual way.

Where a number of regulated facilities are run by a single operator on the same site, a single permit covering them may be applied for; as well as proximity, the extent to which the regulated facilities share a common management system can be taken into account in deciding if they qualify for a single permit. A single permit cannot cover a Part B installation, or a mobile plant, combined with any other category of regulated facility.

Where a range of activities (e.g. A1, A2 and Part B) are carried out in or on the same installation, the highest level of control will apply (Sch.1, para 2 of EPR).

The Secretary of State/Welsh Ministers may direct (reg. 33) that responsibility for either a specific installation or any description of installation be transferred to the other Regulator. They must notify the Environment Agency, local authority and any other affected person. The Environment Agency and local authority must publish the Direction on their websites.

The application form provided by the Regulator must be in a format which enables the applicant to provide sufficient information, including that required by any relevant Directive, to enable the application to be determined. The information to be included is similar to that required for a PPC permit (Table 2.2, page 61). The application should be accompanied by the appropriate fee – for A1 installations fees and charges are set out in the scheme made under s.41 of Environment Act – see 2.6; for Part A2 & B installations, fees and charges are set in accordance with a scheme made under reg. 65 of the EPR.

Schedule 5, Part 1 details the procedure for making an application, which can be submitted online.

Certain installations are exempt from the need for an environmental permit (Sch.1, para 3), and include:

- those carried out in a working museum to demonstrate an industrial activity of historic interest.
- those carried on for educational purpose in a school.
- those carried out an installation or mobile plant used solely for research, development and testing of new products.
- the running on or within an aircraft, hovercraft, mechanically propelled road vehicle, railway locomotive or ship or other vessel of an engine which propels or provides electricity for it.
- the running of an engine in order to test it before installation, or in the course of its development.
- those carried on as a domestic activity in connection with a private dwelling.

Timetable for Determination

Once the Regulator has accepted the application as being "duly made" (i.e. has all information needed and the appropriate fee has been paid), he should let the operator know and indicate the date by which it is expected to determine it – normally four months from the date of receipt for applications subject to public participation, three months for all others, and two months for applications to transfer a permit in whole or in part. More time may be needed if agreement needed on

- whether certain information is commercially confidential;
- for reasons of national security (which the SoS/Welsh Ministers determine);
- where consultation is required with another EU member state.

If he considers it necessary, the Regulator may issue a "further information notice" to enable determination (Sch.5, para 4). The period for determining the application will not begin until the Regulator has received the further information. If the Regulator still considers he does not have enough information to enable determination, he should review the decision to require further information and if he still agrees with it can issue a notice to the applicant stating that the application is deemed to have been withdrawn. There is a right of appeal against the deemed withdrawal.

Once the Regulator has accepted an application for a permit or for a substantial change variation as being "duly made", a copy should be put on the public register "as soon as reasonably practicable" – but see below 2.4.16.

If the application is not determined within the agreed period, the applicant can notify the Regulator that he considers it has been refused and can lodge an appeal – see 2.4.12 below.

Determining Applications

In determining an application for a permit, the Regulator must take into account whether the applicant will operate the installation or mobile plant in accordance with the permit and any conditions attached to it. The Regulator must refuse the application if he considers the applicant is not the operator of the installation, or if he believes the operator will not comply with the conditions attached to the permit. He may also refuse it if he believes the operator is not competent to run the facility in compliance with the permit, its environmental impact would be unacceptable, there is insufficient information on which to determine the application (despite having requested further information), or the additional requirements of relevant EU Directives cannot be met. Regulation 35 gives effect to Schedules 7-18 which outline the requirements in relation to the IPPC and various other Directives to which the Regulators must have regard – setting conditions to ensure compliance with any of the Directives of relevance to the installation. Further guidance on the requirements of each of the Directives has been published by Defra/WAG – see Appendix 2.3.

The Secretary of State/Welsh Ministers may require that certain applications be forwarded to them for determination, such as those of substantial regional or national significance or controversy, or those involving national security or foreign governments (reg. 62). The Regulator should inform the applicant that the application has been forwarded.

Best Available Techniques

One of the main purposes of the Regulations is to achieve a high level of protection for the environment as a whole, by preventing, or if that is not practicable by reducing emissions to air, land and water from Part A installations and mobile plant, and preventing or reducing emissions to air from Part B installations and mobile plant. Thus all permits and conditions attached thereto must therefore ensure that the installation will be operated in such a way as to ensure no significant pollution is caused and that all necessary preventative measures against pollution are taken. Regulators will include a general condition in permits requiring that operators use the Best Available Techniques (see 2.1.4 above) to prevent or reduce emissions. (Note: for existing PPC permits transferring to EP, the condition to use BAT remains an "implied" condition; new EP permits will contain a specific condition.) The BAT for an installation will take account of both technical and operational matters. When considering technical issues, the geographical location and local environmental conditions will be of relevance; when deciding on the best available technique, the cost to the operator of installing a particular technique may be balanced against the benefits to the environment. In determining BAT Regulators are required to have regard to any guidance issued to them; for local authorities this will include Secretary of State sector guidance (SG notes, see Appendix 2.3) and process guidance (PG) notes (see Appendix 2.4).

Defra's *Guidance on the IPPC Directive for Part A Installations and Mobile Plant*, contains guidance on the determination of BAT for various aspects of a plant's operations, as does the *General Guidance Manual on Policy and Procedures for A2 and B Installations*, including determining BAT for energy efficiency. The EU's BREF notes (see Appendix 2.3) also include guidance on BAT. Factors demonstrating operator competence include adequate environmental management systems in place, financial competence, operator's technical competence and his regulatory compliance record.

2.4.5 Consultation and Public Participation (Sch. 5)

Before taking a decision on whether or not to grant a permit or an application for a variation (or regulator-initiated variation) entailing a substantial change, the Regulator must give the public an early and effective opportunity to comment on the application (in accordance with the *Public Participation Directive*, see chapter 1, 1.25.2). This does not apply to mobile plant or Part B standard rules facilities. Nor does it apply to small waste oil burners, dry cleaners, the unloading of petrol into stationary storage tanks, or motor vehicle refuelling activities (Sch.5, part 1, paras 5-8).

Within 30 working days of receiving a duly-made application (and following determination of any issues such as commercial confidentiality, national security etc), the Regulator must ensure that it is brought to the attention of "public consultees" (see also Box), e.g. through local newspaper advertisement, giving details of where the application can be seen, the timescale within which, and the address to which, any representations should be made; it should also be advertised on the Regulator's website. Information given to public consultees should not include anything excluded from the public register on the grounds of national security unless the SoS/Welsh Ministers direct otherwise; similarly information excluded on the grounds of commercial confidentiality should only be given to those public consultees which are public authorities and the information is necessary for the exercise of their functions. Regulators have a duty to take any representations received into account in making their decisions; if they consider it necessary they can request further information and consult again to enable effective public participation in the decision-making process. They must then advertise their draft decision on whether or not to grant the permit, together with reasons for the decision, allowing 20 working days for any further representations. They then have 15 working days to consider further comments and make and publicise their decision. If no further comments are received the permit takes effect five working days later and the applicant notified. Copies of the permit and decision document are placed on the public register.

Public Consultees

The EP Regulations do not refer to statutory consultees, but instead refer to "public consultees", who are defined as "a person who in the regulator's opinion is affected by, is likely to be affected by, or has an interest in an application" (Sch. 5, para 1). Thus, as well as local trade bodies and organisations, NGOs, and individuals, the regulators will also consult appropriate national bodies, such as:

- Natural England/Countryside Council for Wales
- Local Authorities Co-ordinators of Regulatory Services (LACORS)
- Local authorities (Part A1 applications); Environment Agency (Part A2 & B)
- Food Standards Agency
- Health & Safety Executive
- Health Protection Agency/ National Public Health Services for Wales
- Primary care trusts/ Public Health Board in Wales
- sewerage undertakers
- harbour authorities/ local fisheries committee
- petroleum licensing authority

Defra's General Guidance Manual on LA-IPPC and LAPPC provides advice on which national consultees to inform about which applications. Local authorities regulators must also consult the Environment Agency on LA-IPPC applications, as they are responsible for setting water discharge conditions.

Information in applications which is to be excluded from the public register for reasons of industrial or commercial confidentiality or national security is also exempt from consultation, although there may be occasions when it would be appropriate for national consultees to view such information.

Part B and Mobile Plant

For Part B installations, once a copy of the application has been placed on the public register, the local authority should within 30 working days of receiving it take steps to bring it to the attention of public consultees – telling them where it can be viewed, inviting views and the deadline, as well as the address to which any representations should be made. If the local authority approves the application, they should notify the applicant, and place a copy of the permit on the public register.

Consultation is not required for applications for mobile plant or those for a standard rules permit (except if a Part A installation) (see also below); nor does it apply to applications relating to small waste oil burners, to dry cleaners, or to some petrol unloading and vehicle refuelling activities, though petrol licensing authorities must be consulted on service stations applications.

2.4.6 Permits and Conditions

The Regulator must refuse to grant a permit (or its transfer in whole or in part) if he is not satisfied that the applicant will be the operator of the regulated facility, or that the operator will comply with the permit and its conditions (Sch.5, part 1, para 13). If an application for a permit is refused, the applicant should be notified and reasons for refusal given. There is a right of appeal against refusal, also in respect of conditions – see below 2.4.12.

In granting a permit which is specific to a particular installation, the Regulator will include conditions with the aim of achieving a high level of protection for the environment as a whole and based on Best Available Techniques which balances costs to the operator against the benefits to the environment. Conditions must also ensure compliance with any relevant Directive requirements.

For Part A installations and mobile plant, conditions should also ensure

- energy is used efficiently;
- waste production is avoided in accordance with the framework Directive on waste (see chapter 5, 5.6) or where waste is produced it is recovered or, if that is not possible, is disposed of in such a way as to avoid any impact on the environment;
- measures are in place to prevent accidents and to limit their consequences; accidental releases or releases above limits set, and other abnormal occurrences which cause, or may cause, significant pollution must be reported to the Regulator without delay;
- transboundary pollution is minimised;
- protection of soil and groundwater and appropriate management of waste generated by the installation;
- suitable emissions monitoring equipment is in place; measurement methodology, frequency, and evaluation procedure should also be set out; the operator must provide the Regulator with data to enable compliance monitoring and provide regular reports on results of emissions monitoring;
- that when activities cease on site, the necessary measures are taken to avoid any pollution risk and return the site to a satisfactory state.

All permits (A and B installations) will include emission limit values (ELVs) for all pollutants likely to be emitted in significant quantities; these will be based on BAT and in the case of Part A installations will also take into account their potential to transfer pollution to another environmental medium. In setting an ELV account may be taken of the technical characteristics of the installation, its geographical location and local environmental conditions. Conditions should ensure no breach of EU environmental quality standards (Part B – AQS only) and thus where such legislation requires a stricter ELV, then this should be set. Account will also need to be taken of the effect of the installation's emissions on local air quality and the achievement of national air quality objectives.

Local authorities should consult the Environment Agency on permit applications and those for substantial variations for LA-IPPC installations (Part A2); this will enable the Agency to decide on whether to issue a notice under reg. 58 requiring a condition regulating a discharge to water from an A2 installation or mobile plant; the local authority must have regard to any emission limit value specified by the Agency; it may not include a less stringent value, though it may include a stricter one, based on what it considers to be BAT for the entire installation.

Permit conditions for both Part A1 and A2 installations for noise and vibration should reflect the standards of statutory nuisance legislation under the EPA (see chapter 1, 1.29), and thus the "BAT" in this instance should be seen as the "best practicable means" being used to avoid or reduce noise. Activities at an installation regulated by the EP regulations are exempt from statutory nuisance legislation; non-IPPC regulated activities, e.g. barking dogs and burglar alarms, however, are not.

A permit for a Part A installation (reg.15), may include an off-site condition requiring the operator to carry out work which requires consent from another person (e.g. the owner of the land); in such cases (which will have followed consultation) where rights have been granted to carry out the work, compensation may be payable, (Sch.5, Part 2).

Review of Permits

Regulation 34 requires Regulators to review permits periodically, and to make periodic inspections of regulated facilities. As well as reviewing permits to ensure that conditions remain relevant and continue to reflect BAT, conditions may need reviewing to incorporate new or revised EU or national legislation. Conditions reflecting the requirements of the Groundwater Regulations must be reviewed every four years.

2.4.7 Standard Rules (reg. 26-30)

Standard rules – similar to general binding rules – set out requirements which are common to a particular class of activity; they define how the operator must carry out the activity, for example by limiting the types of operation that can be carried on at the site, or place restrictions on such aspects as size or location. In preparing or revising SRs the "rule-making authority" should consult all those it considers are representative both of the communities and of the regulated facilities who will be affected by the rules. SRs should be published on the rule-making authority's website. The rule-making authority in respect of Part A1 installations is the Environment Agency for Part A2 and B, the Secretary of State/Welsh Ministers.

Before publishing any revisions to SRs on the website, the regulator must notify any affected operators of the proposed revisions, and of the date they will be published (which must be at least three months after notification) which is when the revisions will take effect. The rule-making authority may revoke SRs, having consulted any affected persons; the Regulator should then vary environmental permits to delete the condition relating to SRs, replacing it with other appropriate conditions.

The Environment Agency has developed standard rules for a number of low to medium risk activities that required a waste management licence – see Appendix 5.1. Operators of waste management facilities covered by the SRs have the option of applying for either a standard rules permit or a "bespoke permit"; there is no right of appeal against SRs, nor can they be varied; instead if the activities at the site change in a way not covered by SRs, then the operator should apply for a "bespoke permit".

Applications for SR permits are not put out for consultation as there will have been consultation in their development, including risk assessment; any amendment to SRs will be consulted on, also if it is proposed to revoke a set of SRs. Standard rules and any supporting technical guidance are published on the Environment Agency's website.

2.4.8 Variation of Permit (reg. 20, Sch. 5)

Variations to a permit and its conditions may be initiated by the operator or by the Regulator; operator-initiated variations can cover minor changes such as technical alterations or modifications in operational or management practices and can often be notified to the Regulator without the need for a formal application for variation; applications for non-substantial variations do not require consultation or payment of a fee, and will normally be determined within three months.

Operators must however make an application to the Regulator before making a "substantial change" to the installation or its operation. The Regulations define a substantial change as one "which in the Regulator's opinion may have significant negative effects on human beings or the environment, and includes

a) any change in operation which itself meets the thresholds, if any, set out in Part 1 or Schedule 1, and
b) any change in operation of an incineration or co-incineration plant for non hazardous waste which would involve the incineration or co-incineration of hazardous waste."

Applications for substantial changes follow the procedure for making applications for new permits (see above 2.4.4), including arrangements for consultation and public participation, decision-making etc (2.4.5 above), and will usually take four months to determine. There is a right of appeal against refusal to vary a permit. Regulators have the option to issue a consolidated permit reflecting the changes – in such cases the right of appeal only covers the changes to the original permit.

Regulator-initiated variations may result from the need to vary conditions following a periodic review of the permit, revised statutory guidance, the transfer of a permit to another operator, revision of a relevant environmental quality standard, or declaration of an AQMA. As above, non-substantial variations do not need to be publicised or a fee paid, but will be put on the public register. For variations involving a substantial change, the regulator should notify the operator of its intentions and that the changes will be subject to consultation (public participation); the regulator should advise the operator of the fee required and the timescale for making representations. The procedure for determining the proposed variation is similar to that outlined above; the regulator must notify the operator of its decision and reasons for it, the date on which the variation will take effect, and right of appeal with all documentation being placed on the public register.

Where an application for variation is proposing a substantial change in the operation of an installation which is likely to have a significant negative effect on the environment of another Member State, the Secretary of State should send a copy of the application or proposed variation notice to the other Member State at the same time that it is advertised. Such applications will not then be determined until consultations with the Member State concerned have been completed, at which time the Regulator's four month determination period begins.

2.4.9 Transfer of Permit (reg. 21)

Where it is proposed to transfer all or part of a permit to another operator, the current and future operators (the "proposed transferee") should make a joint application to the Regulator, and should enclose the current permit together with the prescribed fee (A1 installations & mobile plant – *Env Act 1995*, s.41; A2 & B – EP reg.65) giving full details of them both, and in the case of a partial transfer, a map or plan of the installation showing which parts are to be transferred.

The Regulator has two months, or longer if agreed with the applicants, to determine the request, which should only be agreed if the Regulator is sure that the proposed transferee will operate the installation (or part installation) in accordance with the permit and its conditions. If the Regulator has neither agreed nor rejected the transfer within the time limit, the applicant(s) can serve notice on the Regulator that they deem the application to have been refused; there is a right of appeal against deemed refusal, which must be made with six months.

Transfers of whole permits will be effected through the endorsement of the current permit with the new operator's details, or they may issue a new permit, if appropriate varying or adding new conditions; in the case of partial transfers, a new permit will be issued to the new operator covering the transferred parts of the installation, with the original permit being returned to the first operator, endorsed to show the extent of the transfer. For partial transfers, it may be necessary for the conditions of the permit to be varied to reflect the new operating responsibilities.

The offence of operating an installation without a permit includes a new operator taking over prior to completion of the transfer procedure; enforcement action may also be taken against both the new and original operators for failing to submit a joint application to transfer.

2.4.10 Revocation of Permit (reg. 22 & 23)

The Regulator may, by notice, revoke all or part of a permit, to take effect not less than 20 working days after the date on which the notice was served; a partial revocation may apply to either the operation of part of the installation or to some of the activities carried on at the installation. As well as stating the date on which the revocation (or partial revocation) is to take effect, it should specify, in the case of partial revocation, the activity or operations covered by the revocation.

If the Regulator considers certain steps need to be taken to avoid a pollution risk or to return the site to a satisfactory condition once operations have ceased or partially ceased because of a revocation notice, the notice should say so and specify the steps (not already required by conditions attached to the permit) to be taken. Those sections of a permit requiring specific action on cessation of operations remain in force until the Regulator is satisfied that the necessary remedial action has been taken and has issued a certificate to this effect.

2.4.11 Surrender of Permit (reg. 24 & 25)

An operator of a mobile plant, Part B installation or a Part A2 installation falling within section 5.1 of Sch.1 (incineration and co-incineration of waste) may surrender his permit, in total or in part, by notifying the Regulator. Such notification should be made on a form provided by the Regulator and will include full contact details for the operator and, in the case of partial surrender, a site map or plan identifying that part affected by the surrender. The date for surrender or partial surrender should be not less than 20 working days from the date of the notification. In the case of a partial surrender, the Regulator may consider it necessary to vary the terms of the permit to take account of the partial surrender; in this case the Regulator will serve a notice on the operator, specifying the variation, and the date that the variation will take effect (with the partial surrender taking effect on the same date as the variation if this is later than the original surrender date).

An application to surrender all or part of a permit for a Part A installation (except for an A2 installation falling with section 5.1 of Sch.1) should be made to the Regulator, with the appropriate fee. It should include full contact details of the operator, and in the case of a partial surrender, a site map or plan identifying the surrender unit. A site report should also be submitted which identifies any changes in the condition of the site from those reported at the time of applying for a permit, and describes the steps taken to avoid any pollution risks or to return the site to a satisfactory state.

The Regulator has three months (or longer as agreed with the operator) in which to determine the application to surrender; if the regulator is satisfied with the condition of the site and that no further steps to avoid any pollution risk need to be taken, it should by notice accept the application, stating the date on which surrender takes effect; if it is not satisfied, it should specify the further steps to be taken to return the site to a satisfactory condition. In the case of a partial surrender, the Regulator may vary the conditions attached to the permit to take account of the surrender. Failure to determine the surrender within the agreed period will be deemed a refusal if the applicant notifies the Regulator that he treats it as such.

2.4.12 Appeals (reg. 23 & Sch. 6)

There is a right of appeal to the Secretary of State [Welsh Ministers] in the following instances – any appeal should be made within six months of the decision being appealed against:

- refusal, or deemed refusal, of permit or disagreement with conditions attached to it;
- refusal, or deemed refusal, to vary conditions of a permit or disagreement with variation notice;
- rejection of transfer application; or disagreement with variation of conditions as a result of the transfer;
- rejection of surrender application, or disagreement with variation of conditions attached to permit following partial surrender;
- non-authorisation of closure procedure in respect of a landfill site.

An appeal may also be made against the following:

- revocation notice – before the date on which the notice takes effect; revocation notices do not take effect until the appeal has been determined or withdrawn;
- variation, enforcement and suspension notices – within two months of the date of the notice; these notices are not suspended pending the outcome of an appeal;
- service of notice that an application is deemed to have been withdrawn (because of no response to earlier request for further information regarding an application) – no later than 15 working days after receiving notice;

- receiving notice that application for information to be kept off the public register has been unsuccessful – no later than 15 working days from receipt of notice.

There is no right of appeal against a notice or decision implementing a Direction from the Secretary of State or Welsh Ministers. The Secretary of State [Welsh Ministers] may determine appeals themselves or appoint the Planning Inspectorate for this purpose (*Env. Act 1995*, s.114 & Sch.20). Appeals relating to the following will normally be determined by the Secretary of State [Welsh Ministers]:

- processes or sites of major importance;
- those giving rise to significant public controversy or legal difficulties;
- those which can only be decided in conjunction with other cases over which inspectors have no jurisdiction;
- those which raise major or novel issues of industrial pollution control which could set a policy precedent;
- those which merit recovery because of the particular circumstances.

Schedule 6, of the EPR outlines the procedure for making and determining appeals. The appellant should send written notice of appeal to the Secretary of State [Welsh Ministers] with the following documentation:

- statement of grounds for the appeal;
- statement as to whether the appellant wishes the matter to be dealt with through written representations or by a hearing.
- copy of the relevant application and permit;
- relevant correspondence between the Regulator and operator;
- copy of the decision which is the subject of the appeal.

A copy of the notice of appeal, and of the documents above should be sent to the Regulator at the same time. An appeal may be withdrawn by notifying both the Secretary of State [Welsh Ministers] and the Regulator.

Within 10 working days of receiving notice of the appeal, the Regulator must send details of it to public consultees and others who have made representations to the Regulator on the subject of the appeal and to any persons who it is thought may have a particular interest in the subject matter of the appeal. As well as giving details of the appeal, the notice should state to whom any representations should be made within 15 working days of the notice and that these will be copied to both the appellant and the Regulator, and that if a hearing is to be held, they will be notified of the date. The notice should also state that any representations received will be put on the public register unless a written request is made that this should not happen. In such cases a statement will be added to the register saying that representations concerning the appeal have been received and are subject to a request to exclude from the register. Within 10 working days of sending out the notice of appeal, the Regulator should notify the Secretary of State [Welsh Ministers] (or their appointee) of all the persons to whom it has been sent.

Where the appeal is to be decided in writing, the Regulator must submit any comments in writing to the Secretary of State [Welsh Ministers] (or appointee), copied to the appellant, within 28 days of receipt of the appeal notice, with the appellant having 17 days (from the date the Regulator submitted its comments) to make further representations, which should be copied to the Regulator. Both sides have 14 days from the date of receipt to comment on any other statements submitted in connection with the appeal.

Where a hearing is to be held, the Secretary of State [Welsh Ministers] (or appointee) will notify the appellant and the Regulator of the time, date and place – a minimum of 28 days notice will be given. If the hearing is to be held partly or wholly in public, all those who have made written representations regarding the appeal matter as well as other consultees should be notified of the date etc, and details of the hearing advertised in a local newspaper – in both cases this should be done at least 21 days in advance of the hearing.

Following determination, copies of the decision and any report (following a hearing) are sent to the appellant, the Regulator and statutory consultees; copies of the decision only are sent to all others who made representations. In those cases where the Secretary of State has recovered an appeal, the Inspector will report to the Secretary of State [Welsh Ministers] who will then make a decision on the appeal. There is a right of appeal to the courts against the appeal decision; if upheld, the Secretary of State [Welsh Ministers] should send all those notified of the original decision details of the further matters to be considered by the appeal, inviting further representations within 28 days and may either re-open or instigate a hearing in which case the procedure for a hearing outlined above will be followed.

Documents relating to the determination of the appeal are put on the public register.

An appellant may withdraw an appeal by written notification to the Secretary of State [Welsh Ministers], copied to the Regulator. The Regulator should then notify all those to whom notice of the appeal was sent that it has been withdrawn.

2.4.13 Inspection and Compliance Monitoring

An EU Recommendation (2001/31/EC) adopted on 4 April 2001 sets minimum criteria for environmental inspections for all industrial installations and other facilities subject to environmental licensing or permitting; the aim is to ensure more consistent application of EU legislation across Member States. This recommends that Member States should draw up plans specifying what environmental inspections should cover and that these should be publicly available. Site inspection reports should be made publicly available within two months of the inspection.

Operators of all regulated facilities are required to carry out their own compliance monitoring to ensure their facility is being operated in accordance with the conditions of its permit and that their activities are not causing pollution of the environment or harm to human health. The regulatory authorities are also required to carry out periodic inspections of all permitted installations and of exempt waste operations. The Regulator also looks at other components of an operator's procedures, including management and competency of personnel and adequacy of monitoring and of the equipment being used. Defra's *General Guidance Manual on Policy and Procedures for A2 and B Installations* summarises the EU recommendation suggesting that it represents "good practice".

In April 2005, Defra published risk assessment methodology for use by local authorities in determining the level of risk posed by Part A2 activities for which they are responsible. This has two elements – environmental impact appraisal (which considers a site's potential environmental impact, level of upgrading to meet regulatory requirements and location), and operator performance appraisal (which assesses how well the operator manages the potential environmental impact of the activity). These impacts will be scored and an operator's overall score will determine the amount of regulatory time the site needs, which in turn will be reflected in the charges levied. Further guidance on this is given in Defra's *General Guidance Manual on Policy and Procedures for A2 and B Installations*, which also provides detailed guidance on monitoring requirements, reporting and record keeping.

To enable the Environment Agency to adopt a common approach to regulation under PPC and other regulatory regimes, as well as target those industries that pose the greatest risk to the environment, it has developed "Environment Protection Operator and Pollution Risk Appraisal" (EP OPRA); this is designed to score operations on the basis of environmental hazard, and operator performance.

The score for environmental hazard will take account of the installation's

- *complexity* – significant releases to one or more media, potential for accidental emissions, whether it is a specified waste activity, amount of regulatory activity needed to assess and maintain compliance and thereby public confidence;
- *location* – presence or absence of key receptors – e.g. schools, hospitals, whether in groundwater protection zone or in AQMA; potential for direct release to controlled water;
- *emissions* – based on the annual load of pollutants which would be emitted if the installation operated at its permitted emission limit values, compared with the emissions threshold which reflects the pollutant's potential to cause environmental harm.

Operator performance will be scored on the basis of whether there is an effective and documented environmental management system in place, with particular attention being paid to operations and maintenance, competence and training, emergency planning, organisation, and auditing, reporting and evaluation. Incidents of non-compliance with permit conditions and the classification (i.e. because of its potential impact) of the breach given by the Agency will also affect the score.

Sites which are particularly complex, have high emissions, are in sensitive locations and have poor or no documented environmental management systems, as well as a poor Compliance Classification Scheme rating (see above) – and thus high OPRA scores – will be targeted by the Agency for more frequent compliance assessments, which will also be reflected in the charges levied. The Agency will review OPRA scores periodically; operators can also request an amendment of their OPRA score demonstrating the improvements and changes made since the assignment of their score.

The Environment Agency's "compliance classification scheme", introduced in April 2004, aims to ensure more consistency in assessing non-compliance with permit conditions and subsequent potential environmental impact, and thus enforcement action. The four categories of non-compliance, and normal prosecution response, are

- Compliance classification scheme (CCS), Category 1 – potentially major environmental effect – normal response will be prosecution;
- CCS Category 2 – potentially significant environmental effect – normal response will be formal caution or prosecution;
- CCS Category 3 – potentially minor environmental effect – normal response will be a warning;
- CCS Category 4 – no potential environmental effect – normal response will be a warning.

An installation's CCS score will be used to assess its overall score under EP OPRA, which in turn affects the amount of regulatory attention received and thus fees and charges to be paid.

2.4.14 Enforcement and Suspension Notices (regs. 36 & 37)

As well as revocation notices (see above), the Regulator may issue an ***enforcement notice*** if he is of the opinion that an operator is not complying with his permit or its conditions. The notice should state the contravention – or potential contravention – the remedial steps to be taken, and the period within which this must be done. The notice may be withdrawn at any time.

If the Regulator is of the opinion that the operation of a regulated facility, or the way in which it is being operated, involves a serious risk of pollution, he can issue a ***suspension notice***. The notice should detail the pollution risk involved in operating the installation, the remedial action and the period within which this should be done. The notice should also state that, until the notice is withdrawn, the permit ceases to authorise the operation of the installation or the carrying out of specified activities; where the permit is to continue to have effect, the notice will authorise the carrying out of certain activities and specify the additional steps to be taken in carrying them out.

A suspension notice will be withdrawn once the steps taken to remove the pollution risk have been taken or in the case of waste management activities, they are back in the hands of a technically competent person.

The Environment Agency's policy statement outlining the general principles which inform its enforcement and prosecution decisions can be accessed on:

www.environmentagency.gov.uk/commondata/acrobat/enfpolicy.pdf

2.4.15 Offences (reg. 38)

It is an offence to operate a regulated facility without a permit, in contravention of a permit or of its conditions, to fail to comply with the terms of an enforcement or suspension notice, or to fail to comply with a court order requiring remedial action following conviction. Summary conviction in these cases may result in a fine not exceeding £20,000 and/or up to 6 months' imprisonment*. Conviction on indictment may result in an unlimited fine and/or up to five years' imprisonment. (Section 154(1) of *the Criminal Justice Act 2003* will enable magistrates to impose a prison sentence of up to 12 months; clause 79 of *the Climate Change Bill* increases the maximum fine to £50,000.)

Other offences, such as failing to comply with an information notice, making false or misleading or false statements to the Regulator, making false entries in any records which are required to be kept, forging documents etc, may result in a fine not exceeding the statutory maximum on summary conviction, or an unlimited fine and/or up to two years' imprisonment on conviction on indictment.

2.4.16 Public Registers (regs. 46-56 & Sch. 19)

The Regulators have a duty to maintain public registers (which can be in electronic form) containing details and all documentation relating to applications for permits, permits, variations, transfers and surrenders of permits, appeals, enforcement, revocation and suspension notices etc, as well as details of any relevant convictions. Local authority registers will also include details of permits etc relating to those regulated facilities in their area for which the Environment Agency is the Regulator.

However, if the operator or someone else who has supplied information for the application – "the information subjects" – object to the inclusion on the Register of information in part of an application on grounds of confidentiality, that information should not be put on the Register until the request ("objection notice") has been determined. The Regulator has 20 working days (or such longer period as is agreed) in which to make a decision on whether to accept the request; if the request is accepted, a statement to this effect should be put on the public register and giving the reason. If the request is not accepted, the Regulator should notify the applicant, allowing 15 working days (in case the applicant lodges an appeal) before putting the application on the public register. If the Regulator considers certain information in an application to be confidential but has not received an objection notice, he should notify the information subject again allowing 15 days for a response as to whether the information is be considered as confidential or not. If the Regulator fails to determine a request for information to be considered confidential within 20 working days (or other agreed period), the applicant should notify the regulator that he deems the request to have been refused, and he can then appeal within 15 working days.

An appeal against a decision of the Regulator with regard to confidentiality should be made to the Secretary of State [Welsh Ministers] within 15 working days of the determination. If the appeal is unsuccessful, information should not be put on the Register until after five working days after determination. If an appeal is withdrawn five working days should be allowed before the information goes on the register.

The information in dispute will not be put on the register until the appeal is determined; if the request is granted a notice must be put on the register to this effect. Exclusion of data from the register on the grounds of commercial confidentiality lasts four years after which it will be added to the register unless a further four year exclusion is applied for and agreed.

The Secretary of State may direct that specific information or information related to specific issues should not be put on the register in the interests of national security. Regulators should notify the Secretary of State of any such information they exclude from the register. He may also direct that information related to certain issues affecting national security be referred to him for decision on inclusion on the register. An operator may also apply to the Secretary of State for specific information to be excluded on grounds of national security (reg. 47(7)), and should notify the regulator that they are doing so; the information remains off the register pending the Secretary of State's determination.

Information will be removed from the register "if it is no longer relevant for the purposes of public participation", and also:

- if an application for permit is withdrawn prior to determination – all particulars to be removed between two to three months of withdrawal;
- if the Regulations cease to apply to an installation or mobile plant due to amendment of Schedule 1 – all particulars to be removed between two and three months from when the installation ceases to be prescribed;
- five years after the date of any formal caution, and convictions in accordance with *the Rehabilitation of Offenders Act 1974*.

The registers must be available for inspection by the public, free of charge, at all reasonable times; a charge may be made for providing a copy of an entry on the register.

2.4.17 Powers to Remedy or Prevent Pollution (reg. 57)

The Regulator may take action to deal with, in his opinion, a serious risk of pollution from the operation of a regulated facility, having given the operator five working days notice. Costs are recoverable from the operator unless he can show that there was no serious risk of pollution, or that the costs were incurred unnecessarily.

2.4.18 Information, Directions and Guidance (regs. 60-64)

The Secretary of State [Welsh Ministers] may, by notice, require the Regulator to supply any information as may be required to enable him to discharge his obligations in relation to European Community or international agreements.

Directions from the Secretary of State [Welsh Ministers] to Regulators may relate to any aspect of their functions under the Regulations; they may be either general or specific and may direct them as to the way in which they are or are not to exercise their powers.

Where the Secretary of State [Welsh Ministers] receives information from another Member State in relation to the operation of an installation outside the UK which is likely to have significant negative effects on the environment in England or Wales, he should direct the Environment Agency to bring it to the attention of anyone likely to be affected, providing those persons with an opportunity to comment.

Regulators have a duty to comply with any Directions, which should be in writing, and which may be varied or revoked by further Direction.

The Environmental Protection (England) (Crematoria Mercury Emissions) Direction 2008 required Regulators to vary the permits for crematoria by 25 July 2008 to insert a condition relating to abatement equipment. The condition requires crematorium operators to provide written details by 31 October 2008 concerning its arrangements for fitting mercury abatement equipment by 31 December 2012.

2.4.19 Fees and Charges (reg. 65)

The Secretary of State (for England) and the Welsh Assembly Government (for Wales) may make a scheme for the fees and charges payable in respect of those installations regulated by local authorities – i.e. Part A2 and B installations; this includes fees for permit applications, variations, transfers and surrenders, as well as subsistence charges. Charges are based on risk, with lower risk installations paying lower charges. Non-payment of subsistence charges may result in revocation of a permit. Where an A2 installation involves a discharge to water, there is an additional charge to cover Environment Agency costs. Separate charges are levied to cover air monitoring. It is also proposed to levy charges for individual installations on the basis of their environmental impact and operator performance.

Service stations, vehicle refinishers, dry cleaners and small waste oil burners under 0.4 MW are eligible for reduced fees.

Charges for 2008-09 were detailed in *the Local Authority Permits for Part A Installations and Mobile Plant (Fees and Charges) (England) Scheme 2008*, and similar scheme for Part B installations and mobile plant. A similar scheme covers Wales.

Fees and charges for A1 installations are levied under a scheme made under the *Environment Act 1995*, s.41 – see 2.6.1.

2.5 POLLUTION PREVENTION & CONTROL REGULATIONS, Scotland & Northern Ireland

The Pollution Prevention and Control (Scotland) Regulations 2000 (SSI 2000/323) came into force on 28 September 2000, with similar regulations for N. Ireland (SR 2003/46) coming into force on 31 March 2003. Implementation of the Regulations belatedly fulfilled the requirement to transpose the EU's IPPC Directive into national legislation, which should have been done by 31 October 1999. All the Regulations have been amended a number of times – see Table 2.1.

The Regulations apply to all installations, including mobile plant, carrying out an activity listed in Annex 1 to the Directive. In ***Scotland*** Schedule 1 to the Regulations lists Part A installations and mobile plant and activities regulated for IPPC and Part B for local air pollution control; the Scottish Environment Protection Agency (SEPA) is the regulating authority for all installations.

In ***Northern Ireland*** Part A and Part B installations and mobile plant (Sch.1 to the NI regulations) are regulated by the Northern Ireland Environment Agency (NIEA) – this was established on 1 July 2008, replacing the Environment and Heritage Service; Part C installations, for local air pollution control are regulated by district councils; Part C mobile plant are normally regulated by the local authority in which the operator has his principal place of business. Schedule 1 also includes the dates within which existing installations had to apply for a permit. Where the principal place of business is outside NI and the mobile plant has not yet been operated in NI, the Regulator will be the local authority in which it first operated, or in whose area it is intended to first operate; or where the plant is already covered by a permit, the local authority which first granted the permit (reg. 7(5)).

Table 2.1: PPC Amendment Regulations

Scotland

SSI 2003/146 & 221, effective 01.04.03 – are largely concerned with clarifying the principal Regulations; they also transfer certain activities from Part B to Part A and amend the date by which existing installations carrying out those activities should apply for a permit.

SSI 2004/110, effective 31.03.04 – make various changes to Sch.1, including: exempting certain small scale chemical plant from PPC; and limit the requirement of a PPC permit for the slaughtering of animals only when carried out in slaughterhouses.

SSI 2004/512, *The Control of Volatile Organic Compounds (Petrol Vapour Recovery) (Scotland) Regulations 2004*, effective 24.12.04 – amend the PPC Regulations to exempt from PPC, controls on VOC emissions from petrol service stations with an annual throughput of less than 500m^3.

SSI 2005/101, effective 01.04.05 – exempts activities involving the use and recovery of chromium, manganese, nickel and zinc from IPPC control (but are still controlled for releases to air); require operators of animal feed incinerators with a capacity exceeding 10 tonnes per day to apply for Part A PPC permit; removes requirement for waste incinerator operators planning to cease operations before 28 December 2005 to apply for WID permit; inserts a definition of animal feed compounding activities into section 6.8 of Schedule 1.

SSI 2005/340, effective 02.07.05 – these insert a number of new activities into Schedule 1 of the Regulations, together with dates for making permit applications; research, development and testing activities of new products and processes are exempted from PPC controls, with other amendments covering mobile plant. A further amendment enables SEPA to issue consolidated permits for a site where, for instance, a permit has been issued covering a variation or an application for a new permit is made.

SSI 2005/510, effective 16.11.05 – implement the public participation provisions of Directive 2003/35 to apply to permits and variations, inserting additional requirements for advertising draft determinations etc. The Regulations also insert a new section 5.4 (waste recovery activities) into Schedule 1, and amend the definition of vehicle coating in section 7 (Sch. 1).

Northern Ireland

SR 2004/507 implement the UK derogation to exempt from PPC controls VOC emissions from petrol service stations with an annual throughput of less than 500 m^3.

SR 2005/285 make various changes to Sch.1 and exempt from PPC VOC emissions from certain service stations.

SR 2006/98 – effective 23.04.06 amend the 2003 Regulations to implement the requirements of Directive 2003/35 (see chapter 1, 1.26.2) to provide for public participation in respect of the drawing up of certain plans and programmes relating to the environment.

SR 2007/245, effective 20.05.07, excludes from the need for a permit landfill gas engines (3-50MW) used to convert landfill gas produced by biological degradation of waste in landfill; they also add motor vehicle refuelling activities (petrol vapour recovery Stage II controls) to those that require a permit.

Reg. 7(6)-(12) enables the DOE to direct that responsibility for a Part C installation be transferred to the NIEA in respect of a specific installation ("specific direction") or a general class of installation ("general direction"). In the former case, the DOE should notify the NIEA, local authority and operator of the installation concerned, and in the latter case the NIEA and all local authorities; the notice, which should also be advertised in the *Belfast Gazette* and locally, should include the date on which the Direction takes effect and, where appropriate, its duration.

The Regulations follow the Directive's definition of an installation as

(i) a stationary technical unit where one or more activities listed in Part 1 of Schedule 1 are carried out; and

(ii) any other location on the same site where any other directly associated activities are carried out which have a technical connection with the activities carried out in the stationary technical unit and which could have an effect on pollution.

New or substantially changed installations in Scotland (i.e. those brought into operation or undergoing substantial change on or after 31 October 1999) had to apply for IPPC permits by 1 April 2001, and in N. Ireland by 1 January 2004.

Existing Part A installations (in operation before 31 October 1999, or for which an authorisation was granted or requested before 31 October 1999 and brought into operation before 31 October 2000) were phased in by sector in Scotland between 2001-2007, and in NI between 2004-2007, in accordance with the timetable in Schedule 3 to the Regulations. Exceptions to this were:

- where it was proposed to make a "substantial change" – defined by the Directive as "a change in operation which in the opinion of the competent authority may have significant negative effects on human beings or the environment"; an application for a permit had to be made ahead of the phase-in date for the sector for that part of the installation affected by the change.
- where an installation is comprised of two IPPC linked activities – i.e. one activity depends on the other for its operation – with different phase-in dates; an application for a permit could be made at the earlier date. However an operator could apply to the Regulator for the phase-in date to be that of the primary activity undertaken (Sch.3, Part 1, para 5).
- where an operator of IPC or LAPC processes on a single site, but with different phase-in dates, wished to bring both within PPC at the same time under a single permit.

The PPC Regulations also apply to, and list (also in Schedule 1), Part B (Sc) and Part C (NI) installations); these largely comprise those formerly prescribed under the EPA regulations (NI: IPCO), but which are not required to be regulated for IPPC; these continue to be regulated for air pollution by SEPA in Scotland and by local authorities in Northern Ireland but under the PPC Regulations, with existing authorised Part B and Part C installations and mobile plant transferring to PPC on the basis of "deemed" applications.

The Regulations do not apply to the following activities (Sch.1, Part 2):

- Part B (& Part C, NI) activity in which releases to air of a prescribed substance are so trivial as to be incapable of causing pollution, or its capacity to cause pollution is insignificant; this exemption does not apply where the release may result in a noticeable offensive odour outside the site.
- An activity carried out at a working museum to demonstrate an industrial activity of historical interest, or is carried out for educational purposes in a school.
- An activity carried out as a domestic activity in connection with a private dwelling.
- Running an engine which provides electricity on or within an aircraft, mechanically propelled vehicle, hovercraft, railway or locomotive, ship or other vessel; running an engine for testing or development purposes.
- Use of a fume cupboard in specified circumstances.
- Research, development and testing of new products and processes if carried out at an installation or mobile plant (or in part of an installation or mobile plant) used solely for that activity (SR 2005/285; SSI 2005/340).

The Offshore Combustion Installations (Prevention and Control of Pollution) Regulations 2001 (SI 1091), which came into force on 19 March 2001, implement the IPPC Directive for offshore combustion installations with an aggregate thermal input exceeding 50 MW. The Regulations, which cover the whole of the United Kingdom, required existing installations to apply for a permit by 30 October 2007. The Regulations are within the remit of the Department for Business, Enterprise and Regulatory Reform (BERR).

Guidance

Practical guidance on IPPC is available from both SEPA and the NIEA, as well as guidance on the application procedure etc – see Appendix 2.3 to the *Pollution Control Handbook*, and also:

- www.sepa.org.uk/ppc/guidance
- www.ni-environment.gov.uk/pollution/ippc
- www.netregs.gov.uk (particularly aimed at helping small and medium sized enterprises businesses understand and comply with environmental laws, as well as providing advice on good environmental practice)

2.5.1 Application for a Permit
(Sc reg. 7-9, NI 6-9; & Sch. 4, Part 1)

An application for a PPC permit to operate an existing installation or mobile plant should be made on the relevant form to the appropriate Regulator during the three month period prescribed in the Regulations (Sch.3 – there are some differences between Scotland & NI), and should be accompanied by the prescribed fee (see 2.5.8). The application should be made by "the person who will have control over the operation of the installation or mobile plant" (reg. 10(3), Sc: 7(3)). The Regulations allow applications to be submitted electronically with the permission of the Regulator. It is an offence to continue to operate an existing installation after the period in which a "duly made" (i.e. able to be legally determined by the Regulator) application should have been made. Schedule 4, Part 1, of the PPC Regulations, details the information required for the application which should also demonstrate that adequate environmental management systems are in place – see Table 2.2.

For new installations, both SEPA's and the NIEA's *IPPC Practical Guide* suggest that an application for a permit be submitted prior to the start of construction, thus ensuring that operational and management techniques meet the requirements of the Regulations. For novel or very complex installations, it is suggested that a "staged" application procedure be used; the use of this procedure must be agreed with the Regulator beforehand, with applications submitted at the design stage of each phase of the construction. The applicant will still need to submit a consolidated version of all the applications for determination. Where an installation comprises various activities with different application periods, then the period for applying for a permit will usually be that for the activity requiring permitting first; if there is more than one operator for the various activities at an installation, then all should apply for a permit at the same time.

In ***Scotland***, SEPA has four months (or longer as agreed) in which to determine the application, beginning with the date on which the application was received; in ***N. Ireland*** 6 months is allowed for new Part A installations and mobile plant and applications for substantial change, and 9 months for existing Part A installations and mobile plant, Sch.4, para 15). The period for determination will normally begin when the Regulator accepts the application as being "duly made" – i.e. it is complete and has all the information the Regulator requires for making a determination; exceptions to this are (Sch.4, para 15)

- applications referred to Scottish Ministers [DOENI] – see below;
- if the Regulator requests additional information, the four month period will begin on the date on which the further information is received;
- where a person has been consulted over off-site conditions, the four month period will begin after the period allowed for representations;
- where issues of commercial confidentiality or national security have been determined, the four month period will begin after the 28 day period allowed for advertising the application (which itself begins 14 days after the issue has been decided).

Within 28 days (beginning 14 days after making the application) the applicant should publish details of the application in one or more local newspapers covering the area in which the installation is located, and for a Part A installation or mobile plant in the *Edinburgh [Belfast] Gazette*, as appropriate, as well; applications for a permit for a non-IPPC landfill site do not have to be advertised in the *Edinburgh [Belfast] Gazette* (see *Landfill Regs*, 5.29.1). The advertisement should state where, how and at what times the application can be inspected by the public; those for Part A installations and mobile plant should also confirm that the application covers all matters required to be included (i.e. complies with public participation Directive: SSI 2005/461; SR 2006/98). The advertisement should also note that any views on the application should be sent in writing (or by email) to the Regulator within 28 days of the date of the advertisement, and that any representations received will be put on the public register unless a request is made that they should not.

Applications for waste oil burners less than 0.4 megawatts, the unloading of petrol at service stations, and those for Part B mobile plant do not have to be advertised (Sch. 4, para 8).

Table 2.2: Information Required for PPC Applications

All Applications

- Full details of the operator (including name, address, ultimate holding company, registered office etc), and for a Part A installation or mobile plant – address, national grid reference, map or plan showing the site – and, for an installation, its location on the site, together with the name and address of any local authority in which the installation is situated.
- Description of the installation or mobile plant – the scheduled activities to be carried out.*
- Nature, quantities and sources of likely emissions to each environmental medium (Part B to air only) and possible significant effects on the environment. For a Part A installation emissions are defined as "direct or indirect release of substances, vibrations, heat or noise ... into the air, water or land"; for a Part B installation emissions are defined as "the direct or indirect release of substances or heat ... into the air" (reg. 2).*
- Technology and techniques to be used to prevent, or if this is not practicable, to reduce, emissions – the regulator needs to be satisfied that the BAT has been chosen.*
- Proposals for monitoring emissions.
- Where the permit is to authorise a specified waste management activity, any additional information which the applicant wishes the Regulator to take into account in considering whether the applicant is a fit and proper person as defined in, and required by, reg. 4.
- Where the permit is to cover an installation or mobile plant for which there are general binding rules (see 2.5.3b below), whether the applicant wishes the permit to be based on these or for specific conditions regarding its operation to be included on the permit.

Part A Applications

- Part A installation and mobile plant: report describing the condition of the site – its environmental and pollution "history", substances on the site with the potential to cause pollution, measures in place to prevent land pollution etc. In some cases the Agency may require further investigation of the site as a condition of granting a permit. Care should be taken in the preparation of the site report as contamination identified on closure of the site and not recorded on the earlier report will be assumed to have occurred during operations on the site. The Agency is to produce guidance on characterisation and assessment of sites.

- Raw and auxiliary materials, other substances and energy to be used or generated in carrying on the activity.*
- Measures to be taken for prevention and recovery of waste generated by the operation of the installation.
- Relevant information obtained as a result of carrying out an environmental assessment in accordance with EU Directive 85/337 on the assessment of the effects of certain public and private projects on the environment (see 1.20).
- Any additional information which the applicant wishes the Regulator to take into account in considering the application.
- Non-technical summary of the application.
- Outline of main alternatives, if any, considered in respect of issues which will arise in determining BAT-based conditions (SSI 2005/510 implementing public participation Directive; SR 2006/98).

* Instead of these details, applications for waste oil burners less than 0.4 MW should give full details of the type of appliance to be used, fuel to be used, height and location of any chimney through which waste gases will be carried away and their efflux velocity, and the location of the fuel storage tanks for the appliance (Sch. 4, para 3).

Part C Mobile Plant: (NI) name and address of local authority in which operator has his principal place of business, or if this is outside NI, the local authority in which the plant was first operated in NI, or if it has not yet been operated in NI, the local authority in which it is first intended to operate it.

Landfill Sites (Additional details required):

- Description of types and quantities of waste to be deposited.
- Proposed capacity of site.
- Description of site, including geological and hydrogeological characteristics.
- Proposed operation, monitoring and control plan.
- Proposed plan for closure and after-care procedures.
- Financial provisions covering after-care and closure procedures.

Copies of all applications will be put on the public register (Sc: reg. 27 & Sch.9; NI: reg. 30 & Sch. 10). An applicant may, however, request that certain information is excluded on the grounds of commercial confidentiality (Sc: reg. 29; NI: reg. 32 & Sch.4, Part 3); the Regulator has 28 days in which to determine the request but if he fails to do so within that period, it is deemed not to be commercially confidential (NI: reg. 32(6); Sc: reg. 29(4)), but will not be entered on the register for 21 days following notification to the applicant. The applicant has a right of appeal during that period with information subject to appeal not entered in the Register until seven days after determination or withdrawal of the appeal. Requests for exclusion on the grounds of national security are determined by Scottish Ministers [DOENI].

2.5.2 Determination of Application
(Sc: reg. 7; NI: reg. 10. Sch. 4, Part 2)

Within 14 days of receiving an application for a permit, the Regulator should send a copy of it to all statutory consultees, who have 28 days in which to make any representations to the Regulator. The following are statutory consultees for Part A installations and mobile plant:

- Health Board (Sc) or Health & Social Services Board (NI), in which the installation or mobile plant is to be operated.
- The Food Standards Agency;
- Where an emission may affect an SSSI, Scottish Natural Heritage, or DOENI, as appropriate;
- Harbour authority, water and sewerage authority), local fisheries committee [NI Dept for Regional Devt, Dept for Agric & Rural Devt or Dept of Culture, Arts & Leisure, NI Loughs Agency or Waterways Ireland], as appropriate – i.e. where the installation may involve the release of a substance affecting their area of responsibility;
- The local authority in which the installation is situated;
- The relevant planning authority, in respect of applications involving a waste management activity (NI);

The Health and Safety Executive should be consulted in respect of permit applications for both Part A and B installations for which a nuclear site licence (see chapter 5, 5.39.1) or major accident prevention policy document (under the COMAH Regulations – see 2.10.2) is required. The Regulations also enable Scottish Ministers [DOENI] to direct that other persons should be consulted.

Scottish Natural Heritage or DOENI should be consulted in respect of applications for a permit covering a Part B (NI, Part C) installation where an emission to air could have an adverse effect on an SSSI. The relevant Petroleum Licensing Authority should be consulted in respect of Part B (NI, Part C) applications covering the unloading of petrol at petrol stations.

Applications for permits in respect of waste oil burners under 0.4 megawatts do not have to be sent to statutory consultees (Sch.4, para 10).

Where it is proposed to issue a permit which includes off-site conditions (Sch.4, para 11), the Regulator should by notice consult the owner, lessee or occupier of the land; again 28 days should be allowed for representations.

Scottish Ministers [DOENI] have the right to call in certain classes of, or individual, applications for determination (Sch.4, para 14); in such cases the Regulator should inform the applicant that the application has been forwarded, together with any representations received. Both the applicant and the Regulator may request a hearing; a request for a hearing should be made in writing within 21 days of being informed that the application has been forwarded. Such applications will then be considered by Scottish Ministers [DOENI] who will then direct the Regulator as to whether a permit should be granted and, if so, with what conditions.

Where the Scottish Ministers [DOENI] become aware that the operation of an installation is likely to have significant negative effects on the environment of another Member State, they should send a copy of the permit application to the Member State concerned to enable bi-lateral consultations to take place. The application should be forwarded to the Member State at the time it is advertised (or as soon as possible after becoming aware of the likely effects of the installation, or when requested by the Member State). Scottish Ministers [DOENI] should advise both the Regulator and the applicant that the permit application has been forwarded to another Member State for bi-lateral consultations. The four month determination period does not then begin until completion of the bi-lateral consultations; any representations made by another Member State must be taken into account when determining the application. (Sch.4, paras 17 & 18). The requirements outlined in the next paragraph also apply to applications with transboundary implications.

As a result of implementation of the public participation Directive requiring the public to be given early and effective opportunities to comment on plans and programmes relating to the environment, regulators of Part A installations will now publish their draft determination (i.e. draft permit) for consultation via their website; this must be accompanied by a "decision document" giving the reasons and matters considered in arriving at the draft determination – these additional requirements apply to all applications for a new Part A permit or for an application to vary a permit due to a substantial change. Regulators should send operators a copy of the draft determination and draft decision document at least 3 days before advertising them on the website. The public should be allowed at least 20 working days to submit any representations and the Regulator must consider them and take a final decision within 15 working days (or such other period agreed with the operator). Where necessary the decision document and the permit should be amended to reflect the outcome of consideration of the representations received and a copy of both put on the public register. The decision must also be advertised on the regulator's website and by any other appropriate means. If no representations are received, then the regulator can make its final decision and inform the operator within 5 working days of the end of the period allowed for representations. The final permit and decision document, with a note stating that no representations were received, will then be put on the public register (Sc: SSI 2005/461; SR 2006/98).

2.5.3 Grant of Permit (Sc: reg. 6-11: NI: reg. 9-15)

(a) Permits & Conditions

In determining an application for a permit, the Regulator must take into account whether the applicant will operate the installation or mobile plant in accordance with the permit and any conditions attached to it. If the permit is to authorise a specified waste management activity, the Regulator must be satisfied that the applicant is a "fit and proper person" (reg. 4) – i.e. that the management of the activity will be in the hands of a technically competent person (with appropriate qualifications), that they have not been convicted of a relevant offence, and that adequate financial provision has been made to deal with after-care and closure obligations (see also chapter 5, 5.26). SSI 2005/510 amends the Scottish PPC Regulations to enable SEPA to grant a permit subject to the applicant obtaining a relevant certificate of technical competence within two years of the grant of permit so long as it is satisfied that the applicant is a fit and proper person to carry out the specified waste management activity. Conditions to be attached to permits for ***landfill*** sites are set out in the *Landfill Regulations 2002* – see also chapter 5, 5.29.1.)

An application for a permit should be refused if the Regulator is of the opinion that

- the installation will have an unacceptable environmental impact;
- lack of information provided by the operator on which to base conditions;
- the operator will be unable to comply with the additional requirements of the *Landfill Regulations, Waste Incineration Regulations*, or the *Solvent Emissions Regulations*.

A permit should also not be granted where the Regulator considers poor management systems, or lack of technical competence, will result in non-compliance with the permit or its conditions or, where the permit covers a specified waste management activity, that this will not be in the hands of a fit and proper person.

All permits and conditions attached thereto must ensure that the installation will be operated in such a way as to ensure no significant pollution is caused and that all necessary preventative measures against pollution are taken; there is an implied condition in all permits that operators will use the Best Available Techniques (see 2.1.4 above) to prevent or reduce emissions. The BAT for an installation will take account of both technical and operational matters. When considering technical issues, the geographical location and local environmental conditions will be of relevance; so far as the operation of the plant is concerned, while not a mandatory requirement, certification to a recognised environmental management system such as ISO 14001 or the EU's Eco-Management and Audit Scheme will be evidence that the installation is likely to be run satisfactorily. Schedule 2 to the PPC Regulations lists the issues to be taken into consideration in defining BAT for the installation, with further guidance, including basic principles for determining BAT, provided in both SEPA's and the DOENI's *IPPC Practical Guide*. Further guidance on BAT is provided in sectoral guidance (see Appendix 2.3).

It should be noted that despite not being explicitly included, as in BAT***NEEC***, the costs and benefits of a particular technique may be taken into account when deciding on the best available technique, in that in selecting the BAT, the cost to the operator installing a particular technique may be balanced against the benefits to the environment.

For Part A installations and mobile plant, conditions should also ensure

- energy is used efficiently;
- waste production is avoided in accordance with the framework Directive on waste (see chapter 5, 5.6) or where waste is produced it is recovered or, if that is not possible, is disposed of in such a way as to avoid any impact on the environment;
- measures are in place to prevent accidents and to limit their consequences; accidental releases or releases above limits set, and other abnormal occurrences which cause, or may cause, significant pollution must be reported to the Regulator without delay;
- transboundary pollution is minimised;
- protection of soil and groundwater and appropriate management of waste generated by the installation;
- suitable emissions monitoring equipment is in place; measurement methodology, frequency, and evaluation procedure should also be set out; the operator must provide the Regulator with data to enable compliance monitoring and provide regular reports on results of emissions monitoring;
- depending on the site report submitted with the application, a requirement to prepare a 'Site Protection and Monitoring Programme'; this should demonstrate the on-going effectiveness of pollution control measures; where a risk is identified, "intrusive investigations" to collect information will be required and the results reported to the Agency within six months of grant of the permit.

All permits will include emission limit values (ELVs) for all pollutants likely to be emitted in significant quantities; these will be based on BAT and in the case of Part A installations will also take into account their potential to transfer pollution to another environmental medium. In setting an ELV account may be taken of the technical characteristics of the installation, its geographical location and local environmental conditions. Conditions should ensure no breach of EU environmental quality standards and thus where such legislation requires a stricter ELV, then this should be set. Account will also need to be taken of the effect of the installation's emissions on local air quality and the achievement of national air quality objectives.

A permit for a Part A installation or mobile plant may include an off-site condition requiring the operator to carry out work which requires consent from another person (e.g. the owner of the land); in such cases (which will have followed consultation) where rights have been granted to carry out the work, compensation may be payable (Sch. 6).

Permits are to be periodically reviewed (NI: reg. 15; Sc: reg. 11) and, in particular, when

- developments in best available techniques make further emissions reductions feasible without imposing excessive costs;
- significant pollution from an installation requires existing emissions limits to be reviewed;
- operational safety necessitates change in techniques to be used.

Activities at an installation regulated by the PPC regulations are exempt from statutory nuisance legislation; non-PPC regulated activities, e.g. barking dogs and burglar alarms, however, are not.

Permit conditions included to ensure compliance with the *Groundwater Regulations 1998* (chapter 6, 6.11.3) should be reviewed every four years.

(b) General Binding Rules (NI: reg. 14; Sc: reg. 10)

The IPPC Directive permits Member States to make "general binding rules" (GBRs) which can be used instead of site-specific conditions, to permit certain types of installation and mobile plant which share a number of similar characteristics. This Regulation enables Scottish Ministers [DOENI] to make GBRs for Part A installations or mobile plant so long as they are satisfied that this will result in the operation of an installation under such rules providing the same "high level of environmental protection and integrated prevention and control of pollution" as if it were permitted using specific conditions. It should however be borne in mind that some aspects of a site's operation may not be suitable for GBRs (e.g. noise and odour) and thus site specific conditions will be developed.

In applying for a permit, an operator of a Part A installation or mobile plant for which GBRs exist may request the regulator to include in the permit a "general binding rules condition" covering the operation of the installation or mobile plant; for regulatory and enforcement purposes, GBRs are treated as though they are site-specific conditions. It should be noted that not all installations within a sector for which GBRs have been developed may be suitable for permitting in this way; for instance if the site is close to an SSSI or other sensitive location, site-specific conditions will need to be included. Sites permitted using GBRs will usually be subject to lower fees and charges.

Scottish Ministers [DOENI] may, by notice, also revoke or vary GBRs. In making, revoking or varying GBRs, they should, as well as serving a copy of the rules or notice of revocation or variation on the Regulators, endeavour to bring them to the attention of all operators in the sector concerned. A notice should also be published in the *Edinburgh [Belfast] Gazette*.

While no GBRs have been made, standard rules and guidance have been developed for the farming industry – see Appendix 2.3.

(c) Low Impact Installations

Guidance from SEPA (2004) outlines the criteria for determining whether an installation might qualify for permit as low impact and thus benefit from lower fees and fewer inspections. Criteria include:

- Discharge of less than 50 m^3 of water per day as waste water or effluent.
- No more than one tonne of Directive waste (or 10 kg of hazardous waste) per day, averaged over a year, with not more than 20 tonnes of Directive waste (or 200 kg of HW) being released in any one day.
- The installation must not have a net rated thermal input of more than 3 MW).
- The inventory of any substance at any one time is not more than 10% of that in the COMAH Directive for lower tier installations.
- Only a low potential for offence due to noise; an installation will not be considered as an LII if it may give rise to noise "noticeable" outside the boundary.
- Only a low potential for offence due to odour; an installation will not be considered as an LII if it may give rise to an offensive smell noticeable outside the boundary.
- No planned or fugitive emission into the ground or any soakaway.
- No likelihood of a release of any substance at a rate greater than that determined as "insignificant" in H1 *Environmental Assessment and Appraisal of BAT* guidance.
- The installation must comply with the criteria for an LII without the need to rely on active abatement.

Enforcement action of any kind – including a formal caution – unless overturned on appeal, will result in the installation no longer being considered as low impact.

In Northern Ireland, *the Pollution Prevention and Control (Industrial Pollution and Radiochemical Inspectorate) Charging Scheme (NI)* defines a low impact installation as one which "...cannot result in emissions or where ... there is no likelihood that it will result in emissions except in a quantity which is so trivial that it is incapable of causing pollution or its capacity to cause pollution is insignificant".

2.5.4 Variations
(NI: reg. 17, Sch. 7; Sc: reg. 13, Sch. 7)

Before making any change to the operation of the installation, reg.16 (Sc: reg. 12) requires the operator to notify the Regulator of the planned change, giving 14 days' notice; this does not however apply if the operator intends to apply for his permit or the conditions to be varied under reg. 17 (Sc: reg. 13). Thus a reg.16 (Sc: reg. 12) notification could be used for minor operational changes which will not result in a breach of the conditions of the permit. While not a requirement of the Regulations, the *IPPC Practical Guides* suggest that Regulators should acknowledge receipt of the notice, either sanctioning the change or requiring the operator to apply for a variation of his permit prior to carrying out the change. However, if nothing is heard from the Regulator, then the operator may at the end of the 14 day period make the planned change.

For more substantial changes – i.e. which may have significant negative effects on human beings or the environment – or if there is any doubt about the effect of the planned change in the operation of the installation, then application should be made to vary the permit.

Conditions attached to a permit may also be varied by the Regulator at any time – e.g. following periodic review of the permit, to update to take account of new developments in BAT, or following notification from the operator that he is intending to make a change to the way in which the installation is operated; or an operator may apply to the Regulator to vary the conditions of a permit.

An application to vary conditions of a permit should be made in writing to the Regulator and should be accompanied by the prescribed fee (see 2.5.8). It should include full details of the operator, the installation or mobile plant and the proposed changes, and an indication of the variations being requested. Where a substantial change is being proposed, a report in compliance with the 1985 EU environmental assessment Directive (as amended) should accompany the application.

Where the Regulator is proposing to vary the conditions of a permit, or an application to do so has been made, it should serve a variation notice on the operator (in the case of an application for variation, within 14 days of receiving it). This should specify the variations to be made and the date or dates on which they are to take effect. Where a variation fee has not already been paid (i.e. Regulator-instigated variations), the notice will also specify the fee to be paid and by when.

The operator should then advertise the application for variation, or proposed variation notice, in a local newspaper and in the case of a Part A installation in the *Edinburgh [Belfast] Gazette* within 28 days of receiving the notification from the Regulator. The advertisement should include the name of the operator, address of the installation, brief details of activities carried on and proposed changes. It should say where the application or variation notice can be seen (i.e. the location of the public register) and that any representations should be made in writing to the Regulator within 28 days; it should also note that such representations will be put on the register unless a specific request is made that this should not be done. The Regulator will also, within 14 days, notify statutory consultees sending a copy of the application or proposed variation notice, inviting representations, also within 28 days.

It should be noted that applications for variations as a result of a "substantial change" or because "the pollution caused by the installation is of such significance that existing emission limit values need to be revised" or new values included, will need to follow the enhanced consultation procedures put in place to implement the public participation Directive (SSI 2005/510, SR 2006/98 and see above 2.5.2).

Applications for variations or proposed variation notices in respect of waste oil burners under 0.4 MW do not have to be notified to statutory consultees or advertised unless the Regulator decides otherwise and notifies the operator of this; in which case the normal consultation procedure will be followed. Applications or proposed variation notices in respect of the unloading of petrol at service stations do not have to be advertised; and neither do those served to comply with a Direction from Scottish Ministers [DOENI], or those modifying a previous proposed variation notice.

Where the Regulator is proposing to include an off-site condition in the variation notice, it should, before serving the notice, notify the owner, lessee or occupier of the land; 28 days should be allowed for representations, which should then be considered by the Regulator before serving the variation notice.

Scottish Ministers [DOENI] may direct the Regulator to forward a specific application, or class of application, for variation for their determination. The Regulator should notify the applicant that the application has been forwarded; both the applicant and the Regulator have 21 days to make a written request for a hearing. Once Scottish Ministers [DOENI] have determined the application, they will notify the Regulator and, if approving the application, direct the Regulator as to the conditions to be included.

Applications for variations, except those forwarded to Scottish Ministers [DOENI], will normally be determined within four months of receipt – three months for those that do not require advertising or notifying to statutory consultees.

Where an application for variation is proposing a substantial change in the operation of an installation which is likely to have a significant negative effect on the environment of another Member State, Scottish Ministers [DOENI] should send a copy of the application or proposed variation notice to the other Member State at the same time that it is advertised. Such applications will not then be determined until consultations with the Member State concerned have been completed, at which time the Regulator's four month determination period begins.

2.5.5 Transfer of Permits (NI: reg. 18; Sc: reg. 14)

Where it is proposed to transfer all or part of a permit to another operator, the current and future operators (the "proposed transferee") should make a joint application to the Regulator, together with the prescribed fee, giving full details of them both, and in the case of a partial transfer, a map or plan of the installation showing which parts are to be transferred.

The Regulator has two months, or longer if agreed with the applicants, to determine the request, which should only be agreed if the Regulator is sure that the proposed transferee will operate the installation (or part installation) in accordance with the permit and its conditions; where the transfer involves a waste management activity, the Regulator should be satisfied that the proposed transferee is a fit and proper person (reg. 4). Transfers of whole permits will be effected through the endorsement of the permit with the new operator's details; in the case of partial transfers, a new permit will be issued to the new operator covering the transferred parts of the installation, with the original permit being returned to the first operator, endorsed to show the extent of the transfer.

Where a partial transfer is effected, it may be necessary for the conditions of the permit to be varied to reflect the new operating responsibilities.

The offence of operating an installation without a permit includes a new operator taking over prior to completion of the transfer procedure; enforcement action may also be taken against both the new and original operators for failing to submit a joint application to transfer.

2.5.6 Application to Surrender a Permit (NI: reg. 19-20; Sc: reg. 15-16)

If an operator ceases, or intends to cease, all or part of his operations at a Part A installation or mobile plant, he should make an application to the Regulator to surrender all or part of the permit. The application (which should be accompanied by the appropriate fee) should include full contact details of the operator and in the case of a partial surrender, a site map or plan identifying the surrender unit. A site report should also be submitted which identifies any changes in the condition of the site from those reported at the time of applying for a permit, and describes the steps taken to avoid any pollution risks or to return the site to a satisfactory state – the Agency has produced guidance (H8, June 2004) for operators on the production of site surrender reports.

The Regulator has three months (or longer as agreed with the operator) in which to determine the application to surrender; if the regulator is satisfied with the condition of the site and that no further steps to avoid any pollution risk need to be taken, it should by notice accept the application, stating the date on which surrender takes effect; if it is not satisfied, it should specify the further steps to be taken to return the site to a satisfactory condition. In the case of a partial surrender, the regulator may vary the conditions attached to the permit to take account of the surrender. Failure to determine the surrender within the agreed period will be deemed a refusal if the applicant notifies the regulator that he treats it as such.

Before accepting the surrender of a permit for a landfill site, the regulator should be satisfied that the site meets the requirements of the Landfill Directive – i.e. that it is not likely to cause a hazard to the environment – and of the IPPC Directive, that the site has been returned to a satisfactory state.

Where an operator of a Part B (NI, Part B or C) installation or mobile plant ceases, or intends to cease, all or part of his activities, he should notify the Regulator of the surrender of all or part of his permit and the date on which it is to take effect. Such notification should include full contact details for the operator and, in the case of partial surrender, a site map or plan identifying that part affected by the surrender. Permits covering Part B (NI, Part B or C) installations or mobile plant surrendered or partially surrendered cease to have effect on the date of the notification, unless the Regulator notifies the applicant that he is of the opinion that it is necessary to vary the conditions of a permit which is only being partially surrendered. In this case the Regulator will serve a variation notice and the partial surrender will take effect from the date on which the varied conditions take effect.

2.5.7 Revocation of Permits
(NI: reg. 21; Sc: reg. 17)

The Regulator may, by notice, revoke all or part of a permit, to take effect not less than 28 days after the date on which the notice was served; a partial revocation may apply to either the operation of part of the installation or to some of the activities carried on at the installation. The Regulator may also serve a revocation notice in relation to a permit authorising specific waste management activities at an installation if it appears that those activities are no longer in the hands of a fit and proper person (because that person has been convicted of a relevant offence – see chapter 5, 5.26.11), or that the activity is no longer managed by a technically competent person.

As well as stating the date on which the revocation (or partial revocation) is to take effect, it should specify, in the case of partial revocation, the activity or operations covered by the revocation.

If the Regulator considers certain steps need to be taken at a Part A installation or mobile plant to avoid a pollution risk or to return the site to a satisfactory condition once operations have ceased or partially ceased because of a revocation notice, the notice should say so and specify the steps (not already required by conditions attached to the permit) to be taken. Those sections of a permit requiring specific action on cessation of operations remain in force until the Regulator is satisfied that the necessary remedial action has been taken and has issued a certificate to this effect.

A revocation notice may be withdrawn before the date on which it takes effect.

2.5.8 Fees and Charges

Regulation 22 of the PPCNI enables the DOENI to make separate schemes covering fees and charges for those installations regulated by either the Chief Inspector (Part A and Part B) and by district councils (Part C). Schemes will cover fees for permit applications, variations, transfers and surrenders, as well as subsistence charges. Charges are based on risk, with lower risk installations paying lower charges. Non-payment of subsistence charges may result in revocation of a permit. *The Pollution Prevention and Control (Industrial Pollution and Radiochemical Inspectorate) Charging Scheme (NI)*, which is usually updated annually covers Part A and B installations.

Fees and charges for Part A and B installations in Scotland are levied under a scheme made under the *Environment Act 1995*, s.41 – see 2.6.1. SEPA was expected to consult in autumn 2008 on proposals to review the way in which charges are levied on permits to take account of risk/compliance factors with a view to implementation from April 2009.

2.5.9 Enforcement (NI: reg. 23-25; Sc: reg. 18-20)

(a) Enforcement & Suspension Notices

As well as revocation notices (see above), the Regulator may issue an ***enforcement notice*** if he is of the opinion that an operator is not complying with his permit or its conditions. The notice should state the contravention – or potential contravention – the remedial steps to be taken, and the period within which this must be done. The notice may be withdrawn at any time.

If the Regulator is of the opinion that the operation of an installation or mobile plant, or the way in which it is being operated, involves a serious risk of pollution, he can issue a ***suspension notice.*** A suspension notice may also be issued to the operator of an installation or mobile plant if the Regulator is of the opinion that he is no longer a fit and proper person on the grounds that waste management activities have ceased to be in the hands of a technically competent person. The notice should detail the pollution risk involved in operating the installation, the remedial action and the period within which this should be done. The notice should also state that, until the notice is withdrawn, the permit ceases to authorise the operation of the installation or the carrying out of specified activities; where the permit is to continue to have effect, the notice will authorise the carrying out of certain activities and specify the additional steps to be taken in carrying them out.

A suspension notice will be withdrawn once the necessary steps to remove the pollution risk have been taken or in the case of waste management activities, they are back in the hands of a technically competent person.

The regulatory agencies have produced policy statements outlining the general principles which inform enforcement and prosecution decisions, which can be accessed on:

- www.sepa.org.uk/pdf/policies/5.pdf
- www.ehsni.gov.uk/pubs/publications/EP_Enforcement_Policy.pdf

(b) Inspection and Compliance Monitoring

Operators of industrial sites are required to carry out their own compliance monitoring, with the regulatory authorities carrying out periodic inspections and reviews to ensure compliance with permits or authorisations and their conditions; they will also look at other component's of an operator's procedures, including management and competency of personnel and adequacy of monitoring and of the equipment being used.

In July 2008, SEPA issued a consultation document on its plans to introduce a new compliance assessment scheme in ***Scotland***, to be phased in from 1 January 2009. The aim is to provide a consistent framework and approach for assessing compliance with permits and licences under various regulatory regimes, thus helping to deliver SEPA's its enforcement policy (see chapter 1, 1.3.2c). Initially the scheme is to apply to

- Registrations and authorisations under *the Radioactive Substances Act 1993* (see chapter 5, 5.39.3).
- Waste management licences under the EPA (see chapter 5, 5.18).
- Permits under the PPC Regulations.
- Licences under *the Water Environment (Controlled Activities) (Scotland) Regulations 2005* (see chapter 6, 6.11.6).

As well as the regulator's inspection, the overall result of assessment will be based on recorded information, monitoring results, complaints and incidents and other relevant information. The use of compliance assessment in setting fees and charges is to be the subject of a separate consultation, though it is SEPA's intention that in the future charges for poorer performing sites should reflect the extra work needed for regulating the site.

2.5.10 Powers to Remedy or Prevent Pollution

(NI: reg. 26; Sc: reg. 21)

The Regulator may, if he is of the opinion that an emergency situation exists, take steps to deal with any serious risk of pollution instead of issuing a suspension notice, recovering the costs from the operator of the installation concerned. If the operator can show that the situation did not warrant emergency action being taken by the regulator, no costs are recoverable.

The Regulator may also take action to deal with any pollution caused by an offence under reg. 30 in Scotland [NI: reg. 33] (see below 2.5.13), having given the operator seven days' notice. Costs are recoverable from the operator.

2.5.11 Appeals

(Sc: reg. 22 & Sch. 8; NI: reg. 28 & Sch. 9)

There is a right of appeal to Scottish Ministers or the Planning Appeals Commission (NI) in the following instances – any appeal should be made within six months of the decision being appealed against:

- Refusal of permit or disagreement with conditions attached to it;
- Refusal to vary conditions of a permit or disagreement with variation notice;
- Rejection of transfer application; or disagreement with variation of conditions as a result of the transfer;
- Rejection of surrender application, or disagreement with variation of conditions attached to permit following partial surrender.

An appeal may also be made against the following:

- Revocation notice – before the date on which the notice takes effect; revocation notices do not take effect until the appeal has been determined or withdrawn;
- Variation, enforcement and suspension notices – within two months of the date of the notice; these notices are not suspended pending the outcome of an appeal.

The *Landfill Regulations* (Sc 2003; NI 2003) amend the PPC Regulations to enable an appeal to be made

- where a request to initiate the closure procedure is refused,
- SEPA [DOENI] has served a closure notice as a result of its decision that, on the basis of the conditioning plan submitted under the *Landfill Regulations*, a site should be closed.

In both cases the closure procedure will not be initiated until the appeal is determined or withdrawn.

There is no right of appeal against a notice or decision implementing a Direction from Scottish Ministers [DOEN]I.

Scottish Ministers may determine the appeal themselves or appoint someone else for this purpose (*Env. Act 1995*, s.114 & Sch. 20).

Schedule 8 to the Scottish Regulations [NI Sch. 9] specifies that written notice of appeal should be sent to Scottish Ministers [NI Planning Appeals Commission], together with the following documentation:

- Statement of grounds for the appeal;
- Statement as to whether the appellant wishes the matter to be dealt with through written representations or by a hearing;
- Copy of the relevant application and permit;
- Relevant correspondence between the regulator and operator;
- Copy of the decision which is the subject of the appeal.

A copy of the notice of appeal, and first two documents above should be sent to the Regulator at the same time.

Within 14 days of receiving notice of the appeal, the Regulator must send details of it to statutory consultees and others who have made representations to the regulator on the subject of the appeal and to any persons who it is thought may have a particular interest in the subject matter of the appeal. As well as giving details of the appeal, the notice should state that any representations should be made to the Scottish Ministers [NI Planning Appeals Commission] within 21 days and that these will be copied to both the appellant and the Regulator, and that if a hearing is to be held, they will be notified of the date. The notice should also state that any representations received will be put on the public register unless a written request is made that this should not happen. In such cases a statement will be added to the register saying that representations concerning the appeal have been received and are subject to a request to exclude from the register. Within 14 days of sending out the notice of appeal, the Regulator should notify Scottish Ministers [NI Planning Appeals Commission] of all the persons to whom it has been sent.

Where the appeal is to be decided in writing, the Regulator must submit any comments in writing to Scottish Ministers [NI Planning Appeals Commission], copied to the appellant, within 28 days of receipt of the appeal notice, with the appellant having 17 days (from the date the Regulator submitted its comments) to make further representations, which should be copied to the Regulator. Both sides have 14 days from the date of receipt to comment on any other statements submitted in connection with the appeal.

Where a hearing is to be held, Scottish Ministers [NI Planning Appeals Commission] will notify the appellant and the Regulator of the time, date and place – a minimum of 28 days notice will be given – and will appoint someone to conduct the hearing. If the hearing is to be held partly or wholly in public, statutory consultees and all those who have made written representations regarding the appeal matter should be notified of the date etc; where the appeal relates to the operation of an installation or Part A mobile plant, details of the hearing should be advertised in a local newspaper – in both cases this should be done at least 21 days in advance of the hearing. Following a hearing the person appointed to conduct the hearing will report to Scottish Ministers [NI Planning Appeals Commission].

Following determination, copies of the decision and any report (following a hearing) are sent to the appellant, the Regulator and statutory consultees; copies of the decision only are sent to all others who made representations There is a right of appeal to the courts against Scottish Ministers [NI Planning Appeals Commission] determination; if upheld, they should send all those notified of the original decision details of the further matters to be considered by the appeal, inviting further representations within 28 days and may either re-open or instigate a hearing in which case the procedure for a hearing outlined above will be followed.

Documents relating to the determination of the appeal are put on the public register.

An appellant may withdraw an appeal by written notification to Scottish Ministers [NI Planning Appeals Commission], copied to the Regulator. The Regulator should then notify all those to whom notice of the appeal was sent that it has been withdrawn.

In Scotland, where Scottish Ministers have determined an appeal both SEPA and the appellant may make an appeal regarding the determination to the Sheriff within 21 days (Sch. 4, part 4, paras 22(5)-(12).

2.5.12 Public Registers
(Sc: reg. 27-29, Sch.9; NI: reg. 30-32, Sch.10)

The Regulators have a duty to maintain public registers containing details and all documentation relating to applications for permits, permits, variations, transfers and surrenders of permits, appeals, enforcement, revocation and suspension notices etc, as well as details of any relevant convictions. *The Landfill Regulations 2003* (both Sc & NI 2003) amend the PPC Regulations to require closure and conditioning plans to be placed on the register. In N. Ireland, local authority registers will also include details of permits etc relating to installations or mobile plant in their area for which the Industrial Pollution Inspectorate is the Regulator.

Where a request has been made that certain information be omitted from the register on the grounds of commercial confidentiality and the Regulator fails to determine the request within 28 days (or an agreed longer period), then the request is deemed to have been refused. If the Regulator refuses a request for confidentiality (or it is deemed to have been refused), the applicant has 21 days in which to appeal to Scottish Ministers [NI Planning Appeals Commission]. The information in dispute will not be put on the register until the appeal is determined; if it is determined as not being commercially confidential, a further seven days must elapse before its entry on the register. Scottish Ministers [NI Planning Appeals Commission] may, however, still require certain information to be put on the register in the public interest even when the Regulator has accepted that it is confidential. Exclusion of data from the register on the grounds of commercial confidentiality lasts four years after which it will be added to the register unless a further four year exclusion is applied for and agreed.

Scottish Ministers [Secretary of State, NI] may direct that specific information or information related to specific issues should not be put on the register in the interests of national security (Sc. reg. 28; NI reg. 31). Regulators should notify Scottish Ministers [Secretary of State, NI] of any such information they exclude from the register. They may also direct that information related to certain issues affecting national security be referred to them for decision on inclusion on the register. In N. Ireland, an operator may also apply to the Secretary of State for specific information to be excluded on grounds of national security (reg. 31(4)), and should notify the regulator that they are doing so; the information remains off the register pending determination.

Information will be removed from the register in the following circumstances:

- Withdrawal of application for permit prior to determination – all particulars to be removed not less than two months but within three months after the date of withdrawal;
- If the Regulations cease to apply to an installation or mobile plant due to amendment of Schedule 1 – all particulars to be removed not less than two months but within three months of the date on which the installation ceases to be prescribed;
- Monitoring information relating to an installation or mobile plant – four years after entering in the register;
- Information relating to an installation or mobile plant which has been superseded – four years after the later information has been entered in the register;
- Details of any relevant formal caution – five years after the date of the caution.

The registers must be available for inspection by the public, free of charge, at all reasonable times; a charge may be made for providing a copy of an entry on the register.

2.5.13 Offences (Sc: reg. 30; NI: reg. 33)

It is an offence to operate an installation without a permit, in contravention of a permit or of its conditions, to fail to comply with the terms of an enforcement or suspension notice, or to fail to comply with a court order requiring remedial action following conviction. Summary conviction in these cases may result in a fine not exceeding £40,000 (NI: £30,000) and/or 6 months' imprisonment; conviction on indictment may result in a fine and/or up to five years' imprisonment, depending on the offence.

Other offences, such as giving misleading or false information to the Regulator, making false entries in any records which are required to be kept, forging documents etc, may result in a fine not exceeding the statutory maximum on summary conviction, or a fine and/or up to two years' imprisonment on conviction on indictment.

2.5.14 Information, Directions and Guidance
(Sc: reg. 26, 23-24; NI: reg. 29, 37-38)

Scottish Ministers [DOENI] may, by notice, require the Regulator to supply any information as may be required to enable them to discharge obligations in relation to European Community or international agreements.

Directions from Scottish Ministers [DOENI] to Regulators may relate to any aspect of their functions under the Regulations; they may be either general or specific and may direct them as to the way in which they are or are not to exercise their powers.

Where the Scottish Ministers [DOENI] receive information from another Member State in relation to the operation of an installation outside the UK which is likely to have significant negative effects on the environment in Scotland [NI], they should direct the Regulator to bring it to the attention of anyone likely to be affected, providing those people with an opportunity to comment.

Regulators have a duty to comply with any Directions, which should be in writing, and which may be varied or revoked by further Direction.

Scottish Ministers [DOE NI] may also issue Guidance to the Regulators relating to the carrying out of any of their functions under the Regulations; Regulators are required to have regard to any such Guidance.

The Pollution Prevention and Control (Solvent Emissions) (Directive) (Scotland) Directions 2002, which came into force on 1 March 2002, require SEPA to include conditions in relevant permits and authorisations ensuring compliance with the Directive (see below 2.6).

2.6 FEES AND CHARGES

2.6.1 England, Wales, Scotland

Sections 41-43 of *the Environment Act 2005*, which came into force on 21 September 1995, enable the Environment Agency and SEPA to make schemes (to be approved by their respective Secretary of State) for the recovery of costs in respect of granting and ensuring compliance with environmental licences. Such costs include:

- Activities relating to issuing authorisations and permits (including applications, variations, transfers and surrender of permits);
- Carrying out compliance monitoring and enforcing controls; enforcement action;
- Sampling and analysis of releases;
- Operation of the public registers of information;
- Administrative expenditure in support of the above functions, including salaries, travel and subsistence costs, office services, etc.

Charges are levied according to the number of components an installation contains. Levels of charges are reviewed each year. Where a permit is issued part way through the year, the charge will be calculated on a pro-rata basis. The financial year runs from 1 April - 31 March.

England & Wales

The Environment Agency has developed a "Unified Charging Framework" (UCF) with the aim of providing a consistent and harmonised approach to charging across the sectors which it regulates.

The Pollution Prevention and Control Charges Scheme, which is usually revised annually, is made under s.41 of *Environment Act 1995*. Charges for individual installations regulated by the Environment Agency (i.e. Part A1) are based on their Environmental Protection Operator and Pollution Risk Appraisal (EP OPRA – see 2.4.13 above) and reflect the amount of regulatory attention/effort needed by the Agency, based on complexity of the operation, emissions and risk posed to the environment, as well as the way in which the installation is operated and managed; those with higher scores and needing more regulatory attention are likely to pay higher fees.

Charges are levied in respect of applications for permits (including pre-application discussions), variations and for substantial changes, annual subsistence (i.e. ongoing compliance assessment and independent monitoring), transfers and surrenders, and landfill closure.

Charges for installations in sectors for which a "standard" permitting regime has been developed (e.g. intensive livestock sector) and for other low impact installations will reflect the reduced amount of regulatory time needed.

Where the Agency becomes the regulatory authority for a Part A2 or Part B activity (because it falls within a Part A1 installation), the charge will be calculated using OPRA. Discharges to water from Part A2 installations will be assessed and charged for by the Agency.

Charges for staged applications will be based on time and materials.

Details of the actual charges which apply can be accessed on the Environment Agency website at http://environment-agency.gov.uk

Scotland

Charges for IPPC installations in Scotland are levied under a scheme made under s.41 of the *Environment Act 1995*, and approved by Scottish Ministers under s.42 of the Act.

The Pollution Prevention and Control Fees and Charges (Scotland) Scheme applies to all operators of Part A installations, with fees being revised annually usually with effect from 1 April each year; the Scheme covers applications for permits and variations; subsistence fees, transfer, renewal, surrender and revocation of permits. While for most sectors standard fees will apply, fees for intensive agricultural installations – both those covered by general binding rules and others – will be calculated on a different basis as will those for low impact installations, inert waste landfill activities and certain waste incineration activities.

Consultation on a new scheme which uses compliance charging factors was due around November 2008.

Details of the actual charges applying can be accessed on SEPA's website at www.sepa.gov.uk

2.6.2 Northern Ireland

The Pollution Prevention and Control (Industrial Pollution and Radiochemical Inspectorate) Charging Scheme (Northern Ireland) is made under reg. 22 of the 2003 NI PPC Regulations; this enables the DOENI to make schemes for fees and charges to be levied by both the Industrial Pollution Inspectorate and district councils.

2.7 POWERS OF ENTRY

Sections 108-109 of *the Environment Act 1995* empower "a person who appears suitable to an enforcing authority to be authorised in writing ..." to enter premises where they have reason to believe a prescribed process is being or has been operated (with or without an authorisation) or where they believe there is a risk of serious pollution. Authorised persons may take photographs, samples etc of substances, and require access to relevant information and records. The authorised person may render harmless any substance or article from which it is considered there is imminent danger of serious harm; a written and signed report of what has been done and why should be given to a responsible person on the premises and a copy of the report served on the owner of the substance or article (if different).

Except in an emergency, a warrant should be obtained if entry has been or is likely to be refused and it is probable that force will be needed to gain entry.

Similar powers for inspectors in Northern Ireland are contained in Articles 17 and 18 of the *Industrial Pollution Control (Northern Ireland) Order 1997*, and in reg. 27 of *The Pollution Prevention and Control (NI) Regulations 2003*.

2.8 LARGE COMBUSTION PLANT

Directions under the PPC Act 1999 and under the EPA 1990 to the Environment Agency in England & Wales and to SEPA in Scotland require them to ensure the requirements of the Directive are fully complied with through IPPC permits, EPA authorisations and conditions attached thereto. A similar (2003) Direction applies in Northern Ireland.

2.8.1 LCP Regulations (Scotland & N. Ireland)

The Large Combustion Plants (Scotland) Regulations 2002 (SSI 2002/493) which came into force on 27 November 2002 partially implement Directive 2001/80/EC on *the limitation of certain pollutants into the air from large combustion plant* (see chapter 3, 3.16.2), with similar Regulations in Northern Ireland, SR 2003/210, which came into force on 25 April 2003. The LCP Regulations for England & Wales, SI 2002/2688, were revoked on 6 April 2008 with the Directive now being implemented through *the Environmental Permitting (England & Wales) Regulations* – see 2.4 above.)

The Regulations provide for "new" plant which have already been authorised to comply with the Directive through "deemed" conditions, pending variation of permits and authorisations. "New" plant not put into operation on or before 27 November 2003 were not allowed to operate until their permit had been varied in line with the requirements for "new-new" plant.

2.8.2 National Emission Reduction Plan

The UK's final National Emission Reduction Plan was submitted to the Commission on 28 February 2006, and sets out the UK's "combined approach" to implementation of the Directive, which allows operators of existing plant to,

- Opt out of the Directive and from 1 January 2008 to operate for a maximum of 20,000 hours, closing down by 31 December 2015; or
- Comply with the Directive through Emission Limit Values (ELV) for individual plant; or
- Be included in the national plan which sets an emissions "bubble" for 2008 for total emissions of SO_2 of 133,445 tonnes per year, 107,720 tonnes of NO_x per year and 9,659 tonnes per year of dust.

Annexed to the NERP are lists of which option individual existing plant in the UK have chosen in order to comply with the Directive.

The NERP enables participating plant to trade emissions allowances for sulphur dioxide, nitrogen oxides and particulate matter, while at the same time preventing them for emitting amounts in excess of their allowances. To facilitate trading of allowances, the Environment Agency holds a UK-wide electronic register, given legal status through *the Large Combustion Plants (National Emission Reduction Plan) Regulations 2007* (SI 2007/2325), which came into force on 10 September 2007; these detail the information to be kept on the register relating to participants, including for the year commencing 1 January 2008 and each subsequent calendar year, their emission allowance allocated by the Secretary of State and the verified annual mass emission for each LCPD pollutant – SEPA and the Chief Inspector (NI), as appropriate are responsible for verifying the annual report of each participating operator and notifying the Environment Agency of the emissions verified by them. The Regulations also detail the rules for transferring emissions allowances, and reporting requirements.

2.9 SOLVENT EMISSIONS

Directive 99/13 *on the limitation of emissions of volatile organic compounds due to the use of organic solvents in certain activities and installations* (see chapter 3, 3.17) has largely been implemented through Directions to the Environment Agency and SEPA requiring them to ensure the Directive's requirements are included in permits and authorisations.

2.9.1 Solvent Emissions Regulations (Scotland & N. Ireland)

The Solvent Emissions (Scotland) Regulations 2004 (SSI 2004/26), which came into force on 28 January 2004, were made under the PPC Act, and extend the PPC Regulations to cover those activities covered by the Directive but outside the scope of either PPC or the EPA; they add a new chapter 7 to Schedule 1 of the Regulations defining the SED activities now covered by the Regulations; SEPA published guidance on the Regulations in June 2004. Similar Regulations for Northern Ireland (SR 2004/36) took effect on 27 February 2004. (The Regulations for England & Wales (SI 2004/107) were revoked on 6 April 2008 by *the Environmental Permitting (England & Wales) Regulations 2007* – see 2.4 above.)

New installations carrying out SED activities are required to apply for a permit prior to being brought into operation, with existing installations required to comply by 31 October 2007 at the latest. Operators carrying out SED activities at the time the Solvent Regulations came into force must apply for a variation to their existing permits (PPC) to incorporate Directive requirements – operators notifying the regulator that they intend to use a reduction scheme to comply with the Directive, had to apply for a permit by, and operate in accordance with that scheme from, 31 October 2005; those using an alternative compliance option had to do so by 31 October 2006. Operators of new installations wishing to carry out a SED activity had to apply (and receive) a permit before bringing the plant into operation; the Regulations prescribe the dates within which previously unregulated operators of installations carrying out SED activities must apply for a permit, with the same dates as above for compliance.

2.9.2 Paints, Varnishes & Vehicle Refinishing Products

A further Directive on the use of solvents in certain paints, varnishes and vehicle refinishing products (2004/42/EC) has been implemented through *the Volatile Organic Compounds in Paints, Varnishes and Vehicle Refinishing Products Regulations 2005* (SI 2005/2773), which came into force on 1 November 2005 and apply throughout the UK. Schedule 1 to the Regulations defines the relevant products covered by the Regulations, with Schedule 2 specifying maximum limit values for VOCs for the various types of paints, varnishes and vehicle refinishing products which may not be exceeded from 1 January 2007 (Phase I), with stricter limits (Phase II) applying from 1 January 2010 – from these dates relevant products exceeding the limits may not be placed on the market within the EU; products must also be labelled with their VOC limit value in g/l, and the maximum content of VOC in g/l of the product in a ready to use condition.

The European Commission has agreed a UK-wide derogation from the Paints Directive for the use of specialist paints which do not comply with the Directive for use in restoring historic buildings and vintage vehicles – the amount of such paint used must be reported to the Commission annually. To fulfil this requirement *the Volatile Organic Compounds (Historic Buildings and Vintage Vehicles) Regulations* (a draft of which was issued for consultation in September 2007) will require business, individual and occasional users to apply for a licence enabling them to obtain specified quantities of non-compliant paints over a specified period. The licensing authorities are:

- England & Wales: local authorities for both historic buildings and vintage vehicles.
- Scotland: local authorities for historic buildings: SEPA in respect of vintage vehicles where the applicant business is subject to the PPC Regulations; other businesses and individuals – the local authority.
- Northern Ireland: NIEA for historic buildings and district councils for vintage vehicles.

2.10 MAJOR ACCIDENT HAZARDS

2.10.1 EU Directives

Following a major explosion at a chemical factory in Seveso, Italy, in 1976, which resulted in health and other environmental problems in the area, the EU adopted a *Directive on the major accident hazards of certain industrial activities* (82/501/EEC). This Directive, commonly known as the Seveso Directive, was amended in 1987 and 1988; on 3 February 1999 it (and the subsequent amendments) were repealed, the date by which Member States had to have implemented its replacement – *Council Directive on the control of major accident hazards involving dangerous substances* (96/82/EC) – Seveso II, which was adopted on 9 December 1996. In the UK the Directive has been implemented through *the Control of Major Accident Hazards Regulations 1999* – see below.

The 1982 Directive aimed, through laying down requirements for notification of major hazard sites, hazard and risk assessments, safety reports and emergency planning, to prevent major accidents which may cause serious damage to the environment or human health. An amendment to the Directive (87/216/EEC) required owners of "top tier" sites – i.e. those storing above specified amounts of hazardous substances and those carrying out particularly toxic or hazardous activities – to provide information to the public on the nature of the hazard and on action to be taken in the event of an accident. The 1996 Directive aims to close some of the loopholes/shortcomings of the earlier Directives which were identified in the light of lessons learned from accidents since the 1982 Directive came into force; many of the accidents were felt to have been preventable given better management and operational procedures.

The 1996 Directive applies mainly to the chemical and petrochemical industries and to those which produce or use substances with flammable, toxic or explosive properties; warehouses and storage facilities for hazardous materials such as fuel oils and gases are included but nuclear and military installations are not. Land-use policies must take major hazard sites into account, and thus more attention must be given to the siting of potentially hazardous installations, ensuring that new installations are sited at an "appropriate distance" from residential or other areas used by the public; these aspects of the Directive have been implemented through the *Planning (Control of Major-Accident Hazards) Regulations 1999* in England and Wales as amended, and similar 2000 Regulations in Scotland – see chapter 1, 1.19.

The Directive defines "hazard" in terms of its potential to damage or harm human health and the environment (flora, fauna, water, soil etc); operators of those establishments where dangerous substances are present above a certain threshold (top tier sites) are required to identify possible hazards and to adopt appropriate management and staff training procedures and policies to prevent accidents or minimise harm caused by accidental releases. This information must be included in the detailed safety report required by the Directive to be submitted to the competent authority (in England, Scotland & Wales – the HSE, the Environment Agency or SEPA); this must also include "main possible major accident scenarios" and how they would be dealt with. The Directive also requires an on-site emergency plan to be prepared and sufficient information given to the competent authority to enable it to prepare an off-site emergency plan. The competent authority must identify those establishments where, because of their proximity or the combinations of substances, the occurrence of an accident becomes even more potentially dangerous.

Safety reports must be publicly available and inspected by the competent authority to ensure that they meet the Directive's requirements. The competent authority must reject the safety report if it is of the opinion that, on the basis of the report, the continued use (or the start up) of an installation would "involve an imminent risk of a major accident". The competent authority must carry out an inspection of the facility (using qualified and trained personnel) and prepare a report of any follow-up measures required, with a follow-up inspection taking place within three months. Safety reports should be reviewed and updated at least every five years in the light of new knowledge about major accident hazards or when changes to the installation or safety management systems could impact on major accident hazards.

Operators of lower tier establishments must notify the competent authority of the hazards and produce a Major Accident Prevention Policy; this should set out aims and principles for preventing major accidents and show that the policy is in place and being implemented.

A further revision of the Directive – 2003/105/EC – was adopted in December 2003 and came into force on 31 December 2003. The revised Directive, which had to be implemented before 1 July 2005, reduces the threshold at which establishments with pyrotechnic and highly explosive substances come within the Directive – this is likely to result in about a further 130 installations in the UK being subject to COMAH for the first time; it also brings certain mining activities within the scope of the Directive, together with landfill waste dumps using chemical, mechanical or thermal processing, and major potassium nitrate processing plants. The Directive also adds to the list of carcinogens, and lowers the threshold at which safety reports and accident prevention policies are required, with the result that some sites will move from being lower tier to top tier sites.

Operators of installations falling within the COMAH regime for the first time had to prepare a major accident prevention policy and notify the authorities by October 2005; those becoming top-tier sites had to prepare their safety report and emergency plans by July 2006 – however inventories of dangerous substances and maps or descriptions showing the areas which could be affected by a major accident do not need to be made publicly available because of the threat of terrorism.

The need to maintain sufficient distance between hazardous facilities and buildings, recreational areas and, where possible, major transport routes, is emphasised. Operators of major hazard sites are required to provide regular information in an appropriate form to the public in their vicinity on the action to take should there be an accident; similar information should also be given to "establishments serving the public" (e.g. schools and hospitals) and to other people who might be affected by an accident.

2.10.2 COMAH Regulations

The Control of Major Accident Hazards Regulations 1999 (COMAH Regulations) (SI 743), which came into force on 1 April 1999, implement the 1996 Directive; 2005 Amendment Regulations (SI 2005/1088) which came into force on 30 June 2005 implement the 2003 Amendment Directive. The Regulations apply in England, Scotland and Wales and replace 1984 Regulations which implemented the 1982 Directive. The Regulations are enforced by the Environment Agency and the HSE in England and Wales and by the Scottish Environment Protection Agency and the HSE in Scotland (the competent authorities). Similar 2000 Regulations (SR 93, effective 1 May, amended by SR 2005/305, effective 25 July 2005) apply in Northern Ireland replacing 1985 Regulations; in Northern Ireland the Environment Agency and the HSE for Northern Ireland are the competent authorities.

The 1999 Regulations (NI: 2000) apply to establishments which keep (or transport) listed dangerous substances in quantities exceeding thresholds (either individually or aggregated) set in the Regulations; thus they apply mainly to the chemical and petrochemical industries and to those which produce or use substances with flammable, toxic or explosive properties, and those which are dangerous for the environment; explosives and chemicals on nuclear sites are included but hazards created by ionising radiation are not. They do not apply to military installations, certain activities of the extractive industries or to waste landfill sites.

Operators of companies covered by the Regulations are placed under a general duty to take all necessary measures to prevent major accidents and to limit their consequences to persons and the environment, and to report any major accidents to the competent authority. They must prepare a "Major Accident Prevention Policy" which should demonstrate that an adequate safety management system is in place. The enforcing authority should be sent details ("notification") of the name and address of the operator, the address of the site concerned and who is in charge, and details and amounts of dangerous substances held at the site; significant increases in the quantities held or in the nature or physical form of the substances held should be notified to the authority.

Site operators holding larger amounts of hazardous substances (top tier sites) must provide the competent authority with a safety report for approval. This should demonstrate that the company's operational, and indeed emergency, procedures are of sufficiently high standard and adequate to prevent accidents and damage both inside and outside plants; the report should include major accident scenarios and safety management systems, using maps, images or equivalent descriptions to support the assessment of the extent and severity of the consequences of identified major accidents; the safety report should also list the organisations involved in drawing up the report. A copy of the safety report will be placed on the public register*, with as agreed with the competent authority, security sensitive, confidential etc information omitted – the 2005 amendment regulations require the operator to provide the competent authority with the amended version of the safety report within three months of agreement on what material may be omitted. The 2005 amendment regulations also require that when operators review their safety reports, any revisions are notified (in writing) to the competent authority; they should also notify the competent authority of any review that has not resulted in a revision, with this notification being placed on the public register.

(*In 2001, safety reports for top tier sites were removed – or were not placed on – public registers to enable "security sensitive" materials to be removed. Information about the location of hazardous substances was permitted to be removed in the light of concern that it could be of use to terrorists – House of Commons Written Answer, 28 November 2001.)

The Regulations also require the operator to prepare an "on-site emergency plan" (which should include details of site remediation and clean up following an accident). The local authority for the area in which the establishment is situated is required to prepare an "off-site emergency plan" and in so doing must consult the competent authority, the emergency services and the health authority, the Environment Agency and SEPA; the local authority must consult the public when the off-site emergency plan is reviewed. Both plans must be reviewed and tested at intervals of not more than three years. The company must also notify all those living in the locality of the fact that there is a hazardous site in the area, and advise what action should be taken in the event of an accident; information and safety reports relating to the establishment are available on public registers held by the competent authorities, subject to issues of commercial confidentiality and national security.

A charging scheme has been introduced to recover the costs of examination of top-tier sites' safety reports, inspection of major accident hazards at top and lower tier sites, enforcement of the Regulations and investigation of major accidents.

Various guidance and background documents on the Regulations are available from the Health and Safety Executive – www.hse.gov.uk/comah.

Separate Regulations – *The Planning (Hazardous Substances) Regulations 1992* – require the storage of certain hazardous substances above a "controlled quantity" to be covered by a consent from the Hazardous Substances Authority (usually the local authority) – see also chapter 1, 1.17. These Regulations have been amended by *The Planning (Control of Major-Accident Hazards) Regulations 1999* (SI 981; Scotland: SSI 179, 2000; NI: SR 101, 2000) to implement the requirements of the Directive that land-use policies must take major hazard sites into account. The Regulations are to be further amended to implement the 2003 Directive requirements for tighter controls on the siting of COMAH sites.

Health and Safety at Work Etc Act 1974

This Act provides a comprehensive and integrated legal and administrative system for securing the health, safety and welfare of persons at work and for protecting other persons against risk to health or safety arising from the activities of persons at work. To this end all the various inspectorates and agencies involved in health and safety issues were brought together in the Health and Safety Executive (established 1 January 1975) under the Health and Safety Commission which had been set up on 1 October 1974. From 1 April 2008, the two bodies were merged with the HSE becoming the single national regulatory body responsible for promoting the cause of better health and safety at work.

In Northern Ireland, the corresponding legislation is the *Health and Safety at Work (Northern Ireland) Order 1978.*

2.11 COSHH REGULATIONS

The Control of Substances Hazardous to Health Regulations 2002 (SI 2677), covering England, Scotland and Wales, came into force on 21 November 2002 (amended by SI 2004/3386, effective 6 April 2005). The Regulations replace and consolidate 1999 Regulations (SI 437) which came into force in March 1999 and subsequent amendments. An Approved Code of Practice (also updated at the same time as the Regulations) provides guidance on the application of the Regulations. Similar 2003 Regulations (SR 34, effective 28 February, as amended SR 2005/165) apply in Northern Ireland; they replace 2000 Regulations (SR 120).

The Regulations apply to all "very toxic, toxic, harmful, corrosive or irritant" substances (defined in *The Chemicals (Hazard Information and Packaging for Supply) Regulations 2002* – SI 1689) and to all places of work; they provide a legal framework for controlling people's exposure to such substances. Hazardous substances may include gases, vapours, liquids, fumes, dusts and solids; they can be components of a mixture of materials including micro-organisms. COSHH also covers substances which have chronic or delayed effects such as those which are carcinogenic, mutagenic or teratogenic. The Regulations also apply to the use of pesticides, for which two codes of practice have been drawn up by the HSE (see chapter 6), to the use of biological agents and to offshore oil and gas installations. Asbestos, lead, materials producing ionising radiations and inhalable dust below ground in coalmines are excluded as they are covered by separate legislation.

As well as implementing the health requirements of EU Directive 98/24/EC on the protection of the health and safety of workers from the risks related to chemical agents at work, the Regulations require

- An assessment of the risk to health of employees arising from their work and what precautions are needed; this assessment must be regularly reviewed, particularly if it is suspected that the assessment is no longer valid or there has been a significant change in the work to which the assessment relates;
- The introduction of appropriate measures to prevent or control the risk;
- That control measures are used and equipment is properly maintained and procedures observed; 2004 amendment Regulations introduce a duty to review, at suitable intervals, control measures other than provision of paint and equipment, including systems of work and supervision.;
- Monitoring the exposure of workers to hazardous substances; in general monitoring should be carried out every 12 months and more frequently for certain processes involving the use of local exhaust ventilation plant (see Schedule 4 to the Regulations); personal exposure monitoring records should be retained for 40 years and general monitoring records for at least 5;
- Health surveillance of employees exposed, or likely to be exposed, to hazardous substances, with records being retained for at least 40 years;
- Informing, instructing and training employees about the risks and the precautions to be taken, and providing them with results of monitoring and health surveillance;
- Compliance with workplace exposure limits for substances hazardous to health and listed in HSE publication EH40, *Workplace Exposure Limits* – these take the place of occupational exposure standards and maximum exposure limits.

HM Factory Inspectorate enforces the Regulations and any breaches are prosecuted under the HSW Act.

Approved Codes of Practice have been published by the Health and Safety Commission to accompany the Regulations. An Approved Code of Practice on the control of occupational asthma aims to ensure that employers are aware of the need to control substances which cause occupational asthma.

Further information is available on www.hse.gov.uk/coshh

2.12 CONTROL OF ASBESTOS AT WORK

2.12.1 EU Directives

EU Directive 98/24/EC relates to the protection of the health and safety of workers from the risks related to chemical agents at work; it requires that all risk assessments must be written down where there is liable to be exposure to asbestos; a new risk assessment must be carried out before any new work involving asbestos is begun; the assessment should include a plan of the work showing the range and use of control measures; the number of workers exposed to asbestos should be reduced to the minimum and details of prevention and protection measures specified, and training for anyone liable to be exposed; immediate steps should be taken to reduce exposure where the control level or action limit is exceeded and information on the location of asbestos should be given to the emergency services.

EU Directive 2003/18/EC *on the protection of workers from the risks related to exposure to asbestos at work*, adopted March 2003, strengthens requirements placed on employers to identify and assess the risk to all employees exposed to asbestos dust and to take appropriate preventative and protective measures. The Directive also requires more detailed information to be provided to the competent authority, and for regular free training on all aspects of working with asbestos to be provided; health assessments of exposed workers must be carried out every three years and medical records kept for at least 40 years. Work that is deemed to be of 'sporadic and low intensity' need not be notified to the enforcing authority and will not require medical surveillance of employees. An exposure limit for airborne concentrations of asbestos of 0.1 fibres per cm^3 as an 8-hour time weighted average is set, above which no worker must be exposed.

2.12.2 UK Regulations

The Control of Asbestos Regulations 2006 (SI 2006/2739), which came into force on 13 November 2006, apply in England, Wales and Scotland and implement the above Directives; they revoke SI 2002/2675. Similar Regulations apply in Northern Ireland – SR 2007/31, which came into force on 6 April 2007, revoking SR 2003/33.

The Regulations prohibit the importation, supply and use of all forms of asbestos, as well as continuing the ban on blue and brown asbestos and on white asbestos (from 1985 & 1999 respectively).

The Regulations cover occupational exposure to asbestos, the removal of which mostly requires to be undertaken by licensed contractors. Work with asbestos will not usually require licensing if employee exposure is "sporadic and of low intensity" – i.e. does not exceed 0.6 fibres per cm^3 of air averaged over any continuous ten minute period – and if the risk assessment shows that the control limit – 0.1 fibres per cm^3 of air measured over four hours – will not be exceeded.

They impose duties on employers, similar to those contained in the COSHH Regulations (see above), for the protection of employees who may be exposed to asbestos at work, and require employers to

- review assessments made under the Regulations at regular intervals;
- prepare a plan of work where the activity concerned entails the removal of asbestos from plant, buildings or other sources;
- put greater emphasis on preventing exposure to asbestos as a carcinogen by using alternative non- or less dangerous substances;
- ensure that only essential workers enter the affected area if there is an unforeseen escape or concentrations are likely to exceed limits.
- employers must assess risk of exposure to asbestos of employees, with extra safeguards to protect those likely to come into contact with asbestos in their normal work (e.g. plumbers, electricians, building workers and carpenters), including training, maintenance and testing of respiratory equipment.

The Regulations also impose a duty to manage asbestos in non-domestic premises; in particular, people in control of premises (i.e. with repair and maintenance responsibilities) are required to

- take reasonable steps to locate materials likely to contain asbestos;
- presume that materials contain asbestos unless there is evidence that they do not;
- keep up-to-date written records of the location of these materials;
- monitor the condition of the materials;
- assess the risk of exposure from asbestos and presumed-asbestos materials;
- prepare and implement a management plan to control the risks.

While the duty will usually fall on the employer, it will include others with a responsibility towards the building, e.g. the owner of the building or managing agent, architects or surveyors. An Approved Code of Practice provides guidance on the management of asbestos in non-domestic premises and is available on HSE's website, together with further guidance on working with asbestos.

2.13 LEAD AT WORK

The Control of Lead at Work Regulations 2002 (SI 2676), together with an Approved Code of Practice, replace and re-enact 1998 Regulations (SI 543). Similar Regulations apply in Northern Ireland – SR 2003/35, which came into force on 28 February 2003, replacing SR 1998/281.

The Regulations require employers, who have the duty under the *Management of Health and Safety at Work Regulations 1992*, to assess risks from exposure to lead in the workplace and to take steps to prevent or adequately control such exposure. If the employer concludes from the assessment that the exposure of employees to lead is likely to be "significant", specific controls must be introduced; these include issuing protective clothing, monitoring of air and medical surveillance. The Regulations also implement the health requirements of the EU Directive on the protection of the health and safety of workers from the risks related to chemical agents at work (98/24/EC).

A revised Approved Code of Practice has been published which reflects the new Regulations, including medical surveillance procedures for employees with elevated blood-lead levels.

2.14 CHIP REGULATIONS

The Chemicals (Hazard Information & Packaging for Supply) Regulations 2002 (SI 2002/1689) as amended, which came into force on 24 July 2002, revoke and re-enact with amendments 1994 Regulations (SI 1994/3247). Similar Regulations took effect in Northern Ireland on 14 November 2002 (SR 2002/301, amended SR 2004/568 & SR 2005/2571). The aim of these Regulations, known as CHIP3, is to protect both human health and the environment from the effects of chemicals by ensuring that chemicals are packaged safely (the 'supply' requirements); suppliers of chemicals are required to identify the hazards associated with the chemicals they supply and to provide adequate information on the chemicals to the people to whom they supply them.

The Regulations also implement EU Directives on the marketing and use of dangerous substances and preparations. The Regulations are to be further amended to implement Directive 2006/8/EC; this amends rules and procedures for classifying and labelling chemical preparations, as well as those containing ozone depleting substances, and adjusts the generic concentration limits to be used for the evaluation of the hazards for the aquatic environment (HSE consultation, April 2008). Separate Regulations cover the transport of chemicals by road.

The Health and Safety Executive has published a number of guides and reference documents in support of the CHIP Regulations, including "The Complete Idiot's Guide to CHIP".

2.15 REACH

EU Regulation 1907/2006 on the *Registration, Evaluation, Authorisation and Restriction of Chemicals* (REACH), was agreed in December 2006, entering into force on 1 June 2007; a corrigendum to 1907/2006 was issued on 27 May 2008. EU Regulations are directly applicable to Member States, with individual Member States required to have an enforcement and penalties regime in place by 1 December 2008.

The European Chemicals Agency is responsible for managing implementation of REACH across the EU – http://echa.europa.eu/

Considerable information on REACH can be found on the HSE's website at www.hse.gov.uk/reach, with detailed guidance also available from the European Chemicals Agency – http://reach.jrc.it/guidance; see also further reading section of the *Pollution Control Handbook*.

2.15.1 EU Regulation

The main elements of REACH are:

- Registration – requires industry to obtain relevant information (technical dossier) on chemical substances produced or supplied above 1 tonne a year and to use that data to manage the chemical safely. A chemical safety report will also be needed for production or use of 10 tonnes or more per year.
- Evaluation – provides the opportunity for regulators to assess whether the information provided by industry is sufficient and that they have applied the right risk management measures.
- Authorisation – controls the use of substances of highest concern. Substances with the most hazardous properties would be banned unless the industry can demonstrate that the risks are properly controlled, or that socio-economic benefits outweigh the risks.

The Regulation applies to chemicals manufacturers, importers, distributors or suppliers and to any business which uses chemicals, who are all required to register with the European Chemicals Agency. For both new and existing chemicals the Agency will require details of characteristics and hazards, if any, associated with a chemical; those that pose a serious hazard may be banned (i.e. restricted) or may be used only following the grant of a specific authorisation. The need to register came into force on 1 June 2008. if a producer pre-registered before 1 December 2008 registration may be spread over the 10 years to 2018, depending on the tonnage produced; pre-registration requires names of the substances being registered, companies on whose behalf the substances are being registered, indication of likely tonnage threshold, names of substances which may have similar properties. Any substances for which pre-registration had not been completed by 1 December 2008, had to be withdrawn from the market pending full registration.

2.15.2 UK Regulations

The REACH Enforcement Regulations 2008 (a draft of which was published for consultation in June 2008) were expected to come into force on 1 December 2008, providing the legislative framework within which the enforcing authorities will operate.

The Regulations designate the Secretary of State (England), Scottish and Welsh Ministers, and in Northern Ireland, the Department of Enterprise, Trade and Investment and the Department of the Environment as the Competent Authorities for the UK. The functions of the Competent Authorities have been delegated to the Health and Safety Executive (and its counterpart in N. Ireland). The HSE and HSENI are to be responsible for enforcing those articles of the EU Regulation relating to registration and the supply chain. Articles relating to use (and environmental breaches) from installations covered by PPC will be enforced by the HSE and HSENI, the Environment Agency and SEPA, DOENI, Secretary of State (BERR). In England and Wales and in Northern Ireland, environmental breaches of the Regulation by non-PPC sites will be dealt with by the Environment Agency and the DOENI (as appropriate); in Scotland such breaches may be dealt with by SEPA with the agreement of the local authority.

The enforcing agencies are given powers to enter premises, ask for information, issue a wide range of enforcement notices and to prosecute non-compliance with the provisions of REACH or the Regulations – Schedule 1 to the Regulations sets out the relevant provisions of REACH and the enforcing authority. Summary conviction may result in a fine of up to £5,000 and/or three months in prison. Conviction on indictment may lead to an unlimited fine and/or two years in prison.

3 CHAPTER

Climate change

Air quality

Depletion of the ozone layer

Control of emissions to air

Agriculture and air pollution

Transport

Radiation

AIR POLLUTION CONTROLS – BACKGROUND

The development of air pollution control legislation in the UK has been a gradual process, beginning with the Industrial Revolution of the late 18th and 19th centuries – when air pollution was the main problem – and continuing to today's legislation which requires an integrated approach to controlling polluting emissions. air pollution was, however, first acknowledged to be a problem many hundreds of years earlier.

3.1 FIRST CONTROLS – UK

Well before the Industrial Revolution, air pollution was becoming a problem; coal was already being used in England by the year 852 and a record of "Seacoales Lane" in London dates from 1228. The use of coal was prohibited in London in 1273 as being prejudicial to health; however the practice must have continued for early in the 14th century a Royal Proclamation prohibited the use of seacoal by artificers (craftsmen) in their furnaces. In 1648, Londoners petitioned Parliament to prohibit the importation of coal from Newcastle because of the injuries they experienced. After complaints of smoke from his neighbour, one baker was ordered to erect a chimney "so high as to convey the smoake clear of the topps of the houses".

In 1661 John Evelyn published his tract *Fumifugium or The Smoake of London Dissipated*. This suggested practical schemes to improve the air over London. These were ignored and it was not until 1819 that Parliament appointed a committee to enquire to what extent persons using steam engines and furnaces could erect them in a manner less prejudicial to the public health and comfort. However, nothing of significance emerged from the committee's deliberations.

3.1.1 Alkali Acts & Best Practicable Means

The Industrial Revolution, which was based on coal as the energy source, brought a worsening of urban squalor and appalling air pollution. Around 1860, discharges to the air from an early alkali works devastated the surrounding country, corroding material and tools and destroying vegetation and crops. Air pollution from the works could no longer be ignored. Following overwhelming public complaint and a Parliamentary Enquiry, the first *Alkali etc Works Act* was passed in 1863. This made no attempt to control smoke, but required that 95% of the offensive emissions should be arrested. The remainder, after adequate dilution, might be allowed to pass to atmosphere. The same Act set up a national inspectorate to enforce the legislation. The improvement in pollution levels was dramatic: acid emissions from alkali works were reduced from an annual rate of almost 14,000 tonnes to about 45 tonnes.

The second *Alkali Act* (1874) required, for the first time, the application of the Best Practicable Means (BPM) to prevent the escape of noxious or offensive gases, whether they arose indirectly from any part of the process or plant, or directly from the exit flues. The requirement also applied to "fugitive" or "uncontained" emissions, including gases or vapours leaking from process storage tanks, pumps, compressors, coking plant or streams of liquid effluent, dust from roadways, stockpiles and even materials carried out of works on the wheels of vehicles. The 1874 Act also introduced the first statutory emission limit – for hydrogen chloride (0.2 grains per cubic foot). This Act was subsequently extended to all the major industries that pollute the air.

A series of Acts followed which were eventually consolidated by the *Alkali etc Works Regulation Act 1906*. This linked together a schedule of carefully defined and chosen processes – works – with an equally carefully chosen list of "noxious and offensive gases". The types of works scheduled were those thought most likely to cause pollution problems and included a substantial part of the chemical industry, petroleum refineries, petrochemicals, electricity generation, coal carbonisation, iron and steel works, non-ferrous metals, mineral processing works etc.

The *Public Health (Smoke Abatement) Act 1926* empowered ministers to make various additions to the list of scheduled works and noxious and offensive gases by an order laid before Parliament for approval and given legislative effect as a statutory instrument. As industrial technology advanced, seven such orders extended control over various processes between 1928 and 1963. These orders were consolidated in 1966 in the *Alkali etc Works Order*.

In 1974 the *Health and Safety at Work etc Act* was passed. It was essentially an enabling measure and was intended to repeal the 1906 Act entirely, replacing it by regulations made under the 1974 Act, e.g. the *Health and Safety (Emissions into the Atmosphere) Regulations 1983*. The remaining sections of the 1906 Alkali Act were repealed in England and Wales on 16 December 1996 following completion of authorisation of all prescribed processes under Part I of the *Environmental Protection Act 1990*, and in Northern Ireland on 3 April 2005. This legislation too has been superseded by controls administered under the integrated pollution prevention and control regime: In Scotland and Northern Ireland, *the Pollution Prevention and Control Regulations* (2000 & 2003, respectively – see chapter 2, 2.5); in England and Wales PPC regulations have been replaced by *the Environmental Permitting (England and Wales) Regulations 2007* (see chapter 2, 2.4).

3.1.2 Smoke Control

While the *Public Health (Smoke Abatement) Act 1926* had enabled the Alkali Inspectorate to extend its influence to industrial smoke, there was no measure to control the far more widespread problem of smoke from domestic chimneys. Other legislation enabled smoke to be dealt with as a possible nuisance, but was largely ineffective because a nuisance had first to be proved in court. Thus the atmosphere of Britain's urban areas was still characterised by the pervading pall of smoke and sulphur fumes from countless stacks, chimneys and funnels. The lack of winter sunshine, the prevalence of pea soup fogs, black buildings and even black snow, had become accepted as the price of progress and bronchitis, exacerbated by smoke and sulphur, was a common illness.

It was not until the infamous London smog of December 1952, which lasted for five days, that any real action was taken. The first casualties of the fog were prize cattle brought to London for the Smithfield Show – one died, 12 had to be slaughtered and some 160 required veterinary treatment. Even an opera performance at the Sadlers Wells Theatre had to be abandoned after the first Act as the audience could not see the stage! More seriously, some 4,000 additional deaths were directly attributed to the episode.

The Government appointed a committee, under the chairmanship of Sir Hugh Beaver, to study the problem of air pollution. Its terms of reference were "to examine the nature, causes and effects of air pollution and the efficacy of present preventive measures; to consider what further measures are practicable; and to make recommendations". The committee issued two reports, the first in 1953 and the second in 1954; these recognised that the constituent parts of the problem were gaseous and particulate emissions, and recommended immediate legislation to reduce smoke, grit and dust.

The eventual result was the *Clean Air Act 1956* which was later amended and extended by the *Clean Air Act 1968*. These Acts (covering England, Scotland and Wales, with a similar 1964 Act for N. Ireland) constituted the operative legislation against pollution by smoke, grit and dust from domestic fires and other commercial and industrial processes not covered by the *Alkali Acts* and other subsequent pollution legislation. They regulated the combustion of solid, liquid and gaseous fuels and controlled the heights of new industrial chimneys not scheduled elsewhere. The legislation also prohibited the emission of "dark" smoke from any chimney, provided for Government funding for the conversion of domestic grates to smokeless operation, and regulated the fuels that could be burned on them.

The 1956 and 1968 Acts have now been consolidated and their provisions re-enacted in the *Clean Air Act 1993* which came into force on 27 August 1993. This applies in England, Scotland and Wales only. (s.30, regulations about motor fuel, also applies in Northern Ireland: see 3.19.4 and 3.27.1 below.) In Northern Ireland, the 1964 Act has been replaced by the *Clean Air (Northern Ireland) Order 1981*.

3.1.3 Air Pollution Monitoring

Measurements of air pollution have been made for many years in the UK, with monitoring (i.e. the systematic measurement of air pollution) begun in 1914 by the voluntary Committee for the Investigation of Atmospheric Pollution. This Committee included, among others, members from a few local authorities, the Coal Smoke Abatement Society (the forerunner of Environmental Protection UK) and the Director of the Meteorological Office. Its work was put on an official footing in 1917, under the direction of the Met. Office with the aim of providing an adequate picture of smoke and sulphur dioxide, grit and dust in towns; this coordinating function was later passed to the Department of Scientific and Industrial Research, followed in turn by the Fuel Research Station, Warren Spring Laboratory and now the National Environmental Technology Centre (NETCEN) (part of AEA Technology).

In 1961, the National Survey of Air Pollution was set up to monitor concentrations of both smoke and sulphur dioxide; measurements were made at about 1,200 urban and rural sites throughout the UK by local authorities and other bodies. This was replaced in 1982 by the UK Smoke and Sulphur Dioxide Monitoring Network, which comprised some 165 sites monitoring black smoke and sulphur dioxide, including sites monitoring in compliance with the EU Directive on air quality limit values and guide values for sulphur dioxide and suspended particulates (see 3.9.3a). The Network was equipped and operated by local authorities, industries and other bodies and coordinated by NETCEN who processed and analysed the data on behalf of Defra. The Network was largely disbanded in 2006, with a few sites being retained to monitor smoke only.

Smoke and sulphur dioxide measurement data are available from the UK National Air Quality Information Archive website – www.airquality.co.uk. The site includes automatic and non-automatic monitoring data and air pollution forecasts, with links to the UK atmospheric emissions inventories. The Environment Agency's *UK Air Pollutants: key facts and monitoring data (UKAP)* (2006) is a digest of ambient air quality monitoring data for 143 pollutants, with further key information for the 51 most important pollutants; it can be downloaded from the Agency's website.

An annual National Atmospheric Emissions Inventory is prepared each year for the Department for Environment, Food and Rural Affairs (Defra) by NETCEN. The Inventory covers black smoke, sulphur dioxide, nitrogen oxides, carbon monoxide, non-methane volatile organics, methane, carbon dioxide, lead, ammonia, metals and halogens; it provides estimates of pollution to atmosphere from all known sources and is used by government for policy formulation and for meeting European legislative requirements and international treaty obligations.

3.2 INTERNATIONAL INITIATIVES

Many pollution problems – depletion of the ozone layer, global warming, export of hazardous wastes and so on – are recognised as global problems requiring global solutions. Here the United Nations Organisation and its agencies, such as the United Nations Environment Programme (UNEP), have an important role in providing a forum for discussion and cooperation on global environmental problems. UNEP itself was founded in 1973 following the UN Conference on the Human Environment held in 1972 where the 113 participating countries expressed concern over the rapidly deteriorating environment; this conference was also important in drawing attention to the concept of transboundary pollution – that pollution does not recognise political or geographical boundaries.

A further conference – the United Nations Conference on Environment and Development (UNCED, or the Earth Summit) – was held in June 1992 in Brazil. This conference was a direct result of the UN Commission on Environment and Development whose report *Our Common Future* (often called the Brundtland Report, after its Chairman Gro Harlem Brundtland), was published in 1987. This identified the importance of "sustainable development" in eradicating poverty and halting further environmental degradation and moving to sustained economic growth.

The Earth Summit in Brazil was attended by world leaders, ministers, officials and representatives of environment non-governmental organisations from over 160 countries. The main aim of the Conference was to find ways to "halt and reverse the effects of environmental degradation" while increasing efforts "to promote sustainable and environmentally sound development in all countries". As well as establishing a Sustainable Development Commission and endorsing a *Declaration of Principles* for the pursuit of sustainable development as the prime means of alleviating Third World poverty and global environmental degradation, two important conventions were agreed, on climate change (3.5 below) and on biological diversity (see chapter 1, 1.9). The conference also agreed a *Statement of Principles* for the sustainable management and conservation of forests and funding for new environmental aid for developing countries.

The UN Commission on Sustainable Development (CSD), established in December 1992, is charged with ensuring effective implementation and follow-up of the agreements reached during the Earth Summit at local, regional, national and international level. It is also responsible for progressing "Agenda 21" – a huge document agreed at the 1992 Earth Summit setting out an "environmental work programme" into the 21st century. It aims to provide guidance for governments in establishing environmental policies that meet the needs of sustainable development; it covers all areas of pollution, energy policy, population and development issues. In the UK, local authorities have been encouraged to adopt their own Agenda 21 – i.e. a strategy for sustainable development at local level – in consultation with business and community groups.

A third conference – "Rio + 10" – was held in Johannesburg in September 2002 at which progress since the Earth Summit was assessed. In re-affirming their commitment to the principles of sustainable development, delegates agreed a "plan of implementation" which includes the following commitments:

- To halve the proportion of people without access to adequate sanitation and safe drinking water by 2015;
- To increase substantially the global share of renewable energy sources and thus increase its contribution to total energy supply; to accelerate the development and dissemination of energy efficiency and energy conservation technologies;
- By 2020, to use and produce chemicals in ways that do not lead to significant adverse effects on human health or the environment;
- By 2020, to achieve a significant reduction in the current loss of biological diversity;
- To facilitate implementation of the Montreal Protocol by ensuring replenishment of funding for this purpose; to assist developing countries to comply with the Protocol by improving access to alternative substances with a less harmful effect on the ozone layer;
- To encourage the application by 2010 of the ecosystem approach for the sustainable development of the oceans; by 2015 to maintain or restore depleted fish stocks.

3.3 EUROPEAN COMMUNITY

The United Kingdom, as a member of the European Community since 1973, is obliged to transpose legislative measures – Directives, Regulations, Decisions, etc – into national legislation. In some cases, EC legislation, such as the adoption in July 1980 of a Directive on health protection standards for levels of sulphur dioxide and suspended particulates in air (80/779/EEC), has resulted in an important change in UK policy. In effect, this Directive – which set limit values and non-mandatory guide values – introduced into the UK the first air quality standards and was thus a break with the UK's tradition of setting non-statutory emission standards, via the pollution inspectorate. Similar Directives for nitrogen oxides (85/203/EEC) and lead (82/884/EEC) followed. Following adoption of Directive 99/30/EC setting limit values for sulphur dioxide, oxides of nitrogen, particulate matter and lead (see below 3.9.3), various articles of the earlier Directives were repealed on 1 January 2000; the remaining Articles, largely relating to the limit values (except for those relating to NO_2) were repealed on 1 January 2005 when the newer limits had to be met.

The EU's 1996 Framework Directive on the Assessment and Management of Air Quality required Member States to assess ambient air quality on the basis of limit values and alert thresholds set in subsequent daughter Directives and to draw up action plans for those areas at risk of exceedence. This Directive, and the subsequent Daughter Directives (except that on heavy metals) have now been merged into a new Directive on Ambient Air Quality and Cleaner Air for Europe (2008/50/EC) adopted on 21 May 2008. The updated Directive includes new targets and limit values for $PM_{2.5}$ as well as a non-binding exposure reduction target to further reduce background levels by 20% between 2010 and 2020. See 3.9.2 & 3 below. The UK has already introduced (in the Air Quality Strategy) a slightly weaker exposure reduction target of 15%.

A brief summary of the European Union, its institutions, legal framework and programmes is given in Appendix 7.

3.4 ENFORCEMENT AGENCIES

3.4.1 Pollution Inspectorates

Prior to April 1996, pollution legislation was administered in England and Wales by HM Inspectorate of Pollution (or its forerunners), with almost identical legislation in Scotland administered by HM Industrial Pollution Inspectorate. The control of scheduled processes by the Inspectorate had three distinct aspects:

- ***Registration:*** Under the *Health and Safety (Emissions into the Atmosphere) Regulations* (SI 943, repealed December 1996) scheduled works had to be registered annually with the Pollution Inspectorate. As an essential prior condition to the first and all subsequent registrations, every scheduled works had to be provided with the Best Practicable Means for preventing the escape of any noxious or offensive substances, and for rendering all such emissions harmless and inoffensive. The *Control of Air Pollution (Registration of Works) Regulations 1989* (SI 318, repealed December 1996) specified the information to be contained in applications for registration.

- ***Inspection:*** As a requirement of registration, the BPM had always to be in use, and pollution control equipment maintained in good and efficient working order. Under the *Health and Safety at Work etc Act 1974*, inspectors were authorised to enter premises for examination or investigation without prior announcement. The emphasis was on prevention rather than cure. If the inspector believed that any part of the legislation was contravened, an improvement notice could be issued specifying the problem, the period during which it should be put right and, if appropriate, the remedial measures to be taken. Where toxic emissions were involved the process would generally be suspended or drastically curtailed. Inspectors could also issue a prohibition notice if there was believed to be a risk of personal injury.

- ***Presumptive Limits:*** Notes on Best Practicable Means specified the amount of pollutant per cubic metre of gas (the "presumptive limit") that could be emitted from the chimney stack. Although non-statutory, there was normally a presumption that if the limit was being met, then the legislation was being complied with in that the BPM was being used. The fact that BPM Notes were non-statutory documents allowed the Inspectorate to tighten emission limits in line with advances in abatement technology or scientific understanding of the effects of pollutants. For many processes there were, in addition, other limits based on subjective assessment, such as smoke emission or other visibility criteria.

The Environment Act 1995 abolished HMIP in England and Wales, replacing it with the Environment Agency; in Scotland, the Scottish Environment Protection Agency replaced HMIPI – see chapter 1, 1.3.1 and 1.3.2 respectively.

3.4.2 Local Authorities

(a) England and Wales

Local authorities have long been responsible for the regulation and control of local air pollution under the *Clean Air Acts*, and for industrial air pollution control first under Part I of the *Environmental Protection Act 1990* (EPA), and now as part of the IPPC regime – *the Environmental Permitting (England and Wales) Regulations 2007* which came into force on 6 April 2008l replacing *the Pollution Prevention and Control (England and Wales) Regulations 2000*. Local authorities also have responsibility for local air quality management and assessment under the *Environment Act 1995* (this chapter, 3.10). Other areas of responsibility include:

- Enforcement of statutory nuisance, and parts of the *Anti-social Behaviour and Clean Neighbourhoods and Environment Acts* (chapter 1, 1.29-1.34);
- Enforcement of noise control legislation (chapter 4);
- Identification of, and ensuring appropriate clean up of, contaminated land under the *Contaminated Land (England) Regulations 2006* (and similar in Wales, 2001) – see chapter 5.

Local authorities in England have also been set a number of environmental performance indicators, including on climate change, air quality and biodiversity – see www.defra.gov.uk/environment/localgovindicators

The *Public Health (Control of Diseases) Act 1984* gives Port Health Authorities in England and Wales similar responsibilities to local authorities in relation to those sections of the *Clean Air Act 1993* and *the Environmental Permitting Regulations 2007* relating to the control of air pollution.

(b) Scotland

Scottish local authorities have similar responsibilities to those in England and Wales for local air quality management, nuisance etc, and for enforcing regulations under *the Clean Air Act*. The Scottish Environment Protection Agency is however responsible for regulating installations controlled for air pollution under *the Pollution Prevention and Control Act 1999* (chapter 2, 2.5).

3.4.3 Northern Ireland

In Northern Ireland, the control of air pollution was the responsibility of the Department of the Environment under the *Alkali Act 1906*, repealed in 2005. The *Industrial Pollution Control (Northern Ireland) Order 1997*, which received Royal Assent on 26 November 1997, established a system of air pollution control similar to that operating in the rest of the UK under *the Environmental Protection Act 1990*. As in other parts of the UK, this regime has been replaced by *The Pollution Prevention and Control Regulations (Northern Ireland) 2003* (SR 2003/46), which came into force on 31 March 2003; all activities subject to IPPC (Part A and Part B installations) are regulated by the Northern Ireland Environment Agency (Industrial Pollution Inspectorate); activities subject to control for emissions to air (Part C) are regulated by local authorities – see chapter 2, 2.5.

District councils in Northern Ireland are also responsible for the review and assessment of air quality in their area, with *the Air Quality (Northern Ireland) Regulations 2003* prescribing the period within which the air quality objectives (in the National Air Quality Strategy) must be achieved.

Local authorities (district councils) are responsible for the control of domestic emissions and for emissions from commercial and industrial premises, for statutory nuisance and noise legislation.

CLIMATE CHANGE

The Intergovernmental Panel on Climate Change (IPCC) was set up in 1988 by the United Nations Environment Programme and the World Meteorological Organization to assess the scientific and technical aspects of climate change. It reported in 1990, 1995 and 2001, with its Fourth Assessment Report being published in three parts during 2007. The "Synthesis Report", published in November 2007, is explicitly targeted at policy makers and integrates and presents the three reports published earlier in the year (see Further Reading section of the Pollution Control Handbook).

The report of Working Group I assessing current scientific knowledge of climate change is unequivocal in its conclusion that the biggest single impact on the climate results from human activities, notably through emissions of carbon dioxide. Working Group II's report on impacts, adaptation and vulnerability concludes that while climate change is already having major impacts, future impacts will result in some areas becoming even drier and others more susceptible to flooding; it forecasts that more than a billion people will face shortages of fresh drinking water by 2050, a halving of crop yields in some countries by 2020, and widespread loss of biodiversity.

3.5 INTERNATIONAL CONVENTIONS & PROTOCOLS

3.5.1 Kyoto Protocol

The Framework Convention on Climate Change came into force on 21 March 1994 following ratification by 50 countries; the

Convention has now (end 2006) been ratified by 192 countries, including both the European Community and the UK. It requires developed countries to take measures aimed at returning emissions of greenhouse gases (in particular carbon dioxide) to 1990 levels by 2000 and to provide assistance to developing countries. Other obligations include compiling inventories of emissions, producing and publishing national programmes of measures to limit emissions and to promote research and public education about climate change.

In December 1997, in Kyoto, Japan, parties to the Climate Treaty agreed to make legally binding cuts in emissions of six greenhouse gases – carbon dioxide, methane, nitrous oxide, hydrofluorocarbons, perfluorocarbons and sulphur hexafluoride. Between 2008-2012 developed countries are to reduce their emissions by an average of 5.2% below 1990 levels (for the first three substances; any year between 1990-1995 for the last three). Different targets have been set for individual countries – e.g. EU 8%, Switzerland 8%, USA 7%, Japan 6%; within the EU, member states are each to reduce CO_2 emissions by an agreed amount (12.5% for the UK) to meet the EU's target of 8%. The Protocol also permits emissions trading, from 2008, between countries as a means of meeting targets. Countries may also use "carbon sinks" – activities which absorb carbon from the atmosphere – as a means of meeting their targets; eligible activities include management of forests, crops and grazing lands and revegetation; the actual amount of target that can be met in this way has been set for each country. The Kyoto Clean Development Mechanisms enable developed countries to contribute to lowering GHG emissions through the joint implementation of projects using clean technologies in developing countries (or other developed countries) where the costs of reducing emissions may be lower; reductions made in the developing country can be "transferred" to the developed country and used to meet its own commitment. Detailed "rules" for the operation of the Protocol are set out in the Marrakesh Accords, drawn up in 2001 and agreed in 2005.

The Protocol came into force on 16 February 2005, following ratification in November 2004 by Russia (accounting for around 17.4% of emissions), thus fulfilling the Protocol's requirement for ratification by 55 countries (on Annex 1 of the Protocol) which together account for 55% of the 1990 carbon dioxide emissions of industrialised countries. As at 12 December 2007, 176 countries plus the EU (and accounting for 63.7% of CO_2 emissions of Annex 1 parties) had ratified the Protocol; the European Community and its 15 (as at that date) member states ratified the Protocol on 31 May 2002. Following Australia's ratification of the Protocol in December 2007, the USA is now the only major developed country that has not ratified it.

Further details on the Climate Convention and Kyoto Protocol can be found on the Convention's website at www.unfccc.int.

3.5.2 Post-Kyoto

December 2007 saw the opening of discussions on a new global agreement to replace the Kyoto Protocol. The Bali Roadmap, agreed by both developing and developed countries, sets out a clear agenda for the negotiations, with a timetable ending in 2009. The aim is that governments would then have time to ratify the new agreement enabling it to come into force at the end of 2012 when existing commitments under Kyoto expire. The new agreement will include new targets for reducing GHG emissions, as well as tackling the issues of adaptation to climate change, technology transfer, financial arrangements, deforestation and carbon markets.

As for specific targets, the document agreed in Bali noted "that achieving the lowest level of emissions in the scenarios assessed by the IPCC would require developed countries to reduce their emissions by 25-40% of 1990 levels by 2020. The EU is proposing that developed countries should collectively cut their emissions by 30% by 2020, and by 60-80% by 2050. Negotiations on a post-Kyoto agreement opened in Bangkok in April 2008 and have continued through 2008.

3.5.3 Greenhouse Gas Emissions from Ships

The International Maritime Organisation (IMO) is developing a mandatory regime to control GHG emissions from international shipping which it hopes to bring into force before the end of 2011. As a first step a "CO_2 Design Index" for new ships is being formulated. As well as enabling the fuel efficiency of different ships to be evaluated at the design stage, the index will contain a required minimum level of fuel efficiency related to baseline established on the basis of the fuel efficiency of ships delivered between 1995-2005. The IMO is also giving consideration to an emissions trading scheme, either one which would require shipping to purchase allowances on the open market, or a closed scheme where trading would be between ships.

3.6 EUROPEAN UNION

The 6th Environment Action Programme covering the period 2001-2010, adopted in June 2002, identifies tackling climate change as one of four priority areas (the others being protection of nature and wildlife, environment and health and sustainable use of natural resources and waste management).

Data from the European Environment Agency (June 2008) show that emissions of greenhouse gases from the EU-27 for 2006 decreased 0.3% between 2005 and 2006 (and are now 7.7% below their 1990 level); emissions of GHGs from the EU-15 over the same period decreased 0.8% and are 2.7% below the base year of 1990.

A Green Paper published by the Commission in June 2007, *Adapting to climate change in Europe – options for EU action* (COM(2007) 354), points to the need for Europe to not only cut its greenhouse gas emissions, but also to develop adaptation measures to lessen impacts of climate change and future warming.

3.6.1 Climate Action & Renewable Energy

At a meeting in March 2007, European Heads of State agreed to cut greenhouse gas emissions by at least 20% by 2020 based on 1990 levels – increasing to 30% when there is an international climate agreement. They also agreed a binding target of 20% of energy to come from renewable sources by 2020, and that 10% of road transport fuels should come from biofuels. (The European Parliament's Environment Committee is suggesting a target of 4% by 2015 and 8-10% by 2020; Opinion, July 2008.)

To meet these targets the Commission published in January 2008 its "Climate action and renewable energy package". The package includes a proposal for a *Directive on the promotion of the use of energy from renewable sources* (COM(2008)19); this sets out the contribution expected from each Member State towards meeting the above targets. The package proposes a series of measures to help achieve the targets, which includes strengthening and expanding the EU emissions trading system (see 3.6.3), and measures to support the development of carbon capture and storage.

For the UK, the Commission proposes

- Reduction of 16% in UK GHG emissions from sectors not covered by the EU-ETS by 2020, based on 2005 levels;
- 15% of energy consumed in the UK should come from renewable sources by 2020;
- 10% of road transport fuels to come from renewable sources (biofuels), subject to them being produced in a sustainable way – see also 3.26.10.

The Commission's package also includes a proposal to amend the 2003 EU-ETS Directive with a view to strengthening, expanding and improving the functioning of the EU-ETS and make it a key tool for cutting emissions cost effectively – see 3.6.3 below.

3.6.2 Climate Change Programme

The EU's second Climate Change Programme was launched in 2005 (the first was published in 2000). Its purpose is to provide a new policy framework for EU climate change policy, looking beyond 2012, and to this end five working groups have been established to

- Propose ways of reducing emissions from aviation – it has already been agreed that emissions from aviation should be included in the European Greenhouse Gas Trading Scheme (EU-ETS) from 2012; all flights (both EU and non-EU carriers) departing from an EU country would be covered – see 3.6.3 below.
- Propose ways of reducing emissions from passenger road transport – based on the working group's report, the Commission has published a revised strategy which envisages mandatory measures for reducing carbon dioxide emissions from cars (see later this chapter, 3.26.8).
- Draw up proposals for geological carbon capture and storage – COM(2006) 843, *Communication on sustainable power generation from fossil fuels: aiming for near zero emissions from coal after 2020* – includes a proposal for 12 experimental carbon capture and storage plants by 2015.
- Examine ways of adapting to climate change.
- Analyse achievements under the first programme.

The Commission's Energy Strategy (COM(2007) 1), *An Energy Policy for Europe*, published in January 2007, focuses on the need for Europe to move to low-carbon energy systems as a means of combating climate change. It sets a target of 21% of electricity to come from renewable sources by 2010, increasing to 34% by 2020, with biofuels to account for 10% of road transport fuels by the same date. The Strategy also proposes that international negotiations for a "post-Kyoto" agreement on emissions reductions should require developed countries to meet a target of 30% below 1990 levels; in advance of this the Commission proposes that the EU should commit to a 20% reduction in atmospheric emissions.

3.6.3 Emissions Trading

In July 2003 the Council of Ministers and European Parliament adopted a *Directive establishing a scheme for greenhouse gas emissions allowance trading within the Community* (2003/87/EC) – EU-ETS; this came into force at the beginning of 2004, with Member States being required to give the Commission details of how emissions quotas were to be distributed by March 2004. It has been implemented in the UK through *The Greenhouse Gas Emissions Trading Scheme Regulations 2005* – see 3.7.2 below.

The scheme, which became effective from January 2005, initially applied to major industrial and power installations and only covers trading in carbon dioxide; installations must have a GHG permit and are allocated a GHG quota (or allowance) enabling the holder to emit a specified amount of GHGs; allowances are transferable, with exceedence of a quota resulting in a penalty. A further Directive enables participants to use "emission credits" from reducing GHG emissions from projects undertaken or the transfer of clean technologies to developing countries and those in transition to meet their annual emissions targets. GHG permits will include requirements for monitoring and reporting, and verification of emissions which should not exceed allowances, with penalties for non-compliance. Between 2005-07 95% of emissions quotas permits were allocated free of charge; from 1 January 2008 until 2012, 90% are free, with the remainder being auctioned.

During the period 2005-2007, Member States may request an exemption for companies or sectors from the EU scheme providing the companies are achieving similar CO_2 reductions through national schemes. As from 1 January 2008, the scheme is binding for all relevant sectors.

In January 2004, *Decision on a Community mechanism for monitoring greenhouse gas emissions and the implementation of the Kyoto Protocol* (280/2004/EC), was formally adopted. This requires Member States to report to the Commission every two years on their actual GHG emissions, forecasts and measures being taken to meet targets. EU Regulation 2216/2004, adopted on 21 December 2004, establishes an electronic registries system which will keep track of all emissions trading; all industrial plant covered by the EU-ETS will have an "account" in their national registry, in which its emissions allowance will be recorded, with all trading being carried out electronically.

The European Parliament has now agreed (July 2008) on the Commission's proposal (COM(2006) 818) for a *Directive amending 2003/87/EC so as to include aviation activities in the scheme for greenhouse gas emission allowance trading within the Community;* the formal agreement of Ministers was expected shortly after. From 1 January 2012 carbon dioxide emissions from all flights starting and/or landing within the EU are to be included in the emissions trading system; the reduction target is to be calculated on the basis of airlines' annual average emissions between 2004-2005; in the first period (2012) emissions are to be cut by 3% and in the second period (2013 onwards) by 5%; 85% of airlines' emissions certificates are to be allocated free and the remainder auctioned. Airlines will be required to begin monitoring and reporting on their emissions from 2010. There are various exemptions including light aircraft with a take-off weight under 5.7 tonnes, emergency and UN-mandated humanitarian flights, military, police, customs and fire-fighting flights, and small airline companies with low emissions.

The Commission's "Climate Action and Renewable Energy Package" published January 2008, includes a proposal to amend the 2003 Directive with the aim of incorporating measures which will see emissions from sectors covered by the system cut by 21% compared with levels in 2005; a single EU-wide cap on ETS emissions will be set, and free allocation of emission allowances will be progressively replaced by auctioning of allowances by 2020. Emissions from sectors not included in the EU-ETS – such as transport, housing, agriculture and waste – will be cut by 10% of 2005 levels by 2020; Member States will contribute to this effort according to relative wealth, with national emission targets ranging from -20% to +2%.

3.6.4 Carbon Dioxide

EU members have agreed that emissions of carbon dioxide (CO_2) – the major greenhouse gas, accounting for 66% of global greenhouse effect – should be stabilised at 1990 levels by 2000. This was in line with a commitment made at the 1992 Earth Summit and confirmed in the Framework Convention on the Climate – see above 3.5.1. In April 1998 the EU (together with individual Member States) signed the Kyoto Protocol, (ratified in February 2005), thus committing EU member states to reducing emissions of six greenhouse gases by 8% based on 1990 levels between 2008-2012. In June 1998 agreement was reached on how to apportion the 8% target cut and, as a result, the UK is to reduce its emissions by 12.5%.

The EU also has in place a number of other measures aimed at stabilising CO_2 emissions.

(a) Energy Efficiency

Directive 2006/32/EC *on Energy End Use Efficiency and Energy Services* was adopted on 5 April 2006, entering into force on 17 May 2006. Most of the Directive had to be transposed by 17 May 2008, with the reporting requirements taking immediate effect. It repeals Directive 93/76 which implemented the Specific Actions for Vigorous Energy Efficiency (SAVE) programme, requiring regular inspections of cars and boilers; energy audits for business; thermal insulation of new buildings; certification of CO_2 emissions related to energy consumption in buildings.

The 2006 Directive requires Member States to adopt, an overall national indicative energy savings target of 9% with the aim of achieving this by 17 May 2017, and to promote energy end-use efficiency and energy services. Member States had to submit to the Commission by 30 June 2007 an Energy Efficiency Action Plan demonstrating the measures planned to meet the energy savings target; the first EEAP should include an intermediate energy savings target to be achieved by 17 May 2011 and the measures planned for its achievement. Subsequent EEAPs, which should include an analysis and evaluation of the previous EEAP, should be submitted to the Commission by 30 June 2011 and 30 June 2014.

Directive 2002/91/EC on Energy Performance in Buildings, adopted in December 2002, and which came into force in January 2003, requires new buildings to meet minimum energy performance standards; for certificates of energy performance to be available to prospective buyers or tenants; for regular inspections of boilers and air conditioning plant; and for a common methodology for measuring energy performance to be adopted. Existing buildings should, if feasible, be upgraded to meet the new standards at the time they undergo major renovation. These measures will, the Commission believes, result in potential energy savings in buildings of 22% by 2010. The Directive has been implemented by the UK and Devolved Administrations – see 3.7.6 below.

(b) Monitoring CO_2 Emissions

Council Decision 99/296/EC, which amends Decision 93/389, outlines *"A Monitoring Mechanism for Community CO_2 and other Greenhouse Gas Emissions"*; it extends the scope of the earlier Decision to cover all emissions of greenhouse gases, including those not covered by the Kyoto Protocol; it requires stabilisation of CO_2 emissions at 1990 levels by 2000, the drawing up of national programmes for meeting EU commitments under the Kyoto Protocol, monitoring in this respect and reporting to the Commission. National programmes should include actions to increase energy conservation and efficiency, encourage a switch to low or no carbon fuels and implement Community legislation. Member States are required to report on and publicise their policies and actions to limit emissions of CO_2 and other greenhouse gases and to evaluate the impact of these measures.

In January 2004, *Decision 280/2004/EC on a monitoring mechanism of community greenhouse gas emissions and the implementation of the Kyoto Protocol* was formally adopted. This replaces Decision 93/389 and requires Member States to report to the Commission every two years actual GHG emissions, forecasts and measures being taken to meet targets, using either 1990 or 1995 as the reference year.

(c) Reducing CO_2 Emissions from Cars

The Commission has brokered agreements with European, Japanese and South Korean car manufacturers to reduce emissions from new cars, but is now proposing mandatory measures – see 3.26.8.

3.6.5 Methane

Methane (CH_4) is the second most important greenhouse gas (after carbon dioxide), accounting for 18% of the global greenhouse effect. In November 1996, the Commission published a Strategy Paper (COM(96) 557) for reducing methane emissions from the three sectors chiefly responsible for emissions: agriculture (45% of EU emissions), waste (32%) and energy (23%). It is hoped that the measures proposed for these sectors (see below) will lead to a reduction in methane emissions of 30% by 2005 and 41% by 2010.

- **Agriculture:** promotion of research and incentives to reduce emissions from, in particular, cattle and sheep; animal manure management, e.g. anaerobic digesters or covered lagoons.
- **Waste:** promotion of measures to encourage separate collection of organic wastes, composting and recycling; new anaerobic landfills to be equipped with methane recovery and use systems; wherever possible, existing landfills to be retrofitted with methane collection and use systems; otherwise encourage flaring.
- **Energy:** CH_4 reduction schemes promoting best available recovery techniques in coal mines. For natural gas, a

minimum leakages standard and increased control frequency of pipelines.

In 2006 total UK methane emissions were estimated at 2339.8 thousand tonnes, with waste management accounting for 41% and agriculture for 38%. (Defra, 2007 Greenhouse Gas Emissions, provisional figures, March 2008).

3.6.6 Fluorinated Greenhouse Gases

Two legislative measures have been adopted which aim to reduce emissions of fluorinated greenhouse gases (hydrofluorocarbons, perfluorocarbons and sulphur hexafluoride) by 20 million tonnes of CO_2 eq. per year until 2012, and by 40-50 m. tonnes of CO_2 eq. per year after that.

Regulation 842/2006, adopted on 17 May 2006, and which applies from 4 July 2007, requires improved containment of the gases, and sets leakage inspection standards for refrigeration, air conditioning and fire fighting equipment, as well as requiring recovery of the gases when such equipment reaches the end of its life. It provides for monitoring and reporting of emissions and labelling of products and equipment; the Regulation also establishes minimum standards for training and certification of personnel for working with both F gases and ozone depleting substances, which must be established by 4 July 2008. Where containment is not feasible or the use of certain fluorinated gases is inappropriate, marketing and use will be banned – examples include non-refillable containers, self-chilling drinks cans, tyres, magnesium die-casting, footwear, double glazing systems, new fire protection systems and fire extinguishers containing such gases. A further Regulation (1497/2007) adopted on 18 December 2007 sets out standard leakage requirements for fire protection systems containing 3kg or more of FGHGs. Regulation 1493/2007 of 17 December 2007 establishes the format of the reports to be submitted by producers, importers and exporters.

Guidance on compliance with the EU Regulation was published by Defra (on behalf of the UK Government and the Devolved Administrations) in August 2006. *The Fluorinated Greenhouse Gases Regulations 2008* (SI 2008/41) detail the sanctions and penalties for non-compliance with the EU Regulation. The Regulations apply in England, Wales and Scotland, to offshore installations and to Northern Ireland for the purposes of importations; the Secretary of State is the competent authority for the purposes of the EU Regulation. Two further sets of Regulations are due to come into force in early 2009, and on which Defra consulted in July 2008: *the Fluorinated Greenhouse Gases Regulations 2009* will set out certification and qualification requirements for personnel working in five industry sectors covered by EU Regulation 842/2006, as well as deal with other requirements relating to leakage checking, reporting and labelling. *The Ozone-Depleting Substances (Qualifications) Regulations 2009* will revoke SI 2006/1510 (see below 3.14) to take the newly proposed qualifications into account for working with ozone depleting substances. Defra has issued (August 2008) a series of guidance notes, both general and sector specific, on how the Regulations apply to manufacturers, sellers and users in different sectors.

Directive 2006/40/EC, also adopted on 17 May 2006, requires the phasing out of HFC-134a (which has a global warming potential of 1,300) in vehicle air-conditioning systems. With effect from 12 months from the date of adoption of a harmonised leakage detection test (i.e. 4 July 2008), type approval may no longer be granted to new vehicle models fitted with an air conditioning system designed to contain FGHGs with a global warming potential higher than 150, unless the rate of leakage does not exceed 40g of FGHG per year for a single evaporator or 60g for a dual evaporator system; this will apply to all new vehicles from 4 July 2009. From 1 January 2011, type approval may not be granted to new vehicle models fitted with an air conditioning system designed to contain FGHGs with a global warming potential higher than 150; this will apply from 1 January 2017 to all new vehicles. National legislative measures implementing the Directive must apply from 5 January 2008.

3.7 CLIMATE CHANGE & THE UK

The Department of Energy and Climate Change established in early October 2008 brings together much of Defra's climate change responsibilities with the energy component from BERR, to focus on solving the challenges of climate change and energy supply. Policy regarding adaptation to climate change remains with Defra.

3.7.1 Climate Change Programmes

In the UK, emissions of the 'basket' of six greenhouse gases in 2007 were provisionally estimated at 639.4 million tonnes carbon dioxide equivalent (2% lower than the 2006 figure of 652.3 million tonnes). Net emissions of CO_2 – the main greenhouse gas – in 2007 were provisionally estimated at 543.7 million tonnes – 2% lower than 2006 (554.5 million tonnes) (Defra press release, 27.03.08).

In its 2003 Energy White Paper the Government proposed a target of cutting emissions of CO_2 by 60% by 2050, a target reiterated in subsequent White Papers. Central to the achievement of the target is the introduction of a carbon emissions trading scheme – see below. Other measures include action to cut emissions in all sectors of the UK (including business, public sector, transport, domestic, agriculture, forestry and land use), and improved energy efficiency in the home.

The UK's first *Climate Change Programme*, published November 2000, outlined various policies and measures to meet the UK's Kyoto target of reducing total greenhouse gas emissions by 12.5% over the period 2008-2012, relative to the base year, and to meet the Government's aim of a reduction in UK emissions of greenhouse gases of 23% by 2010 compared to 1990, with a reduction in carbon dioxide emissions of 20% by 2010 compared to 1990 and a reduction of 60% by 2050. The Scottish Government published its own climate change programme at the same time which included a range of measures and policies aimed at achieving emissions reductions in those areas devolved to it. The 2000 programme outlined the actions which various sectors (e.g. individuals, industry and commerce and the public sector) could take to reduce energy consumption and also government measures. It also took account of already agreed policies which aimed at reducing carbon dioxide emissions such as the climate change levy (introduced from 1 April 2001), and negotiated agreements with the energy intensive industries; in return for commitments to improving energy efficiency and reducing their environmental impact, energy intensive industries are eligible for an 80% discount in the climate change levy. All non-domestic energy users are liable for the climate change levy and are taxed on the energy (electricity, gas, coal and LPG) supplied to their business.

The UK's second *Climate Change Programme*, published March 2006, builds on the earlier programme and outlines plans for tackling climate change at global, national and individual level, with the aim of reducing the UK's emissions of greenhouse gases to 23-25% below the base year levels and to reduce CO_2 emissions to 15-18% below 1990 levels by 2010. The programme includes the following plans:

- Energy efficiency standards for new and refurbished buildings are being raised (see 3.7.6), with further measures aimed at encouraging householders to improve energy efficiency, both in the private and public sector.
- Additional action by local authorities – a new £4 million local government best practice support programme will aim to proactively benchmark the performance of local authorities on climate change and sustainable energy and target those who need the most help to raise their performance.
- A new planning policy statement, *Planning and Climate Change*, published as a supplement to PPS 1 in December 2007. This requires regional planning bodies and planning authorities to have regard to the need to limit carbon dioxide emissions in decisions on the location and design of new development (see also chapter 1, 1.16.2).
- Measures to encourage the uptake of biofuels in the transport sector; by 2010 5% of transport fuel sold in the UK will have to come from renewable sources.
- Finalisation of the National Allocation Plan for the second phase of the EU emissions trading scheme (see below).
- Working with both EU member states and international agencies to help developing countries evolve as low-carbon economies

Voluntary agreements with car manufacturers to cut engine emissions and to improve the fuel efficiency of new cars (see this chapter, 3.27.8) are also an important element in the UK programme, as is action to cut emissions in all sectors of the UK (including business, public, transport, domestic, agriculture, forestry and land use), and improved energy efficiency in the home.

London

The Mayor of London's climate change action plan (*Action today to protect tomorrow*), published March 2007, proposes a target reduction of 60% in London's CO_2 emissions by 2025; this is to be achieved through switching supplies to combined heat and power schemes and local sources such as solar panels and micro wind turbines; particular emphasis will be placed on measures promoting public transport use and encouraging Londoners to switch to low-carbon vehicles and fuels. *The Greater London Authority Act 2007* (s.42) extends the 1999 Act to place a duty on the Mayor and London Assembly to take account of government policies and any guidance issued, with regard to climate change or its consequences. The Mayor is also required to publish a London climate change mitigation and energy strategy and an adaptation to climate change strategy for London – the Act details what both strategies should contain.

3.7.2 Emissions Trading

The Greenhouse Gas Emissions Trading Scheme Regulations 2005 (SI 2005/925), which came into force on 21 April 2005, as amended by SI 2006/737, implement EU Directive 2003/87/EC establishing a scheme for GHG emission allowance trading within the Community; they also implement Commission Decision 2004/156/EC establishing guidelines for monitoring and reporting GHGs, and Regulation 2216/2004 regarding the electronic registry. The Regulations apply throughout the UK. Emissions of carbon dioxide from listed activities are controlled and installations carrying out such activities were required to apply for a permit from the relevant Agency before 31 January 2004 to ensure the permit was issued by 31 March 2004. Operators of activities included in the second phase of the scheme announced early 2006 had to apply for a permit from the Environment Agency by 1 June 2006; those with a permit but with other activities covered by the expanded scheme had to apply for a variation by 31 December 2007. Conditions of permits must ensure that emissions are properly monitored and reported and that the operator surrenders within four months of the end of each scheme year allowances equal to the annual reportable emissions from the installation during that year. The Regulations make provision for variation, transfer, surrender and revocation of permits and for fees to be paid. Exceedence of an allowance will result in both a financial penalty and the following year's allowance being reduced by the amount of exceedence.

In March 2007, Defra published the approved National Allocation Plan and final list of installation-level allocations for industry covering the second phase of the EU Emissions Trading Scheme which runs from 2008-2012; this has been given effect through 2007 amendment Regulations (SI 2007/3433). A web-based Emissions Trading Registry, established in the UK in May 2005, allows operators to access their carbon accounts and to trade their allowances both within the UK and within other participating countries.

As allowed by Phase II of the EU-ETS, Defra is proposing to set up a scheme enabling 7% of allowances to be auctioned (Phase II of the EU scheme allows up to 10% to be auctioned).

Defra has published (February 2006) an *Operator's Guide to the EU Emissions Trading Scheme – The Steps to Compliance.*

The Community Emissions Trading Scheme (Allocation of Allowances for Payment) Regulations 2008 (SI 2008/1825 as amended), and Scheme 2008 (issued by the Treasury), sets out the conduct and terms of allocation of allowances in the UK's auctions in Phase II of the EU-ETS.

The Waste and Emissions Trading Act 2003 will enable penalties to be levied on direct participants in the UK greenhouse gas emissions trading scheme (or any future emissions schemes) who fail to comply with emissions reduction targets.

3.7.3 Climate Change Bills

Introduced to Parliament in November 2007, the Bill sets a statutory target of reducing UK carbon dioxide emissions by 60% by 2050 (subsequently increased to 80%; see below) and by at least 26% by 2020 (based on 1990 levels).

Legally binding five year "carbon budgets", to be set at least 15 years ahead, will set a limit on CO_2 emissions for that five year period, with the Government publishing its proposals and policies for meeting each budget at the same time. A independent Committee on Climate Change has been established to provide independent expert advice and guidance to Government on achieving targets and staying within budgets – the Committee will be legally established once the Bill has been enacted.

The Committee is charged with providing advice on the first three carbon budgets covering the period 2008-2022 by 1 December 2008; the Committee is also given a statutory duty to review the 60% target, advising on whether it should be increased, possibly to 80%; the Committee is also to advise on whether other greenhouse gases and emissions from international shipping and aviation should be included in the targets. The Committee will also be required to provide an annual progress report to Parliament to which the Government must respond. The Bill includes powers for the Government to introduce new measures to facilitate emission reductions measures without the need for primary legislation. (The Committee has recommended that the target be increased to 80% by 2050, to include emissions of all greenhouse gases from all sectors of the UK economy, including shipping and aviation; however because of problems in allocating emissions from international transport to the UK, they should not be included in the five yearly carbon budgets. In a statement to MPs, 16.10.08, the Secretary of State for Energy and Climate Change has accepted the Committee's recommendations.)

A Strategy published with the Bill shows how the UK can move to a low carbon economy through

- Investment in low carbon fuels and technologies, such as carbon capture and storage, wind, wave and solar power.
- Significantly more efficient use of energy.
- Encouraging energy suppliers to focus on reducing demand, rather than supplying as much energy as possible.
- Encouraging consumers to become producers as well as consumers of energy.

The Bill was expected to receive Royal Assent in autumn 2008.

Scotland

Following consultation in January 2008, The Scottish Government plans to bring forward a Climate Change Bill during its 2008-09 session. As well as measures to reduce emissions over the period to 2011, the Bill will include

- A longer term target of reducing emissions by 80% by 2050;
- Create a framework of carbon budgets for Scottish emissions to set out a pathway towards the 2050 target;
- Provide for annual scrutiny and reporting of progress;
- Publication of transparent, robust and independent advice on when, and to what level, cuts in emissions should be made.

Data published in September 2007 show that emissions of greenhouse gases in Scotland have fallen over 15% since 1990 (News release, 18/09/07).

3.7.4 Climate Change & Sustainable Energy Act 2006

The principal purpose of this Act, which received Royal Assent on 21 June 2006, is to enhance the UK's contribution to combating climate change. As such it requires the Secretary of State to report annually to Parliament on levels of emissions of GHGs in the UK, including any increases or decreases, and on the steps taken by government departments to reduce emissions during the previous calendar year. It also requires the Secretary of State to publish "energy measures reports" outlining measures which local authorities can take to improve efficiency of energy use; increase the amount of electricity generated by micro-generation or other low emissions sources or technologies; reduce GHG emissions; and reduce the number of households living in fuel poverty.

The Act (which does not cover N. Ireland) also gives the Secretary of State powers to set national targets for micro-generation and adds a new section to the *Building Act 1984* regarding breach of the building regulations on conservation of fuel and power; places a duty on the Secretary of State to promote community energy projects and to promote the use of heat from renewable sources. *The Climate Change and Sustainable Energy Act 2006 (Sources of Energy and Technologies) Order 2008* (SI 2008/1767) amends s.26(2) of the Act to add heat from air, water or the ground to the technologies or sources which constitute microgeneration.

3.7.5 Home Energy Conservation Act 1995

This Act requires local authorities to report to the Secretary of State on how "significant energy savings" – initially 30% – can be made from residential properties, including homes in multiple occupation, in their area, and also to report progress. (*The Local Authorities' Plans and Strategies (Disapplication) (England) Order 2005* removes the requirement to report from those LAs classified as "excellent".)

3.7.6 Energy Efficiency in Buildings

EU Directive 2002/91/EC on Energy Performance in Buildings requiring new buildings to meet minimum energy performance standards, and for certificates of energy performance to be available (see 3.6.4a).

From 2010, large non-energy intensive private and public sector organisations whose half-hourly metered electricity use exceeds 6,000 MWh will be required to participate in the Carbon Reduction Commitment Scheme being developed by Defra, and on which consultation is expected in December 2008. The aim is to encourage such organisations to reduce their energy use (based on 2008) and thus their CO_2 emissions through energy efficiency measures. The Ethical Corporation has published (September 2008) a guide for companies entitled *How to manage carbon reduction and make it pay.*

England and Wales

The Building and Approved Inspectors (Amendment) Regulations 2006 (SI 2006/652), which came into force on 6 April 2006, implement the above EU Directive; the Regulations amend Part L (conservation of fuel and power) setting maximum carbon dioxide emissions (based on heating, ventilation, hot water and lighting) for whole buildings with a view to increasing their energy efficiency by 20% for dwellings and up to 27% for other buildings. Existing buildings with a floor area over 1000m[2] undergoing major renovation should upgrade to the higher standards as far as is feasible. Transitional arrangements enabled buildings for which approval had been received before 6 April 2006 to comply with the 2002 Part L requirements so long as building work was started before 1 April 2007; those for which full building plans approval was not required but for which a contract had been entered into before 6 April 2006 had to begin work before 1 October 2006.

The Energy Performance of Buildings (Certificates and Inspections) (England and Wales) Regulations 2007 (SI 2007/991) require energy performance certificates (EPCs) for all buildings when they are constructed, sold or rented; EPCs became mandatory for transactions related to commercial buildings over 500 m^2 from April 2008, and for the sale or rental of all homes and commercial buildings from October 2008. EPCs are valid for 10 years.

Also, from 1 October 2008, all public buildings over 1000 m^2 must display a Yearly Display Energy Certificate showing actual annual energy consumption and the resulting CO_2 emissions. Guidance on the *Requirements for Energy Performance Certificates when marketing homes for sale or let* was published by the Communities and Local Government Department in October 2008.

The Welsh Assembly Government requires that all new buildings which it funds must be built to BREEAM "excellent" standards (the highest level of environmental performance; The WAG is aiming for all new buildings in Wales to be zero carbon from 2011.

Other measures aimed at promoting energy efficiency announced by the Office of the Deputy Prime Minister (now Department for Communities & Local Government) in March 2006, include:

- New Planning Policy Statement setting out how participants in the planning process should work towards the reduction of carbon emissions in the location and design of new development (now published as supplement to PPS 1, see chapter 1, 1.16.2).
- Code for Sustainable Homes setting higher standards for energy and water efficiency; a draft revision of the sustainable construction strategy (Department for Business, Enterprise & Regulatory Reform, July 2007) proposes that all new homes should be zero carbon by 2016, with the energy efficiency of all residential homes in England to be increased by at least 20% by 2010.
- Continued action to upgrade the energy efficiency of social and rented homes.
- Review of implementation of planning policy to ensure that local authorities are taking sufficient action to promote on-site renewable technology.

Scotland

In Scotland, the Energy Efficiency Directive has been implemented using *the Building (Scotland) Act 2003*; an amendment to the 2004 Building Regulations (SSI 2008/310) inserts a definition of "energy performance certificate" into the main regulations. *The Energy Performance of Buildings (Scotland) Regulations 2008* (SSI 2008/309), which came into force on 4 January 2009 require the owner of a building to make an EPC available to prospective buyers or tenants on all buildings to be sold or let. They also require the owner or occupier of public buildings to ensure that the EPC is displayed in a place clearly visible to members of the public.

Scotland's *Planning Policy on Renewable energy* (PPS 6; March 2007) sets out how the planning system should manage the process of encouraging, approving and implementing renewable energy proposals when preparing development plans and determining planning applications, with a view to generating 40% of Scotland's electricity from renewable sources by 2020. It also expects all new housing and commercial developments in Scotland with a floorspace of 500 sq metres or more to deliver a 15% reduction in CO_2 emissions above that required by building regulations.

Northern Ireland

The Energy Performance of Buildings (Certificates and Inspections) Regulations (Northern Ireland) 2008 (SR2008/170, as amended SR 2008/241), which came into force on 30 June 2008, implement the EU's Directive on Energy Performance in Buildings. They lay down requirements for the production of energy performance certificates when buildings are constructed, sold or rented out, and for the display of annual display energy certificates in large public buildings providing a public service. Sellers and prospective landlords are required to make energy performance certificates available to prospective buyers and tenants at the earliest opportunity. The person responsible for having construction work carried out, when a building is constructed or modified, must provide an energy performance certificate to the owner of the building not more than five days after the work has been completed. Energy performance certificates shall be accompanied by recommendation reports for the improvement of the energy performance of the building.

EPCs were required for new newly built buildings from 30 September 2008, and for those for rental or sale from 30 December 2008. Certificates may be no more than 10 years old.

The NI Regulations also provide for regular inspection and reports on air-conditioning systems.

AIR QUALITY

The effects of air pollution on health have been recognised for more than three centuries – in 1648 Londoners petitioned Parliament to ban the importation of coal from Newcastle because of its injurious effects. More recently, the European Commission in its *Thematic Strategy on Air Pollution* (September 2005) suggests that the premature deaths of up to 370,000 Europeans a year can be attributed to air pollution; more than 90% of these premature deaths are caused by fine particulates and most of the remainder by ground-level ozone.

In the UK a 1998 report from the Committee on the Medical Effects of Air Pollutants (*Quantification of the Effects of Air Pollution on Health*) suggested that the short term impact of air pollution on health results in the premature death of between 12,000-24,000 vulnerable (i.e. the very old, very young or sick and those with respiratory diseases) people in Great Britain each year; the Committee also estimated that between 14,000 and 24,000 hospital admissions and readmissions each year may be associated with short term air pollution. A 2006 report from COMEAP (*Cardiovascular Disease and Air Pollution*) found that outdoor air pollutants are likely to be associated with increased deaths and hospital admissions (though this association is not as large as family history, smoking and hypertension).

3.8 WHO AIR QUALITY GUIDELINES

The World Health Organisation's recommended air quality guidelines are used as a basis for setting EU standards and were also taken into account by the UK Expert Panel on Air Quality Standards when making recommendations for UK air quality standards (see this chapter, 3.11.8). Originally published in 1987, as *Air Quality Guidelines for Europe*, WHO's recommendations are not mandatory but are generally accepted as being levels not to be exceeded if healthy air quality is to be maintained. In 1993 WHO began reviewing the guidelines in the light of new scientific data and at a Consultation meeting in October 1996 revised guidelines were agreed; a fully revised version of *Air Quality Guidelines for Europe* was published in 2000. Further review of the Guidelines in the light of new scientific data on the health effects of air pollution has resulted in revised guidelines, together with interim targets, for ozone, nitrogen dioxide, sulphur dioxide and particulate matter (see Table 3.1 below). The revised Guidelines (*Global Update 2005*), published in October 2006, are intended to be relevant globally.

3.9 AIR QUALITY & THE EU

This section covers EU legislation and programmes to improve local air quality and reduce air pollution from all sources, including industry and transport, and measures for meeting global commitments under the Climate Change and other Conventions.

A summary of the institutions of the European Union, its legislative procedures and a list of the Directives on air pollution are given in Appendix 7 and 7.3, respectively.

3.9.1 CAFÉ & the Thematic Strategy on Air Pollution

In May 2001, the European Commission formally adopted the Clean Air For Europe, or CAFÉ, programme (COM(2001) 245), in a move which hopes to integrate the various strands of air pollution policy under the 6th Environmental Action Programme.

Table 3.1: WHO Air Quality Guidelines

(This table includes only those pollutants for which EU standards exist or are being proposed)

Substances	*Average*	*Time-weighted Averaging time*
Classical air pollutants		
Nitrogen dioxide		
(unchanged, 2005	200 µg/m³	1 hour mean
(unchanged, 2005)	40 µg/m³	annual mean
Ozone (revised, 2005)	100 µg/m³	daily max. 8 hour mean
Sulphur dioxide		
(unchanged, 2005)	500 µg/m³	10 minutes mean
(revised, 2005)	20 µg/m³	24 hour mean
Particulate matter (new, 2005)		
PM_{10}	50 µg/m³	24 hour mean*
	20 µg/m³	annual mean
$PM_{2.5}$	25 µg/m³	24 hour mean*
	10 µg/m³	annual mean
Carbon monoxide	100 mg/m³	15 minutes
	60 mg/m³	30 minutes
	30 mg/m³	1 hour
	10 mg/m³	8 hour
Organic pollutants		
PAH (benzo-a-pyrene)	8.7×10^{-5} (ng/m³)$^{-1}$	UR** / lifetime
Benzene	6×10^{-6} (µg/m³)$^{-1}$	UR / lifetime
1.3 butadiene	no guideline	
Inorganic pollutants		
Lead	0.5 µg/m³	annual
Arsenic	1.5×10^{-3} (µg/m³)$^{-1}$	UR / lifetime
Cadmium	5 ng/m³	annual
Mercury	1.0 µg/m³	annual

* 99th percentile (3 days/year)
** UR is the excess risk of dying from cancer following lifetime exposure.
More information: www.who.int/peh/air/airqualitygd.htm

CAFÉ will effectively form both an air quality strategy for Europe, and an active framework within which air pollution measures, such as the Auto Oil programme, national emissions ceilings Directives and the air quality Daughter Directives can be coordinated.

The Commission has now published (September 2005) a *Thematic Strategy on Air Pollution* (COM(2005) 446) – one of seven arising from the 6th Environmental Action Programme (see Appendix 7); its aim is to cut the annual number of premature deaths caused by air pollution by 40% by 2020 from the 2000 level and to reduce the continuing damage to Europe's ecosystems. To do this the Strategy says that emissions of sulphur dioxide will need to be reduced by 82%, nitrogen oxide by 60%, volatile organic compounds by 51%, ammonia by 27% and fine particulate matter by 59% (compared to their 2000 levels).

In accordance with the Thematic Strategy, EU air quality legislation has been streamlined with the adoption on 21 May 2008 of Directive 2008/50/EC on *Ambient Air Quality and Cleaner Air for Europe*; this will replace the 1996 Air Quality Framework Directive and three of its Daughter Directives (see below 3.9.3); other measures include review of the *National Emission Ceilings Directive* (see below 3.16.1); tighter (Euro V) emission limits for cars and Euro 6 for heavy goods vehicles (3.26.2 & 5); extending the IPPC Directive to cover small combustion plant; reducing VOC emissions from fuel stations, setting NOx emission limit values for ships; and reducing nitrogen use for animal feedstuffs and fertilisers.

Full details of the CAFÉ programme and of the Thematic Strategy can be found on the DG Environment website, at http://europa.eu.int/comm/environment/air/cafe/index.htm

3.9.2 Ambient Air Quality Directive

Directive 2008/50/EC *Ambient Air Quality and Cleaner Air for Europe* was adopted on 21 May 2008, entering into force on 11 June 2008; Member States are required to incorporate its provisions into national legislation before 11 June 2010. The new Directive consolidates the 1996 Framework Directive, and the first three of its Daughter Directives (i.e. excluding that on heavy metals, which will be merged at a later date), which are to be repealed on 11 June 2010; Council Decision 97/101/EC on the exchange of air pollution data (see 3.9.4 below) will be repealed two years after the Commission has established a more streamlined way for the collection and reporting of data, and for the reciprocal exchange of information and data from networks. Existing air quality limit values are retained (see Table 3.2, p.92), while the Directive strengthens the provisions which oblige Member States to prepare and implement plans and programmes in the case of non-compliance. Where in a particular zone or agglomeration Member States can demonstrate that they have taken all reasonable steps to attain compliance by the due date (e.g by establishing air quality action plans), they will be allowed to extend the deadline as follows:

- nitrogen dioxide, benzene: extend by maximum of 5 years (i.e. to January 2015).
- PM_{10}: 11June 2011.

The UK is seeking derogations to both the particulates (PM_{10}) and nitrogen dioxide deadlines.

The Directive requires Member States "to take all necessary measures not entailing disproportionate cost to reduce exposure to $PM_{2.5}$, and sets an annual concentration limit for $PM_{2.5}$ of 25 μg/m³ to be to be attained by 1 January 2015, or where possible by 2010; Member States are required to reduce human exposure by 20% between 2010 and 2019 – these provisions are to be reviewed five years after adoption of the Directive with a view to making specific legally binding exposure reduction targets for each Member State.

Monitoring and reporting requirements are simplified by moving towards a shared information system and electronic reporting, with some reporting requirements repealed.

3.9.3 Assessment and Management

The EU's Framework *Directive on ambient air quality assessment and management* (96/62/EC) was formally adopted on 27 September 1996; it came into force on 21 November 1996, and had to be implemented by Member States by 21 May 1998. The main aim of the Directive, which is to be repealed on 11 June 2010 (see above), is to protect human health and the environment by avoiding, reducing or preventing harmful concentrations of air pollutants; this is to be achieved through

- the definition and fixing of objectives for air quality and setting of limit values and/or alert thresholds (and/or target values for ozone);
- assessing air quality in a uniform manner;
- making information available to the public;
- maintaining or improving ambient air quality.

The framework Directive required the Commission to propose "Daughter" Directives setting air quality objectives, limit values, alert thresholds, guidance on monitoring, siting and measurement methods for individual pollutants, with daughter Directives subsequently adopted on sulphur dioxide, nitrogen dioxide, fine particulate matter (PM_{10}), suspended particulate matter & lead; benzene & carbon monoxide; ozone; and heavy metals – see below.

Factors to be taken into account when setting limit values and alert thresholds include: degree of exposure of population (including sensitive groups); climatic conditions; sensitivity of flora and fauna and of historic heritage; economic and technical feasibility; and long-range transmission of pollutants. The World Health Organisation's guidelines (see above, 3.8) are also taken into account when setting air quality objectives.

Criteria for monitoring, modelling and estimation techniques will be fixed for individual pollutants. Monitoring will be mandatory in population centres of more than 250,000, in those areas where the level of the pollutant is above a certain proportion of its limit value, and in other areas where limit values are exceeded. Member States must designate the national, regional or local authorities to be responsible for implementation of the Directive and also accredit laboratories for quality assurance and assessment.

Member States must assess ambient air quality on the basis of the limit values and alert thresholds set, drawing up action plans indicating measures to be taken in the short term where there is a risk that these may be exceeded. Certain measures are required depending on measured air quality:

Table 3.2: EU Air Quality Limit Values, 99/30 and 2008/50

[2008 additions/amendments in square brackets]

Sulphur dioxide

- *Hourly limit value for the protection of human health:* 350 μg/m³, not to be exceeded more than 24 times per calendar year (pcy) – should have been met by 1.1.2005.
- *Daily limit value for the protection of human health:* 125 μg/m³, not to be exceeded more than 3 times pcy – should have been met by 1.1.2005.
- *Alert threshold:* 500 μg/m³ measured over 3 hours.
- *Critical level for the for the protection of vegetation:* 20 μg/m³, averaged over calendar year and winter (1 October - 31 March)

Nitrogen dioxide

- *Hourly limit value for the protection of human health:* 200 μg/m³, not to be exceeded more than 18 times pcy; target date: 1.1.2010. [derogation permitted to 1.1.2015]
- *Annual limit value for the protection of human health:* 40 μg/m³; target date: 1.1.10. [derogation permitted to 1.1.2015]
- *Alert threshold:* 400 μg/m³ measured over 3 hours.
- *Critical level for the protection of vegetation:* 30 μg/m³, averaged over a calendar year.

Benzene

- *Annual limit value for the protection of human health:* 5 μg/m³; target date 1.1.2010 [derogation permitted to 01.01.2015].

Carbon monoxide

- Maximum daily 8 hour mean for the protection of human health: 10 mg/m³; should have been met by 1.1.2005.

Lead

- *Annual limit value for the protection of human health:* 0.5 μg/m³; should have been met by 1.1.2005. Limit value to be met by 1.1.2010 in immediate vicinity of the specific industrial sources situated on sites contaminated by decades of industrial activities; until then limit value is 1.0 μg/m³.

Particulate matter (PM_{10})

Stage 1:

- *Daily limit value for the protection of human health:* 50 μg/m³, not to be exceeded more than 35 times pcy; should have been met by 1. 1.2005; [derogation permitted to 11.6.2011]
- *Annual limit value for the protection of human health:* 40 μg/m³; should have been met by 1.1.2005. [derogation permitted to 11.6.2011]

[Stage 2 is to be repealed

- *Indicative daily limit value for the protection of human health:* 50 μg/m³, not to be exceeded more than 7 times pcy; target date: 1.1.2010.
- *Indicative annual limit value for the protection of human health:* 20 μg/m³; target date: 1.1.2010.]

[Particulate matter $PM_{2.5}$

- Annual mean target value of 25 μg/m³ to be met by 1.1.2010, with derogation permitted to 1.1.2015.
- Annual mean limit value of 25 μg/m³ to be met by 1.1.2015
- Indicative limit value of 20 μg/m³ to be met by 1.1.2020
- Average human exposure reduction of 20% on 2010 value to be achieved by 2020; 20 μg/m³ exposure concentration to be met as an average across urban background by 2015]

Ozone – see (c) below.

- in zones where levels exceed or are likely to exceed the limit value, plus the margin of tolerance (i.e. the percentage of the limit value by which the value may be exceeded) – plans and programmes to achieve the limit value within the specific time limit must be drawn up and implemented and the public given details of them;
- in zones where the levels are lower than the limit value – these levels should be maintained and the best ambient air quality compatible with sustainable development maintained;
- where the alert threshold is exceeded, steps must be taken to inform the public and a report sent to the Commission within three months of the episode, with details such as duration and levels of pollutants recorded.

This Directive is now implemented through separate *Air Quality Standards Regulations 2007* in England, Wales, Scotland and Northern Ireland, replacing and revoking previous *Air Quality Limit Values Regulations* – see 3.11.6 below.

(a) Sulphur Dioxide, Oxides of Nitrogen, Particulate Matter & Lead

The first Daughter Directive under the *Air Quality Assessment and Management Directive – Directive Relating to Limit Values for Sulphur Dioxide, Oxides of Nitrogen, Particulate Matter and Lead* (99/30) – was formally adopted on 22 April 1999, was brought into force on 19 July 1999 and had to be implemented by 19 July 2001. It is now implemented in each country of the UK through separate 2007 *Air Quality Standards Regulations* – see 3.11.6 below. This Directive has been subsumed into Directive 2008/50/EC on ambient air quality and is to be repealed on 11 June 2010 – see above – 3.9.2 and Table 3.2.

The Directive sets limit values for the above pollutants with the aim of avoiding, preventing or reducing their harmful effects on human health and on the environment as a whole (Table 3.2). Where the limit value is likely to be exceeded on more than the permitted number of days per year then action programmes setting out plans to meet the limit value by the target date must be drawn up; where the limit value is exceeded, steps must be taken to ensure it is met by the target date; where air quality is below the limit value, it should not be allowed to deteriorate. When the alert threshold for sulphur dioxide or nitrogen dioxide is exceeded, specified information must be made available to the public. Information (updated hourly, or every three months for lead) on ambient concentrations of all the pollutants should be made available to the public, including exceedences of the public indicator threshold (i.e. the hourly or daily limit values below). The Directive recognises that some Member States may not be able to meet some of the limit values because of climatic or other special circumstances.

(b) Carbon Monoxide & Benzene

The second "daughter Directive" under the AQMA Directive was formally adopted in December 2000 (2000/69/EC), and had to be implemented by December 2002. The Directive establishes a limit value of 10 mg/m³ (8.5 ppm) averaged over 8 hours for carbon monoxide to be achieved by 1 January 2005; for benzene a limit value of 5 μg/m³ (about 1.66 ppb) averaged over a calendar year is to be achieved by 1 January 2010; for those countries which would have difficulty in meeting the limit, a five year derogation is allowed but benzene concentrations must not be allowed to exceed a yearly average of 10 µg/m³. This Directive has been subsumed into Directive 2008/50/EC on Ambient Air Quality – see above 3.9.2 and Table 3.2.

The carbon monoxide limit value accords with WHO Guidelines, and also with EU policy on the environment and on protecting human health. The benzene limit has been decided on the basis of what is achievable but is also considered to offer a high level of protection for human health.

The Directive specifies the number of, and siting of, measurement stations, as well as reference methods for air quality measurements. Member States are required to supply regular and up-to-date information to the public and appropriate organisations, and to draw up action plans where limit values are exceeded. The Commission is expected to review the benzene limit shortly.

This Directive is now implemented in each country of the UK through separate 2007 *Air Quality Standards Regulations* – see 3.11.6 below.

(c) Monitoring & Information – Ozone

The third Daughter Directive to be proposed under the 1996 AQMA Directive was adopted in February 2002; Directive 2002/3/EC *relating to ozone in ambient air*, which also forms part of the Commission's acidification strategy (see 3.16), came into force on 9 March 2002. Member States were required to bring in the measures necessary to implement the Directive by 9 September 2003, when the 1992 Directive on Air Pollution by Ozone (92/72/EEC)1992 was repealed. This Directive has been subsumed into Directive 2008/50/EC on Ambient Air Quality and is to be repealed on 11 June 2010 – see above 3.9.2 and Table 3.2.

As well as setting target values for the protection of health and of vegetation (see Table 3.3), the ozone Directive specifies the information to be made available to the public when levels reach an information threshold of 180 µg/m³, 1 hour average, and an alert threshold (aimed at sensitive sections of the population) of 240 µg/m³, 1 hour average measured over 3 consecutive hours. Such information should include the area affected, forecast of length of episode and whether ozone levels are likely to increase further; information on what health precautions are advisable (e.g. the need to avoid strenuous exercise) should also be given. Monitoring results are also to be made publicly available.

Member States must identify zones and agglomerations where either the target values or long term objectives are unlikely to be met within the specified period, and draw up action plans or programmes in accordance with the Directive. Member States should also identify those zones and agglomerations in which the long term objective is already met and should ensure that air quality is maintained in those areas.

This Directive is now implemented in each country of the UK through separate 2007 *Air Quality Standards Regulations* – see 3.11.6 below. Information on levels of ozone is made available to the public on a regular basis – see below 3.11.9.

A report from the European Environment Agency (Technical report 5/2008) notes levels of ozone across Europe during the summer of 2007 were the lowest recorded for eleven years; however, the EU's long-term objective for the protection of human health was exceeded in all countries in almost every month of the summer, with 83% registering one or more exceedence.

Table 3.3: Ozone in Ambient Air (2002/3/EC) & 2008/50/EC

Target values for the protection of human health

- 120 µg/m³, maximum daily 8 hour mean to be achieved by 01.01.2010, not to be exceeded on more than 25 days per calendar year averaged over 3 years.
- Long term objective of 120 µg/m³ (daily maximum for an 8 hour mean within a calendar year), [to be met by 2020 – date not defined in 2008 Directive]

Target values for the protection of vegetation

- 18000 µg/m³.h, (AOT40, calculated from 1 h values from May-July), averaged over 5 years, to be achieved by 01.01.2010.
- Long term objective of 6000 µg/m³.h (AOT40, calculated from 1 hour values from May-July) [to be met by 2020 - date not defined in 2008 Directive]

(d) Heavy Metals in Ambient Air

The fourth Daughter Directive under the *Directive on Air Quality Management and Assessment* – 2004/107/EC relating to arsenic, cadmium, mercury, nickel and polycyclic aromatic hydrocarbons in ambient air – was adopted on 15 December 2004 and entered into force 15 February 2005; it had to be transposed into national legislation by 15 February 2007. Consideration is to be given to subsuming this Directive into Directive 2008/50/EC on Ambient Air Quality once sufficient experience of implementation of the former has been gained.

The Directive sets target values (Table 3.4) for four heavy metals, with Member States required to "take all necessary measures, not entailing disproportionate cost" to meet the target values by 31 December 2012. Installations regulated for IPPC will be expected to use "best available techniques" to reduce emissions.

Mandatory monitoring will be required in zones and agglomerations where concentrations exceed upper assessment thresholds (on the basis of concentrations during the previous five years); where concentrations are between the thresholds a combination of monitoring and modelling may be used, and modelling used where concentrations are below the lower threshold. The Directive provides for the siting and number of monitoring sites, for Member States to report to the Commission and for the public to be given regular information on air quality and concentrations of the pollutants.

This Directive has been implemented in each country of the UK through 2007 Air Quality Standards Regulations – see 3.11.6 below.

Table 3.4: Heavy Metals

	*Target Value** ng/m³	*Upper Assessment* ng/m³	*Lower Threshold* ng/m³
Arsenic	6	3.6	2.4
Cadmium	5	3	2
Nickel	20	14	10
Benzo(a)pyrene	1	0.6	0.4

*for the total content in the PM_{10} fraction averaged over a calendar year

3.9.4 Exchange of Air Pollution Data

Decision 97/101/EC approved in January 1997 establishes a "reciprocal exchange of information and data from networks and individual stations measuring ambient air pollution within Member States". This supersedes an earlier 1982 Decision which aimed to achieve equality throughout the Community and harmonise measurements and, subsequently, techniques. The current scheme covers all the pollutants listed in the air quality assessment and management Directive (see above); Member States are required to submit to the Commission data collected and validated in accordance with the Decision; data for the year must be sent by October of the following year. The data will be stored by the European Environment Agency; a request for access to the data may be made to the EEA or to a Member State.

This Decision is to be repealed once revised reporting and exchange of data procedures have been finalised in accordance with Directive 2008/50/EC on Ambient air Quality (see 3.9.2 above).

3.10 AIR QUALITY & THE UK

National and EU legislation (as well as international agreements) over the years has tended to focus on reducing pollution from a particular sector, e.g. from industry or from motor vehicles. Now, however, the emphasis is on maintaining or, where necessary, improving local air quality to protect public health: targets or limit values are set for individual pollutants and action plans drawn up for those areas where there is a risk that standards might be, or are being, exceeded – see above, 3.9.3, the EU *Ambient Air Quality Assessment and Management Directive*, and 3.11.1 below, the Air Quality Strategy for England, Scotland, Wales and Northern Ireland. The UNECE Convention on Long-Range Transboundary Air Pollution (see 3.15), whose signatories include most European countries, aims to reduce levels of certain pollutants to protect, and to prevent further damage to, sensitive ecosystems.

Data on current concentrations for a wide number of air pollutants, air quality objectives etc, can be found on www.airquality.co.uk.

3.11 AIR QUALITY MANAGEMENT

Part IV of *the Environment Act 1995* requires the Secretary of State to draw up a National Air Quality Strategy, and for local authorities to review and assess local air quality against standards and objectives in the National Air Quality Strategy and where necessary to declare an air quality management area.

3.11.1 National Air Quality Strategy

Section 80 of the Act was brought into force on 1 February 1996; it requires the Secretary of State to formulate a national air quality strategy in consultation with the environment agencies, representatives of local authorities and of industry and other relevant bodies or persons considered appropriate. Local authorities must take account of the objectives in the Air Quality Strategy when reviewing local air quality and, where necessary, draw up an air quality management plan if standards are being breached or are at risk.

Following a period of consultation in 1996 the first AQ Strategy, *The United Kingdom National Air Quality Strategy* was published in March 1997. Objectives for the eight main health-threatening pollutants were based on the recommendations of the Government's Expert Panel on Air Quality Standards (see 3.11.8 below). A second Strategy, *The Air Quality Strategy for England, Scotland, Wales and Northern Ireland – Working Together for Clean Air*, published in January 2000 (and addendum, 2003), built on the earlier Strategy, outlining a national framework for reducing hazards to health from air pollution in the UK. The objectives in both strategies were given statutory force through Regulations.

Proposals for revising the AQ Strategy were published by Defra and the Devolved Administrations in April 2006, with the final version published on 17 July 2007. It aims to map out as far as possible current and future ambient air quality policy in the medium term, providing the best practicable protection to human health and the environment by setting evidence-based objectives for the main air pollutants. The revised Strategy retains the air quality objectives from the 2000 Strategy and its Addendum, and introduces an exposure reduction approach to reducing emissions of certain pollutants – such as particulate matter – for which no threshold for adverse health effects has been identified, similar to that included EU's 2008 Ambient Air Quality Directive (see above 3.9.2).

As well as setting AQ standards and objectives, the Strategy includes:

- an assessment of current air quality, and describes the international and European framework within which the Strategy has been drawn up;
- what the Government is doing to achieve the Strategy's objectives;
- the contribution which other sectors – including industry, transport, local government, the environment agencies – can make towards achieving the objectives of the National Strategy.

The Environment Agency and SEPA are required to take account of the Strategy in carrying out their pollution control functions (s.81). Section 7(2) & (12) of the *Environmental Protection Act 1990*, as amended by the *Environment Act*, Sch. 22, para 49(2) place the same obligation on local authorities in England and Wales. The Environment Agency's document *Regulating to Improve Air Quality* (AQPG3, version 1, 14/7/08) outlines the Agency's duties related to air quality and the way in which it works with local authorities to deliver LAQM – it can be accessed via Defra's air quality website pages.

Greater London

The Greater London Authority Act 1999 (GLAA 1999) requires the Mayor of London to publish an air quality strategy for London containing proposals and policies for implementing the National Air Quality Strategy and achieving the statutory objectives in Greater London; it should also include an assessment of current and future air quality in Greater London. In drawing up the London Air Quality Strategy the Mayor should consult London local authorities and the Environment Agency. In carrying out their duties under the *Environment Act* London local authorities will need to have regard to the London Strategy as well as the National Strategy, Defra Guidance and any Directions from the Secretary of State.

Table 3.5: National Air Quality Objectives

Objectives and Target Values for the Protection of Human Health *(note: objectives are the same in all areas of the UK unless otherwise stated)*	
Benzene	16.25 μg.m^{-3} running annual mean to be achieved by 31.12.03; supplemented by ***E & W:*** 5 μg.m^{-3} annual average to be achieved by 31.12.10. ***Scotland & N. Ireland:*** 3.25 μg.m^{-3} running annual mean to be achieved by 31.12.10.
1,3 Butadiene	2.25 μg.m^{-3} running annual mean to be achieved by 31.12.03.
Carbon monoxide	10 mg.m^{-3} (E&W, NI max. daily running 8-hour mean) (Sc running 8-hr mean) to be achieved by 31.12.03.
Lead	0.5 μg.m^{-3} annual mean to be achieved by 31.12.04; 0.25 μg.m^{-3} annual mean by 31.12.08.
Nitrogen dioxide	200 μg.m^{-3} 1 hour mean to be achieved by 31.12.05, not to be exceeded more than 18 times a year; 40 μg.m^{-3} annual mean to be achieved by 31.12.05.
Particles (PM_{10})	50 μg.m^{-3} 24 hour mean to be achieved by 31.12.04, not to be exceeded more than 35 times a year; 40 μg.m^{-3} annual mean to be achieved by 31.12.04. ***Scotland*** 50 μg.m^{-3} 24-hour mean not to be exceeded more than 7 times a year 18 μg.m^{-3} annual mean - both to be achieved by 31.12.10.
Particles ($PM_{2.5}$) (exposure reduction)	***E, W & NI:*** 25 μg.m^{-3} annual mean to be achieved by 2020. ***Scotland:*** 12 μg.m^{-3} annual mean to be achieved by 2020. ***UK urban areas:*** target of 15% reduction in annual mean concentrations at urban background, to be achieved between 2010-2020.
PAH	0.25 ng.m^{-3} B[a]P, as annual average to be achieved by 31.12.10.
Ozone	100 μg.m^{-3} 8-hour mean, to be achieved by 31.12.05; not to be exceeded more than 10 times per year.
Sulphur dioxide	266 μg.m^{-3} 15 minute mean to be achieved by 31.12.05, not to be exceeded more than 35 times per year. 350 μg.m^{-3} 1 hour mean, to be achieved by 31.12.04, not to be exceeded more than 24 times per year; 125 μg.m^{-3} 24 hour mean, to be achieved by 31.12.04, not to be exceeded more than 3 times per year.
Objectives and Target Values for the Protection of Vegetation and Ecoystems	
Nitrogen oxides	30 μg.m^{-3} annual mean, to be achieved by 31.12.00.
Sulphur dioxide	20 μg.m^{-3} annual mean, to be achieved by 31.12.00. 20 μg.m^{-3} winter average, to be achieved by 31.12.00.
Ozone	Target value of 18,000 μg.m^{-3}, average over 5 years, based on AOT40 to be calculated from 1 hour values from May to July, and to be achieved, so far as possible, by 2010.

The final version of the Strategy, *Cleaning London's Air*, published in September 2002, highlights traffic as the main cause of air pollution in London and proposes a number of measures aimed at reducing emissions from road vehicles, particularly the most polluting, by for example reducing the amount of traffic in London (e.g. through the introduction in February 2003 of the congestion charge) and investing in public transport, encouraging use of cleaner vehicles and investigation of introducing one or more low emission zones in London (see p. 133). The Mayor also proposes to reduce emissions as a result of poor energy use in buildings, encouraging improved fuel and energy efficiency. The GLA has published (November 2006) a Construction Best Practice Guide which aims to reduce emissions of dust and other pollutants and improve air quality around construction sites.

Northern Ireland

Part III of *The Environment (Northern Ireland) Order 2002* (SI 2002/3153) requires district councils in Northern Ireland to review air quality in their area, with *The Air Quality (Northern Ireland) Regulations 2007* (SR 2007/265, replacing SR 2003/342) prescribing the period within which the air quality objectives (in the National Air Quality Strategy) must be achieved. The Regulations implement a similar system of review and assessment to that which applies in the rest of the UK. District councils are responsible for review and assessment and for declaring AQ management areas and for drawing up action plans as necessary.

Prior to the formal introduction of the LAQM regime in Northern Ireland many district councils had already carried out Stage 1 review and assessment. Statutory guidance (LAQM.PG NI (03)) published in 2003 required that those authorities which had identified a potential problem should complete Stages 2 and 3 reports by 31 December 2003, with consultation on these reports completed in March 2004, and declaration of air quality management areas in April 2004; action plans were to be drafted by July 2004 and completed by April 2005, together with progress reports. Dates for the second round of review and assessment follow the pattern outlined for the rest of the UK – see next section. Guidance on the preparation of AQ progress reports (LAQM.PRGNI(04)) was published in November 2004.

3.11.2 Local Authority Functions & Duties

Sections 82-84 of the *Environment Act* (E, S, W), which were brought into force on 23 December 1997 (NI Art.13b of 2002 Order), place a duty on local authorities to review air quality in their area. The review should include the likely future air quality within the relevant period and an assessment of whether air quality standards and objectives are being achieved, or are likely to be achieved measured against the National Air Quality Strategy. Initial reviews and assessments should have been completed by June 2000, and air quality management areas – i.e. any part of the authority's area in which standards or objectives were not being met, or were unlikely to be met within the relevant period set out in the *Air Quality Regulations 2000* – designated by October 2000.

The current timetable for review and assessment, which is carried out on a three year cycle is set out in LAQM policy guidance; in E, S & W, LAQM.PG(03), and in NI LAQM.PG NI (03). Draft revised policy guidance for England was issued for consultation in July 2008, and was expected to be finalised by the end of 2008:

- All authorities – updating and screening assessment (USA) to be completed by 30 April 2009 (and again 2012 and 2015)
- Detailed assessment if USA shows possible breach of an AQO, with view to establishing whether a new AQMA is needed or an existing one needs amending – to be completed by 30 April 2010 (and again 2013 and 2016)
- Authorities not required to carry out detailed assessment – progress report by 30 April 2010 (and again 2013 and 2016)
- All authorities – progress report by 30 April 2011 (and again 2014 and 2017).

Assessments and reports are submitted to Defra, Welsh Assembly, Scottish Government or NI Environment Agency, as appropriate. From 2009 Local authorities in England will be able to submit updating and screening assessments and progress reports online – see revised policy guidance expected end 2008.

Within 12 months of designation, the local authority should draw up an action plan* for each AQMA setting out the various measures, together with target dates, by which it aims to meet air quality standards. Following consultation on the plan, and its revision as appropriate, local authorities should aim to have the action plan in place within 12-18 months of designation of the AQMA. The plan may be revised as appropriate. AQM Orders may be varied by subsequent Order or revoked (where standards and objectives are being achieved or likely to be during the relevant period), but only following a further air quality review.

(**The Local Authorities' Plans and Strategies (Disapplication) (England) Order 2005* (SI 2005/157), which came into force on 1 February 2005, removes the duty for those local authorities categorised as "excellent" to produce an AQ action plan; "excellent" authorities who had produced an action plan prior to the Order taking effect will, however, need to continue to keep it under review and report progress on its implementation. If a local authority loses its "excellent" classification, it will have one year in which to regain it before the need to prepare an action plan is reinstated – it will then have a year in which to produce the plan.)

In carrying out its functions in respect of an air quality review, assessment or preparing an action plan, the local authority must consult:

- the Secretary of State,
- the appropriate Environment Agency,
- in England and Wales, any highway authority affected by the review or plan,
- any local authority with whom it shares a border,
- in England, any relevant county council,
- appropriate National Parks authority,

and, as the local authority considers appropriate, any public authorities with functions within the area, representatives of business interests, or any other bodies or persons (Sch.11 to the Act). *The Greater London Authority Act 1999* amends Sch.11 to require local authorities in Greater London and those adjoining a Greater London authority to, in addition, consult the Mayor. In Northern Ireland, district councils should consult the DOE, neighbouring district councils, other relevant authorities, bodies representing local business interests, and such other bodies or persons as considered appropriate (Sch.2 to the NI 2002 Env. Order).

Schedule 11 also specifies that the local authority should ensure the public have free access to the following documents, with copies obtainable at a reasonable charge:

- results of air quality review
- results of any assessment made
- any Order designating an Air Quality Management Area
- any action plan
- any Directions from the Secretary of State
- in England, proposals and statements submitted by a county council under s.86 or any Secretary of State Direction.

Guidance for local authorities

Guidance is issued by the Secretary of State under s.88(1) of the Environment Act 2005; section 88(2) of the *Environment Act* requires local authorities to have regard to any guidance issued by the Secretary of State in carrying out their duties under this Part of the Act. London local authorities are also required to have regard to the London air quality strategy (*Greater London Authority Act 1999*, s.364).

In February 2003 Defra and the National Assembly for Wales published Local Air Quality Management Policy Guidance LAQM.PG(03), with similar guidance published by the Scottish Government and the DOE NI. This replaced earlier guidance documents issued in 2000, and covers:

- The statutory background and the legislative framework within which LAs have to work;
- New principles governing review and assessment up to 2010 and recommended steps that LAs should take;
- The designation of AQMAs and implementation of action plans;
- Taking forward the development of local and regional AQ strategies;
- Consultation and liaison;
- Possible local transport measures;
- General principles behind AQ and land-use planning.

An Addendum (LAQM.PGA(05), published March 2005, updates LAQM.PG(03) for English authorities and provides guidance on the integration of AQ action plans into local transport plans. This arises from the Government's "Freedoms and Flexibilities" agenda under which local authorities who have declared an AQMA solely because of local transport pollution no longer have to prepare an action plan, but can instead include measures for dealing with the problem in their local transport plan. Local authorities in England classified as "excellent" do not have to prepare an LTP (*The Local Authorities' Plans and Strategies (Disapplication (England) Order 2005*).

Following consultation in July 2008, revised policy guidance for England is expected towards the end of 2008, together with new practice guidance covering

- Economic principles for the assessment of local measures to improve air quality;
- Low emission zones;
- Measures to encourage the uptake of low emission vehicles;

- Measures to encourage the update of retro-fitted abatement equipment on vehicles;
- Using the planning system to reduce transport emissions.

The Technical Guidance (LAQM.TG(03)), published jointly by Defra and the Devolved Administrations in February 2003, sets out the approach to be taken for review and assessment, based on up to date understanding of pollutant concentrations and sources and methods to predict future levels; appendices cover monitoring, estimating emissions and selection and use of dispersion models. The Guidance was partially updated at the end of 2005 to provide guidance on the Updating and Screening Assessments (USAs) required as part of the third round of review and assessment and which had to be submitted in April 2006. Revised technical guidance was due for publication towards the end of 2008, following consultation in July 2008.

LAQM.PRG(03), published December 2003, for England, Scotland and Wales (NI: LAQM.PRGNI(04)) outlines the information to be included in progress reports, which includes: monitoring data which shows progress towards achieving objectives, and factors which could impact on future air quality (e.g. new industrial installations, other planned new developments). It is also suggested that progress reports incorporate progress on implementing action plans, though local authorities can submit this report separately; progress reports on action plans should, where possible, also detail the impact measures are having on local air quality and a recommended format for doing so is included in the Guidance.

In addition, Annex 1 (pollution control, air and water quality) of PPS23 on *Planning and Pollution Control*, published 2004, notes the need for planners and air quality staff to work together in carrying out reviews and assessments, in drawing up action plans, and on any other matters where development decisions may have an indirect or direct effect on air quality.

Defra and the Devolved Administrations fund a number of helpdesks providing assistance to local authorities on various aspects of the LAQM regime and procedures – see list of websites at the back of the of the *Pollution Control Handbook* and www.laqmsupport.org.uk. They have also produced (2006) a *guide for local authorities purchasing air quality monitoring equipment*.

All guidance is available via the websites of Defra and the Devolved Administrations, as appropriate.

3.11.3 Reserve Powers of Secretary of State

Section 85 of the Act (brought into force on 23 December 1997) provides powers for the Secretary of State (in E & W) and SEPA (in Scotland, acting with the approval of the Secretary of State) to carry out or to have carried out a review of air quality within the area of a local authority; to assess whether air quality objectives and standards are being met; to identify those parts of a local authority's area where it seems objectives and standards are not being met; and to make an assessment of the possible reasons why they are not being met either within the area of the LA or within a designated area.

If it appears to the Secretary of State/SEPA that a local authority is not meeting its obligations under ss.82-84, or that AQSs or objectives are unlikely to be achieved, the Secretary of State/SEPA may direct the local authority to:

- arrange for an air quality review to be carried out as specified by the Directions;
- arrange for a new air quality review to be carried out, covering such area as directed or in the way specified by the Directions;
- to revoke or modify an Order made under s.81 of the Act;
- to prepare or modify its action plan or to implement any measures in its action plan.

The Greater London Authority Act 1999 amends *the Environment Act 1995* to give similar powers to the Mayor of London with regard to London local authorities; however such powers may only be exercised following consultation with the authority concerned and having regard to any Guidance issued under s.88(1) by the Secretary of State.

The Secretary of State/SEPA may also direct local authorities (other than those in Greater London, s.367(4) GLAA 1999 amendment to s.85(5) of Env. Act 1995) to take account of European or international commitments in their action plan. The *Air Quality Standards (England) Regulations 2007* provide a similar power for the Secretary of State to direct the Mayor (and for the Mayor to direct London local authorities) for the purposes of meeting his obligations under the 1996 and 1999 EU air quality Directives (see 3.9.3 above and 3.11.6 below).

3.11.4 County Council Functions

Section 86 of the Act, which came into force on 23 December 1997, applies in those districts in England covered by a district council which are also in an area comprising a county council.

The relevant county council may make recommendations to the district council with regard to the district council's duties under this part of the Act, which the district council must take into account.

Where the district council is preparing an action plan, the county council should submit its proposals, together with a statement outlining the timescale, for meeting air quality standards and objectives within the designated area. The *Air Quality (England) Regulations 2000* require the county council to submit its proposals within nine months of first being consulted by the district council on its action plan.

If the Secretary of State is of the opinion that the council is not meeting its obligations under this section of the Act, he may through Directions require the county council to

- submit its proposals and statement of timescale for meeting AQSs and objectives;
- modify its proposals or statement in accordance with the Directions and to submit this to the district council;
- implement any measures in an action plan as required by the Direction.

Directions may also require county councils to take steps to implement international or European obligations with regard to air quality.

Schedule 11 to the Act requires county and district councils in England to provide each other with information "as is reasonably requested" in pursuit of their respective duties and functions.

Where a local authority in London is drawing up an action plan in respect of a designated area, the Mayor of London should submit his proposals for the area to the authority, together with a statement outlining when he proposes to implement each of his proposals (new s.86A to *Env. Act 1995*, added by s.368 of *GLAA 1999*).

3.11.5 Regulations

Section 87 (brought into force on 1 February 1996) provides for regulations covering all aspects of Part IV of the Act, including air quality standards; objectives for restricting levels of particular substances in the air; prohibiting or restricting specified activities, vehicle or mobile equipment access in general or in certain circumstances; communicating AQ information to the public; regulations may also relate to air quality reviews, assessments, orders designating AQ management areas and action plans; vehicle emission spot checks; offences, penalties (including fixed penalty offences) and appeals.

(a) Air Quality Regulations

These Regulations incorporate the objectives contained in the Air Quality Strategy for England, Scotland, Wales and Northern Ireland, against which local authorities must review and assess air quality (see above, s.80 of the Act); they also specify the period within which county councils in England must submit proposals to district councils for inclusion in an air quality action plan (see above, s.86 of the Act).

The Air Quality (England) Regulations 2000 (SI 2000/928) came into force on 6 April 2000, with similar Regulations (SSI 2000/97) taking effect in Scotland on 7 April and in Wales (SI 2000/1940, W.138) on 1 August 2000. The 1997 Regulations (SI 1997/3043) have been revoked. Similar Regulations took effect in Northern Ireland on 1 September 2003 – *The Air Quality Regulations (Northern Ireland) 2003* (SR 2003/342, as amended SR 2003/543).

In Scotland 2002 Amendment Regulations (SSI 297), which came into force on 12 June 2002, prescribe new additional objectives for benzene and particles and a new tighter objective for carbon monoxide; in England the 2002 Amendment Regulations (SI 3043), which came into force on 11 December 2002, cover the objectives for benzene and carbon monoxide (see Table 3.5).

(Separate Regulations have been made transposing the EU Directives relating to limit values for sulphur dioxide, oxides of nitrogen, particulate matter and lead (3.9.3a), carbon monoxide and benzene (3.9.3b), and target values for ozone (3.9.3c) – see below 3.11.6.

(b) Vehicle Emissions (Fixed Penalty) Regulations

The Road Traffic (Vehicle Emissions) (Fixed Penalty) (England) Regulations 2002 (SI 2002/1808) came into force on 18 July 2002, with similar Regulations applying in Wales (SI 2003/300, W.42, effective 1 May 2003), and in Scotland (SSI 2003/212, effective 21 March 2003). They replace and revoke earlier 1997 Regulations (SI 3058) which applied in specified local authority areas – see also 3.27.8.

Local authorities in England or Wales who have declared an air quality management area can apply to the Secretary of State for designation to enable authorised officers to undertake roadside emissions testing (in Scotland this applies to all local authorities, regardless of designation); the regulations also enable all local authorities in England to issue fixed penalty notices to drivers who leave vehicle engines running unnecessarily, having been asked to switch it off. For further details of the Regulations see later this chapter 3.27.8.

3.11.6 Air Quality Standards Regulations

The Air Quality Standards Regulations 2007 (SI 2007/64) consolidate and replace *the Air Quality Limit Values Regulations 2003*, as amended; they came into force on 15 February 2007 and implement the EU's 1996 Directive on air Quality Management and Assessment and subsequent "Daughter Directives" – see 3.9.3 above. These Regulations apply in England only, except for the Regulations covering the Secretary of State's duties to inform other Member States regarding transboundary pollution which apply throughout the UK. These Regulations will also repeal, on 1 January 2010, the last remaining part (reg. 6) of *the Air Quality Standards Regulations 1989* (SI 317) prescribing a limit value for nitrogen dioxide.

The corresponding Regulations in the Devolved Administrations are:

- *The Air Quality Standards (Scotland) Regulations 2007* (SSI 2007/182, effective 29 March 2007), replacing SSI 2003/547, as amended, with Scottish Ministers responsible for implementation.
- *The Air Quality Standards (Wales) Regulations 2007* (SI 2007/717, W.63, effective 15 March 2007), which replace SI 2002/3183, W.299; the National Assembly for Wales is responsible for implementation.
- *The Air Quality Standards Regulations (Northern Ireland) 2007* (SR 2007/265), which came into force on 28 May 2007, replacing and revoking SR 2002/94 & SR 2003/240.

The competent authorities designated for the purposes of Art.3 of the 1996 Directive on Air Quality Assessment and Management are the Secretary of State, Scottish Ministers, National Assembly of Wales, and Department for Environment (NI) as appropriate.

(a) Attainment of Air Quality Standards

The Regulations require the competent authority to take the necessary measures to ensure that in each zone – or agglomeration – air quality standards are attained; for Group B pollutants such measures should not entail disproportionate cost and where emissions arise from an installation controlled under IPPC, the best available techniques should be used to limit emissions; in the case of ozone, measures should be proportionate and in relation to the long-term objective cost-effective. Schedule 1 to the Regulations sets out the limit values for Group A pollutants (benzene, carbon monoxide, lead, nitrogen dioxide and oxides of nitrogen, PM_{10} and sulphur dioxide), target values for Group B pollutants (arsenic, benzo(a)pyrene, cadmium and nickel and their compounds), and target values and long-term objectives for ozone – see Tables 3.2, 3.3 and 3.4 above.

A zone is classified as an agglomeration if it has a population of more than 250,000 inhabitants, or one with less but with a population density per sq km, which it is felt justifies the need for separate assessment and management. Different zones may be established for different pollutants.

(b) Improvement Plans

For each zone in which concentrations of benzene or nitrogen dioxide exceed the limit value (plus the margin of tolerance) prior to 1 January 2010, or concentrations of ozone exceed the target value, the competent authority must prepare and implement an improvement plan aimed at meeting the attainment date. Schedule 2 to the Regulations specifies the information to be included, such as extent of pollution and reasons for it, authorities responsible for development and implementation of improvement plans, details of measures or projects for improvement which existed both before and after 21 November 1996.

Where concentrations of Group B pollutants exceed the relevant target value or those of ozone are equal to or below the target value but exceed the long term objective, the competent authority should identify the cause of the pollution and ensure that improvement measures (for Group B pollutants) do not entail disproportionate cost or accord with BAT for emissions from IPPC installations; and in the case of ozone are cost-effective.

In accordance with EU Directive 2003/35 ("public participation directive"), the competent authority is required to consult the public where it is proposed to prepare, modify or review an improvement plan. The public should be given all relevant background information and informed of their right to participate in the decision making process. The means by which the public can participate should be outlined and a reasonable time allowed for consultation. The competent authority is required to take account of the results of the consultation process in coming to a decision, and to inform the public and provide information as to the reasons and considerations on which the decision was taken.

(c) Action Plans

In each zone where concentrations of any Group A pollutants are below the limit value, or for Group B below target values, or long term objective for ozone is met, the Regulations require the competent authority to ensure compliance is maintained.

Where the competent authority consider there is a risk of limit values of any of the relevant pollutants being exceeded, or of the alert thresholds for sulphur dioxide, nitrogen dioxide or ozone being exceeded, they should prepare and implement action plans; these should aim to reduce the risk of exceedence or if it is not possible to prevent exceedence, to limit the duration and severity of the occurrence. In the case of ozone the obligation to prepare an action plan applies in so far as the competent authority considers that there is significant potential to achieve the objectives of the action plan given geographical, meteorological and economic conditions.

(d) Duty to Assess

The competent authority has a duty to assess concentrations of Group A and Group B pollutants and ozone within each zone – the Regulations detail assessment methods and sampling points.

(e) Other Pollutants – Background Monitoring

Additionally the Regulations require the measurement of concentrations of $PM_{2.5}$ and of ozone precursor substances, and the monitoring of concentrations of polycyclic aromatic hydrocarbons. Background monitoring is also required for Group B pollutants, basic PAHs and of total gaseous mercury.

(f) Public Information

The competent authority is required to ensure that up-to-date information on all the pollutants is routinely made available to the public in a clear, comprehensive and accessible form – the public includes health-care bodies, other organisations with an interest in ambient air quality or those representing the interests of sensitive people or the environment. This includes information on:

- *Zones* – lists of those which have attained air quality standards and those which have not and the pollutants concerned; until 31 December 2009 list of zones in which concentrations of benzene or nitrogen dioxide either exceed or are between the limit value and margin of tolerance; lists classifying each zone in relation to the method by which Group A and B pollutants have been assessed.
- *Concentrations* – of Group A and B pollutants and of ozone; and of concentrations of mercury, $PM_{2.5}$, PAHs and deposition rates of Group B pollutants, mercury and basic PAHs. Information on benzene (average value over the preceding 12 months) should be updated every three months and where practicable, monthly; for carbon monoxide (max. running average over 8 hours), it should be updated at least daily and if practicable, hourly; information on lead should be updated every three months. Information on NO_2, SO_2 and PM_{10} should be updated daily and in the case of hourly values for NO_2 and SO_2, hourly where practicable. Information on concentrations of ozone should be updated daily and where practicable hourly.
- *Breach of alert threshold* for NO_2 or SO_2 – information to be made available as soon as possible on date, hour, place of occurrence and reasons, forecast of any change in concentrations, duration or location, type of population potentially sensitive and precautions to take.
- *Breach of alert and information thresholds* for ozone (predicted or actual) – information to be made available as soon as possible on location, forecast for following afternoon, day or days, type of population at risk, possible health effects and recommended action to minimise effects.
- *Breach of air quality standards* – for Group A pollutants, the extent to which any limit values (and the alert thresholds for NO_2 and SO_2) have been exceeded, short assessment of the occurrences and effects on health. For Group B pollutants, occasions on which any target values have been exceeded and the area within each zone affected, reasons, predominant source of emissions, short assessment of effect of exceedence on overall compliance with the target value and prospects for attainment.
- *Ozone* – information should be given on any occasions on which the long-term objective for the protection of human health, or the information or alert thresholds, are exceeded, short assessment of occurrence, its extent and effects on health. The Secretary of State is also required to produce an annual report on ozone with details of exceedences of target values, alert and information thresholds for both human health and vegetation and effects.
- *Action and Improvement Plans.*

(g) Transboundary Pollution

Regulation 29 of *the Air Quality Standards Regulations 2007* requires the Secretary of State to consult another Member State when

- Concentrations of any Group A pollutants (or until 31 December 2009 for NO_2 and benzene, the limit value plus margin of tolerance) or the alert threshold for NO_2 or SO_2 is exceeded or likely to be exceeded anywhere in the UK as a result of significant pollution in another Member State;
- Concentrations of ozone anywhere in the UK exceed the target value or long term objective, with the predominant reason being the emission of ozone precursors in another Member State;
- Concentrations of ozone in another Member State exceed the target value or long term objective, with the predominant reason being the emission of ozone precursors in the UK.

Where the pollution occurs from a Group A pollutant in Scotland, Wales or Northern Ireland or in the case of ozone affects another Member State, the relevant administration should inform the Secretary of State who will then consult the Member State with regard to any remedial action. Where the transboundary pollution is caused by ozone precursors and the Secretary of State considers that the target value or long term objective can be achieved through proportionate measures, he and the Member State concerned should cooperate in the preparation of joint plans and programmes to this end.

3.11.7 Monitoring Networks

In January 1992 the Enhanced Urban Network (EUN) – now the Automatic Urban Network (AUN) – funded by Defra, became operational, and a number of local authority sites were integrated into the Network. Sulphur dioxide, nitrogen oxides, carbon monoxide, ozone and particulate matter (PM_{10}) are monitored continuously using automatic instruments. Defra also funds the Automatic Rural Network which monitors ozone by automatic UV absorption analysers. The main purpose of the sites is to provide up to date information on air quality which can be used to increase understanding of air quality problems, to assess personal exposure to air pollution and to assess the extent to which air quality standards and targets are met.

In April 1998, the Urban and Rural Networks were combined under a single Central Management and Coordination Unit, with Bureau Veritas (environmental consultants) being contracted to carry out day to day management and control of the combined Networks – the UK Automatic Urban and Rural Air Quality Monitoring Networks (AURN), which now consists of some 127 sites. Responsibilities include specifying the monitoring equipment, site housings, data handling and maintenance requirements; sites are operated mainly by trained staff within local authorities and data from the sites coordinated by NETCEN on behalf of Defra. NETCEN is responsible for quality assurance and quality control in respect of the AUN and the National Physical Laboratory for the rural and London networks. Air quality information is logged on a central computer system which gathers and analyses data from all the sites.

The AURN can be accessed at www.bv-aurnsiteinfo.co.uk.

Nitrogen dioxide is measured (using diffusion tubes) at approximately 1,200 urban background and roadside sites. In compliance with EU Directives, ground level concentrations of nitrogen dioxide and airborne lead are also monitored. Acid deposition and toxic organic micropollutants (dioxins, PCBs and PAHs) are also measured at a number of sites.

The Hydrocarbons Network, also funded by Defra, monitors 25 VOCs, including benzene and 1,3 butadiene. The National Physical Laboratory is the Central Management and Coordination Unit for this Network (as well as being responsible for quality assurance/quality control).

More information: www.airquality.co.uk

3.11.8 Research & Advice

There are a wide range of independent and government sponsored bodies carrying out research in all areas relating to air pollution and air quality.

The *Expert Panel on Air Quality Standards* (EPAQS), was set up in 1991 under the auspices of the Department of Environment (now Defra) to advise the Government on health-based air quality standards for the UK. More recently, EPAQS has looked at industrial emissions, providing guidance values for various pollutants (such as arsenic and chromium) regulated by the Environment Agency and which have been shown to affect human health. EPAQS is being dissolved and that part of its role of "providing independent UK-based expertise in specifying, interpreting and using air quality standards" subsumed in COMEAP.

The Committee on the Medical Effects of Air Pollutants (COMEAP) is an advisory committee of experts set up (under the Department of Health) in 1992 to advise the government on the effects of environmental factors on health. COMEAP also provides advice on the type of information which should be provided to the public during air pollution episodes. COMEAP has also drafted guidance (2000) on assessing the health impacts of industrial emissions to enable regulators to respond, in particular, to concerns from nearby residents. This outlines a procedure for assessing such impacts – including looking at statistics for illness, mortality, monitoring and modelling and studies to determine causes of clusters of illness. In 2005 COMEAP published guidelines for a number of indoor air pollutants (see 3.12 below).

In 2001 Defra and the Scottish Government established the Air Quality Expert Group (AQEG) to advise the Government on levels, sources and characteristics of air pollutants; to assess the extent of exceedences of the Air Quality Strategy objectives, proposed objectives, proposed and actual EU limit values; to analyse trends in pollutant concentrations; to assess current and future ambient concentrations of air pollutants in the UK; and to suggest potential priority areas for future research.

See Further Reading section (air pollution research and air pollution & health) of the *Pollution Control Handbook* for list of reports.

Defra and the Devolved Administrations also commission research from various independent consultancies.

3.11.9 Air Quality Information

The former Department of the Environment first established an air quality information service in 1990 to advise the public both on the quality of the air and on any precautions which should be taken during pollution episodes. This was replaced by a new system in 1997 with air pollution bands reflecting effects on health. Following consultation by the Department for

Environment, Food and Rural Affairs and the Devolved Administrations, a new banding system was launched in August 2001. The Air Pollution Bulletin system is now based on a numerical index (1-10) to enable the general public to interpret pollution warnings with more ease – see Table 3.6. When sulphur dioxide and nitrogen dioxide levels exceed the "alert thresholds" specified in the EU's Air Quality Directives (see 3.9.3 above), relevant information will be added to the bulletins.

Hourly-updated information on air quality in 16 regions and 16 urban areas, with regionalised 24 hour forecasts on levels of air pollution can be accessed on www.airquality.co.uk and is also available on freephone 0800 556677 and Teletext, page 155, with health advice on Teletext, page 169. The information is also sent to the media for inclusion in weather bulletins. Data from automatic monitoring networks and from non-automatic networks (the National Air Quality Archive), as well as data from the National Atmospheric Emissions Inventory can also be accessed from this site. Information on air quality in the Devolved Administrations is available at:

- Wales www.welshairquality.co.uk
- Northern Ireland www.airqualityni.co.uk
- Scotland: www.scottishairquality.co.uk

A new web-based Air Pollution Information System (APIS) – www.apis.ac.uk – supported by the UK's statutory environmental organisations, was launched in 2004. APIS is primarily aimed at government agency staff, for example in helping them evaluate local environmental impact assessments. It gives users a breakdown of the polluting agents and shows the impact of air pollution on habitats and species across the UK.

3.12 INDOOR AIR QUALITY

Most people spend around 90% of their time indoors and the increasing use of double glazing and other energy conservation measures will generally mean less natural ventilation and the build up of a range of potentially harmful pollutants. Indoor pollution can arise from a number of sources: heating, ventilation or cooking appliances can generate carbon monoxide, nitrogen dioxide and VOCs; the solvents used in many cleaning products, household sprays and some paints; tobacco smoke which gives off a range of harmful chemicals including formaldehyde, nicotine, tar, carbon monoxide and oxides of nitrogen; wood products such as chipboard and some glues contain formaldehyde. Naturally occurring uranium, present in certain rocks such as granite, produces radon, a naturally radioactive gas, which if it seeps upwards can build up to dangerously high concentrations in houses – see 3.30.1 and 3.30.4.

The presence of pollutants in the home environment – apart from radon – is however largely in the control of the householder and there are no regulations covering indoor air pollution; there are of course regulations and/or guidance relating to matters such as ventilation in homes (e.g. in areas where radon might be a problem) and the siting and servicing of appliances such as gas cookers and heaters; householders too should ensure adequate ventilation.

In 2004, the Committee on the Medical Effects of Air Pollutants (COMEAP) published *Guidance on the Effects on Health of Indoor Air Pollutants*; these provide guidance on how to minimise the production of air pollutants within the home, as well as guidelines – see Table 3.7. These are non-statutory guidelines intended to help manufacturers of products that

Table 3.6: Air Pollution Information Bands

Band/Index	Ozone 8 hourly or hourly mean* μgm^{-3}/ppb	Nitrogen Dioxide hourly mean μgm^{-3}/ppb	Sulphur Dioxide 15 minute mean μgm^{-3}/ppb	Carbon Monoxide 8 hour mean mgm^{-3}/ppm	PM_{10} Particles 24 hour mean μgm^{-3}
Low					
1.	0-32 (0-16)	0-95 (0-49)	0-88 (0-32)	0-3.8 (0.0-3.2)	0-16
2.	33-66 (17-32)	96-190 (50-99)	89-176 (33-66)	3.9-7.6 (3.3-6.6)	17-32
3.	67-99 (33-49)	191-286 (100-149)	177-265 (67-99)	7.7-11.5 (6.7-9.9)	33-49
Moderate					
4.	100-126 (50-62)	287-381 (150-199)	266-354 (100-132)	11.6-13.4 (10.0-11.5)	50-57
5.	127-152 (63-76)	382-476 (200-249)	355-442 (133-166)	13.5-15.4 (11.6-13.2)	58-66
6.	153-179 (77-89)	477-572 (250-299)	443-531 (167-199)	15.5-17.3 (13.3-14.9)	67-74
High					
7.	180-239 (90-119)	573-635 (300-332)	532-708 (200-266)	17.4-19.2 (15.0-16.5)	75-82
8.	240-299 (120-149)	636-700 (333-366)	709-886 (267-332)	19.3-21.2 (16.6-18.2)	83-91
9.	300-359 (150-179)	701-763 (367-399)	887-1063 (333-399)	21.3-23.1 (18.3-19.9)	92-99
Very High					
10.	360 (180) or more	764 (400) or more	1064 (400) or more	23.2 (20) or more	100 or more

* The maximum of the 8 hourly and hourly mean is used to calculate the index value.

Health Descriptor

Low:	Effects are unlikely to be noticed even by individuals who know they are sensitive to air pollutants.
Moderate:	Mild effects, unlikely to require action, may be noticed amongst sensitive individuals.
High:	Significant effects may be noticed by sensitive individuals and action to avoid or reduce these effects may be needed (e.g. reducing exposure by spending less time in polluted areas outdoors). Asthmatics will find that their 'reliever' inhaler is likely to reverse the effects on the lung.
Very High:	The effects on sensitive individuals described for "High" levels of pollution may worsen.

Source: www.airquality.co.uk

release pollutants into the domestic environment, architects and building engineers wishing to ensure adequate ventilation, and individuals concerned about air pollutants in their home; the guidelines have been set at a level at which the majority of people would not suffer "significant health effects".

Also of relevance to indoor air quality is the legislation banning smoking in all enclosed and substantially closed public places and work places, including in work vehicles used by more than one person. The ban came into force in Scotland on 26 March 2006, 1 July 2007 in England, 2 April 2007 in Wales and 30 April 2007 in Northern Ireland.

In the workplace, however, the *Control of Substances Hazardous to Health Regulations*, which are regularly updated, place strict controls on the use and storage of hazardous or toxic substances to ensure risks to health are prevented or minimised – see chapter 2, 2.11.

Table 3.7: Guideline Values for Indoor Air Pollutants

Pollutant	*Concentration*	*Averaging Time*
Nitrogen dioxide	150 ppb (300 µg/m³)	1 hour
	20 ppb (40 µg/m³)	Annual (provisional)
Carbon monoxide	90 ppm (100 mg/m³)	15 minutes
	50 ppm (60 mg/m³)	30 minutes
	25 ppm (30 mg/m³)	1 hour
	10 ppm (10 mg/m³)	8 hours
Formaldehyde	0.1 mg/m³ (0.1 ppm)	30 minutes
Benzene	1.6 ppb (5.0 µg/m³)	Annual
Benzo[a]pyrene	0.25 ng/m³	Annual (provisional)

DEPLETION OF OZONE LAYER

The ozone layer which forms part of the earth's stratosphere, lies about 15-20 km above the earth's surface. Ozone molecules absorb some of the sun's ultraviolet radiation, thus acting as a protective filter. Research shows that increased ultraviolet light on earth as a result of ozone depletion can increase the risk of skin cancer and eye cataracts, depress the human immune system, and harm aquatic systems and crops.

Concern over depletion of the stratospheric ozone layer, particularly over Antarctica, was first raised in the 1970s, with scientific evidence that chlorine based chemicals – such as chlorofluorocarbons – appeared to be a major cause of continuing depletion. More recently, preliminary results for research carried out by the European Arctic Stratospheric Ozone Experiment revealed that ozone amounts in the Northern Hemisphere were also falling, thus confirming the need for urgent measures if further damage is to be halted. The World Meteorological Organisation has reported (October 2006) that the 2006 hole in the Antarctic ozone layer is the worst on record "registering the largest depletion ever measured of the naturally occurring gas ..." NASA instruments showed that on 25 September 2006, the area of the hole reached 29.5 million km². The latest report from the Scientific Assessment Panel, published 2007, predicts that average and Arctic ozone levels are now expected to recover by 2050 and the Antarctic ozone hole between 2060-2075.

More information http://ozone.unep.org and www.wmo.int

3.13 MONTREAL PROTOCOL

A series of meetings held under the auspices of the UN Environment Programme culminated in the adoption in March 1985 of the *Vienna Convention for the Protection of the Ozone Layer*. While this called for cooperation on monitoring and research data, it was not until the signing in Montreal in September 1987 of the *Agreement on Substances that deplete the Ozone Layer* – the Montreal Protocol – that any measures for reducing use etc of ozone depleting substances were introduced. The 1987 Protocol, which came into force on 1 January 1989, has now been ratified by over 190 countries worldwide, although ratification of subsequent amendments is not so high. The international ozone regime now covers some 100 ozone-depleting chemicals, some of which also contribute to global warming.

The 1987 Protocol committed industrialised nations to reduce consumption of CFCs by 50% by 1999 and to freeze production of halons in 1992. In 1992 methyl bromide, hydrochlorofluorocarbons and hydrobromofluorocarbons were included in the Protocol, with further controls on the first two substances being agreed in 1995, and the timetable for methyl bromide being tightened further in 1997. Further restrictions on HCFCs and methyl bromide were agreed in December 1999 and bromochloromethane added to the list of restricted substances – these came into force in February 2002. The November 2003 meeting of Parties to the Montreal Protocol discussed revisions to the control timetable for the use of methyl bromide in developing countries, critical use nominations for methyl bromide and conditions for critical use exemptions in developed countries, which were finally agreed at a subsequent meeting in March 2004. At the 2007 meeting an accelerated programme for phasing-out HCFCs was agreed.

The Protocol, as amended, currently provides for the following:

- **Chlorofluorocarbons:** developed countries required to phase out by 1 January 1996. Developing countries to freeze production and consumption at 1995/7 average levels, 50% cut in 2005, phase out by 1 January 2010.
- **Halons:** developed countries required to phase out by 1 January 1994. Developing countries – as for CFCs.
- **Carbon Tetrachloride:** developed countries required to phase out by 1 January 1996. Developing countries – phase out by 1 January 2010.
- **1,1,1-trichloroethane:** based on 1989 levels, 50% reduction by 1 January 1994; phase out by 1 January 1996.
- **Methyl Bromide:** for developed countries, freeze on use at 1991 levels from 1 January 1995; 25% reduction by 1 January 1999, followed by 50% reduction by 1 January 2001, 70% by 1 January 2003 and phase out by 1 January 2005, but may request temporary exemptions from this deadline if there are no technical or economically feasible alternatives. Developing countries are to reduce use by

20% by 2005 using a baseline of the average of 1995-98 consumption and phase out by 2015. Quarantine and pre-shipment fumigations to be limited to 21 days before export, with obligation to report quantity used for such purposes; imports and exports to any country not a party to the 1994 amendments on methyl bromide are prohibited.

In March 2004, parties to the protocol agreed to cap production of methyl bromide at 30% of their 1991 level. Critical use exemptions for 2005 were agreed with the UK being allowed 129 tonnes.

- **Hydrochlorofluorocarbons:** at the September 2007 meeting of parties to the Protocol, developed countries agreed to reduce production and consumption by 75% by 2010, by 90% by 2015, 99.5%* phase-out by 2020 and total phase out by 2030. (*0.5% relates to servicing of refrigeration and air conditioning equipment existing between 2020-2030).

  Developing countries are to freeze production and consumption of HCFCs at average of 2009/2010 levels in 2013, reducing production and consumption by 10% in 2015, 35% by 2020, 67.5% by 2025, with final phase-out in 2030.

  For developed countries trade in HCFCs with countries not party to the Protocol has been banned since 1 January 2004; this ban takes effect in developing countries from 1 January 2016.

- **Hydrobromofluorocarbons:** developed and developing countries required to phase out by 1 January 1996.
- **Methyl chloroform:** developed countries required to phase out by 1 January 1996. Developing countries to phase out by 1 January 2015.
- **Bromochloromethane:** developed and developing countries required to phase out by 1 January 2002.

In ratifying the original Protocol, which came into force in August 1992, governments committed themselves to its requirements. The European Community has implemented the latest revisions to the Protocol through Regulation 2037/2000/EC – see next section.

In the case of the first four substances listed above, limited production for essential uses will be allowed after the phase out dates. An annual list of critical uses in developed countries is to be published by UNEP from 2004. Criteria to be taken into account in considering whether a use is essential include:

- the use must be necessary for the health and safety of society or critical for its functioning;
- unavailability of technically or economically feasible alternatives which are acceptable on health and safety grounds;
- insufficient recycled material, both in terms of quantity and quality, available.

In 1997, parties to the Protocol also agreed that with effect from 1 January 2000, imports and exports of new and recycled CFCs and halons, methyl bromide and HCFCs would require licensing, which it is hoped will curb the illegal trade in these substances. In February 2007 the Commission published a proposal for a Directive (COM(2007) 51) which will require Member States to make the unlawful trade in ozone depleting substances a criminal offence – see chapter 1, 1.15 for more details.

3.14 EU & UK REGULATIONS

The Council of Environment Ministers agreed on a Decision to ratify the Montreal Protocol in June 1988 and has implemented it, and subsequent amendments through Regulations, the latest being Regulation 2037/2000, which came into force on 1 October 2000. This repealed an earlier Regulation (3093/1994) to implement, and in some cases go further than, the latest revisions to the Montreal Protocol. DTI (now BERR) Guidance published on 25 October 2000 summarises the requirements of this Regulation. Regulation 1804/2003, adopted on 22 September 2003, amends 2037/2000 as regards the critical uses and export of halons, the export of products and equipment containing CFCs and controls on bromochloromethane.

Following a review of the 2000 Regulation, the Commission has published proposals for recasting and updating the Regulation to ensure compliance with the 2007 revisions to the Montreal Protocol, delete obsolete provisions, and ensure future challenges can be addressed.

The main provisions of COM(2008) 505, *Proposed Regulation on Substances that Deplete the Ozone Layer*, include

- bringing forward the date for phasing-out HCFCs to 2020;
- strengthening rules on the recovery and destruction of ozone depleting substances contained in products, and extending the list of products and equipment subject to the rules, and specifying the technology to be used.
- reducing the cap on the use of methyl bromide for fumigating crops to 200 tonnes (from 600 tonnes), with full phase-out proposed for 2015.
- strengthening provisions relating to enforcement of, and prevention of, the illegal trade of ozone depleting substances.

The current EU schedule, leading to the phase-out of the main ozone depleting chemicals, is as follows:

- **Chlorofluorocarbons (11, 12, 113, 114, 115):** production banned from 1 January 1995. Supply and use banned from 1 October 2000. Ban on use to maintain existing equipment from 1 January 2001; exemptions permitted for use in certain existing military applications until 31 December 2008.
- **Other Fully Halogenated CFCs (13, 111, 112, 211, 212, 213, 214, 215, 216, 217):** as for other CFCs.
- **Halons:** production banned from 1 January 1994. Halons 1211 and 1301 (used in firefighting) phased out by 31 December 2002, with total ban on use in firefighting equipment from 31 December 2003. The supply and use of all halons, including halon 1011 (bromochloromethane) banned from 1 October 2000.
- **Carbon Tetrachloride:** supply and use banned from 1 October 2000.
- **1,1,1-trichloroethane:** supply and use banned from 1 October 2000.
- **Methyl Bromide:** using 1991 as a base level, 60% reduction on production and consumption from 1 January 2001, 75% reduction from 1 January 2003, with prohibition on production and supply by 1 January 2005. The amounts do not include methyl bromide produced for quarantine and pre-shipment allocations; further exemptions may be provided from 1.1.05 for any uses agreed to be critical.

- **Hydrochlorofluorocarbons:** 2.8% cap on consumption, leading to phase out in 2020. Use of HCFCs in new air conditioning equipment and refrigeration equipment banned from 1 January 2001. Use of virgin HCFCs is banned in existing refrigeration and air conditioning equipment from 1 January 2010, and the use of recycled HCFCs from 1 January 2015; their use in solvents banned from 1 January 2002 and for polyurethane foams from 1 January 2003; the use of HCFCs in foam blowing banned from 1 January 2004. Exports to countries who have not signed the Montreal Protocol banned from 1 January 2004.
- **Hydrobromofluorocarbons:** production banned from 1 January 1996; supply and use banned from 1 October 2000.

While the production of most ozone depleting substances is generally prohibited, limited exceptions (to be updated annually, together with the amount that can be produced) are permitted where substitutes have not been found for "essential uses", e.g. medical uses, in particular metered dose inhalers for asthma and other lung diseases; in a communication published in November 1998, the Commission advised that essential use status for MDIs containing CFCs was to be phased out by 2003. Other essential uses include laboratory and analytical uses, and cardio-vascular surgical materials. "All practicable precautionary measures" have to be taken to ensure no leakages of ozone depleting substances from various commercial and industrial installations and manufacturing processes. A Commission Decision published each year allocates the quantities of controlled substances allowed for essential use under Regulation 2037/2000.

The Commission granted the UK a waiver to use certain CFCs for military uses until the end of 2004, and in existing navy vessels, submarines and aircraft until the end of 2008.

EU Regulation 2037/2000 also requires that from October 2000 all CFCs and HCFCs had to be recovered from industrial and commercial equipment before being recycled or disposed of, and from domestic refrigerators from 1 January 2002.

An export licensing system has been introduced, similar to that included in the Montreal Protocol, as a means of halting the illegal trade in CFCs etc; the import of products containing ozone depleting substances is banned (except for strictly controlled essential uses for which annual import quotas are specified in a Decision from the Commission) as is the export of ozone depleting substances and products containing them.

The *Environmental Protection (Controls on Ozone Depleting Substances) Regulations 2002* (SI 528) which came into force on 31 March 2002 implement the requirements of the 2000 EU Regulation; those sections of the Regulations relating to imports also cover Northern Ireland, with separate Regulations (SR 2003/97) implementing other requirements of the EU Regulation. 2008 Amendment Regulations (SI 2008/91), which came into force on 15 February 2008, extend the 2002 Regulations to offshore installations; they also give the Secretary of State powers to require disposal of substances, products and equipment imported, placed on the market or exported in contravention of the 2002 Regulations; the written prior authorisation of the local authority or port health authority is required for use of methyl bromide for fumigation for quarantine or pre-shipment.

The Ozone Depleting Substances (Qualifications) Regulations 2006 (SI 2006/1510, as amended by SI 2008/97), which came into force on 10 July 2006, relate to minimum qualifications for those working on the recovery, recycling, reclamation or destruction of controlled substances and the prevention and minimising of leakages. These are to revoked by new Regulations to take the newly proposed qualifications into account for working with ozone depleting substances. Defra has issued (August 2008) a series of guidance notes, both general and sector specific, on how the Regulations apply to manufacturers, sellers and users in different sectors – see 3.6.5 above. In Northern Ireland, SR 2006/321, which came into force on 31 August 2006, is similar to SI 2006/1510.

For EU and UK controls on the use of hydrofluorocarbons (HFCs), see earlier this chapter, 3.6.5.

CONTROL OF EMISSIONS TO AIR

3.15 TRANSBOUNDARY AIR POLLUTION

The Convention on Long-Range Transboundary Air Pollution was adopted in Geneva in 1979 and came into force in 1983; it was drawn up under the auspices of the UN Economic Commission for Europe which now comprises some 56 countries in Eastern and Western Europe and North America. There are now 51 parties to the Convention which, was initially drawn up as a result of concern – particularly from Norway and Sweden – of the damage to the environment caused by the long-range transport of certain pollutants (mainly sulphur dioxide and nitrogen oxides). The Convention calls on countries to "endeavour to limit and, as far as possible, gradually reduce and prevent air pollution, including long range transboundary air pollution". This should be achieved through the "use of best available technology that is economically feasible".

The Convention also deals with the long-range transport of nitrogen and chlorine compounds, polycyclic aromatic hydrocarbons, heavy metals and particles of various sizes.

For up to date information and background on the Convention and its protocols, visit www.acidrain.org, the website of The Swedish NGO Secretariat on Acid Rain or the Convention Secretariat at www.unece.org/env/lrtap

Critical Loads

Following a report to the Executive Body of the Convention in the early 1990s, it was agreed that reductions in emissions of both sulphur dioxide and nitrogen oxides should be negotiated "taking into account the best available scientific and technical developments . . . and internationally accepted critical loads". The "critical loads" approach takes account of the level of pollutant that a receptor – e.g. ecosystem, human being, plant or material – can tolerate without suffering long term adverse effect according to current knowledge.

Critical loads maps covering the whole of Europe are being drawn up by the European Evaluation and Monitoring Programme (EMEP) under the LRTAP Convention. In the UK the Department for Environment, Food and Rural Affairs (Defra) is responsible for preparing maps showing the levels of deposition at which soils in the UK are vulnerable to acidity. Other maps show actual estimates of deposition and where this is likely to exceed the critical load.

In the Second Edition of its *Air Quality Guidelines for Europe* (WHO, 2000), the World Health Organisation recommended the following critical levels and loads:

- *Sulphur dioxide:* critical level annual and winter means – 30 µg/m³ (agricultural crops); 15-20 µg/m³, depending on temperature (forests and natural vegetation). Lichens 10 µg/m³ annual mean; forests 1.0 µg/m³ (sulphate particulate) annual mean, where ground level cloud present 10% or more of time. Critical load 250-1500 eqS/ha/yr depending on the type of soil and ecosystem.
- *Nitrogen oxides:* critical level for NO_x (NO and NO_2, added in ppb and expressed as NO_2) 30 µg/m³ (annual); critical level for NH_3 8 µg/m³ (annual). Critical load 5-35 kgN/ha/yr depending on the type of soil and ecosystem.
- *Ozone:* critical level, (Accumulated exposure Over a Threshold of 40 ppb, AOT 40) – crops (yield) 3 ppm.h (3 months); crops (visible injury) 0.2 ppm.h (5 days, humid air), 0.5 ppm.h (5 days, dry air); forests 10 ppm.h (6 months); semi-natural vegetation 3 ppm.h (3 months).

3.15.1 Sulphur Protocols

The Helsinki Protocol adopted in 1985 came into force in 1987; it required signatories to reduce national sulphur emissions, or their transboundary fluxes by 30% on 1980 levels, by 1993. This Protocol was ratified by 23 countries, but not the UK or the European Community (although some individual Member States did do so). All signatories to the Protocol achieved reductions of more than 30%, as did a number of non-signatories including the UK with a reduction of 37% by the end of 1993 (45% by the end of 1994).

The second Sulphur Protocol, which was officially signed in Oslo in June 1994, and came into force on 5 August 1998, was negotiated using data based on critical loads assessments. This required countries to reduce, by the year 2000, their sulphur emissions to meet a UNECE-wide target of 60% of the gap between sulphur emissions and the critical load. Particularly sensitive areas of Scandinavia, Germany and The Netherlands where natural and unattributable emissions exceed the critical load were excluded from the calculations. Individual countries' target reductions are based on their contribution to acid deposition over the areas included in the calculations. To meet the 60% target, the UK agreed to reduce its own sulphur dioxide emissions by 50% by 2000, 70% by 2005 and 80% by 2010 (on 1980 levels). (To meet the target by the year 2000 would have required the UK to reduce its emissions by 79%, a target to which the UK would not agree.)

Parties to the Protocol had to submit their national strategies for meeting their targets to the UNECE monitoring committee within six months of ratification, which the UK Government did at the time of ratification. Briefly, this suggested that current emissions reduction and control programmes would ensure the 2000 and 2005 targets were met, with advances in technology helping to meet the 2010 target. In 2006 UK emissions of sulphur dioxide were 86% below 1980 levels. (www.defra.gov.uk/environment/statistics)

The Second Sulphur Protocol has been ratified by 28 countries, including the UK who did so on 17 December 1996, and the European Community (April 1998). A report to the LRTAP Executive Body in December 2006 estimated total releases of sulphur dioxide across Europe at less than 15 million tonnes in 2004 – a decrease of 65% from the 1990 figure.

3.15.2 Nitrogen Oxides Protocol

The 1988 Sofia Protocol, which came into effect in 1991, freezes nitrogen oxides emissions, or their transboundary fluxes, by 1994 using a 1987 baseline. As at June 2006, this Protocol had been ratified by 32 countries including the UK (1990) and the European Community (1993). In 1994 UK NO_x emissions were 11% lower than in 1987, and by the end of 1996 had fallen a further 10% (*Digest of Environmental Statistics*, No. 20, 1998). Between 1990-2005 UK NO_x emissions fell 45% (44% between 1990-2004). (www.defra.gov.uk/environment/statistics)

A report to the LRTAP Executive Body in December 2006 estimated total emissions of nitrogen oxides across Europe at less than 18 million tonnes in 2004 – a decrease of 30% from the 1990 figure.

3.15.3 "Multi-Effect" Protocol

The Protocol to Abate Acidification, Eutrophication and Ground Level Ozone ("multi-effect protocol") covering nitrogen oxides, sulphur dioxide, VOCs and ammonia was finalised in Gothenburg, Sweden in November 1999; it entered into force on 17 May 2005, 90 days after the 16th ratification. It has now been ratified by 24 states, including the European Community (June 2003), and the UK (December 2005).

Critical loads have been used as a basis for setting national emissions ceilings, to be met by 2010, based on 1990, to combat acidification, eutrophication and tropospheric ozone formation. Parties to the Protocol have signed up to a national emissions ceiling, which requires them to reduce total emissions of each of the pollutants to their ceiling level or below by the end of 2010. Emissions ceilings for the UK are

- sulphur dioxide – 625 kilotonnes/year
- nitrogen oxides – 1181 kilotonnes/year
- VOCs – 1200 kilotonnes/year
- ammonia – 297 kilotonnes/year

The Protocol also sets emission limits for both stationary and mobile sources (including combustion plant, electricity plant, dry cleaning, cars and lorries) and requires the use of best available techniques to meet limits. Parties to the Protocol "may apply different emission reduction strategies that achieve equivalent overall emission levels for all source categories together".

The requirements of the Protocol have largely been incorporated in *The National Emissions Ceilings Regulations 2003* (see 3.16.1 below), with compliance with other measures, including the Large Combustion Plants Directive (see below 3.16.2), the Solvents Directive (3.17), fuel quality Directives (3.26.9) and the IPPC regime, also contributing to achievement of the targets.

UK emissions of sulphur dioxide decreased 1.8% in 2006 to 0.68 million tonnes (2005: 0.69 million tonnes), with emissions of nitrogen oxides decreasing 1.5% to 1.59 million tonnes in 2006 (2005: 1.62 million tonnes). Emissions of non-methane volatile organic compounds have fallen 62% between 1990-2006, and from 0.96 million tonnes in 2005 to 0.91 million tonnes in 2006. Emissions of ammonia in 2006 remained much

the same as 2005, at 315 thousand tonnes – an 18% reduction on 1990 (383 thousand tonnes).
(www.defra.gov.uk/environment/statistics)

3.15.4 VOC Protocol

This Protocol, signed in Geneva in 1991, aims to control emissions of volatile organic compounds (VOCs) or their transboundary fluxes. It came into force on 29 September 1997, and has now been ratified by 23 countries, including the UK which did so in June 1994.

VOCs are defined as "all organic compounds of anthropogenic nature, other than methane, that are capable of producing photochemical oxidants by reactions with nitrogen oxides in the presence of sunlight". VOCs are involved in the formation of ground level ozone and in depletion of the ozone layer. They also contribute to the greenhouse effect in that methane and photochemical oxidants produced from the use of VOCs are both greenhouse gases. Thus they have both local and regional/transboundary effects.

The Protocol obliged most parties to secure a 30% overall reduction in their VOC emissions by 1999, using 1988 as a base (or another year between 1984 and 1990 to be specified when acceding to the Protocol). Other basic obligations required

- By September 1999 – national or international emission standards to be applied to new sources of VOCs, taking account of guidance on control technologies given in the Protocol, two years after the Protocol entered into force; and national or international measures to be applied to products that contained solvents; to promote the use of products with low or nil content; and to foster public participation in VOC emission control.
- By September 2002 – to apply economically feasible best available technologies to existing stationary sources in those areas where ozone standards are exceeded or transboundary fluxes originate; and for techniques to reduce VOC emissions from petrol distribution and motor vehicle refuellings and to reduce the volatility of petrol to be implemented.

In November 1993, the UK Government published its strategy for meeting its commitments under the Protocol, which it was hoped would result in a 36% reduction in emissions of VOCs by 1999. The strategy included setting emission limits in conditions attached to authorisations for both Part A and Part B processes regulated under the *Environmental Protection Act 1990*; implementation of EU measures to reduce evaporative emissions during petrol storage and distribution; information to the public on reducing usage – and thus emissions – of household products containing VOCs. Between 1988-1999 UK VOC emissions fell 41%; between 1990-2004 emissions fell by 57% to 1.0 million tonnes.
(www.defra.gov.uk/environment/statistics)

The Protocol was implemented in the EU by Directive 99/13/EC limiting emissions of VOCs due to the use of organic solvents in certain activities and installations (see 3.17), and by Directive 94/63/EC controlling VOC emissions during petrol storage and distribution (see 3.26.11). Both Directives have been transposed into UK legislation – see 2.4 & 2.5 and 3.27.4 respectively.

3.15.5 Heavy Metals

This Protocol, which aims to reduce airborne emissions of three transboundary pollutants, entered into force on 29 December 2003, and has now been ratified by 29 states, including the UK (6 July 2005).

The Protocol requires emissions of cadmium, lead and mercury to be reduced to below their 1990 levels (or an agreed year between 1985 and 1995), and for emission inventories to be maintained. The Protocol also aims to cut emissions from new and existing industrial sources and waste incinerators by laying down strict limit values based on the use of best available techniques.

In the UK annual emissions of all three pollutants in 1999 were well below their 1990 levels:

- cadmium – 6 tonnes (1990: 21 tonnes – 71% reduction)
- lead – 553 tonnes (1990: 2831 tonnes – 80% reduction)
- mercury – 9 tonnes (1990: 30 tonnes – 70% reduction)
  (UK National Atmospheric Emissions Inventory)

In March 2002 both Defra and the Scottish Government published consultation documents for ensuring the necessary strategies for compliance are in place to enable ratification (the Welsh and Northern Ireland Assemblies will need to consult on ratification). Most of the requirements of the Protocol are already met under existing legislation; Secretary of State Directions under s7(3) of *The Environmental Protection Act 1990* and under *The Pollution Prevention and Control Regulations 2000* will be necessary to ensure emission limit values for lead, mercury and general particulate matter specified in the Protocol are met (the Devolved Administrations will need to issue their own Directions).

3.15.6 Persistent Organic Pollutants

This Protocol entered into force on 23 October 2003 and has now been ratified by 29 states, including the UK (on 2 September 2005). Its aim is to phase out production and use of a defined list of substances as well as imposing requirements to eliminate discharges, emissions and losses, and to ensure safe disposal methods. The list covers 16 substances in three categories:

- pesticides (aldrin, dieldrin, endrin, chlordane, DDT, hexachlorocyclohexane, heptachlor, chlordecone, mirex and toxaphene);
- industrial chemicals (PCBs, hexachlorobenzene and hexabromobiphenyl);
- by-products and contaminants (dioxins, furans and PAHs).

Annexes to the Protocol set emission limit values, as well as outline best available techniques for the various industrial sectors which emit substances covered by the Protocol.

A UNEP Convention banning or, in some cases, restricting the production and use of POPs was agreed in Johannesburg in December 2000, and adopted and signed in Stockholm in May 2001. The Convention, which is global rather than just the member countries of the UNECE, came into force on 17 May 2004, 90 days after the 50th ratification; as at October 2006, it had been ratified by 133 parties. The Convention covers 12 substances known to be linked with cancer, birth defects and abnormalities in infants: hexachlorobenzene, endrin, mirex, toxaphene, chlordane, heptachlor, DDT, aldrin, dieldrin, polychlorobiphenyls, dioxins and furans. An exception is made for the use of DDT where it is used as an insecticide to combat malaria; some countries are allowed to continue using electrical transformers with PCBs until 2025. Technical and financial assistance will be available for developing countries to enable them to switch to substitute products.

At the first meeting of the parties to the Convention in May 2005, it was agreed to consider the inclusion of four more compounds in the Convention: penta-bromo diphenyl ether, hexabromobiphenyl, chlordecone and hexachlorocyclohexanes.

The European Community ratified the UNECE Protocol on 29 April 2004, and the Stockholm Convention on 18 November 2004. EU Regulation 850/2004 (as amended by Regulation 172/2007) transposing them both into Community law came into force on 20 May 2004; the Regulation, which is directly applicable in all Member States,

- bans the production, use and placing on the market of all POPs listed on both the Protocol and the Convention;
- requires holders of stockpiles to manage them in an environmentally sound manner, ensuring that the POP content is destroyed or irreversibly transformed;
- prohibits waste disposal or recovery operations that could lead to recycling or re-use of any of the banned substances;
- requires Member States and the Commission to establish appropriate programmes and mechanisms for the environmental monitoring of unintentionally produced POPs, including comprehensive release inventories for dioxins, furans, PCBs and PAHs and action plans to minimise total releases;
- allows Member States to grant a phase-out period for certain uses of HCHs until the end of 2007.

The European Commission has published (April 2007) a *Community Implementation Plan for the Stockholm Convention on POPs*; this envisages compliance with the Convention largely through better enforcement and amendment of existing legislation, including the IPPC, pesticides and biocides Directives, as well as waste legislation.

The UK's National Implementation Plan for the Stockholm Convention, published by Defra and the Devolved Administrations in April 2007, outlines the measures already in place, as well as the additional steps to be taken, to implement the Convention. It also includes the UK Dioxins Action Plan which summarises the measures taken to control unintentional emissions of dioxins, PCBs and HCB and human exposure via food, and further planned measures.

The Persistent Organic Pollutants Regulations 2007 (SI 2007/3106), which came into force on 3 December 2007, enforce the EU Regulation and designate the Environment Agency, SEPA and the DOE NI as both the competent and enforcing authorities for the purposes of the Regulation. As well as monitoring the use and management of stockpiles containing POPs, the Agencies will have powers of entry enabling them to investigate compliance with the Regulations and will be able to issue enforcement notices and to deal with requests for derogations in the way in which the waste is to be dealt with. It is an offence to contravene the EU Regulation punishable by a fine not exceeding the statutory maximum and/or three months' imprisonment on summary conviction and a fine and/or up to two years' imprisonment on conviction on indictment.

3.15.7 Evaluation and Monitoring Programme

This 1984 Protocol came into force on 28 January 1988 and has been ratified by 42 states, including the UK in 1985 and the European Community in 1986; it provides funding for the programme of monitoring and evaluation of the long-range transport of air pollution (EMEP). EMEP comprises some 100 monitoring stations in 25 countries which provide data on all Convention substances.

3.16 EU STRATEGY TO COMBAT ACIDIFICATION

There has been considerable concern about the effects of acid deposition on parts of the EU for many years and various measures aimed at reducing discharges of pollutants thought to contribute to the problem have been put in place. These provide a framework for reduction of industrial air pollution and propose specific emission limits for pollutants such as sulphur dioxide, nitrogen oxides and suspended particulates. The EU's long-term objective, as stated in the Fifth Environmental Action Programme is that there should be no exceedence of critical loads or levels for acidification anywhere in the EU. On the basis of measures in place prior to 1997, however, the Commission estimates that about nine million hectares of the Community will still exceed critical loads in 2010. In an effort to cut this deficit to 4.5 million hectares, the Commission published in March 1997 a *Communication on a Community Strategy to Combat Acidification* (COM(97) 88); the main elements of the Strategy are outlined below. As part of its Strategy, the European Community also ratified the UNECE *Sulphur Protocol* in April 1998 (Decision 98/686) (see also 3.15.1 above).

Implementation of both the *National Emissions Ceilings Directive* and the latest *Directive on Large Combustion Plant* should enable the EU to meet the emissions ceilings set in the UNECE *Protocol to Abate Acidification, Eutrophication and Ground Level Ozone* (see 3.15.3 above).

3.16.1 National Emission Ceilings

Directive 2001/81/EC setting *national emission ceilings for certain atmospheric pollutants* was adopted on 23 October 2001, and came into force on 27 November 2001; Member States had to have the necessary implementing measures in place by 27 November 2002.

The Directive sets national emission ceilings for four pollutants for each Member State, to be achieved by 2010; the UK targets (actual emissions for 2006, www.defra.gov.uk/environment/statistics, in brackets) are:

- Sulphur dioxide: 585 thousand tonnes (2006: 0.68 million tonnes)
- Nitrogen oxides: 1,167 thousand tonnes (2006: 1.59 million tonnes)
- VOCs: 1.20 million tonnes (2006: 0.91 million tonnes)
- Ammonia: 297 thousand tonnes (2006: approx 315 th. tonnes)

The national ceilings have been set with a view to meeting the following interim environmental objectives for the Community as a whole by 2010:

- Acidification: that the areas where critical loads are exceeded be reduced by at least 50% (in each grid cell), compared to 1990;

- Health related ground level ozone exposure: that the ozone load above the critical level for human health be reduced by two-thirds in all grid cells compared to 1990; and
- Vegetation related ground level ozone exposure; that the ozone load above the critical level for crops and semi-natural vegetation be reduced by one-third in all grid cells compared to 1990.

The Directive required Member States to draw up programmes by October 2002 for reducing emissions of the pollutants from all sources (including energy, transport, industry and agriculture) and to send these to the Commission; national programmes had to be reviewed, and where necessary revised, by 1 October 2006. Member States are required to submit annual emission inventories to the Commission by 31 December each year, as well as projections to 2010.

The Directive requires progress towards meeting national targets to be reviewed in 2004 and 2008; these should take account of progress towards meeting the following emissions ceilings for the Community as a whole which are aimed at meeting the interim environmental objectives by 2010:

- Sulphur dioxide: 3,634 kilotonnes
- Nitrogen oxides: 5,923 kilotonnes
- VOCs: 5,581 kilotonnes

The review should include an evaluation of the extent to which further emissions reductions might be necessary to meet the interim objectives and will also consider the measures needed to meet long-term objectives of no exceedence of critical loads and levels by 2020. The NECD is to be revised to reflect the objectives of the EU's *Thematic Strategy on Air Pollution* to reduce the number of air pollution related premature deaths by 2020.

Emissions from international maritime traffic and from aircraft beyond the landing and take-off cycles are excluded from the scope of the Directive. However the Commission was to report to the European Parliament on the contribution of these sources to acidification, eutrophication and the formation of ground-level ozone within the Community by the end of 2002 and 2004 respectively.

The National Emission Ceilings Regulations 2002 (SI 2002/3118), which came into force on 10 January 2003, require the Secretary of State to ensure that in 2010 and thereafter, emissions of SO_2, NO_x, VOCs and ammonia do not exceed the specified amount in the UK. The Secretary of State is also required to prepare a national programme for progressively reducing emissions of these pollutants, to be revised and updated as necessary by 1 October 2006 (a revised plan was published for consultation late 2006). The programme should be made available to the public and to the Regulators which should have regard to it; a *National Emissions Reduction Plan*, which includes proposals for trading of SO_2 and nitrous oxides, was submitted to the Commission in November 2003. Finally, the Secretary of State is required to prepare annual emissions inventories, with emissions projections to 2010, which should also be made available to the public and reported to the European Commission.

3.16.2 Large Combustion Plant

The first *Large Combustion Plant Directive* (88/609/EEC) committed Member States to specific reductions in emissions of sulphur dioxide and nitrogen oxides from large fossil fuel burning plant (50 MW or more) – mainly power stations. The UK agreed to reduce sulphur dioxide emissions from existing plant (i.e. licensed prior to 1 July 1987) in steps of 20, 40 and 60% of the 1980 baseline by 1993, 1998 and 2003 respectively, and reductions in nitrogen oxides emissions of 15% by 1993 and 30% by 1998. Emission limits for sulphur dioxide and nitrogen oxides for new plant were included in an annex to the Directive. In the UK emissions of sulphur dioxide from LCP were 60% below 1980 levels by 1998, 79% below in 2003 and 89% below in 2006; emissions of nitrogen oxides in 1998 were 60% lower than in 1980 and 59% below in 2006 (Defra environment statistics). During the period 1980-93, total EU emissions of SO_2 and NO_x reduced by 52.5% and 45.8% respectively. See also 3.15.1 & 2 for the UK's commitments under the sulphur and nitrogen oxides protocols to the UNECE Convention on Long-Range Transboundary Air Pollution.

The 1988 Directive did not cover sulphur dioxide emissions from new solid fuel plant between 50-100 MW as agreement could not be reached on the limit to be applied to these plant. However in December 1994 an amendment Directive (94/66) covering such plant was adopted; this came into force on 24 June 1995 and set an emission limit of 2000 mg/m^3 for such plant (i.e. those authorised to start after 1 July 1987).

Directive 2001/80/EC on the *limitation of emissions of certain pollutants into the air from large combustion plant* was adopted on 23 October 2001; it came into force on 27 November 2001, and had to be implemented by 27 November 2002, when the 1988 Directive was repealed. The Directive covers both new and existing LCP with a rated thermal input equal to, or greater than 50 MWth, irrespective of the fuel used – solid, liquid or gaseous. The Directive defines three categories of plant – with a group of boilers discharging through a single stack being counted as a single combustion plant:

- "New-new" plant which are subject to request for a licence on or after 27 November 2002; these must comply with the Directive and meet emission limit values (ELVs) from when they are brought into operation.
- "New" plant – licensed on or after 1 July 1987, but before 27 November 2002, or subject to a full licence request before 27 November 2002 and which came into operation before 27 November 2003; these must comply with the Directive and meet ELVs from 27 November 2002.
- "Existing" plant, licensed before 1 July 1987; for this category Member States may choose to meet required emissions reductions by 1 January 2008, either through ELVs or a national plan (to be communicated to the Commission by 27 November 2003) – see also chapter 2, 2.8.

Individual operators of existing plant may alternatively choose not to be included in the ELV or national plan approach and instead undertake to close down after 20,000 operational hours beginning 1 January 2008 and ending 31 December 2015. Operators had to inform the competent authority (i.e. the Regulators) of their decision before 30 June 2004 (see note below, under existing plant). The operator must submit annual reports to the competent authority showing the time used and unused time allowed for the rest of the plant's operational life.

As well as taking account of technical developments since the adoption of the 1988 Directive, one of the main objectives of the new Directive is to contribute to the goals of the acidification strategy through the setting of strict emission limit values and operating procedures. Emission limit values to be met by new plant are shown in Table 3.8. It has been implemented in the UK through Regulations and Directions under the *Pollution Prevention and Control Act 1999* (see chapter 2, 2.2.1).

The Directive required the Commission to report to the European Parliament and Council of Ministers by December 2004 on the need for further measures.

Table 3.8 : ELVs New Large Combustion Plant

Sulphur dioxide *(mg/Nm³)*			
Plant size (MW_{th}):	*50-100*	*100-300*	*>300*
Solid fuels*			
- general	850	200	200
- biomass	200	200	200
Liquid fuels*	850	400-200	200
Gaseous fuels			
- general		35	
- liquefied gas		5	
- low calorific gases from coke oven		400	
- low calorific gases from blast furnace		200	
* except gas turbines			
Nitrogen oxides *(mg/Nm³)*			
Plant size (MW_{th})	*50-100*	*100-300*	*>300*
Solid fuels			
- general	400	200	200
- biomass	400	300	200
Liquid fuels	400	200	200
Gaseous fuels			
- natural	150	150	100
- other	200	200	200
Gas turbines (> 50 MW_{th}): natural gas – 50; liquid fuels – 120; gaseous fuels – 120.			
Dust *(mg/Nm³) (with the exception of gas turbines)*			
Plant size (MW_{th})	*50-100*	*>100*	
Solid fuels	50	30	
Liquid fuels	50	30	
Gaseous fuels : as a rule 5; for blast furnaces 10; and for gases produced by the steel industry which can be used elsewhere 30.			

Existing Plant

Following lengthy negotiations, the Commission has now accepted (February 2006) the UK Government's "combined approach", which will allow operators of existing plant to,

- Opt out of the Directive and from 1 January 2008 to operate for a maximum of 20,000 hours, closing down by 31 December 2015; or
- Comply with the Directive through Emission Limit Values (ELV) for individual plant; or
- Be included in the national plan which sets an emissions "bubble" for 2008 for total emissions of SO_2 of 133,445 tonnes per year, 107,720 tonnes of NO_x per year and 9,659 tonnes per year of dust.

Nine power stations have chosen to close down by 31 December 2015, with 13 others choosing one of the other two options.

3.16.3 Sulphur Content of Liquid Fuels

A *Directive on the sulphur content of certain liquid fuels* (99/32) was formally adopted on 26 April 1999 and had to be implemented by 1 July 2000. It maintains the limit for the sulphur content of gas oil at 0.2% by weight (as set in Directive 93/12, see 3.26.9) as a minimum standard, reducing to 0.1% from 1 January 2008. The sulphur content of heavy fuel oils (used mainly in refineries, power stations and industry) has been limited to 1% by weight from 1 January 2003. Operators of combustion plant burning heavy fuel oil with a sulphur content of more than 1% may continue to do so as long as total SO_2 emissions do not exceed 1700 mg/m³. Member States which contribute very little (or nothing) to acidification have a derogation to use gas oils with a sulphur content of between 0.1-0.2%, and heavy fuel oils with a sulphur content up to 3% until 1 January 2013. Refineries may also continue to burn high sulphur fuels so long as total emissions do not exceed 1700 mg/m³ (existing plant) or 1000 mg/m³ (new plant).

An amendment to the 1999 Directive, 2005/33/EC, adopted on 12 July 2005, extends the Directive's requirements to limit the sulphur content of marine fuels used and marketed in the EU and used on board ships operating in Member States' waters; it also makes other minor amendments for the purpose of separating the land-based and marine provisions of the 1999 Directive – see this chapter, 3.29.2.

The 1999 and 2005 Directives have been implemented in England and Wales, and in Scotland and Northern Ireland through Regulations made under the *European Communities Act 1972.* As well as making it an offence to use heavy fuel oils and gas oils exceeding the sulphur limits in the Directive, the Regulations detail sampling requirements to confirm compliance – see this chapter, 3.20.

3.16.4 Sulphur Dioxide Control Areas

The Commission is negotiating the inclusion of "*sulphur dioxide control areas*" in the current revision of the IMO's MARPOL Convention controlling maritime pollution (see also 3.29); in such areas ships would be required to use marine bunker fuel oil with a sulphur content not exceeding 1.5%. Both the Baltic and North Seas have now been declared "SOx Emission Control Areas", although the reduced sulphur limit will not be enforced until countries together representing at least half the world's tonnage have ratified the decisions.

Directive on the Sulphur Content of Marine Fuels (2005/33/EC), which was formally adopted by the European Parliament and the Council on 12 July 2005, places a limit of 1.5% sulphur for marine fuels used by all seagoing vessels in the Baltic Sea from 11 August 2006 and in the North Sea and the Channel from autumn 2007.

This, and other requirements of the Directive (and of Annex VI to MARPOL) are to be implemented by the *Merchant Shipping (Prevention of Air Pollution from Ships) Regulations*, a draft of which was issued for consultation by the Marine and Coastguard Agency in July 2008 – see this chapter, 3.29.3.

3.17 Solvent Emissions

In March 1999 Environment Ministers adopted Council Directive 99/13 on the *limitation of emissions of volatile organic compounds due to the use of organic solvents in certain activities and installations*. The aim of the Directive, which will mainly apply to the paint coatings and pharmaceuticals industries, is to reduce emissions of VOCs by approximately 50% by 2010, based on 1990 levels, from various activities listed in an Annex to the Directive.

Annex III specifies annual thresholds for solvent consumption for the activities and emission limits. The Directive gives Member States the option of meeting the overall reduction target through the establishment of a national plan for existing installations: this would enable Member States to set their own reduction targets for sectors taking into account national patterns of emissions, while still meeting the overall objectives of the Directive – existing installations using solvents which are carcinogens, mutagens or reproductive toxicants cannot be included in a national plan. New installations were also excluded and are required to comply with emission limits for the sector concerned and to be registered with, or authorised by the appropriate authority upon implementation of the Directive by 1 April 2001; existing installations had until 31 October 2007 to comply. Installations requiring a permit under the IPPC Directive do not also require authorisation under this Directive although they must comply with its requirements. Member States who already have measures reducing VOC emissions but which are incompatible with the Directive will have until 2010 to comply.

The 1999 Directive has been implemented in both England and Wales, and in Scotland through Directions and Regulations under the *Pollution Prevention and Control Act 1999* and the *Environmental Protection Act 1990* – see chapter 2. In Northern Ireland, it has been implemented through *the Solvent Emissions Regulations (Northern Ireland) 2004* (SR 2004/36), which came into force on 27 February 2004.

Directive 2004/42/EC on the limitation of emissions of volatile organic compounds due to the use of organic solvents in certain paints and varnishes and vehicle refinishing products, and amending Directive 1999/13/EC, was adopted in April 2004; it came into force on 30 April 2004 and must be implemented by 30 October 2005. This Directive, which does not apply to products sold for exclusive use in installations covered by the 1999 Directive, places a limit on the amount of solvent per litre to between 50 and 750 grams (depending on the product) from 1 January 2007, with further reductions from 2010. The Directive also requires the Commission to report to the Council and European Parliament on the scope for extending the Directive to other products and whether further reductions in VOC limits are feasible. This Directive has been implemented through *the Volatile Organic Compounds in Paints, Varnishes and Vehicle Refinishing Products Regulations 2005* (SI 2005/2773), which came into force on 1 November 2005 and apply throughout the UK – see chapter 2, 2.9.2.

3.18 LOCAL AIR POLLUTION

Processes regulated for local air pollution control, in England, Scotland and Wales under *the Environmental Protection Act 1990* are now regulated under the pollution prevention and control regime (see chapter 2, 2.4 and 2.5). In Northern Ireland the transfer of Part B processes from regulation under *the Industrial Pollution Control Order 1997* to control under the PPC regime was due to be completed in April 2008 (see chapter 2, 2.3 and 2.5).

Much of the principal legislation for dealing with industrial smoke pollution is contained in the *Clean Air Act 1993* which came into force on 27 August 1993 and extends to England, Scotland and Wales only – 3.19 below. The provisions of this Act enabling local authorities to designate "Smoke Control Areas" also apply to residential areas; the content of motor fuel is regulated by Regulations made under s.30 of the Act (which also applies in N. Ireland). The *Clean Air Act 1993* is a consolidating Act and replaced the *Clean Air Acts 1956 and 1968*, Part IV of the *Control of Pollution Act 1974*, *the Control of Smoke Pollution Act 1989*, as well as various sections of a number of other Acts. The *Clean Air (Northern Ireland) Order 1981* provides similar controls for Northern Ireland.

3.19 CLEAN AIR ACT 1993

This Act covers England, Scotland and Wales, with s.30 (regulations on motor fuel) applying also to Northern Ireland. It came into force in August 1993 and consolidates the 1956 and 1968 *Clean Air Acts* (which were repealed); it also incorporates clean air legislation contained in other Acts such as the *Control of Pollution Act 1974*, the *Control of Smoke Pollution Act 1989* (which is also repealed) and the EPA. Regulations made under the earlier Acts still apply.

Similar controls in Northern Ireland are provided in the *Clean Air (Northern Ireland) Order 1981*.

It should be noted that Parts I, II and III of the Act do not apply to installations regulated under the *Pollution Prevention and Control Act 1999* from the date on which a permit has been granted, refused or refusal confirmed following an appeal (s.41, as amended by the *Pollution Prevention and Control (Scotland) Regulations 2000* or *the Environmental Permitting Regulations (England & Wales) 2007*); e.g small-scale incinerators or combustion plant burning waste materials.

Offences

Prosecutions for most offences under the *Clean Air Act 1993* are dealt with in the Magistrates' Court (Sheriff Court, Scotland); offences are subject to a fine on the standard scale as specified in the *Criminal Justice Act 1982/Criminal Procedure (Scotland) Act 1975*, both as amended; for offences under ss.10, 42 and 43, where conviction relates to an offence which is substantially a repetition or continuation of an offence for which the defendant has already been convicted, a cumulative fine of £50 per day may be substituted (s.50 of CAA).

3.19.1 Part I: Dark Smoke

Prohibition of Dark Smoke from Chimneys (s.1)

This section prohibits the emission of dark smoke from a chimney of any building; it also applies to chimneys not attached to a building serving furnaces of fixed boilers or industrial plant, and could include incinerators and crematoria. The Secretary of State may, by regulation, exempt prescribed lengths of emission from action under this section. Section 43 of the Act brings dark smoke emissions from railway engines within the scope of s.1.

There are four defences available in any proceedings for dark/black smoke emission. These are that the alleged emission was:

a) solely due to lighting a furnace from cold and all practicable steps had been taken to minimise emissions;

b) solely due to unavoidable mechanical failure of part of the plant, that this could not reasonably have been foreseen or if foreseen could not reasonably have been provided for and that the emission could not have been prevented after failure occurred;

c) solely due to unavoidable use of unsuitable fuel, suitable fuel not being available and the best available fuel being used; and all practical steps were taken to minimise the emission;

d) due to any combination of a, b and c.

It should be noted that these are not absolute defences and are available only if every practical effort is made to avoid and/or minimise emissions.

Prohibition of Dark Smoke from Industrial or Trade Premises (s.2)

Subject to certain exemptions, it is an offence to cause or permit the emission of dark smoke from industrial or trade premises (as distinct from chimneys). Unless the contrary is proved, an emission of dark smoke is deemed to have taken place if material is burned on those premises in circumstances where the burning would be likely to give rise to the emission of dark smoke. This can include night-time burning of cable and vehicles on open ground by removing the necessity for a local authority to prove by direct observation that dark smoke has been emitted.

"Industrial or trade premises" means premises normally used for industrial or trade purposes, or premises not normally so used, but which, at the time of the offence, were being used for industrial or trade burning, e.g. demolition sites. For the purposes of this section, land being used for commercial agriculture or horticulture constitutes trade premises (Code of Good Agricultural Practice for the Protection of Air, MAFF, 1998). Thus, for example, emitting dark smoke from the burning of tyres on a commercial farm would probably be an offence.

An offence under this section of the Act may result in a fine not exceeding £20,000 (*Environment Act 1995*, Sch. 22, para 195).

Meaning of "Dark Smoke" (s.3)

Where legal standards of emission are prescribed for smoke, they refer to "dark" and "black" smoke.

Dark smoke is defined by reference to a shade on the British Standard Ringelmann Chart (see Appendix 3.1) and means smoke which if compared ... with the Ringelmann Chart would appear to be as dark as, or darker than shade 2 on the chart.

Black smoke (defined in the *Dark Smoke (Permitted Periods) Regulations 1958* and in the *Dark Smoke (Permitted Periods) (Vessels) Regulations 1958)* means smoke which, if compared ... with the Ringelmann Chart, would appear to be as dark as, or darker than shade 4 on the chart.

Although legislation defines dark and black smoke by reference to colour shades on the Ringelmann Chart, the use of the Chart is not compulsory. The 1993 Act says that "for the avoidance of doubt, it is hereby declared that ... the court may be satisfied that smoke is or is not dark smoke ... [even if] there has been no actual comparison of the smoke with a chart of the type mentioned".

For an illustration and description of how to use the Ringelmann chart see Appendix 3.1.

3.19.2 Part II: Smoke, Grit, Dust, and Fumes

Installation of Furnaces (s.4)

Before installing a furnace (except a domestic furnace) in a building or fixed boiler, the local authority must be informed; any such furnace must be capable of being operated continuously without emitting smoke when burning fuel of a type for which the furnace was designed.

There is no definition of "furnace", but a practical interpretation of this word whenever it appears in clean air legislation is usually taken as "any enclosed or partly enclosed

Table 3.9: The Dark Smoke (Permitted Periods) Regulations 1958 (SI 498)

These Regulations were made under s.1(2) of the 1956 Act. Similar 1958 Regulations apply in Scotland and 1965 Regulations in Northern Ireland.

Specified permitted periods of emission are as follows:

a) Aggregate emissions of dark smoke

Number of furnaces served by the chimney	Permitted emission of dark smoke in any period of eight hours	
	If not soot blowing during period	If soot blowing during period
One	10 mins	14 mins
Two	18 mins	25 mins
Three	24 mins	34 mins
Four or more	29 mins	41 mins

Where a single boiler or unit of industrial plant is fired by more than one furnace discharging into the same chimney, those furnaces shall be deemed to be one furnace.

b) Continuous emission of dark smoke

The continuous emission of dark smoke in excess of four minutes, caused otherwise than by soot blowing, is prohibited.

c) Aggregate emission of black smoke

No emission of black smoke exceeding two minutes aggregate in any period of thirty minutes is allowed. There are thus three standards of emission which are acceptable in relation to any one chimney.

(Soot blowing is a method of cleaning deposited carbon from the internal surfaces of large industrial boilers. It includes the use of a jet of steam onto heat exchange surfaces on a regular, usually daily basis.)

space in which liquid, solid or gaseous matter is burned, or in which heat is produced". Domestic furnaces are defined as those with a maximum heating capacity of less than 16.12 kilowatts.

Limits on Rate of Emission of Grit and Dust (s.5)

The Secretary of State may by regulation prescribe limits on the rates of emission of grit and dust from the chimneys of all furnaces (except domestic furnaces); it is a defence to prove that the best practicable means were used to prevent any emission in excess of the permitted rates.

The Clean Air Act (Emission of Grit and Dust from Furnaces) Regulations 1971 (SI 162)

These Regulations (which do not cover N. Ireland) were made under s.2(1) of the 1968 Act and apply to:

a) boilers;
b) indirect heating appliances (being heating appliances in which the combustion gases are not in contact with the material being heated);
c) furnaces in which the combustion gases are in contact with the material being heated but that material does not in itself contribute to the grit and dust in the combustion gases.

The Regulations do not apply to incinerators burning refuse or waste matter (solid or liquid) whether or not the resulting heat is used for any purpose; incinerators are regulated for PPC – see chapter 2. Schedules 1 and 2 of the Regulations tabulate the quantities of grit and dust which may be emitted, and where the rating of the boiler or furnace is intermediate between the tabulated values, the minimum permitted emission is obtained by interpolation (see Appendix 3.2).

Arrestment Plant for Furnaces (ss.6-9)

Section 6 requires all furnaces (except domestic furnaces) to be equipped with grit and dust arrestment plant approved by the local authority and the arrestors to be properly maintained and used if used for the following purposes:

a) to burn pulverised fuel; or
b) to burn, at a rate of 45.4 kg or more an hour, any other solid matter; or
c) to burn, at a rate equivalent to 366.4 kilowatts or more, any liquid or gaseous matter.

The Secretary of State may by regulation prescribe limits on the rates of emission of grit and dust from the chimneys of furnaces, and prescribe different limits for different cases and according to different circumstances. Such regulations would not however apply to furnaces begun, or an agreement for the purchase or installation entered into, before the implementation of the regulations.

Section 7 of the 1993 Act specifies under what circumstances exemptions from s.6 will be permitted:

a) The Secretary of State has power to provide in Regulations that furnaces of any particular class, whilst used for a prescribed purpose, should be exempted. Here the *Clean Air (Arrestment Plant) (Exemption) Regulations 1969* are relevant (see below).

b) The local authority may, on application, exempt a specific furnace installation providing that it is satisfied that the emissions will not be prejudicial to health or a nuisance. The local authority must give a written decision within eight weeks, and an aggrieved applicant may appeal within a further 28 days, to the Secretary of State.

Table 3.10: Clean Air (Emission of Dark Smoke) (Exemption) Regulations 1969 (SI 1263)

These Regulations (made under s.1(3) of the 1968 Act) apply to England and Wales. They exempt the emission of dark smoke caused by the burning of certain materials subject to specified conditions. The corresponding Regulations for Scotland are the Clean Air (Emission of Dark Smoke) (Exemption) (Scotland) Regulations 1969. In Northern Ireland, the Clean Air (Emission of Dark Smoke) (Exemption) Regulations 1981 apply.

Exempted Matter	Conditions*
1. Timber and any other waste matter (other than natural or synthetic rubber or flock or feathers) which results from the demolition of a building or clearance of a site in connection with any building operation or work of engineering construction (within the meaning of section 176 of the Factories Act 1961).	A, B and C
2. Explosive (within the meaning of the Explosives Act 1975) which has become waste; and matter which has been contaminated by such explosive.	A and C
3. Matter which is burnt in connection with: a) research into the cause or control of fire or b) training in fire fighting.	C
4. Tar, pitch, asphalt and other matter which is burnt in connection with the preparation and laying of any surface, or which is burnt off any surface in connection with re-surfacing, together with any fuel used for any such purpose.	C
5. Carcasses of animals or poultry which a) have died, or are reasonably believed to have died, because of disease; b) have been slaughtered because of disease; or c) have been required to be slaughtered pursuant to the Diseases of Animals Act 1950	A and C unless the burning is carried out by or on behalf of an inspector (within the meaning of section 84 of the Diseases of Animals Act 1950)
6. Containers which are contaminated by any pesticide or by any toxic substances used for veterinary or agricultural purposes; and in this paragraph "container" includes any sack, box or receptacle of any kind. (Scotland; revoked by SI 2007/1156 in England & Wales and SR 2008/18 in NI.)	A, B and C

* Condition A	That there is no other reasonable safe and practicable method of disposing of the matter.
* Condition B	That the burning is carried out in such a manner as to minimise the emission of dark smoke.
* Condition C	That the burning is carried out under the direct and continuous supervision of the occupier of the premises concerned or a person authorised to act on his behalf.

An offence occurs if a furnace is exempt and is then used for any purpose other than that prescribed in the exemption.

When considering an application for approval of a proposed arrestor plant (or an application for exemption from the need to install an arrestor plant) the local authority will have regard to the probable grit and dust burden in the flue gases, and to the known efficiency of the proposed arrestor plant so that it can compare the probable emissions with any standard which might at the time be in force or which is under consideration. It should be noted that arrestor plant, unless very well maintained at all times, rarely operate at full design efficiency after a few years in use.

Measurement of Grit, Dust and Fumes (ss.10-11)

Grit is defined in the *Clean Air (Emission of Grit and Dust from Furnaces) Regulations 1971* as particles exceeding 76 µm in diameter; the *Clean Air Act 1993* (s.64) defines fumes as any airborne solid matter smaller than dust. British Standard BS 3405 defines dust as small solid particles between 1-75 µm in diameter.

Under s.10 if a furnace (or range of furnaces served by one chimney) is used to burn:

a) pulverised fuel,
b) any other solid matter at a rate of 45.4 kg, or more per hour, or
c) any liquid or gaseous matter, at a rate equivalent to 366.4 kilowatts or more,

the local authority may serve a notice on the owner of the plant or occupier of the building, requiring grit, dust and fume emissions to be measured from time to time, to adapt the chimney for that purpose, to provide and maintain the necessary measuring apparatus, and to inform the local authority of measuring results. The occupier of the building is under a duty to permit the local authority to be present during the measuring and recording of emissions. The local authority may revoke its notice at any time.

If the plant burns at a rate of:

a) less than 1.02 tonnes/hour of solid matter other than pulverised fuel, or
b) less than 8.21 Megawatts of liquid or gaseous matter,

the industrialist may serve a notice on the local authority requiring the authority from time to time to measure and record emissions at its own expense until such notice is withdrawn in writing.

The Clean Air (Measurement of Grit and Dust from Furnaces) Regulations 1971 (SI 161)

These Regulations were made under s.7(2) of the 1956 Act and prescribe the administrative process to invoke the above provisions (similar Regulations apply in Scotland):

a) A local authority must give at least six weeks' notice in writing to an industrialist requiring:
   - adaptations to be made to the chimney or flues; and
   - the provision of the necessary equipment, to enable measurements to be made in accordance with BS 3405 (Simplified) Method for Measurement of Particulate Emission Including Grit and Dust. The British Standard states the principles to adopt, sets out requirements to be met in designing apparatus and indicates basic sampling procedures.
b) When the sampling points have been installed and the equipment provided (or it is known that the firm are engaging a consultant who possesses the equipment), the local authority must give at least 28 days' notice in writing, requiring the tests to be carried out in accordance with the methods detailed in *Measurements of Solids in Flue Gases* (published by the Institute of Energy).
c) Before making the measurement the industrialist must give the local authority at least 48 hours' notice in writing of the date and time of the commencement of the tests.

Table 3.11: Clean Air (Arrestment Plant) (Exemption) Regulations 1969 (SI 1262)

These Regulations were made under s.4(1) of the 1968 Act.

Class of Furnace

1. Mobile or transportable furnaces.

Purpose

a) providing a temporary source of heat or power during any building operation or work of engineering construction (within the meaning of section 176 of the Factories Act 1961);
b) providing a temporary source of heat or power for investigation or research;
c) providing heat or power for the purpose of agriculture (within the meaning of section 109(3) of the Agriculture Act 1947).

Class of Furnace

2. Furnaces other than furnaces designed to burn solid matter at a rate of 1.02 tonnes an hour or more, which fall within any of the following descriptions and in which the matter being heated does not contribute to the emission of grit and dust:

a) furnaces burning liquid matter, gas, or liquid matter and gas;
b) hand-fired sectional furnaces designed to burn solid matter at a rate of not more than 11.3 kg an hour for each square foot of grate surface;
c) magazine type gravity-fed furnaces designed to burn solid matter at a rate of not more than 11.3 kg an hour for each square foot of grate surface;
c) furnaces fitted with an underfeed stoker designed to burn solid matter at a rate of not more than 11.3 kg an hour for each square foot of the plan area of the combustion chamber;
e) furnaces fitted with a chain grate stoker designed to burn solid matter at a rate of not more than 11.3 kg an hour for each square foot of the grate surface;
f) furnaces fitted with a coking stoker designed to burn solid matter at a rate of not more than 11.3 kg an hour for each square foot of area covered by the fire bars excluding the solid coking plate.

Purpose

Any purpose except the incineration of refuse.

d) The result of the test must be sent to the local authority within 14 days from the making of the measurements and the reports shall include:
   - the date(s) of the test(s);
   - the number of furnaces discharging into the chimney on that date;
   - the results of the measurements expressed in lbs/hour of grit and dust emitted and, in the case of solid fuel fired plant, the percentage of grit in the solids.

A notice may require the making of measurements from time to time or at stated intervals, but not at intervals of less than three months unless, in the opinion of the local authority, the true level of emission cannot be determined without further measurement being made.

If an industrialist serves a "counter notice" on the local authority, he need only make adaptations to the chimney to enable the measurements to be made, including the provision of scaffolding where necessary, and the provision of facilities such as electrical connections to enable the measurements to be made. The local authority must provide the sampling equipment and conduct the test entirely at its own expense.

Information about Furnaces and Fuel Consumed (s.12)

Local authorities may serve a notice on the occupier requesting information about furnaces in a building and the fuel or waste being burned to be given to them within 14 days or longer.

Outdoor Furnaces (s.13)

This section notes that ss.5-12 apply to boilers or industrial plant furnaces attached to buildings or installed on land.

Height of Chimneys (ss.14-16)

At whatever height smoke and flue gases are discharged, gravity will eventually bring the larger particles of grit, dust and soot to the ground. Additionally, because of the natural turbulence of the atmosphere, a proportion of the gases and of the freely suspended fine particles will reach the ground although not affected by gravity. The higher the point of discharge and the greater the total heat content of the discharged gases, the more widespread and diluted will be the fine particles and gases by the time they reach ground level. The control of chimney heights enables local authorities to take into account a number of relevant factors in determining the height of a chimney. These include the need to avoid downdraught or downwash created by the chimney itself, or by buildings or topographical features; to avoid the ground level concentration of combustion products becoming prejudicial to health or a nuisance; in the case of smaller units to prevent the flue gases from entering nearby buildings in too high a concentration.

Under s.14 of the Act, unless the local authority has approved the height of the chimney and any conditions attached to approval adhered to, it is an offence to cause or knowingly permit a furnace to be used to

- burn pulverised fuel;
- burn at a rate of 45.4 kg or more an hour any other solid matter; or
- burn at a rate equivalent to 366.4 kilowatts or more any liquid or gaseous matter.

An application for chimney height approval must contain adequate information to enable the necessary calculations to be carried out. Additional information on the calculation of chimney heights is given in Appendix 3.3. Supplementary technical assistance for estimating the minimum permissible chimney height for small boilers emitting sulphur dioxide was also prepared by Stanger Science and Environment at the request of the Department for Environment, and was published in *Clean Air*, Vol. 31, Autumn 2001, pp89-95; email admin@environmental-protection.org.uk to request a copy.

The local authority must consider an application for approval for a chimney height for a furnace and give a written decision within 28 days of its receipt, unless it is agreed in writing between the local authority and the applicant that a longer period may be allowed. Should the local authority fail to deal with the application within the agreed period, then approval without qualification will be deemed to have been given.

The local authority, however, must not approve the proposed chimney height unless it is satisfied that it will be sufficient to prevent, so far as is practicable, the smoke, grit, dust, gases or fumes emitted from the chimney from becoming prejudicial to health or a nuisance, having regard to:

a) the purpose of the chimney;
b) the position and descriptions of buildings near to it;
c) the levels of the neighbouring ground;
d) any other matters requiring consideration in the circumstances.

Any approval of the height of a chimney may be granted unconditionally or subject to conditions as to the rate and/or quality of emissions from the chimney. "Rate of emissions" is defined as the quantities of any specified substance which may be emitted in a period specified in the conditions attached to the approval, e.g. kg/hr of sulphur dioxide. Many other miscellaneous emissions may, of necessity, be considered for inclusion in conditions as to the rate of emission, but there will be cases where a limitation on "rate" alone will not sufficiently ensure against excessive pollution by materials other than normal combustion products; in those cases it might be appropriate to impose conditions as to the "quality" (i.e. concentration) of the emission. A limitation on the quality of emission would be appropriate if the products of combustion contain gases or acid mists resulting from abnormal reactions or breakdown of materials derived from chlorinated or sulphur bearing compounds etc, or if products of combustion are mixed with waste gases from noncombustion processes.

If the local authority decides not to approve the height of a chimney or to attach conditions to its approval, it must give the applicant a written notification of its decision, stating its reasons; it must additionally specify the lowest height, if any, which it is prepared to approve unconditionally and/or the lowest height it is prepared to approve with any specified conditions. The applicant may appeal against the decision to the Secretary of State within 28 days. The Secretary of State may confirm the local authority's decision, or may amend the height or conditions if this is thought to be appropriate. The Secretary of State must also give a written notification of the decision stating the reasons and, where a chimney height is not approved, specifying what height, with and/or without conditions would be approved.

Section 16 applies where plans have been submitted in accordance with building regulations to the local authority to erect or extend a building outside Greater London or in an outer London borough, and the plans include a chimney outside the scope of s.15 above (i.e. one not serving a furnace). It does not cover residences, shops or offices. Again, the local authority must reject the plans if it is thought that the proposed height of chimney will not be sufficient, so far as practicable, to prevent emissions of smoke, grit, dust or gases from becoming prejudicial to health or a nuisance. The factors to be taken into account are the same as listed above. There is a right of appeal against the local authority's decision to the Secretary of State.

The Secretary of State may by regulation exempt boiler or plant to be used for certain prescribed purposes from the above requirements so long as the chimney height has been approved for that purpose.

Clean Air (Heights of Chimneys) (Exemption) Regulations 1969 (SI 411)

These Regulations were made under s.6(1) of the 1968 Act. It is not necessary to apply for chimney height approval for any exempted boiler or plant specified in these Regulations. Such boiler or plant include those used:

a) as a temporary replacement for a boiler or plant which is under inspection or being maintained, repaired, rebuilt or replaced by another permanent boiler or plant;
b) as a temporary source of heat or power for building or engineering construction work;
c) as a temporary source of heat or power for investigation or research;
d) as an auxiliary plant used to bring other plant to an operating temperature;
e) as a mobile or transportable source of heat or power for agricultural operations.

Two other pieces of legislation are also of relevance where approval of chimney height is needed. These are

- **Building Act 1984:** Section 73 of this Act applies when a new building is erected which will over-reach the chimney of an adjoining building. Providing that the chimney of the lower building is in the party wall between it and the taller building, or is six feet or less from the taller building, the local authority may by notice require the person who erects the taller building to raise the height of the chimneys of the adjoining building if it is reasonably practicable to do so, so that the chimneys are the same height as the taller building or its chimneys, whichever is the higher. The notice will also require the owner or occupier of the adjoining building to allow the work to be done, but he may elect to do the work himself and charge the reasonable cost of so doing to the person on whom the notice was served. In the event of non-compliance with the notice, the local authority may do the work in default.

- **The Building Regulations 2000:** The *Clean Air Act 1993* controls chimney heights only if the fuel consumption exceeds 45.4 kg or 366.4 kilowatts or more per hour. *The Building Regulations 2000* (SI 2531, effective 1 January 2001) consolidate and replace the 1991 Regulations and subsequent revisions; they have been amended by 2001 Regulations which came into force on 1 April 2002. The Regulations set standards for various aspects of building work, including use of appropriate materials, when building plans are required. Schedule 1 to the Regulations outlines standards to be met for such aspects as fire protection, insulation, water and drainage services, heat producing appliances etc. Standards for this latter category are detailed in Part J of Schedule 1 (of the 2001 Regulations); this covers fixed heat producing appliances designed to burn solid fuel, oil or gas, or are incinerators, as well as chimneys, irrespective of the capacity of the fireplace, furnace or type of building. Such appliances must have adequate provision for the discharge of the products of combustion to outside air, and be so constructed as to reduce to a reasonable level the risk of the building catching fire.

The relevant Regulations in Scotland are the *Building Standards (Scotland) Regulations 1990*. Part F deals with heat producing installations and storage of liquid and gaseous fuels.

Smoke Nuisances in Scotland (s.17)

This section was repealed on 1 April 1996 following implementation of s.107 and Schedule 17 of the *Environment Act 1995*. It enabled action to abate certain nuisances from smoke to be taken under the *Public Health (Scotland) Act 1897*.

3.19.3 Part III: Smoke Control

Creation of Smoke Control Areas (ss.18-19)

Section 18 (and Sch. 1) allows a local authority to declare the whole or any part of its district a smoke control area by making a "Smoke Control Order". Before doing so it must publicise its intention, and the effect of the Order, in the *London Gazette* (or *Edinburgh Gazette*) and a local newspaper on two successive weeks. Unless postponed through resolution, the Order must come into effect not later than six months after it was made. The Order may specify to what area the Order applies, the types of buildings covered, and any exemptions (e.g. specified buildings or fireplaces).

Section 19 enables the Secretary of State (or in Scotland, the Scottish Environment Protection Agency, *Environment Act 1995*, Sch. 22, para 196) to direct a local authority to bring forward, within six months, proposals for making a Smoke Control Order for the purposes of abating smoke pollution in its area.

Local authorities may themselves subsequently revoke or vary Orders; Smoke Control Orders originally confirmed by the Secretary of State (Scotland, SEPA) can only be revoked or varied by a similarly confirmed Order.

Full details of the procedure for applying for a fuel to be authorised or for an appliance to be exempted, as well as lists of authorised fuels and exempted fireplaces can be found on www.uksmokecontrolareas.co.uk

Prohibition on Emission of Smoke in Smoke Control Area (ss.20-22)

When operative, it is an offence for an occupier of premises within a smoke control area to allow smoke emission from a chimney, unless the smoke is caused by the use of an "authorised fuel" (listed in Regulations – see Appendix 3.4a). It is recognised that there may be times when certain of these fuels may cause short periods of light smoke emission.

Appliances that may be used in a smoke control area are defined in *Smoke Control (Exempted Fireplaces) Orders*. These conditionally exempt the emission of smoke from chimneys serving the various fireplaces (see Appendix 3.4b). It should be noted that some appliances, exempted under clean air legislation, e.g. small incinerators and boilers used in industrial and commercial premises, may now be subject to PPC as a result of regulations implementing the *Waste Incineration Directive*, and thus require a permit.

The implications of the law are that coal, oil and wood cannot be used as fuel in a smoke control area unless burnt on an exempted fireplace (unless they can be burnt without any emission of smoke). Refuse cannot be burnt in an incinerator in a smoke control area unless there is a specific exemption written into the Order, a general exemption for incinerators, or unless smokeless combustion can always be achieved. This prohibition would even apply to the simple incinerator consisting of a dustbin with a chimney.

The local authority may also include in the Smoke Control Order other exemptions, which it may feel are necessary, subject to reasonable conditions. One of these is to allow the lighting of domestic fires with sticks and paper if there is no gas in the dwelling.

Dealings with Unauthorised Fuel (s.23)

It is an offence to acquire for use or to sell by retail for delivery in a smoke control area any fuel other than an authorised fuel, unless in each case the premises or fireplace are exempt. It is a defence to prove that the sale for delivery of unauthorised fuel was carried out in the belief that a Smoke Control Order did not cover either the building or the fireplace/boiler or plant.

In Northern Ireland, the sale or delivery of unauthorised fuels in smoke control areas is prohibited by *The Smoke Control Areas (Sale or Delivery of Unauthorised Fuel) Regulations (Northern Ireland) 1998* (SR 328). *The Sulphur Content of Solid Fuel Regulations (Northern Ireland)* 1998 (SR 329) prohibit the sale or delivery within Northern Ireland of solid fuel with a sulphur content exceeding 2%.

Adaptation of Fireplaces (ss.24-29)

The local authority may serve a notice on the owner of a private dwelling in a smoke control area requiring work to be carried out to ensure compliance with a Smoke Control Order. So long as the work has been carried out to the local authority's satisfaction, it should repay seven-tenths of the reasonable cost of the adaptation, and it may repay all or any part of the remaining three-tenths. Grants are not available for "new dwellings" begun on or after 16 August 1964.

A local authority may designate as unsuitable any appliances for which suitable fuels are not fully available, and in those circumstances a grant is not available for installation of those appliances. The principles which local authorities must follow in assessing eligibility of works of adaptation for grant are contained in the Memorandum on Smoke Control Areas, as revised. The Department of Environment (now Department for Environment, Food and Rural Affairs – Defra) also sets cost limits for local authority guidance in assessing "reasonable costs".

The Act enabled the Secretary of State to repay to the local authority a proportion of their expenditure in implementing a Smoke Control Order. Exchequer contributions for smoke control orders in England and Wales made after 31 March 1995 are not normally available. Claims for payments could be made up to 31 December 1996 for payment by 31 March 1997. In Scotland such grants were no longer available from 1 April 1993; applications for approval of schemes had to be submitted to the Scottish Office Environment Department by 31 March 1993 and contributions for all approved schemes claimed by 28 February 1995.

In Northern Ireland, grant-aid is apportioned as follows: DOE(NI) – 40% of approved costs; the council – 30%; and the householder/occupier – 30%.

3.19.4 Part IV: Control of Certain Forms of Air Pollution

Regulations about Motor Fuel (s.30)

This section enables the Secretary of State to make regulations limiting the composition and content of motor fuel and preventing or restricting the production, treatment, distribution, import, sale or use of any fuel which fails to comply with the requirements and which is for use in the United Kingdom. Before making any regulations the Secretary of State must consult representatives of manufacturers, users of vehicles, producers and users of fuel for motor vehicles as well as air pollution experts. *The Motor Fuel (Composition and Content) Regulations 1999* (SI 3107), as amended, are summarised below at 3.27.1.

Regulations about the Sulphur Content of Oil Fuel for Furnaces or Engines (s.31)

The Secretary of State is empowered to make regulations limiting the sulphur content of a wide range of fuel oils used in industry, commerce and in domestic boilers. The Secretary of State has a duty to consult similar to that outlined for s.30 above. 1994 Regulations on the marketing of gas oil (SI2249) were revoked in 2000 (NI in 2002) and replaced by regulations on the sulphur content of liquid fuels – see below 3.20.

Supplementary Provisions (s.32)

This section notes that anyone contravening or failing to comply with the provisions of Regulations made under ss.30 and 31 shall be guilty of an offence and liable to a fine, subject to any exclusion for liability included in the Regulations. Regulations made under ss.30 and 31 apply to fuel used in the public service of the Crown.

Cable Burning (s.33)

Cable burning for the purposes of recovering the metal is illegal unless the burning is carried out as part of a process subject to permit under the *Pollution Prevention and Control Act 1999*.

3.19.5 Part V: Information about Air Pollution

Research and Publicity (s.34)

This enables local authorities to carry out or contribute towards the cost of, investigations and research relevant to air pollution. It can arrange lectures, addresses and discussions and publish information; organise exhibitions or other displays including films; prepare or contribute towards the cost of making a film or exhibition, etc. Care should be taken to ensure that any material published does not breach trade secrets, unless the written consent of the person providing the information has been obtained.

Obtaining Information (s.35)

In order to obtain information about emissions into the air – gaseous, liquid or solid or any combination (s.40) – the local authority may issue a notice under s.36 (see below); it can enter premises to measure and record emissions (either by agreement or using powers under s.56 of the Act which confers rights of entry and inspection); or it can agree with the occupiers of premises that they will measure and record emissions for the local authority. Private dwellings and caravans are excluded from this section.

The local authority may only use its powers of entry to measure and record emissions after serving and expiry of a 21 day notice on the occupier which

- specified the kind of emissions in question and how it was proposed to measure and record them;
- stated that it would monitor and record the emissions itself unless the occupier requested the local authority to serve a section 36 notice.

Investigation of emissions from processes subject to Part I of the *Environmental Protection Act 1990* or an activity regulated by permit under the *Pollution Prevention and Control Act 1999* may only be carried out following service of a s.36 notice or by contributing to the cost of such investigation, as provided for in s.34, without entering the premises.

In exercising its powers to obtain information, the local authority should set up a consultation committee of locally representative people with knowledge of air pollution problems or an interest in the local amenity; this should meet not less than twice a year to discuss the extent and way in which information collected should be made publicly available.

Notices Requiring Information about Air Pollution (s.36)

The local authority may by notice require an occupier of premises in its area to provide certain information about emissions of pollutants into the air at specified intervals. If the notice is served on a process for which a permit is required, only information certified by the appropriate enforcing inspector as being required under PPC need be supplied. The person on whom a notice is served has six weeks within which to comply, or longer as agreed with the local authority.

Notices may require periodical returns or single event emission data; but if periodical returns are requested, the interval between returns shall be not less than three months, and no one return is to cover a period of more than one year.

Section 36 applies to all premises, including Crown premises, unless specifically exempted by the Secretary of State. *The Control of Atmospheric Pollution (Exempted Premises) Regulations 1977* (SI 18) covering England and Wales list government research establishments to which this section does not apply.

Failure to comply with a notice, without reasonable excuse is an offence, as is the furnishing of false information.

Appeal against Notice (s.37)

A person on whom a s.36 notice is served may appeal to the Secretary of State on the grounds that

- disclosure would unreasonably prejudice some private interest, or be contrary to public interest, or
- the information is not readily available and cannot readily be collected without incurring undue expenditure.

Following consultation with representatives of local authorities, industry and air pollution experts, the Secretary of State may make regulations relating to the appeals process.

Regulations under the *Control of Pollution Act 1974* – the *Control of Atmospheric Pollution (Appeals) Regulations 1977* (SI 17) – set out the information to be submitted with the appeal, allow the Secretary of State to withhold certain, secret information from the local authority and indicate how the appeal should be determined. If an appeal is upheld, the Secretary of State may direct the local authority to withdraw or to modify the notice, or to take steps to ensure that the prejudicial information is not disclosed to the public. Similar Regulations (1982) apply in Scotland.

Regulations about Local Authority Functions under ss.34, 35 and 36 (s.38)

Following consultation with representatives of local authorities, industry and air pollution experts, the Secretary of State may make regulations prescribing emissions for which a s.36 notice may be issued, the information which such a notice may require and the way in which such notice is to be given. Regulations may also specify under what circumstances local authorities may arrange for the occupier of premises to measure and record emissions and the apparatus which local authorities may provide and use for measuring and recording emissions. Regulations would also provide for the setting up of a public register.

The Control of Atmospheric Pollution (Research and Publicity) Regulations 1977 (SI 19) limit the information to be required. Notices may only relate to sulphur dioxide or particulates from processes where stock does not contribute to the emission, or gas or particulates where stock does contribute to the emission, or gas or particulates from non-combustion processes. Also laid down is the information which can be required in respect of such emissions. The Regulations contain provisions relating to the service of notices and the maintenance of a register of information under these Regulations. Similar Regulations (1982) apply in Scotland.

Provision by Local Authorities of Information for Secretary of State (s.39)

This section enables the Secretary of State, after consultation with a local authority, to direct it to monitor air pollution and to send the information to the Secretary of State. Any capital expenditure incurred by the local authority in providing and installing apparatus to be able to comply with a direction may be reclaimed from the Secretary of State.

3.19.6 Part VI: Special Cases

Relation to Environmental Protection Act 1990 (s.41)

Parts I-III of the *Clean Air Act 1993* (dark smoke; smoke, grit, dust and fumes; smoke control areas) do not apply to a process prescribed for control under Part I of the EPA from the date of authorisation, refusal of authorisation or confirmation of refusal following an appeal. This section is to be repealed by the *Pollution Prevention and Control Act 1999* (Sch.3).

Colliery Spoilbanks (s.42)

The owner of a mine or quarry from which coal or shale is, has been, or will be obtained must use all practicable means to prevent combustion of refuse from the mine or quarry and to prevent or minimise emissions of smoke and fumes from the refuse.

The nuisance provisions of the *Environmental Protection Act 1990* and Parts I-III of the *Clean Air Act 1993* may not be applied to smoke, grit or dust arising from combustion of mine refuse.

This section of the Act does not apply to refuse from a mine or quarry deposited before 5 July 1956 if it was no longer in use and ownership had passed from the mine or quarry owner.

Railway Engines (s.43)

Emissions from railway engines are subject to the dark smoke provisions of s.1 of the 1993 Act and owners of railway engines are required to use any practicable means for minimising emissions of dark smoke. The remainder of Part I, together with Parts II and III of the 1993 Act do not apply to smoke, grit or dust from railway engines. However emissions could be the subject of a statutory nuisance action under s.79 of the EPA (see chapter 1, 1.29.1).

The Dark Smoke (Permitted Periods) Regulations 1958 made under s.19 of the 1956 Act are also applicable.

Vessels (s.44)

Vessels in water not navigable by sea-going ships and in waters within the seaward limits of the territorial waters (including docks, ports, harbours and rivers) of the UK, must comply with the dark smoke provisions of s.1 of the 1993 Act. This section does not apply to ships in HM Navy or "Government Shipping" (e.g. merchant shipping) being used by the Navy.

The Dark Smoke (Permitted Periods) (Vessels) Regulations 1958 made under s.1(2) of the 1956 Act specify the types of emissions which will be permissible and the length of time. There are similar Regulations for Scotland (1958) and for Northern Ireland (1965).

A vessel is not under way when it is at anchor or made fast to the shore or bottom. However a vessel which is aground is deemed to be under way. The Regulations include a proviso that continuous emissions of dark smoke caused otherwise than by the soot blowing of a water tube boiler shall not exceed:

a) in the case of classes 1 and 2: 4 minutes;
b) in the case of natural draught oil-fired boiler furnaces in class 4: 10 minutes; and
c) in no case shall black smoke be emitted for more than 3 minutes in the aggregate in any period of 30 minutes.

Exemption for Purposes of Investigations and Research (s.45)

The local authority may exempt a particular chimney, boiler or furnace etc or the acquisition or sale of a specified fuel from various provisions of the Act, for example to enable investigations or research relevant to air pollution to be carried out without making the applicant liable to proceedings. If the local authority refuses the exemption, the applicant has a right of appeal to the Secretary of State.

Crown Premises (s.46)

In the event of excessive emissions of smoke, grit and dust from Crown premises, the local authority should notify the Minister responsible for the premises who should take appropriate action to prevent or minimise the emission. This includes premises within a smoke control area and those being used by visiting forces (except for Parts IV and V); ships in the Navy or Government Ships (e.g. merchant shipping) being used by the Navy are also subject to the Act.

Table 3.12: The Dark Smoke (Permitted Periods) (Vessels) Regulations 1958 (SI 878)

Class of Case	*Permitted Period for Emission of Dark Smoke*
1. Emissions from a forced draught oil-fired boiler furnace, or an oil engine.	10 minutes in the aggregate in any period of two hours.
2. Emissions from a natural draught oil-fired boiler furnace, (except in the case falling within class 4 below).	10 minutes in the aggregate in any period of one hour.
3. Emissions from a coal-fired boiler furnace:–	
a) when the vessel is not under way (except in the cases falling within class 4 below).	10 minutes in the aggregate in any period of one hour.
b) when the vessel is under way.	20 minutes in the aggregate in any period of one hour.
4. Emission from a natural draught oil-fired boiler furnace or a coal-fired boiler furnace in the following cases:– a) a vessel with funnels shortened for the purpose of navigating the Manchester Ship Canal; b) a tug not under way, but preparing to get under way or supplying power to other vessels or to shore installations; c) a vessel not under way but using main power for the purpose of dredging, lifting pumping or performing some other special operation for which the vessel is designed.	20 minutes in the aggregate in any period of one hour.
5.Emission from any other source.	5 minutes in the aggregate in any period of one hour.

3.19.7 Part VII: Miscellaneous and General

Application of Certain Provisions to Fumes and Gases (s.47)

This enables the Secretary of State to make regulations extending the scope of ss.4, 5, 6, 7, 11, 42, 43, 44 and 46 (covering variously grit, smoke and dust) to fumes or prescribed gases.

Power to Give Effect to International Agreements (s.48)

The Secretary of State may make Regulations relating to any of the provisions of Parts IV, V and VII of the 1993 Act to enable compliance with international agreements.

Administration and Enforcement (ss.49-62)

These sections of the Act are mainly administrative, relating to such matters as improper disclosure of information by an officer (s.49), penalties for offences under the Act (s.50), county court powers to authorise works and order payments (s.54) and general provisions as to enforcement (s.55).

Duty to Notify Occupiers of Offences (s.51)

Where an offence has been or is being committed under ss.1, 2 or 20 (prohibition of certain smoke emissions), the local authority's authorised officer must notify the occupier of the premises, either verbally or in writing "as soon as may be". If notification is given verbally this should be confirmed in writing before the end of four days following that on which the officer became aware of the offence. In any subsequent proceedings it is a defence to prove that the notification procedure was not complied with.

Rights of Entry and Inspection (ss.56-57)

In order to carry out their duties under the Act, authorised officers of a local authority are allowed access to land (which includes premises) and vessels at any reasonable time; they may also carry out any necessary measurements and tests and remove samples for analysis. This section does not apply to private dwellings except in relation to work required to effect compliance with a smoke control order.

The local authority may apply to a Magistrate (Sheriff in Scotland) for a warrant to enter (if necessary by force) in the following circumstances:

- if refused entry despite seven days notice having been given;
- the land/vessel is unoccupied or the occupier absent;
- there is an emergency;
- to apply for admission would defeat the object of the exercise.

Power of Local Authority to Obtain Information (s.58)

The local authority may serve a notice specifying what information it requires and by when in relation to Parts IV and V of the Act (Control of Certain Forms of Air Pollution and Information about Air Pollution). Non-compliance with such a notice is an offence, as is furnishing false information.

Inquiries (s.59)

The Secretary of State may cause an inquiry to be held in connection with any part of the *Clean Air Act* or for preventing or dealing with air pollution.

3.20 SULPHUR CONTENT OF LIQUID FUELS

The *Sulphur Content of Liquid Fuels (England and Wales) Regulations 2007* (SI 2007/79) which came into force on 16 February 2007 implement EU Directive 1999/32/EC as amended by 2005/33/EC on the sulphur content of certain liquid fuels – see 3.16.3. The Regulations were made under the *European Communities Act 1972*, and replace SI 2000/1460. There are similar Regulations for Scotland (SSI 2007/27, which came into force on 12 March 2007, replacing SSI 2000/169), and Northern Ireland, SR 2007/272, which came into force on 31 May 2007, replacing SR 2002/28. The enforcing authority for England and Wales is the local authority, SEPA in Scotland, and in Northern Ireland, the NI Environment Agency or local authority.

The 2007 Regulations are not significantly different from the earlier Regulations but incorporate the land-based elements of the 2005 Directive.

Since 1 January 2003 it has been an offence to use any heavy fuel oil with a sulphur content exceeding 1% by mass. This regulation does not apply to operators of "new" large combustion plant, or to "existing" plant (from 1 January 2008) covered by a PPC permit which contains conditions requiring the plant to meet emission limits for SO_2 which are at least as strict as those in Annex IV of the 2001 Large Combustion Plant Directive (see 3.16.2).

Operators of other combustion plant who do not require a PPC permit (those below 20 MW) may choose to comply with the 2003 limit or will need to have a "sulphur content of liquid fuels permit" from the relevant enforcing authority. As well as full details of the applicant, address of the plant and fuel used, the application should give details of the condition to be included in the permit which will satisfy the Regulations – i.e. ensure emissions of SO_2 do not exceed 1,700 mg/Nm^3 at an oxygen content in the flue gas of 3% by volume on a dry basis; where the combustion plant is used for combustion in a refinery, the monthly average of all SO_2 emissions from all over the plant (excluding those from new large combustion plant) shall not exceed 1,700 mg/Nm^3. A permit may be transferred to another person, with the person to whom the permit is transferred notifying the relevant enforcing authority within 21 days of the date of the transfer. A permit may be surrendered by notifying the enforcing authority which granted the permit.

The Secretary of State (Scottish Ministers, DOENI) is responsible for ensuring that heavy fuel oil and gas oil is sampled on a regular basis to ensure compliance with the limits.

AGRICULTURE & AIR POLLUTION

3.21 OVERVIEW

Agricultural practices can cause a wide range of pollution problems; these are regulated either explicitly, or by implication, in a number of statutes and codes of good practice, which are summarised in this and other chapters of the Pollution Control Handbook, as follows.

3.21.1 Air Pollution and Nuisance

Clean Air Act 1993

- Part I: s.2 prohibits the emission of dark smoke from industrial or trade premises; this includes land being used for commercial agriculture or horticulture (this chapter, 3.19.1).
- Part III: farms falling within a smoke control area must comply with ss.18-29 of the Act which prohibit the emission of smoke from any building or chimney and require the use only of authorised fuels (this chapter, 3.19.3).
- *Clean Air (Emission of Dark Smoke) (Exemption) Regulations 1969*: these permit the burning, in certain circumstances, of various farm wastes; these include the burning of animal or poultry carcases and pesticide containers (this chapter, 3.19.3).

Environmental Protection Act 1990

- Part III: this enables local authorities and individuals to take action against statutory nuisance; statutory nuisances include smoke, odour, noise, accumulations or deposits and animals kept in such a way as to be prejudicial to health or a nuisance (see chapter 1, 1.29.1).
- *Highways (Amendment) Act 1986*: it is an offence to light a fire, permit or direct one to be lit, the smoke from which injures, interrupts or endangers anyone using a highway. This provision could also be used for bonfire smoke.

3.21.2 Industrial Pollution and Waste Management

- ***Pollution Prevention and Control Regulations*** *(Sc, 2000; NI, 2003);* ***Environmental Permitting Regulations*** *(E & W, 2007):* operators of new (brought into operation on or after 31 October 1999) intensive farming installations require a permit covering their activities; existing installations were required to apply for a permit between 1.11.06 and 31.01.07. The Regulations apply to facilities with 40,000 places for poultry, 2,000 for production pigs and 750 places for sows. The Environment Agency has also drafted for consultation guidance on noise management at intensive livestock installations.

- ***Ammonia emissions*** – farming activities, particularly cattle, pig and poultry farming, are responsible for a significant proportion of ammonia emissions in the UK. Both the UNECE Protocol to Abate Acidification, Eutrophication and Ground Level Ozone (see 3.15.3) and the EU's Directive on National Emissions Ceilings (see 3.16.1) require the reduction of emissions of ammonia (and other pollutants) below a specified ceiling by 2010. One of the main ways of meeting the UK's obligations so far as agricultural emissions are concerned will be through conditions attached to PPC permits. Defra has published a booklet (*Ammonia in the UK*, October 2002), which describes the problem and the environmental damage caused, as well as methods of curbing emissions.

- ***Environmental Protection Act 1990***, Part II: the disposal of certain organic wastes to land requires a waste management licence from the waste regulation authority (Environment Agency or SEPA) – see chapters 2 and 5.

- ***Waste Management Regulations 2006 (E&W); Waste (Scotland) Regulations 2005:*** These Regulations apply waste management, including licensing, controls to waste from agricultural premises – these include premises used for all types of farming, horticulture, fruit growing, seed growing, market gardens, nursery grounds and osier land. Farm dumps or tips may no longer be used to dispose of waste. There are similar Regulations for Northern Ireland. See chapter 5, 5.19.

- ***The Sludge (Use in Agriculture) Regulations 1989*** (SI 1263) (as amended and Code of Practice dated 1996) control the spreading of sewage sludge on agricultural land, and restrict the planting, grazing and harvesting of some crops following application. The main aim of the controls is to prevent the accumulation of hazardous concentrations of heavy metals in soil and to prevent bacteriological contamination of crops; the Code of Practice recommends maximum concentrations for contaminants to limit risks to crops and animals, utilisation practices on the farm and monitoring requirements for soils receiving sludge. Similar Regulations came into force in N. Ireland in July 1990.

3.21.3 Water Pollution

The control of water pollution from agricultural practices, in particular the use of nitrates and pesticides is covered in chapter 6. A statutory *Code of Practice on Using Plant Protection Products* (February 2006) covering England and Wales (replacing 1998 Code of practice on the safe use of pesticides on farms & holdings) sets out best practice for the use of pesticides which are controlled under the *Control of Pesticides Regulations* and other plant protection regulations. It provides advice on the laws regarding protection of groundwater and waste management, and on training and certification. Advice on the new legal requirement to keep spray records is given, and on what should be done if members of the public report being affected by spraying.

3.22 CODES OF PRACTICE

3.22.1 England and Wales

A *Code of Good Agricultural Practice for the Protection of Air* was published in 1992, and a revised version in 1998. It covers England and Wales and summarises the various pieces of legislation which farmers should be aware of; it gives practical guidance on reducing the problem of odours from housed livestock systems, slurry and manure storage, production of compost for mushrooms and land spreading of livestock wastes. There is also guidance on minimising the need to burn waste and other materials and alternative uses for straw. The Code describes ways of reducing emissions of ammonia and of the greenhouse gases. It does not cover noise nuisance (action for which can be taken under Part III of the *Environmental Protection Act 1990*, see chapter 1, 1.29.1; nor does it cover pesticide use for which there are separate codes of practice (see chapter 6).

The Code, which was prepared by the Ministry of Agriculture, Fisheries and Foods (now Defra) and the Welsh Office Agriculture Department, has no legal status; adherence to the Code is not a defence in a legal action.

A revised version of the 1993 *Code of Good Agricultural Practice for the Protection of Soil* was published in 1998. Like the Code covering the protection of the air. It summarises the legislative controls farmers should be aware of and suggests ways of preventing soil erosion and the loss of organic matter. The code gives a reminder that wastes and pesticides should not be applied in such quantities that they are likely to accumulate in the soil. It also provides advice on protecting soil from contamination and acidification.

3.22.2 Scotland

In Scotland there is a single Code of Practice covering the prevention of air, soil and water pollution – *The Code of Good Practice for the Prevention of Environmental Pollution from Agricultural Activity*, published in 2005 (replacing the 1997 code). Those parts of the Code covering water pollution have statutory backing (SSI 2005/63) under s.5(1) of the *Control of Pollution Act 1974*, as amended by the *Water Act 1989* (see chapter 6).

The Code summarises all the various pieces of legislation in Scotland of relevance to the control of agricultural pollution. As well as providing advice and information on the prevention and control of emissions to air, the code covers diffuse agricultural pollution; soil protection and sustainability; collection, storage and application to land of livestock slurries and manures; nitrogen and phosphorus; silos and silage effluent; sheep dip; pesticides; disposal of animal carcasses; agricultural fuel oil; waste management and minimisation. The various pollution problems caused by incorrect – or inadvisable – treatment and disposal practices are summarised, together with the advantages and disadvantages of various recommended good practices and action necessary to comply with legislation.

In Scotland a 'Four Point Plan' (available on the Scottish Government website) developed by the farming industry and environmentalists provides best practice guidance on the management of pollution from livestock farming. It describes a range of measures for the minimisation of slurry and dirty water, better nutrient planning using slurry and manure, and risk assessment for manure and slurry spreading, and the protection of water margins.

3.22.3 Northern Ireland

The 1994 codes for the protection of air, water and soil for Northern Ireland are similar to those for the rest of the UK.

3.23 STRAW AND STUBBLE BURNING

Section 152 of the *Environmental Protection Act 1990*, Part VIII enables Regulations to be made prohibiting or restricting the burning of crop residues on agricultural land and for exemptions relating to specified areas, specified crop residues, or in specified circumstances. The Regulations may impose requirements to be complied with before or after burning, and also make contravention of any Regulations an offence which on conviction is subject to a fine. Straw and stubble burning in England and Wales was previously controlled through local authority byelaws; these were all repealed with effect from 2 April 1992.

3.23.1 Crop Burning Regulations

The Crop Residues (Burning) Regulations 1993 (SI 1366) came into effect on 29 June 1993 and extend to England and Wales only, revoking earlier 1991 Regulations. The burning of cereal straw and stubble, and the residues of field beans and peas harvested dry and oilseed rape is banned. They may, however, be burnt in the following circumstances:

- for education and research purposes;
- in compliance with a notice served under the *Plant Health (Great Britain) Order 1993* (e.g. to eliminate pests);
- to dispose of broken bales and the remains of straw stacks.

The burning of linseed residues is currently exempted from the ban, but may subsequently be included in the light of the availability and practicability of alternative methods of disposal and the level of public complaint.

Any burning for education and research, in compliance with a plant health order or of linseed residues must be carried out in compliance with the conditions contained in Schedule 2 of the Regulations; these are:

a) No burning at weekends, bank holidays, or from one hour before sunset until the following sunrise.

b) No more than an area of ten hectares of straw and stubble and 20 hectares of other crop residues may be burnt in a single operation. Additional guidance from MAFF (now Defra) states that if sufficient fire-fighting equipment and manpower are available and fire-break requirements are observed, two or more burning operations may be undertaken simultaneously, so long as they are at least 150 metres apart.

c) No burning within certain distances of various "vulnerable" objects; these are:

   - 15 metres for straw and stubble burn and 5 metres for other residues: tree trunks (including coppices or scrubland); fence belonging to another property; telegraph or telephone poles; electricity pole, pylon or substation.

   - 50 metres for straw and stubble burn and 15 metres for other residues: residential buildings; thatched roofs; building, structure, fixed plant or machinery which could be set alight by heat from a fire; scheduled monument which could be set alight; hay or straw stack; accumulation of combustible material (other than crop residues) removed in making of fire-break; mature standing crop; woodland or land managed as a nature reserve; above ground gas or oil installation.

   - 100 metres for burning of any residue: motorways, dual carriageways; A roads; and railway lines.

d) The area to be burnt must be completely surrounded by a fire-break. This must be 10 metres wide for cereal straw and stubble and 5 metres wide for other crop residues. Precautions should be taken to ensure that the fire will not cross the fire-break.

e) At least two adults who are familiar with the Regulations must be present during burning, one of who must be in control of the operation and must be experienced in burning of crop residues.

f) At least one hour's notice (but not more than 24 hours' notice) must be given to the following: local environmental health department; occupiers of all premises adjacent to the area to be burned; air traffic control of any aerodrome within 800 metres.

g) At least 1000 litres of water must be available on the burning site, together with equipment to dispense the water in a spray or jet at a rate of 100 litres per minute; there must be five implements suitable for fire beating available and all vehicles used in connection with burning must be equipped with fire extinguishers.

h) Ash remaining from cereal straw and stubble burning should normally be incorporated into the soil within 24 hours of burning; if windy conditions mean that incorporating the ash might cause a nuisance, it should be incorporated as soon as conditions allow.

The maximum fine for each breach of the Regulations is currently £5,000.

The burning of broken bales and straw stack remains is still permitted, as is the burning of other crop residues not specifically mentioned, e.g. herbage seeds, reeds, lavender, hop bines and potato haulms. The local environmental health department must be notified of any permitted burning.

The Regulations do not ban burning when winds exceed certain speeds. However, guidance notes issued by MAFF (now part of Defra) suggest farmers should not burn when winds exceed, or are forecast to exceed, force 3 (8-12 mph). Farmers are also advised to pay attention to wind direction in order to minimise nuisance and hazard.

Action for burning on agricultural or horticultural land may also be taken under s.2 of the *Clean Air Act 1993* (see section 3.19.1 above), or under the nuisance provisions of the *Environmental Protection Act* (see chapter 1, 1.29.1 above).

In Scotland, there are no specific regulations covering straw and stubble burning as only a small percentage of stubble is burnt. A voluntary Code of Practice (*Straw and Stubble Burning and Muirburn Practice*) has however been agreed between the Scottish NFU and the Scottish Office Agriculture and Fisheries Department. Additional guidance is contained in the 1997 *Code of Good Practice on Prevention of Environmental Pollution from Agricultural Activity*. The advice given is similar to that contained in the Regulations for England and Wales.

Under the "set aside" arrangements, the burning of "set aside" is prohibited.

3.23.2 Heather and Grass Burning

The Heather and Grass etc Burning (England) Regulations 2007 (SI 2007/2003), which came into force on 1 October 2007 revoke 1986 Regulations in England. There are similar Regulations for Wales – SI 2008/1081, W.115, which came into force on 6 May 2008, revoking 1986 Regulations. The Regulations do not apply to land cultivated as private gardens or allotment gardens or to the burning of "specified vegetation" on railway land or under the authority of Network Rail. "Specified vegetation" means heather, rough grass, bracken, gorse or vaccinium.

The Regulations prohibit the burning of specified vegetation on any land between sunset and sunrise, and no burning can be carried out unless a burning plan has been prepared, there are sufficient people and equipment available to control the burning, and all reasonable precautions taken to prevent injury or damage to adjacent land, or to any person or thing on that land. Not less than 24 hours or more than 72 hours notice (including extent of burn, date and time etc) prior to the burn must be given in writing to any other person with an interest in the land either as owner or occupier.

No burning of specified vegetation may be carried out outside the burning season except under licence from Natural England [Welsh Ministers] – and indeed the burning of larger areas or separate burns close to each other within the burning season will also require a licence. In England the burning season in relation to land within an upland area is 1 October to 15 April (inclusive) of the following year, and land outside an upland area 1 November to 31 March (inclusive) the following year; for Wales the corresponding dates are 1 October to 31 March and 1 November to 15 March.

The licence application should be made not less than 28 days before the proposed burn; in the case of a proposed burn during a burning season the application may not be made before the end of the preceding burning season; in the case of a proposed burn outside the burning season, an application may not be made more than 56 days before the proposed burn. A licence to burn may only be granted if Natural England [Welsh Ministers] are of the opinion that

- the burn will be carried out in accordance with the provisions of the burning plan;
- in the case of railway land, it is expedient for good maintenance or pest control;
- it is necessary or expedient for the conservation, enhancement or management of the natural environment.

if it is decided not to issue a licence or to issue one with conditions, the applicant should be notified in writing. If Natural England [Welsh Ministers] believe burning has been carried out in contravention of the Regulations, they may issue a "Burning Notice" requiring the occupier of the land to provide prior notification of any future burns. Any appeal against such a notice should be made within 28 days.

The Regulations also amend *the Common Agriculture Policy Single Payment and Support Schemes (Cross-Compliance) (England) Regulations 2005* (SI 2005/3459) – Wales, SI 2004/3280, W.284) – to impose similar requirements on farmers as regards the burning of heather, rough grass, bracken, gorse or vaccinium. A licence is required for burning on upland areas in England from 16 April to 30 September (inclusive) or from 1 April to 31 October (inclusive) on all other land [Wales, 1 April to 30 September, and 16 March to 31 October].

3.24 GENETICALLY MODIFIED ORGANISMS

3.24.1 Contained Use Regulations

The Genetically Modified Organisms (Contained Use) Regulations 2000 (SI 2831), made under *The Health & Safety at Work Act 1974*, cover England, Scotland and Wales; they came into force on 15 November 2000, replacing 1992

Regulations (SI 3217) and subsequent amendments. Similar Regulations for Northern Ireland (SR 295/2001) took effect on 25 September 2001. The Regulations implement EU Directive 90/219/EEC, as amended by 98/81/EC; this required the setting up of a notification, consent and emergency system to protect human health and the environment (implemented through the 1992 Regulations). The 1998 Directive aims to streamline administrative arrangements, specify minimum containment and control measures for each risk class and introduces the possibility of exemptions for safe GMOs. "Contained Use" refers to any work with GMOs which takes place in conditions which are intended to prevent any escape of the organisms into the environment. The Commission has now published (COM(2007) 736, 29.11.07) a proposal for a Directive codifying the 1990 Directive and amendments into a single instrument.

The competent authority as regards England and Wales is the Secretary of State for Environment, Food and Rural Affairs and the Health and Safety Executive, in Scotland the HSE and Scottish Ministers, and in Northern Ireland the Health and Safety Executive for Northern Ireland.

The main provisions of the Regulations are:

- Prior to carrying out any activity involving genetically modified micro-organisms (GMMOs) or GMOs, an assessment of risks to human health and the environment should be carried out; a genetic modification safety committee should be established to advise on the risk assessment.
- Risk assessments should be recorded and reviewed if there is reason to suspect it is no longer valid or that activities have changed significantly; records of assessments to be kept for 10 years.
- Notification of intention to use premises for the first time for activities involving genetic modification, and notification to HSE of an intention to carry out an activity and, in certain cases, to await consent.
- Classification of operations and the organisms used according to a prescribed scheme:
  - Class 1: activities of no or negligible risk for which containment level 1 is appropriate to protect human health and the environment;
  - Class 2: activities of low risk, for which containment level 2 is appropriate to protect human health and the environment;
  - Class 3: activities of moderate risk, for which containment level 3 is appropriate to protect human health and the environment;
  - Class 4: activities of high risk, for which containment level 4 is appropriate to protect human health and the environment.
- Notification to the competent authority of individual activities of classes 2, 3 and 4 involving GMMs; formal consent to be obtained from the authorities before starting activities of classes 3 and 4. The Regulations detail the notification process and fees payable.
- Adopt controls, including suitable containment measures – these relate to measures to be taken in the laboratory, equipment, system of work, waste, etc.
- If the risk assessment shows that there is a risk to the health and safety of persons outside the premises or of serious damage to the environment as a result of the activity, an emergency plan of the measures to be taken in the event of an accident should be drawn up. The HSE should be notified of accidents involving GMOs.
- Inclusion of information from notifications on a public register, with provision for confidentiality of commercially sensitive information and personal data. Information may be removed 10 years after notification that the activity has ceased or 10 years after notification that a premises is no longer used for the activity.

Legislation governing the environmental aspects of control over GMOs is responsibility of the Devolved Administrations for Scotland, Wales and Northern Ireland.

An amendment to the Regulations – SI 2002/63, effective 8 February 2002 – provides for information relating to GMO facilities to be kept confidential in the interests of national security; it also enables the exclusion of information from the public registers which, in the opinion of the Secretary of State, is not in the interests of national security. Further 2005 Amendments (SI 2005/2466) update the confidentially requirements to accord with the *Environmental Information Regulations 2004*, update references to other legislation which has been amended since the 2000 Regulations, remove references to MAFF and clarify the powers of the National Assembly for Wales to issue deliberate release consents under the EPA.

Defra has issued proposals (England only, July 2006) for ensuring that any growing of genetically modified crops does not disadvantage neighbouring farmers. As well as the imposition of a strict separation distance between GM and conventional crops, Defra is seeking views on whether special rules should apply for coexistence between GM and organic crops; whether there should be a public GM crop register; compensation scheme for non GM-farmers whose crops have GM material in them; and on guidance to farmers on voluntary GM-free zones.

3.24.2 Deliberate Release

Part VI of the *Environmental Protection Act 1990*, ss.106-127, aims to prevent or minimise damage to the environment resulting from the escape or release (i.e. deliberate release) from human control of GMOs. This part of the Act also applies to Northern Ireland.

The Act defines "organism" as any "acellular, unicellular or multicellular entity (in any form) other than humans or human embryos; and ... any article or substance consisting of or including biological matter". "Genetically modified" means "if any of the genes or other genetic material in the organism have been modified by means of an artificial technique ... or are inherited or otherwise derived, through any number of replications, from genes or other genetic material which were so modified".

The Act places certain duties on persons proposing to import, acquire, keep or release or market GMOs. These include the need to carry out a risk assessment of any potential damage to the environment as a result of the activity; notifying the Secretary of State of the intention to carry out the activity and in most instances applying for a consent; this may include conditions and other limits or restrictions on the activities permitted. Implicit in all consents is a requirement to use BATNEEC to control organisms and prevent damage to the environment; all reasonable steps should be taken to remain informed about any damage which the activity might have caused and to notify the Secretary of State if the risk appears more serious than was first apparent; if it appears that despite the use of BATNEEC for keeping the GMOs under control, damage might occur, then the activity should be ceased.

Other requirements of this part of the Act relate to enforcement and prohibition notices, offences and fees and charges, and are similar to those in Part I of the EPA.

The *Genetically Modified Organisms (Deliberate Release) Regulations 1992* (SI 3280), as amended by 1995 Regulations (SI 304) came into force on 1 February 1993 and implemented EU Directive 90/220/EEC on the Release of Genetically Modified Organisms into the Environment. They also implemented the requirements of the EPA relating to Secretary of State consents and a charging scheme, risk assessments and make certain information publicly available subject to various exclusions on the grounds of confidentiality. Similar provisions for Northern Ireland are contained in the *Genetically Modified Organisms (Northern Ireland) Order 1991* (SR 1991/755).

The 1992 Regulations have been revoked and replaced by Regulations implementing Directive 2001/18/EC (which also repealed the 1990 Directive with effect from 17 October 2002); the replacement Regulations are: England – *the Genetically Modified Organisms (Deliberate Release) England Regulations 2002* (SI 2443) which came into force on 17 October 2002; Scotland – SSI 2002/541, 5 December 2002; Wales – SI 2002/3188, 31 December 2002, amended by SI 2005/1913; Northern Ireland – SR 2003/167, 15 April 2003, amended SR 2005/272.

The 2001 Directive defines a GMO as "an organism, with the exception of human beings, in which the genetic material has been altered in a way that does not occur naturally by mating and/or natural recombination"; it requires marketing consents to be renewed every ten years and mandatory monitoring for environmental impact of products containing GMOs, with renewal of authorisation dependent on monitoring results. Other provisions relate to harmonisation of risk assessment and classification of risks, labelling and streamlining of administrative procedures, with a simplified authorisation procedure for certain categories of release (defined in an Annex to the Directive). Authorisation authorities are required to hold public consultations for a minimum of 48 days on all GMO research trial applications, and to take any views into account in decision making. The location of GM crops must be notified to the competent authority and details kept on a public register.

TRANSPORT EMISSIONS

This section of the *Pollution Control Handbook* looks at the control of emissions of air pollutants primarily from road transport, with summaries covering emissions from rail, aviation and shipping. It also covers fuel quality and petrol storage and distribution. Noise pollution from all forms of transport is covered in chapter 4, and pollution of the marine environment from shipping in chapter 6.

3.25 ROAD TRANSPORT – OVERVIEW

Early legislation in both the UK and EU limiting vehicle emissions was often based on model standards developed by the UN Economic Commission for Europe (which includes both EU and non-EU countries), with the first European Directives limiting emissions from cars dating from the early 1970s. In the UK EU Directives limiting exhaust emissions from road vehicles and technical standards are implemented through the *Road Vehicles (Construction and Use) Regulations* and the *Motor Vehicles (Type Approval) (Great Britain) Regulations* made under the *Road Traffic Act 1988*. Similar regulations made under the *Road Traffic (Northern Ireland) Order 1981* apply in Northern Ireland. The *Transport Act 1982* also specifies requirements for the manufacture of vehicles; Regulations made under the *Clean Air Act 1993* impose requirements as to the composition, content and marketing of fuel used in motor vehicles – see 3.27.1.

The Future of Transport White Paper (Cm 6234, published July 2004) sets out the Government's strategy over the next 30 years for providing a coherent transport network that both meets the increasing demand for travel and satisfies environmental and economic objectives. This is to be achieved through sustained investment in the transport networks, improvements in transport management and planning ahead.

Scotland's *National Transport Strategy* launched in December 2006, sets out a vision of "an accessible Scotland with safe, integrated and reliable transport that supports economic growth, provides opportunities for all and is easy to use; ... that meets everyone's needs, respects the environment, and contributes to health..." To achieve this vision, the Scottish Government will focus on reducing vehicle emissions, improving public transport and better integration between different transport modes.

The *Transport (Wales) Act 2006* requires the National Assembly to prepare and publish a Transport Strategy which sets out the Assembly's "policies for the promotion and encouragement of safe, integrated, sustainable, efficient and economic transport facilities and services to, from and within Wales. The aim of *One Wales: Connecting the Nation* (April 2008) is to promote sustainable transport networks that safeguard the environment while strengthening Wales's economic and social life. Reducing greenhouse gas emissions and other environmental impacts, as well as integrating local transport systems are just two of the goals of the Strategy which are to be delivered through the development of a National Transport Plan and four regional transport plans.

Northern Ireland's Regional Transportation Strategy 2002-2012 (RTS), published by the Department for Regional Development, identifies strategic transportation investment priorities. A Sub-Regional Transport Plan (SRTP), published in 2007 takes forward the initiatives of the RTS, and identifies measures for the period 2002-2015 by mode of transport (walking & cycling; bus; rail; and highways).

In June 2007, the Council of the EU supported the further development of the Commission's Energy Strategy for Transport, with the priorities for such a strategy aimed at

- Improving the energy efficiency of all modes of transport;
- Increasing the use of alternative and renewable fuels along with efficient drive trains;
- Designing instruments to promote transport user behaviour that is energy conscious and climate change conscious;
- Promoting integrated transport systems and planning to minimise energy use in transport.

3.25.1 Traffic Growth

Despite advances in technology, emissions from road transport remain a major source of air pollution. The European Environment Bureau's *EU Environmental Policy Handbook* (2005) notes that in 2001 road traffic in the EU-15 accounted for nearly half of all emissions of nitrogen oxides and one third of VOC emissions, as well as contributing significantly to emissions of fine particulates and of carbon dioxide. The problem is exacerbated by the ever increasing number of vehicles on our roads. In its Green Paper on the Impact of Transport on the Environment published in 1992, the European Commission predicted a probable growth in private cars in the European Union (EU) of 45% between 1987-2010 (i.e. from 115 million to a total of 167 million). Goods traffic is expected to increase by 42% between 1990-2010; total vehicle kilometres are predicted to grow 25% throughout the same period.

Statistics published by the Department for Transport in February 2008 provisionally estimate that compared with 2006, car traffic in Great Britain decreased by 1% in 2007, although light van traffic was 9% higher and heavy goods vehicle traffic 1% higher. Based on the provisional estimates cars accounted for 78% of all motor vehicles in 2007, vans for 14%, good vehicles for 6% and other vehicles for 2%. The number of licensed vehicles increased by 74% between 1980 and 2006, from 19.2 to 33.4 million (DfT Transport Trends 2007).

3.25.2 Planning Policy Statements

England

Planning Policy Statements (which are replacing planning policy guidance) are advice from the Secretary of State to local authorities and others on policies and the operation of the planning system. PPG 13, *Transport*, a revised version of which was published in 2001, replacing the 1994 version, looks at the needs of all road users (including pedestrians and cyclists), and the way in which land use planning and transport planning can be integrated at the national, regional, strategic and local level to promote sustainable travel choices and to make optimum use of the transport system thus reducing the need to use the car, and indeed the length of journeys. To this end, the guidance suggests that

- Developers should submit a "transport assessment" with their planning application, outlining any significant transport implications which the development may have; this should illustrate access to the site by all modes and proposed measures (such as travel plans) to improve access by walking, cycling and public transport;
- Parking policies, alongside other transport and planning measures, should be used to promote sustainable travel choices and reduce reliance on the car for travel to work;
- Attention be given to the need to ensure developments are not dependent on travel by car for access, and to providing alternative modes of travel, such as bus services; the guidance does, however, acknowledge that in many rural areas there will be no viable alternative to the car;
- The use of green belt land may in some circumstances be appropriate for "park and ride" schemes; further guidance on this is given in PPG2 on Green Belts;
- The needs of disabled people both as pedestrians and users of public transport should be properly taken account of in the implementation of planning policies and transport.

Scotland

Scottish Planning Policy guidance, *Planning for Transport* (SPP 17), published in 2005, looks at the integration of land-use and transport planning, and notes the need for issues surrounding the interaction of accessibility, transport and the development strategy to be considered early in the planning process. Development should also be in accordance with the National Planning Framework and SPP1 (*The Planning System*). Other issues to be taken into account include:

- Development plans should be co-ordinated with regional and local transport strategies;
- Land allocations and regeneration projects should be related to transport constraints and opportunities – i.e. locating new development so as to encourage and maximise use of public transport, while reducing private car use; the needs of the mobility impaired should be taken into account;
- The location of roadside facilities should take account of land use, environmental, economic and transport objectives.

Wales

Planning Guidance (TAN 18) issued in 2007 by the Welsh Assembly Government contains advice on the integration between land use and transport, with advice on assessing impacts and managing implementation, planning for transport infrastructure, outlining how local authorities should use the planning system to increase sustainable transport provision and reduce the impact of transport on climate change and air quality.

Northern Ireland

Planning Policy Statement 13 (PPS 13) on Transportation and Land Use (2005) provides guidance on integrating land use planning and transport through the promotion of sustainable transport choices, promoting accessibility for all, and reducing the need to travel, especially by private car. (From January 2008, the Department of the Environment took over responsibility for PPSs from the Department for Regional Development.)

3.26 EU CONTROLS – ROAD TRANSPORT

In 1992 the European Commission adopted a Green Paper on the *Impact of Transport on the Environment* (COM(92) 46 final). This outlines an EU strategy for "sustainable mobility" and highlights the expected growth in all forms of transport with a consequent rise in levels of pollution. Between 1990 and 2010, the Commission forecast a growth in road/goods transport of 42%, in rail/goods transport of 33%, and in air/passenger transport of 74%. Private cars were predicted to increase 45% over the period 1987-2010 and vehicle kilometres 25% between 1990 and 2010.

The Green Paper while noting the EU's current and future legislative programmes on emission limits and standards says that these will not be enough to contain likely environmental damage from increased transport. The Commission suggests the need for a more integrated approach to transport policy and policy in other areas on which transport has an impact, such as planning and economic policy.

In addition, the Commission's "Climate action and renewable energy package" published in January 2008 calls for 10% of road transport fuels to come from renewable sources (biofuels), subject to them being produced in a sustainable way – see earlier this chapter 3.5.1, and below 3.26.10.

EU Directives covering pollution from road transport are listed in Appendix 7.3.

3.26.1 The Auto/Oil Programme

In 1992, the European Commission initiated a tripartite research programme with the European oil and motor industries. The aim of the programme was to identify technical measures to help achieve health-based air quality targets at least cost to society. The Auto/Oil programme was intended to consider not just motor and oil industry solutions but also the contributions of traffic management, public transport, new technologies and vehicle maintenance programmes.

Based on the results of the research, the Commission published proposals in June 1996 aimed at reducing emissions from road transport by up to 70% by the year 2010. Published as a Communication (COM(96) 248) "on a future strategy for the control of atmospheric emissions from road transport", the proposals take into account EU air quality targets, technological advances and improvements in fuel quality; the Communication also discusses the practicalities of measures such as fiscal incentives, traffic management and encouraging urban public transport as a means of reducing overall emissions.

A second Auto-Oil Programme, launched in 1997, as well as resulting in further legislative initiatives, has evaluated the role of the auto-oil programmes in reducing emissions from road transport and their contribution to overall reduction of emissions. This shows that emissions from road transport of CO, benzene, SO_2, NO_x and particulates are expected to fall to 20% of their 1995 levels; however, CO_2 emissions from road transport are continuing to rise. And as emissions from road transport will continue to be a major, albeit declining, contributor to poor air quality, there is a need to address sources other than road transport if further improvements in air quality are to be achieved.

A number of Directives arising out of the Auto/Oil Programme have now been formally adopted, and are summarised below.

3.26.2 Light Duty Vehicles

Light duty vehicles are cars and vans under 3.5 tonnes. The first EU Directive (70/220/EEC) relating to measures to be taken against air pollution from positive ignition engines continues to be the basis for current Directives and proposals and has thus been amended and adapted a number of times; it applies to vehicles with no more than six seating positions and a maximum permissible mass of not more than 2500 kg.

Euro 1

Two Directives adopted in 1988 amending 70/220/EEC and progressively tightening emissions limits for CO, HC and NO_x and later particulates were 88/76EEC – "the Luxembourg Agreement" – which imposed emission limits according to engine size: more than 2 litres, 1.4-2 litres and less than 1.4 litres, and 88/436/EEC, which imposed limits on particulate emissions. These Directives were largely superseded by Directive 91/441/EEC – "the Consolidated Directive".

This Directive, adopted in June 1991, implemented new stricter limits (Euro 1) for all passenger cars (except direct injection diesel vehicles) with no more than six seats, and under 2500 kg (see Table 3.13). It also contained an improved driving cycle with a high speed element as the basis for tests. Emission limits are expressed in grams per kilometre instead of grams per test (i.e. the amount of emissions allowable per km averaged over the whole test cycle). The new standards were mandatory and applied to all new cars from 31 December 1992, and to new models from 1 July 1992, and were implemented through a 1992 amendment (SI 1992/2137) to the *Road Vehicles (Construction and Use) Regulations 1986* (see 3.27.2).

Table 3.13: 91/441/EEC

	Emission limit (g/km)		
	CO	$HC+NO_x$	Particulates
Type approval	2.72	0.97	0.14
Conformity of production	3.16	1.13	0.18

From 1 July 1994 the above limits also applied to new model light duty direct injection diesel vehicles and to new models from 31 December 1994.

To meet the standards, new petrol driven cars registered after 1 January 1993 had to be fitted with three-way catalytic converters; these fit into the car exhaust and act by converting CO, HC and NO_x into CO_2, water and nitrogen. Diesel engined vehicles also required "state of the art" technology. These requirements were implemented through an amendment (SI 1992/1341) to the *Type Approval Regulations* (see 3.27.3 below). The 1991 Directive also contains provisions for the introduction of tax incentives to promote the introduction of cleaner vehicles.

Euro 2

In 1994 a further Directive (94/12/EC) amending the original 1970 Directive on vehicle emission limits was agreed (Euro 2). Different limits are set for petrol and diesel engined vehicles and for the first time no distinction is made between standards for type approval and conformity of production (Table 3.14).

Table 3.14: 94/12/EC

	Emission limit (g/km)		
	CO	$HC+NO_x$	Particulates
Petrol engines	2.2	0.5	
Diesel engines			
- indirect injection	1.1	0.7	0.08
- direct injection	1.0	0.9	0.10
DI from 1.10.99	1.0	0.7	0.08

The new limits were mandatory from 1 January 1996 for new model vehicles and from 1 January 1997 for new registrations. The Directive has been implemented through a 1995 amendment (SI 1995/2210) to the 1986 *Construction and Use Regulations*. As with the 1991 Directive it was permissible to offer fiscal incentives to encourage manufacturers to meet the new standards before the deadline. Any such schemes had to be approved by the Commission prior to introduction and meet a number of criteria, including compatibility with the functioning of the internal market and be non-discriminatory.

Euro 3 & 4

Current emission limits – Euro 3 and 4 – (Table 3.15) for passenger vehicles (up to 2,500 kg) are contained in Directive 98/69/EC which originated from the Auto/Oil Programme, and was formally adopted in September 1998; this Directive also covers light commercial vehicles. The 1 January 2000 and 2005 compliance dates are those for new model types, with compliance by new cars a year later in each case.

Table 3.15: 98/69/EC

Petrol fuelled vehicles to meet following limits (g/km) by		
	2000 "Euro 3"	2005 "Euro 4"
CO	2.3	1.0
HC	0.20	0.10
NOx	0.15	0.08
Diesel fuelled vehicles to meet following limits (g/km) by		
	2000	2005
CO	0.64	0.50
HC/NOx	0.56	0.30
NOx	0.50	0.25
Particles	0.05	0.025

New model type petrol-engined passenger cars and light commercial vehicles (category I) up to 2,500 kg with electronically controlled catalytic converters are required to be fitted with an onboard diagnostic (OBD) system from 1 January 2000 (new vehicles from 1 January 2001) and diesel engined vehicles from 2005 (OBD systems show if vehicle emissions are above permissible levels). Tax incentives will be permissible for vehicles already meeting the 2005 standards from 2000. Directive 2001/1/EC, adopted on 22 January 2001, amends the dates by which different categories of vehicles had to be fitted with OBD systems as follows:

- Petrol fuelled engines: as above and passenger cars and light commercial vehicles (categories II and III) exceeding 2,500 kg – 1 January 2001 (new model types) and 1 January 2002 (new models);
- LPG and natural gas fuelled vehicles: passenger cars and light commercial vehicles (category I) up to 2,500 kg – 1 January 2003 (new model types) and 1 January 2004 (new models); passenger cars and light commercial vehicles (categories II and III) exceeding 2,500 kg – 1 January 2006 (new model types) and 1 January 2007 (new models).

Directives 96/69/EC, 98/69/EC, 99/96/EC, 99/102/EC have been implemented through an amendment to the *Construction and Use Regulations*, SI 2000/3197, and the 2001 Directive by SI 2001/1825.

Euro 5 & 6

Regulation 715/2007 of 20 June 2007 *on type approval of motor vehicles with respect to emissions from light passenger vehicles and commercial vehicles (Euro 5 and Euro 6) and on access to vehicle repair and maintenance information* imposes stricter emission limits to begin coming into force on 1 September 2009 (Euro 5) when all new model diesel vehicles will have to be fitted with a particle filter; new existing models will need to be fitted with a particle filter from 1 January 2011. Passenger cars over 2,500 kg (e.g. sports utility vehicles) will be required to meet Euro 5 emission limits for passenger cars (rather than those for light commercial vehicles as at present) from 1 September 2012. From 1 September 2014 (new type approvals) and 1 September 2015 (new existing models) diesel cars will need to meet Euro 6 standards for NO_x emissions.

Manufacturers will also be required to guarantee the durability of emission control devices (e.g. catalytic converters and particulate traps) for 160,000 km instead of the current 80,000 km. Information on vehicle repairs will need to be made easily available to independent repairers.

Commission Regulation 692/2008 of 18 July 2008 implements and amends Regulation 715/2007, establishing administrative provisions for EC type approval for light duty vehicles, including provisions for conformity of production and in service conformity.

Among the Directives to be consolidated into the Regulation and repealed on 2 January 2013 are: 70/220/EEC; 80/1268/EEC; 89/458/EEC; 91/441/EEC; 93/59/EEC; 94/12/EEC; 96/69/EEC; 98/69/EC and 2004/3/EEC.

Table 3.16: Regulation 715/2007: Euro 5 and 6 Standards

	Euro 5 Petrol mg/km	*Diesel mg/km*	*Euro 6 Diesel mg/km*
Hydrocarbons	100		
Nitrogen oxides	60	180	80
HC + NOx	-	230	170
Particulates	5*	5	
Carbon monoxide	1000	500	

* lean-burn direct injection engines only

Design & Manufacture

Directive 2005/64/EC, adopted in November 2005, covers the reusability, recylability and recoverability of motor vehicles with the aim of ensuring that vehicles are designed and manufactured in such a way as to ensure compliance with the requirements of the Directive on End of Life Vehicles (2000/53/EC) – see also chapter 5, 5.11 and 5.31. It had to be transposed into national legislation by 15 December 2006. From 15 December 2008, member states must refuse the registration, sale or entry into service (type approval) of new vehicles which are not reusable and/or recyclable to a minimum of 85% by mass, and reusable and/or recoverable to a minimum of 95%; from 15 July 2010 the sale of vehicles not meeting type approval will be banned. The Directive does not apply to special purpose vehicles, multi-stage built vehicles and those produced in small series.

3.26.3 Light Commercial Vehicles

A further amendment to Directive 70/220/EEC (Directive 93/59) adopted by Environment Ministers in June 1993 set emission limits for light commercial vehicles which were as strict as those for private vehicles (Euro 1); vehicles were categorised according to their reference mass, i.e. the mass of the unladen vehicle in running order, as follows:

- Category I: commercial vehicles with reference mass up to 1250 kg and vans/cars designed to carry up to nine people;
- Category II: commercial vehicles with reference mass between 1250-1700 kg;
- Category III: commercial vehicles with reference mass between 1701-3500 kg.

In October 1996 the Council of Ministers formally adopted Directive 96/69/EC (Table 3.17) – Euro 2; this brought emission limits in line with those of Directive 94/12/EC. For Category I the limits applied to new models from 1 January 1997 and to new vehicles from 1 October 1997; for Categories II and III the respective dates were 1 January 1998 and 1 October 1998. Member States could use tax incentives as a means of encouraging the introduction of vehicles meeting the new emission limits before the mandatory dates.

These Directives have been implemented by amendments to the *Construction and Use Regulations* (SI 1993/2199 & 1995/2210) and Type Approval Regulations (see below, 3.37.2 and 3.37.3).

Directive 98/69 which tightens emission limits for passenger vehicles (see above and Table 3.15) also covers limits for light commercial vehicles (Euro 3/4).

These Directives will be repealed on 2 January 2013 by Regulation 715/2007 implementing Euro 5 emission limits for both light duty and light commercial vehicles, see above and Table 3.16.

3.26.4 Motorcycles and Mopeds

Directive 97/24/EC, introduced limits for CO and HC and NO_x emissions from motorcycles, mopeds, tricycles and quadricycles; limits have now been tightened through the adoption in September 2002 of a *Directive on the reduction of the level of pollutant emissions from two and three-wheel motor vehicles* (2002/51/EC) which also amends the 1997 Directive. New limits (Stage 1) applied to new vehicle types from 1 April 2003 and to all new vehicles from 1 July 2004 (trial and enduro motorcycles – 1 January 2004 and 1 July 2005), with further reductions (Stage 2) applying from 1 January 2006/2007 – see Table 3.18. The Directive, which had to be implemented by 1 April 2003, also includes testing procedures and permits tax incentives for vehicles meeting the limits ahead of the specified dates. Directive 2003/77/EC defines the test cycles.

Directive 2002/24/EC, as amended by 2005/30/EC relating to type approval of two or three wheeled vehicles was implemented through an amendment to the 1999 *Motor Cycle (EC Type Approval) Regulations* (SI 2006/2935).

3.26.5 Heavy Duty Vehicles

Directive 2005/55/EC which was adopted on 28 September 2005 consolidates various Directives relating to emissions from, and emissions testing of, heavy duty vehicles (lorries and buses over 3.5 tonnes); it came into force on 9 November 2005, with Member States required to implement it by 9 November 2006, when the earlier Directives were repealed – i.e. 88/77/EEC, 91/542/EEC, 96/1/EC; 1999/96/EC and 2001/27/EC. Directive 2005/78/EC contains the technical measures necessary to implement Directive 2005/55/EC. Directive 2006/51/EC of 6 June 2006 amends both the 2005 Directives to introduce improved requirements relating to the verification of operational conditions, failures and demonstration of emission control monitoring systems, with further amendments contained in Directive 2008/74/EC of 18 July 2008.

The Directive outlines the measures to be taken against the emission of gaseous and particulate pollution from compression-ignition engines for use in vehicles and the emission of gaseous pollution from positive ignition engines fuelled with natural gas or liquefied petroleum gas for use in vehicles. Emission limits from Directive 1999/96/EC are restated (Euro IV & V – see Table 3.19), with the Commission required to review the need for emission limits for as yet unregulated pollutants based on the introduction of new alternative fuels and on the introduction of new additive enabled exhaust emission control systems. The current emission limits (Table 3.19) applied to new vehicle and engine types from 1 October 2000 and a year later to new engines and vehicles (Euro III), with further cuts in emission limits to be met by 1 October 2005 (Euro IV). A further reduction in NO_x emissions to 2.0 g/kWh has been agreed for 2008 (Euro V). The limit values to be applied depend on the test cycle used: ESC and ELR test cycles for conventional diesel engines, including those fitted

Table 3.18: Limits for Motorcycles etc (2002/51/EC)

	Emission limit (g/km)		
	CO	HC	NOx
Motorcycles & scooters (97/24)			
Two-stroke	8.0	4.0	0.1
Four-stroke	13.0	3.0	0.3
Motorcycles & scooters (2002/51)			
Stage 1 (from 1.4.03/1.7.04)			
Two-stroke & Four-stroke	5.5	1.2	0.3
Stage 2 (from 1.1.06/1.1.07			
Two-stroke & Four-stroke	2.0	0.8	0.15
Tri- & quadricycles (from 1.4.03/1.7.04)			
Petrol engined	7.0	1.5	0.4
Diesel engined	2.0	1.0	0.65

Table 3.17: 96/69/EC

	Emission limit (g/km)				
	CO		HC+NOx		Particulates
	P*	D*	P	D**	D**
Category I	2.2	1.0	0.5	0.9/0.7	0.10/0.08
Category II	4.0	1.25	0.6	1.3/1.0	0.14/0.12
Category III	5.0	1.5	0.7	1.6/1.2	0.20/0.17

* P = petrol; D = diesel
** the first limit value applies to direct injection vehicles until 30.9.99

with electronic fuel injection equipment, exhaust gas recirculation and/or oxidation catalysts; ESC, ELR and ETC test cycles for diesel engines fitted with advance emission control systems, including de-NOx catalysts and/or particulate traps, and also gas engines. Tax incentives may be used to encourage early compliance with emission limits. In addition all new type approval vehicles have been required to carry onboard diagnostic systems since 1 October 2005, with all other new vehicles required to carry them from 1 October 2006. The same dates applied for manufacturers being required to demonstrate the durability of emission control systems, with the length of time or kilometres depending on the type of vehicle.

In December 2005 the Commission published proposals for a Directive (COM(2005) 634) which will require public bodies – state, regional, and local authorities, and operators contracted to provide a public service – to allocate a minimum of 25% of their annual procurement (both purchase and leasing) of heavy duty vehicles to "enhanced environmentally friendly vehicles (EEV); the aim of the Directive is to encourage the development of vehicles adapted to high blends of biofuels, and to other technologies such as natural gas, LPG, hydrogen, electric and hybrid vehicles – this has been replaced by a proposal covering procurement of all vehicles used by public bodies (COM(2007)817) – see 3.26.8 below.

In December 2007, the Commission published a *Proposal for a Regulation on type approval of motor vehicles and engines with respect to emissions from heavy duty vehicles and on access to vehicle repair and maintenance information* (COM(2007) 851; the Regulation sets rules for heavy vehicles over 2,610 kg (Euro VI standards) to apply from 1 April 2013 for new types of engine or vehicle and from 1 October 2014 for new vehicle registrations – see Table 3.20 below; The proposed emission limits would result in a reduction of 80% in NO_x emissions and 67% reduction in particulates, compared with Euro V standards. The Regulation includes a requirement that vehicle onboard diagnostic information be made available on websites in a standardised format, and for type-approval of exhaust after-treatment components. Tax incentives for new vehicles put on the market which comply with the Regulation early may be offered until 1 October 2014. Directives 2005/78/EC and 2005/55/EC, and 2006/51/EC are all to be repealed from 1 October 2014.

The 2005 and 2006 Directives have been implemented by amendment of the Construction & Use and Type Approval Regulations (SI 2006/2565).

The Commission's *Greening Transport Package* adopted on 8 July 2008 includes a proposal for a Directive (COM(2008) 436) which would enable Member States to levy charges on heavy duty vehicles for costs related to air pollution, noise and traffic congestion.

3.26.6 Non-Road Mobile Machinery

In December 1997 Environment Ministers adopted Directive 97/68 (subsequently amended by Directive 2002/88/EC) relating to emissions of gaseous and particulate pollutants from non-road mobile machinery (NRMM). This Directive applies to agricultural and forestry equipment, snow ploughs, fork lift trucks, mobile cranes and construction equipment. Stage I limits were phased in between 30 September 1998 to 31 March 1999 and Stage II limits between December 2000 and 31 December 2003.

Directive 2002/88/EC, adopted at the end of 2002 and brought into force on 11 February 2003, amends Directive 97/68 to limit emissions from both small (< 19 kW) non-road petrol engines, such as lawn mowers, and other garden equipment. A first stage of emissions reductions came into force on 11 August 2004, with a second stage of reductions coming into force between 1 August 2004 and 1 February 2009, depending on the type and size of engine.

Table 3.20: Proposed Euro VI Emission Limits

	ESC/CI[1]	ETC/CI[2]	ETC/PI[3]
CO (mg/kWh)	1500	4000	4000
THC (mg/kWh)	130	160	
NMHC (mg/kWh)			160
CH_4 (mg/kWh)			500
NO_x (mg/kWh)	400	400	400
NH_3 (ppm)	10	10	10
PM mass (mg/kWh)	10	10	10

1. European steady state [test] cycle, compression ignition engines
2. European transient [test] cycle, compression ignition engines
3. European transient [test]cycle], positive ignition engines

Table 3.19: 2005/55/EC

Limit values for engines tested on ESC and ELR test cycles (g/kWh, except smoke)

	2000	2005	2008
Mass of CO	2.1	1.5	
Mass of HC	0.66	0.46	
Mass of NOx	5.0	3.5	2.0
Mass of Particulates	0.10*	0.02	
Smoke (m^{-1})	0.8	0.5	

(*Derogation to 1.10.05 - 0.13 for engines with swept vol < $0.7dm^3$ per cylinder and rated power > 3,000 min^{-1})

Limit values for engines tested on ESC, ELR & ETC test cycles (g/kWh)

	2000	2005	2008
Mass of CO	5.45	4.0	
Mass of NHMC	0.78	0.55	
Mass of methane	1.6*	1.1	
Mass of NOx	5.0	3.5	2.0
Mass of Particulates	0.16**	0.03	

(*for natural gas engines only; **Derogation to 1.10.05 - 0.21 for engines with swept volume < 0.7 dm^3 per cylinder and rated power of > 3,000 min^{-1} (diesel engines only))

These Directives have been implemented through *The Non-Road Mobile Machinery (Emission of Gaseous and Particulate Pollutants) Regulations 1999* (SI 1999/1053), as amended by 2002 and 2004 Regulations (SI 2002/1649 and 2004/2034).

Directive 2004/26/EC, adopted in April 2004, amends 97/68; it applies to diesel fuel engines from 19-560 kW, and tightens emission limits for NOx+HC by 30-40% (compared to Stage II) from between 30 June 2005 and 31 December 2010 (depending on engine size) – Stage IIIA; Stage IIIB, which will come into force between 1 January 2010 and 31 December 2014 (except for inland waterway vessels), will tighten particulate limits by about 90%. The Directive extends the definition of NRMM to new engines in railcars and sets emission standards for inland waterway vessel engines, but will not apply to machinery already in use at the time of the Directive's agreement even if the engine needs replacing. The Commission is to review the second stage standards before 31.12.07 in case they need adjustment because of technical difficulties.

The 2004 Directive has been implemented by a further amendment to the 1999 Regulations – SI 2006/29; SI 2008/2011 amends the 1999 Regulations to take account of Directive 2006/105/EC on the accession of Bulgaria and Romania to the EU.

3.26.7 Agricultural & Forestry Tractors

In May 2000, Directive 2000/25/EC on action to be taken against the emission of gaseous and particulate pollutants by engines intended to power agricultural or forestry tractors was adopted. Emission limits are similar to those for non-road mobile machinery covered by Directive 97/68/EC. Stage I limits became mandatory for all new tractor engines on 1 July 2001 and Stage II took effect between 1 January 2002 and 1 January 2004 depending on engine size.

The Directive has been implemented through *The Agricultural or Forestry Tractors (Emissions of Gaseous and Particulate Pollutants) Regulations 2002*; these Regulations, which came into force on 12 August 2002, apply in England, Wales and Northern Ireland. Directive 2005/13/EC adopted on 1 March 2005 implements standards similar to those agreed for NRMM to come into force, depending on engine category in 2006-2008 (Stage IIIA), 2010-2012 (Stage IIIB), and 2014 (Stage IV). 2006 Amendment Regulations, SI 2006/2393, implement the 2005 Directive.

3.26.8 Carbon Dioxide Emissions

In June 1996 Environment Ministers endorsed the Commission's proposal for *A Community strategy to reduce carbon dioxide emissions from passenger cars and improve fuel economy* – Communication COM(95) 689. This forms part of the overall strategy to stabilise emissions of carbon dioxide at 1990 levels by the year 2000 – see this chapter 3.6.3. An important element of the strategy included trying to reach agreement with the auto industry to reach a specific CO_2 emission target for cars within an agreed timeframe: the Commission suggested a 33% reduction (186g/km to 120 g/km) in the average of CO_2 emissions of new cars sold in the EU by 2010 at the latest compared with 1990; the European Car Manufacturers Association (ACEA) agreed that average emissions should be reduced to 140 g/km by 2008 (2006 average was 160 g/km). In 1999, Japanese and South Korean car manufacturers agreed to a target of 140g/km by 2009 (2006 averages 161 g/km and 164 g/km, respectively).

As well as supporting research and development programmes to improve the performance of cars and the promotion of alternative modes of transport, the Strategy included a proposal for monitoring CO_2 emissions from passenger cars (given effect through Decision 1753/2000) and for the promotion of fuel efficient passenger cars and fuel economy labelling – Directive 99/94/EC requires new passenger cars to carry a label on fuel consumption and CO_2 emissions and for information on fuel economy to be available at the point of sale. Directive 2003/73/EC amends Annex III to the 1999 Directive specifying the format of such information at the point-of-sale, on both posters and electronic screens. In June 2008, the Commission initiated an online consultation seeking views on how the Directive could be amended to provide information on fuel consumption and CO_2 emissions in a more informative way.

The Strategy also envisages further research aimed at reducing emissions from new cars to an average of 95g CO_2/km by 2020, measures to promote fuel efficient vehicles (notably through improved labelling, and encouraging car taxes based on CO_2 emissions. Finally car manufacturers were invited to sign up, by mid-2007, to a voluntary EU code of good practice on car marketing and advertising aimed at promoting more sustainable consumption patterns.

The Commission has now reviewed the Strategy and in a Communication (COM(2007) 10) published 24 January 2007 concluded that mandatory targets would be required if its target of 120 g/km CO_2 is to be met by 2010. Accordingly, in December 2007, it published a *proposal for a Regulation setting performance standards for new passenger cars as part of the Community's integrated approach to reducing CO_2 emissions from light duty vehicles.* COM(2007)856) will require average emissions from new cars sold in the EU-27 to be reduced to 120 g CO_2/km by 2012; a reduction to 130 g/km is to be achieved through improvements in technology, and the remaining 10 g/km through efficiency measures such as tyres and air conditioning systems, and reduction in the carbon content of road fuels through increased use of biofuels. The legislation defines a "limit curve" of CO_2 emissions allowed by new vehicles according to the mass of the vehicle; manufacturers whose measured fleet average emissions are above the curve in 2012 will be charged an "excess emissions premium".

UK initiatives aimed at reducing carbon dioxide emissions from vehicles are summarised below at 3.27.4.

Clean and Efficient Vehicles

COM(2007) 817, published in December 2007, is a *proposal for a Directive on the promotion of clean and energy efficient road transport vehicles.* It will require public authorities to include operational lifetime costs of energy consumption, CO_2 and other pollutant emissions in award criteria for all procurement of road transport vehicles by public authorities and by operators providing services under contract, including passenger transport services.

Car Safety Package

Published in May 2008 by the Commission, this includes a proposal for a *Regulation concerning type-approval requirements for the general safety of motor vehicles* – COM(2008) 316. Among the measures proposed are requirements for the type approval of tyres with regard to their CO_2 and noise emissions. Low rolling resistance tyres are to become mandatory in 2012; this will reduce fuel consumption and CO_2 by up to 7gr per km. See also chapter 4, 4.19.3 with regard to tyre noise.

3.26.9 Fuel Quality

Directive 85/210/EEC, set a maximum of 0.15 grammes per litre of lead in petrol, and limited benzene to 5% by volume; it also required that unleaded petrol was to be made available in Member States by 1989, new model cars designed to run on it from 1989, and all new cars to run on unleaded petrol from 1990. An amendment to this Directive (87/416/EEC) permitted Member States to ban the sale of regular grade leaded petrol with six months notice.

Directive 93/12/EEC limits the sulphur content of all gas oils (except aviation kerosene) to 0.2% by weight as from 1 October 1994, with a further reduction in the sulphur content of diesel fuel to 0.05% by weight as from 1 October 1996. The Directive also required Member States to ensure that diesel fuel with a sulphur content of 0.05% or less became available from 1 October 1995, thus enabling compliance with EU Directive 91/542/EEC on emissions from heavy duty vehicles (see 3.26.5). (As part of its strategy to combat acidification (see 3.16), Directive 99/32 adopted on 26 April 1999 amends Directive 93/12/EC to reduce the sulphur content of gas oil to 0.1% by weight from 1 January 2008.) Directive 93/12 is to be repealed following adoption of a new Directive on the specification of petrol, diesel and gas oil – see below.

In September 1998, a Directive relating to the quality of petrol and diesel fuels and amending Directive 93/12/EEC was formally adopted. This Directive, 98/70/EC, arises out of the Auto/Oil Programme, and includes the following provisions:

- A ban on the marketing of leaded petrol from 1 January 2000; five year derogation for those Member States where this timetable may result in "severe socio-economic problems or would not lead to overall environmental or health benefits".
- Reducing the amount of benzene in petrol from 5% to 1% by volume by 2000, and olefins to 18%; aromatics to be limited to 42% by volume and sulphur to 150 mg/kg (with a three year derogation for those Member States where this would cause "severe socio-economic problems"). For diesel, polycyclic aromatic hydrocarbons are to be limited to 11% by volume and sulphur to 350 mg/kg. Sulphur in petrol must be reduced to 50 mg/kg from 2005 (to be phased in for diesel) and a limit of 35% aromatics. The summertime (June - August) Reid Vapour pressure is limited to 70 kPa; distillation evap/ted at 100C is limited to 46 v/v and at 150C to 75% v/v.
- From 2000 Member States could give tax incentives for fuels already meeting the 2005 limits.

On 3 March 2003, Directive 2003/17/EC amending Directive 98/70/EC was adopted. From 1 January 2009 all petrol and diesel road fuel must be "sulphur free" (i.e. max. 10 ppm) – this date was to be confirmed for diesel fuels no later than 31 December 2005 (see next para). Sulphur free petrol and diesel had to be available "on an appropriately balanced geographical basis" by 1 January 2005. The sulphur content of diesel fuels for use by non-road mobile machinery and agricultural and forestry tractors should not exceed 2000 ppm and by 1 January 2008 should not exceed 1000 ppm.

A further amendment to the 1998 Directive (COM(2007) 18) published in January 2007 covers the specification of petrol, diesel and gas oil and introduces a mechanism to monitor and reduce greenhouse gas (GHG) emissions from the use of road transport fuels; it also amends Directive 1999/32/EC on the specification of fuel to be used by inland waterway vessels (see 3.16.3 and chapter 6, 6.23.2). This Directive will require fuel suppliers to reduce GHG emissions caused by the production, transport and use of fuels for road transport and non-road mobile machinery by 10% between 2011-2020 (from 2010 baseline) – mandatory monitoring and reporting of the lifecycle GHG emissions from those fuels is to be required from 1 January 2009. The PAH content of diesel is to be cut from 11% by volume to 8%; the proposed Directive also confirms that from 1 January 2009 all diesel fuels must be sulphur free (i.e. max. 10 ppm). The sulphur content of gas oil for non-road mobile machinery and agricultural and forestry tractors must not exceed 1000 ppm from 1 January 2008, reducing to 10 ppm from 1 January 2010; the sulphur content of gas oils for inland waterway vessels must not exceed 1000 ppm from 1 January 2008, 300 ppm from 1 January 2010 and 10 ppm from 1 January 2012.

To enable a higher volume of biofuels to be used in petrol, the Directive proposes establishing a petrol blend with a higher permitted content of oxygen-containing additives ("oxygenates"), including up to 10% ethanol. Petrol pumps selling such blends would have to be clearly marked as they may not be suitable for all vehicles. In addition the Commission will bring forward proposals for mandatory vapour recovery equipment at filling stations because of the increase in emissions of polluting vapours resulting from greater use of ethanol.

Directive 98/70/EC has been implemented in the UK through the *Motor Fuel (Composition and Content) Regulations 1999*, made under the *Clean Air Act 1993*, Part IV, with 2003 amendment Regulations (SI 2003/3078) implementing the 2003 Directive. Further amendment Regulations (SI 2007/1608) the Directive's requirement for sulphur-free fuel to be available on a geographically balanced basis prior to 1 January 2009. – see below, 3.27.1.

3.26.10 Biofuels

Directive 2003/30 on the *Promotion of the Use of Biofuels and other Renewable Fuels for Transport*, adopted in May 2003, lays down targets for the progressive introduction of biofuels derived from agricultural, forestry and organic waste products. Member States must set targets for the market share of biofuels using the Directive's benchmarks of 2% by December 2005, rising to 5.75% by December 2010. Member States setting lower targets must justify their decision to the Commission. The Directive, which had to be transposed into national legislation by 31 December 2004, also requires Member States to ensure that all road fuels comprise a minimum of 1% biofuel by 2009, rising to 1.75% in 2010.

The Biofuel (Labelling) Regulations 2004 (SI 2004/3349), which came into force on 1 February 2005 and extend to the whole of the UK, require a warning label on pumps dispensing biofuels or fuel blends containing more than 5% of biofuel by volume to show that the fuel may not be suitable for all vehicles. As from 15 April 2008, the "Renewable Transport Fuels Obligation" (RTFO) has required that 2.5% of all road fuels sold in the UK must come from a renewable source, rising to 5% in 2013/14 (previously 2010; DfT announcement 7 July 2008). The RTFO includes a requirement for fuel suppliers to submit reports on the biofuels they deliver; this should cover matters such as the carbon saving compared with the fossil fuels replaced, where the fuel comes from and whether the feedstock is grown to a recognised sustainability standard.

In February 2006, the Commission published an *EU Strategy for Biofuels* (COM(2006) 34); this aims to promote biofuels in both the EU and developing countries; prepare for large-scale use of biofuels by improving their cost competitiveness and increasing research into second-generation fuels; and support developing countries where biofuel production could stimulate economic growth.

The Commission's "Climate Action and Renewable Energy Package", published January 2008, includes a proposal for a *Directive on the promotion of the use of energy from renewable sources* (COM(2008)19) – this suggests a target of 10% biofuel content for transport fuels by 2020 – see earlier this chapter 3.6.1.

3.26.11 Petrol Storage & Distribution

In December 1994, Ministers formally adopted a Directive (94/63/EC) on controlling VOC emissions resulting from the storage of petrol and its distribution from terminals to service stations ("Stage I" controls). The Directive, which came into force on 31 December 1995, also covers the loading of petrol on to rail or new oil tanker lorries and inland waterways vessels. The Directive aims to reduce emissions from these sources by 90% over ten years through the use of technical measures such as vapour recovery, painting storage tanks with reflectance paint and bottom loading of tankers. Design and operating standards for installations and equipment have been set so as to reduce vapour losses to below a "target reference value" of annual throughput.

New storage installations, the loading and unloading of new road or rail tankers at new terminals or new petrol stations (i.e. not authorised or operated before 31 December 1995) had to comply with the new standards by 31 December 1995. Existing storage installations with an annual throughput of more than 50,000 tonnes, loading and unloading of rail tankers and barges at terminals with an annual throughput of more than 150,000 tonnes had a further three years to meet the Directive, as did the loading and unloading at petrol stations with an annual throughput of more than 1,000 m^3; where the annual throughput exceeds 25,000 tonnes at existing storage installations or terminals loading and unloading road and rail tankers and petrol stations with an annual throughput of more than 500 m^3, the compliance date was 31 December 2001; remaining installations, terminals and petrol stations had a further three years. Where vapour emissions at storage installations, terminals or petrol stations could cause adverse health or environmental effects, stricter requirements may be imposed.

Small service stations (below 1,000 m^3 annual throughput) have been given longer to meet the new standards; small rural service stations (100-500 m^3 annual throughput) have a limited derogation from controls where VOC emissions are likely to pose an insignificant threat to the environment; all service stations below 100 m^3 annual throughput have been given an absolute derogation.

In early 2008, the Commission issued a consultation document seeking views on the mandatory introduction of Stage 2 petrol vapour controls at service stations (to control emissions during refuelling of passenger cars).

3.26.12 Roadworthiness Testing

The EU Roadworthiness Directive (96/96/EC) was formally adopted on 20 December 1996 and had to be implemented by Member States by 9 March 1998. It consolidates and repeals Directive 77/143/EEC and subsequent amendments and applies to all motor vehicles and their trailers. Member States may, however, exclude vehicles of historic interest from testing. The Directive lays down the periodicity of testing and the items to be checked, including exhaust emissions, and requires Member States to ensure that vehicles be issued with proof that they have complied with testing requirements. This Directive, and the in-service emission requirements, have been implemented through amendments to the Construction and Use Regulations – see 3.27.2c.

Directive 2001/9/EC amends the 1996 Directive in line with technical progress enabling Member States to use On Board Diagnostics (OBD) in lieu of part of the roadworthiness inspection. Directive 2003/27/EC sets new limit values for testing for petrol and diesel vehicles in line with the more recent Directives on vehicle emission limits; and 2003/26/EC introduces the same requirements for "unexpected" roadside inspections for commercial vehicles. Both were implemented on 1 January 2004, through a further amendment to the Construction and Use Regulations (SI 2003/3145).

3.27 UK REGULATIONS – ROAD TRANSPORT

3.27.1 Composition of Fuels

The composition of motor fuel is governed by *the Motor Fuel (Composition and Content) Regulations 1999* (SI 1999/3107), made under s.30 of *the Clean Air Act 1993* (see above 3.19.4). These Regulations, which came into force on 1 January 2000, replace 1994 Regulations (SI 2295) and apply throughout the UK. They implement Directive 98/70/EC relating to the quality of petrol and diesel fuels, which also bans the marketing of leaded petrol from 1 January 2000 (see 3.26.9).

The Regulations apply to both the distribution of fuel to, and the sale of fuel from, filling stations; the limits to be met are summarised earlier at 3.26.9. Petrol stations with low throughput will not have to meet the summertime volatility requirements laid down in the Regulations until three deliveries of such fuel have been made. Fuel sold for tests and experimental purposes is exempt from the Regulations.

2003 Amendment Regulations (SI 2003/3078), brought into force on 31 December 2003, implement the requirements of Directive 2003/17/EC which reduces the maximum permissible amount of sulphur in petrol and diesel fuels to 10 ppm from 1 January 2009; the sulphur content of diesel fuels for non-road mobile machinery and agricultural tractors is restricted to no more than 2000 mg/kg, reducing to 1000 mg/kg from 1 January 2008. 2007 Amendment Regulations (SI 2007/1608) implement the Directive's requirement for sulphur-free fuel to be available on a geographically balanced basis prior to 1 January 2009, by specifying that the level of sulphur in diesel fuel and super unleaded petrol sold before 1 January 2009 must not exceed 10 mg/kg at a filling station at which at least 3 million litres of motor fuel were sold in the preceding calendar year. To satisfy summer petrol requirements, the vapour pressure should not exceed 70 kPa.

The 1999 Regulations enable the Secretary of State to grant temporary exemptions from compliance with one more of the requirements for petrol and diesel in the event of exceptional events resulting in refineries being unable to meet any of the standards, so long as this has been authorised by the Commission. Further amendment Regulations (draft published early 2008) will allow the Secretary of State to grant temporary exemptions from one or more of the standards for gas oil components.

Leaded Petrol

Leaded petrol was withdrawn from general sale on 1 January 2000; the Directive allows for a limited amount of leaded petrol to be sold for use in old vehicles – i.e. classic and historic vehicles – and this is to be made available through a permit system under the Regulations. Producers of petrol for use in the UK and importers of petrol into the UK, who are also members of the Federation of British Historic Vehicle Clubs, may apply to the Secretary of State for a permit (or permits). Initially, permits are to be valid for one year to enable amendments to the system to deal with any problems; thereafter permits will be valid for three years.

An application for a permit should include full details of the applicant, evidence showing that he is a producer or importer of petrol into the UK and a registered member of the FBHVC. (A 2001 amendment to the Regulations (SI 2001/3896) removed the requirement for permit holders to provide detailed information, both when applying for a permit, and subsequently monthly, on the distribution of leaded petrol to the nominated filling stations.) Permit holders must agree to sell leaded petrol for the duration of the permit. In common with other permit and environmental licence systems, the Regulations provide for permits to be varied, revoked or surrendered, as well as providing for appeals to the Secretary of State on any matters relating to a permit. It should be noted that permits may normally be surrendered on one month's notice; holders of single permits must however give six months' notice of their intention to cease the distribution and sale of leaded petrol – this is to enable the Secretary of State to find another permit holder.

3.27.2 Construction and Use

The main objectives of *The Road Vehicles (Construction and Use) Regulations 1986* (SI 1078) (N. Ireland SR 1999/454), as amended, are to ensure vehicles are manufactured to high standards, and to ensure such standards are maintained while vehicles are in-use; they also implement EU Directives on emission limits and roadworthiness testing – details of the relevant SI are given in the text with the Directives (see previous section 3.26).

(a) Use of Unleaded Petrol

The *Road Vehicles (Construction and Use) (Amendment No. 6) Regulations 1988, Motor Vehicles (Type Approval) (Great Britain) (Amendment) Regulations 1988* and *Motor Vehicles (Type Approval for Goods Vehicles) (Great Britain) (Amendment) Regulations 1988* which came into force on 1 October 1988 required that all new vehicles should be capable of running on unleaded petrol in the 1990s. Implementation dates were 1 October 1989 for new type approvals and 1 October 1990 for existing model types. Special arrangements covered the small number of models that would need major engineering changes to comply.

(b) Compliance with Emission Standards

Under Regulation 61(5) of the *Road Vehicles (Construction and Use) Regulations 1986* (SI 1078), as amended, it is an offence to use a vehicle if it is emitting "smoke, visible vapour, grit, sparks, ashes, cinders or oily substances" in such a way as is likely to cause "damage to any property or injury or danger to any person". The Regulations also require vehicle users to keep engines in tune and any emission control equipment, such as catalysts, working efficiently.

These Regulations are regularly amended to give effect to EU Directives on vehicle emissions and emissions testing, and to update the Department for Transport publication on In-Service Exhaust Emission Standards for Road Vehicles.

The V*ehicle Excise Duty (Reduced Pollution) Regulations 1998*, effective 1 January 1999, enable HGVs and buses meeting particulate standards in EU Directive 99/96/EC (replaced by 2005/55/EC) to qualify for a Reduced Pollution Certificate, renewable annually when the vehicle undergoes its roadworthiness test, and thus a reduction in vehicle excise duty of up to £500. 1998 Type Approval Regulations prescribe requirements for reduced pollution devices.

The EU Directives on emissions from non-road mobile machinery (see 3.26.6 above) have been implemented through *the Non-Road Mobile Machinery (Emissions of Gaseous and Particulate Pollutants) Regulations 1999* (SI 1999/1053, as amended).

London

Proposals for introducing a Low Emission Zone covering Greater London were published by the Mayor of London in October 2005 and the LEZ established on 4 February 2008. Diesel engined vehicles with a gross vehicle weight of over 3.5 tonnes are required to meet Euro III standards for particulates in order to drive through the LEZ at no charge; this covers lorries, motor caravans, motorised horseboxes, buses and coaches, and minibuses with more than 8 seats. Operators of non-compliant vehicles will need to pay a charge of £200 for each charging day they are driven in the zone. From January 2012, all these vehicles will be required to meet Euro IV standards. From October 2010, light goods vehicles (1.205 tonnes unladen, and 3.5 tonnes gvw) and passenger vehicles with more than 8 seats and below 5 tonnes gvw will be required to meet Euro 3/III standards to be able to drive through the LEZ free of charge.

The Mayor of London's Taxi Emissions Strategy will require taxis to fit abatement technology in order to meet Euro 3 Standards by 1 January 2008, (see 3.26.2 and Table 3.15). This requirement is being implemented progressively with the oldest taxis (i.e. pre-Euro emissions standards) required to meet the standard from 1 July 2006, those currently meeting Euro 1 from 1 January 2007, and those currently meeting Euro 2 from 1 January 2008.

A further scheme launched by the Mayor in March 2008 aims to reduce exhaust emissions from machinery used on construction and demolition sites. An accreditation scheme run by the Energy Saving Trust provides companies with a list of suitable devices for fitting to exhausts of the machinery. The programme offers independent certification of after-treatment equipment to ensure that it meets the requirements of the London-wide Best Practice Guide to cutting air pollution from construction sites (launched in 2006).

(c) Vehicle Testing

In-service emission requirements for all vehicles in line with EU Directives are specified in amendments to the Construction and Use Regulations.

Petrol engined vehicles first used before 1 August 1975 are tested visually and checks on vehicles first used after this date are by metered test; diesel engined vehicles first used before 1 August 1979 are tested visually and those first used after that date by metered smoke test, or may comply with a "relevant instrument" – e.g. EU Directive 72/306/EEC as amended, which prescribes limit values for the pollutants contained in the smoke emitted by vehicles. The standards are enforced through Type Approval Regulations and checked at the annual MOT (or roadworthiness check for HGVs) and at roadside checks.

From 1 January 1996 vehicles with advanced emissions control systems (e.g. catalytic converters) and other newer vehicles listed in Department for Transport (DfT) publication *In-service Exhaust Emission Standards for Road Vehicles* (this is regularly updated to include new models coming on to the market) had to meet emission limits specified by the manufacturer of the particular model. Since 1 August 1997, vehicles with more than five seats (plus the driver) or vans or other vehicles of 3.5 tonnes or less, first used from 1 August 1994 have also been required to meet the standards for vehicles with advanced emission control systems.

For all vehicles the Regulations require that "no excessive smoke" (dense blue or clearly visible black smoke) should be emitted from the exhaust after 5 seconds at idling.

Vehicle owners have 14 days in which to have the necessary repairs carried out and the vehicle rechecked, or face prosecution with a possible fine. Owners who have neglected to maintain their vehicles are prosecuted and face a fine as well as the cost of putting their vehicle right.

London taxis are tested annually at the Public Carriage Office and have to meet requirements which are equal to those of the MOT.

The Regulations are regularly amended to include new model vehicles coming onto the market, and the MOT updated to reflect new emission requirements.

(d) Steam Road Vehicles

Steam road vehicles – such as traction engines and steamrollers – used on the public highway are not liable to statutory nuisance action under the *Environmental Protection Act 1990*. They are however, subject to the *Road Vehicles (Construction and Use) Regulations*.

3.27.3 Type Approval

Before a new vehicle model is introduced on to the market it must be tested by the Type Approval Authority to ensure it complies with relevant EU Directives. All new vehicles produced must then meet Conformity of Production limits. These are set at 15-30% above type approval limits, depending on the pollutant.

Passenger cars, including three wheeled cars, first used on or after 1 August 1978 (and not manufactured before 1 October 1977) are subject to the *Motor Vehicles (Type Approval) (Great Britain) Regulations 1984* as amended. Goods vehicles are subject to the *Motor Vehicles (Type Approval for Goods Vehicles) (Great Britain) Regulations 1982*, as amended. The main effect of *Type Approval Regulations* is the enforcement of construction requirements and of construction requirements necessitated by EU and other legislation – e.g. to enable compliance with emission limits. 2007 Regulations implement Directive 2005/64/EC imposing requirements relating to reusability, recyclability and recoverability (see also chapter 5, 5.11). Specimen vehicles are tested by the DfT to ensure compliance with the international standards referred to in *Construction and Use Regulations*. No routine testing of new vehicles for this purpose is carried out under the Regulations.

3.27.4 Carbon Dioxide Emissions

In 2006 average CO_2 emissions from new cars bought in the UK were 167.2 g/km (2005: 169.4 g/km) (Society of Motor Manufacturers & Traders).

The Passenger Car (Fuel Consumption and CO_2 Emissions Information) Regulations 2001 (SI 2001/3523), which took effect on 21 November 2001 and apply throughout the UK, implement EU Directive 99/94/EC requiring new passenger cars to carry a label on fuel consumption. 2004 amendment Regulations (SI 2004/1661) implement Directive 2003/73/EC which specifies the format of such information at the point-of-sale, on both posters and electronic screens. The DfT has issued guidance (November 2001, revised 2003) for enforcement officers, car dealers and car manufacturers on complying with the Regulations.

In the UK, a voluntary energy labelling scheme for new cars was launched by the Government in 2005; the label to be displayed in car showrooms will enable consumers to compare energy ratings of different vehicles.

From April 2008, vehicle excise duty for Band G petrol and diesel vehicles (CO_2 emissions of 226 g/km is to rise to £300; for Band A vehicles (100 g/km CO_2 and below) will be exempt, and VED for Band B vehicles (100-120 g/km CO_2) will be £35 (Budget Statement, April 2007).

3.27.5 Petrol Storage and Distribution

EU Directive 94/63/EC "on controlling VOC emissions resulting from the storage of petrol and its distribution from terminals to service stations" ("Stage I" controls) is now implemented in England and Wales through *the Environmental Permitting (England and Wales) Regulations 2007* (Sch.18 refers), and in Scotland and Northern Ireland through *the Pollution Prevention and Control Regulations* (Scotland SSI 2000/323; Northern Ireland: SR 2003/46). These Regulations replace, respectively, controls under the 1990 *Environmental Protection Act* and, in NI the 1997 *Industrial Pollution Control Order.*

Existing installations with an annual throughput of between 100-500 m^3 (i.e. in operation or for which planning permission had been received before the end of 1995) require an authorisation, depending on annual throughput, by the end of 1998, 2001 or 2004; new service stations with an annual throughput of between 100-500 m^3 in rural areas of Northern Scotland (where vapour emissions are unlikely to have an adverse environmental or health effect) do not require an authorisation.

The derogation permitted by the 1994 Directive to exempt existing petrol stations with an annual throughput of less than 500 m^3 from the need to install vapour control equipment has been implemented in England and Wales through *the Environmental Permitting Regulations* (NI: amendment to PPC Regulations, SR 2004/507). The exemption applies in those areas where the service station is located in a geographical area, or on a site where vapour emissions are unlikely to contribute to environmental or health problems. *The Control of Volatile Organic Compounds (Petrol Vapour Recovery) (Scotland) Regulations 2004* (SSI 2004/512) have the same effect in Scotland.

Petrol vapour recovery stage II controls (PVR II) to reduce emissions of VOCs during refuelling of vehicles at petrol stations have also been implemented in England and in Wales through *the Environmental Permitting Regulations* – all service stations with annual petrol throughput greater that 3500 million litres to fit equipment to capture the fumes by 1 January 2010.

An Approved Code of Practice published by the Health and Safety Commission [now Executive] in 2003 aims to reduce the risk of accidents from road tankers unloading petrol at filling stations.

3.27.6 Road Traffic Reduction Acts

The *Road Traffic Reduction Act 1997* was brought into force in England and Wales on 10 March 2000 and in Scotland on 21 April 2000. The Act puts local traffic authorities under a duty to prepare a report assessing local traffic levels and forecast growth; it should specify targets for reducing local traffic levels or for reducing the rate of growth – these can be different for the different areas within an authority (e.g. specific to congested areas or rural areas) and how these are to be achieved. The report must also contain any other information or proposals relating to levels of local traffic or which the Secretary of State may specify in guidance. The Act enables the Secretary of State to issue guidance relating to preparation of the report and on those who should be consulted during its preparation. All reports on traffic levels should be sent to the Secretary of State and published.

The *Road Traffic Reduction (National Targets) Act 1998* requires the Secretary of State and the devolved administrations for Scotland and Wales to prepare and publish reports which include targets for road traffic reduction in England, Scotland and Wales, aimed at reducing the adverse environmental, social and economic impacts of road traffic. If it is considered that traffic reduction targets are not the most appropriate way for reducing the adverse impacts of traffic, including congestion and pollution, the report should explain why and outline alternative measures or targets and their expected impact on traffic reduction. A report, covering England only, published in early 2000, proposes that targets aimed at reducing the adverse impacts of traffic, rather than actual traffic reduction targets, should be set.

Neither of the above Acts extends to Northern Ireland.

3.27.7 The Road Traffic Regulation Act 1984

This Act enables a local authority to make a traffic regulation order for preserving or improving the amenities of an area through which a road runs. The *Environment Act 1995* amends the Act to ensure that such an order could be made for the purpose of meeting statutory requirements in respect of air quality – see this chapter 3.11.

3.27.8 Vehicle Emissions Regulations

The Road Traffic (Vehicle Emissions) (Fixed Penalty) (England) Regulations 2002 (SI 1808) came into force on 18 July 2002, replacing and revoking earlier 1997 Regulations. The Regulations enable certain local authorities in England to stop vehicles for the purpose of checking emissions and enable all local authorities in England to request drivers who are stationary to switch off their engine.

Similar Regulations apply in Wales – SI 2003/300, W.42, which came into force on 1 May 2003 – and in Scotland – SSI 2003/212 which came into force on 21 March 2003.

The Road Traffic (Vehicle Emissions) (Fixed Penalty) Regulations 1997 (SI 3058), which piloted emissions testing in seven local authority areas, have now been revoked.

Emissions Testing

A local authority in England and Wales who has declared a traffic related air quality management area (AQMA) under s.83 of *the Environment Act 1995* (see this chapter 3.11) may apply to the Secretary of State to be designated for the purposes of these Regulations (in Scotland any local authority may apply for designation); adequately trained authorised persons of the authority, or persons authorised by the authority, may then carry out an emissions test, or require an emissions test to be carried out, on a vehicle being driven within the AQMA, or which is about to pass through it, or has passed through it. Guidance on conducting roadside emissions tests has been published by Defra (April 2003).

In those instances where the authorised person has carried out an emissions test on-the-spot and considers that an offence (under reg.61 or 61A of the 1986 Construction & Use Regulations – emissions of smoke etc) has been committed, he should issue a fixed penalty notice (see below); this should be issued "as soon as reasonably practicable", but not more than 24 hours after the test. This is also the case where the vehicle's driver is required to have an emissions test carried out by an authorised person at a specified place and time (within 14 days).

Alternatively, an authorised person may require the vehicle to be presented for examination under s.45 (tests of satisfactory condition of vehicles) of the *Road Traffic Act 1988*. The driver will also be required to produce a test certificate (or notice of failure) on a specified date (within 21 days), time, place, and to a named person. Failure to do so is an offence resulting in a fine not exceeding level 3 on the standard scale. (Production of a test certificate dated before the requirement to produce another one is not satisfactory.)

Stationary Idling Offence

An authorised person may request a driver to switch off their engine if he believes that a stationary idling offence (under reg.98 of the 1986 C & U Regulations – stopping of engine when stationary) is being committed. Failure to comply is an offence resulting in a fine not exceeding level 3 on the standard scale. A fixed penalty notice may also be issued as soon as reasonably practicable but within 24 hours of the offence (see below). The Scottish Government has issued guidance on the use of this power (April 2003).

Fixed Penalty Notices

The amount of the fixed penalty in respect of an emissions offence is £60 and for a stationary idling offence £20, increasing to £90 and £40, respectively, if not paid within the time allowed and no request for a hearing has been made.

The notice should state: the offence and the date on which it was committed; name and address of the person to whom the notice is issued and registration number of vehicle; and amount of penalty. It should also give details of to whom and where payment should be made and any correspondence sent, and the period (not less than 28 days beginning with the date of issue of the notice) within which the penalty should be paid. The notice will also state to whom any request for a hearing should be made, or reduction or waiver of the penalty.

A request may be made for a fixed penalty in respect of an emissions offence to be reduced to £30 (Scotland – waived completely) if the emissions defect is corrected within 14 days (and corroborated by MOT test), the vehicle had passed an MOT within the preceding six months, or it can be demonstrated that all reasonable steps had been taken prior to the emissions test to maintain the vehicle's emissions performance; the penalty may be waived completely if the emissions defect is corrected within 14 days and the vehicle had passed an MOT within the preceding six months, or if the emissions defect is corrected within 14 days and it can be demonstrated that all reasonable steps had been taken prior to the emissions test to maintain the vehicle's emissions performance.

3.28 AVIATION EMISSIONS

Emission standards for aircraft are set by the International Civil Aviation Organisation which, since 1981 has issued progressively stricter standards; the ICAO also looks at ways of reducing emissions through operational and market-based measures.

Relatively few complaints relating to air pollution at airports are recorded and research so far shows that compared with other domestic, transport and industrial sources, aircraft make a negligible overall contribution to air pollution. However, because pollutants are mainly emitted at high altitudes, where they take longer to disperse, their contribution to global problems (such as ozone depletion and the greenhouse effect) is more serious. This is an area that is now coming under increasing investigation; tighter emission controls in other areas and environmental pressures are likely to have an effect on air traffic movements in the future. In the UK studies are being carried out around various airports to establish levels of pollution.

Fuel combustion in aircraft engines is relatively efficient and jet exhausts are now virtually smoke free. Current jet engines do however produce significant quantities of nitrogen oxides which react with sunlight to form ozone, a greenhouse gas. Other emissions of note from aircraft include carbon monoxide, hydrocarbons, nitrogen dioxide, water, sulphur dioxide and carbon dioxide. Improvements in fuel efficiency and aircraft design are however unlikely to offset the effect of aircraft on climate due to the forecast increase in air traffic of between 2-4% by 2050. According to the European Commission, carbon dioxide emissions from aviation grew by nearly 70% between 1990 and 2002. Estimates from the National Environmental Technology Centre suggest that in 2000 UK civil passenger aeroplanes emitted 30 million tonnes of carbon dioxide (5% of total UK emissions); this is forecast to rise to 55 million tonnes in 2020 and about 70 million tonnes by 2030.

European Union Ministers have now agreed to bring aircraft emissions contributing to climate change within the greenhouse gas emissions trading scheme (see 3.6.3 above).

The Air Navigation (Environmental Standards) Order 2002 (SI 2002/798), which came into force 5 April 2002, covers both noise and emissions certification. The Order revokes the 1990 noise certification Order (see chapter 4, 4.22.6) and 1986 and 1988 Orders on aeroplane and aircraft emissions. With regard to emissions certification, the Order bans aircraft and aeroplanes from taking-off or landing at UK airports unless they comply with fuel venting requirements and are fitted with the engines specified in the emissions certificate which comply with emissions requirements for smoke, unburned hydrocarbons, carbon monoxide and nitrogen. Emissions certificates for aircraft registered in the UK are issued by the Civil Aviation Authority, and for aircraft registered elsewhere by the competent authority of the state concerned.

In December 2003 the Government published a White Paper – *The Future of Air Transport;* this sets out a strategic framework for the next 30 years for managing the increased demand for air travel and the resulting need for additional airports or the expansion of current airports, while doing more to reduce the environmental impacts of aviation. *The Civil Aviation Act 2006*, which received Royal Assent on 8 November 2006 (see also chapter 4, 4.22.3) will enable airport operators to set charges for aircraft by reference to their emissions to air (as well as noise levels). (More information: http://www.aef.org.uk)

3.29 EMISSIONS FROM SHIPPING

A report published in 2007 suggests that worldwide 60,000 deaths from heart and lung disease in 2002 can be linked to emissions from international shipping; 27,000 of the deaths were in Europe. Almost 70% of ship emissions occur within 400 km of land; pollutants of most concern are fine particulate matter, accounting for approx. 1.6 million tonnes of global emissions, sulphur oxides 6.5 million tonnes and nitrogen oxides 6.9 million tonnes. (Mortality from Ship Emissions: a Global Assessment, published online in the American Chemical Society journal, *Environmental Science and Technology*, 07.11.07).

A UN study (reported in *The Guardian*, 13.02.08) calculates annual emissions of carbon dioxide from international shipping at almost 1.12 billion tonnes (4.5% of global CO_2 emissions).

For regulations etc controlling oil pollution and waste from ships, see chapter 6.

3.29.1 International Maritime Organization

The International Convention for the Prevention of Pollution from Ships – MARPOL – was adopted following the 1973 international conference on marine pollution held under the auspices of the International Maritime Organisation. Included were articles dealing with oil, chemical, sewage and other pollution from ships, and requiring provision of adequate port reception facilities to deal with such wastes. Further measures aimed at reducing pollution from ships were incorporated in MARPOL for implementation from January 1992, including reducing pollutants in ships' exhaust gases, banning the use of CFCs in refrigeration units and of halons in new fire extinguishers.

A new Annex VI to the MARPOL Convention was signed in London in September 1997; this aims to control emissions of air pollutants from shipping. It finally came into force on 19 May 2005 following ratification by 15 states representing about 54% of the world's merchant tonnage; before it could come into force, the Annex required ratification by 15 (out of a possible 100) member states who together accounted for 50% of the world's merchant tonnage. The sulphur content of bunker oil is limited to 4.50%, with a lower limit of 1.50% to apply in the Baltic Sea, which has been designated a "SOx Emission Control Area", from May 2006; in April 2000 it was agreed that the North Sea should have similar status – this was formally agreed in July 2005, with the lower limit coming into force on 21 November 2006 and full implementation 12 months later. Limits on emissions of nitrogen oxides from diesel engines are also set and the incineration on board ships of contaminated packaging materials and PCBs is banned. Deliberate emissions of ozone depleting gases are also banned as are new installations with ozone depleting substances on board ships although those containing hydrochlorofluorocarbons are permitted until 1 January 2020. Annex VI is to be implemented in the UK through regulations to be made under *the Merchant Shipping (Pollution) Act 2006*.

In October 2008, the IMO's Marine Environment Protection Committee agreed some revisions to Annex VI to come into force on 1 July 2010. The sulphur content of all marine fuels worldwide is to be capped at 0.50% from 2020 (with an interim limit of 3.50% from 1 January 2012. The agreement provides for the 2020 limit to be delayed to 2025 if an IMO review to be completed no later than 2018 of the availability of low sulphur fuels shows that the 2020 limit cannot be met. In the SOx emission control areas (the Baltic and North Seas), the limits have been set at 1% from 1 July 2010 and 0.10% from 1 January 2015. Emissions of nitrogen oxides from the engines of new ships are to be reduced by between 16-22% (based on 2000 levels) by 2011, with those installed on ships constructed on or after 1 January 2016 and operating in an Emission Control Area required to reduce emissions by 80%. The revised Annex also allows for an Emission Control Area to be designated for SO_x and particulate matter, or NO_x, or for all three, subject to a proposal from a Party or Parties to the Annex, for approval by the IMO.

3.29.2 EU Strategy on Reducing Emissions from Ships

In November 2002 the European Commission published a Communication on *A European Union Strategy on Reducing Atmospheric Emissions from Seagoing Ships* (COM(2002) 595). Pollutants of concern include sulphur dioxide, nitrogen oxides, VOCs and particles, as well as greenhouse gases and ozone depleting substances. The main aims of the Strategy are to reduce the impact of emissions on local air quality as well as their contribution to acidification and eutrophication, and to promote shipping as an environmentally friendly mode of transport. The Strategy's main proposals, as amended and confirmed by the *Directive on the Sulphur Content of Marine Fuels* (2005/33/EC), which was formally adopted by the European Parliament and the Council on 12 July 2005, and entered into force on 11 August 2005, are:

- A limit of 1.5% sulphur for marine fuels used by all seagoing vessels in the Baltic Sea from 11 August 2006 and in the North Sea and the Channel from autumn 2007; this will reduce the effect of ship emissions on acidification and on air quality in Northern Europe. As part of a review of the Directive in 2008, the Commission is to consider extending the 1.5% sulphur limit to cover all EU sea areas and to establish a second phase reducing the sulphur limit to 0.5%.
- A limit of 1.5% sulphur for marine fuels used by passenger vessels on regular services to or from any port within the EU from 11 August 2006; this will improve air quality around ports and coasts.
- A limit of 0.1% sulphur on fuel used by ships using inland waterways or while they are at berth in ports inside the EU from 1 January 2010; this will reduce local emissions of sulphur dioxide and particulate matter, and improve local air quality.

This Directive amends the marine fuels aspects of Directive 1999/32/EC – see chapter 3, 3.18.3 and 3.23 and is to be implemented through *Sulphur Content of Marine Fuels Regulations*.

The Commission's Strategy also proposed the following actions:

- Extending Directive 1997/68 setting emission limits for non-road mobile machinery to cover engines marketed and for use on board vessels operating on inland waterways (see chapter 3, 3.26.6).
- If by the end of 2006, the IMO has not proposed tighter NOx standards (through an amendment to Annex VI of MARPOL), the Commission will consider bringing forward its own proposal.
- By 2010, to remove the exemption under regulation EC 2037/00 on substances that deplete the ozone layer which permits the use of halons on board existing cargo ships operating in EU waters (see chapter 3, 3.14).
- To look at the possibility of regulating the abatement of VOCs from ship-loading.

3.29.3 UK Regulations

The requirements of the above Directive (and of MARPOL) are to be implemented by the *Merchant Shipping (Prevention of Air Pollution from Ships) Regulations*, a draft of which was issued for consultation by the Marine and Coastguard Agency in July 2008.

(a) Merchant Shipping (Pollution) Act 2006

Section 1 of the Act enables the UK to implement revisions which have been made to international arrangements relating to compensation for oil pollution from ships, thus ensuring that victims of oil pollution from tankers are properly and promptly compensated. Section 2 of the Act amends s.128 of the *Merchant Shipping Act 1995* to make reference to Annex VI of MARPOL which limits atmospheric emissions from ships, thus enabling Regulations to be made (see above).

(b) Offshore Combustion Installations

The Offshore Combustion Installations (Prevention & Control of Pollution) Regulations 2001 (SI 1091), cover off-shore combustion installations with over 50 MW capacity and took effect on 19 March 2001. They apply in relation to sea adjacent to England and Wales, and to UK territorial sea apart from those areas in Scottish controlled waters. The DTI is the permitting authority; new installations were required to apply for a permit as soon as the Regulations came into force and existing ones by October 2007. The Regulations include similar requirements to those for on-shore installations, although only permits and monitoring data will need to be put on the public register (see chapter 2, 2.4.1). 2007 Amendment Regulations (SI 2007/938), which came into force on 16 April 2007, implement the public participation requirements of Directive 2003/35/EC (see chapter 1, 1.26.2), as well as outlining consultation requirements between the Secretary of State and another EEA state covering installations with transboundary implications.

(b) Offshore Chemicals

The Offshore Chemical Regulations 2002 (SI 1355) were made under the *Pollution Prevention and Control Act 1999*; they implement UK obligations under the OSPAR Convention (see chapter 6, 6.20.4) for a harmonised mandatory control system for the use and reduction of the discharge of offshore chemicals in relation to offshore activities (Decision 2000/2). The Regulations took effect on 15 May 2002 and apply throughout the UK. Operators of offshore oil and gas installations within UK territorial waters are required to apply for a permit from the Secretary of State BERR, covering their use and discharge of chemicals into the marine environment. Among the conditions which may be attached to permits are requirements relating to:

- quantity, frequency, location and duration of any permitted use or discharge of offshore chemicals;
- a requirement to take all appropriate measures to prevent pollution through the use of appropriate technology to limit discharges, emissions and waste;
- the need to take necessary measures to prevent accidents affecting the environment or, where they do occur, to limit their effects on the environment;
- the minimisation of long-distance or transboundary pollution.

For more details on these Regulations, see chapter 6, 6.24.4a.

3.30 RECREATIONAL CRAFT

Directive 2003/44/EC, adopted on 16 June 2003, limits both exhaust emissions and noise from recreational craft. It came into force on 26 August 2003 and must be transposed into national legislation by 30 June 2004, and the required measures applied by 1 January 2005. With regard to exhaust emissions it applies to propulsion engines installed, or intended for installation, on or in recreational craft (2.5-24 metres) and personal watercraft (e.g. jetskis), and to those installed on these craft which are subject to "major engine modification". Craft built for own use, original historical craft and individual replicas based on pre-1950 designs are exempt. Limits are set for CO, HC, NOx and particulates; the Commission estimate that the Directive will reduce CO emissions from spark ignition engines on relevant craft by 42% and HC by 89% and from compression ignition engines by 5% and 23% respectively; NOx and particulates from compression ignition engines are expected to be reduced by 31% and 29% respectively.

This Directive has been implemented through *The Recreational Craft Regulations 2004* (SI 2004/1464, as amended SI 2004/3201), which partly came into force on 30 June 2004, with implementation being completed on 1 January 2005. They revoke 1996 Regulations (which implemented Directive 1994/25/EC), except in respect of craft on the market prior to 1 January 2005.

The Directive also covers noise from recreational craft and personal watercraft – see chapter 4.

RADIATION

Radiation of natural origin has always been present in our environment, and always will be. Natural radioactive materials are present in the earth, in the buildings we inhabit, in the food and water we consume, and in the air we breathe. Even our bodies contain natural radioactivity. Radiation of artificial origin has been used since the beginning of the century. There is no difference between the effects of radiation whether it is emitted from natural or from artificial radioactive materials. Natural radiation accounts for most of the public's exposure (about 86% of the total). Most of the exposure from artificial radiation is due to medical uses (about 14% of the total).

3.31 SOURCES AND EFFECTS

3.31.1 Natural Sources

(a) Cosmic Rays

A part of natural radiation comes from the sun and outer space in the form of cosmic rays. The atmosphere acts as a partial shield against these rays, but those that reach the earth's surface easily penetrate buildings. About 10% of our exposure comes from this source.

(b) Radiation from the Ground and Building Materials

Another part of natural radiation comes from the earth. Soils and rocks contain radioactive materials such as uranium and potassium-40, a radioactive form of the element potassium. Most building materials are extracted from the earth, so these are also somewhat radioactive. About 13% of our exposure comes from this source, principally from building materials, but with a small contribution directly from the ground itself. The level of exposure therefore varies mainly according to the type of building materials used in the construction of houses. Exposures in granite buildings, for instance, are generally higher than those in brick buildings. The overall variation in exposure is as large as the average exposure to radiation of artificial origin.

(c) Internal Exposure

Potassium-40 and other natural radioactive materials (from the ground or produced by cosmic rays in the atmosphere) are present in our diet and thus cause some radiation exposure inside our bodies. About 12% of our exposure arises from these radioactive materials, with some variation depending on the individual's build and bodily processes.

(d) Radon

Naturally-occurring uranium, present in certain rocks, such as granite, and building materials, produces a gas called radon; it is a colourless, odourless and almost chemically inert gas. A similar gas (thoron) is produced from naturally-occurring thorium, which is also present in rocks and building materials. Together radon and thoron contribute about 50% of total exposure to radiation of a typical member of the UK population. In the open air, radon is diluted to very low levels and poses no danger. However, it can also seep upwards and become trapped in houses where it can build up to dangerously high concentrations.

The level of exposure received mainly depends on the type of rock underneath homes and how well ventilated they are. When radon is breathed in, it is mainly people's lungs that are exposed to radiation. A study, funded by the European Commission and Cancer Research UK and published late 2004, suggests that almost 20,000 deaths a year across the EU can be attributed to lung cancer caused by radon, with nearly a 1,000 deaths a year in Britain.

3.31.2 Artificial Sources

(a) Medical Procedures

Radiation is used in a wide range of diagnostic procedures, as in chest and dental x-rays. Radiation is also used for treatment purposes, for example, in killing off cancerous cells. Actual exposure in any year depends on how many examinations an individual is subjected to, and the type of examination, but an average figure would be 14%.

(b) Fallout

Most tests of nuclear weapons now take place underground. But in the 1950s and early 1960s many tests took place in the atmosphere. Fallout is the term used to describe the radioactive particles that are released and which eventually drop to earth from the atmosphere. This radioactivity can add to the radiation from the ground. It can also be breathed in, and eaten when taken up in food. In the UK, exposure due to fallout from weapons tests is now less than one-tenth of the highest level experienced, which occurred in the mid 1960s.

Some of the radioactivity released by the Chernobyl accident in 1986 was deposited on the UK, the deposition being heaviest in the areas where it was raining at the time that the radioactive cloud passed overhead. Some of this radioactivity was taken up in foodstuffs. In the first year after the Chernobyl accident the total exposure of a typical individual in the UK was increased by about one per cent. Some people might have received increases of up to 40%, depending upon where they lived and what they ate.

Fallout from weapons tests and Chernobyl now contributes about 0.2% of our exposure on average. There are some variations in this average due, for example, to differences in rainfall between areas.

(c) Discharges from the Nuclear Industry

These occur routinely from nuclear power stations and plant that manufacture or reprocess the fuel. On average less than 0.1% (one-thousandth) of total radiation exposure is due to such discharges. Some people will get higher exposures than this if they live near nuclear installations. In some cases, these higher exposures are due to the concentration of radioactivity in locally produced foodstuffs such as shellfish. In the mid-1980s the most exposed individuals were getting up to one-third of their total exposure from the nuclear industry, but by the late 1980s this had fallen to about onetenth.

(d) Occupational Exposure

This covers workers who deal with radiation in the nuclear power industry, medicine, dentistry, research laboratories and general industry. There are also underground miners and aircrew who are exposed to increased levels of natural radiation. Miners in non-coal mines receive the highest average exposures of all workers. They typically get more than ten times the exposure of a worker in a nuclear power station. Overall, all these workers receive an average additional exposure amounting to nearly 50% of typical total exposure. Averaged out over the whole population, working with radiation contributes about 0.3% of the total population exposure.

(e) Miscellaneous Sources

This covers enhanced cosmic radiation exposure of aeroplane passengers, release of radioactive material through coal burning (mainly in power stations), and consumer products containing small amounts of radioactivity for various purposes (e.g. smoke alarms, and radioluminous watches and clocks). These combined sources are responsible for less than 0.1% of exposure.

3.31.3 Effects of Radiation

If the exposure of the whole body to radiation is very high, death may occur within a matter of weeks. In this context "very high" means about 2000 times higher than a typical annual total exposure. Even if death does not occur, high exposures may soon lead to other effects such as reddening of the skin, cataracts and sterility. These same doses of radiation are given to individual organs in radiotherapy to kill cancer cells and prolong patients' lives.

If the radiation exposure is lower, or is received over a long period of time, immediate effects will not occur, but there may be risks that the exposure will lead to cancer later in life or to hereditary defects in future generations. A small increase in cancers has been observed in a number of population groups exposed to relatively high exposures. For instance, for the survivors of the Hiroshima and Nagasaki bombs, the cancer death rate is about 10% higher than that expected in a similar Japanese population.

For the purposes of controlling exposures it is assumed that the cancer risk increases in proportion to the exposure, even at very low levels of radiation. This is a cautious assumption, and deliberately so. It ignores the possibilities that very low levels of radiation may be incapable of causing damage or that the body may be able to repair any damage caused by low exposures. For the average person, radiation poses a very small risk and most of that risk is due to natural radiation.

3.31.4 Radiation Exposure Limits

The basic assumptions about the risks from radiation exposure given in the previous section have led the International Commission on Radiological Protection (ICRP) to state three key recommendations for radiation protection. ICRP is the principle standard setting body for radiation protection and many countries throughout the world, including the UK, have adopted their recommendations. Their three key recommendations formulated in 1990 are:

a) *Justification:* no practice involving exposure to radiation should be adopted unless it produces sufficient benefit to the exposed individuals or to society to offset the radiation detriment it causes.

b) *Optimisation:* radiation doses and risks should be kept as low as reasonably achievable, economic and social factors being taken into account; constraints on risk or dose should be used to provide upper bounds to the optimisation process.

c) *Limits:* the exposure of individuals should be subject to dose or risk limits above which the radiation risk would be unacceptable.

All of these recommendations apply to all sources of radiation, except that exposure to normal levels of natural radiation and exposure due to medical practices are not subject to exposure limits, and thus only the first two recommendations should be applied. For example, public exposure due to both routine discharges of radioactivity to the environment and solid radioactive waste disposal is constrained by the application of all three principles. It should be noted that it is not sufficient merely to comply with the limits placed on exposure. The second principle ensures that efforts will be made to keep exposure well below these limits.

Nothing can sensibly be done about normal levels of exposure from natural radiation. But human activities can result in increased exposure to natural radiation. Examples of this include flying at high altitude resulting in elevated exposure to cosmic rays, and living in a house where there is a high concentration of radon. Where control of such exposures is feasible, it is given full consideration. This is the case with the relatively high levels of radon exposure which occur in some parts of the UK. Government-funded surveys are being carried out to identify buildings with high radon concentrations.

The National Radiological Protection Board (since April 2005, part of the Health Protection Agency) set an "action level" of 200 Bq m^{-3} average annual concentration above which it recommends measures to exclude radon from houses. Current estimates are that more than 1 in 100 homes in Cornwall, Devon, Somerset, Northamptonshire and Derbyshire are above this recommended action level – 80% of England's radon-affected homes are in these counties; the remaining 20% are in Avon, Cumbria, Dorset, Gloucestershire, Leicestershire, Lincolnshire, Northumberland, North Yorkshire, Nottinghamshire, Oxfordshire, Shropshire, Staffordshire and Wiltshire. The Health Protection Agency, in collaboration with the British Geological Survey, has produced an online map (www.ukradon.org) of England and Wales which identifies those areas most at risk from exposure to radon. For a small fee householders and businesses can check by postcode whether their premises are in an affected area and should therefore be comprehensively tested. Properties with elevated levels of radon can then take advice on remedial measures. The Health Protection Agency has also recommended that UK Building Regulations be amended to require all new property and extensions to be constructed in a way which prevents the build up of radon.

In new buildings exposure to radon gas can be prevented by the use of suspended concrete floors. In existing buildings, such measures may involve improved underfloor ventilation by the installation of a fan and sealing of the floor to prevent the upward flow of radon gas. Home improvement grants are available towards the cost of carrying out remedial work, although these are means-tested. With regard to new buildings in radon-affected areas, *BR211 Radon: guidance on protective measures for new dwellings*, (effective from 14 February 2000, published by the Building Research Establishment) provides technical guidance on measures to be taken as well as details of areas where they may be necessary. This guidance is also referred to in Approved Document C to the *Building Regulations 2000* (SI 2531).

The Building (Scotland) Regulations 2004 (SSI 2004/406), which came into force on 1 May 2005, require all buildings to be designed and constructed in such a way that the emission and containment of radon gas will not be a health threat to people in or around the building.

The Health Protection Agency carries out scientific research on radiation hazards and advises the Government on whether or not recommendations made by bodies such as the ICRP should be applied in the UK. It provides information and advice about radiation protection to UK government departments and other interested parties.

3.32 REGULATORY CONTROLS

In the UK several government departments have responsibility for protecting the public from radiation exposure. No significant amount of radioactive waste can be discharged without a specific authorisation from the Environment Agency or the Scottish Environment Protection Agency (as appropriate). Discharges from major nuclear sites in England also require authorisation from the Secretary of State for Environment, Food and Rural Affairs. The Secretaries of State for Scotland and Wales grant authorisations within their respective countries. The Department of the Environment is the regulatory body for Northern Ireland. Authorisations specify discharge limits, and inspectors check that the conditions of the authorisations are adhered to. Unless the level of discharge is extremely small, this includes a requirement to monitor the local environment (see chapter 5).

The UK Strategy for Radioactive Discharges 2001-2020, published in July 2002, reaffirms the Government's commitment to the ALARA principles, whereby "radiological doses and risks are reduced to a level that represents a balance between radiological and other factors, including social and economic factors".

3.32.1 Ionising Radiations

Responsibility for the control of radiation hazards at work falls on the Health and Safety Executive which employs inspectors to enforce the *Ionising Radiations Regulations 1999* (made under the *Health and Safety at Work Act 1974*) – SI 1999/3232; these Regulations came into force on 1 January 2000 replacing and consolidating 1985 and 1993 Regulations. The 1999 Regulations reflect EU Directive 96/29 "laying down the basic safety standards for the protection of the health of workers and the general public against dangers arising from ionising radiation".

The Regulations (which cover England, Scotland and Wales) provide protection for the workforce from ionising radiations, including the handling and storage of radioactive waste; exposure to radiation must be kept as low as reasonably practicable, whether the source is artificial (e.g. nuclear reactor) or naturally occurring (e.g. radon). A schedule to the Regulations sets out the annual dose limits which should not be exceeded for both occupational exposure and exposure of the general public. The Regulations require continuous monitoring to be carried out and designation as "classified persons" and records kept of any employee likely to receive an effective dose in excess of 6mSv per year; employers must make an assessment of the potential radiation hazard in the event of a possible accidental release or other accident. Employers conducting certain types of work, including the use of electrical equipment intended to produce x-rays for various purposes, must obtain prior authorisation from the HSE; 28 days notice must be given to the HSE of any work with ionising radiation to be carried out for the first time and an assessment of its risk must be made to ensure that suitable protective measures are in place to restrict exposure of employees and other persons. Similar 2000 Regulations covering Northern Ireland (SR 375) were brought into force from 8 January 2001.

The "Justification" principle – see 3.31.4 above – has been given statutory force through *the Justification of Practices Involving Ionising Radiation Regulations 2004* (SI 2004/1769). These require an assessment to be carried out of all new activities (i.e. carried out since 13 May 2000) looking at both the benefits and disadvantages of the practice.

The Ionising Radiation (Medical Exposure) Regulations 2000 (SI 2000/1059), as amended by SI 2006/2253) implement Directive 97/43/Eurotom on health protection of individuals against the dangers of ionising radiation in relation to medical exposure.

The Commission has published proposals for a Directive (COM(2007)51) which will require Member States to make the intentional discharge of ionising radiation into air, soil or water a criminal offence (see also chapter 1, 1.15 for further details of the proposal).

3.32.2 Emergency Preparedness & Public Information

The Radiation (Emergency Preparedness and Public Information) Regulations 2001, (REPPIR) (SI 2001/2975), covering England, Scotland and Wales, came into force on 20 September 2001; they replace and consolidate the 1992 *Public Information for Radiation Emergency Regulations* (SI 2997). Similar Regulations for Northern Ireland (SR 2001/436) came into force on 4 February 2002. The Regulations, which are enforced by the Health and Safety Executive, implement those articles of Euratom Directive 96/29 covering emergency planning and information to the public in the event of an accident or an emergency at a major radiation site – other aspects of the Directive have been implemented through *The Ionising Radiations Regulations 1999*.

The REPPIR Regulations apply to all employers, including carriers, who work with ionising radiation involving specified quantities of radionuclides; all those covered by the Regulations have a duty to identify hazards and evaluate risks for all (big and small) radiation accidents, to prevent accidents occurring and to limit the consequences if they do occur; this hazard assessment, which should be reviewed at least every three years, should be sent to the HSE.

Operators at those premises where it is reasonably foreseeable that a radiation emergency may occur have to prepare an emergency plan; this should aim to restrict exposure and should take account of steps taken to prevent and limit the consequences of potential emergencies (as identified in the hazard assessment). In drawing up the plan the operator should consult those working on the site, the HSE, emergency services, the local authority and the health authority, and any other appropriate persons. Local authorities covering those operators required to prepare an emergency plan, have themselves to prepare an off-site emergency plan; this has the same aim as the operator's emergency plan and should be drawn up in consultation with the HSE, the operator – who must ensure the local authority has access to adequate information to enable it to draw up the plan – the emergency services, health authorities and members of the public. Operators should also provide members of the public likely to be affected by a radiation emergency with information about what they should do in the event of an emergency; should an emergency occur, the local authority should ensure that all members of the public within the vicinity of the plant receive information of the protective measures they should take.

Carriers, too, must prepare an emergency plan if it is reasonably foreseeable that an emergency might occur during transport of radionuclides, and should consult the HSE and the Radioactive Materials Transport Division; as it is intended that the plan should be generic – i.e. could be used on any route – the carrier should consult representatives of the local authorities, health authorities and emergency services.

The Regulations make provision for local authorities to charge for preparation of the off-site emergency plan, and for the review and testing of emergency plans at least every three years. Operators and carriers should have in place adequate systems to ensure that emergency plans are implemented as soon as possible after an accident. Operators and carriers should advise the HSE of the dose level at which certain personnel may receive emergency exposure (e.g. in putting an emergency plan into effect); such personnel should be identified and given appropriate information, training, equipment and medical surveillance.

Copies of hazard assessments and emergency plans should be made available to the public, subject to exclusions for reasons of industrial, commercial or personal confidentiality, public security or national defence. The REPPIR Regulations are consistent with the relevant sections of *The Control of Major Accident Hazards Regulations 1999* dealing with emergency plans etc – see 2.10.

4 NOISE

NOISE POLLUTION – GENERAL

The Pollution Control Handbook is indebted to Colin English, President of the Institute of Acoustics* (2007) for bringing to our attention a book called *Filth, Noise and Stench in England* which demonstrates that noise nuisance has been a problem for at least as long as air pollution. Legislative controls on noise pollution are, however, of more recent origin than controls on other forms of pollution, and prior to 1960 noise was largely dealt with locally as a nuisance and under local bye-laws.

The first recorded case of noise nuisance was heard in 1378 and concerned the case of a noisy armourer who had set up a forge in his house which shook the party walls. There was no concept of industrial zones and craftsmen often established their workshops in their homes, causing considerable noise nuisance to their neighbours. In the 15th century coopers were the subject of many actions, while in 1611 a smith was forced to remove a forge that he had erected in a residential street. A century later coppersmiths were the trade regularly appearing in court to answer noise nuisance charges.

Neighbour noise was also a significant problem as it is today. In 1595 a bylaw in London prevented any man from causing a sudden outcry after nine at night; offences listed included singing, making an affray, and beating a wife or servant. Traffic noise, too, was a problem in towns and cities; a 1596 Act in London enabled fines to be imposed for any coach or cart heard to creak or 'pype' through want of oil.

*www.ioa.org.uk

4.1 SOURCES & DEFINITION

In its widest sense, neighbourhood noise might be defined as any unwanted sound in the vicinity of the home or its locality. That definition might embrace industrial noise, noise from transport, as well as noise from domestic premises, which is the biggest source of noise nuisance and complaints. During 2004/05 there were 286,872 separate complaints about noise made to environmental health departments in England and Wales; this figure breaks down into 206,086 relating to noise from domestic premises; 7,522 from industrial premises; 35,784 from commercial and leisure premises; 11,305 complaints about vehicles, machinery, equipment in streets; and 11,714 about construction and demolition noise; a further 14,461 complaints related to miscellaneous sources (e.g. public buildings, places of worship, military facilities and traffic) (Chartered Institute of Environmental Health annual survey into local authority noise enforcement action).

In Scotland, local authorities received 5,225 complaints about noise and vibration from domestic premises in 2004/05, 2,750 about noise and vibration from commercial premises and 433 complaints relating to industrial premises. (Environmental Health in Scotland 2004/05, Royal Environmental Health Institute of Scotland)

For the year 2006/07, 11,923 (2005/06: 11,337) complaints about noise were made to local authorities in Northern Ireland, of which 85% related to noise from domestic premises (including barking dogs and loud music); commercial and leisure facilities accounted for 6% of complaints, and industrial activities (including agriculture) for 2%. (Noise Complaint Statistics for NI 2006/07, DOE NI)

4.2 EFFECTS OF NOISE

Noise can cause annoyance, interfere with communication and sleep, cause fatigue, reduce efficiency and damage hearing. Physiological effects of exposure to noise include constriction of blood vessels, tightening of muscles, increased heart rate and blood pressure and changes in stomach and abdomen movement. Although hearing sensitivity varies and the effects of exposure are therefore personal, exposure to constant or very loud noise – either occupational or leisure-associated – can cause temporary or permanent damage to hearing.

The World Health Organisation's *Guidelines for Community Noise* (2000) includes the following recommendations for guideline levels to prevent critical health effects:

- Outdoor living area – 55 dB LA_{eq} measured over 16 hours, to prevent serious annoyance and 50 dB LA_{eq} over 16 hours, to prevent moderate annoyance (daytime and evening);
- Dwelling indoors – 35 dB LA_{eq} measured over 16 hours, to prevent moderate annoyance (daytime and evening).
- Inside bedrooms – 30 dB LA_{eq} measured over 8 hours (nighttimes), for undisturbed sleep;
- Impulse sounds from toys, fireworks and firearms – peak sound pressure of 140 dB (adults) and 120 dB (children), measured 100 mm from the ear, to prevent hearing impairment.

New guidelines are likely as a result of research being carried out by the WHO's Working Group on the Noise Environmental Burden of Disease. Preliminary findings published in 2007 suggest that long-term exposure to traffic noise accounts for up to 3% of cases of heart disease in the EU, as well as 3% of cases of tinnitus worldwide; up to 1.7% of 7-19 year olds in the EU have hearing loss as a result of "leisure noise" (e.g. listening to MP3 players etc). The Working Group also note that 2% of EU citizens suffer "severely disturbed" sleep as a result of noise and up to 15% suffer "severe annoyance". (More: Night noise guidelines for Europe, Martin van den Berg, paper to Inter-Noise 2007).

Recent research has focused on the impacts of noise on health in both the city and the countryside. A report[1] commissioned by the Greater London Authority, and published in 2008 links persistent high levels of noise from road traffic to significant health impacts, finding that up to 108 heart attacks and nearly 500 cases of heart disease in the Greater London area could be caused by prolonged exposure to high levels of road traffic noise. A report[2] issued by the European Federation for Transport and the Environment, also in 2008, estimated that across Europe 50,000 fatal heart attacks could be attributed to noise.

1. *The Effect of Noise on Physical Health Risk in London*, Berry Environmental Ltd
2. *Traffic noise reduction in Europe: health effects social costs and technical and policy options to reduce road and rail traffic noise.*

So far as low frequency noise is concerned, it is not clear at what level it may be physically damaging; however, the unpleasant symptoms it can induce are sufficient to cause disruption and significant social and economic penalties to sufferers.

4.3 EU PROGRAMMES

The European Union's 1996 Green Paper on *the Future of Noise Policy* (COM(96) 540) notes that the data available on noise exposure is generally poor in comparison to that collected to measure other environmental problems. However it estimated that around 20% of the then EU's population, or close to 80 million people, suffer from noise levels that scientists and health experts consider to be unacceptable, where most people become annoyed, where sleep is disturbed and where adverse health effects are to be feared. An additional 170 million citizens were living in so-called "grey areas" where the noise levels were such to cause serious annoyance during the daytime.

The Commission's Fifth Action Programme on the Environment, *Towards Sustainability* (1993-2000) (see also Appendix 7) proposed a number of measures aimed at reducing people's exposure to night-time noise, including

- exposure to more than 65 dB(A) should be phased out and at no time should 85 dB(A) be exceeded;
- those exposed to levels of between 65-55 dB(A), and those currently exposed to less than 55 dB(A), should not suffer any increase.

To achieve these levels the Commission proposes noise abatement programmes; standardisation of noise measurement and rating (see 4.5 below); Directives further reducing noise from transportation; measures related to infrastructure and planning (e.g. better zoning around airports, industrial areas, main roads and railways).

In 2006, the EU adopted a renewed Strategy on Sustainable Development which aims at "reducing transport noise both at source and through mitigation measures to ensure overall exposure levels minimise impacts on health".

As in other areas of environmental policy, the European Union (EU) has adopted a number of Directives aimed at reducing noise in the environment. As well as the Environmental Noise Directive (see below 4.5), there are Directives establishing maximum permitted noise levels for motor vehicles (cars, trucks, buses, motorcycles and agricultural tractors) and outdoor equipment and subsonic jet aircraft, all of which are summarised later in this chapter. Exposure to occupational noise is limited through a 2003 Directive (see 4.28).

A list of Directives and other EU measures relating to noise is at Appendix 7.4. Brief details of the institutions of the EU and of the legislative process are given in Appendix 7.

4.4 REGULATORY AUTHORITIES

Much of the responsibility for practical action to prevent or abate noise rests with local authorities and in particular environmental health departments under the general direction of central government departments.

At central government level, environmental noise and noise nuisance is the responsibility of the Devolved Administrations, with the UK Government ultimately responsible for ensuring compliance with EU Directives and reporting to the Commission.

The Department for Environment, Food and Rural Affairs (Defra) is responsible for the preparation of noise maps and action plans in England arising from implementation of the Environmental Noise Directive, and for policy on the control of noise from fixed sources and statutory noise nuisance. The Department for Communities and Local Government is responsible for the preparation of guidance on the use of planning controls to prevent noise problems; the Department for Transport deals with legislation and policy on road traffic noise and noise from civil aircraft. Occupational noise is the responsibility of the Health and Safety Executive.

In England and Wales the police, while having no specific powers to deal with noise nuisance, provide some support to environmental health officers and other officials in carrying out their duties; they also have powers to noise-test road vehicles. In Scotland, the police have powers to deal with noise both under *the Civic Government (Scotland Act) 1982* and under Part 5 of *the Anti-Social Behaviour etc. (Scotland) Act 2004.*

Guidance drawn up by the Chartered Institute of Environmental Health and the Association of Chief Police Officers – *Good practice guidance for police and local authority cooperation* (CIEH, 1997) – provides guidance on what is considered to be a minimum standard of cooperation between the police and LAs when dealing with noise complaints. It sets out the roles and responsibilities of each, giving examples of effective local liaison arrangements, and suggests that both local authorities and the police in each area should draw up an agreement covering the way in which they will cooperate.

The CIEH has also published (2006) *Neighbourhood Noise Policies and Practice for Local Authorities – a Management Guide*. This replaces, extends and updates 1997 guidance, aiming to encourage consistency in the way local authorities deliver noise control services, while still enabling them to respond to local circumstances and needs. It provides advice on dealing with noise produced by a person's neighbours, noise in the street (from vehicles, machinery and equipment), noise from pubs, clubs and other recreational and leisure sources, and noise from commercial, local industrial and construction sites. Appendices provide examples of some practices and procedures sourced from local authorities. Similar Guidance, but reflecting Scottish law and practice, was also published in 2006 by the Scottish Government.

ENVIRONMENTAL NOISE

Directive 2002/49/EC relating to the *Assessment and Management of Environmental Noise* was adopted on 25 June 2002, and came into effect on 18 July 2002 (the day on which it was published in the Official Journal). Regulations implementing the Directive should have been in place by 18 July 2004, but were brought into force throughout the UK in October 2006 (see 4.6 below).

4.5 ENVIRONMENTAL NOISE DIRECTIVE

The Directive is the result of the Commission's 1996 Green Paper on Noise Policy (COM(96) 540) which recognised that current measures for reducing people's exposure to unacceptable levels of noise – particularly traffic noise – were not enough. While acknowledging that noise is essentially a local issue, the Commission suggested that there were a number of areas where a Community-wide approach was needed and it is on these areas that the Directive concentrates.

The Directive seeks to limit people's exposure to environmental noise, in particular in built-up areas, public parks, or other quiet areas, and in noise sensitive buildings such as schools and hospitals. Environmental noise is defined as "unwanted or harmful outdoor sound created by human activities, including noise emitted by means of transport, road traffic, rail traffic, air traffic, and from sites of industrial activity..." to which humans are exposed in the domestic environment (e.g. in and near the home, in public parks, in schools and hospitals etc). It does not cover noise from domestic activities, noise created by neighbours, noise at workplaces, noise inside means of transport or due to military activities in military areas. The Directive requires Member States to provide the Commission with data on exposure to environmental noise (through noise mapping), to ensure that information on exposure to environmental noise and its effects is available to the public, and to adopt action plans, based on the results of noise mapping with a view to preventing or reducing noise exposure and preserving environmental noise quality where it is currently good.

4.5.1 Noise Indicators

L_{den} and L_{night}* are to be used as indicators in EU noise policy, in new legislation on noise mapping, and in any revision to existing legislation (other indicators may be used for acoustical planning and noise zoning). However until the Commission has established common assessment methods for the determination of L_{den} and L_{night}, member states may use existing national indicators converted to L_{den} and L_{night} using the assessment methods in Annex II – this sets out assessment methods for both indicators and is to be revised by the Commission. L_{den} is the day-evening-night level in decibels and is an indicator for "annoyance"*; L_{night} is the "overall night-time noise indicator", a reduction in which would reduce sleep disturbance.

* "Annoyance" is the scientific expression for non-specific disturbance by noise. When associated with dose-effect relations, L_{den} and L_{night} are both indicators that are able to predict the average response of a population that is subject to long-term noise exposure. [Thus they] are suitable for planning purposes and for an integral approach for a residential area, a city and larger areas. The value of noise indicators can be determined either by measurement or by computation. *Extracted from The Noise Policy of the European Union, Year 2* (1999-2000).

4.5.2 Noise Maps

Noise maps – using common indicators and assessment methods – covering agglomerations with more than 250,000 inhabitants, major roads with more than 6 million vehicle movements per year, major railways with more than 60,000 movements per year, and major civil airports (with over 50,000 movements per year) should have been prepared and approved by 30 June 2007 covering the previous calendar year. No later than 30 June 2005, and then every five years, Member States should advise the Commission of all major roads, major railways and major airports and agglomerations, noted above.

For agglomerations between 100,000 and 250,000 inhabitants, roads with more than 3 million vehicle movements per year and railways with more than 30,000 movements per year, noise maps will need to be prepared and approved by 30 June 2012. No later than 31 December 2008, Member States must advise the Commission of all roads, railways and agglomerations covered by this paragraph.

Noise maps should include data on noise levels, number of citizens affected and to what extent; noise maps are to be reviewed at least every five years after their preparation. Member States should cooperate on strategic noise mapping near borders. Noise maps are to be published and made publicly available and thus should be in a form which is easily understood.

4.5.3 Action Plans

By 18 July 2008, action plans should have been drawn up for places near major roads (more than 6 million vehicle movements per year), major railways (with more than 60,000 train passages a year), major airports and major agglomerations (with more than 250,000 inhabitants). These should include, where necessary, measures for reducing noise exposure and for protecting quiet areas from any increase in noise. By 18 July 2013 action plans should be drawn up for agglomerations (100,000-250,000), major roads (over 3 million vehicle movements), and major railways (over 30,000 train movements per year). Action plans are to be reviewed and revised, as appropriate, should there be a major development that affects the noise situation, and at least every five years after their approval. Action plans should be publicly available.

The public should be consulted about proposals for action plans and should be given early and effective opportunities to participate in their preparation and review.

4.5.4 Reporting

Member States should, within 6 months of the relevant date, forward information from their noise maps and summaries of action plans to the Commission for their noise data bank. The Commission is to publish a summary report of data from the strategic noise maps and action plans every five years, with the first report due 18 July 2009. The Directive also includes a requirement for the Commission to submit to the Parliament and the Council a report, no later than 18 July 2009, summarising the effects of the Directive and, if appropriate, to propose environmental noise quality objectives and strategies for implementing them; the report should also include an assessment of the need for further legislative proposals for reducing noise from major sources, including road and rail vehicles, aircraft, outdoor and industrial equipment and mobile machinery.

4.6 ENVIRONMENTAL NOISE REGULATIONS

The EU's Environmental Noise Directive has been implemented in England and each of the Devolved Administrations by separate, but similar, Regulations: England, SI 2006/2238, which came into force on 1 October 2006, as amended by SI 2008/375; N. Ireland – SR 2006/387, effective 20.10.06; Scotland – SSI 2006/465, effective 5.10.06; Wales – WSI 2006/2629 (W.225), effective 4.10.06. The Regulations set out the timetable and process by which noise mapping and the action planning process, including arrangements for consulting the public, is to be taken forward, to meet the requirements of the Directive.

4.6.1 Competent Authorities

(a) England, Wales, Scotland

The Secretary of State has been designated as the competent authority, responsible for identification of noise sources and preparation of noise maps for all major agglomerations, major roads and railways; the Secretary of State is also responsible for mapping of the designated airports (Heathrow, Gatwick and Stansted), with airport operators responsible for the non-designated airports. The Secretary of State is also the competent authority with regard to preparation of action plans, except for airports, where the airport authorities are responsible for their preparation for all airports.

Similar arrangements apply in Scotland and Wales, where Scottish Ministers and the National Assembly for Wales (NAW), as appropriate, are the competent authorities.

(b) Northern Ireland

The Department of Environment has been designated the competent authority responsible for identification of noise sources, for the preparation of noise maps for industrial sources within agglomerations, and for the preparation of a consolidated strategic noise map and action plan for all sources within agglomerations. The Department of Regional Development is the competent authority responsible for the preparation of strategic noise maps and action plans for major roads, the Northern Ireland Transport Holding Company for those covering major railways, and airport operators in respect of airports.

4.6.2 Identification of Noise Sources

The competent authority was required to identify by 31 December 2006* the noise sources for which strategic noise maps will be required for:

- large agglomerations (> 250,000 inhabitants and a population density equal to or greater than 500 people per km^2) ("first round agglomerations");
- major roads (> 6 million vehicle passages per year);
- major railways (> 60,000 passages per year);
- major civil airports (> 50,000 movements per year).

Schedules to *the Environmental Noise (Identification of Noise Sources) (England) Regulations 2007* (SI 2007/415, amended by SI 2007/2458) identify first round agglomerations, major roads and major railways shown in maps deposited at Defra, and list major airports (see Appendix 4.1). The maps can be accessed via Defra's website or inspected at its 123 Victoria Street, London SW1 office.

Regulations 4, 5, and 6 of *the Environmental Noise (Identification of Noise Sources) (Wales) Regulations 2007* (WSI 2007/3519, W.311) identify first round agglomerations, major roads and railways. The relevant maps have been deposited at the offices of the Welsh Assembly Government (Cathays Park, Cardiff CF10 2NQ) and can also be accessed online. No major airports in Wales have been identified.

By 31 December 2011*, Regulations identifying noise sources for all agglomerations with a population of more than 100,000 (and population density equal to or more than 500 people per km^2, roads with more than 3 million vehicle passages per year and railways with more than 30,000 passages per year and major airports (> 50,000 movements per year) should be made. The Regulations should be reviewed, and revised as appropriate, in 2016 and every five years thereafter to ensure they are up to date as regards major developments that may affect the noise climate.

*In ***Northern Ireland***, where the competent authority is not the DOE, the relevant dates are 31 October 2006 and 31 October 2011.

4.6.3 Strategic Noise Maps

Strategic noise maps, to be drawn up by the competent authority covering noise sources identified for 2006 for major agglomerations, roads, railways and airports (first para of 4.6.2 above) should have been made and adopted no later than 30 June 2007. Noise maps covering sources identified in the second round of the process must be made and adopted no later than 30 June 2012, and thereafter every five years. Strategic noise maps should be reviewed and revised from time to time to take account of any major development that may affect the noise situation.

Airport operators for non-designated major airports were required to submit to the Secretary of State (NAW, Scottish Ministers) their first strategic noise map covering the preceding year no later than 31 March 2007, and thereafter every five years.

Strategic noise maps covering the preceding year for "non-designated other airports" in an agglomeration with a population of more than 250,000 should also have been submitted to the Secretary of State (NAW, Scottish Ministers) by 31 March 2007; this covers those airports in which air traffic results in air traffic noise of an L_{den} value of 55 dB(A) or more, or an L_{night} value of 50 dB(A) or more anywhere within the agglomeration. Similar strategic noise maps for agglomerations with a population of more than 100,000 should be submitted to the Secretary of State (NAW, Scottish Ministers) by 31 March 2012 and thereafter every five years. Maps should be reviewed and revised from time to time to take account of any major development affecting the noise situation.

Strategic noise maps, must meet the requirements of Annex IV of the Environmental Noise Directive:

- Presentation of data on one or more of the following aspects – an existing, a previous or predicted noise situation in terms of a noise indicator; the exceeding of a limit value; estimated number of buildings, schools and hospitals in a certain area exposed to specific value of a noise indicator.
- Estimated number of people both within and outside agglomerations exposed to noise from road, rail and air

traffic and industrial sources, at various noise levels for both day and night and how many have special insulation against the noise in question (Annex VI to the Directive) – this data should also be sent to the Commission.

- Maps may be presented to the public as graphical plots, numerical data in tables or numerical data in electronic form.
- For agglomerations special emphasis should be put on noise from road and rail traffic, airports and industrial activities, including ports.
- For major roads, railways and airports, total area in km^2 exposed to values of L_{den} higher than 55, 65, 75 dB, including estimated number of dwellings and total number of people living in each area.
- Once adopted, for the purposes of informing the public, more detailed information should be made available, such as: a graphical presentation; maps disclosing the exceeding of a limit value; difference maps (comparing the existing situation to a possible future situation); maps showing the value of a noise indicator at a height other than 4m where appropriate.

The Secretary of State (NAW, Scottish Ministers) should adopt strategic noise maps submitted to him (and those made or revised by him) if they meet the requirements of the Regulations; if he considers that a submitted strategic noise map does not, he may either amend and adopt it or reject it. If the latter he should notify the competent authority of the reasons for rejection and the date by which it should be revised and resubmitted.

The Secretary of State (Scottish Ministers) is required to compile and publish a consolidated noise map from all strategic noise maps by 30 October 2007.

In ***Northern Ireland*** the competent authorities were required to draw up and submit their strategic noise maps to the DOE no later than 31 March 2007 and for the second round, by 31 March 2012, and thereafter every five years. By the same dates, the relevant competent authority is also required to draw up and submit to the DOE a strategic noise map of those sources within an agglomeration for which it is the competent authority. The DOE, as the competent authority, will make a consolidated strategic noise map of all sources within the agglomeration.

First round maps have now been published for each of the areas of the UK and are available online at:

- Scotland www.scottishnoisemapping.org – maps of road, rail, aviation and industrial sources; an interactive viewer contains maps that shows calculated levels of noise sources across the Glasgow and Edinburgh agglomerations and the main transportation and rail links.
- Northern Ireland www.noiseni.co.uk – road, rail, aviation and industrial sources can be viewed on individual maps or on a consolidated map; the maps also show how many people are affected by noise, with the consolidated map for the Belfast agglomeration identifying candidate quiet areas.
- England www.defra.gov.uk/noisemapping – show levels of environmental noise from major industries, road and rail networks in 23 urban areas and for 18 airports; the site also includes information on the number of people exposed to various noise levels. (See also Appendix 4.1)
- Wales www.wales.gov.uk/environment – maps show environmental noise on major roads and railways, as well as from transport and industry in two large urban areas (Cardiff and Swansea-Neath Port Talbot) (there are no qualifying airports).

4.6.4 Identification of Quiet Areas

Under the Directive a "quiet area" in an agglomeration is a space in an urban area that is delimited as such by a competent authority and, for example, is not exposed to noise levels above a certain limit.

The Secretary of State (NAW, Scottish Ministers, DOE NI) was required to identify in Regulations "quiet areas" – for first round agglomerations by 30 September 2007 and for other agglomerations by 30 September 2012. The Regulations should be revised to ensure they remain appropriate no later than 30 September 2017 and every five years thereafter.

The Welsh Assembly Government has published proposals for amending the *Environmental Noise Regulations* in Wales to require quiet areas in first round agglomerations and agglomerations to be identified in the action plans for those agglomerations. They will also require that from 2011 onwards maps are published identifying noise sources listed in reg. 3. Similar proposals are expected for England.

4.6.5 Action Plans

The Secretary of State (NAW, Scottish Ministers) as the competent authority is required by 18 July 2008 to draw up action plans for places near first round major roads, first round major railways and first round agglomerations. Action plans covering other agglomerations, and places near other major roads and railways covered by the Regulations are required by 18 July 2013.

In ***Northern Ireland*** the relevant competent authority had to draw up and submit its action plan to the DOE, for the first round by 30 April 2008, and subsequently by 30 April 2013, and thereafter every five years. By the same dates each one is to draw up an action plan for places near railways, roads and airports within agglomerations for submission to the DOE. A consolidated action plan for all noise within agglomerations was to be drawn up by DOENI by 18 July 2008 and thereafter every five years.

For major airports and non-designated other airports where air traffic noise results in an L_{den} value of 55 dB(A) or more or an L_{night} value of 50 d(B)A or more anywhere in a first round agglomeration or agglomeration, and the competent authority is the airport operator, an action plan should be drawn up and submitted to the Secretary of State by 30 June 2008 (NAW, Scottish Ministers by 30 April 2008). Where the airport operator in England was not the competent authority on or before 30 June 2008 (Scotland, Wales 30 April 2008), the date for drawing up and submission of the action plan is 30 June 2013 (Scotland, Wales 30 April 2013.) (Date changed for England by SI 2008/375.)

The Regulations required the Secretary of State (NAW, Scottish Ministers) to publish Guidance setting out limit values or other criteria for the identification of priorities for action plans by 18 July 2007

Action plans should show how noise issues and effects, are to be managed and what measures are to be taken to prevent and reduce environmental noise where necessary, and particularly where exposure levels can induce harmful effects on human health; they should also aim to protect quiet areas identified in all agglomerations against an increase in noise and address priorities identified in any guidance. As a minimum, action plans should include:

- A description of the agglomeration, major roads, railways or major airports and other noise sources taken into account, and an evaluation of the number of people exposed to noise.
- The authority responsible and the actions which they are intending to take, e.g. traffic planning, land-use planning, technical measures, regulatory or economic incentives, etc.
- Limit values in place.
- Summary of the results of noise mapping.
- Long term strategy, financial information (e.g. cost effectiveness/benefit assessments), and provisions for evaluating the implementation of the action plan.
- Estimate in terms of the reduction of the number of people affected (annoyed, sleep disturbance, or other).

Action plans may identify public authorities (including any person who exercises functions of a public nature) as being responsible for implementing part of the action plan; where this is the case the authority should treat the action plan as its policy. If the authority wishes to depart from the policy (action plan) it should give the Secretary of State (NAW, Scottish Ministers) and/or competent authority which made the plan its reasons, and publish its reasons. A summary of the action plan (not exceeding 10 pages) covering those aspects referred to in Annex V (Minimum Requirements for Action Plans) should be prepared.

The public should be given early and effective opportunities to participate in the preparation and review of action plans, and should be kept informed of any decisions taken.

The Secretary of State (NAW, Scottish Ministers, DOE NI) should adopt action plans submitted to him (and those made or revised by him) if they meet the requirements of the Regulations; if he considers that a submitted action plan does not, he may either amend and adopt it or reject it. If the latter he should notify the competent authority of the reasons for rejection and the date by which it should be revised and resubmitted.

Action plans should be reviewed and revised every five years after adoption or whenever a major development occurs affecting the noise situation.

Scotland – Draft noise action plans for Edinburgh and Glasgow, and for road and rail noise, were put online for consultation at the end of May, with a deadline for comments of 21 July. Using available knowledge, the criteria used for potential noise exposure effects which might be used in Action Plans for Scotland include annoyance. The draft plans for the cities include proposals for designating Quiet Areas, using a noise level filter of less than 55 dB L_{day} and a minimum area filter of 9 hectares, and minimum area 4.5 hectares within an area. For assessing which areas are candidates for action to manage noise a prioritisation a matrix system has been used – areas identified for potential action are those affected mostly by road, and some by rail, noise. For Glasgow 42 candidate management areas affected by road noise and 11 by rail have been identified in this first round of mapping and exposure assessment, and for Edinburgh 19 road and 6 rail. Draft action plans for the three airports (Glasgow, Edinburgh, Aberdeen) are also online on the airport websites.

Wales – Draft Environmental Noise Action Planning Guidance published in August 2008 states that environmental noise will be considered in the context of sustainable development to contribute to improvements in health and quality of life.

England – Draft noise action planning guidance for civil airports in England was published by Defra for comment in early September 2008. The proposed guidance includes actions for airport operators to take account of when drafting their plans. It also acknowledges potential conflicts of interest that will need to be resolved – requiring consideration of the protection of National Parks and Areas of Outstanding Natural Beauty from increased noise, provided this does not add to the environmental burdens on more densely populated areas.

4.6.6 Information to the Public

Strategic noise maps and action plans made available to the public prior to their adoption should include a prominent notice identifying it as a draft subject to adoption by the Secretary of State (NAW, Scottish Ministers, DOE NI).

The Secretary of State (NAW, Scottish Ministers, DOE, NI) is required to publish all adopted strategic noise maps, action plans and consolidated noise maps and to provide a summary of the most important points.

4.7 NATIONAL NOISE STRATEGY – ENGLAND

Consultation on a noise strategy for England covering both ambient and environmental noise was expected during 2008.

London

The *Greater London Authority Act 1999* requires the Mayor to produce a noise strategy for London. The "London Ambient Noise Strategy" should include information about ambient noise levels related to road and rail traffic, aircraft and water transport in London and their impact on people living and working there. It should include an assessment of the impact on ambient noise levels of the other Strategies that the Mayor is required to prepare and a summary of the action proposed or taken to reduce ambient noise levels. In preparing or revising the Strategy, the Mayor is required to consult the Environment Agency. The final version of the Strategy, *Sounder City*, was published in March 2004; its aim is to "minimise the adverse impacts of noise on people ... using the best available practices and technology within a sustainable development framework". Among the proposals for achieving its aim, are

- Quieter roads, through use of quieter vehicles, better street repairs and use of low-noise road surfaces;
- Traffic – improved traffic management and encouraging better driving;
- Railways – use of quieter trains, improved railway track maintenance and control of noise and vibration; better trackside screening;
- Aircraft – support a ban on night flights, financial incentives to operators to phase out noisier aircraft, and for aviation to pay its environmental costs;
- Planning building design – buildings to be designed to screen housing and other uses from noise; creation of new quiet outdoor spaces; promotion of tranquil spaces.

4.8 PLANNING CONTROLS

Effective use of planning controls and related guidance will go a long way towards preventing many noise-related problems arising in the first place. Before any development can take place, planning permission under the *Town and Country Planning Act 1990* (Scotland – 1997 Act) must be obtained. In some cases an environmental impact assessment may be required (see also chapter 1). In either case noise will be one of the criteria to be considered.

The *Town and Country Planning General Development Order 1988* permits the use of a specific site for an activity (e.g. clay pigeon shooting, helicopter landing site) for up to 28 days a year without the need for planning permission. There is no limit as to the amount of shootings or number of take-offs and landings in any one day.

The Town and Country Planning (General Permitted Development) (Amendment) (England) Order 2008 (SI 2008/675), which came into force on 6 April 2008, amends Part 2 (Sch.2) of the 1995 GDO to allow for the installation of solar photovoltaics, solar thermal, ground and water source heat pumps, and combined heat and power systems within the curtilage of a dwelling house without the need for planning permission. For the moment the amendment to the GDO in respect of heat pumps and micro wind turbines only applies to detached dwellings, and indeed will only apply when the European Commission has approved the Microgeneration Certification Scheme; this has been established to ensure high standards for all microgeneration technologies and in particular that neighbours are not exposed to noise levels exceeding 45 decibels – this limit is to be reviewed in 2010 (see British Wind Energy Association website, www.bwea.com.

4.8.1 Planning Policy Guidance

PPG 24, Planning Policy Guidance: Planning and Noise (1994) replaced the 1973 *Planning and Noise Circular* (10/73); it is now itself under review and will, in common with other PPGs, become a "Planning Policy Statement" (PPS) – consultation on the revised PPS is expected. PPG 24, builds on the principles established in the earlier circular and suggests new mechanisms and guidelines for local authorities to adopt. General principles of how noise should be taken into account in the planning process are suggested, including the need for special attention to be given to proposed developments which could affect "quiet enjoyment" of areas of outstanding natural beauty, national parks etc. Examples of planning conditions are given and the various statutory and other noise controls listed. The PPG deals with noise from road traffic; aircraft; railways; industrial and commercial developments; construction and waste disposal sites; sporting, entertainment and recreational activities; and mixed noise sources. An Appendix to the PPG lists four categories of noise exposure for the assessment of new residential developments near a noise source. The categories are:

- A: noise need not be considered as a determining factor in granting planning permission, although the noise level at the high end of the category should not be regarded as a desirable level;
- B: noise should be taken into account when determining planning applications and, where appropriate, conditions imposed to ensure an adequate level of protection against noise;
- C: planning permission should not normally be granted. Where it is considered permission should be given, for example because there are no alternative quieter sites available, conditions should be imposed to ensure a commensurate level of protection against noise;
- D: planning permission should normally be refused.

For each of these categories the PPG recommends limits for both day (0700-2300) and night (2300-0700) time exposure for new dwellings which it is planned to develop near one of the following existing noise sources – road traffic, rail traffic, air traffic, or mixed source.

PPGs, and now PPSs, are produced by the Department for Communities and Local Government. Minerals Policy Guidance – also produced by the DCLG and being revised as Minerals Policy Statements – are also of relevance to noise control. MPS2, *Controlling and Mitigating the Environmental Effects of Minerals Extraction in England* (2005) includes an annex outlining the policy considerations in relation to noise from mineral workings and how they should be dealt with in local development documents etc.

In Scotland, SODD Circular 10.1999, *Planning and Noise*, and Planning Advice Note (PAN) 56 (2000) provide guidance on how planners should deal with noise in considering plans for new developments and the issues to be taken into account. The PAN introduces the use of noise exposure categories, outlines ways of mitigating the adverse impacts of noise and provides specific guidance on noisy and noise sensitive developments. In Wales TAN 11 (1997) covers similar concerns.

4.8.2 British Standard 4142

BS 4142 (1997) describes a "Method for Rating Industrial Noise Affecting Mixed Residential and Industrial Areas" and is thus of particular use when considering planning applications. It can be used to determine the likelihood of a complaint from an industrial source under normal operating conditions using a mixture of calculations and measurements, including measurement of background noise levels. It can be used to assess noise levels from both existing and new or modified premises in the vicinity of existing housing and other noise sensitive uses.

NEIGHBOURHOOD NOISE

Anyone living in the United Kingdom has a legal right to be protected against noise nuisance. Nuisance itself may be defined as "an unlawful interference with a person's use or enjoyment of land or of some right over, or in connection with it". Nuisance at law is a tort, i.e. a civil wrong, for which courts can provide a remedy, usually damages. In any action for noise nuisance the key factors are the need for an aggrieved person to establish that his occupation of land or property is affected by noise, and that each case will be considered on its merit.

Under *the Noise Abatement Act 1960*, noise became a statutory nuisance for the purposes of the *Public Health Act 1936*, thus enabling local authorities to take action for noise nuisance; it also introduced a complaints procedure whereby three or more private individuals, each of whom occupied land or premises affected by a noise nuisance, could complain direct to a magistrate; the "three neighbour rule" was to deter malicious and frivolous complaints, but was not re-enacted in either the *Control of Pollution Act 1974* or the *Environmental Protection Act 1990*, which require only one person to be affected for action to be initiated for statutory nuisance. The 1960 Act was repealed by the *Control of Pollution Act 1974*, Part III. As well as dealing with statutory noise nuisance, it introduced provisions relating to noise from construction sites, street noise, the setting up of noise abatement zones and on drawing up codes of practice on specific noise problems (e.g. burglar alarms). The sections dealing with statutory noise nuisance were repealed and re-enacted in the *Environmental Protection Act 1990* (see chapter 1, 1.29).

Minor noise nuisance is often restricted by local authority bye-laws, which the local authority enforces (although an individual may also prosecute). Bye-laws can be used to control such noise sources as: noisy conduct at night; noisy street trading; noisy animals; loud music; radios, television and other audio equipment; seaside pleasure boats; model aircraft; fireworks. Other regulations cover the use of loudspeakers and chimes.

Independent mediation services can also be an effective means of solving neighbour noise disputes where a statutory noise nuisance has not been established; in many cases mediation can be a quicker and more cost-effective way of tackling a range of neighbour nuisance disputes at an early stage and thus avoiding formal action. Independent or community based mediation services are now available in many local authorities. Mediation is not, however, a substitute for formal action where a statutory nuisance has been established: in such instances an abatement notice must be issued – but see below, 4.9.

Excessive noise from council dwellings can often be dealt with as a breach of tenancy agreement where this includes a clause restricting noise, and the tenants taken to court. The *Housing Act 1996* also enables both private and council landlords to evict tenants for anti-social behaviour – e.g. for causing a nuisance.

Section 1 of the *Crime and Disorder Act 1998* enables an anti-social behaviour order to be served on those responsible for various types of anti-social behaviour, including excessive noise.

Scotland

In addition to the *Environmental Protection Act 1990*, action in Scotland can also be taken under the *Civic Government (Scotland) Act 1982*, which covers many aspects of minor (or neighbourhood noise).

Northern Ireland

The control of noise and noise nuisance is regulated under the *Pollution Control and Local Government (Northern Ireland) Order 1978* and the *Noise Act 1996*.

Article 38 of the 1978 Order gives district councils powers to deal with noise from premises, including land, which they consider to be a statutory nuisance; a noise abatement notice should be served placing a prohibition or restriction on the noise; an appeal against such a notice should be lodged within 21 days. Article 39 enables individuals to complain direct to the Magistrates Court if the council is unwilling or unable to take action on their behalf. For noise caused in the course of trade or business, it is a defence to prove that the best practicable means have been used to prevent the noise. The Order also covers construction site noise and the establishing of noise abatement zones and codes of practice.

4.9 ENVIRONMENTAL PROTECTION ACT 1990

General noise nuisance may be dealt with under ss.79-81 of the *Environmental Protection Act 1990* (as amended), which empowers local authorities to deal with noise from fixed premises: factories, shops, discos and dwellings. Part III of the Act deals with statutory nuisance (including noise) in England, Wales and Scotland. Section 79 of the EPA places a duty on local authorities to require its area to be inspected from time to time to detect whether a statutory nuisance exists; this includes "noise emitted from premises so as to be prejudicial to health or a nuisance". Premises include land and vessels (except those powered by "steam reciprocating machinery") and noise includes vibration. Noise from aircraft other than from model aircraft is excluded. Similar powers to investigate noise in the street are provided through an amendment to s.79 of the EPA in the *Noise and Statutory Nuisance Act 1993* (see below, 4.11).

Local authorities must also take such steps as are reasonably practicable to investigate any complaints of statutory nuisance in their area. Where the local authority is satisfied that a statutory nuisance exists, or is likely to occur or recur, it must serve an abatement notice (s.80) requiring

- the abatement of the nuisance or prohibiting or restricting its occurrence or recurrence; and/or
- the carrying out of such works or other action necessary to abate the nuisance.

Section 86 of *the Clean Neighbourhoods and Environment Act 2005*, brought into force in England on 6 April 2006 and in Wales on 27 October 2006, amends s.80 of the EPA 1990 to allow a local authority in England and Wales, if it so wishes, to delay the issuing of an abatement notice in respect of noise for seven days while it takes other appropriate steps to persuade the person on whom such a notice would be served to abate the nuisance – if a local authority decides to use this power, it should record its reasons for doing so. However if the noise is not abated within seven days or the local authority concludes that it will not be, then it should proceed to serve an abatement notice in the normal way. Guidance on this section is available on Defra's website at www.defra.gov.uk/environment/localenv/legislation/cnea/index.htm

The London Local Authorities Act 2004 enables local authorities in London to issue a fixed penalty notice in respect of a breach of, or failure to comply with, an abatement notice served under the EPA. The 2007 LLA allows a council (or two or more acting together) to apply to the Secretary of State to designate an area of land as an "enforcement action zone" where higher fines can be imposed for a range of environmental offences, including noise nuisance.

Individuals may also take action against statutory nuisance. Information about this, together with details of local authority powers with regard to statutory nuisance are covered in chapter 1, 1.29.2.

4.10 CONTROL OF POLLUTION ACT 1974

Part III of COPA is specifically concerned with the control of noise pollution ***(England, Scotland & Wales)***. Sections 57-59, which covered local authorities' duty to inspect, noise abatement notices and individuals' rights to take action in respect of a noise nuisance were repealed on 1 April 1996. On that date the statutory nuisance provisions of the EPA were extended to Scotland through implementation of s.107 and Sch. 17 of the *Environment Act 1995* (see chapter 1, 1.29).

In ***Northern Ireland***, the *Pollution Control and Local Government (Northern Ireland) Order 1978* covers the same areas as the remaining sections of Part III of COPA.

4.10.1 Construction Sites

Construction activities are inherently noisy and often take place in areas that are normally quiet. During 2004/05, environmental health officers in England and Wales received 11,714 complaints relating to construction and demolition noise (CIEH annual survey into local authority noise enforcement action). In Scotland there were 1,114 complaints regarding noise and vibration from building sites in 2004-05 (Environmental Health in Scotland 2004-05, REHIS). In Northern Ireland there were 426 complaints about noise from construction sites in 2006/07 (Noise Complaint Statistics 2006/07, DOE NI)

Section 60 (Art.40, 1978 NI Order) gives local authorities the power to serve a notice imposing requirements as to the way in which construction works are to be carried out.

British Standard 5228, *Noise Control on Construction and Open Sites*, provides guidance to enable compliance with s.60 and is applicable throughout the UK. It is in five parts, with Part 1 (1997) being a code of practice for basic information and procedures for noise and vibration control; Part 2 (1997) provides a guide to noise and vibration control legislation for construction and demolition including road construction and maintenance; Part 3 (1997) is a code of practice applicable to surface coal extraction by opencast methods; Part 4 (1992) is a code of practice for noise and vibration control applicable to piling operations; and Part 5 (1997) is a code of practice applicable to surface mineral extraction (except coal) sites. Parts 1, 3, 4 and 5 have been given formal approval under s.71 of COPA – *The Control of Noise (Codes of Practice for Construction and Open Sites) (England) Order 2002*, SI 2002/461, effective 25 March 2002; Scotland: SSI 2002/104, 29 March 2002; Wales: SI 2002/1795 (W.170), 1 August 2002; NI: SR 2002/303, 1 November 2002.

Neither COPA nor the related codes of practice set down specific limits for construction site noise, on the basis that the local authority knows its own locality best and would have a better idea of suitable noise limits. The person served with the notice may appeal within 21 days to the Magistrates' Court (Sheriff in Scotland; N. Ireland – Court of Summary Jurisdiction) and is guilty of an offence under the Act if, without reasonable excuse, any requirement of the notice is contravened.

Under s.61 (Art.41, 1978 NI Order), the person intending to carry out works may apply in advance for a consent as to the methods by which the works are to be carried out, with the local authority having 28 days in which to make a decision. If it does not give consent to a building contractor/developer's own scheme of noise control, or does but with conditions or other limitations with which the contractor does not agree, the contractor may appeal to the court within 21 days. Once a consent has been given, the contractor is effectively immune from action on noise grounds taken by the local authority so long as the terms of the consent are complied with.

The activities covered by these sections of the Act are:

a) the erection, construction, alteration, repair or maintenance of buildings, structures or roads;
b) breaking up, opening or boring under any road or adjacent land in connection with the construction, inspection, maintenance or removal of works;
c) demolition or dredging work; and
d) (whether or not also comprised in paras a, b, or c) any work of engineering construction.

Individual occupiers of premises (England, Scotland and Wales) may apply to a Magistrates' Court (Sheriff in Scotland) for an order under s.82 of the *Environmental Protection Act 1990*, on the grounds that the noise from the site nevertheless amounts to a statutory nuisance.

Construction work undertaken by statutory operators (i.e. on behalf of the government) is usually regarded as "emergency" and therefore exempt from COPA.

Noise Control: the Law and its Enforcement, 3rd edition (Christopher Penn, Shaw & Sons, 2002) has a detailed chapter on noise control at construction sites and methods of reducing noise from construction plant. See also 4.20 below summarising further legislation on industrial and construction site noise.

4.10.2 Noise in Streets

Sub-section 62(1) of COPA bans the use of a loudspeaker in a street (Scotland – road) between the hours of 9.00 pm and 8.00 am (Art.42, 1978 NI Order). A loudspeaker includes a megaphone or other amplifying instrument. Under the *Noise and Statutory Nuisance Act 1993* (NASNA), the Secretary of State may vary these times but not so as to permit the use of loudspeakers between these times; additionally, local authorities may adopt powers enabling them to grant a Consent for a loudspeaker to be used later than 9.00 pm but not for advertising or electioneering purposes (see below, 4.11.2).

Under COPA, it is an offence to use loudspeakers in the street at any hour for advertising any entertainment, trade or business. Sub-section 62(2) exempts the emergency services (fire, police, ambulance) as well as the water authorities and local authorities in certain circumstances. Also exempt are loudspeakers on ground being used by pleasure fairs, and loudspeakers that form part of a public telephone system.

Vehicles which sell perishable foodstuffs (such as ice-cream) are exempted from this section, but may only use loudspeakers between noon and 7.00 pm; again, the loudspeaker must be operated in such a way as not to give reasonable cause for annoyance to persons in the vicinity. These times may be varied by local authority Consent under the *Noise and Statutory Nuisance Act 1993* (see below, 4.11.2).

In 2006/07, there were 253 complaints in N. Ireland about noise in streets (this includes complaints about loudspeakers, ice-cream van chimes and street traders) (Noise Complaint Statistics 2006/07, DOENI).

From 1997/98 complaints under this section of the Act, including roadworks, in England and Wales have been included in complaints related to vehicles, machinery and equipment in the street and are dealt with under the *Noise and Statutory Nuisance Act 1993* – see 4.11.1

4.10.3 Noise Abatement Zones

A local authority may designate all or part of its area as a noise abatement zone (NAZ) (COPA, s.63; Art.43 NI 1978 Order); the purpose of a NAZ is the long-term control of noise from fixed premises in order to prevent any further increases in existing levels of neighbourhood or community noise levels, and to achieve reduction of those levels wherever possible.

Following the implementation of a noise abatement zone order, the local authority measures noise levels from those types of premises (specified in the order) within the zone (s.64). These are recorded in a register kept by the local authority for the purpose and open to public inspection. Details of methods of measurement and maintenance of the register are given in the *Control of Noise (Measurement and Registers) Regulations 1976* (SI 1996/37; Sc: *Noise Levels (Measurement and Registers) (Scotland) Regulations 1982* (SI 1982/600). Appeals in respect of entries on the noise level register, refusal to allow exceedence of a registered level, and in respect of proposed noise level from a new building are determined by the Secretary of State – the *Control of Noise (Appeal) Regulations 1975* (SI 1975/2116) refer.

Once the noise level has been registered, it may not be exceeded except with the local authority's consent (s.65). Over a period of time, the local authority may seek to achieve a reduction in the initially registered levels of noise by serving a reduction notice under s.66 of the Act, but only if it is satisfied that a reduction is practicable at a reasonable cost and would afford a public benefit. Section 67 empowers the local authority to determine the acceptance level of noise for a proposed new building, which when constructed will be subject to a noise abatement order. Section 68 allows the Secretary of State to make regulations for the reduction of noise caused by plant or machinery, whether or not in a noise abatement zone and s.69 gives the local authority default powers in the case of failure to comply with a noise abatement or noise reduction notice. There is a right of appeal (s.70) to the Magistrates' Court for three months from the date on which a Noise Reduction Notice is served. It is a defence to prove that the best practicable means were used to prevent or counteract the effects of the noise.

4.10.4 Codes of Practice

Section 71 of COPA gives the Secretary of State powers to prepare and approve, and issue codes of practice for minimising noise, or to approve such codes issued by non-government bodies, such as the Code of Practice on the control of noise on construction and demolition sites (BS 5228) prepared by the British Standards Institution (see 4.10.1 above).

The Department of Environment (now Defra) issued three codes, all in 1982 (with similar codes in Northern Ireland issued under Art.51 of the 1978 Order):

- Noise from ice cream chimes, etc (limits the use of chimes to between 12 noon and 7.00 pm).
- Noise from burglar intruder alarms (see also below, 4.11.3 & 4.14).
- Noise from model aircraft.

Such codes, although having statutory recognition, do not have the force of regulations, and infringement does not constitute an offence. Non-compliance, however, will usually be taken into account in any proceedings for noise nuisance.

A review of these codes, commissioned by Defra, was published in September 2005; the review team has made a number of recommendations for updating the codes – many to take account of technical and legislative developments since their publication – and also notes that most of those consulted in the course of the review felt that the codes' statutory status did enhance their effectiveness.

Other Codes

There are a number of other draft and finalised codes of practice which have been prepared by various interest groups but which have not received approval under s.71, although some have been submitted for such approval. These codes can, however, be used as guidance in determining what constitutes good practice. Such codes include:

- *Noise from Organised Off-Road Motor Cycle Sport:* prepared by the Noise Council in association with various other bodies (1994) – provides guidance on noise control and limits for various types of event.
- *Clay Target Shooting: Guidance on the Control of Noise:* CIEH (2003) – this guidance which is intended primarily for local authorities, will also be of use to those organising shoots or involved in their management. It provides practical guidance on the ways in which noise from such shoots can be controlled to minimise or prevent intrusion. A recommended method for noise measurement and subsequent rating, produced by the Building Research Establishment, is also given. Available on the CIEH website www.cieh.org
- *Water Skiing and Noise* (British Water Ski Federation, 1996): provides guidance for clubs on avoiding significant impact from noise on the surrounding community for both existing and new sites. Methods for assessing noise from sites, advice on course layout, hours of operation, screening etc is also provided, as well as noise limits for specific activities and events. (Note: *the Recreational Craft Regulations 2004* include limits on both emissions to air and noise from recreational boats and jet skis, see 4.24.)
- *Environmental Noise Control at Concerts* (Noise Council, 1994): it provides assistance for those planning concerts, for licensing authorities and for local authorities enforcing nuisance legislation. It includes music noise levels for both indoor and outdoor events and single as well as multiple events. (CIEH is to "refresh and revise" this code, Environmental Protection UK Noise Update, March 2008.)
- *Guide to Health, Safety and Welfare at Pop Concerts and Similar Events* (HSE, Home Office, Scottish Office, 1993): this includes guidance on noise levels, covering the risks to health and safety (including noise) both to those at the concerts and to those living near the concert venue.

- *Short Oval Circuit Motor Racing* (NSCA's National Noise Committee, 1996): this covers stock car, hot rod, banger racing and similar events; it aims to help local authorities and organisers to minimise the noise impact of such events on nearby premises; it also outlines legal controls.
- *Good Practice Guide on the Control of Noise from Pubs and Clubs* (Institute of Acoustics, March 2003): provides guidance on control of noise affecting noise sensitive premises from the public and private use of pubs, clubs, hotels, discos, restaurants, cafés, community and village halls etc; noises covered include public address systems, children's play areas and people in general, but not noise at live sporting events. Further work to be done on criteria for measurement units and techniques.

In August 2002 Defra published Guidance on preparing codes of practice for minimising noise in England, with similar guidance published by the Scottish Government in March 2002 and by the Welsh Assembly also in August 2002. As well as providing guidance on developing such codes of practice and the issues to be covered, it sets out the procedures to be followed – consultation etc – if it is intended to submit a code for approval under s.71 of COPA.

4.10.5 Best Practicable Means

It is a defence in any summary proceedings relating to noise caused by a trade or business to prove that the Best Practicable Means (BPM) have been used to prevent or counteract the effect of noise. Section 72 (NI: Art 52 of the 1978 Order) defines practicable as "reasonably practicable, having regard among other things to local conditions and circumstances, to the current state of technical knowledge and to the financial implications". Means includes "the design, installation, maintenance and manner and periods of operation of plant and machinery, and the design, construction and maintenance of buildings and acoustic structures".

4.10.6 Other Sections

Section 73 deals with the interpretation of some of the terms used in Part III of COPA, and s.74 with penalties for offences under Part III. (NI: Arts 53 & 54 of the 1978 Order)

4.11 NOISE & STATUTORY NUISANCE ACT 1993

This Act (apart from the burglar alarm provisions, which have been repealed in E & W by *the Clean Neighbourhoods & Environment Act 2005*), came into force on 5 January 1994; it covers England, Scotland and Wales. It amends ss.79-82 of the *Environmental Protection Act 1990* to make noise in a street (Scotland – road) a statutory nuisance. Section 62 of COPA dealing with the hours in which loudspeakers may be used is also amended and provision is made for regulating burglar alarms. The Act is enforced by local authorities; the procedure for taking statutory nuisance action and the appeals procedure are outlined in chapter 1, 1.29.

Department of the Environment, Transport and the Regions Circular (Environment Circular 9/97, WO Circular 42/97) provides guidance to local authorities on implementation of the Act and their duties; it also explains the operational role of the police.

4.11.1 Statutory Nuisance – Noise in Street

Sections 2-5 of NASNA add an additional subsection to s.79 of the *Environmental Protection Act 1990* so as to make noise emitted from or caused by a vehicle, machinery or equipment in a street (Scotland – road) a statutory nuisance. This could include, for example, noisy car repairs, car alarms, car radios and parked refrigerator vehicles, as well as (in England and Wales) roadworks. In 2004/05 there were 11,305 complaints to local authorities in England and Wales under these sections of the Act (CIEH annual survey into local authority noise enforcement action). Traffic noise, noise by military forces or from political or other campaigning demonstrations (but not noise or disturbance as result of picketing) is excluded. Having satisfied itself that a statutory nuisance exists or is likely to occur or recur, the local authority must serve an abatement notice; the procedure for this is set out in chapter 1, 1.29.

The *Noise and Statutory Nuisance Act 1993* does, however, amend the EPA procedure for serving an abatement notice in respect of a nuisance from or caused by a vehicle, machinery or equipment (VME) on the street; if possible the notice should be served on the person responsible, but if the VME is unattended then the notice may be affixed directly to the VME. The enforcing officer should then spend one hour endeavouring to trace the person responsible for the VME to serve a copy of the notice on them; where the person is found additional time should be allowed before the notice takes effect to enable them to abate the nuisance; if the person cannot be found the enforcing officer may then take the necessary steps to abate the nuisance.

Authorised persons are empowered to gain entry to a vehicle in order to carry out the terms of an abatement notice (e.g. to turn off an alarm); before doing so the police should be notified; the vehicle should be left as secure as before entry or, if this is not possible, should be immobilised or removed to a secure place and the police notified of where it has been taken.

4.11.2 Loudspeakers

Section 7 of the Act gives the Secretary of State powers to further restrict the hours in which loudspeakers may be used (8.00 am - 9.00 pm) in a street (Scotland – road) under s.62 of COPA (see above, 4.7.2).

Under s.8 local authorities may adopt powers enabling them to grant a Consent for a loudspeaker to be used outside the hours of 9.00 pm and 8.00 am (except for advertising or electioneering). Schedule 2 to the Act sets out the procedure for granting a Consent: anybody wanting to use a loudspeaker at a different time must apply for a Consent specifying the purpose, date, location and duration of use. The local authority must deal with the application within 21 days and may attach any conditions it feels appropriate to the Consent. A charge may be made and the local authority may require details of the Consent to be published in a local newspaper.

4.11.3 Audible Intruder Alarms

Section 9 and Schedule 3 of NASNA, which were never brought into force, were repealed in England on 6 April 2006 and in March 2007 in Wales by *the Clean Neighbourhoods and Environment Act 2005* – see below 4.14. This section has not been brought into force in Scotland.

4.11.4 Recovery of Expenses

Section 10 of the Act applies in England and Wales only. Its effect is to add two additional sections (81A and 81B) to the *Environmental Protection Act 1990* enabling local authorities to recover expenses incurred in abating a nuisance by placing a charge on the property (see chapter 1, 1.29).

4.12 NOISE ACT 1996

This Act, covering England, Wales and Northern Ireland, was introduced into Parliament as a Private Member's Bill in February 1996 and received Royal Assent on 18 July 1996. It deals with noise nuisance – in particular night-time noise – from domestic premises. In Northern Ireland, ss.2-9 relating to the procedure for dealing with night-time noise nuisance are adoptive – i.e. the district council must resolve to implement these sections for application in its area; alternatively, the Secretary of State may, by order, require a district council to implement these sections. *The Anti-Social Behaviour Act 2003* amends ss.2-9 to enable all local authorities in England and Wales to investigate and deal with complaints about night-time noise – this amendment took effect on 31 March 2004. Section 10 of *the Noise Act* and the Schedule to the Act – confiscation of noise-making equipment – apply to all local authorities in England, Wales and N. Ireland.

Guidance on *the Noise Act* taking account of the extension of the Act to licensed premises is available from Defra (March 2008), with separate Guidance available in Wales, (NAFWC 18/07).

In Scotland, similar measures are contained in Part 5 of *the Anti-Social Behaviour etc (Scotland) Act 2004* – see 4.13 below.

4.12.1 Night-time Noise Nuisance

Section 2(6) defines night as 11.00 pm to 7.00 am the following morning.

These sections of the Act were brought into force on 23 July 1997, and could be used from 23 October 1997. *The Anti-Social Behaviour Act 2003* (s.42) has amended ss.2-9 in England and Wales, with effect from 31 March 2004, so that all local authorities may use the powers without first having to adopt those sections. *The Clean Neighbourhoods and Environment Act 2005* (s.84) extends the powers of local authorities in England & Wales to take action against noise at night from licensed premises or from premises where a temporary event notice (under *the Licensing Act 2003*) has effect.

In ***Northern Ireland***, a district council wishing to use these powers must pass a resolution stating from when the offence is to apply in its area – this should not be less than three months from the date of the resolution. It must publish a notice in a local newspaper on two consecutive weeks ending at least two months before the commencement date. The notice should state that a resolution has been passed, give the commencement date and summarise the effect of the resolution. An order from the Secretary of State requiring implementation of these sections must also allow three months from the date of the order before they are brought into force. Once brought into force a district council has a duty to take reasonable steps to investigate a complaint from an individual about noise being emitted from another dwelling during night hours. To date only Belfast City Council has implemented these provisions.

In ***England and Wales*** ss.2-9 apply in all local authorities; where a local authority receives a complaint from an individual in a dwelling about noise being emitted from another dwelling during night hours, it may arrange for an officer to take reasonable steps to investigate a complaint (rather than having a duty to do so) (*ASB Act, s.42*). Section 84 & Sch.1 of the CN&EA enable local authorities to deal with complaints of noise at night from licensed premises, including those with a temporary event notice – the procedure is similar to that for private dwellings.

Section 2(7) enables a local authority to investigate a complaint about night-time noise from a dwelling (E & W: or licensed premises) outside its area, (NI: whether or not these sections of the Act apply in that area). Section 11 defines dwelling as any building or part of a building used or intended to be used as a dwelling, or any garden, yard, outhouse or other appurtenance belonging to or enjoyed with the dwelling.

If it is thought that the noise exceeds, or might exceed, the permitted level as measured within the complainant's dwelling then a warning notice may be served on the person appearing to be responsible for the noise. (Note – it is not a requirement of the Act that the noise should be measured; for instance, there may be circumstances when it is obvious that the noise exceeds the permitted level.)

It should be noted that there may be circumstances that while the noise does not exceed the night-time permitted level, the investigating officer considers that it does constitute a statutory nuisance under s.80 of the *Environmental Protection Act 1990*; in such cases an abatement notice must be served, although the investigating officer (England & Wales) has discretion to defer doing so for seven days (see chapter 1, 1.29.1a). If however the investigating officer considers that the noise does exceed the permitted level and constitutes a statutory nuisance, he can issue both an abatement notice and a warning notice; in any subsequent enforcement action, a decision should be taken under which Act it would be most appropriate to proceed.

If a warning notice is to be served, it should state (s.3(1)) that

- in the opinion of the local authority officer, noise is being emitted during night hours and that the noise exceeds, or might exceed, the permitted level as measured from within the complainant's dwelling;
- any person responsible for the noise being emitted during the time specified in the notice (beginning 10 minutes after service of the notice until 7.00 am) may be guilty of an offence.

If it is not possible to identify the person responsible for the noise then the notice should be left at the offending dwelling (E & W: or premises). The notice should state the time at which it was served or left.

The "permitted level" of noise is set by the Secretary of State (National Assembly of Wales/NAW) in Directions made under s.5 (*The Permitted Level of Noise (England) Directions 2008* and similar 2007 Directions in Wales); s.6 provides powers to approve measuring devices and to specify the manner and purpose for which such devices are to be used *(The Measuring Devices (Noise Act 1996) (England) Approval 2008* and similar 2007 Approval in Wales); the England

Directions and Approval came into force on 28 February 2008, and those in Wales on 5 April 2007. In both cases the earlier the earlier Directions and Approval are revoked.

The permitted level of noise from dwellings and licensed premises in both England and Wales is 34 dB(A) where the background level does not exceed 24 dB(A), or 10 dB(A) above the background level where this exceeds 24 dB(A); the noise level should be measured over a period of not less than one minute and no more than five.

Detailed guidance (March 2008) on the powers available to local authorities in England to deal with night noise from dwellings and licensed premises which exceeds the permitted level is available from Defra, with similar guidance in Wales ((NAFWC 18/07).

An offence is committed (s.4) if the noise, when measured from inside the complainant's dwelling, continues to exceed the permitted level during the time specified in the warning notice; if convicted the offender will become liable to a fine not exceeding level 3 on the standard scale (currently £1,000). However the local authority authorised officer may, if he thinks that an offence has or is being committed, issue a fixed penalty notice (s.8), set at £100 (unless the local authority has set a different amount (new ss.8A and 8B, inserted by s.82(3) of the CN&EA); the offender has 14 days to pay the penalty and in which case will not be prosecuted further. The amount of the fixed penalty in respect of noise from a licensed premises is £500 (Sch.1 CN&EA). If the noise continues after the fixed penalty notice has been issued, then the local authority officer can prosecute the offender who becomes liable for a maximum fine of £5,000 if convicted. In any proceedings, evidence (s.7 as amended by the CN&EA) may be given by the production of a document signed by the local authority officer stating he had identified the specified dwelling/premises as being the source of the noise – a copy of this document must be served on the offender not less than seven days before the hearing or trial.

In England and Wales, *the Anti-Social Behaviour Act 2003* amends *the Noise Act* to enable local authorities to retain penalty receipts for use in carrying out its functions under the Act (or other functions as specified in Regulations).

4.12.2 Seizure of Noise-making Equipment

Section 10 and the schedule to the Act apply where a warning notice has been served and the noise continues to exceed the permitted level; it was brought into force on 23 July 1997. (Section 10(7) clarifies that the powers of a local authority to abate a noise nuisance under s.81(3) of the *Environmental Protection Act 1990* include being able to seize and remove the noise-making equipment; this was brought into force on 19 September 1996 – see chapter 1, 1.29.1a.)

A local authority officer, or other authorised officer, may enter the offending dwelling and seize and remove – confiscate – any equipment which it is thought is or has been used to emit noise. If entry is refused, likely to be refused, or request for admission to the dwelling would defeat the object of the entry, the local authority authorised officer may obtain a warrant from a justice of the peace authorising entrance (s.10(4)). The authorised officer may take with him such other people or equipment as is necessary (s.10(5)). Prosecution for obstructing confiscation may result in a maximum fine of £1,000.

The seized equipment may be retained for 28 days or, if court proceedings are instigated, until the case has been dealt with or discontinued. If no court proceedings are begun, then the equipment should be returned to any person who appears to be the owner, or who makes a claim for the equipment within six months of the expiry of the 28 day period. The local authority should take reasonable steps to advise a potential claimant of their right to do so. If no claim is made within six months the local authority may dispose of the equipment.

Equipment seized at the time that a fixed penalty notice has been served under s.8 of the Act may not be retained if the fixed penalty is paid within the allowed period (14 days).

The court may make a forfeiture order if it convicts someone of a noise offence and is satisfied that the seized equipment was actually used for that offence; such an order can be made as the sole means of punishment for the offence. An offender loses all right to the equipment forfeited.

Where equipment has been forfeited, a claimant (other than the offender) may apply to the court within six months of the date of the forfeiture order for return of the equipment. If the court is satisfied that

- the claimant had not given the offender permission to use the equipment; or
- did not know, or had no reason to suspect, that the equipment would be used in committing an offence, and
- is satisfied that the claimant is the owner of the forfeited equipment

it will make an order requiring the local authority to deliver the equipment to that person. Local authorities also have a duty to take reasonable steps to bring to the attention of any person entitled to claim the forfeited equipment their right to do so. If the equipment has not been claimed on the expiry of six months from the date of the forfeiture order, the local authority may dispose of it.

Regulations will provide for the retention and safekeeping of any equipment seized and the charges the local authority may recover. The local authority will be able to keep any equipment until all charges have been paid; a person claiming seized or forfeited equipment will not be liable for charges if the local authority is satisfied that the person did not know and had no reason to suspect that their equipment was likely to be used to commit an offence.

4.13 NOISE NUISANCE (SCOTLAND)

Part 5 of *the Antisocial Behaviour etc (Scotland) Act 2004*, which came into force on 28 October 2004, enables local authorities, who resolve to do so, to apply noise controls in specific areas and/or covering specific times. The provisions in this part of the Act are similar to those for England and Wales for dealing with night-time noise nuisance (see above): local authority officers may investigate noise complaints; maximum permitted noise levels and approved measuring devices will be specified in Regulations; if a local authority officer considers noise from a domestic dwelling exceeds the permitted level a warning notice may be issued; if the noise continues to exceed the permitted level during the specified time of the warning notice, an offence is committed and may lead to prosecution; alternatively the local authority officer or the police may issue a fixed penalty notice of £100, which if not paid will lead to prosecution; local authority officers may also obtain a warrant to seize noisemaking equipment.

The Antisocial Behaviour (Noise Control) (Scotland) Regulations 2005 (SSI 2005/43), which came into force on 28 February 2005, specify the permitted levels of noise which apply at different times:

- 31 dB between the hours of 23.00 and 07.00 (night) where the underlying level does not exceed 21 dB;
- 37 dB between the hours of 19.00 and 23.00 (evening) where the underlying level does not exceed 27 dB;
- 41 dB between the hours of 07.00 and 19.00 (day) where the underlying level does not exceed 31 dB.

Where the underlying level exceeds that specified for the particular period, then the permitted level is 10 dB in excess of the underlying level.

The Regulations also prescribe approved measuring devices, how noise is to be measured and verification and testing of approved devices.

The *Civic Government (Scotland) Act 1982* also provides the police in Scotland with powers to abate noise nuisance from, among other things, musical instruments and sound-producing devices; see this chapter 4.15.2.

4.14 AUDIBLE INTRUDER ALARMS (E&W)

Sections 69-76 of *the Clean Neighbourhoods & Environment Act 2005*, brought into force in England on 6 April 2006 and on 15 March 2007 in Wales, provide powers for local authorities to deal with faulty burglar alarms. Guidance for England on these sections of the CN&EA is available on Defra's website at www.defra.gov.uk/environment/localenv/legislation/cnea/index.htm

Section 69 of the CN&EA enables local authorities to designate all or part of their area as an "alarm notification area"; before doing so they must publicise their plans in a local newspaper, allowing 28 days for representations; a decision to continue with a designation must also be publicised in a local newspaper and written notification sent to all affected premises, stating the date (which must be at least 28 days later) when the area will become designated. If it is decided not to go ahead with designation this decision must be advertised in a local newspaper. For all premises with an audible intruder alarm, a responsible person – defined as the occupier or owner of the premises – must nominate a key-holder for the premises and give the local authority that person's contact details (s.71) within 28 days of the area becoming designated or within 28 days of the alarm being fitted. Key-holders must have appropriate keys for the premises and knowledge of the alarm system to enable them to silence it (s.72). Failure either to nominate a key-holder or notify the local authority is an offence and if convicted may result in a fine of up to £1,000 level 3; alternatively, the local authority may offer an offender the option of discharging the offence through payment of a fixed penalty fine of £75 (or other amount between £50-£80 fixed by the local authority). Section 75 of the Act enables local authorities to retain receipts from fixed penalty notices for use in carrying out its functions with respect to audible intruder alarms.

Section 77 of the Act provides authorised officers of the local authority with powers to enter premises (without force), whether or not they are in a designated alarm notification area, to silence an alarm if it has been sounding for at least 20 minutes continuously or intermittently for more than an hour; if the premises are in a designated area "reasonable steps" must be taken to get the key-holder to silence the alarm. If it is not possible to gain entry to the premises without the use of "reasonable force", a warrant must first be obtained from a justice of the peace (s.78). Section 79 enables a local authority to recover expenses incurred in silencing an alarm.

The 1982 Code of Practice on Noise from Audible Intruder Alarms (approved under s.71 of COPA) gives guidance on minimising nuisance from faulty alarms. It suggests that alarms should be fitted with a 20 minute cut-out device and that key-holders of properties with alarms should be notified to the police.

Action to deal with faulty alarms could also be taken under the nuisance provisions of the EPA (see chapter 1, 1.29).

Section 23 of *the London Local Authorities Act 1991*, which empowered local authorities to prosecute owners or occupiers of buildings whose audible intruder alarms contravened specified requirements or caused annoyance to people in the vicinity was repealed on 6 April 2006 following the introduction of the CN&EA provisions on alarms.

ENTERTAINMENT NOISE

The provision of entertainment or entertainment facilities "regulated entertainment", or alcohol now requires a "premises licence" under *the Licensing Act 2003* (E & W). This replaces the previous system under which separate licences were required for providing entertainment and alcohol from, respectively the local authority and the magistrates' court. In Scotland *the Civic Government (Scotland) Act 1982* requires all "places of public entertainment" to be licensed, with similar provisions for Northern Ireland included in *the Local Government (Miscellaneous Provisions) (NI) Order 1985*.

4.15 LICENSING CONTROLS

4.15.1 Licensing Act 2003 (E & W)

Premises licences are granted by the Licensing Authority (i.e. of the local authority) and may include conditions tailored to the particular premises; conditions should not duplicate the requirements of existing legislation, including health and safety law and noise nuisance legislation. However, guidance issued under s.182 of the Act notes that conditions aimed at preventing public nuisance – e.g. noise which affects either a few people locally or a major disturbance affecting the whole community – may be appropriate; and says that "conditions relating to noise will normally concern steps necessary to control the levels of noise emanating from the premises...ensuring that doors and windows are kept closed after a particular time..."; it would also be permissible to include a condition requiring the licence holder to put up signs by exits encouraging departing customers to respect the rights of people living nearby to a peaceful night.

Licensable activities are

- the sale by retail of alcohol
- the supply of alcohol by or on behalf of a club
- the provision of late night refreshment
- the provision of regulated entertainment.

Regulated entertainment is any that takes place in front of an audience or is provided for the purpose of entertaining an audience, and usually with a view to profit; it includes

- the performance of a play
- an exhibition of a film
- indoor sporting event; boxing or wrestling entertainment
- performance of dance
- performance of live music; any playing of recorded music; or entertainment of a similar description to live or recorded music or dance.

There are a number of exempted activities including: religious services, morris dancing or similar, incidental music, garden fetes, film exhibitions at museums and galleries, or for the purposes of advertisement, information, education etc. However if alcohol is supplied or late night refreshment provided a licence will be required.

It is an offence to carry out a licensable activity without a licence, the penalty on summary conviction being imprisonment for a maximum of six months or a fine not exceeding £20,000, or both. The person performing the entertainment is not liable.

Organisers of "one-off" events – or events taking place only irregularly – at which entertainment, alcohol or late night refreshments are provided should notify the licensing authority and the police. Such notifications – Temporary Event Notices (TENs) – are limited to events lasting for up to 96 hours and cannot involve more than 499 people at any one time; the same premises may only be used for a maximum of 12 occasions each year, subject to an overall aggregate of 15 days. A TEN may be used to cover anything from a pop festival or rave to events organised in schools and colleges or village halls. Failure to adhere to any of the requirements for TENs would render the event unauthorised, and the premises user liable to prosecution. Premises where a TEN is in effect will also be subject to the night noise provisions of *the Noise Act 1996*, as amended – see above.

See also Closure of Licensed Premises, below.

4.15.2 Civic Government (Scotland) Act 1982

All places of "public entertainment" require licensing; these are defined as "a place where on payment of money or money's worth, members of the public are admitted or may use any facilities for the purposes of entertainment or recreation." Such places include educational establishments, athletic and sports grounds, and church halls – in each case when not being used for their prime purpose – community centres, barn dances, discos, leisure and sports centres, bowling alleys, dance halls, marquees, circuses and fairs etc, licensed premises (outside their permitted hours). Applications may be made for a three year, one year or temporary licence (for a one-off event). Licences may include a condition specifying the days and times when the licensed activity may take place. (The provision of alcohol at an event will usually require a separate licence.)

Section 54 of the *Civic Government (Scotland) Act 1982* empowers uniformed police officers to request that noise from, for example, sound-producing devices be reduced if it is giving any other person "reasonable cause for annoyance". If the request is ignored then the police may confiscate the offending equipment as evidence for any prosecution. This section of the Act also covers annoyance caused by singing, playing musical instruments, televisions, radios etc. Loudspeakers used by the emergency services are exempt.

The Licensing (Scotland) Act 1976 enables the Licensing Board to include conditions limiting noise.

Part 5 of *the Anti-Social Behaviour Act etc (Scotland) 2004* enables local authorities, who resolve to do so, to implement noise controls covering specific areas and/or times of day – see 4.10 above.

4.15.3 Local Government (Miscellaneous Provisions) (NI) Order 1985

All indoor and outdoor entertainment for which a charge is made (whether paid by money or money's worth) requires a licence. Indoor entertainment includes theatrical performances, music, singing and dancing, circuses, public contests, matches, exhibitions etc, and machines for entertainment or amusement. Outdoor entertainment is defined as that taking place in the open on private land and is of a musical nature. Exemptions may be made for religious events or in some circumstances those taking place in schools. Application may be made for an annual licence, a 14 day occasional unspecified dates or 14 day occasional specified dates. Licensing authorities may take into account the effect of the entertainment on neighbouring properties. Action for noise nuisance can be taken under *the Pollution Control and Local Government (NI) Order 1978*.

4.16 STATUTORY NUISANCE

Action for noise nuisance arising from entertainment and parties can be taken under Part III of the *Environmental Protection Act 1990*. Under s.80 of the EPA, an abatement notice can be served requiring abatement of the nuisance; the notice can be served on anyone involved with the organisation of the party, e.g. host, disc jockey, sound engineer; the notice can be addressed to the occupier of the premises or indeed simply affixed to the premises. An abatement notice can also be served in anticipation of a nuisance – in such cases the notice could prohibit the delivery of audio equipment to the party site or require its removal. Contravention of an abatement notice is an offence.

The EPA provides local authorities with powers under s.80(3) to take whatever action is necessary to abate a nuisance themselves. This might include confiscating the offending noise-making equipment, a power clarified under s.10(7) of *the Noise Act 1996* (see above, 4.12.2). Schedule 2 of the Act provides powers of entry to the premises, although where these are mainly or wholly used for residential purposes 24 hours' notice should be given or where entry is likely to be refused, a warrant obtained.

Section 43 of the *Powers of Criminal Courts Act 1973* also enables local authorities in England and Wales to seek a deprivation order from the court for the permanent confiscation of noise making equipment following a prosecution.

For further details regarding procedures under nuisance legislation, see chapter 1, 1.29.1.

Statutory nuisance legislation may also be used to deal with noisy parties – or alternatively *The Noise Act 1996* (England, Wales and Northern Ireland) sets a limit on the level of noise which can be made between 11.00 pm and 7.00 am from a dwelling; disobeying a warning notice is a criminal offence – see above 4.12.1.

4.17 CLOSURE OF LICENSED PREMISES (E & W)

Part 8 of *the Licensing Act 2003* enables the police to seek a court order to close for a period not exceeding 24 hours all licensed premises in a geographical area that is experiencing or likely to experience disorder – these powers are also available in respect of premises licensed for the provision of regulated entertainment or for the provision of late night refreshment, also for premises covered by a Temporary Events Notice.

A senior police officer may also order the immediate closure for up to 24 hours a licensed premises – including premises for which a Temporary Event Notice is held – which is disorderly, likely to become disorderly or causing a public nuisance owing to the noise emanating from the premises. A court order should be applied for as soon as possible; in certain circumstances a closure notice may be extended for a further 24 hours. Guidance issued under s.182 of *the Licensing Act* explains the powers of the police in more detail.

The Anti-Social Behaviour Act 2003, which received Royal Assent on 20 November 2003, enables local authorities in England and Wales to serve a notice on a licensed premises deemed to be causing a public nuisance through noise, requiring its immediate closure for 24 hours. Failure to comply may result in a maximum fine of £20,000 or a prison sentence of three months or both. This applies whether a premises licence is held or a temporary event notice – this provision came into force on 31 March 2004.

The Clean Neighbourhoods and Environment Act 2005, s.84 and Sch.1 extends the powers of local authorities to deal with noise at night under ss.8-9 of *the Noise Act 1996* to licensed premises and those with temporary event notices – see above 4.12.

4.18 ILLEGAL RAVES

The *Criminal Justice and Public Order Act 1994*, as amended by s.58 of *the Anti-Social Behaviour Act 2003*, enables the police to require people whom it is reasonably thought are preparing for, or waiting to attend an illegal rave to leave the land, whether or not they are trespassing, and ultimately to confiscate sound equipment. It applies to gatherings of 20 or more people, not necessarily in the open air, at which amplified music is played which is "wholly or predominantly characterised by ... repetitive beats".

4.19 FIREWORKS

The Fireworks Act 2003, which received Royal Assent on 18 September 2003, applies in England, Wales and Scotland and aims to reduce the nuisance, noise and injuries caused by fireworks and their anti-social use. The Act enables the Secretary of State for Trade and Industry to make Regulations regarding the supply of fireworks as well as their use, or misuse.

Directive 2007/23/EC, adopted on 23 May 2007, on *the placing on the market of pyrotechnic articles*, aims to improve the safety of fireworks, restricting their use and possession depending on their strength:

- *Category 1* – very low hazard and negligible noise level for use in confined areas and those for use inside domestic buildings
- *Category 2* – low hazard and low noise level for outdoor use in confined areas
- *Category 3* – medium hazard for outdoor use in large open area and whose noise level is not harmful to human health
- *Category 4* – high hazard for professional use only and noise is not harmful to human health.

The Directive, which must be implemented by 4 January 2010, also requires that fireworks should be able to be disposed of with minimum environmental impact.

The fireworks industry has agreed a voluntary code of practice which, among other things, limits the sale of fireworks to the general public to the three weeks before 5 November and for a few days after that date, and for a similar period over new year.

4.19.1 Fireworks Regulations

The Fireworks Regulations 2004 (SI 2004/1836) apply in England, Scotland and Wales, replacing SI 2003/3085; most of the Regulations came into force on 7 August 2004, with the remainder in force from 1 January 2005, with further amendment Regulations (SI 2004/3262) also taking effect on 1 January 2005.

Under the Regulations it is an offence for anyone under the age of 18 to possess fireworks (except items such as sparklers or party poppers) in a public place; the possession of category 4 fireworks (i.e. professional display fireworks) except by "fireworks professionals" is also prohibited. Fireworks professionals include professional organisers or operators of firework displays, and manufacturers and suppliers or their employees. The Regulations also

- ban the use of fireworks except for various official displays between 11 pm and 7 am, with the following exemptions: the first day of the Chinese New Year until 1.00 am the following day; 5 November 11.00 pm ending at midnight; 11.00pm on the day of Diwali until 1.00 am the following day; and 31 December 11.00 pm until 1.00 am on 1 January (this reg. does not apply in Scotland – see below*);
- prohibit the supply to the public of category 3 fireworks that exceed 120 dB;
- require persons intending to supply fireworks to the public all year round, or the premises from which the fireworks are to be supplied, to be licensed by their local authority – the annual fee for a licence is £500; unlicensed traders may only supply fireworks on the first day of the Chinese New Year and Diwali and the three days preceding them; between 15 October and 10 November; and between 26-31 December;
- require suppliers of any adult fireworks, including sparklers, to the public, whether by internet, mail-order, retail or wholesale, to display a notice informing consumers of the law regarding under-age sales and possession; suppliers are required to keep records for three years of both their supplier and to whom the fireworks were supplied of any with a total explosive content exceeding 50 kg;

- require fireworks importers to provide Customs & Excise with details as to the destination of their products.

**The Fireworks (Scotland) Regulations 2004* (SSI 2004/393), which came into force on 7 October 2004, extend the use of fireworks on 5 November to midnight, and to 0100 hours the following day on Hogmanay, Chinese New Year and Diwali.

INDUSTRIAL & CONSTRUCTION SITE NOISE

In 2004/05 there were 7,522 complaints about noise from industrial premises in England and Wales and 11,714 complaints about noise from construction and demolition sites (CIEH annual survey into local authority noise enforcement action). In Scotland local authorities received 433 complaints about noise and vibration from industrial premises in 2004/05 (REHIS Environmental Health in Scotland 2004/05). In Northern Ireland there were 223 complaints about noise from industrial premises (including agriculture) in 2006/07 and 426 complaints of noise from construction sites (Noise Complaints Statistics for NI, 2006/07, DOENI).

The use of the planning system in controlling noise has already been summarised (see this chapter 4.8).

The Noise Emission in the Environment by Equipment for Use Outdoors Regulations 2001 require various types of construction site equipment, including concrete breakers and excavators, welding and power generators, to meet specified noise levels – see 4.26 below.

4.20 REGULATORY CONTROLS

4.20.1 Control of Pollution Act 1974

Sections 60-61 cover noise from construction sites and enable local authorities to impose requirements, including noise limits on the way in which work is carried out. British Standard 5228, *Noise Control on Construction and Open Sites*, provides guidance to enable compliance with s.60 (see 4.10.1 above).

Under ss.63-67 of COPA a local authority may by order designate all or part of its area a noise abatement zone; premises classified under the order – usually industrial premises – may not exceed their registered noise level (details of which are kept in a public register) – see 4.10.3 above. It is a defence to prove that the best practicable means have been used to counteract the effects of noise caused in the course of trade or business from premises in a NAZ.

In Northern Ireland, similar controls are contained in *the Pollution Control and Local Government (NI) Order 1978* (Art.40).

4.20.2 Environmental Protection Act 1990

Statutory Nuisance

Action against noise (including vibration) from industrial or trade premises may be taken using Part III of the EPA. It is a defence to prove that the best practicable means were used to prevent or counteract the effects of the nuisance. See chapter 1, 1.29.1.

Waste Management Licensing

The EU's Waste Framework Directive (see chapter 5, 5.6) requires waste disposal processes to be regulated to ensure that they do not cause noise nuisance; relevant conditions are included in waste management licences. This, however, does not preclude action under statutory nuisance legislation.

Internal Guidance for the Regulation of Noise at Waste Management Facilities (Environment Agency, version 3.0, July 2002) includes guidance on appropriate noise limits in relation to the location of the facility. Conditions will be included in waste management licences ((E & W, permits) to ensure the adverse noise impacts of waste management activities – including noise from vehicle movements, equipment on site etc – are minimised. Where the site is close to dwellings and private gardens, it is suggested that noise from sites (except landfills) should be kept below 10 dB above background levels and that operators should draw up noise action plans where noise levels exceed 5 dB above background; if levels exceed 10 dB above background, enforcement action will be considered. See also chapter 5, 5.18.11.

4.20.3 Integrated Pollution Prevention & Control

The *Pollution Prevention and Control (Scotland) Regulations 2000* (SSI 2000/323), which came into force on 28 September 2000 and in Northern Ireland (SR 2003/46) of 31 March 2003, implement a 1996 EU Directive which includes noise and vibration among the pollutants which operators of processes covered by the Directive are required to control – see chapter 2, 2.1 and 2.5. In England and Wales *the Environmental Permitting (E & W) Regulations 2007* (SI 2007/3538) now implement the IPPC Directive, replacing 2000 PPC Regulations (see chapter 2, 2.4).

The Directive defines pollution from noise and vibration as that "which may be harmful to human health or the quality of the environment, result in damage to material property or impair or interfere with amenities and other legitimate uses of the environment". The Best Available Techniques must be used to control noise sources.

The Environment Agency, together with SEPA and the EHS (now NIEA) has published technical guidance on assessing and controlling noise from A1 IPPC installations – H3 Horizontal Noise Guidance. Part I outlines the main considerations relating to the permitting and regulation of noise and vibration and is aimed at regulators. Part II describes the principles of noise measurement, prediction, control and management and will help to determine BAT for the installation. The Guidance, which is applicable throughout the UK, adopts the following broad approach:

- where appropriate planning conditions covering noise from the installation exist, are enforceable and represent BAT, these should be mirrored in the IPPC permit;

- where there are no existing planning conditions covering noise, or they are no longer appropriate, then conditions should be developed to ensure compliance with IPPC legislation.

In applying for a permit, the operator will need to provide information on all sources of noise and vibration, assessing their impact on both the environment and local population. The operator will need to demonstrate that BAT is being used to control noise and vibration. The guidance suggests that noise from the installation should not exceed 50 dB during the day and 45 dB during the night. It should also be noted that non-PPC regulated activities, such as barking dogs and burglar alarms at an installation regulated for PPC, will still be covered by statutory nuisance legislation.

The Guidance notes the expertise of local authorities in noise matters and suggests that while the Agency may deviate from local authority advice, it will need to provide a robust justification for doing so. The Agency is consulting on the drafting of separate guidance on mitigating noise at intensive livestock installations. This will outline sources of noise and ways of controlling it, and the expected standards of control. In applying for a permit, farmers would need to provide a noise management plan, which if not complied with may lead to enforcement action.

The Health Protection Agency has produced *IPPC: A practical guide for health authorities* (version 1.1, February 2002) to help them in their role as statutory consultees for IPPC; this includes a chapter on noise and lists the points health authorities should consider in commenting on IPPC applications, such as

- whether the application has identified all the likely noise sources from on-site activities;
- whether it identifies all noise sensitive premises likely to be affected by noise;
- whether it details the methodology used to model the predicted noise levels;
- whether it details all the noise control measures that will be in place;
- if maximum noise levels are recommended, are these acceptable.

The Guidance can be viewed on www.doh.gov.uk/-pdfs/ippchag.pdf. (The HPA has also produced *IPPC: A guide for primary care trusts and local health boards – responding to IPPC applications* (September 2004).

Guidance for local authorities in England and Wales on noise and vibration issues relevant to A2 (LA-IPPC) installations is included in Defra's *General Guidance Manual on Policy and Procedures for A2 an B Installations* (January 2008).

Defra has also produced guidance (February 2005) on the control of noise and odour for operators of commercial kitchens (e.g. restaurants, takeaways, pubs); while not statutory it provides guidance on best techniques for minimising nuisance odours and noise from kitchen exhaust systems.

See also earlier this chapter, 4.8.1, on planning guidance available of relevance in minimising noise problems.

TRANSPORT NOISE

The need to address increasing levels of transport noise has gained importance with the publication of various reports linking ill health to ongoing persistent exposure to such noise.

Research[1] carried out for Transport and the Environment (T&E) and published in 2008, found that almost half of Europeans are regularly exposed to potentially unhealthy transport noise. It links noise from road and rail to 50,000 fatal heart attacks and 200,000 cases of cardiovascular disease in the EU every year. The study found that around 210 million EU citizens are regularly exposed to 55 decibels or more of road noise and 35 million to the same level of rail noise – WHO's threshold for "serious annoyance" and negative health effects from environmental noise is 55 decibels. The HYENA[2] report, also published in 2008, found that exposure to noise from night flights and daily road traffic increases the risk of hypertension and heart attacks.

Meanwhile, a report[3] commissioned by the Greater London Authority, and also published in 2008 links persistent high levels of noise from road traffic to significant health impacts, finding that up to 108 heart attacks and nearly 500 cases of heart disease in the Greater London area could be caused by prolonged exposure to high levels of road traffic noise.

For the year 2003/04, local authorities in England and Wales reported 1,058 complaints relating to traffic noise, 3,949 relating to aircraft noise and 684 relating to railway noise (CIEH annual survey into local authority noise enforcement action). In Northern Ireland, road traffic noise accounted for 48% of all transport noise complaints in 2006/07 (Noise Complaint Statistics for NI, 2006/07, DOENI).

1. *Traffic noise reduction in Europe: health effects social costs and technical and policy options to reduce road and rail traffic noise.*
2. *Hypertension and Exposure to Noise Near Airports*; www.hyena.eu.com
3. *The Effect of Noise on Physical Health Risk in London*, Berry Environmental Ltd

4.21 ROAD VEHICLES

4.21.1 Introduction

Road traffic noise has two main components: the mechanical noise associated with the engine and transmission system of the vehicle, and the "rolling noise" of the vehicle – this latter is due to factors such as the frictional contact between the tyres and the road and the aerodynamic noise caused by the passage of the vehicle through the air. The noise levels caused by traffic on a road are dependent on the number of vehicles using the road, their speed and the proportion of heavy vehicles. Heavy vehicles are of course much noisier than cars and hence have a greater influence on overall traffic noise levels.

Other factors affecting traffic noise levels are road surface and the immediate topography of the street. Rough road surfaces cause more noise than smooth ones; asphalt surfaces, and in particular porous asphalt, tend to generate less noise than other conventional surfaces; nearby high walls or buildings also reflect noise causing an increase in the overall level. Noise from smoothly flowing traffic is less than noise from an interrupted flow because changes in engine speed and the use of low gears, necessary at traffic lights, junctions and hills, produce more noise than a steady cruising speed.

4.21.2 Cars, Light Goods Vehicles

Maximum noise limits from the engines and exhausts of motor vehicles when new and in use are set out in regulations made under the *Road Traffic Act 1972; the Road Vehicles (Construction & Use) Regulations 1986* (SI 1078) (and amendments), and the *Motor Vehicle (Type Approval) (Great Britain) Regulations. Type Approval Regulations* set requirements to meet noise limits in EU Directives. The latest Directive, 92/97/EEC, adopted in 1992 has reduced the limits set in a 1984 Directive (84/424/EEC) by between 2-5 dB(A) – see Table 4.1 below; it was implemented by SI 1996/2329. The new limits applied to new model vehicles, cars, dual purpose vehicles and light goods vehicles with petrol engines from 1 October 1995 and to all new vehicles from 1 October 1996. Subject to the agreement of the European Commission, Member States were permitted to offer tax incentives until October 1996 for all vehicles meeting the new limits before the deadline. Directive 92/97/EEC has been implemented in the UK through amendments to the type approval and construction and use regulations which came into force on 1 October 1996 (see also chapter 3, 3.27.2). These regulations also implement Noise Directive 96/20/EC which makes certain changes to the method of noise level testing set out in the 1992 Directive.

The Commission is expected to make proposals for further reductions in noise emissions in the near future.

Table 4.1: Maximum Noise Limits for Road Vehicles

Vehicle	*dB(A)*
Goods vehicles over 3.5 tonnes	
- over 150 kW engine power	80
- 75-150 kW engine power	78
- below 75 kW engine power	77
Goods vehicles below 3.5 tonnes	
- goods vehicles 2-3.5 tonnes	77
- goods vehicles below 2 tonnes	76
Buses	
- with over 9 seats, over 3.5 tonnes GVW and over 150 kW engine power	80
- with over 9 seats, over 3.5 tonnes GVW but below 150 kW engine power	78
Cars	74

The Road Vehicles (Construction and Use) Regulations 1986 also make it an offence to use a motor vehicle in such a way as to cause excessive noise; the Regulations do not cover off-road uses, such as motorcycle scrambling. Maximum permitted levels for vehicles in use on the road are generally 3 dB(A) higher than the construction noise limits specified in Regulations, because of differences in measurement. Higher limits apply to older vehicles. Similar Regulations apply in Northern Ireland.

In addition the Regulations specify that motor horns may not be sounded in a restricted road between 11.30 pm and 7.00 am and not at all when a vehicle is stationary unless there is danger to another moving vehicle. Private vehicles must not be fitted with a gong, bell, siren or two-tone horn. For vehicles first used on or after 1 August 1973, no other multi-tone horns are permitted.

Alarms fitted to vehicles to prevent theft of the vehicle or its contents must have a five minute cut-out device; if the alarm continues to sound then the police may prosecute the vehicle's owner (reg.37(7) & (8)).

4.21.3 Motorcycles

The first Directive to limit motorcycle noise was adopted in 1978 (78/1015/EEC) and set optional limits of between 86 dB(A) for motorcycles over 500 cc and 78 dB(A) for those less than 80 cc; these limits were reduced in two stages by Directive 87/56/EEC to 80 dB(A) for motorcycles over 175 cc, 77 dB(A) for those between 80-175 cc and 75 dB(A) for those under 80 cc. The 1978 Directive and its amendment have been repealed following implementation of Directive 97/24 which was adopted in mid-1997; Directive 2006/72/EC amends the requirements on marking of original silencers. The 1997 Directive, which also covers 2 and 3 wheel mopeds and tricycles, had to be complied with by 18 December 1998 and the limits observed by 17 June 1999 – see Table 4.2.

Table 4.2: Motorcycle Noise Limits

Vehicle	*dB(A)*
Motorcycles	
- < 80 cc	75
- 80 - 175 cc	77
- > 175 cc	80
Tricycles	80
Two Wheel Mopeds	
- up to 25 km/h	66
- > 25 km/h	71
Three Wheel Mopeds	76

A 1989 Directive (89/235 EEC) laid down standards for replacement exhausts and silencers and prohibited the marketing and use of those that did not meet the required standards from October 1990. Implementation was optional. The Commission is however expected to bring forward proposals for a Directive which would make implementation obligatory and require all replacement exhausts to meet EU standards.

In the UK, the construction and use requirements of the Directives are implemented through the *Road Vehicles (Construction and Use) Regulations 1986* and subsequent amendments (see also chapter 3, 3.27.2). So far as noise from motorcycles is concerned, the Regulations require silencers to be maintained in good and efficient working order; the silencer should not be altered so as to increase the noise made by the escape of exhaust gases. A silencer which forms part of the exhaust system of a motorcycle first used on or after 1 January 1985 shall be either the original one, or one that meets the relevant British Standard (and is so marked), or is of a make and type specified by the manufacturer. Older motorcycles (used after 1 April 1970) must be able to meet noise standards specified in the Regulations.

Regulations made under *the Motor Cycle Noise Act 1987*, the *Motor Cycle Silencer and Exhaust Systems Regulations 1995* (SI 1995/2370), effective 1 August 1996, make it an offence to sell, supply, offer to supply, or have available for supply, exhaust systems and silencers that are not marked BS AU 193/T2, BS AU 193a: 1990/T2 or BS AU 193a: 1990/T3. The Regulations (which also apply to mopeds and scooters) do not apply to silencers and exhausts to be used by off-road motorcycles but must be clearly marked "Not for Road Use"; nor do they apply to motorcycles first used before 1 January 1985 and the silencer is marked "Pre 1985 MC only".

The Motorcycles (Sound Level Measurement Certificates) (Amendment) Regulations 1989 ensure that a certificate may only be issued when the relevant standards are met.

4.21.4 Tyre Noise

Directive 92/23/EC covers the approval of new tyres as components, and includes requirements ensuring new vehicles are equipped with suitable tyres. The 1992 Directive was amended by 2001/43/EC to insert additional requirements to limit tyre rolling noise emissions; it set noise emission limit values between 72-76 dB(A), depending on type of vehicle and tyre width, to apply until 30 June 2007, with a further 1 dB(A) reduction applying from 1 July 2007.

As part of its *Car Safety Package*, the Commission has issued a proposal for a *Regulation concerning type-approval requirements for the general safety of motor vehicles* (COM(2008) 316). As well as requiring the mandatory fitting of various safety features, it also sets more stringent noise rolling noise limit values of between 70-75 dB(A), depending on type of vehicle and tyre width. It also introduces requirements on tyres regarding rolling resistance and foresees the mandatory fitting of tyre pressure monitoring systems to passenger cars. The 1992 Directive will be repealed.

4.21.5 Charging for Noise

The European Commission's *Greening Transport Package* adopted in July 2008, includes a proposal for a Directive *on the charging of heavy goods vehicles for the use of certain infrastructures* (COM(2008)436. This would enable Member States to charge operators of heavy goods vehicles for noise, congestion and air pollution caused by their lorries. Member States would not have to impose the tolls but could choose to do so for vehicles weighing more than 3.5 tonnes; charges would be capped, with lower tolls for less polluting lorries; there would also be an incentive to use motorways and avoid suburban routes. Any revenues raised would have to be used to finance efforts to reduce the environmental impact from transport, for example developing more energy-efficient vehicles.

4.21.6 Other Controls – Traffic Noise

A number of other legislative measures are of relevance for controlling and abating noise pollution from road traffic.

The Road Traffic Regulation Act 1984 gives local authorities powers to make a traffic regulation order for preserving or improving the amenities of an area through which a road runs.

The Police Reform Act 2002 gives the police powers to seize a motor vehicle if they have reasonable grounds to believe it is being used "in a manner likely to cause harassment, alarm or distress" to members of the public; this could include noise nuisance.

Part I of the *Land Compensation Act 1973* provides for compensation in cases where property values can be shown to have fallen as a direct result of increased noise levels (and other specific factors). Part II, enables householders to claim a grant from their local authority to provide noise insulation if:

a) their property is within 300 metres of a new road or substantially improved roads and they experience increased noise levels as a result; and

b) the noise level reaches or exceeds 68 dB(A) $L_{10(18\ hour)}$ with at least 1 dB(A) resulting from the increase in traffic.

The Noise Insulation Regulations 1975, SI 1975/1763, (and amendments), made under Part II of the *Land Compensation Act 1973*, apply in England and Wales, and provide for grants to cover the costs of insulation. Grants are available for secondary glazing of eligible rooms of dwellings where it is estimated that maximum noise levels during the 15 years after the road opening will be at least 68 dB(a) $L_{10(18\ hour)}$, with a contribution of at least 1 dB(A) from traffic using the new road. However grants are not available to householders affected by increased traffic along existing roads resulting from re-routing or traffic management schemes, or from a general increase in traffic flow.

Similar legislation – *The Land Compensation (Scotland) Act 1973* and Regulations made under it, the *Noise Insulation (Scotland) Regulations 1975*, SI 1095/460 – applies in Scotland.

4.22 AIRCRAFT NOISE

4.22.1 Introduction

In common with other forms of transport, air travel has an environmental impact, and is increasing rapidly. In the UK we have two of the world's busiest international airports – Heathrow and Gatwick. Heathrow is currently home to around 90 airlines, flying over 67 million passengers to more than 180 destinations; Gatwick is also used by about 90 airlines flying 34 million passengers to over 200 destinations – Gatwick's development strategy is based on expansion to handle 40 million passengers by 2010/11 (source: www.BAA.com, 29.10.07).

Following seven regional consultation documents setting out proposals for the future development of air transport across the UK over the next 30 years, the Department for Transport published a White Paper in 2003 – *The Future of Air Transport*. This sets out a strategic framework for the development of airport capacity in the UK to meet a projected near tripling in passenger numbers to 2030. So far as noise is concerned, the White Paper proposed a number of measures which have been included in *the Civil Aviation Act 2006* – see below.

The most obvious environmental threat from aircraft is noise – ranging from the regular rumble of large jets to the buzz of microlights and other light aircraft. In 2003/04 local authorities in England and Wales received 3,949 complaints about aircraft noise (CIEH annual survey of local authority noise enforcement action). The impact of emissions to atmosphere from aircraft is briefly summarised in chapter 3.28.

The noise generated by an aircraft is related to air velocity. Thus any fast moving components – propellers, compressor blades – as well as jet exhaust gases, will be efficient generators of noise.

Aircraft noise comprises a broadband background noise with the periodic components of rotating machinery noise superimposed on it. In newer aircraft with high by-pass ratio turbo-fan engines, these components operate at much lower speeds and are significantly quieter as a result.

Different aircraft produce different noise problems; for instance, jets produce more periodic component noise on landing than on take-off, when the broadband exhaust noise predominates; noise from the hull on landing tends to predominate in quiet-engined aircraft.

Jets currently in service are Chapter 3 aircraft – the term comes from annex 16 of the Chicago Convention of 1944 which sets standards. (Chapter 1 aircraft included the DC8 and VC10; Chapter 2 aircraft, which included the BAe 1-11, DC9 and Boeing 737-200, have been banned in Europe since April 2002.) Chapter 3 aircraft include the Boeing 747, 757 and 767, DC10 and airbus. In June 2001, ICAO adopted a new standard – Chapter 4 – to be met by all new aircraft submitted for certification from 1 January 2006. This increases the stringency of the current Chapter 3 cumulative levels by 10 EPNd, with the sum of the certification level margins at any two of the three measurement points (take-off, approach and a sideline point) at least 2 dB; for helicopters the new limits will require them to be 3 dB quieter at take-off, 4 dB quieter at flyover and 1 dB quieter at approach, and apply to most new helicopters submitted for certification after January 2002. The ICAO Assembly has endorsed the concept of a "balanced approach" to noise control at airports (resolution 14/1, 33rd Assembly, 2001); this enables airports to address noise problems in the most cost-effective manner, e.g. through noise reduction at source (quieter aircraft), land-use planning and management, and noise abatement operational procedures and operating restrictions.

Noise standards are drawn up by the International Civil Aviation Organisation (ICAO) – a UN body – and by the European Civil Aviation Conference, whose members are drawn from throughout Europe, including the European Community. Their standards are then incorporated into European and national legislation. ICAO is also responsible for developing operating procedures, for example to minimise take-off and landing noise. It also carries out research and is currently looking at air quality around airports and emissions at high altitude.

4.22.2 The EU & Aircraft Noise

The European Union's first Directive on aeroplane noise (80/51/EEC) prevented the addition of any further non-noise certificated aeroplanes (i.e. Chapter 1 aircraft) to the civil registers of Member States, and required the removal of any such aeroplanes by 31 December 1986. A small number of exemptions enabled some of these aeroplanes to continue flying until 31 December 1988. An amendment to this Directive (83/206/EEC) prevented non-noise certificated aircraft registered outside the EU from landing within it from 1 January 1988 (with some exemptions until 31 December 1989). A 1989 Directive (89/629/EEC) limited noise from subsonic aircraft and permitted only Chapter 3 aircraft able to meet the tighter limits to be added to Registers.

A further Directive (92/14/EEC) limiting aircraft noise was agreed in March 1992, and amendments, 98/20/EC and 1999/28/EC, adopted in March 1998 and April 1999 respectively. These Directives have all been repealed and their provisions codified into Directive 2006/93/EC on *the regulation of the operation of aeroplanes covered by Part II, Chapter 3, Volume 1 of Annex 16* of the Chicago Convention. All aircraft unable to meet Chapter 3 standards or Chapter 2 standards and were less than 25 years old, have been banned from operating in the EU after 1 April 1995; all Chapter 2 aircraft have been banned since 1 April 2002. Various exemptions to the Directive cover such matters as extensions to an aircraft's life and aircraft from developing countries.

The above Directives, as codified, have been implemented through *The Aeroplane Noise Regulations 1999* (SI 1452) which came into force on 27 May 1999 and 1999 Amendment Regulations (SI 2253) which came into force on 16 August 1999.

Directive 2002/30/EC *on the establishment of rules and procedures with regard to the introduction of noise-related operating restrictions at Community airports* was adopted on 26 March 2002; it entered into force on 28 March 2002, when Regulation 925/99 banning the use of hushkits for meeting Chapter 3 standards was repealed. Member States had to implement the new Directive by 28 September 2003. The Directive implements the ICAO Chapter 4 standards and applies to civil airports with more than 50,000 movements per year of civil subsonic jet aeroplanes and to certain city airports, listed in an annex to the Directive, which may apply more stringent measures if these are identified as having a particular noise problem. The Directive requires that a range of options be assessed in order to find the most appropriate measures for managing noise at an individual airport, and that certain information (such as currently in place noise mitigation measures) be taken into account where operating restrictions appear necessary. The Directive sets out a timescale and conditions for enabling the withdrawal of "marginally compliant" (Chapter 3) aircraft and for exemptions for aircraft from developing countries. This Directive has been implemented through *The Aerodromes (Noise Restrictions) (Rules and Procedures) Regulations 2003* (see below 4.22.4).

EU Regulation 1702/2003, as amended by 335/2007, concerns the environmental certification of aircraft and related products, as well as implementing the standards and guidelines for the administration of noise certification documents in accordance with the Chicago Convention.

4.22.3 Civil Aviation Acts

The principal legislation covering aircraft operations in the United Kingdom is the *Civil Aviation Act 1982*, as amended by *the Civil Aviation Act 2006* which received Royal Assent on 8 November 2006.

The regulation of aviation, including policy on the control of aircraft noise, and of air transport is the responsibility of the UK Government; in Scotland, the Scottish Government has various powers relating to aerodromes including environmental and planning issues; the National Assembly for Wales also has planning powers, including those that affect aerodromes; in Northern Ireland, the NI Assembly and the Executive have powers relating to aerodromes and can legislate on civil aviation matters, with the consent of the Secretary of State.

The 1982 Act contains three main powers aimed at aircraft noise control at particular airports. Section 76 does, however, exclude the properly controlled flight of aircraft from action in respect of trespass or nuisance provided that rules and regulations such as the *Rules of the Air and Air Traffic Control Regulations* have been observed. And while powers to control environmental noise are contained in Part III of the *Environmental Protection Act 1990*, s.79(6) specifically excludes noise from aircraft (other than model aircraft) from these controls which are designed to deal with noise from fixed rather than moving sources.

Section 78 of the *Civil Aviation Act 1982* gives the Secretary of State wide powers to enforce noise standards on aircraft, apply operational controls and restrictions and give directions to owners of designated airports in relation to operational requirements and noise insulation grant schemes. The policy is concerned with containing and effectively reducing aircraft noise by restrictions on aircraft movements, control of flight paths and use of certain operational procedures. Section 2 of the 2006 Act (brought into force on 1 March 2007) amends s.78 to enable the Secretary of State to give directions for the purpose of avoiding, limiting or mitigating the effect of aircraft noise and vibration either generally or in a particular area or areas.

(a) Environmental Considerations

In licensing or relicensing an aerodrome, the Civil Aviation Authority is primarily concerned with safety issues, although under s.5 of the 1982 Act the Secretary of State can require it to take account of environmental issues, such as noise and atmospheric pollution.

In March 2005, the Secretary of State for Transport re-issued Guidance to the Civil Aviation Authority (under s.70(2) of *the Transport Act 2000*) on the environmental objectives to which the CAA should have regard in exercising its air navigation functions (this was first issued by the then DTLR in January 2002). Advice on best practice and noise mitigation procedures includes a requirement that the CAA should consult relevant local authorities on any proposals to change air navigation procedures which might have a significant effect on the environment in their area. The guidance also reiterates Directions given to the CAA under s.66(1) of *the Transport Act* on the "need to reduce, control and mitigate as far as possible the environmental impacts of civil aviation operations, and in particular the annoyance and disturbance caused to the general public arising from aircraft noise and vibration, and emissions from aircraft engines". Guidance is also given on the Government's sustainable development strategy and how this relates to the CAA's functions.

(b) Consultation Facilities

Where the Secretary of State designates an airport under s.35 of the 1982 Act, the airport management is required to establish suitable consultation facilities with users, nearby local authorities, and local representative organisations. Forty-seven airports have been designated under s.35, covering all the national and regional airports and some of the general aviation airfields.

(c) Aerodrome Charges

Section 1 of the 2006 Act, brought into force on 1 March 2007, substitutes s.38 of the 1982 Act to enable airport operators to set charges that reflect local noise and atmospheric emissions from aircraft. The Secretary of State is given powers to direct airport operators to levy such charges.

(d) Noise Control Schemes

Section 4 of the 2006 Act (brought into force on 1 March 2007) inserts ss.38A-38C into the 1982 Act; they apply in England, Scotland and Wales and enable airport operators at non-designated airports to establish and enforce measures to avoid, limit or mitigate the effects of noise (within and outside their boundaries) from aircraft taking off or landing. Such schemes may

- require aircraft operators to comply with specific requirements during take off or landing;
- prohibit certain aircraft from take off or landing during specified periods;
- specify the maximum number of occasions aircraft of a particular description may take off or land during a specified period;
- impose other restrictions to limit the cumulative amount of noise caused by specified aircraft during specified periods.

The area to be covered in any scheme by requirements relating to take off or landing will either be specified by Order by the Secretary of State; if no area is specified the requirements imposed by the scheme will cover an area represented by a circle with a radius of 40 km from the centre of the longest runway at the aerodrome. Before making an Order, the Secretary of State should consult the CAA, aerodome operator, organisations representing aircraft operators at the aerodrome, relevant local authorities and any other body representing people in the locality.

The aerodrome operator should specify in the Noise Control Schemes the amount payable by the aircraft operator for any breach of the noise scheme. The aerodrome operator should make payments equal to the amount of penalties "for purposes ... likely to be of benefit to persons who live in the area..."

The CAA, together with the Department for Transport, BAA and a number of airlines have agreed a voluntary code of practice for reducing noise from arriving aircraft (2nd ed, November 2006). This provides technical advice, primarily for pilots and air traffic controllers, but also includes advice of relevance to airports and the regulators.

(e) Designated Airports

Section 80 of the 1982 Act enables the Secretary of State to designate an aerodrome for the purposes of ss.78 and 79. Heathrow, Gatwick and Stansted are designated for both sections.

The Secretary of State is directly responsible for noise abatement measures for aircraft landing and taking off (including during the night) at airports designated for the purposes of s.78; he can by notice require aircraft operators to adopt procedures limiting noise and vibration. In so doing, restrictions can be placed on the numbers or types of particular aircraft using the airport at specified times (see below for current restrictions on day and night time use).

Section 79 enables the Secretary of State to make noise insulation grant schemes for airports designated for the purposes of s.79.

Daytime Noise Limits

Daytime noise limits were first introduced at Heathrow in 1959, at Gatwick in 1968 and at Stansted in 1993, and cover the period 0700-2300. The current limit is 94 dB(A). The airport companies are responsible for monitoring compliance with noise limits with breaches of the limit subject to financial penalty – currently £500 if the limit is exceeded by 3dB(A) or less, £1,000 if it is exceeded by more than 3dB(A).

Certain aircraft were exempt from the limit under the provisions of the EU Directive 92/14 until 31 March 2002 (e.g. on grounds of economic hardship or those registered in developing countries).

Night-time Noise Limits & Flight Restrictions

Night flights have been restricted at Heathrow since 1962, at Gatwick since 1971 and at Stansted since 1978, and cover the period 2300-0700. Revised restrictions covering the period until the end of October 2012 came into effect on 29 October 2006; they are set out in the document *Night Restrictions at Heathrow, Gatwick and Stansted* (DfT, June 2006).

All three airports are allocated both a movements limit (take-offs and landings) and noise quota for night flights covering both the summer and winter seasons (demarcated by when the clocks change). The current night period runs from 2300-0700 and the night quota period from 2330-0600. During the night period the noisiest aircraft (i.e. those with a QC of 8 or 16 – see below) are banned from both landing and take off and may only do so in exceptional circumstances during the night quota period; aircraft with a QC rating of 4 may also not be scheduled to operate during the night quota period. The movements limit applies between 2330-0600. In addition, during the night quota period movements by most types of aircraft (including those with a QC of 0.25) will be restricted by the movements limit and noise quota.

Quota Counts range from 0.25 for aircraft with a certificated noise performance of 84-86 EPNdB and 0.5 (less than 90 EPNdB (79 dB(A)) to QC/4 (EPNdB of 96-98.9), QC/8 for those between 99-101 EPNdB (88 dB(A) or more) and QC/16 for those above 101.9 EPNdB. The noisier the aircraft the more points will be deducted from the airport's allocation. Only aircraft with a certificated noise performance below 84 EPNdB (QC/0) do not count against either the movements limit or noise quota.

The quota are divided between summer and winter seasons, with airports allowed to carry forward up to 10% of the quota to the next season or to use 10% of the next season's quota to deal with any overrun. Any exceedence of the quota over 10% results in a loss of quota double the exceedence. The airport companies are responsible for monitoring compliance, with the same financial penalties applying as for daytime exceedences of the limit.

Specific noise abatement objectives for all three designated airports require them to:

- Minimise sleep disturbance resulting from the overflight of the noisiest aircraft;
- Mitigate the effects of noise (in particular sleep disturbance effects) by encouraging the adoption of appropriate night-noise-related criteria for domestic and other noise-sensitive premises, to determine which residents should be offered sound insulation paid for or contributed to by the airport.

4.22.4 Noise Restrictions' Regulations

The Aerodromes (Noise Restrictions) (Rules and Procedures) Regulations 2003 (SI 2003/1742), which came into force on 6 August 2003, implement EU Directive 2002/30 (see above 4.22.2). They apply to the following UK airports: Birmingham, Gatwick, Heathrow, Luton, Stansted and Manchester; Edinburgh and Glasgow; and also to London City and Belfast City airports. The competent authorities – i.e. the Secretary of State at the designated airports (Heathrow, Gatwick and Stansted), and airport operators at the other airports – are responsible for carrying out the assessments required by the Directive and for any consequent decisions on operating restrictions.

In addressing noise problems at an airport, the competent authority should publish an environmental objective for the airport which "adopts a balanced approach" but may consider economic incentives as a noise management measure. If on the basis of assessment, it is decided that the only way of meeting the environmental objective for the airport is through the imposition of operating restrictions, marginally compliant aircraft should be allowed to operate for 6 months on the same basis as the corresponding period in the previous year (i.e. the same number of services); 12 months from the decision to impose operating restrictions, the competent authority may require the aircraft operator to reduce the number of movements by marginally compliant aircraft by up to 20% of that at the time of introduction of operating restrictions. City airport operators may introduce more stringent measures if they wish provided these do not affect the operation of civil subsonic jets complying with Chapter 4 noise standards. Marginally compliant aircraft from developing countries are exempt from any operating restrictions imposed under the Regulations until 28 March 2012, providing they have been used at the airport between 1 January 1996 and 31 December 2001 and are still on the same register as they were between those dates, and meet Chapter 3 standards.

As required by the Regulations, environmental objectives for Heathrow, Gatwick and Stansted have now been finalised (and published in *Night Flying Restrictions and Heathrow, Gatwick and Stansted*, DfT, 2006), as follows:

- Progressively to encourage the use of quieter aircraft by day and by night (Heathrow & Gatwick), and at Stansted at night while allowing overall growth of the airport as envisaged by the White Paper;
- To avoid allowing the overall noise from aircraft during the night quota period to increase above what was permitted in 2002-03 (Heathrow & Gatwick); and at Stansted to limit the overall noise from aircraft during the night quota period close to existing levels while permitting expansion of the airport's overall traffic in line with White Paper objectives;
- At all three airports, to meet noise abatement objectives as adopted from time to time.

4.22.5 Rules of the Air

These cover general flight rules; usually, aircraft or helicopters may not fly over any congested area of a town or settlement below the height of 1,500 feet above the highest fixed object and within 2,000 feet of it, or at such height as would not permit the aircraft to glide clear of the area in the event of a failure of a power unit. Landing and take-off routes are prescribed by air traffic control procedures in Regulations and usually include a requirement as to the rate of height gain or loss and turning on track with the intention of minimising noise over built up areas. Aircraft are prohibited from flying within 500 feet of persons, vessels, vehicles or structures. The Civil Aviation Authority enforces air Traffic Control Regulations, which include Northern Ireland.

International airports often have formal "minimum noise" routes, which are to be observed as far as possible.

4.22.6 Noise Certification

Relevant standards, drawn up by the International Civil Aviation Organisation, are enacted in the UK, and domestic regulations tightened as the ICAO standards are revised. They are incorporated in *Air Navigation Orders* and prohibit non-certificated aircraft from taking off or landing in the UK, as well as implementing the latest international recommendations in respect of the control of aircraft noise at source.

The Air Navigation (Environmental Standards) Order 2002 (SI 798), which consolidated and revoked earlier Orders, also covers engine emissions (see chapter 3, 3.28); it came into effect on 5 April 2002. Specified supersonic aeroplanes, including microlights and helicopters, are prohibited from taking off or landing at UK airports unless they have a valid noise certificate from the CAA, or if registered outside the UK, from the competent authority for the state concerned. Schedules to the Order specify the standards to be met, test and measurement procedures.

In 1984 the British Microlight Association drew up a voluntary code of practice which gives advice on minimising the noise impact of microlights.

4.22.7 Helicopters

While the provisions of the *Civil Aviation Act 1982* apply to helicopters, their ability to land almost anywhere can be a particular source of noise nuisance in residential areas; currently very little can be done to alleviate the problem, except by use of planning controls (see below, 4.22.9). Under s.76 of *the Civil Aviation Act* the properly controlled flight of aircraft is excluded from action in respect of trespass and nuisance provided that rules and regulations such as the *Rules of the Air* and *Air Traffic Control Regulations* have been observed.

4.22.8 Military Aircraft

In most cases legal liability in respect of nuisance by military aircraft is excluded by statute. Military airfields usually have their own schemes to limit disturbance, including sound insulation and compensation schemes.

4.22.9 Planning Controls

New airport developments are subject to planning permission under the *Town and Country Planning Act 1990*. However, airports which have established use operate without planning controls and increasing use does not require planning permission. Permanent helicopter landing sites require planning permission, although temporary use of land for up to 28 days per year is permitted under the *General Development Order 1988*. In this instance, there is no limit to the number of take-offs and landings in one day.

Planning Policy Guidance Notes (being redrafted and reissued as Planning Policy Statements) – in particular PPG 24, Planning and Noise – provide guidance for planning authorities on the use of planning powers to minimise the noise impact of new development, including transport; PPG 24 specifies noise exposure categories for residential and other various noise sensitive developments.

Private Aerodromes

Where individuals and companies, establish private aerodromes normal planning procedures will apply except on those vested in or controlled by the British Airports Authority. Planning permission may include conditions to secure the abatement of noise and local authority bye-laws may also be used to restrict operations so as to limit or mitigate the effect of noise. The operating authority of the aerodrome may also make such bye-laws.

Compensation is payable in certain circumstances where there is a new runway or if there have been major changes to an existing runway/other features at the airport. Compensation will be paid where the value of an interest in land is depreciated by more than £50 as a result of noise and other physical factors. Compensation will only be payable where the authority responsible for the public works involved can claim statutory immunity from an action for nuisance in respect of the work.

4.22.10 Monitoring Aircraft Noise

Since September 1990 the Department of Transport (now Department for Transport) has used "equivalent continuous sound level" (L_{Aeq}) to measure the extent and intensity of noise disturbance around airports for the daytime period (0700-2300). This can be defined as the sound level of a steady sound having the same energy as a fluctuating sound over a specified measuring period (Batho Report, 1990).

4.23 RAIL NOISE

4.23.1 EU Communication on Rail Noise

The European Commission's *Greening Transport Package* adopted on 8 July 2008, includes a *Communication on Rail Noise* (COM(2008) 432. It says that estimates show that 10% of the EU's population is exposed to high rail noise levels. Noise emissions from new rolling stock are already limited – Commission Decision 2006/66/EC required new and renewed rolling stock, including freight wagons, to be equipped with low-noise brake blocks, reducing the noise emission by 50%. It is proposed to bring forward legislation to encourage the fitting of low noise brakes to existing rail freight trains; the Commission's strategy combines noise emissions ceilings, voluntary commitments and legislation setting financial incentives – existing rules on track access charging would be adapted to enable quieter wagons to be charged less than noisier ones.

4.23.2 Noise Insulation

Under the *Noise Insulation (Railways and Other Guided Transport Systems) Regulations 1996* (SI 428), an authority constructing a new railway (or other guided transport system) or additional track has a duty to provide insulation or a grant towards the cost of insulation for homes affected by the noise. Only residential buildings within 300 metres of the works for the new, altered or extended system are eligible for grant or insulation. The Regulations also include discretionary powers to provide grants for homes affected by altered lines or the noise of construction work. The level at which insulation should be provided or a grant paid towards its cost is 68 dB $L_{Aeq\ 18h}$ during the day (0600-midnight) and 63 dB $L_{Aeq\ 6h}$ at night (midnight-0600). 1998 Amendment Regulations (SI 1701) amend the 1996 Regulations to include procedures for predicting noise from Eurostar trains. Noise levels should be calculated using *Calculation of Railway Noise 1995* (as amended, and available from The Stationery Office), and based on expected normal traffic flows within a 15 year period.

4.23.3 Train Horns

The use of train horns is governed by regulations and instructions set out in the "Master Rule Book" with which all rail staff must comply. Complaints in 2002/4 about the noise generated by horns on new rolling stock resulted in a review by the Rail Safety and Standards Board of both the level of noise and the circumstances when horns must be used. The RSSB has approved a reduction in the minimum volume of horns of up to 8 decibels (which should reduce the noise emitted by between 33-50%). However, it is for train operators to decide whether and how to modify train horn equipment. Train drivers are also no longer required, to sound their horn when entering or passing through a long tunnel, unless there are clear indications of work taking place in the tunnel; they need only sound the horn when approaching the tunnel exit between 0800-1700 if they cannot see that the line immediately ahead is clear of people working. The use of horns during the night-time between 2330-0700 is generally banned except in an emergency. For more information visit www.rssb.co.uk

4.24 RECREATIONAL CRAFT

Directive 2003/44/EC, adopted in June 2003 and which entered into force on 26 August 2003, amends an earlier Directive (94/25/EC) on exhaust and noise emissions from recreational craft. It had to be transposed into national legislation by 30 June 2004, and the measures implemented from 1 January 2005. The Directive covers inboard and outboard propulsion engines on recreational craft (of 2.4-24 metres in length) and personal watercraft under 4 metres in length (e.g. jet skis). It does not apply to experimental craft, those specifically to be crewed or which carry passengers for commercial purposes, submersibles, air cushion vehicles, hydrofoils or original and individual replicas of historical propulsion engines based on a pre-1950 design. Noise emission levels must not exceed a maximum sound pressure level of between 67 and 75dB, depending on rated engine power. The Directive sets out conformity of production, testing and measurement requirements.

The Directive has been implemented through *The Recreational Craft Regulations 2004* (SI 2004/1464) which came into force on 30 June 2004, with most of the provisions taking effect on 1 January 2005. These Regulations largely revoke earlier 1996 Regulations, except for craft on the market prior to 1 January 2005.

See chapter 3, 3.29 for exhaust emission requirements.

EQUIPMENT NOISE

A number of EU Directives establish maximum noise levels for different types of machinery and equipment, including construction machinery and lawn mowers, and have been transposed into UK legislation. The aim is to reduce the noise level of individual machines and thus bring about an overall reduction in neighbourhood noise.

4.25 EU DIRECTIVES

In May 2000 the European Parliament and Council of Ministers formally adopted Directive 2000/14/EC relating to the noise emission in the environment by equipment for use outdoors. As well as introducing consistent noise emission requirements and noise labelling for some equipment, the Directive also aims to reduce overall noise exposure to and nuisance from outdoor equipment. The Directive covers 57 types of equipment, of which 22 are subject to mandatory noise limits and the remainder required to carry a noise label. The equipment covered ranges from concrete breakers, dumpers and excavators, to lawn mowers, hedge trimmers and leaf blowers. Manufacturers must guarantee that the noise level of a product conforms to the specific level for that category of equipment, and for a label confirming this to be affixed to it. For equipment for which a limit has not been set, there is a requirement for it to be labelled with its sound output. The Directive came into force on 3 July 2000, with Regulations implementing it taking effect throughout the UK on 3 July 2001 – see below. On 3 January 2002, the noise limits of the new Directive became mandatory and all the earlier Directives repealed. A second stage of tighter limits took effect on 3 January 2006; however the Commission has agreed (August 2005), through an amending Directive, 2005/88/EC, that eight types of plant may be exempted from the tighter limits "for the time being" as they are unable to meet the standard; these are: steel tracked bulldozers and loaders of specified size; compacting screed paver finishers, vibratory plates and rammers, walk behind vibratory rollers, industrial and rough terrain lift trucks, and hand held motorised concrete breakers and picks.

4.26 OUTDOOR EQUIPMENT

The Noise Emission in the Environment by Equipment for Use Outdoors Regulations 2001 (SI 1701) implement EU Directive 2000/14/EC – see above. The Regulations, which cover equipment ranging from concrete breakers and excavators, and welding and power generators, to lawn mowers, came into force on 3 July 2001 and had to be complied with by 2 January 2002, when earlier Regulations were repealed. Manufacturers of 35 categories of outdoor equipment are required to affix a label to their product which shows its "guaranteed sound power level". In addition manufacturers of a further 22 categories of outdoor equipment are required to ensure (and also guarantee by means of a label fixed to the product) that the equipment meets specified noise levels – a second stage of stricter noise levels came into force on 3 January 2006 although eight types of plant have been exempted from the tighter limits "for the time being" as they are unable to meet the standard (Directive 2005/88/EC and amendment Regulations SI 2005/3325). Equipment put on the market before 2 January 2002 is exempt from the Regulations.

The Department of Trade and Industry has appointed the Vehicle Certification Agency (VCA) to monitor conformity with the requirements of the Directive – i.e. that manufacturers are providing accurate information on sound levels. It has also appointed a number of "Notified Bodies" to monitor conformity with noise levels for the 22 categories of equipment for which levels have been set.

Equipment covered by these Regulations ranges from concrete breakers, dumpers and excavators, to lawn mowers and leaf blowers.

4.27 HOUSEHOLD APPLIANCES

The Household Appliances (Noise Emission) Regulations 1990 (SI 1990/161) establish general principles for information on noise from household appliances, including those for cleaning, preparation and storage of foodstuffs and those for production and distribution of heat and cold and air conditioning; the Regulations detail measurement methods for determination of noise levels and arrangements for monitoring. Manufacturers do not have to supply information on the noise emitted by their appliances, but if they do they must comply with the Regulations. The Regulations implement EU Directive 86/594/EEC on airborne noise emitted by household appliances.

NOISE AT WORK

It is estimated that over 1.1 million people in the UK are at risk from high levels of noise at work (HSE, 2006). People typically at risk are shipbuilders and steel workers using pneumatic hammers; workers engaged in riveting or drop forging; workers in canning and bottling plants; and those who operate wood-working machinery, as well as those working in the music and entertainment industry (including pubs and clubs).

Occupational noise is controlled under statute by the *Health and Safety at Work Act 1974*. Part I provides a comprehensive and integrated system of law to deal with the health, safety and welfare of persons at work and the health and safety of the public as they may be affected by work activities, including protection from occupational noise exposure.

4.28 EU DIRECTIVES

Directive 2003/10/EC on the minimum health and safety requirements regarding exposure of workers to the risks arising from physical agents (noise), was adopted and brought into force in February 2003. It had to be transposed into national legislation by 15 February 2006 when the 1986 *Directive on the protection of workers from the risks related to exposure to noise at work* (86/188/EEC) was repealed. Member States were allowed an additional transitional period of two years before the legislation applies to the music and entertainment industry to allow the code of conduct to be drawn up (see below) and until 15 February 2011 to apply the Directive to the shipping industry. Member States are required to draw up a code of conduct providing practical guidelines to enable employers and workers in the music and entertainment industry to comply with the Directive.

The 2003 Directive requires that the risk of noise be eliminated at source or reduced to a minimum; where risks cannot be prevented, individual hearing protection must be made available. Noise exposure must be assessed at the points where employees are, or are likely to be, exposed to risk; assessments, which are to be carried out within the framework of risk assessment procedures, are to be carried out over 8 hours except where the noise varies from day to day when they may be carried out over a period of a week.

Noise exposure limit values, which must not be exceeded, and the action required by the employer at each level are as follows:

- Lower exposure action values for daily or weekly exposure: 80 dB(A) and a peak sound pressure of 135 dB(C) – hearing protection to be made available (protection to be selected to eliminate risk or reduce it to a minimum); information and training to be provided.
- Upper exposure action values for daily or weekly exposure: 85 dB(A) and a peak sound pressure of 137 dB(C) – a programme of control measures must be instigated for exposure at this level, which must be reduced to as low as reasonably practicable, and access to such areas restricted; hearing protection must be worn; appropriate health surveillance required where noise assessment indicates a risk to health; employees have a right to hearing checks.
- Daily or weekly personal noise exposure limits: 87 dB(A) and a peak sound pressure of 140 dB(C).

In determining exposure, the use of hearing protection provided by the employer may be taken into account when assessing personal exposure limits, but not when assessing either the upper or lower action levels.

Where an employee's hearing damage can be attributed to exposure at work, the risk assessment must be reviewed with a view to taking further measures to limit exposure and consideration given to moving the employee to alternative work to prevent further exposure.

Member States may apply, after consulting both sides of the industry concerned, a derogation from the Directive for those sectors where the wearing of hearing protection could put the employee at greater risk; such derogations to be reviewed every four years, and the employees concerned subject to increased health surveillance.

(The European Parliament and the Council of Ministers have also agreed the text of a Directive *laying down minimum standards for the health and safety of workers exposed to hand-arm and whole-body vibration*; this was formally adopted in June 2002 as Directive 2002/44/EC. Member States have three years in which to implement the Directive which requires employers to assess and measure exposure to mechanical vibration and to establish programmes to reduce exposure where action levels are exceeded. The Directive is implemented through *the Control of Vibration at Work Regulations 2005*, which came into force on 6 July 2005, and *the Merchant Shipping and Fishing Vessels (Control of Vibration at Work) Regulations 2007* (SI 2007/3077), which came into force on 23 February 2008)

4.29 UK REGULATIONS

The 2003 Directive has been implemented through *the Control of Noise at Work Regulations 2005* (SI 2005/1643) which came into effect on 6 April 2006, replacing and revoking the earlier 1989 Regulations (SI 1989/1970), with the music and entertainment industry being required to comply from 6 April 2008. Enforcement of the Regulations is mainly the responsibility of the Health and Safety Executive, and of local authorities. The Regulations apply in E, W & Sc, with similar 2006 Regulations (SR 2006/1) applying in Northern Ireland.

Separate Regulations apply the Directive to UK shipping: *the Merchant Shipping and Fishing Vessels (Control of Noise at Work) Regulations 2007* (SI 2007/3075), which came into force on 31 October 2007, with the regulations covering the provision of music and entertainment on ships taking effect on 6 April 2008. Regulations 7(4) and (5) – exposure to daily or weekly personal noise exposure limits of 87 dB(A) and a peak sound pressure of 140 dB(C), and the need to take action to reduce exposure, identify reasons for limit being exceeded and amend programme of measures – do not come into force for sea-going ships until 6 April 2011. The Regulations apply to UK shipping wherever they may be and to non-UK shipping in UK waters.

The Regulations put the emphasis on identifying measures to eliminate or reduce risks from exposure to noise at work, with the exposure action values (see above) being the points at which employers must take specific action. Exposure should be assessed over a working day, or a week if exposure varies markedly from day to day. In summary the Regulations require employers to

- Assess the risks to their employees from noise at work
- Take action to reduce the noise exposure that produces those risks
- Provide employees with hearing protection if they can't reduce the noise exposure through other methods
- Make hearing protection available on request at 80 dB, and ensure it is worn at 85 dB
- Ensure the legal limits on noise exposure – 87 dB daily or weekly exposure or peak sound pressure of 140 dB taking account of hearing protection – are not exceeded
- Provide employees with information, instruction and training
- Carry out health surveillance where there is a risk to health.

Exposure to members of the public from their non-work activities is not covered by the Regulations.

A guide to the Regulations and advice for employers on reducing exposure is available on HSE's website www.hse.gov.uk. A draft Sound Advice web guide, containing practical advice on controlling noise at work in the music and entertainment sectors is currently being finalised, with publication of both printed and online versions due July 2008.

SOUND INSULATION

4.30 BUILDING REGULATIONS

Part E of Schedule 1 of the *Building Regulations 2000*, revised through an amendment to the Regulations (SI 2002/2871), which came into force on 1 July 2003, requires floors and walls in new built dwellings for residential use – including houses, flats, hotels, student accommodation, residential homes for the elderly – and buildings where there is a material change of use (e.g. flat conversions) to meet specified performance targets to improve sound insulation; from 1 July 2003 the Regulations have required pre-completion testing on a sample of separating walls and floors on new dwellings formed from flat conversions to check compliance. This requirement was extended to new flats and houses from 1 July 2004. An amendment to the Regulations (SI 2004/1465) which took effect on 1 July 2004, does not require pre-completion testing so long as the person carrying out the building work notifies the local authority that one or more design details approved by Robust Details Limited are being used – i.e. building components and construction methods which have been tested and found to exceed the insulation standards which would be required under pre-completion testing. *Approved Document E – Resistance to the Passage of Sound* describes ways of meeting the requirements, with examples of widely used "satisfactory" constructions and test results.

In Scotland, Schedule 5 (section 5) of *the Building (Scotland) Regulations 2004* (SSI 2004/406) which came into force on 1 May 2005 requires buildings to be designed and constructed in such a way to minimise the transmission of sound between dwellings.

British Standard 8233 (1999) is a code of practice providing guidance on how to insulate buildings to reduce sound and general noise reduction advice. BS 9142 (2007) provides guidance on choosing noise impact assessment methods for projects. The Code for Sustainable Homes, published by the Government in December 2006 sets out a number of standards and measures aimed at encouraging housing developers to reduce the environmental impact (including noise) of new homes, building to standards that go beyond Building Regulations. It can be accessed via www.planningportal.gov.uk

LOW FREQUENCY NOISE

4.31 AN OVERVIEW

A low frequency noise is very often characterised by a hum or rumble and may be confused with tinnitus. These sounds often have no obvious source though in some areas a number of people may complain about the same unidentified "noise". While many of these noises remain unidentified, sources of LFN can include ventilation and air conditioning systems in large buildings; domestic boilers; diesel powered transport, such as ships, locomotives and lorries; and even the vibration of high rise buildings in the wind. These can cause complaints for two reasons:

a) primary noise (low frequency or infrasonic) radiated from source and entering houses either through their structure or through open windows;

b) secondary noise, e.g. rattling windows and doors. This either causes disturbance in itself or wakes people up, whereupon they become conscious of the low frequency pulsation.

Although there is no convenient measurement unit to describe it, low frequency sound may be defined as having a frequency below 150 Hz, with most problems reported within the 40-60 Hz range. Low frequency sound has a long wavelength which radiates in a spherical manner, rather like the ripples caused by a stone being dropped into a pond. In this way it can cover a vast area.

It can be particularly difficult to track down the source of LFN and assess its magnitude – ordinary sound level meters may not be able to detect it on the decibel scale as its decibel level is often lower than background noise. Practical methods of control are technically difficult and often prohibitive in cost. Sound proofing in buildings is usually impracticable as the design – particularly of modern buildings – may enhance the effect. Enclosing the noise source is a better option and will provide a more comprehensive solution. This too is often difficult and expensive as it involves enclosing the source in a combination of massive structures to reduce sound transmission. LFN from machinery can sometimes be reduced by the use of vibration absorbing mountings.

An update on low frequency noise and flow chart for its investigation has been prepared by Casella Stanger under contract from Defra and the Devolved Administrations – see bibliography. The Noise Council's 1995 *Code of Practice on Environmental Noise Control at Concerts* provides some guidance, as well as guideline limits, on minimising low frequency noise at outdoor concert venues. This notes that because "it is the frequency imbalance which causes disturbance ... there is less of a problem from the low frequency content of the music noise near to an open air venue than further away" [i.e. more than 2 km from a venue]. It goes on to suggest that "a level up to 70 dB in either of the 63 Hz or 125 Hz octave frequency band is satisfactory: a level of 80 dB or more in either of those octave frequency bands causes a significant disturbance."

The most recent studies carried out at the University of Salford and published in 2005, include guidance for local authorities in investigating complaints of LFN (with criteria and methodology).

A review of published research on low frequency noise and its effects can be downloaded from Defra's website – www.defra.gov.uk/environment/noise/lowfrequency/index.htm.

MEASUREMENT OF SOUND

4.32 AN OVERVIEW

The simple definition of noise as "sound which is undesired by the recipient" (Wilson Committee on the Problem of Noise, Cmnd 2056, 1963) requires that three conditions obtain: there must be a sound source, a transmission path, and a receiver who would have some adverse subjective reaction to the sound. It is this area of subjectivity which creates the major difficulty in the measurement and assessment of noise sources. There is no (nor is it likely there ever will be) equipment available which can measure "noise", but there are a wide range of instruments which measure sound levels and there are more sophisticated derivatives capable of carrying out statistical evaluations of total noise climates. Thus it is relatively easy, given suitable equipment, to measure sound but it is much more difficult to allow for the subjective reaction to noise, which varies widely within an exposed population.

For a more detailed explanation of noise measurement and taking sound level measurements, as well as methods of noise reduction, see *Noise Control*, (3rd edition), Christopher Penn (Shaw & Sons, 2002).

4.32.1 The Decibel

The logarithmic scale on which sound levels are measured is called the decibel (dB). The fundamental unit is the Bel and this is the logarithm of the ratio of the intensity of the given sound and the reference level; 1 Bel equals 10 dB or a factor of 10 in intensity – a change of 10 dB roughly corresponds to a doubling or halving of the loudness of a sound, i.e. a sound of 80 dB sounds about twice as loud as one of 70 dB. 0 dB is the threshold of hearing and 140 dB the threshold of pain.

4.32.2 Effective Perceived Noise Level

EPNdB and PNdB are used to measure aircraft noise with 3 EPNdB representing a doubling of noise energy. The quietest aircraft currently generate less than 90 EPNdB (about 75 dB(A)). It includes corrections for the duration and tonal quality of aircraft noise which are dependent on the rotational frequency of propellers or fans. The measure is used internationally for noise certification of aircraft.

4.32.3 Frequency

Another physical parameter which it is often necessary to measure is the frequency of the sound. The ear has a nominal range of between 20 hertz (Hz) (formerly called cycles per second) and 20 kHz (20,000 Hz), although at very high levels sound outside this range may be sensed. This sensitivity obtains only for the young, healthy, adult ear; as one gets older, the upper limit is reduced progressively perhaps to around 10 kHz. Because the ear does not respond equally to sound of different frequencies, being most sensitive around the 1,000 Hz region and least sensitive at very low frequencies, a measurement of sound without a frequency "weighting" would not indicate the likely response which the auditory mechanism might have to that sound.

In order that a sound level meter may approximate the response of the human ear, certain standardised filter networks are built into instruments and the most commonly utilised network is termed the "A" weighting. Measurements made when this circuit is switched on are referred to as dB(A) – decibels on the A weighted scale. Other weightings are occasionally used, of which the most important is the "D" network used for certain (approximate) aircraft noise measurements.

5 WASTE MANAGEMENT

WASTE – AN OVERVIEW

Section 75(2) of *the Environmental Protection Act 1990*, as amended by *the Environment Act 1995* (Sch. 22, paras 88 & 95), implements the 1991 European Union (EU) Framework Directive on Waste (91/156/EEC)* definition of waste:

"waste shall mean any substance or object in the categories set out in Schedule 2B to this Act [i.e. EPA] which the holder discards or intends or is required to discard;"

"Holder" is defined as the producer of the waste or the person who is in possession of it; and the "producer" is anyone whose activities produce waste or anyone who carries out pre-processing, mixing or other operations resulting in a change in the nature or composition of this waste.

(*the 1991 Directive has been repealed and its provisions codified in Directive 2006/12/EC – see below 5.6)

5.1 WHAT IS WASTE – DEFINITIONS

The above definition of waste was brought into force in England and Wales on 15 May 2006 by *the Environment Act 1995 (Commencement Order No. 23) (England and Wales) Order 2006* (SI 2006/934). It was brought into force in Scotland on 1 April 2003 for the purposes of *the Landfill (Scotland) Regulations 2003 – Env. Act 1995 (Commencement No. 21) (Scotland) Order 2003* (SSI 2003/206), and by Commencement Order no. 22 (SSI 2004/541) on 1 January 2005, and in Northern Ireland by SR 1998/390 of 1 December 1998.

Department of the Environment Guidance (Circular 11/94, SOED 10/94, WO 26/94) suggests the following be used as a starting point in deciding whether or not something is waste:

a) [has it been] discarded, disposed of or got rid of by the holder; or
b) [is it] intended to be discarded, disposed of or got rid of by the holder; or
c) [is it] required to be discarded, disposed of or got rid of by the holder.

A further consideration is whether "the substance or object [has] been discarded so that it is no longer part of the normal commercial cycle or chain of utility". Thus some items although eventually recycled will be treated as waste (discarded) because they need to be reprocessed before they can be brought back into re-use.

Waste subject to the provisions of the EPA is known as "controlled waste" and includes wastes arising from domestic, industrial and commercial premises, as well as hazardous waste (formerly "special waste") for which there are additional regulations (see this chapter 5.23). The types of waste to be treated under each of these classifications for the purposes of Part II of the EPA are defined in the *Controlled Waste Regulations 1992*, which took effect on 1 April 1992 and apply in England, Scotland and Wales (Northern Ireland: SR 2002/248):

- **Household waste** includes waste from: premises occupied by a charity; land belonging to domestic property, caravan or residential home, or a private garage for a car; private (domestic) premises; moored houseboat; camp sites; prisons and penal institutions; public meeting halls; royal palaces; litter collected under s.89 of the EPA. Local authorities may charge for the collection of certain types of household waste – see 5.22 below.
- **Industrial waste** includes waste from: commercial garages/maintenance premises (for vehicles, vessels, aircraft); laboratories and scientific research associations; workshops; dredging and tunnelling waste; clinical waste (other than from domestic property, residential home or houseboat); aircraft, vehicles, vessels not used for domestic purposes; leachate; poisonous or noxious waste from certain processes (e.g. dry cleaning, paint mixing/selling, pesticide sales); premises for breeding, boarding, stabling or exhibiting animals; waste oils, waste solvent, scrap metals (except from domestic premises); waste imports and waste from ships.
- **Commercial waste** includes waste from: offices; showrooms; hotels; private garages (more than 25 sq m); club/social premises; markets or fairs; courts, government departments, local and central government premises; corporate bodies; tents on land other than camp sites.

The *Waste Management Licensing Regulations 1994* [NI: 2003 Regulations] (see below, 5.18) apply to disposal and recovery activities relating to "Directive Waste" – that is waste covered by the EU Framework Directive (see below, 5.6). In England and Wales waste management activities are regulated under *the Environmental Permitting (England and Wales) Regulations 2007* (see chapter 2, 2.4).

Table 5.1: EU Framework Directive on Waste: Categories of Waste

Annex I to Directive 2006/12/EC; Schedule 2B to *Environmental Protection Act 1990*

1. Production or consumption residues not otherwise specified below.
2. Off-specification products.
3. Products whose date for appropriate use has expired.
4. Materials spilled, lost or having undergone other mishap, including any materials, equipment, etc, contaminated as a result of the mishap.
5. Materials contaminated or soiled as a result of planned actions (e.g. residues from cleaning operations, packing materials, containers, etc).
6. Unusable parts (e.g. reject batteries, exhausted catalysts, etc).
7. Substances which no longer perform satisfactorily (e.g. contaminated acids, contaminated solvents, exhausted tempering salts, etc).
8. Residues of industrial processes (e.g. slags, still bottoms, etc).
9. Residues from pollution abatement processes (e.g. scrubber sludges, baghouse dusts, spent filters, etc).
10. Machining/finishing residues (e.g. lathe turnings, mill scales, etc).
11. Residues from raw materials extraction and processing (e.g. mining residues, oil field slops, etc).
12. Adulterated materials (e.g. oils contaminated with PCBs etc).
13. Any materials, substances or products the use of which has been banned by law.
14. Products for which the holder has no further use (e.g. agricultural, household, office, commercial and shop discards, etc).
15. Contaminated materials, substances or products resulting from remedial action with respect to land.
16. Any materials, substances or products which are not contained in the above mentioned categories.

Until 2005, in contravention of the EU Framework Directive, non-mineral waste from a mine or quarry and waste from premises used for agriculture had not been classified as "controlled waste" under the *Controlled Waste Regulations*, and thus did not require a waste management licence. In Scotland, this was rectified through *the Waste (Scotland) Regulations 2005* (SSI 2005/22) which came into force on 21 January 2005 – these also categorise waste from a mine or quarry as industrial waste and that from premises used for agriculture as commercial waste. The *Waste Management (England & Wales) Regulations 2006*, which came into force on 15 May 2006 and *the Waste Management Regulations (Northern Ireland) 2006* (SR 2006/280), which came into force on 31 July 2006, implement similar controls. See below 5.19.

A Department of Environment Circular, 14/92, (WO 30/92; SOED 24/92) defines in detail wastes to be treated as "controlled waste". Further guidance is contained in DOE Circular 11/94 (WO 26/94; SOED 10/94) on waste management licensing and the Framework Directive on waste.

Clinical Waste

Clinical waste – also called, healthcare waste or healthcare risk waste – other than that from a private dwelling or residential home, is classified as industrial waste for legislative purposes. Handlers of clinical waste are also subject to the duty of care provisions of s.34 of the EPA: this places a duty on producers, handlers and disposers of waste to ensure it is disposed of safely and legally (see 5.17 below). Most clinical waste (except that from households) is also subject to the *Hazardous Waste Regulations 2005* (E&W and NI) and in Scotland, the *Special Waste Regulations 1996*, as amended (see 5.23.2 & 3 below). The *Controlled Waste Regulations 1992* (NI: 2002) define clinical waste as

- any waste which consists wholly or partly of human or animal tissue, blood or other body fluids, excretions, drugs or other pharmaceutical products, swabs or dressings, or syringes, needles or other sharp instruments, being waste which unless rendered safe may prove hazardous to any person coming into contact with it; and
- any other waste arising from medical, nursing, dental, veterinary, pharmaceutical or similar practice, investigation, treatment, care, teaching or research, or the collection of blood for transfusion, being waste which may cause infection to any persons coming into contact with it.

Technical guidance on clinical waste facilities was issued by the Environment Agency in August 2003; this covers matters such as waste acceptance, storage and handling, monitoring and criteria to be met before surrender of a licence. Appendix 5 to IPPC S5.06 (issued for consultation in March 2006) deals with the alternative treatment of clinical waste, laying down indicative standards that include waste pre-acceptance, waste receipt and acceptance, treatment efficacy and monitoring.

Clinical waste arisings from NHS Trusts were estimated at around 200,000 tonnes a year, with about the same amount being generated by private hospitals, GPs, dentists, nursing homes etc (MEL Research, 1998, for Environment Agency). HSC Guidance (*Safe Disposal of Clinical Waste*, 1999) provides guidelines on the disposal of the different types of clinical waste.

5.2 WASTE ARISINGS

Total waste arisings for the UK for 2004 were estimated to be around 335 million tonnes of which nearly 220 million tonnes was controlled waste (household, industrial & commercial and construction and demolition), and nearly 100 million tonnes mining and quarrying waste. Special waste arisings for each of the five years to 2003 remained constant at around 5 million tonnes a year. (Defra website)

5.2.1 Waste Arisings – England

Total municipal waste arisings in England increased by 1.4% to 29.1 million tonnes in 2006/07 (2005/06: 28.7 m. tonnes); 31% of household waste was recycled or composted in 2006/07 (2005/06: 27%). 16.9 m. tonnes of municipal waste was disposed of to landfill in 2006/07 (2005/06: 17.9 m. tonnes). (*Municipal Waste Management Statistics 2006/07*, Defra)

5.2.2 Waste Arisings – Scotland

Total controlled waste arisings for Scotland for the year 2005/06 were estimated at 22.22 million tonnes; this is made up of: household waste 2.89 m. tonnes (13%); commercial & industrial waste 8.41 m. tonnes (38%); construction & demolition 10.60 m. tonnes (48%); and agricultural waste 0.32 m. tonnes (1%); data on mines and quarries waste is not available. Around 3.54 m. tonnes of controlled waste was collected by or on behalf of local authorities in 2005/06, of which 955,005 (26.96%) was collected for recycling and composting. (Waste Data Digest 7, SEPA)

5.2.3 Waste Arisings – Wales

Municipal waste arisings in Wales for 2004/05 totalled 1.93 million tonnes (2003/04: 1.82 million tonnes) of which 78% (2003/4: 82%) was landfilled and 21.7% (2003/04: 17.7%) recycled or composted; waste from household sources (1.58 million tonnes) accounted for 82% of municipal waste in 2004/05 (2003/04: 84%); 286,455 tonnes (18.1%) of household waste was collected for recycling or composting (2003/04: 16.5%). (Municipal Waste Management Survey for Wales 2004/05, www.wales.gov.uk).

5.2.4 Waste Arisings – Northern Ireland

In Northern Ireland total municipal waste arisings increased 0.05% in 2006/07 to 1,064,090 tonnes (2005/06: 1,063,510 tonnes), of which 25.5% was recycled and composted (2005/06: 23.02%). 76.5% (813,253 tonnes) of municipal waste was landfilled in 2006/07 (2005/06: 76.98% - 818,699 tonnes). Household waste accounted for 88.22% of total municipal waste in 2006/07 (2005/06: 88.14%), of which 27.7% was collected for recycling or composting (2005/06: 24.5%). (*Municipal Waste Management Northern Ireland, 2006/07*, NI Environment Agency)

Total commercial and industrial waste arisings for 2004/05 were estimated to be 1,560,371 tonnes of which it is thought 45% is produced by industry and 55% by commerce (RPS Consulting Engineers survey for EHSNI (now NIEA), 2007)

5.3 REGULATORY AGENCIES

Waste management policy is the responsibility of the Department for Environment, Food and Rural Affairs (Defra). In Scotland, Wales and Northern Ireland most environmental and planning issues are the responsibility of the Devolved Administrations.

5.3.1 Environment Agency, England & Wales

As from 1 April 1996 the Environment Agency took over responsibility for waste regulation from the local authority waste regulation authorities. As well as ensuring high standards of waste management and implementing the National Waste Strategy, the Environment Agency's responsibilities include

- regulation of waste management activities, landfill sites, waste incineration, WEEE, etc under *the Environmental Permitting (England and Wales) Regulations 2007*) (chapter 2, 2.4).;
- ensuring compliance with the 1991 duty of care regulations (5.17 below);
- registration of waste carriers and brokers (5.34 below).

The Agency is also responsible for those areas of contaminated land designated as "special sites" because of the nature of contamination – see below 5.41.

5.3.2 Scottish Environment Protection Agency

SEPA is responsible for the regulation of waste management activities under *the Waste Management Licensing Regulations 1994* (as amended); landfill sites and waste incineration are regulated under *the Pollution Prevention and Control (Scotland) Regulations 2000*, and 2003 landfill and waste incineration Regulations (see chapter 2, 2.5 and this chapter 5.29 and 5.30. WEEE is regulated under 2006/2007 Regulations (5.32).

Details of SEPA's objectives and other responsibilities are given in chapter 1, 1.3.2.

5.3.3 Northern Ireland Environment Agency

The Northern Ireland Environment Agency (NIEA) is responsible for enforcement of waste management legislation in NI. It was established on 1 July 2008 as an executive agency of the Department of Environment, replacing the Environment and Heritage Service

The Waste and Contaminated Land (Northern Ireland) Order 1997 (WCLO), which received Royal Assent on 26 November 1997 and came into operation in March 1998 introduces controls on waste management and contaminated land similar to those in the rest of the United Kingdom – these have been given effect through *the Waste Management Licensing Regulations (Northern Ireland) Order 2003. The Waste (Amendment) (Northern Ireland) Order 2007* (SI 2007/611, NI3) makes a number of amendments to the WCLO which have been reflected in the text of this chapter. A statutory waste strategy was published in 2006 (replacing that published in April 2000) (see 5.20.3) and waste licensing regulations came into force on 19 December 2003.

The Pollution Prevention and Control Regulations (Northern Ireland) 2003 (see chapter 2, 2.5.1) and *The Landfill Regulations (Northern Ireland) 2003* (see 5.29) transpose the IPPC and landfill Directives into Northern Ireland legislation.

5.3.4 Local Authorities

Waste disposal authorities – i.e. county or district councils or unitary authorities as appropriate – are responsible for arranging for the disposal of all controlled waste in their area collected by the waste collection authority. Section 32 of the EPA requiring local authorities to contract out their waste disposal functions was repealed in England and Wales by s.47 of *the Clean Neighbourhoods & Environment Act 2005*, with effect from 18.10.05 (E) and 16.03.06 (W).

Waste collection is the responsibility of district, metropolitan or city councils, or of the appropriate unitary authority. Waste collection authorities have a duty to arrange for the collection of household and, if requested, commercial waste. WCAs are also responsible for drawing up and implementing recycling plans. *The Climate Change Bill* (see chapter 3, 3.7.3 and 5.22.1 below) includes powers for the Secretary to State to permit a maximum of five local authorities in England to pilot incentives for people to reduce, reuse and recycle waste.

The Local Government and Public Involvement in Health Act 2007 makes provision for groups of two or more local authorities in England or Wales to form Joint Waste Authorities to which they can transfer waste disposal, collection and/or street cleaning functions; these powers are due to come into force in 2009. In England, *the [draft] Joint Waste Authorities (Proposals) Regulations 2008*, issued for consultation in March 2008 together with draft guidance, outline the information to be included in an application to the Secretary of State to establish a JWA.

Under the *Pollution Control and Local Government (NI) Order 1978* district and borough councils in Northern Ireland have a duty to arrange for the collection and disposal of controlled waste.

Local authorities also have responsibilities under contaminated land legislation (see later this chapter 5.41), and are also responsible for enforcing controls on litter (chapter 1, 1.35).

Municipal Waste Management Strategies

The Waste and Emissions Trading Act 2003, s.32, requires waste disposal authorities in two tier areas in England (i.e. where disposal is the responsibility of the county and collection of the district council) to prepare joint municipal waste management strategies; these should outline plans for diverting biodegradable waste away from landfill and increasing recycling rates in line with Waste Strategy 2000. Initial strategies had to be drawn up by 13 May 2005. The Secretary of State may disapply the requirement to those authorities whose performance in carrying out their functions is at or above a specified level. (*The Joint Municipal Waste Management Strategies (Disapplication of Duties) (England) Regulations 2007* (SI 2007/63) which came into force on 20 February 2007 (revoking SI 2004/3242), implement this for WDAs or WCAs categorised as excellent under *the Local Government Act 2003*, or who have met specified standards under the 1999 LGA; the Regulations also specify the circumstances under which such disapplication may be withdrawn.) Guidance (Defra, July 2005) on the preparation of municipal waste management strategies and outlining local authorities' central role in delivering sustainable waste management, is available on Defra's website.

The Waste and Emissions Trading Act also gives the National Assembly for Wales powers enabling it to make regulations requiring local authorities in Wales to produce municipal waste management strategies

5.4 PLANNING CONTROLS

5.4.1 Planning Permission

New sites for waste disposal will require planning permission, and this will usually contain conditions on the use of the site such as the limits of the area involved in the operations and, where relevant, the provisions to be made for the restoration of the land after operations have ceased. For land which may already be contaminated, the planning permission may be granted subject to conditions relating to the remedial action required. Specific details as to how a site should be operated will be included in the conditions attached to a waste management licence issued under *Waste Management Licensing Regulations* (Scotland & NI, see below 5.18), or *the Environmental Permitting Regulations* (E & W, chapter 2, 2.4).

Annex I of the landfill Directive requires that the location of a proposed landfill site be taken into consideration when granting planning permission for a site. *The Landfill (Scotland) Regulations 2003* (SSI 2003/235), and similar Regulations in NI (SR 2003/496 – (see also later this chapter, 5.29). In England and Wales, requirements of the Directive are effected through *Environmental Permitting (E & W) Regulations 2007* (see chapter 2, 2.4). Factors to be taken into account include:

- distances from the site boundary to residential and recreational areas, waterways, water bodies and other agricultural or urban sites;
- existence of groundwater, coastal water or nature protection zones in the area;
- geological and hydrological conditions;
- risk of flooding, subsidence, landslides or avalanches on the site;
- protection of the natural or cultural heritage in the area.

5.4.2 Planning Policy

Planning Policy Statements set out the Government's policies on different aspects of land-use planning, and as such must be taken into account by waste planning authorities and also by regional planning bodies and the Mayor of London in the preparation of spatial strategies, and local planning authorities in the preparation of local development documents. Regional planning bodies will also need to take account of the objectives and requirements for managing waste included in documents such as the National Waste Strategy, and indeed requirements for carrying out strategic environmental assessments and sustainability appraisals – see chapter 1, 1.20.

(a) England

PPG 10, *Planning and Waste Management*, published in September 1999, replaced the pollution and waste management sections of PPG 23 (now PPS 23), and has itself been replaced in England by Planning Policy Statement 10 *Planning for Sustainable Waste Management*, published July 2005. PPS 10 puts the waste hierarchy at the centre of policy, with greater emphasis on waste as a resource; the need to deliver a better match between the waste communities generate, and the [local] facilities needed to manage that waste is emphasised. Thus regional planning bodies and local authorities (through their planning strategies) will need to plan for managing their waste, and ensuring that it is disposed of as near as possible to its place of production. A "Companion Guide" to PPS 10, published June 2006, provides further advice for planning authorities, developers and communities on the delivery of appropriate waste management facilities.

Annex 2 of Planning Policy Statement 23, *Planning and Pollution Control* (PPS 23, 2004), looks at the relationship of the contaminated land regime under Part IIA of the *Environmental Protection Act 1990* to the planning system and explains the legislation relevant to the consideration of the development of land affected by contamination.

PPSs and PPGs can both be accessed via the website of the Department for Communities and Local Government – www.communities.gov.uk.

The Town and Country Planning (Development Plan) (England) Regulations 1999 (SI 3280) which came into force on 4 January 2000 require local planning authorities to have regard to the National Waste Strategy and to the land use planning requirements of the COMAH Directive when preparing structure plans; among the latter's requirements are the need to ensure "appropriate" distance between residential areas and hazardous installations; this would include hazardous waste facilities.

In London, *the Greater London Authority Act 2007* requires London's waste authorities to act in general conformity with the Mayor's Municipal Waste Management Strategy; it also includes powers for the Mayor to call in applications on "strategically important" development proposals, such as waste management facilities

(b) Scotland

National Planning Policy Guidance 10, Planning and Waste Management (March 1996) reminds planning authorities of the need to take account of the Framework Directive on Waste (see below, 5.6) – i.e. that waste should be recovered, or disposed of in the nearest suitable facility, without risk to the environment or human health. Planners are urged to safeguard groundwater when developing plans for all new waste facilities and to consult SEPA at appropriate stages. Guidance is also given on which decisions belong to planning and which to pollution control.

Scottish Planning Policy 10: Planning for Waste Management; SPP 10, published August 2007 (replacing) NPPG 10), provides guidance to planning authorities on their role in helping to further the objectives of the National Waste Plan in relation to sustainable waste management, as well as providing guidance on compliance with both Scottish and EU waste legislation. It emphasises that a sustainable approach to waste management planning should reflect a number of factors including the waste hierarchy, the need to reduce reliance on landfill as well as principles such as the polluter pays and issues relating to proximity. It reminds planning authorities of the need to involve communities in the development planning process. The SPP focuses on industrial land, and promotes a model policy for planning authorities to adopt in their development plans. Planning authorities are also required to encourage the provision for waste separation and kerbside collection of recyclable materials when considering proposals for new housing.

(c) Northern Ireland

In Northern Ireland, *Planning Policy Statement 11, Planning and Waste Management* (December 2002) outlines policies for the development of waste management facilities, aiming to promote the "highest environmental standards" in such developments. The PPS also outlines the relationship between the planning system and the authorities responsible for waste regulation and management.

(d) Wales

Planning Policy Wales (March 2002) sets out the land use planning policies of the Welsh Assembly Government, and is supplemented by a series of Technical Advice Notes. TAN 21 (2001) deals with waste and is intended to facilitate the introduction of a comprehensive and sustainable land use planning framework for waste management in Wales. It looks at planning considerations in waste issues, with the various types of waste considered in more detail. More information on www.planningportal.gov.uk/Wales

THE EU AND WASTE MANAGEMENT

The European Commission's 6th Environment Action Programme (2002-2012) – see also Appendix 7 – which came into force in July 2002, identifies sustainable use of resources and waste management as one of four priority areas (the others being climate change, environment and health and biodiversity). The Programme reiterates that the Community's approach to waste management is that priority should be given to prevention of waste; thereafter the priority reuse of materials and products, material recycling, incineration with energy recovery, and as a last resort, landfilling. The Programme set as an objective to reduce the quantity of waste going to final disposal by 20% on 2000 levels by 2010 and by around 50% by 2050; it also set a target of reducing hazardous waste arisings by 20% by 2010 and 50% by 2020.

A list of EU measures relating to waste is at Appendix 7.5. Brief details of the institutions of the European Union, and of the legislative process are given in Appendix 7.

5.5 THEMATIC STRATEGY

As required by the 6th Action Programme, the European Commission has formulated a *Thematic Strategy on the Prevention and Recycling of Waste* – one of seven thematic strategies required under the 6th Environment Action Programme – which was adopted in December 2005. The Strategy includes a Proposal for a Directive to amend the Waste Framework Directive. The full documents are available from the European Commission's waste strategy website: http://europa.eu.int/comm/environment/waste/strategy.htm

The Thematic Strategy (COM(2005) 666) sets out the overall objectives and outlines the means through which the EU can improve its waste management practices. It reiterates the current basic objectives of EU waste policy which are: "to prevent waste and promote re-use, recycling and recovery so as to reduce [its] negative environmental impact". The Thematic Strategy notes that it is recognised that there are problems in waste management across the EU that relate to differences in interpretation of previous legislation, and intends to address this through removing the ambiguities, resolving disputed interpretations and amending ineffective legislation. To achieve this, the Commission's Waste Management Committee will be used as a forum to exchange information and best practice and to bring implementation difficulties to its attention.

In order to implement changes to waste management practices across Europe with a view to developing a comprehensive approach to waste prevention and recycling, and secure a higher level of environmental protection, the Strategy included a proposal for revising the Waste Framework Directive, which has now been agreed – see below 5.6.2.

EU Waste Statistics

1.8 billion tonnes of waste is generated each year in Europe – this consists mainly of waste from households, commercial activities (e.g. shops, restaurants, hospitals etc), industry (e.g. pharmaceutical companies, clothes manufacturers etc), agriculture (e.g. slurry), construction and demolition projects, mining and quarrying activities, and from energy generation. In 2005 49% of EU municipal waste was disposed of through landfill, 18% incinerated and 28% recycled or composted. *European Parliament press release announcing agreement on revised Waste Framework Directive, 16.06.08*

5.6 FRAMEWORK DIRECTIVE ON WASTE

5.6.1 Framework Directive 2006/12/EC

Among the most important of the Directives adopted on waste was the 1975 "Framework" Directive 75/442/EEC, as amended by Directive 91/156/EEC, which established general rules for waste management. These Directives were repealed on 5 April 2006 and their provisions codified into Directive 2006/12/EC which came into force on 17 May 2006 – this Directive does not amend the earlier Directives, but combines their provisions into a single instrument. This Directive is due to be repealed following implementation of the new Framework Directive, on which agreement was reached in June 2008 – see below.

Directive 2006/12/EC identifies 16 specific categories of waste, which is defined as "any substance or object which the holder discards or intends or is required to discard" (see 5.1 above), with further annexes defining disposal and recovery operations. The Commission Decision 2000/532 adopted on 3 May 2000 (amended 22 January 2001, 2001/118/EC) lists materials to be treated as waste, replacing the *European Waste Catalogue* (Decision, 94/3/EC, January 1994) and the Hazardous Waste list (Decision 94/904/EC, December 1994). The List categorises waste by industry sector or process, and classifies materials as hazardous, non-hazardous or potentially hazardous according to their physical, chemical and biological properties.

An important objective of the Framework Directive (Article 4) is to ensure "that waste is recovered* or disposed of without endangering human health and without using processes or methods which could harm the environment and in particular

- Without risk to water, air or soil, or to plants or animals;
- Without causing a nuisance through noise or odours;
- Without adversely affecting the countryside or places of special interest".

*The European Court of Justice has defined "recovery" of waste in the following terms: "the essential characteristic of a waste recovery operation is that its principal objective is that the waste serve a useful purpose in replacing other materials which would have had to be used for that purpose, thereby conserving natural resources". (Case C-6/00, ruling published 27.02.02)

Considerable emphasis is placed on the prevention, reduction, re-use and recycling of waste, and on the use of waste as a source of energy. With the ultimate aim of the EU becoming self-sufficient in waste disposal, the Directive calls on Member States to establish a network of disposal facilities.

Member States are required to establish competent authorities who must draw up waste disposal plans in collaboration with other Member States and the Commission; the designated competent authorities are also to be responsible for issuing authorisations and waste management licences. Specified waste recovery operations and businesses which dispose of their waste on site do not require an authorisation but must be registered with the relevant authority; before permitting exemption from authorisation, Member States must have appropriate Regulations specifying the types of activity exempted, quantities of waste, disposal and recovery methods permitted and other conditions.

The Directive is largely implemented through *Waste Management Licensing Regulations* in Scotland and Northern Ireland (see later this chapter, 5.18) and *the Environmental Permitting Regulations* in England and Wales (see chapter 2, 2.4).

5.6.2 Revised Framework Directive

In June 2008 the European Parliament and Council of Members reached agreement on a new waste Directive (COM(2005) 667); following its formal adoption and publication in the Official Journal, Member States will have two years to transpose its provisions into national law. The existing Waste Framework Directive, the 1991 Hazardous Waste Directive and the 1975 Waste Oils Directive are to be repealed.

The new Directive sets a clear five-step hierarchy of waste management options, with prevention being the preferred option, followed by reuse, recycling, other forms of recovery, and, as a last resort, safe and environmentally sound disposal. Within five years of the Directive coming into force Member States must develop national waste management plans and waste prevention programmes, together with waste prevention objectives. The Directive calls on the Commission to propose (if appropriate) by the end of 2014, waste prevention and decoupling objectives for 2020.

The Directive clarifies a number of important definitions – such as recycling, recovery and waste itself; in particular it draws a line between waste and by-products and defines when waste has been recovered enough – through recycling or other treatment – to cease being waste. Incineration of waste is to be categorised as a recovery operation, rather than disposal, providing it meets a specified energy efficiency standard (Annex II to the Directive includes a formula for this).

The Directive obliges Member States to take the necessary measures to achieve a target of 50% for the reuse and recycling of waste materials such as paper, metal and glass from household and similar waste streams by 2020; the target for non-hazardous construction and demolition waste is 70%. Member States are to establish separate collections for paper, glass and metals by 2015 so long as this is technically, environmentally and economically feasible.

5.7 LANDFILL

In April 1999 Environment Ministers formally adopted Directive 99/31/EC on the landfill of waste; its aim is to reduce the amount of waste landfilled, to promote recycling and recovery, and to establish high standards of landfill practice across the EU and, through the harmonising of standards, to prevent the shipping of waste from one country to another.

The Directive, which had to be implemented by May 2001, required Member States to have a strategy in place by July 2003 which will reduce the amount of biodegradable municipal waste sent to landfill to 75% of the total generated in 1995 by 2006, 50% by 2009 and 35% by 2016. The final target will be reviewed in 2014. Member States (including the UK) who currently landfill a majority of waste may apply for a derogation giving an extra four years to meet each target – in a statement to the House of Commons (April 2003), the then Environment Minister said that a decision would be taken one year prior to each target as to whether the derogation will be made use of. The Directive defines "municipal waste" as "waste from households, as well as other waste which because, of its nature or composition, is similar to waste from households". "Biodegradable waste" is defined as "any waste that is capable of undergoing anaerobic or aerobic decomposition, such as food and garden waste and paper and cardboard".

Three types of landfill site are defined – hazardous waste, non-hazardous waste and inert waste – with the Directive also covering underground and mine storage. Internal waste disposal sites (i.e. where a producer of waste is carrying out his own disposal at the place of production) are subject to the Directive, as are sites used for more than one year for the temporary storage of waste. A Council Decision (2003/33/EC) establishes acceptance criteria and procedures for allowing waste into landfill sites. Facilities where waste is unloaded to permit its preparation for further transport for recovery or treatment or disposal elsewhere are not covered by the Directive, and nor is the storage of waste prior to recovery or treatment for less than three years, or its storage prior to disposal for less than one year.

All waste will need to be pretreated prior to landfilling unless this would have no environmental benefit or is not feasible (e.g. for inert wastes); co-disposal of hazardous and non-hazardous waste is also banned from 2004. The disposal of the following wastes to landfill will be banned: liquid waste; explosive, corrosive, oxidising or inflammable waste; infectious hospital waste; used whole and shredded tyres (two years and five years respectively after the Directive takes effect); and other waste not meeting certain criteria set out in an Annex to the Directive. Monitoring schemes to establish the amounts of waste being sent to landfill, and its biodegradable content, had to be set up by 2001. Charges for landfilling must reflect the full costs of management of the site, as well as closure and aftercare for up to 30 years. A system of operating permits, some of the features of which are similar to the UK's current waste management licensing procedures, must be implemented. All sites will require licensing; existing sites not conforming to the Directive will need to do so by 2009 or close down.

Member States may exempt from the Directive landfills with a total capacity of less than 10,000 tonnes serving islands; also exempt are landfills serving isolated settlements – i.e. those with fewer than 500 inhabitants per municipality, and no more than 5 per sq. km, and more than 50 km from the nearest urban area with more than 250 inhabitants per sq. km.

In England and Wales all landfill sites require a permit under *the Environmental Permitting Regulations (England and Wales) Regulations 2007* which also incorporate the requirements of the Landfill Directive (see chapter 2, 2.4). In Scotland and Northern Ireland, landfill sites require a permit under *Pollution Prevention and Control Regulations* (Sc: 2000; NI: 2003), with additional requirements of the Directive being implemented through 2003 *Landfill Regulations* – see chapter 2, 2.5 and this chapter, 5.29.

The Waste and Emissions Trading Act 2003, which received Royal Assent on 13 November 2003, requires the Secretary of State to set targets for reducing the amount of biodegradable municipal waste (BMW) landfilled – this has been done through Regulations specifying the maximum amount of municipal waste which may be sent to landfill for the UK, and for each of the four countries in the target years, and also for each of the four countries the maximum amount in non-target years. Further Regulations in England and each of the Devolved Administrations allocate landfill allowances to waste disposal authorities, with the amounts being progressively reduced to meet each of the Directive's target dates – see later this chapter 5.29.2. The Act also amends the *Environmental Protection Act 1990* to empower county councils to direct district councils (waste collection authorities) as to the form (e.g. segregating waste) in which waste should be delivered to landfill sites to help meet targets.

5.8 WASTE INCINERATION

Directive 2000/76/EC on waste incineration was adopted on 4 December 2000; it entered into force on 28 December 2000 and had to be implemented by Member States by 28 December 2002. New plant had to comply by 28 December 2002, and stationary and mobile plant used for the generation of energy by 28 December 2004. Existing plant (i.e. in operation and complying with national legislation before 28 December 2002) were required to comply with the Directive by 28 December 2005; on this date the two 1989 Municipal Waste Incineration Directives (89/369 & 429/EEC) and the 1994 Hazardous Waste Incineration Directive (94/67/EC) were repealed.

Incineration plant are defined as any mobile or stationary technical unit and equipment dedicated to the thermal treatment of waste with or without the recovery of the combustion heat generated.

The Directive covers the incineration of both hazardous and non-hazardous wastes, requiring strict emission limits, based on best available techniques, as well as compliance with the IPPC Directive. A priority of the Directive is, through stricter controls, to reduce air, water and ground pollution and thus improve human health. Limit values have been set for hydrogen chloride, hydrogen fluoride, heavy metals, sulphur oxides, nitrogen oxides, dioxins and furans (Table 5.2), and for discharges into water (Table 5.3) and leachate from residues, see p.182; monitoring requirements for both hazardous waste and non-hazardous waste incinerators are specified. The Directive introduces strict requirements for the co-incineration of waste (i.e. the incineration of waste as an additional fuel in plants for energy generation) to ensure that operational standards are as high as for other plant. Operators of incineration plant have a duty "to take all necessary precautions concerning the delivery and reception of waste to prevent negative effects to the environment", and require a permit from the competent authority which will include measures to ensure that

- the plant is operated in accordance with the Directive;
- heat generated during incineration is, as far as possible, recovered;
- residues are prevented, reduced or recycled as far as possible;
- the disposal of residues which cannot be prevented, reduced or recycled is carried out in conformity with Community legislation.

All waste waters discharged from an incineration plant also require authorisation with strict conditions based on BAT, and operational controls parameters must be set for pH, temperature, flow and turbidity. The Directive requires instantaneous daily measurement for total suspended solids at the point of discharge, daily and monthly measurement of mercury, cadmium and thallium and daily measurements of the other heavy metals. For the first 12 months of operation of the incinerator, heavy metals, dioxins and furans must be measured every three months, and thereafter at least two measurements must be taken annually.

The following incineration plant are excluded from this Directive:

- vegetable waste from agriculture* and forestry;
- vegetable waste from the food processing industry, so long as heat is recovered;
- the treatment of agriculture and forest residues (except where they may contain halogenic organic compounds or heavy metals as a result of treatment);
- animal carcase incinerators, including pet crematoria;
- waste resulting from the exploration of, or exploitation of, oil and gas resources from off-shore installations, and incinerated on board;
- cork waste;
- radioactive waste.

(*other agricultural waste incineration processes, e.g. the burning of plastic film and containers, are subject to the Directive)

Also excluded are experimental plant used for research, development and testing to improve the incineration process and which treat less than 40 tonnes of waste a year.

All incineration plant burning more than 2 tonnes/hour must publish an annual report which includes information on emissions from the plant. Lists of smaller plant must be publicly available.

In England and Wales, the Directive is now implemented through *the Environmental Permitting (England and Wales) Regulations 2007* (see chapter 2, 2.4). In Scotland *the Waste Incineration (Scotland) Regulations 2003* (SSI 2003170), which came into force on 1 April 2003, and Directions from the Secretary of State, incorporate the requirements of the Directive, to ensure relevant waste incineration operations have a permit under *the Pollution Prevention and Control Regulations* by the dates set in the Incineration Directive; in Northern Ireland the relevant Regulations (SR 2003/290) took effect on 22 September 2003 (see this chapter 5.30.2 and chapter 2, 2.5).

Table 5.2: Air Emission Limit Values from Incinerators

(a) Daily average values (mg/m³)

Total dust	10[a]
Total organic carbon	10
Hydrogen chloride	10
Hydrogen fluoride	1
Sulphur dioxide	50

[a] exemption may be authorised for existing plant until 1.1.08, provided permit foresees no exceedence of daily average of 20 mg/m³

NO & NO_2 expressed as NO_2	
- existing plant with capacity exceeding 6 tonnes/hr, or new plant*[b]	200
- existing plant with capacity of 6 tonnes/hr or less*[b]	400

* Until 1.1.07, does not apply to plant incinerating hazardous waste only
[b] Exemptions for NOx emissions from existing plant may be authorised up to 1.1.08; exemptions are based on capacity of plant.

(b) Half-hourly average values (mg/m³)

	100% compliance	97% compliance
Total dust	30	10
Total organic carbon	20	10
Hydrogen chloride	60	10
Hydrogen fluoride	4	2
Sulphur dioxide	200	50
NO & NO_2 expressed as NO_2		
- existing plant with capacity exceeding 3 tonnes/hr, or new plant*[c]	400	200

* Until 1.1.07, does not apply to plant incinerating hazardous waste only
[c] Exemptions may be authorised at existing plant with capacity between 6-16 tonnes per hour until 1.1.10 providing permit foresees no exceedence of 600 mg/m³ (100%) or 400 mg/m³ (97%)

(c) Average values over sample period, min. 30 mins, max. 8 hrs (mg/m³)

Hg	0.05	0.1*
Cd + Tl	0.05 (total)	0.1 (total)*
Sb+As+Pb+Cr+Co+Cu+Mn+Ni+V	0.5 (total)	1 (total)*

(*Until 1.1.07 average values for existing plant for which permit granted before 31.12.96 and which incinerate hazardous waste only.)

(d) Average values over sample period, min 6 hrs, max 8 hrs

Dioxins and furans	0.1 ng/m³ (total)

(e) Carbon monoxide

Emission limit values not to be exceeded in combustion gas (excluding start up and shut down)
50 mg/m³ as daily average value
150 mg/m³ (95% of measurements) as 10 minute average value, or
100 mg/m³ (all measurements) as half-hourly average values taken in any 24 hr period.

Table 5.3: Emission Limit Values for Discharges of Waste Water from Cleaning of Exhaust Gases

Emission limit values expressed in mass concentrations of unfiltered samples

Total suspended solids*	30 mg/l (95%)	45 mg/l (100%)
Mercury & its compounds	0.03 mg/l	
Cadmium & its compounds	0.05 mg/l	
Thallium & its compounds	0.05 mg/l	
Arsenic & its compounds	0.15 mg/l	
Lead & its compounds	0.2 mg/l	
Copper & its compounds	0.5 mg/l	
Chromium & its compounds	0.5 mg/l	
Nickel & its compounds	0.5 mg/l	
Zinc & its compounds	1.5 mg/l	
Dioxins and furans	0.3 ng/l	

* Exemptions may be authorised until 1.1.08 for existing plant provided that permit foresees 80% of measured values do not exceed 30 mg/l and none exceed 45 mg/l

5.9 HAZARDOUS WASTE

5.9.1 Hazardous Waste Management

Directive 91/689/EEC on Hazardous Waste was adopted in December 1991 but implementation was deferred until 27 June 1995 (when Directive 78/319/EEC on toxic and dangerous waste was repealed) owing to difficulties in drawing up a definitive list of hazardous wastes. This was finally adopted in December 1994 (Council Decision 94/904/EC), and included 200 types of waste split into 20 categories. This Decision (together with Decision 94/3/EC which established the European Waste Catalogue) has been replaced by Decision 2000/532, adopted on 3 May 2000; this is a list of hazardous and non-hazardous waste pursuant to both the Framework Directive on Waste (see above) and to the Hazardous Waste Directive and had to be implemented by 1 January 2002. A further Decision (2001/118/EC) adopted on 22 January 2001 adds end-of-life vehicles to the List. This Directive has been incorporated into the revised Framework Directive on Waste and is to be repealed – see 5.6.2 above.

The 1991 Directive redefines "hazardous waste" and aims to achieve greater harmonisation in the management of hazardous waste. All wastes with characteristics which render them hazardous, e.g. corrosive, infectious or ecotoxic substances, are covered. Annexes to the Directive list the categories of waste to be covered, the various components which would make it hazardous; a third annex lists 15 characteristics of which one or more must be present for the waste to be hazardous. Domestic waste from domestic premises is exempt from this Directive.

The Directive prohibits the mixing of hazardous waste with other waste except where it is a necessary part of the disposal operation and places stricter controls on carriers of hazardous waste. Hazardous waste producers will be subject to periodic inspection and they and waste carriers must keep records of waste transactions for one and three years respectively; waste transfers must comply with the EU's 2006 Shipments of Waste Regulation which came into force in July 2007 – see below.

Compliance with the Directive in England is effected through *the Hazardous Waste (England and Wales) Regulations 2005* (SI 2005/894), with similar Regulations in Wales (SI 2005/1806, W.138) and Northern Ireland (SR 2005/300); in Scotland, *the Special Waste (Amendment) (Scotland) Regulations 2004* (SSI 2004/112) implement the Directive – see 5.23.2 & 3.

5.9.2 Transfrontier Shipments

A 1984 Directive on the Transfrontier Shipment of Hazardous Wastes (84/631/EEC), amended in 1986 (86/279/EEC), required pre-notification of transfrontier shipments of hazardous waste; they had to be sent only to facilities that could dispose of them without danger to human life or the environment, their passage documented, and exports of hazardous waste outside the European Union prohibited unless the receiving country had given its prior consent. The 1984 Directive was replaced by Regulation 259/93 on the Supervision and Control of Shipments of Waste within, into and out of the EU, which came into force in May 1994 – this covered the movement of all wastes (i.e. not just hazardous waste), and enabled the Community to ratify the 1989 Basel Convention on the Control of Transboundary Movements of Hazardous Wastes and their Disposal. This Regulation was repealed on 12 July 2007 when Regulation 1013/2006 on *shipments of waste*, which was adopted on 14 June 2006, had to be applied.

The current Regulation aims to clarify and simplify procedures for transfrontier shipments of waste, thus ensuring that waste is properly handled from the time it is shipped to the time it is disposed of or recovered at destination. The Regulation also fully implements the Basel Convention banning the export of hazardous waste from the EU to developing countries and the ban on the export of waste for disposal to countries outside the EU and EFTA – see below.

The Regulation places an obligation on the producer, notifier and other undertakings involved in a shipment of waste, and/or its recovery or disposal, to take the necessary steps to ensure that any waste they ship is managed in an environmentally sound manner and without endangering human health throughout the shipment and during its recovery and disposal.

All shipments of waste for disposal (listed in Annex III) and of hazardous and semi-hazardous waste for disposal or recovery (listed in Annex IV) require written prior notification to the authority responsible for the shipment; the notifier has to make a contract with the recipient of the waste, as well as a financial guarantee covering the shipment up to its recovery or disposal – there are different requirements depending on whether the shipment is between, into or outside (including transit through) Member States. The authority responsible for the waste will send a copy of the notification to the destination authority who should acknowledge receipt. There will then be a 30-day period to enable shipping, transit and destination authorities to notify their consent or objections to the shipment. The shipment may not take place until the notifier has the written consent of the despatch and destination authorities, and must be accompanied by all relevant notification and shipment documents. All waste shipments will remain under the control of the authorities until final disposal or recovery is effected, including while awaiting treatment or disposal. EU Regulation 669/2008 of 15 July 2008 contains the specific instructions for completing notification and movement documents in accordance with Annex IC of Regulation 1013/2006.

For "Green List" wastes (Annex III) a movement form must be completed and must accompany each shipment of waste; this should include details of the composition, quantity and destination of the waste, and a declaration that a contract with the consignee has been entered into. EU Regulations 1418/2007 and 740/2008 take account of the rules of non-OECD countries on the import of green-listed non-hazardous wastes for recovery and the wastes they would like to receive.

EU Regulations are directly applicable on Member States and thus do not require transposition into national legislation; national legislation is however necessary to specify the competent authorities and to set out offences and penalties for contravening the Regulations. *The Transfrontier Shipment of Waste Regulations 2007* (SI 2007/1711) (as amended), which came into force on 12 July 2007 apply, revoking earlier 1994 Regulations and subsequent amendments (see 5.37 below).

The transfrontier shipment of radioactive waste is regulated by a European Atomic Energy Union Directive 92/3/EURATOM, adopted in February 1992; this is to be repealed by Directive 2006/117/EURATOM on 25 December 2008 – see this chapter 5.39.5. The shipment of radioactive wastes within the EU is regulated by Euratom Regulation 1493/93 of June 1993.

In February 2007, the Commission published a *proposal for a Directive on the protection of the environment through criminal law* (COM(2007) 51). Among the environmental crimes to be covered by the Directive is the unlawful treatment, transport, export or import of waste, including hazardous waste – see chapter 1, 1.15 for more details.

Basel Convention

The *Convention on the Control of Transboundary Movements of Hazardous Wastes and their Disposal*, prepared under the auspices of the UN Environment Programme, came into force on 5 May 1992. As at November 2006, there were 168 parties to the Convention, including the European Community and the UK, which both ratified it in February 1994.

The Convention takes as a basic principle the need to reduce both the generation of hazardous wastes and their transboundary movement to a minimum. It says that all states have the right to ban the import of hazardous waste and there are extensive pre-notification requirements for all transfrontier movements of household and industrial wastes. The exporting state has a duty to ensure that all wastes, whether for recovery or disposal, can be dealt with in an environmentally sound manner; it also has a duty to arrange for the return of any wastes failing to go to an appropriate treatment or disposal plant. The Convention permits parties to it to agree bi-lateral arrangements for trade in hazardous waste and prohibits such arrangements with others. Hazardous wastes which are to be exported must be packaged, labelled and transported in accordance with recognised international standards.

In September 1995, parties to the Convention agreed an amendment to the Convention which has now been ratified by 63 (as at November 2006) of the parties that agreed to the amendment (including the European Community and the UK). It has however not yet come into force owing to disagreement on the number of ratifications needed (either three-quarters of those agreeing the amendment, or three-quarters of all parties to the Convention) – the issue remained unresolved following discussion at the 9th meeting of parties to the Convention in June 2008. The amendment provides for an immediate ban on exports of hazardous waste for disposal from OECD to non-OECD countries and a similar ban on exports for recovery or recycling from 1 January 1998, although such exports may be continued by agreement between an OECD and non-OECD country. Non-OECD countries that wish to continue importing hazardous waste for recycling or recovery must notify the Convention Secretariat of the types and quantities of waste they will accept, processes to be used and method of final disposal of the residues. Lists of hazardous and non-hazardous wastes covered by the ban to be annexed to the Convention were agreed by Parties to the Convention in February 1998. List A includes over 50 categories of waste defined as hazardous; List B covers those waste categories which the Basel Convention does not define as hazardous but which nevertheless may contain hazardous characteristics and thus be covered by the ban. A further list – List C – covers wastes "awaiting classification". This latter amendment to the Convention was implemented in the UK in November 1998 (Command Paper 4061).

A Protocol to the Convention signed in December 1999 establishes a liability and compensation system for damage for accidents resulting from the transfrontier shipment of waste; it applies primarily to hazardous wastes defined by the Convention and to other wastes notified to the Convention Secretariat which have been defined or are considered to be hazardous by the importing or exporting country. Compensation covers injuries, death and loss of income for individuals and environmental damage. The person initiating the transboundary movement will be strictly liable for any damage caused until the waste is delivered to the person recycling or disposing of that waste, when they become liable until completion of the operation. All those who are strictly liable will be required to carry insurance up to the limit of their strict liability. Where liability cannot be established or the liable party is unable to pay, compensation will be paid from a Fund to be administered by the Convention Secretariat.

The Protocol requires ratification by 20 countries before it can come into force 90 days later; as at November 2006, there had been seven ratifications. The UK signed the Protocol on 7 December 2000 but will need to introduce new legislation before being able to ratify it.

Further information on the Convention and its Protocols can be found on http://basel.int

5.10 DANGEROUS SUBSTANCES

Directives have been adopted on polychlorinated biphenyls (PCBs) and polychlorinated terphenyls (PCTs), 96/59/EC which repeals a 1976 Directive (76/403/EEC); on wastes from the titanium dioxide industry (1978 and 1982); and on the use of sewage sludge in agriculture (86/278/EEC, proposal for revision due 2007). This latter was implemented in the UK in September 1989 through the *Sludge (Use in Agriculture) Regulations* (amended 1990) – see chapter 3, 3.21.

The 1996 Directive on PCBs and PCTs, adopted in September 1996, aims to reduce risks to human health and the environment through stricter controls on their disposal; it includes requirements for inventories, labelling and treatment of all significant PCB holdings and tighter regulation of all treatment facilities. The Directive has been implemented through *The Environmental Protection (Disposal of Polychlorinated Biphenyls and other Dangerous Substances) (England and Wales) Regulations 2000* (SI 1043) and similar in Scotland (SSI 95) and Northern Ireland (SR 232). Both sets of Regulations were made under the *European Communities Act 1972* – see this chapter 5.35.

A number of Directives dealing with the discharge of hazardous substances to water have been adopted and information about these is contained in chapter 6 on water pollution.

5.10.1 Batteries

Directive 2006/66/EC *on batteries and accumulators and spent batteries and accumulators* (and corrigenda dated 10.11.06. OJ L311/58) was formally adopted on 6 September 2006, entering into force on 26 September 2006. It replaces Directive 91/157/EEC, which was repealed on 26 September 2008, the date by which its replacement had to be transposed – see later this chapter 5.35.

COM(2008) 211 is a proposal to amend Article 6(2) of the 2006 Directive to clarify that batteries or accumulators legally placed on the market before 26 September 2008 but which do not comply with the 2006 Directive do not have to be withdrawn from the market or relabelled.

The main elements of the Directive, which will require the collection and recycling of all batteries and accumulators sold within the EU and the metals in them to be recovered and reused, are as follows:.

- landfilling or incineration of automotive and industrial batteries is banned.
- Collection systems for spent portable batteries and accumulators, as well as accessible collection facilities, enabling consumers to return spent batteries free of charge, must be set up by 26 September 2008.
- Producers of industrial batteries and accumulators, or third parties acting on their behalf, will be required to take back spent batteries and accumulators from end users. However, producers of very small quantities may in certain circumstances be exempted from the financing regime to be established. All producers will need to be registered.
- Producers of automotive batteries and accumulators, or third parties acting on their behalf, will be required to set up collection systems for spent batteries and accumulators.
- The sale of all batteries and accumulators with more than 0.0005% of mercury by weight and 0.002% of cadmium by weight is prohibited; those used in emergency and alarm systems, medical equipment and cordless power tools will be exempt, with the latter exemption being reviewed by the Commission and a report submitted to the European Parliament by 26 September 2010; button cells with less than 2% mercury will also be exempt. The exempted batteries will be collected under the collection systems to be set up under this Directive or, if incorporated in appliances, under the WEEE Directive.
- Appliances must be designed to ensure that batteries and accumulators can be readily removed, thus facilitating their collection, treatment and recycling.
- Information to consumers relating to the potential effects of batteries on the environment, the need to dispose of them properly in collection and recycling facilities, and the meaning of the various symbols on batteries and packaging – producers are to assume the net costs of public information campaigns. From 26 September 2009 labels on batteries and accumulators must show their actual capacity.

The Directive sets a target for collection of 25% and 45% of portable batteries to be achieved by 26 September 2012 and 2016 respectively, and all industrial and commercial batteries. As a general principle all batteries are to be sent to recycling facilities, and the "best available treatment and recycling techniques" used – these facilities must be established by September 2009; the Directive requires the recovery of all the lead and 65% of the average weight of lead-acid batteries; for nickel-cadmium batteries, all the cadmium should be recovered and at least 75% of the average weight of the batteries; and for all other batteries at least 50% of their average weight – these targets to be achieved by 26 September 2011.

5.11 END OF LIFE VEHICLES

This Directive (2000/53/EC), which came into force on 21 October 2000, covers the disposal of end-of-life vehicles (ELVs); it aims to prevent polluting discharges as a result of the scrapping of such vehicles and to encourage the recycling and recovery of vehicle parts. Decision 2001/18/EC adds ELVs to the Hazardous Waste List; thus ELV collection and treatment facilities will also be required to comply with the Hazardous Waste Directive (see 5.9 above).

To encourage the recovery and re-use of vehicle components, as well as recycling, the Directive sets a target of 85% of ELVs (by average weight of vehicle per year) to be reused or recovered and 80% recycled by 1 January 2006. For pre-1980 vehicles the targets are 75% and 70% respectively. By 1 January 2015 95% of all vehicles should be reused or recovered and 85% recycled. Following a review taking account of technical developments the Commission is recommending (January 07 report) that these targets should be retained.

New cars placed on the market after 1 July 2003 may not contain lead, mercury, cadmium or hexavalent chromium, subject to certain exemptions (Annex II of the Directive); these include lead in batteries and mercury in light bulbs. A Council Decision of June 2002 bans the sale of cadmium for use in batteries for electric vehicles from 31 December 2005, except as replacement parts for vehicles already on the market at that date. A maximum 2% lead in alloys was permitted until 1 July 2005, reducing to 1% from 1 July 2008; lead in wheel balancing weights was permitted until 1 July 2003 and its use will continue to be allowed in batteries and in certain electrical components. A further Council Decision (2005/673/EC) of January 2005 amends Annex II to exempt from the Directive's ban on heavy metals spare parts for use in vehicles placed on the market before 1 July 2003.

Member States were required to ensure that there are adequate collection and treatment facilities for ELVs by 1 July 2002; vehicles may only be de-registered when transferred to an authorised treatment facility and a certificate of destruction obtained – this latter requirement should have been implemented in April 2002, together with a requirement for impermeable drainage and storage facilities for liquids to be licensed. Certificates of destruction must contain detailed information identifying the vehicle, as well as details of the scrapyard and of the competent authority issuing the certificate. Treatment facilities must be registered, or have a permit from the competent authority. From 1 January 2007, last owners of vehicles must be able to return their vehicles to a treatment facility free of charge.

Those aspects of the Directive covering restrictions on the use of heavy metals, certificates of destruction, marking of components to aid recycling, free take-back of ELVs put on the market from 1 July 2002 and licensing of authorised treatment facilities have been implemented through *The End of Life Vehicles Regulations 2003* in England and Wales, and similar in Scotland and Northern Ireland; further 2005 Regulations cover "free take back" of ELVs of any age from 1 January 2007 and recycling targets for 2006 and 2015 – see this chapter 5.31.

Directive 2005/64/EC, adopted in November 2005, covers the reusability, recylability and recoverability of motor vehicles and must be transposed into national legislation by 15 December 2006. From 15 December 2008, member states must refuse the registration, sale or entry into service (type approval) of new vehicles which are not reusable and/or recyclable to a minimum of 85% by mass, and reusable and/or recoverable to a minimum of 95%; from 15 July 2010 the sale of vehicles not meeting type approval will be banned. The

Directive, which does not apply to special purpose vehicles, multi-stage built vehicles and those produced in small series, has been implemented through an amendment to *the Motor Vehicles (EC Type Approval) (Amendment) Regulations 2007* (SI 2007/855).

5.12 ELECTRICAL & ELECTRONIC EQUIPMENT

In January 2003, two Directives relating to waste electrical and electronic equipment (WEEE), were adopted. Directive 2002/96/EC seeks to prevent WEEE and to encourage its recycling and recovery. Directive 2002/95/EC aims to restrict the use of certain hazardous substances in EEE and thus contribute to the protection of human health and encourage environmentally sound recovery and disposal of WEEE.

5.12.1 Waste EEE

The *Directive on Waste Electrical and Electronic Equipment* – 2002/96/EC, which came into force on 13 February 2003 – requires manufacturers to recover and recycle electrical and electronic equipment, thus reducing the amount sent for final disposal to landfill or incineration without processing. Member States had until 13 August 2004 to implement this Directive, which covers: large and small household appliances, IT and telecommunications equipment, consumer equipment (radios, TVs etc), lighting equipment, electrical and electronic tools, toys, leisure and sports equipment, medical devices, monitoring and control instruments (smoke detectors, thermostats etc), and automatic dispensers.

From 13 August 2005 Member States are required to provide convenient facilities for householders to return old electrical equipment free of charge; when supplying a new product, distributors are responsible for taking back the old (equivalent) product free of charge, although Member States need not adhere to this provision so long as this does not make it more difficult for the householder to return the old product to a convenient facility. Also by 13 August 2005, producers (i.e. manufacturers) must provide for the financing of the collection, treatment, recovery and environmentally sound disposal of WEEE from householders deposited at collection facilities and must finance these operations for their own products which are put on the market after 13 August 2005 – as with packaging waste, producers may either set up collective schemes to ensure compliance or be responsible individually.

The costs of dealing with "historic" waste (i.e. equipment placed on the market prior to 13 August 2005) is to be financed by all producers. Producers are to be responsible for financing the collection and treatment etc of commercial WEEE from 13 August 2005, and also for financing the costs of managing "historic" commercial WEEE, though Member States may impose this cost on other sectors (apart from householders). An amendment to the Directive adopted on 25 November 2003 (2003/108/EC) limits producers' responsibility to take back "historic" commercial WEEE only to the equivalent amount of new EEE being supplied; where a commercial user is not replacing EEE, then he becomes liable for the disposal costs.

The Directive includes a mandatory target for the separate collection of 4kg of 'electro-scrap', on average, per person per year by 31 December 2006. A minimum recovery target of 80% by average weight per appliance is set, and recycling of 75% by weight, for large household appliances and automatic dispensers; for IT, telecom and consumer equipment the targets are 75% and 65% respectively; and for small household appliances, lighting, tools, toys, monitoring and control instruments 70% and 50% respectively.

To ensure the proper disposal of WEEE, all equipment is to be marked with a symbol (crossed out bin), by 13 August 2005, showing it should not be put in the household bin, and consumers are to be given adequate information to ensure they know what to do with such waste.

Registers of producers had to be established in 2005, and the Commission provided with details of the quantities of products put on the market, collected, reused, recycled or otherwise recovered every two years. Every three years, beginning 2007, Member States must report to the Commission on their implementation of the Directive.

Following review of the Directive, the Commission was expected to bring forward proposals for its amendment before the end of 2008.

The Directive has been transposed in the UK through *the Waste Electrical and Electronic Equipment Regulations 2006* (SI 2006/3289) which came into force on 2 January 2007 – 5.32.1.

5.12.2 Hazardous Substances in EEE (RoHS)

The *Directive on the Restriction of the Use of Certain Hazardous Substances in Electrical and Electronic Equipment* – 2002/95/EC – also entered into force on 13 February 2003 and should have been implemented by 13 August 2004. With effect from 1 July 2006, new EEE put on the market must not, except for certain listed exemptions, contain lead, mercury, cadmium or hexavalent chromium, or two brominated flame retardants – polybrominated biphenyls (PBB) and polybrominated diphenyl ethers (PBDE). This Directive applies to the use of these substances in all the product categories covered by the WEEE Directive (except medical devices and monitoring and control instruments); it applies to electric light bulbs, and luminaires in households. It does not apply to spare parts for the repair or reuse of EEE put on the market before 1 July 2006. Exemptions from the RoHS Directive are contained in various Commission Decisions covering exemptions for uses and applications of lead (2006/310/EC, 2006/690/EC, 2006/691/EC, 2008/385/EC), lead oxide (2008/385/EC), cadmium (2006/691/EC, 2008/385/EC), and hexavalent chromium (2006/692/EC).

The Directive is to be reviewed by the Commission with any proposals for amendment expected in 2008.

The 2002 Directive has been implemented by *the Restriction of the Use of Certain Hazardous Substances in Electrical and Electronic Equipment Regulations 2006* – see 5.32.3.

5.13 MINE AND QUARRY WASTES

Directive 2006/21/EC *on the Management of Waste from extractive Industries* (i.e. mine and quarry wastes) was adopted on 15 March 2006; it entered into force on 1 May 2006, with Member States required to transpose it before 1 May 2008. Waste facilities covered by the Directive which have been granted a permit or are already in operation by 1 May 2008 must comply with the Directive by 1 May 2012.

The Directive applies to waste from land-based minerals extraction, treatment and storage and to the working of quarries, with the aim of preventing water and soil pollution and accidents arising from long-term storage of extractive wastes. Operators of such sites will be under a general duty to "prevent or reduce as far as possible" the adverse effects on the environment of their operations. The Directive will require operators to have a permit, to include a waste management plan, for their facility, preparation of a major accident prevention policy and a safety management system, and financial guarantees covering obligations under their permit, including closure and after care. Member States are required to ensure that the public are given an early opportunity to participate in decisions regarding permits. The EU's IPPC Bureau is finalising a draft BREF (Best Available Techniques Reference document) outlining the requirements with which mine and quarry operators would need to have regard when applying for a permit.

Provisions regarding waste management plans, drafting of a major accident prevention policy, permits and public participation do not apply to waste facilities which stopped accepting waste before 1 May 2006, are completing a closure procedure in accordance with national or EU legislation and will be effectively closed by 31 December 2010.

A consultation paper published by the Department for Communities and Local Government in April 2008 proposes implementing the Directive in England and Wales through *the Environmental Permitting (E & W) Regulations 2007*, with the Environment Agency as the principal competent authority. Operators of existing facilities in operation at 1 May 2008 would be required to apply for an environmental permit by May 2010, with the Environment Agency issuing permits by 1 May 2012. The Scottish Government is proposing (April 2008) to implement the Directive by modifying existing planning application procedures so that they explicitly incorporate the requirements of the Directive.

5.14 PACKAGING WASTE

The 1994 Directive on Packaging and Packaging Waste (94/62/EC), which came into force on 31 December 1994, applied to all waste packaging – household, commercial and industrial – and had to be implemented by 30 June 1996, when Directive 83/339/EEC (containers of liquids for human consumption) was repealed. The Directive placed a strong emphasis on the prevention of packaging waste and on reducing the amount of packaging used (while safeguarding consumer safety and product quality); in promoting the use of recyclable and reusable materials and the use of packaging likely to have the least impact on the environment, an annex to the Directive laid down standards for manufacture, composition, etc to be complied with by the end of 1997. The composition by weight of lead, cadmium, hexavalent chromium and mercury in packaging and packaging waste had to be reduced over five years.

The Directive required the recovery of 50-65% of packaging materials in waste by June 2001, and 25-45% recycled with a minimum of 15% of each material. Recovery of packaging waste includes incineration with energy recovery; recycling includes composting. Recycling of packaging waste does not take place until the material has been put back into productive use. Also by June 2001, the Directive required all packaging to be labelled with its recyclable or recoverable content; Member States had to set up suitable systems for the collection of packaging and packaging waste and its recycling or reuse and should ensure that consumers have adequate information regarding the marking of packaging, the benefits of reusable packaging and about return systems.

The Directive permits the use of economic instruments as a means of achieving its objectives but only "in accordance with the principles governing Community policy on the environment, such as the polluter pays principle, and with full respect for obligations stemming from the Treaty".

Decision 97/129/EC established an identification system for packaging materials as required by the Directive; Decision 97/138/EC relates to the formats of databases on packaging and packing waste established under the Directive with the aim of making these compatible throughout the EU.

A new Directive – 2004/12/EC, as amended by 2005/20/EC – amending and updating the 1994 Directive was formally adopted on February 2004 and came into force on 18 February 2004. The Directive includes new targets to be met by 31 December 2008 (with a longer period allowed for the countries which joined the EU in May 2004):

- Minimum of 60% by weight of packaging to be recovered or incinerated at waste incineration plants with energy recovery;
- Recycling of a minimum of 55% and maximum 80% by weight of packaging waste;
- Materials specific recycling targets, by weight: glass 60%; paper & board 60%; metals 50%; plastics 22.5%; wood 15%.

See 5.33 below for implementation of the Directives in the UK.

UK REGULATORY CONTROLS

Until 1972, when public concern about the dumping of toxic waste in the Midlands resulted in the *Deposit of Poisonous Wastes Act*, there had been no specific legislation in the UK dealing with the management of wastes on land, although local authorities had for many years had powers to control waste in the interests of public health.

The *Public Health Act 1936* had given local authorities powers to remove house and trade refuse and to require removal of any accumulation of noxious matter. It also placed a duty on them to inspect their areas "for accumulation or deposit which is prejudicial to health or a nuisance", with powers to serve abatement notices and to prosecute offenders. *The Town and Country Planning Act 1947* included a requirement for waste disposal sites to have planning permission.

The Deposit of Poisonous Wastes Act 1972 had made it an offence to deposit on land any poisonous, noxious or polluting waste in a manner likely to create an environmental hazard. The Act also required that the waste disposal authority or water authority be given prior notification of the composition, quantity and destination of the waste. This Act was repealed in 1981 and replaced by the *Control of Pollution (Special Waste) Regulations 1980* made under s.17 of the *Control of Pollution Act 1974* (superseded by 1996 and then 2005 Regulations, see 5.23).

Stricter controls governing the management and disposal of waste were put in place through Part I of the *Control of Pollution Act 1974*, which introduced licensing with the main objective being to ensure that licensed activities did not cause pollution of water, danger to public health or detriment to local amenities. This in turn was superseded by Part II of *the Environmental Protection Act 1990* (in England, Scotland & Wales) and similar Regulations in Northern Ireland made under *the Waste and Contaminated Land (Northern Ireland) Order 1997*. These introduced stricter licensing controls and other provisions aimed at ensuring waste handling, disposal and recovery operations do not harm the environment. It also covers hazardous or "special" waste – i.e. waste that is difficult or dangerous to dispose of – for which special provisions are made. Part II of the EPA (NI, WCLO, Part II) also introduced a duty of care for producers and handlers of waste; the separation of local authority regulatory and disposal functions; and amended waste disposal planning requirements and licensing controls.

In this chapter of the *Pollution Control Handbook*, unless otherwise stated, the term "regulatory authority", means the Environment Agency in England and Wales, the Scottish Environment Protection Agency (SEPA), or the Waste and Contaminated Land Inspectorate of the Northern Ireland Environment Agency.

5.15 OVERVIEW OF REGULATORY CONTROLS

5.15.1 England & Wales

In England and Wales the management of waste operations has from 6 April 2008 been regulated through *the Environmental Permitting (England & Wales) Regulations 2007*; these Regulations also provide for implementation of various waste-related Directives, including those on landfill, waste incineration, end-of-life vehicles and waste electrical and electronic equipment – see chapter 2, 2.4.

Sections 35-43 (waste management licensing, ss.64-66 (public registers) and s.74 (fit and proper persons) of the EPA, no longer apply in E & W (EPR, Sch. 21, paras, 10 and 17).

The Waste Management (England and Wales) Regulations 2006 (SI 2006/937), which came into force on 15 May 2006, amend the EPA to define agricultural waste and non-mineral waste from mines and quarries as controlled waste, and confirm their classification as industrial waste, thus applying waste management controls to these wastes – see 5.19 below. They also amend s.33 of the EPA to require householders to dispose of their waste in a way which does not harm the environment, and s.34 is amended to extend the duty of care to householders to hand over their waste only to authorised persons.

The Environment Agency has prime responsibility for enforcing waste management activities through the environmental permitting regime (see chapter 2, 2.4), as well as for controls on the handling, transfer and transport of waste, and packaging waste, which are all covered in this chapter.

5.15.2 Scotland

Part II of the *Environmental Protection Act 1990* (EPA) (ss.29-78), provides the framework for the control of waste management activities in Scotland, with the detailed implementation set out in *the Waste Management Licensing Regulations 1994*, Responsibility for waste rests on the person who produces it and everyone who handles it, right through to final disposal or reclamation. Only "fit and proper" persons may run waste sites and responsibility for a closed landfill site continues until all risks of pollution or harm to human health and safety are past. Certain waste activities (including incineration) require an IPPC permit as a result of implementation of the *EU Directive on Integrated Pollution Prevention and Control* (see chapter 2, 2.1 and 2.5). All landfill activities require a permit to comply with both the *IPPC Directive and the Landfill Directive* (see 5.7 and 5.29), and incineration plant will also need to comply with the *Waste Incineration Directive* (see 5.8 and 5.30).

The Waste (Scotland) Regulations 2005 (SSI 2005/22), which came into force on 21 January 2005, amend the EPA to define agricultural waste and non-mineral waste from mines and quarries as controlled waste, and confirm their classification as industrial waste, thus applying waste management controls to these wastes – see 5.19 below; s.33 of the EPA is amended to require householders to dispose of their waste in a way which does not harm the environment, and s.34 is amended to extend the duty of care to householders to hand over their waste only to authorised persons. The Regulations also make various amendments to the waste management licensing regulations and amend the *Groundwater Regulations 1998* to bring waste management licensing into line with other types of authorisations for the purposes of complying with the Groundwater Directive.

The Scottish Government and SEPA issued a joint consultation paper – *Better Waste Regulation* – in March 2007 seeking views on whether current regulations governing waste management need to be improved, in particular to ensure Scotland can meet its obligations under various EU Directives and to meet its targets for reducing landfill and increase recycling rates. This initial consultation closed on 22 June 2007.

The Scottish Environment Protection Agency is responsibility for the regulation of waste management operations.

5.15.3 Northern Ireland

Under *the Pollution Control and Local Government (NI) Order 1978* district councils were responsible for licensing private waste disposal facilities, monitoring special waste and transfrontier shipment of waste movements and for drawing up ten year waste disposal plans for their area, as well as for the collection and disposal of waste.

The Waste and Contaminated Land (Northern Ireland) Order 1997 introduced similar controls to those applying in the rest of the UK. Reference to the appropriate article of the Northern Ireland legislation is given throughout this chapter and is shown as follows: (NI: WCLO, Art.0). These articles have now been given effect through *The Waste Management Licensing Regulations (Northern Ireland) 2003* (SR 2003/493), which came into force on 19 December 2003. The regulatory authority for Northern Ireland is the Department of the Environment (through the Northern Ireland Environment Agency, which was established 1 July 2008, replacing the Environment and Heritage Service).

The Waste (Amendment) (Northern Ireland) Order 2007 (SI 2007/611, NI 3), made on 6 March 2007, strengthens the powers of the regulatory authority to deal with the illegal transport and disposal of waste; its provisions, many of which have now been brought into force by Order, have been incorporated into the text below.

The Waste Management (Northern Ireland) Regulations 2006 (SR 2006/280) amend the WCLO to define agricultural waste and non-mineral waste from mines and quarries as controlled waste and are similar to Regulations applying elsewhere in the UK – see 5.19.

District councils are responsible for the collection and disposal of waste in accordance with Part 2 of *the Pollution Control and Local Government (NI) Order 1978*, (SI 1978/1049) – *the Waste Collection and Disposal Regulations (NI) 1992* (SR 1992/254 as amended, SR 1997/52). Article 30 of the WCLO requires district councils to draw up waste management plans for their area (see 5.22.2).

5.16 OFFENCES

Prohibition on Unauthorised or Harmful Depositing, Treatment or Disposal of Waste

Under section 33(1) of the EPA (NI: WCLO, Art.4)

(a) it is an offence to deposit, knowingly cause or permit the disposal of controlled waste on land without a waste management licence;

(b) any waste to be deposited must be treated, kept or disposed of in accordance with the terms of the waste management licence, and

(c) waste must be dealt with in such a way that it is unlikely to cause pollution or harm to human health.

The Waste Management (E&W) Regulations 2006 (SI 2006/937) and *The Waste (Scotland) Regulations 2005* (SSI 2005/22) amend the EPA to apply sub-section 33(1)(c) to household waste which is treated, kept or disposed of by a private individual within an area attached to a domestic property (sub-s.33(1)(c) already applied to clinical waste, asbestos, or any mineral or synthetic oil or grease by virtue of the *Controlled Waste Regulations 1992*). However, the whole of sub-section 33(1) applies to the treatment, keeping or disposal of household waste by an establishment or undertaking. An offence under this section is punishable on summary conviction by a fine not exceeding the statutory maximum (currently £5,000) and on conviction on indictment a fine; the maximum amount that a Magistrates Court may order a private individual (i.e. a person who is not an establishment or undertaking) to pay in clean-up costs is £5,000.

It is a defence to prove that an act (e.g. unauthorised deposit, treatment or disposal etc) in relation to the waste was done in an emergency and appropriate steps taken to minimise environmental pollution and harm to human health, and that full details of the act were given to the waste regulation authority as soon as practicable (*Env. Act 1995*, Sch. 22, para 64). Section 40 (effective from 7.06.05) of *the Clean Neighbourhoods & Environment Act 2005* (CN&EA – E&W), removed the defence of acting under an employer's instructions (s.33(7)(b)); Article 3 of *the Waste (Amendment) (NI) Order 2007*, brought into force 25.06.07, makes similar provision.

Dumping of any waste (including hazardous waste) or treating it without a licence may result in a maximum fine of £50,000 (E&W, CN&EA; NI, Art.4, 2007 Order/SR 2007/294); Sc. £40,000, ASB Act 2004) and/or six months in prison on summary conviction or an unlimited fine and/or a maximum five years in prison on conviction on indictment. Where a person has been convicted of an offence under s.33, s.33A (inserted by s.42 of the CN&EA) enables the court to make an order requiring the offender to pay the enforcing authorities' investigation and enforcement costs, including those resulting from seizure of vehicles involved in the offence; s.33B (inserted by s.43 of the CN&EA) enables the court to make an order requiring an offender to pay the regulatory authority, or the owner or occupier of the land for any costs involved in removing the illegal waste and for any action taken to eliminate or reduce the consequences of the illegal deposit. Section 33C (inserted by s.44 of the CN&EA) enables the court to make an order depriving an offender of his rights to the vehicle used for the offence. Sections 33A-C were brought into force on 18 October 2005. Defra Guidance on the CN&EA (2006) explains the new powers. Articles 6-9 of the 2007 NI Order insert new provisions into the WCLO 1997 to make similar provision in NI; Arts. 6, 7 & 8 were brought into force by SR 2007/294.

Section 33 has been amended in E & W by *the Environmental Permitting Regulations (England and Wales) 2007* to reflect those Regulations (EPR, Sch.21, paras 4-7).

In Scotland, *The Controlled Waste (Fixed Penalty Notices) (Scotland) Order 2004* (SI 2004/426), which came into force on 5 November 2004, enables authorised officers of the local authority, a constable, or an authorised officer of a waste regulation authority to issue a fixed penalty notice in respect of an offence under s.33a. This allows the offender 14 days to pay the fine stipulated in the notice, during which time proceedings in relation to the offence will not be commenced.

5.17 DUTY OF CARE

Section 34 of the EPA, aimed at curbing the illegal disposal of controlled waste, has been implemented through Regulations which came into force on 1 April 1992 (E, S & W). Regulations (SR 2002/271) and a code of practice, implementing the duty of care in Northern Ireland (WCLA, Art.5) were brought into operation on 1 October 2002; they are similar to those in force in the rest of the UK.

A legal duty of care is imposed on anyone – from producers, to carriers and disposers of waste – to ensure that

- waste is not illegally disposed of or dealt with without a licence or in breach of a licence or in a way that causes pollution or harm;
- waste does not escape from a person's control;
- waste is transferred only to an "authorised person", such as a local authority, a registered carrier or a licensed disposer;
- when the waste is transferred, it is accompanied by a full written description so that each person who has it knows enough to deal with it properly and thus avoids committing an offence under s.33 of the Act.

New s.34(A) of the EPA – inserted by s.45 of the CN&EA and brought into force in both England & Wales in 2006 (SI 2006/1361; WSI 2006/768) – enables the Environment Agency and waste collection authorities to issue a fixed penalty notice to a person who has not complied with a request from the enforcement authority to furnish it with any relevant documents; the fixed penalty is set at £300 though the enforcing authority may substitute a different amount; it may also make provision for accepting a lesser sum in discharge of the offence if paid within a specified shorter period. Section 46 of the CN&EA inserts new sections 34B & C, which will enable enforcement authorities to stop, search and seize vehicles which they believe have committed an offence under either ss. 33 or 34 – the procedures are the same as those under s.38 of the CN&EA which replace s.5 of *the Control of Pollution (Amendment) Act 1989* – see 5.23 below. Defra Guidance on the CN&EA (2006) also explains the new powers.

Articles 5, 8 and 9 *the Waste (Amendment) (NI) Order 2007* insert new Articles 5A, D, E & F into the WCLO to make similar provisions for NI. (Art.8 was brought into force by SR 2007/294, 25.06.07)

The duty of care applies to the Crown and since 1 October 1995 has applied to scrap metal yards. Hazardous waste (Sc. special waste) is covered by the duty of care; such waste must of course also comply with the requirements of the *Hazardous Waste Regulations 2005* (Sc. *Special Waste Regulations 1996*, as amended) (see 5.15.9 below).

5.17.1 Duty of Care Regulations

The Environmental Protection (Duty of Care Regulations 1991 (SI 1991/2839) cover England*, Wales* and Scotland and came into force in April 1992 (Northern Ireland, SR 2002/271, effective 1.10.02, as amended SR 2004/277, effective 9.8.04). They provide for a mandatory system of signed transfer notes and require all those subject to the duty of care to keep records of waste received and transferred. Information to be recorded on the transfer note must include a full description of the waste – its special features, hazardous substances, source, how it was produced etc. Other details to be recorded include quantity of waste (by weight if possible), whether/how containerised, names and addresses of those involved in the transfer, and place and date of transfer. In summary, sufficient information should be given to ensure that anybody coming into contact with the waste has enough detail to deal with it properly and in compliance with the duty of care.

A transfer note, together with the full description of the waste must accompany each consignment of waste, using the relevant code from *the List of Waste Regulations 2005* (E: SI 2005/895; W: SI2005/1820; NI: SR 2005/301); Sc: amended Special Waste Regulations SSI 2004/112). Regular or multiple transfers of waste between the same people need not be individually documented provided that the details on the transfer note remain unchanged (amendment to EPA by s.33 of *Deregulation and Contracting Out Act 1994*); in the case of regular transfers, the transfer note should be renewed at least once a year. Records of all waste must be kept for at least two years.

Before transferring waste to a waste carrier it is important to check that a valid registration certificate is held; waste holders should assure themselves that their waste falls within the scope of the licence or exemption at the site to be used.

Where a waste regulation authority (i.e. Environment Agency or SEPA) wishes to assure itself that the duty of care is being complied with, or to follow up a suspected unlicensed dumping of waste, it may serve notice on a waste holder requesting copies of relevant documents within seven days. *The Environmental Protection (Duty of Care) (England) (Amendment) Regulations 2003* (SI 2003/63), which came into force on 20 February 2003, give waste collection authorities a similar power, which will enable them to investigate fly-tipping incidents more easily. Similar Regulations (SI 2003/1720, W.187) came into force in Wales on 31 July 2003, and in Scotland (SSI 2003/533) on 1 December 2003.

The Waste (Household Waste Duty of Care) (England & Wales) Regulations 2005 (SI 2005/2900), which came into force on 21 November 2005, require occupiers of domestic properties in England to take reasonable measures to ensure that their household waste is transferred only to an authorised person (e.g. registered waste carrier); failure to do so may result in a fine up to £5,000. Despite their name, these Regulations do in fact only apply to domestic properties in England, with WSI 2006/123, W.16, which came into force on 26 January 2006, covering Wales; *The Waste (Scotland) Regulations 2005* (SSI 2005/22), which came into force on 21 January 2005, include a similar provision. This extends the previous duty which only applied to householders who disposed of household waste from a neighbour's property and to builders disposing of rubble etc from a house in which they are working.

The Environmental Permitting (England and Wales) Regulations 2007 amend the 1991 Duty of Care Regulations to apply the duty of care to waste activities regulated under the EPR (EPR, Sch. 21, para 32).

**The 1991 Duty of Care Regulations* are to be revoked in England and Wales, together with the 2003 Amendment Regulations (England) once the proposed *Waste Controls (England and Wales) Regulations 2009* have been brought into force – see 5.34.3 below. These Regulations will provide for the registration of waste brokers and dealers, and of waste carriers, transfer notes for controlled waste, and seizure of vehicles used for the illegal transport of waste.

5.17.2 Duty of Care Code of Practice

A statutory code of practice (1991, updated 1996) giving practical guidance on how to discharge the duty of care has been drawn up by the Department of Environment (now Defra) and the Welsh Office, and a similar code for Scotland by the Scottish Development Department (NI: 2002, modified 2004). These give step by step advice on following the duty of care and what to do if it is suspected that the waste does not correspond to its description or is being handled incorrectly. The code also outlines the law with regard to the duty of care and the regulations on keeping records.

Breach of the code is not an offence (although breach of the duty of care is). It is however admissible in court as evidence in deciding if a breach of the duty has occurred. An offence under this section of the Act may result in a £5,000 fine on summary conviction, or an unlimited fine on conviction on indictment.

A joint circular from the Department of Environment (19/91), Welsh Office (63/91) and Scottish Office (25/91), gives advice on the operation of the duty of care. It complements the advice for producers and holders of waste given in the code of practice.

5.18 WASTE MANAGEMENT LICENSING

> **England and Wales:** From April 2008, waste management activities have been regulated through *the Environmental Permitting (England and Wales) Regulations 2007*, which came into force on 6 April 2008 – see chapter 2, 2.4. Existing waste management licences automatically became environmental permits under the new Regulations without the need for a new application. Applications for permits or waste management licences made prior to the new Regulations taking effect are determined under the "old" system, and once determined will then be treated as an environmental permit under the "new" system. Transitional arrangements also provide for a "holder" of an existing waste management licence to be considered as the "operator" under the EPR – on the occasion of a licence (i.e. permit) transfer, it will need to be transferred to the person who has control of the facility (the operator). Sections 35-43 of the 1990 EPA have been repealed in England and Wales (EPR, Sch. 23), and regs. 2-19 and Sch. 1-3 of the *Waste Management Licensing Regulations 1994* omitted (EPR, Sch. 21, para 34).

Sections 35-43 of *the Environmental Protection Act 1990*, relating to waste management licensing now only apply in **Scotland**, and have been implemented through the *Waste Management Licensing Regulations 1994* (as amended) which came into force on 1 May 1994. They replace and repeal ss.3-11 of the *Control of Pollution Act 1974* which covered site licensing, and under which almost anybody could get a site licence provided planning permission had been obtained for use of the land. Changes to procedures and the regulation of waste management licensing as a result of amendments to the EPA in Schedule 22 to the *Environment Act 1995* are reflected in the text below.

In **Northern Ireland**, Articles 6-18 of the *Waste and Contaminated Land (Northern Ireland) Order 1997* were brought into force on 27 November 2003 (SR 2003/489), and given effect through *the Waste Management Licensing Regulations (Northern Ireland) 2003* (SR 2003/493), which came into force on 19 December 2003.

The Regulations set out the procedure for obtaining a licence and also deal with revocations and suspensions of licences, appeals, public registers (EPA: ss.64-66; NI: WCLO, Art.34-36) and the definition of fit and proper persons (EPA: s.74; NI: WCLO, Art.3). The Regulations list the offences under which if a person has been convicted they would not be considered a fit and proper person to be a licence holder. They also introduce a system for the registration of waste brokers – see below 5.34. The Northern Ireland Regulations also partly transpose the End-of Life-Vehicles Directive (see 5.31).

The Regulations have been amended a number of times:

- ***SI 1995/288 (E, W, Sc):*** extend the licensing regulations to the scrap metal industry.
- ***SI 1996/634, SI 1997/2203 (Sc only):*** largely concerned with amending the provisions relating to qualifications for fit and proper persons (see below, 5.26).
- ***SI 1998/606, (E, W, Sc):*** relate to registration of waste brokers and charging schemes for registration of scrap metal recovery and vehicle dismantling sites (see 5.34 below).
- ***SSI 2003/171 (Sc):*** amend the definition of scrap metal dealer and amend the qualification requirements for persons to be regarded as technically competent for the purposes of s.74 of the EPA (see 5.26 below).
- ***SI 2006/937 (E&W):*** amend the Regulations to apply controls to agricultural waste and to waste from mines and quarries (see 5.19 below); reg.15 regarding authorisation of activities impacting on groundwater is revoked. SI 2007/1156 amend the 1994 Regulations to add further exemptions for agriculture wastes.
- ***SSI 2003/171 (Sc):*** also amend the definition of "environmental licence" in the *Environment Act 1995* and extend powers of SEPA to charge for the registration of certain exempt activities under s.41 – see also (c) below; as from 1 October 2003, 21 days' notice had to be given before land spreading of wastes, together with details of the types and amounts of waste.
- ***SSI 2004/275 (Sc):*** make further amendments to the definition of environmental licence to charge for the registration of certain exempt activities and impose thresholds on some activities which will now require a full waste management licence.
- ***SSI 2005/22 (Sc):*** amend the WMLR to include a definition of mines and quarries waste, revoke reg.15 which made provision for authorisation of WML activities impacting on groundwater, and provide for an exemption from WML for certain agricultural activities.
- ***SSI 2006/541 (Sc):*** amend the 1994 regulations to insert definitions of the various incineration plant and of WEEE; re-enact with various amendments the regulation relating to exempt activities and substitute a new Sch. 3 (exempt activities) and Sch.3A (plans and documents needed for registration of certain exempt activities) into the Regulations. They also amend the Regulations to provide that the deemed authorisation for the purposes of the *Controlled Activities Regulations* (see chapter 6) of an exempt activity which is a controlled activity does not apply to certain works carried out in or around surface waters.
- ***SSI 2007/172 (Sc):*** amend the 1994 regulations to facilitate the granting or variation of WMLs in respect of WEEE and to require SEPA to ensure that the requirements of the WEEE Directive for the storage and treatment of WEEE are complied with.
- ***SR 2006/519 (NI):*** transpose the permit requirements of the WEEE Directive and amend the 2003 WML(NI) Regulations to provide for exemptions from permit requirements for treatment of WEEE; they also amend reg.3 and Sch.1 of the 1994 Regulations to provide for new exemptions for the treatment and storage of WEEE.

The 1997 *Waste Management (Miscellaneous Provisions) Regulations* (SI 1997/351) make evasion of landfill tax a "relevant offence" for the purposes of whether someone is "fit and proper" to hold a waste management licence (EPA, s.74; see 5.26 below).

Fees and charges in relation to waste management licences in Scotland and NI are fixed in accordance with ss.41-43 of the *Environment Act 1995* (NI: WCLO, Art.15) – see 5.27 below.

Future Developments – Scotland

In August 2008 SEPA published proposals for revising controls on waste management in Scotland. SEPA's *Better Waste Regulation Action Programme* outlines a number of actions for consideration, including

- consolidation of the *1994 Waste Management Licensing Regulations* and subsequent amendments;
- review of exemptions from waste management licensing, and faster system for adding new exemptions;
- revisions to various elements of the fit and proper persons test (financial provisions; certificates of technical competence and relevant convictions);
- review of charging scheme
- review of special waste regulations and their implementation;
- review of duty of care regime and registration of waste carriers.

5.18.1 Requirement for a Licence (Sc, NI)

The deposit, keeping, treatment or disposal of controlled waste (i.e. Directive Waste – see above 5.6 – in or on land (including treatment or disposal in mobile plant) requires a waste management licence. There are various exemptions (see below). It should be noted that certain waste installations – including all landfill sites receiving controlled waste – will fall within the scope of the IPPC Directive and will thus require a permit under *Pollution Prevention and Control Regulations* (SSI 2000/323; SR 2003/46 (see chapter 2, 2.5); landfill sites also need to comply with the *Landfill Regulations* (see 5.29). From 1 January 2006 waste management licences for facilities wishing to continue to treat waste electrical and electronic equipment had to comply with the requirements of the WEEE Directive (see 5.12 and 5.32), with conditions ensuring that WEEE and its parts and components are recovered as far as possible; operators wishing to register an exempt activity had to do so by 31 March 2006.

Since April 2006 operators of mobile plant in Scotland have only needed a single licence authorising several pieces of mobile plant, used either singly or in combination, for remediation operations carried out at the same time but at different sites.

The licence is issued by the waste regulation authority within which the waste disposal facility is situated (i.e. the appropriate office of SEPA or NIEA). In the case of mobile plant, this will be the office in which the operator's principal place of business is located.

5.18.2 Exclusions from Licensing (Sc, NI)

Certain activities do not require a waste management licence where these are part of a process registered or authorised under other legislation. These are:

- recovery or disposal of waste from an installation permitted under PPC;
- disposal of waste from an incinerator permitted under PPC;
- disposal of liquid waste under a consent for discharge to water (but not to a sewer) granted under Part III of the *Water Resources Act 1991*;
- recovery or disposal of waste where activity forms part of operation requiring licence under Part II of the *Food and Environment Protection Act 1985* (dumping at sea).
- operators of waste effluent treatment plant where disposal is subject to a discharge consent under Part III of the *Water Resources Act 1991*, or a discharge consent under Part IV of the *Water Industry Act 1991*;
- discharges of effluent covered by the *Urban Waste Water Treatment Regulations 1995* (in both Scotland and Northern Ireland) – see chapter 6, 6.20.

The Waste Management Licensing (Water Environment) (Scotland) Regulations 2006 (SSI 2006/128) amend the 1994 WML Regulations to exclude from the need for a waste management licence the disposal of agricultural waste in or under land granted, or deemed to have been granted, an authorisation under *the Controlled Activities Regulations 2005* (SSI 2005/348). (SR 2006/280 amend WMLNI 2003 to similar effect.)

These exclusions, which do not have to be registered (see next section) also apply where the activity involves hazardous (Sc. special) waste; however a waste management licence will be required if the activity involves the final disposal of waste in or on land.

5.18.3 Exemptions from licensing

Schedule 3 to the WML Regulations, as amended, lists the activities for which a waste management licence will not normally be required. These mainly relate to small-scale waste storage and waste recovery operations, though it should be noted that an activity falling into one of the exempted categories will not necessarily be exempt and thus the Schedule to the Regulations, as amended, should be consulted for details of any limitations or other requirements affecting the exemption. To qualify for exemption the waste operation must be carried out without endangering human health, and without using processes or methods likely to harm the environment; it must not cause noise or odour nuisance or adversely affect the countryside or places of special interest. In most cases the exemptions do not apply to hazardous (Sc. special) waste and the person carrying on the exempt activity must have the permission of the occupier of the land or be entitled to use the land for the activity.

All establishments or undertakings (i.e. not private individuals) carrying on an exempted activity must register with the appropriate registration authority (SEPA); non-registration of an exempt activity is an offence, subject to a fine. The relevant authority usually inspects registered sites once a year. The register, which is open to the public, has details of the establishment or undertaking, the exempt activity and where it takes place.

Registered establishments or undertakings must renew their registration annually by providing the registration authority with a notice confirming that they wish to continue to carry on the exempt activity; details of the quantity of waste to be disposed of or recovered should be provided, as well as any other information required by the registration authority.

In Scotland, 2003 Amendment Regulations (SSI 2003/171), which came into force on 1 April 2003, require that from 1 October 2003 establishments or undertakings proposing to recover waste for the purposes of land treatment, or for use in land reclamation and construction, must notify SEPA at least 21 days before the activity is due to begin; the notice should include details of the types and amounts of waste to be recovered and be accompanied by the appropriate fee. Where relevant, evidence must be provided that the activity will lead to agricultural benefit or ecological improvement. SEPA also has powers, under certain circumstances, to refuse to register an exemption or to remove an exemption from the register. 2004 Amendment Regulations (SSI 2004/275) make further amendments to the schedule of exempt activities now requiring notification to SEPA and registration, and alter the thresholds at which some will now require a full waste management licence.

*The Waste Management (**England and Wales**) Regulations 2006* (SI 2006/937) amend Sch.3 of the 1994 Regulations to include those activities in relation to agricultural waste exempt from WML (with further amendments made by SI 2007/1156); the Regulations are also amended to require records to be kept of the quantity, nature, origin and, where relevant, destination and treatment method of all such waste recovered. 2006 Amendment Regulations in ***Scotland*** (SSI 2006/541) re-enact with various amendments the registration requirements for both existing and new exempted activities and substitute a new Sch.3 of exempted activities and Sch.3A (plans and documents needed for the registration of certain exempted activities).

In ***Northern Ireland***, waste operators were required to register exemptions by 19 July 2004; registrations must be renewed either annually or every three years (depending on the activity concerned), with a fee being charged according to the activities exempted. The regulator may refuse or revoke a registration; where appropriate a certificate must be submitted showing how the activity will lead to agricultural or ecological improvement. *The Waste Management Regulations (NI) 2006* (SR 2006/280) amend the 2003 WMLNI Regulations to cover agricultural waste. *The Waste Management (Miscellaneous Provisions) Regulations (NI) 2008* (SR 2008/18) provide exemptions for the disposal of various agricultural wastes and for the treatment of land with those agricultural wastes subject to certain conditions and evidence that such treatment will result in benefit to agriculture or ecological improvement.

Future Developments – Scotland

A joint consultation paper published by the Scottish Government and SEPA in March 2007 – *Better Waste Regulation* – seeks views on giving SEPA powers to make new exemptions from WML (rather than currently via the Scottish Government and Parliamentary process). This, it is suggested, would enable the system for amending existing exemptions or adding new ones to be both simplified and made considerably quicker. It is also proposed to enable operators to appeal against a decision to refuse an exemption or to remove an existing exemption from the register for the more complex exemptions (e.g. those relating to composting and recovery to land and building operations). Views are also sought on whether to relax the existing requirement for SEPA to inspect exempted activities annually to require "periodic" inspection.

5.18.4 Licence Applications (Sc, NI)

Applications must be made on a form provided by the appropriate Agency (*Env. Act 1995*, Sch. 22, para 68(2), brought into effect 1 April 1998; WCLO, Art. 8(1)); as well as general details of the site and of the operator, the application should include:

- details of the nature and quantities of wastes to be dealt with;
- details relating to the operation and management of the site;
- site specific risk assessment (see Environment Agency publication, *A practical guide to environmental risk assessment for waste management facilities*, 2000);
- risk assessment relating to odour from the site (see Environment Agency *Guidance for regulation of odour at waste management facilities*, v.3.0, July 2002);
- any other information required by the regulatory authority.

The licence application should demonstrate that the applicant is a "fit and proper" person (EPA, s.74 – 5.26 below), and should be accompanied by the appropriate fee (*Env. Act 1995*, s.41 – see 5.27 below).

5.18.5 Consideration of Applications (Sc, NI)

The regulatory authority may refuse to deal with an application if the information which it requires to accompany the application has not been provided, or until the information is provided (s.36(1A), see *Env. Act 1995*, Sch. 22, para 68(2); WCLO, Art.3) The regulatory authority has four months (or longer if mutually agreed) in which to consider the application – the four months beginning once all the information has been received (ss.36(9) & 36(9A) – *Env. Act*, Sch. 22, para 68(5); WCLO, Art.8(6)). If at the end of that time the regulatory authority has neither advised the applicant of acceptance of the application, nor that it intends to reject it, then the application is deemed to have been refused.

The regulatory authority may only reject a licence in order to prevent pollution of the environment, harm to human health or serious detriment to local amenities – thus meeting a prime objective of the EU Framework Directive, included in Sch. 4 of the Regulations, that waste must be disposed of without endangering human health and "without risk to water, air soil and plants and animals". A licence application may also be rejected if the regulatory authority is of the opinion that the applicant is not a fit and proper person or has insufficient financial resources to comply fully with licence conditions. The applicant has a right of appeal to Scottish Ministers [NI: Planning Appeals Commission] with regard to any refusal of licence (EPA, s.43, & *Env. Act 1995*, s.114; NI: WCLO, Art.17 – see 5.18.10 below).

If SEPA plans to grant a licence it must first consult both the appropriate planning authority and the Health and Safety Executive and, if any of the land falls within a Site of Special Scientific Interest, Scottish Natural Heritage. All consultees have 28 days from the day on which the regulatory authority received the application in which to submit any comments (*Env. Act 1995*, Sch. 22, para 68(6)). In Northern Ireland the regulatory authority is required to consult such bodies or persons it considers appropriate (WCLO, Art.8(5).

Licences must ensure compliance with the EU's Framework Directive on Waste; if granted, the licence will include conditions relating to types and quantities of waste covered by the licence, treatment methods, operation and management, hours of operation, record keeping and security provisions. The conditions will also cover standards to be achieved and measures to ensure site activities cause no pollution or other harm to the environment or human health. If necessary conditions to control noise impacts on the surrounding environment will also be included.

Licence conditions will remain in force even after closure of the site and until a certificate of completion has been issued. In addition site operators will also be required to provide financial guarantees covering care of the site following its closure and until surrender of the licence is accepted – see also 5.18.8 below: surrender of site licence.

It is an offence to breach any of the conditions of a waste management licence. Summary conviction may result in a fine of £20,000 and/or six months in prison, or an unlimited fine and/or two years in prison on conviction on indictment.

Details of licence applications, and of any comments on the application, as well as full details (subject to certain exemptions) of all licences, conditions, etc must be kept by the regulatory authority on a register open to public inspection (see below, 5.24).

Licence conditions will normally be reviewed annually to ensure they remain appropriate to the activities carried out at the site. Unless otherwise revoked or surrendered, the waste management licence and any accompanying conditions apply to closed disposal sites, with operators retaining a duty of care for them.

The *Environment Act 1995*, Sch. 22, para 69 (brought into force on 1 April 1998), adds new s.36A to the EPA: if the regulatory authority plans to issue a licence which includes a condition requiring work which the applicant may not be entitled to do, it should notify all those with an interest in the land (e.g. the owner, lessee or occupier) who will be required to grant rights to the licence holder to carry out such works; the notice should set out the condition, the work required in connection with it, by when it has to be done and the date by which any representations must be received (the length of time allowed for such representations to be set in Regulations); the regulatory authority must take any such representations into account. A similar requirement is imposed by Article 9 of the *Waste and Contaminated Land (Northern Ireland) Order 1997* (brought into force 03.04.08, SR 2008/138). The *Waste Management Licences (Consultation and Compensation) Regulations 1999* (SI 481), made under ss.35A, 36A & 37A (paras 67, 69 & 71 of *Env. Act* Sch. 22), came into force in Scotland on 1 April 1999 (revoked in E & W 06.06.08 by *Environmental Permitting Regulations*). They specify a consultation period of three months in the case of a new licence and six weeks (see below) in the case of a variation. The Regulations also specify the circumstances under which compensation may be payable for loss or damage to those granting rights; a claim should be made within 12 months of the date in which entitlement to compensation arises or within six months of the licence holder exercising their rights. Similar Regulations apply in NI – SR 2008/160, which came into force on 19 May 2008.

5.18.6 Variation of Licence (Sc, NI)

Section 37 of the EPA (NI: WCLO, Art.10) places a duty on the regulatory authority to take the necessary steps to ensure that licence conditions are being met and that licensed activities are not causing any pollution or harm to human health; if necessary the regulatory authority may by notice modify the licence conditions, stating when the modification is to be effected. Licence holders may also apply (with the appropriate fee) for licence conditions to be varied. However, if after two months (or longer as mutually agreed), the regulatory authority has neither approved nor rejected the application for modification, it must be assumed that it has been rejected. In all instances there is a right of appeal to Scottish Ministers [NI: Planning Appeals Commission].

New s.37A (*Env. Act 1995*, Sch. 22, para 71, brought into force on 1 April 1998) (NI: WCLO, Art.11, brought into force 03.04.08) requires the regulatory authority to consult anyone with an interest in the land where it proposes to vary a licence by adding a new condition requiring certain works which the site operator may not be entitled to do. In such instances the regulatory authority should serve a notice on the interested persons setting out the new condition and the work which might be required to meet it. The notice should specify the date by which any representations should be received (the period allowed for such representations to be set in Regulations); any representations received must be taken into account before finally issuing the licence variation.

5.18.7 Revocation and Suspension of Licence (Sc, NI)

Where the regulatory authority has reason to believe that the licence holder has ceased to be a fit and proper person, that licence conditions are being breached or that the activities covered by the licence are, or are likely to, cause serious harm to human health or the environment, it may serve notice that it intends to suspend or partially revoke the licence (EPA: s.38; NI: WCLO, Art.12). In suspending, or partially suspending a licence, the regulatory authority may require the licence holder to take the necessary steps to deal with or avert the pollution; s.36A of the EPA (see above) – consulting those with an interest in the land – applies in this instance, though such consultation or notification may be postponed in an emergency (EPA s.38(9A-C), *Env. Act 1995*, Sch. 22, para 71(1)). If emergency remedial action is required, the regulatory authority may carry out the necessary work and recover the expenditure from the licence holder, or in the case of a surrendered licence from the former licence holder.

Notices suspending, or partially or totally revoking a licence, or modifying its conditions to prevent pollution or harm take immediate effect, whether or not an appeal has been lodged (see below). Failure to pay fees or charges also results in immediate revocation.

Where a licence has been suspended or revoked, and action required by the regulatory authority for remedying a situation has not been carried out, it may prosecute the licence holder, who on summary conviction may be liable to a maximum fine of £20,000 (and/or six months' imprisonment if special waste involved); on conviction on indictment, the offender becomes liable to imprisonment for up to two years (five years in relation to special waste) and/or an unlimited fine.

Copies of all enforcement notices and convictions will be put on the public register.

5.18.8 Surrender of Site Licence (Sc, NI)

Section 39 of the EPA, as amended by the *Environment Act 1995* (Sch. 22, para 73), In (NI, WCLO, Art. 13) applies to all sites for which a waste management licence is required. An application to surrender a licence must be made on a form provided by the regulatory authority, together with any information and evidence required by them, and the appropriate fee.

Prior to making such an application, however, and despite the fact that the site is no longer in use, the site operator will need to maintain any pollution control systems and continue regular monitoring until the site has stabilised – this may take 30 years or more. Once this state has been reached, and in agreement with the regulatory authority, it is considered that the site no longer requires a pollution control system, a period of completion monitoring may be commenced to confirm the site's stability and safety. The regulatory authority will decide the period of completion monitoring – probably two years if regular monitoring data exists or five years in other instances.

Schedule 1 to the *Waste Management Licensing Regulations* (Sc.) specifies the information to be given on the application: this includes full details of the site location, licence number, activities carried out on the site and when and estimated quantities of each type of waste dealt with; details should also be given on landfill gas and leachate production and quality of surface water and groundwater; in the case of special wastes, plans should be included showing where the waste was deposited and information about possible contamination.

In considering whether to accept the surrender of a licence, the regulatory authority has a duty to inspect the site to determine whether the condition of the land is likely to cause pollution or harm to human health on the basis of deposits made during the lifetime of the licence. If the regulatory authority plans to accept an application for surrender, it must refer the proposal to the appropriate planning authority which has 28 days from the day on which the regulatory authority receives the application (*Env. Act 1995*, Sch. 22, paras 68(6) & 73(3)) to make any comments, or longer if mutually agreed. In NI, the DOE should consult such body or person it considers appropriate (WCLO, Art.13(6)). Where the regulatory authority has neither accepted nor rejected an application for surrender within three months of its receipt (or longer if mutually agreed), then it is deemed to have been rejected. If the appropriate planning authority wishes the application to be rejected because it is of the opinion that there is a risk of pollution, then either it or the regulatory authority may refer the matter to the Secretary of State for determination. Surrender of a licence for a closed site will only be accepted after the site has been certified safe and a "certificate of completion" issued.

Full details of applications to surrender a licence etc must be put on the public register kept by the regulatory authorities.

5.18.9 Transfer of Licences (Sc, NI)

Where it is proposed to transfer a licence (EPA, s.40 & Sch.2 to WML; NI: WCLO, Art.14), the current and proposed licence holders must make a joint application to the regulatory authority (on the form provided), together with the relevant fee. Information to be given should include full details of the licence and information to demonstrate that the proposed licence holder is a fit and proper person (EPA, s.74, 5.26 below). If within two months (or longer as mutually agreed) from the date of receiving the transfer application, the regulatory authority has neither agreed to nor rejected the transfer, then it is deemed to have been refused. Again, full details will be put on the public register.

5.18.10 Appeals (Sc, NI)

There is a right of appeal (EPA: s.43; NI: WCLO, Art.17) to Scottish Ministers (NI: Planning Appeals Commission) against a decision of the regulatory authority in the following instances:

- an application for a licence or modification of licence conditions is rejected;
- disagreement about licence conditions;
- disagreement over modification of licence conditions;
- a licence is suspended;
- a licence is revoked because it is thought the licence holder is no longer a fit and proper person or the licensed activities may cause environmental pollution or harm to human health and subsequent requirements to remedy the situation have not been carried out;
- rejection of an application for surrender of a licence;
- rejection of an application to transfer a licence.

There is also a right of appeal under s.66 of the EPA (NI: WCLO, Art.36(5)) where the regulatory authority rejects a request for information to be kept off the public register on the grounds of commercial confidentiality.

The written statement should include the reasons for the appeal, copies of relevant documents and a statement as to whether the appellant wishes the appeal to be determined by means of a hearing or by correspondence. Appeals relating to licences should be made within six months of the action precipitating the appeal; those relating to information for the public register should be made within 21 days of the regulatory authority's initial decision.

Details of the appeal etc should be copied to the regulatory authority who will put a copy on the public register. Copies of all appeal decisions will also be put on the public register.

5.18.11 Supervision of Licensed Activities (Sc, NI)

Section 42 of the EPA (as amended by the *Env. Act 1995*, Sch. 22, para 76) (NI: WCLO, Art.16) places a duty on regulatory authorities to ensure that licences and their conditions are being complied with and that authorised activities are not causing pollution, harm to human health or having a detrimental effect on local amenities.

If the regulatory authority is of the opinion that emergency work is needed to prevent pollution etc, it may carry out the work itself, recovering any expenditure from the licence holder or as the case may be the former holder of the licence.

Where the regulatory authority is of the opinion that a licence condition is not being complied with, or is likely not to be complied with, it should serve a notice on the licence holder stating this; the notice will detail the activities constituting the non-compliance or likely non-compliance, the remedial steps to be taken and the time within which those steps must be taken. Non-compliance with the notice may result in partial or total revocation of the licence or in its suspension; if it is felt that this would be an ineffectual remedy, further action may be taken through the Courts to secure compliance (EPA, s.42(6A), see *Env. Act 1995*, Sch. 22, para 76(7)). See also above, Revocation and Suspension of Licences and Appeals.

5.19 AGRICULTURAL WASTE

The disposal of agricultural waste is now controlled through *the Waste Management (England and Wales) Regulations 2006* (SI 2006/937) – also known as *the Agricultural Waste Regulations* – which came into force on 15 May 2006. There are similar regulations for Scotland – *The Waste (Scotland) Regulations 2005* (SSI 2005/22), which came into force on 21 January 2005 – and for Northern Ireland, *the Waste Management Regulations (Northern Ireland) 2006* (SR 2006/280), which came into force on 31 July 2006. The Regulations are enforced by the Environment Agency, SEPA and the Northern Ireland Environment Agency, as appropriate.

The main effect of the Regulations is to apply waste management, including licensing, controls to waste from agricultural premises – these include premises used for all types of farming, horticulture, fruit growing, seed growing, market gardens, nursery grounds and osier land. As such, they amend the EPA (NI WCLO) to define agricultural waste and non-mineral waste from mines and quarries as controlled waste, and confirm their classification as industrial waste.

They also amend s.33 of the EPA (Art.4, NI WCLO) to require householders to dispose of their waste in a way which does not harm the environment; s.34 (Art.5, NI WCLO) is amended to extend the duty of care to householders to hand over their waste only to authorised persons. The Regulations also make various amendments to the *Waste Management Licensing Regulations 1994* (NI 2003) and amend the *Groundwater Regulations 1998* (NI 1998) to bring waste management licensing into line with other types of authorisations for the purposes of complying with the Groundwater Directive. The landfill and hazardous waste regulations are also amended to bring agricultural and mines and quarries waste within their remit (with effect from 15 May 2007) – see 5.29 and 5.23.

With regard to the application of *the Hazardous Waste Regulations* to agricultural waste in England and Wales, the Environment Agency has produced guidance for farmers and growers, which includes examples of when their waste may be hazardous. All farmers and growers producing more than 500 kg or 500 litres of hazardous waste a year need to register with the Agency. Producers of smaller amounts, while not having to register, will still need to prepare a consignment note before any hazardous waste is removed from site.

Farm dumps or tips may no longer be used to dispose of waste; burning or burying farm plastics and other materials such as tyres is also banned, with the Regulations providing the following options (which can be used in combination) for dealing with farm waste:

1. Storage of waste on-farm in compliance with the duty of care (see 5.17 above) for up to 12 months – then recover or dispose of it using one of the options below.
2. Take the waste for recycling or disposal off-farm at a licensed site; farmers and growers transporting their own waste do not need to register as waste carriers. However if they wish to collect waste from another farm for recovery at their own farm or to take to a waste disposal facility, they will need to register with the Agency (free of charge) as a waste carrier. (In Scotland waste carriers, agents and brokers who deal only with agricultural waste do not have to register with SEPA).
3. Get an authorised waste contractor to take the waste away.
4. Register licence exemptions with the Agency to recycle or dispose of the waste on-farm from existing activities (required by 15 May 2007); exemption for new activities should be registered before commencing the activity. The Regulations amend Sch.3 of the 1994 Regulations to define those agricultural activities which are exempt from the licensing regime, with further exemptions listed in SI 2007/1156; registration of exempt activities is not subject to a fee (reg.3 amends s.41 of the Environment Act); records of any such waste recovered – including quantity, nature, origin and, where relevant, destination and treatment method – should be kept for at least two years; records do not however have to be kept in respect of activities carried out on land subject to an action programme under the Action Programme for Nitrate Vulnerable Zones (chapter 6, 6.12.1).
5. Apply to the regulatory authority for a waste management licence, or landfill permit, to recover or dispose of waste on-farm, in accordance with the procedures outlined at, in E & W, chapter 2, 2.4, or Sc. & NI, this chapter, 5.18 and 5.29.

A statutory producer responsibility scheme is to be introduced covering the collection and recovery/recycling of non-packaging farm plastics, such as silage wrap and horticultural film. Regulations will cover England, Scotland and Wales; the DOE, N. Ireland, is consulting on whether to set up a similar scheme for NI to or to join a UK-wide scheme. In the meantime the Agricultural Waste Plastics Collection and Recovery Progamme has issued general guidance to farmers on the handling, segregation and storage of agricultural waste plastics (December 2007) – available for download from www.agwasteplastics.org.uk.

The disposal of sheep dip and pesticide washings to land requires an authorisation under *1998 Groundwater Regulations* in E & W and NI (see chapter 6, 6.11.3), and in Scotland under *the Controlled Activities Regulations* (chapter 6, 6.11.6).

Further information and guidance on the Regulations and on dealing with agricultural waste is available via www.netregs.gov.uk.

5.20 NATIONAL WASTE STRATEGIES (E, W, NI)

In England and Wales, ss.44A-B of the EPA, together with Schedule 2A, were added by the *Environment Act 1995*, s.92 and Schedule 12. (It should be noted that the requirement for waste regulation authorities to prepare waste disposal plans under s.50 of the EPA was revoked by the *Env. Act*, Sch. 22, para 78 with effect from 1 April 1996.) Section 44A requires the Secretary of State to prepare a National Waste Strategy for England and Wales.

In Northern Ireland, the Department of the Environment has a similar obligation under Article 19 and Schedule 3 of *the Waste and Contaminated Land (Northern Ireland) Order 1997*, which was brought into force on 1 December 1998.

[Section 44B of the EPA, which placed a similar obligation on the Scottish Environment Protection Agency as regards Scotland, has been repealed by *the National Waste Management Plan for Scotland Regulations 2007* (SSI 2007/251) which came into force on 23 March 2007 – see 5.21 below.]

The Strategies, the preparation of which implemented Article 7 of the EU's Waste Framework Directive, should include policies for meeting the following objectives:

- ensuring that waste is recovered or disposed of without endangering human health or using processes or methods which could harm the environment (including causing noise or odour nuisance).
- establishing an integrated and adequate network of waste disposal installations taking account of BATNEEC; the network should also aim to help the EU meet its objective of becoming self-sufficient in waste disposal, and that waste is disposed of in the nearest appropriate installation.
- encouraging the prevention or reduction of waste through the development of clean technologies, the technical development and marketing of products with the least impact on the waste stream, and the development of techniques for the final disposal of dangerous substances in waste for recovery.
- encouraging recovery, reuse, reclamation, recycling etc and the use of waste as a source of energy.

The Strategy should also include details of the amounts of waste to be recovered or disposed of, general technical requirements and requirements for specific types of waste.

The Environment Agency must be consulted in the preparation or modification of the Strategy, together with representatives of local government and industry; others may be consulted as considered appropriate. DOE(NI) should consult similar bodies.

The Secretary of State may require the Environment Agency to advise on the policies to be included in the Strategy and to carry out an investigation and report on the amount and types of likely waste arisings, and the facilities which are or may be needed to deal with them. Before carrying out its investigation, the Agency should consult representatives of local planning authorities and of industry, and should report its findings to them. *Waste Management Planning - Principles and Practice*, (DOE, 1995), sets out good practice for the conduct and design of surveys to obtain key information on waste arisings and waste management facilities; the guidance suggests that surveys should be carried out every three years.

5.20.1 Waste Strategy for England

In May 2007, Defra published *Waste Strategy for England 2007*. This summarises progress since *Waste Strategy 2000*, sets new objectives and targets and outlines key proposals for action and for implementing the strategy. A key aim is to reduce waste by making products with fewer natural resources, with most products being able to be reused or their materials recycled – such measures will also help to reduce emissions of carbon dioxide and methane emissions from landfill sites, both major greenhouse gases contributing to climate change.

The waste hierarchy is: waste prevention – reuse – recycle/compost – energy recovery – disposal.

The Government's key objectives are to:

- decouple waste growth from economic growth, putting more emphasis on waste prevention and reuse;
- greater emphasis on waste reduction – home composting, food waste reduction, less packaging;
- meet and exceed targets for reducing the amount of biodegradable municipal waste landfilled;
- increase diversion of non-municipal waste from landfill and secure better integration of treatment of municipal and non-municipal waste;
- secure the investment in infrastructure which is needed to divert waste from landfill and for the management of hazardous waste; and
- get the most environmental benefit from that investment, through increased recycling of resources and recovery of energy from residual waste.

Specific targets include:

- reducing the amount of household waste not reused, recycled or composted from over 22.2 million tonnes in 2000 by 29% to 15.8 million tonnes in 2010, with aspirational targets of 14.3 million tonnes in 2015 and 12.2 million tonnes in 2020 (a reduction of 45%);
- to recycle and compost at least 40% of household waste by 2010, 45% by 2015 and 50% by 2020;
- to recover 53% of municipal waste by 2010, 67% by 2015 and 75% by 2020.

A new target is to be set for reducing the amount of commercial and industrial waste going to landfill – the 2000 Strategy included a target of reducing by 2005, the amount landfilled to 85% of its 1998 levels. The Government is also considering a target of halving the amount of construction, demolition and excavation wastes going to landfill by 2012 as a result of waste reduction, reuse and recycling.

As in the earlier Strategy, all sectors – business, retailers, consumers, local authorities and the waste management industry – are seen as having a role in delivering the Strategy. A new Waste Strategy Board (to be led by Defra) is being established which will be responsible for developing the policy actions necessary to achieve its aims. These are summarised as

- informing consumers, retailers and businesses on how to reduce, reuse and recycle waste;
- set up voluntary agreements with producers and retailers;
- use government procurement to accelerate development of products which use less natural resources;
- invest in local government waste collection and disposal;
- incentivise producers, consumers and disposers to reduce, reuse and recycle waste, and divert from landfill through tax or trading. (The landfill tax is to be increased by £8 per tonne from 2008 until 2010/2011; it is also proposed to consult on removing the ban on local authorities introducing financial incentives for households who reduce and recycle their waste.)

In addition, the regulatory regime is to be simplified through reform of the permitting and exemption regime, allowing for more targeted risk-based enforcement, and enforcement action to deal with fly-tippers and illegal waste exports strengthened.

Greater London

Section 353 of the Greater *London Authority Act 1999* requires the Mayor of London to prepare and publish a "Municipal Waste Management Strategy". This should contain proposals and policies for the recovery, treatment and disposal of municipal waste in Greater London; in preparing (or revising) the Strategy, the Mayor should have regard to recycling plans prepared by London local authorities and to the National Waste Strategy (ss.49 and 44A respectively of the EPA), as well as to any guidance or Directions from the Secretary of State. The Mayor is required to consult: the Environment Agency; waste disposal authorities in Greater London and any whose area adjoins Greater London; local authorities in whose area municipal waste is disposed of by WDAs in London; and other bodies concerned with waste minimisation, recovery, treatment or disposal of municipal waste which the Mayor considers it appropriate to consult. Waste disposal authorities and waste collection authorities in London must have regard to the Mayor's Waste Strategy.

In September 2003, the Mayor published his municipal waste management strategy – *Rethinking Waste in London*. It includes the following main policy proposals aimed at promoting waste minimisation, increasing the amount of waste recycled and ensuring that all waste is handled in the most sustainable manner, with minimum impact on the environment:

- Seek legislative changes to enable the recycling and composting of 50% of municipal waste by 2010 and 60% by 2015.
- Require all London Boroughs to introduce collection systems for recyclables ("except where impracticable") by September 2004, with 'bring' facilities for household recycling being maintained and extended.
- Request all London Boroughs to prepare fully costed feasibility studies for collecting separated kitchen and garden wastes to help meet the targets for reducing the amount of biodegradable waste sent to landfill.
- Further consideration (following a pilot study in two London Boroughs) of the impact of offering financial incentives to householders to encourage recycling, as well as a London-wide programme promoting recycling and other sustainable waste practices.
- To encourage the development of new and emerging advanced conversion technologies for non-recyclable residual waste and new waste treatment methods such as mechanical biological treatment.
- Seek legislative changes allowing the creation of a single waste authority to coordinate disposal and recycling services.

The Greater London Authority Act 2007 requires London's waste authorities to exercise their waste functions in general conformity with the Mayor's municipal waste management strategy.

The Mayor's draft strategy aimed at managing waste from London's businesses (75% of London's rubbish) was published for consultation in early 2008. It includes targets for London's businesses to reuse, recycle or compost at least 70% of their waste by 2020, with a target of 95% recycling and reuse in construction, demolition and excavation waste by 2020.

The Greater London Authority Act 2007 inserts s. 356A in to the GLA Act 1999 to establish the London Waste and Recycling Board. The purpose of the Board (as set out in the *London Waste and Recycling Board Order 2008* (SI 2008/2038) is to "promote and facilitate an increase in environmentally sustainable waste management across London".

5.20.2 Wise About Waste, Wales

Wise About Waste, published on 14 June 2002, outlines the measures and targets necessary to enable Wales to move towards achieving sustainable waste management. The Strategy aims to encourage all sectors to limit the amount of waste produced and to manage it better – with particular emphasis on recycling and composting of unavoidable waste, and reducing to a minimum the use of landfill or incineration.

The Strategy sets the following targets for recycling and composting of municipal waste:

- by 2003/04 achieve at least 15% recycling/composting, with minimum each 5% composting and recycling;
- by 2006/07 achieve at least 25% recycling/composting, with minimum each 10% composting and recycling;
- by 2009/10 (and beyond) achieve at least 40% recycling/composting with minimum each 15% composting and recycling;
- Local authorities are to work towards a target that by 2009/10, waste arisings per household should be no greater than for 1997/98; by 2020 waste arisings per person should be less than 300 kg per person;
- by 2005 to reduce the amount of industrial and commercial waste landfilled to 85% of its 1998 levels, and 80% by 2010.

The Welsh Assembly Government is currently reviewing its waste strategy and is expected to consult on its proposals in Autumn 2008. Among the revisions under discussion is a proposal for local authorities to work towards a long-term target of recycling and composting 70% of municipal waste by 2025. The revised strategy is also expected to include proposals for waste prevention, better collection sites and recycling centres, multi-material kerbside collections, separate collection of food wastes and improved public awareness.

The regional waste plans prepared by local authorities, as well as including data on waste arisings and existing waste disposal facilities, include proposals for meeting the Strategy targets and for providing sufficient waste facilities.

The Household Waste Recycling Act 2003 (see 5.22.3) will enable the Welsh Assembly to make an Order requiring Welsh councils to provide kerbside recycling facilities for at least two types of recyclables by 2010.

The Wales Environment Strategy (published at the end of May 2007) includes a proposal to give local authorities powers to charge householders for collecting non-recyclable waste.

5.20.3 Towards Resource Management, Northern Ireland

Northern Ireland's first Waste Management Strategy was published in March 2000 and included as a key objective the need to reduce the amount of waste sent to landfill by two-thirds. Other key objectives were to fully achieve sustainable waste management, maximising opportunities for the re-use, recycling and recovery of waste, with landfill disposal as a last resort, and to minimise the risk of environmental damage or harm to human health. To facilitate the preparation of area waste plans aimed at meeting the Strategy's objectives and deciding on the types of waste management facilities needed, three sub-regional groupings of local authorities were established. Planning Policy Statement 11 (Planning and Waste Management), published by the DOENI in December 2002, set out the Department's policies for the development of waste management facilities and included guidance on the issues likely to be considered in the determination of planning applications.

A revised Strategy *"Towards Resource Management"* was published in March 2006, together with a programme for delivering its targets. The new Strategy builds on progress made since 2000 with the primary aim of "increasing resource efficiency through the promotion of recycling and recovery of waste, based on a lifecycle approach which balances consumption and production", and includes the following recycling targets

- 35% of household waste by 2010, 40% by 2015, and 45% by 2020;
- 60% of commercial and industrial waste by 2020;
- 75% of construction, demolition and excavation waste by 2020.

The NI Strategy also includes a commitment to draw up proposals to enable local authorities to charge householders for collection of waste, and for major construction projects to be required to draw up site waste management plans. Consideration is also to be given to requiring local councils to collect at least two materials for recycling and recovery (as is required in England under *the Household Waste Recycling Act 2003* – see 5.22.3 below).

5.21 NATIONAL WASTE MANAGEMENT PLAN, SCOTLAND

Section 44B of the EPA, which required the Scottish Environment Protection Agency to prepare a national waste strategy for Scotland, has been repealed by *the National Waste Management Plan for Scotland Regulations 2007* (SSI 2007/251) which came into force on 23 March 2007. These implement the relevant requirements of the 2006 EU Framework Directive on Waste, as well as those on hazardous waste, packaging and Directive 2003/35/EC allowing for public participation in the drawing up of certain plans and programmes.

The 2007 Regulations require Scottish Ministers to have a National Waste Management Plan (NWMP) which contains their policies in relation to the recovery and disposal of waste in Scotland; it should include provisions relating to

- the type, quantity and origin of waste to be recovered or disposed of;
- general technical requirements
- any special arrangements for particular wastes
- suitable disposal sites or installations
- the management of hazardous waste.

The Schedule to the Regulations details the objectives which the NWMP should meet (i.e. the objectives of the EU's Framework Directive). In preparing the plan, or any subsequent modification, Scottish Ministers must consult SEPA, Scottish Natural Heritage, other representative bodies or persons and any other persons as they think fit. In line with the public participation Directive, steps must be taken to bring the draft plan to the attention of any persons or bodies who are affected by it, or may be affected by it or have an interest in it, stating where comments should be sent. Any comments should be taken into account and as soon as practicable after taking decisions on the plan, Scottish Ministers should inform statutory and public consultees. A copy of the plan should be made available for public inspection free of charge.

The Regulations enable Ministers to give Directions to SEPA requiring it to provide advice on policies to be included in the plan or to carry out surveys or investigations in connection with it.

5.21.1 Zero Waste Scotland

A new National Waste Plan for Scotland is being developed during 2008/09. In the meantime the Scottish Government announced (24.01.08) its intention to consult on new targets aimed at increasing recycling and reducing the amount of waste sent to landfill. Proposed new targets include:

- the amount of municipal waste recycled or composted to be increased to 40% by 2010, 50% by 2013, 60% by 2020, and 70% by 2025.
- the percentage of municipal waste going to landfill to be reduced to 56% by 2010, 36% by 2013, 15% by 2020 and to 5% by 2025.
- from 2025, no more than 25% of municipal waste may be used to generate energy.

In August 2008 the Scottish Government published its proposals for achieving its waste and recycling targets, with a view to including including the following legislative measures in the Scottish Climate Change Bill:

- Public bodies and businesses to have a duty to provide recycling facilities.
- Powers for the Scottish Government to make regulations setting statutory targets for retailers to make packaging reductions.
- Powers to enable the Scottish Government to place a duty on public bodies and businesses to specify a minimum percentage of materials made from recyclate in procurement contracts.
- Public bodies and businesses to have a duty to prepare waste prevention plans.
- Powers to enable the Government to introduce deposit and return systems for some glass containers.
- Businesses to be required to send waste data returns to SEPA.

5.21.2 Current Waste Strategy

Prior to preparation of the national waste management plan, the Regulations allow for the following documents to be taken as constituting the NWMP: statement under s.44B of the EPA (Scotland's waste strategy), the National Waste Plan, National Planning Policy Guideline 10, Planning Advice Note 63, the Waste Data Digest and the National Planning Framework.

Scotland's National Waste Strategy was adopted by the Scottish Government and published by SEPA in December 1999. It outlines programmes aimed at ensuring all those involved in waste planning and management put in place the policies needed to achieve sustainable waste management. It envisages a need for up to 11 new waste treatment facilities – including recycling, composting and energy to waste plant – to meet the EU targets for reducing the amount of biodegradable waste sent to landfill (see 5.7). To facilitate achievement of the Strategy's objectives and to encourage locally based waste management proposals, eleven regional waste strategy groups have been established – their area waste plans, together with a National Waste Plan (Scottish Government/SEPA) which sets targets for sustainable waste management until 2020, were published in February 2003.

Key measures for reducing waste and implementing sustainable waste management policies in Scotland include:

- To provide kerbside recycling collections to 85% of households by 2010 and to 90% by 2020;
- Halt growth in amount of municipal waste produced by 2010**;
- Achieve 25% recycling and composting of municipal waste by 2006*, 30% by 2008, 38% by 2010 and 55% by 2020 (35% recycling and 20% composting);
- Recover energy from 14% of municipal waste;
- Reduce landfilling of municipal waste from around 90% to 30%*;
- Providing advice about waste minimisation to businesses;
- Develop markets for recycled materials.

*for the year 2003/04, 85% of municipal waste was landfilled, and 12.3% recycling/composting was achieved; www.audit-scotland.gov.uk

**The Scottish Government is consulting on whether "direct variable charging" for household waste would help to achieve this aim.

Additional targets in the National Waste Strategy include:

- industrial waste arisings (excluding construction and demolition waste) to be reduced by 3-5% by 2005, 6-9% by 2010 and 10-12% by 2015;
- targets for recovery and recycling of construction and demolition waste, and other industrial waste, to be established by 2002.

The Business Waste Framework, published jointly by the Scottish Government and SEPA in March 2007, aims to reduce the amount of business and industrial waste by at least 200,000 tonnes a year. Key points in the Framework include advice to business on waste minimisation, further work to boost recycling services for small and medium sized enterprises and encouraging sustainable waste management.

National Waste Plan

The National Waste Plan 2003 also sets out how Scotland is to achieve the EU targets in the Landfill Directive reducing the amount of biodegradable waste sent to landfill (see 5.7).

The Local Government in Scotland Act 2003 amends the EPA to add new sections 44ZA-44ZD which place a duty on local authorities to prepare integrated waste management plans for approval by Scottish Ministers; this plan will set out

- how the local authority plans to carry out its waste management functions in relation to policies contained in the national waste management and the relevant Area Waste Plan;
- how the local authority plans to comply with Directions from Scottish Ministers specifying what other matters are to be included, such as statements setting out performance targets and steps the local authority plans to take "in endeavouring to meet" such targets.

Directions from Scottish Ministers will also specify when such plans are to be prepared.

5.22 COLLECTION, DISPOSAL OR TREATMENT OF CONTROLLED WASTE

5.22.1 Collection of Controlled Waste

Sections 45-48 and 51 detail the duties of waste collection authorities for collecting controlled waste, and the provision of bins for both household and commercial or industrial waste; they largely came into effect on 1 April 1992 (E, S & W). Similar provisions for Northern Ireland are contained in Article 20 of the *Waste and Contaminated Land (Northern Ireland) Order 1997*, which came into effect on 6.10.99; the waste collection authority is the district council.

The WCA may charge for the collection of commercial and industrial waste; household waste is collected free of charge, although under the *Controlled Waste Regulations 1992* (E, S & W), a local authority may charge for the collection of certain types of household waste or from those premises whose waste is classified as household waste (see 5.1 above). Examples of when a charge may be made include: garden waste; clinical waste from a domestic property; waste from camp sites or self-catering accommodation; waste from residential hostels, homes, educational establishments or premises forming part of a hospital or nursing home; asbestos; dead domestic pets; waste from halls used for public meetings; and waste from royal palaces.

The Climate Change Bill (see chapter 3, 3.7.3) includes powers to enable up to five local authorities in ***England*** to pilot incentives for householders to reduce, reuse and recycle waste. Under the scheme local authorities may choose between a "rebate scheme" offering incentives to those householders throwing away the least waste, or a "rebate and charge scheme" whereby those throwing away the most could be expected to pay a charge. Local authorities wishing to take part in the pilot would need, as a minimum, to meet the following criteria:

- a recycling and composting rate of at least 20%;
- free kerbside collection of at least two waste streams, of which at least one must be a dry recyclable (i.e. not food or garden waste).

Schemes should take account of the needs of any potentially disadvantaged groups. Draft guidance for local authorities was published by Defra in June 2008; it is hoped that the first pilot scheme will start in April 2009.

The Household Waste Recycling Act 2003, which received Royal Assent in November 2003, amends s.45 of the EPA to require local authorities in *England** to provide kerbside recycling for at least two different recyclable or compostable wastes by 31 December 2010 (or 2015 subject to permission from the Secretary of State). A local authority may apply for exemption if it can show that the cost of providing facilities would be unacceptably high or that there is a "comparative alternative arrangement" (e.g. bring banks close to housing). The Act required the Secretary of State to report on WCAs' progress on meeting the target by 31 October 2004. Defra has published guidance for waste collection authorities on the Act (April 2005). (*The Act enables the Welsh Assembly to make an order applying its requirements to Welsh councils.)

Seventy-nine per cent of households in England are now served by kerbside collection schemes, and during 2003/4 1.9 million tonnes of waste was collected through such schemes –

a rise of 52% over the previous year. In Wales 59% of households were served by kerbside collection schemes in 2003/4 (2002/3: 47%) (*Municipal Waste Management Surveys* 2003/4, England & Wales).

Section 31 of *the Waste and Emissions Trading Act 2003*, brought into force on 1 January 2005 (SI 2004/3319) amends s.48 to require WCAs in ***England*** to discharge their duties with regard to separation of waste in accordance with any directions from the WDA for its area. Section 51 is amended to enable a WDA to give a direction to a WCA concerning requirements for separating waste.

Section 48 of *the Clean Neighbourhoods and Environment Act 2005*, brought into force in 2006, inserts new sections 47ZA and B into the EPA: authorised officers of WCAs ***(E & W)*** may issue a fixed penalty notice to any householder or business who fails to comply with notices regarding, for instance, where and when they should put waste receptacles out for collection. Article 10 of *the Waste (Amendment) (**Northern Ireland**) Order 2007* (brought into force 07.04.08, SR 2008/75) makes similar provision with regard to commercial and industrial waste.

5.22.2 Waste Management Plans

(a) Northern Ireland

Article 23 of the WCLO was brought into force on 6 October 1999 through Commencement Order (SR 373). Each district council in Northern Ireland is required to carry out an investigation with a view to deciding what arrangements are needed for the purpose of treating, recovering or disposing of controlled waste within its area; the need to prevent or minimise environmental pollution and harm to health is a prime objective. In drawing up an appropriate plan the council should take account of the objectives and targets contained in the Waste Strategy (Art.19, and see 5.20.3 above). The Strategy also suggests that councils should cooperate regionally in producing their plans which, it says, "is essential to improving the environmental performance of waste management facilities ... and meeting recycling and recovery targets".

The plan should include information on types and quantities of waste to be disposed of; methods of disposal and treatment with costs and savings attributable to each method; quantities of waste which the district council expects to be disposed of or treated in its area, including waste to be brought into its area or taken out of it. Wherever possible priority should be given to recovering waste. In preparing the plan, the district council should consult the DOE(NI), any other district council which might be affected by its plans as well as others (e.g. local business or industry) who might have an interest in it. Before finalising the plan, the district council must publicise it locally to give the public an opportunity to comment on it. The draft had to be sent to the DOE(NI) by 31 January 2001, and a copy of the final version by 30 June 2001, and publicised locally. The district council should keep a copy of the plan, to which members of the public should have free access; a charge may be made for copies of the plan.

The NI Waste Strategy (see above, 5.20.3) includes a commitment to make it a statutory requirement for developers and contractors of major construction and demolition projects to provide a written Site Waste Management Plan identifying the volume and types of material to be demolished and /or excavated. The Plan would need to demonstrate how off-site disposal of wastes would be minimised and managed.

Section 50 of the EPA, which was repealed by the *Environment Act 1995* on 1 April 1996, contained similar provisions for England, Scotland and Wales. Locally based waste disposal plans have been replaced by the waste strategies developed in England, Wales and Scotland – see ss.44A & 44B of the EPA, above.

(b) England & Wales

The Clean Neighbourhoods & Environment Act 2005, s.54 (brought into force 7 June 2005), provides powers to enable regulations to be made requiring developers and contractors of construction and demolition projects to prepare site waste management plans (SWMP) outlining how it is intended to manage and dispose of waste generated during the project.

The Site Waste Management Plans Regulations 2008 (SI 2008/314), came into force on 6 April 2008 and cover England only. The Regulations apply to all new construction (and demolition) projects on which construction work (or demolition) began on or after 1 July 2008 worth more than £300,000 excluding VAT. Projects relating to Part A installations as defined in the *Environmental Permitting (England and Wales) Regulations 2007* are exempt from these Regulations.

A site waste management plan (SWMP) must be prepared for all projects covered by the Regulations giving a breakdown of the waste that will be produced and detail how it will be reused, recycled or disposed of. The plan should be updated throughout the project to show what actually happens to the waste, with more detailed requirements covering projects over £500,000. The duty of care lies with the main contractor and the person in charge of the project, who are also responsible for ensuring materials are handled efficiently and waste managed appropriately. The Environment Agency and local authorities are responsible for enforcing the Regulations and as such have powers to enter premises to check on compliance.

It is an offence not to make a SWMP, keep it up to date or produce it on request; it is also an offence to forge a plan. Summary conviction may result in a fine not exceeding the statutory maximum; conviction on indictment may result in a fine. If the enforcing authority (Environment Agency and local authorities) has reason to believe that no SWMP has been produced it may issue a notice to the contractor (or person in charge of the project) offering him an opportunity to discharge his liability by payment of a fixed penalty of £300. The notice will outline the offence, the period (14 days) during which further proceedings will not be taken so long as the fixed penalty is paid before the expiry of that period, and the name of the person and address to which payment should be made.

Non-statutory guidance on the Regulations has been produced by Defra. In addition CL:AIRE (Contaminated Land: Applications in Real Environments) has developed a voluntary code of practice for the development industry on the definition of waste; this sets out good practice for assessing whether materials are classified as waste or not, and determining when treated waste can cease to be waste for a particular use. One element of good practice is ensuring that an adequate Materials Management Plan is in place covering the use of materials on a specific site; an MMP is like to form part of the SWMP for a development site. The code is available at www.claire.co.uk.

5.22.3 Waste Recycling

Section 49 of the EPA requiring waste collection authorities to draw up plans for recycling household and commercial waste has been repealed in Scotland by *The Local Government in Scotland Act 2003*, and in England and Wales by *The Waste and Emissions Trading Act 2003* (W: 25.06.04, SI 2004/1488, W.153; E: 1.01.05, SI 2004/3321).

Section 52 of the Act (which mainly applies in England only), introduces a system of "recycling credits". Where the WDA avoids disposal costs due to recycling initiatives by the WCA, it must pay the WCA a sum equivalent to that saved. Similarly, if the WDA retains waste for recycling, thus saving the WCA collection costs, the WCA must pay the WDA an equivalent sum. Both the WCA and WDA may pay a third party (e.g. a charity) to recycle waste on their behalf (this latter point applies in Scotland and Wales also); s.52(8A), inserted by s.49 of the CN&EA enables the Secretary of State to produce guidance to assist English WDAs and WCAs in determining whether to make such payments.

The Waste and Emissions Trading Act 2003 (s.31, brought into force on 1.01.05, SI 2004/3321) adds s.52A to the EPA requiring WDAs in England to financially compensate a WCA for costs incurred in carrying out a direction under s.51 of the Act to separate waste.

Article 24 of the *Waste and Contaminated Land (Northern Ireland) Order 1997* (brought into force 27.11.03) makes provision for the district council to pay another person in respect of waste collected by them for recycling; the payment should represent the council's net saving in not collecting and disposing of the waste itself.

Waste collection and disposal authorities may buy and sell waste for the purpose of recycling and make arrangements for the production of heat and electricity from waste (EPA: s.55; NI: WCLO, Art.26 – effective 6.10.99).

Section 49 of *the Clean Neighbourhoods and Environment Act 2005*, brought into force in March and April 2006, amends the arrangements for paying recycling credits. Section 52 of the EPA is amended to enable the Secretary of State to make regulations in England setting the method of calculating payments; new s.52(1A) enables him by Order to disapply this duty, and new s.52(1B) exempts a WDA in England from the duty to make payments to a WCA in respect of waste retained for recycling where the two authorities agree to alternative arrangements. In England these arrangements have been incorporated in *the Environmental Protection (Waste Recycling Payments) Regulations 2006* (SI 2006/743) which came into force on 6 April 2006. Guidance on the Recycling Credit Scheme is available on Defra's website.

Under *the Joint Waste Disposal Authorities (Recycling Payments) (Disapplication) (England) Order 2006* (SI 2006/651), which came into force on 1 April 2006, a joint WDA is no longer required to make payments to a WCA for the waste retained and recycled by the WCA in the WDA's area.

Waste Policy Guidance: Preparing and Revising Local Authority Recycling Strategies and Recycling Plans (DETR, 1998) suggested ways in which local authorities could plan to achieve the target of recovering value from 40% of municipal waste by 2005. It also shows how waste disposal and collection authorities can together develop a coordinated approach to municipal waste management in their area, which integrates the various treatment and disposal options for municipal waste. DETR (now Defra) has also published (1999) guidance for local authorities on monitoring and evaluating their recycling, composting and recovery programmes; this outlines a single methodology for calculating recycling, composting and recovery rates and indicators for assessing a programme's performance.

Local authorities in England have also been given statutory performance standards for recycling and are required to draw up municipal waste management strategies incorporating these – see 5.20.1.

An EU Directive on packaging waste adopted in 1994 (amended and updated by a 2004 Directive) also has as a prime purpose the recycling of such waste; this was originally implemented in the UK through the *Producer Responsibility Obligations (Packaging Waste) Regulations 1997* (now replaced by 2007 Regulations) which place certain obligations on producers to recover and recycle packaging waste – see this chapter 5.14 and 5.33.1.

5.22.4 Acceptance of Waste

Sections 57-58 (NI: WCLO, Art.27) empower the Secretary of State (NI: DOE) to require holders of waste management licences (and in Scotland a WDA) to accept and keep/treat or dispose of controlled waste as directed by the Secretary of State. *The Environmental Protection Act 1990 (Amendment of Section 57) (England and Wales) Regulations 2005* (SI 2005/3026), which came into force on 22 November 2005, extend the Secretary of State's powers of direction to include permits under the PPC Regulations which authorise disposal or recovery of waste.

5.22.5 Removal of Unlawfully Deposited Waste

Under s.59 of the Act (NI: WCLO, Art.28), waste regulation authorities and waste collection authorities may by notice on the occupier of land require the removal or treatment within 21 days of waste deposited unlawfully (fly-tipped) on any land in their area. Section 50 of *the Clean Neighbourhoods and Environment Act 2005*, brought into force on 6 April 2006, inserts s.59ZA into the EPA and enables the regulatory authority (E & W) to serve a notice on the owner of the land if the occupier is unknown, cannot be found or has successfully appealed against the notice.

The occupier or owner of the land may appeal against the notice to the appropriate Magistrates' Court (Sheriff Court in Scotland; Court of Summary Jurisdiction in NI). The court may quash the notice if the occupier/owner proves that the waste was not deposited by them or that they did not knowingly cause or knowingly permit its deposit.

Failure to comply with the notice without adequate excuse is an offence. Where there is a danger of pollution or harm to health, the regulatory authority or WCA may arrange for removal of the waste and recover the costs from the occupier or owner of the land.

Guidance on the CN&EA (2006) includes guidance on when to use their powers to serve a notice on a landowner. Defra has also published a good practice guide (July 2006) offering councils advice on preventing fly-tipping – e.g. longer opening hours at local tips and publicising potential fines for fly-tipping.

Article 12 of *the Waste (Amendment) (N. Ireland) Order 2007* (brought into force 25.06.07) adds to Article 28 of the WCLO to make similar provision in NI.

5.22.6 Interference with Waste & Waste Sites

Under s.60 of the Act (NI: WCLO, Art.29) it is an offence for an unauthorised person to interfere with waste put out for collection or at a disposal site, unless consent has been obtained from the WCA, WDA or waste disposal company.

5.23 SPECIAL WASTE & NON-CONTROLLED WASTE

Section 62 of the EPA (which came into force on 11 August 1995), as amended by the *Environment Act 1995* (Sch. 22, para 80) empowers the Secretary of State to make regulations for the treatment, keeping or disposal of particularly dangerous or difficult wastes. Article 30 of the NI WCLO is similar and took effect on 17 September 1998. The regulations may impose, among other things, requirements on consignors and consignees relating to their handling of the waste, as well as provisions relating to record keeping and public registers and the penalty for contravening the regulations. Provision may also be made relating to the supervision by regulatory authorities of the activities authorised by the regulations or of the persons carrying out the activity, for the recovery of costs incurred by the regulatory authorities in carrying out their regulatory functions and for appeals in respect of their decisions.

It should be noted that all producers, handlers and disposers of hazardous waste are subject to the duty of care provisions (EPA: s.34; NI: WCLO, Art.5).

New legislation in England, in Wales and in Northern Ireland (and in Scotland amended legislation – see 5.23.3 below) implements EU requirements for classifying hazardous waste in line with the European Waste Catalogue and for dealing with HW in accordance with the Hazardous Waste Directive (see 5.9).

5.23.1 List of Wastes Regulations (E, W, NI)

The List of Waste (England) Regulations 2005 (SI 2005/895, as amended SI 2005/1673), which also came into force on 16 July 2005, implement Commission Decision 2000/532/EC, as amended, which adopted the List of Wastes. This replaced the European Waste Catalogue and provides for the classification of wastes, including the classification of wastes as hazardous or non-hazardous. The EU Decision also provides for a member state to determine that a specific batch of HW may be treated as non-hazardous, or that a specific batch of non-HW should be treated as hazardous based on its particular properties; any such determinations should be notified annually to the Commission. This part of the EU Decision has been transposed through the HW Regulations and guidance on its application is available on the Environment Agency's website.

There are similar Regulations for Wales (SI 2005/1820, W.148), and Northern Ireland (SR 2005/301, as amended SR 2005/462).

5.23.2 Hazardous Waste Regulations (E, W, NI)

The Hazardous Waste (England and Wales) Regulations 2005 (SI 2005/894) came into force on 16 July 2005, revoking the 1996 Special Waste Regulations in England. It should be noted that only Part 11 of the 2005 Regulations which replaces references to special waste in the EPA with hazardous waste, applies in Wales. *The Hazardous Waste (Wales) Regulations 2005* (SI 2005/1806, W.138) revoke the 1996 Regulations in Wales and otherwise implement similar controls to those applying in England. In Northern Ireland, SR 2005/300 (as amended SR 2005/461) are also similar and revoke their 1998 Special Waste Regulations. (Scotland, see page 205 below)

The Regulations are enforced by the Environment Agency in England and Wales and by the Department of Environment (NI Environment Agency) in Northern Ireland.

The HW Regulations apply to all wastes listed as hazardous in the *List of Wastes Regulations* – waste is hazardous if it contains substances or has properties that might make it harmful to human health or the environment; waste may be hazardous only if concentrations of a dangerous substance exceed a specific threshold, but some will be classified as hazardous at whatever level they contain a dangerous substance. While most wastes that were "special" under the 1996 regulations will be hazardous under the 2005 Regulations, there are a number of differences to note, including:

- only cytostatic and cytotoxic medicines are covered by the HW Regulations (instead of all prescription only medicines as in the 1996 Regulations);
- Hazardous wastes (but which were not "special") include discarded single-use cameras, untreated end-of-life-vehicles; dental amalgam; some discarded electronic & electrical equipment such as TVs and computer monitors, and some waste wood.

Hazardous waste from domestic dwellings (i.e. non-commercial properties), and disposed of via normal refuse collections, is outside the scope of the Regulations. However where such waste is collected separately (or collected at the same time but not mixed with non-hazardous waste), the Regulations will apply to the waste from the point at which it reaches the central collection point – e.g. waste transfer station or civic amenity site – which is then treated as the waste producer. The Regulations apply to domestic asbestos waste produced by a person working for a householder, who becomes both the producer and consignor (unless he has engaged another person to remove the waste). A householder who takes his domestic asbestos waste to a civic amenity site, is not obligated by the Regulations, nor is a friend or neighbour who helps without payment.

Also excluded is most radioactive waste which is regulated under the *Radioactive Substances Act 1993*; however the disposal of radioactive waste items such as some clocks and watches, illuminants, indicators and smoke detectors, which does not have to be authorised under the RSA, but which has one or more hazardous properties arising other than from its radioactive nature would be.

The Regulations ban the mixing of hazardous wastes with other categories of HW or with non-HW unless specifically allowed by permit or registered exemption; where HW has been mixed other than in accordance with a permit or registered exemption, the Regulations require the holder to separate the waste wherever technically and economically feasible. Guidance from Defra also says that different categories of HW may be collected on a single vehicle so long as they are kept separate and not mixed – the Guidance also offers further clarification on what is meant by "categories of hazardous waste" and when "mixing" of HW may be permissible and when it would definitely not be.

Notification

As from 16 July 2005, all premises (subject to certain exemptions – see box) where hazardous waste is to be produced or removed, including those where it is disposed of or recovered at the place of production, must be notified to the Environment Agency in England or Wales – this requirement does not apply in Northern Ireland. The duty to notify premises rests with the producer – usually the owner or occupier of the premises – who should do so before the production of any HW. Where the waste is being produced at premises by a mobile service (i.e. carrying out construction, maintenance or repair work), responsibility for notification will usually lie with the operator of the mobile service. However if they are visiting a number of sites in any one year, producing only a little hazardous waste at each (e.g double glazing fitter, roadside breakdown service), then only the service operator's premises or principal place of business need be notified – if these premises themselves need to be notified because of the amount of hazardous waste produced, the mobile service operator may submit a single notification covering his premises and the mobile service. If more than 200kg of hazardous waste (agricultural premises until April 2008 – 500 kg) is to be produced at a site at which a mobile service is operating, then that site will need to be notified in its own right.

Premises may be notified to the Agency in writing (paper or electronic) or by telephone by either the producer or consignor, or by a third party on behalf of either, not more than one month before the "effective time" (i.e. the commencement date of production of HW, expiration of a previous notification, or 4th business day following that on which notification is given). If the producer of the HW is unknown, or cannot be found, or the waste was not produced on the premises, the Regulations enable notification of the premises to be effected by the consignor (person who proposes to remove the HW, or cause it to be removed). Premises on which waste has been fly-tipped in contravention of s.33 of the EPA do not have to be notified, though the person removing the waste will have to complete consignment notes in the usual way (see below).

All notifications should include: the name, address and contact details of the person providing the notification and where different of the person for whom the notification is being carried out; the address of the premises to which the notification applies and its SIC classification (UK Standard Industrial Classification of Economic Activities 2003, Office of National Statistics, HMSO, 2002); information required by the Agency to enable it to fulfil its inspection and monitoring functions. The appropriate fee – £28 for each premises notified in writing, £23 by telephone, or £18 in electronic form – should be submitted with the notification. The Agency will then issue a "Premises Code", which is valid for 12 months, and which should be used on all documentation relating to the movement of any HW from the premises. Guidance on site premises notification is available from the Environment Agency (version 2, May 2005). It is an offence to remove waste from premises which have not been notified and are not exempt from notification.

Premises exempt from Notification (E&W)

Certain premises do not need to be notified – but must still comply with requirements regarding consignment notes – to the Environment Agency if they produce less than 200 kg of hazardous waste in any twelve month period: these are:

- Office and Shop premises
- Agricultural premises producing less than 200kg of HW per year (500 kg per year until April 2008)
- Premises listed in s.75(5) of the EPA: caravans; residential and nursing homes; universities, schools and other educational establishments; hospitals; dental, doctor and veterinary surgeries; health clinics; ships.

Note:

1. The exemption for notification applies only if removal of hazardous waste from the premises is carried out by a registered waste carrier or by a waste carrier exempt from registration (e.g. WCA, WDA, WRA, charity or voluntary organisation or waste producer carrying his own waste);
2. All the above premises – except ships – will need to notify the Agency immediately if it becomes clear the 200 kg limit of HW is to be exceeded;
3. All industrial premises producing HW, regardless of the amount produced, will need to be notified to the Agency.
4. EA guidance suggests that 200 kg roughly equates to: 10 small TVs, or 14 lead acid batteries, or 500 fluorescent tubes, or 5 small domestic fridges.
5. Defra guidance for dentists requires all practices to install amalgam separators to prevent mercury releases to drains.

Consignment

Prior to the removal of any HW from premises, the producer or holder as appropriate must prepare a copy of a consignment note for each party involved (i.e. producer or holder where different from the consignor, the consignor, the carrier and the consignee); each consignment must be allocated a unique consignment code in accordance with the Agency's coding standard. Following completion of relevant parts, the producer/holder and/or consignor retain one copy, and the carrier has the remaining copies which he must ensure remain with the consignment; following acceptance of the waste and completion of the remainder of the documentation by the consignee, one copy should be returned to the carrier.

If more than one carrier is to be used to transport a consignment of waste, then the consignor should prepare a schedule of carriers, with each carrier retaining one copy before handing over the waste, together with the rest of the copies of the schedule to the next carrier.

A carrier may collect more than one consignment of waste in a journey, so long as all are from premises in England and/or Wales (or NI, as appropriate) and are destined for the same consignee. In such cases the carrier should use a "multiple collection consignment note"; copies should be prepared for both the carrier and the consignee and for each waste producer or holder (or consignor where different) from whom waste is to be collected (and a unique consignment note allocated for each premises where HW is to be collected). An annex to the note must also be prepared for each producer or holder (or consignor) and when completed must be signed by both the carrier and consignor confirming the accuracy of the details of the consignment.

Where a consignee has rejected all or part of a consignment of HW, he should indicate, together with his reason (e.g. does not have a permit or registered exemption for the recovery or disposal of the waste; incorrect description of the waste) on the consignment note, retaining one copy, giving one to the carrier and sending one to the consignor (or producer or holder if known). The carrier should inform the Agency that a consignment has been rejected and get instructions from the producer or holder as to what action to take with regard to transferring the waste to another consignee. If an alternative cannot be found within five business days, the HW producer or holder as noted on the consignment note must make arrangements for the waste to be returned to the originating premises and stored in accordance with the Waste Directive until a suitable consignee can be found. Before rejected HW is moved from the consignee, the producer or holder must ensure that a new consignment note is prepared, with the letter "R" being added to the consignment code. Where two or more consignments from a multiple collection are rejected, the carrier should prepare a multiple consignment collection note covering the rejected consignments which must be completed and signed by all parties prior to removal of the waste from the original delivery premises.

The Regulations allow for cross-border recognition of consignment notes; the NI and Scottish Regulations still require pre-notification of hazardous waste consignments to the NI Environment Agency or SEPA, as appropriate. For HW consignments from England and Wales to Scotland or Northern Ireland, an additional copy of the consignment note must be prepared and sent to the appropriate regulator at least 72 hours before the waste is moved; the consignee will also need an extra copy of the consignment note for sending to the regulator in compliance with the NI and Scottish Regulations.

Record Keeping

The Regulations require detailed records to be kept by all parties, including consignment notes, carriers' schedules etc; producers, holders and consignors' records must include details of quantity, nature, origin, and where relevant destination, frequency of collection, mode of transport (and details sufficient to identify the carrier) and treatment method, and should be kept while the waste remains in their possession, and for at least three years from when the waste is transferred to another person. Carriers' records should include the same information and should be retained for at least 12 months' following delivery of the waste to its destination. All parties must also let the Agency have any other information as it may reasonably require for the purposes of carrying out its functions to monitor the production, movement, storage, treatment, recovery and disposal of HW. The Agency is also under a duty to carry out periodic inspections of hazardous waste producers.

All records or registers (as well as those held under the 1980 or 1996 *Special Waste Regulations* should be sent to the Agency when a waste permit is revoked or surrendered; the Agency is required to keep them for not less than three years following receipt.

Quarterly Returns

Consignees are required to make quarterly returns to the Agency of all consignments of hazardous waste received, rejected consignments, HW delivered by pipeline and of individual consignments received as part of multiple collections. A quarterly return should cover the quarter ending 31 March, 30 June, 30 September, 31 December, as appropriate, and should be sent to the Agency no later than the end of the month following that to which the return applies. A fee of £10 for each consignment forming part of a multiple collection or £19 for any other consignment applies for returns made in writing (£5 and £10, respectively, if submitted electronically). Producers disposing of their HW within their own premises are also required to make quarterly returns to the Agency with a fee of £19 applying if made in writing and £10 if made electronically.

Consignees are also required to send producers, holders or consignors a return detailing consignments accepted on their behalf within one month of the end of the quarter in which the waste was accepted; if preferred, the consignee may instead send a copy of the completed consignment note, together with a description of the method of disposal or recovery of the waste, within the same timescale.

Non-compliance with the Regulations relating to requirements for notification of premises, consignment codes and notes, cross-border consignments and consignee returns to both the Agency and producers etc is an offence and on summary conviction may lead to a fine not exceeding the statutory maximum. The Agency may instead offer the offender the opportunity of discharging liability for the offence by payment of a fixed penalty fine of £300, to be paid within 28 days of the issuing of the fixed penalty notice. If however, non-compliance is connected with non-compliance with any other requirements of the Regulations, this may lead to a fine not exceeding the statutory maximum or on indictment to a fine and/or imprisonment for up to two years.

Guidance

The Environment Agency provides information on the legislation, updates, guidance etc; this includes guidance on *what is a hazardous waste* (HWR01, v4, July 2008), obtaining and use of consignment notes, and Best Practice Guidance on recovery and disposal – more information on www.environment-agency.gov.uk and see Appendix 5.1 to the *Pollution Control Handbook*. Joint guidance (November 2005) from the Environment Agency, SEPA and NIEA on the definition and classification of hazardous waste can be found on SEPA's website

5.23.3 Special Waste Regulations – Scotland

The Special Waste Regulations 1996 (SI 972), which came into effect on 1 September 1996 (together with amending regulations SI 1996/2019 and SI 1997/257) now apply only in Scotland. They were revoked in England and Wales (as were similar 1998 Regulations in Northern Ireland) on 16 July 2005 following implementation of new Hazardous Waste Regulations – see 5.23.2 above. In August 2008 SEPA published proposals for revising controls on waste management in Scotland. SEPA's *Better Waste Regulation Action Programme* outlines a number of actions for consideration, including review of the special waste regulations and their implementation.

In Scotland, *the Special Waste Regulations* have been further amended by SSI 2004/112 (as amended by SSI 2004/204) which came into force on 1 July 2004, to implement the 1991 Hazardous Waste Directive. These amend the definition of special waste to that in the hazardous waste Directive and also give effect to the EU's List of Hazardous and Non-Hazardous Waste (Decision 2000/532, as amended), thus widening the range of wastes now falling under special waste controls. The Regulations replace the exemption for household waste with one for domestic waste; domestic asbestos waste when removed by anyone other than the householder is not exempt from control. All special waste must be packaged and labelled in accordance with the *Carriage of Dangerous Goods (Classification, Packaging and Labelling) and Use of Transportable Pressures Receptacles Regulations 1996* (SI 1996/2092).

The Regulations impose a duty on establishments or undertakings which dispose of or recover special waste, collect or transport special waste, which is already mixed with other substances or materials to separate it as soon as reasonably practicable so long as this is technically and economically feasible, and necessary to comply with Article 4 of the Framework Directive (see 5.6 above).

The Regulations require SEPA to assign to each consignment or carrier's round of special waste a unique code which must be used on all documentation; waste should be identified by reference to the European Waste Catalogue. (A carrier's round is defined as a journey made by a carrier during which special waste is collected from more than one producer for transportation to the same consignee; each type of waste must be listed on a schedule to the consignment note; the round must normally be completed within 24 hours. SEPA need not issue a code until the relevant fee has been paid and, in any event, fees must be paid within two months of requesting a code.

Consignment Notes

All movements of special waste (unless exempt) must be prenotified through completion of the relevant part of the standard consignment note to the SEPA office in the area to which the special waste is being sent; this should be done at least three working days, but not more than one month, in advance of any movement of the waste. Repeat consignments of the same waste to the same destination can be notified at the time of the first consignment with no further notification being required for 12 months, although the consignment note should indicate that there are to be a number of consignments and estimate the quantities of waste to be involved in each; waste carriers collecting similar consignments of special waste from various premises to be transported to the same destination need only complete one prenotification which again will last for 12 months.

Certain types of special waste movements are exempt from prenotification, although the standard consignment note procedure still applies; these include:

- special waste movements (except by waste management businesses) within a company where the waste is to be stored prior to recovery or disposal;
- special waste products or materials being returned to their originator because they do not meet the required specifications;
- lead acid batteries.

The consignment note has five sections – Parts A and B are completed by the consignor or producer of the waste and include details of the waste (description, quantity etc), of the consignor and of where the waste is being transported from and to – one copy should be sent to the local SEPA office by way of pre-notification not more than one month, but not less than 72 hours, before removal of the consignment; Part C is completed by the carrier, Part D by the consignor (who confirms that the information in Parts B & C is correct), and who also retains one copy of the consignment note. The final Part, E, is completed by the consignee confirming receipt of the waste, date, quantity, registration number of vehicle, and waste management operation. If for any reason the consignee does not accept the consignment, he should give his reasons on Part E of the consignment note.) If there is no consignment note, the consignee should send SEPA a written explanation of his reasons for refusing the waste.) One copy of the completed consignment note is retained by the consignee, one given to the carrier and one sent to the local SEPA office – this should be done within one day of receipt of the special waste.

A similar procedure is followed for the second and subsequent removals of repeat consignments or carrier's rounds (for which a schedule in quadruplet must also be prepared) of special waste, and for removals of ships' waste to reception facilities.

Where a delivery has been refused, the carrier should also inform both the consignor and SEPA of the fact and get instructions from the consignor as to where the waste should be taken. Before delivery to the new premises, four copies of a new consignment note should be prepared; following completion of Parts A and B, on which should be entered a new code and the previous code; Part C should be completed by the carrier and Part D by the consignor (though the carrier may complete Part D on behalf of the consignor if he has received written instructions to this effect, sending the consignor's copy to him); a copy of the carrier's schedule, annotated to show the consignment which was refused should be attached to each copy of the consignment note. Three copies of the consignment note should accompany the consignment and be given to the consignee; he should complete Part E of the consignment note, retain one copy, give one to the carrier and send one to the local office of the Environment Agency.

The Regulations allow for cross-border recognition of consignment notes.

Fees

A fee applies to most consignments and carrier's rounds of special waste, payable when SEPA assigns a code to the consignment. No fee is payable for:

- second or subsequent carrier's rounds, only one consignment is collected from any consignor during the succession, the total weight collected in each round in the succession does not exceed 400 kg, or there is less than one week between collection of the first consignment and delivery of the last consignment in a round;

- the removal of a single consignment of special waste if it does not meet the specifications of the person to whom it was supplied;
- the removal of special waste from a ship in a harbour area to a conveyance for transportation outside the area, to reception facilities within the harbour area, or by pipeline to reception facilities outside the harbour area – such waste is also exempt from pre-notification though a consignment note must be completed and a code number obtained from SEPA.

Registers

Registers containing a copy of all consignment notes and where relevant a copy of the carrier's schedule must be kept by consignors, carriers and consignees, and should include details of the quantity, nature, origin and, where appropriate, the destination, frequency of collection and mode of transport of the special waste produced; where the waste is transported other than by the producer, the producer's register must record sufficient particulars of the other person. Records of special waste should be kept by the producer while he remains the holder of that waste and for three years following its transfer. Copies of consignment notes, carrier's schedules in respect of each consignment, etc should also be retained. SEPA is required to inspect the registers at appropriate intervals. Data must be retained on the registers of consignors and carriers for not less than three years. Consignees must retain all data on their registers until their waste management licence is revoked or surrendered; they should then send their register for the site to SEPA who should keep it for not less than three years.

Site Records

Regulation 16 requires that consignees should maintain full site records of all deposits of special waste; this should include a site plan marked with a grid, or with overlays on which deposits are shown. Descriptions of deposits should accord with those on the register of consignment notes; where the waste is disposed of by pipeline or at the place of production, a full record of the quantity, composition and date of disposal should be kept. A written statement of quantity, composition and date of disposal should be prepared of any liquid waste discharged without containers into underground strata or disused workings. All site records must be kept until a waste management licence is surrendered or revoked, when they should be sent to SEPA.

It should be noted that the consignor, carrier and consignee registers and the consignee site records are not publicly accessible; only when a permit is surrendered or revoked does SEPA put the consignee register and site records on the public register, subject to the usual constraints of confidentiality.

It is an offence not to comply with the Regulations; it is a defence to prove that non-compliance was due to an emergency or grave danger and that all steps were taken to minimise any threat to the public or the environment and that subsequent compliance took place as soon as reasonably practicable. Summary conviction renders the offender liable to a fine not exceeding level 5 on the standard scale; conviction on indictment may result in a fine and/or imprisonment for up to two years.

A Guide to Consigning Special Waste was published by SEPA in June 2006 (updated from January 2005) and is available on SEPA's website.

5.23.4 Non-Controlled Waste

Under s.63 of the EPA (WCLO, Art.31), it is an offence to deposit or knowingly cause or permit the deposit of non-controlled waste which, if it were controlled, would be special (or hazardous) waste. Such a deposit done in accordance with a condition of a waste management licence or other consent is not an offence. The penalty for an illegal deposit is the same as that for an illegal deposit of special (hazardous) waste – see above s.33 of EPA.

5.23.5 Waste Minimisation

Section 63A of the EPA was added by the *Waste Minimisation Act 1998*, which received Royal Assent in November 1998. It enables local authorities to include targets for reducing and recycling waste in waste contracts, to include strategies for minimising waste in waste plans and to promote schemes which encourage consumers to reuse or repair products, e.g. laundry services and household appliance repair services.

5.24 PUBLIC REGISTERS (Sc & NI)

Sections 64-66 of the EPA (as amended by the *Env. Act 1995*, Sch. 22, paras 82-83) have been implemented through *the Waste Management Licensing Regulations 1994*; Arts.34-36 of the WCLO are implemented through *the Waste Management Licensing Regulations (Northern Ireland) 2003*. [Sections 64-66 have been revoked in England and Wales as a result of implementation of *the Environmental Permitting (E & W) Regulations 2007* – see chapter 2.2.4.]

Waste regulation authorities must maintain public registers containing copies of applications for waste management licences and supporting documentation, as well as copies of the actual licences. In Northern Ireland the register will be maintained by the DOE(NI), with district councils having registers relating to the keeping, treatment and disposal of waste in their areas. The registers must also include details of variations of licence conditions and notices relating to breaches of licence conditions; revocations, suspensions and convictions; monitoring data and inspectors' site reports; appeals and other written representations etc; applications for surrender of a licence and completion certificates (issued once a closed landfill site has been declared safe). Information relating to the Hazardous Waste Regulations 2005 in NI and in Scotland, *the Special Waste Regulations* will also be put on the register. *The Landfill Regulations 2003* (sc) (see 5.29.1 below) amend both the Scottish and NI *WML* and *PPC Regulations* to require copies of the following documents to be placed on the register: landfill site conditioning plans; operator notifications regarding intention of ceasing to accept waste after 16 July 2002; Agency [DOENI] notices advising of decision that a site should close; and closure notices issued by the Agency [DOENI].

Exemption may be sought for exclusion of information from the register on the grounds of commercial confidentiality or national security (EPA, ss.65 & 66, WCLO, Arts.35 & 36). In the latter case Scottish Ministers [Secretary of State, NI] may give directions as to the types of information to be excluded or types of information which they will determine whether or not should be excluded. A person may give notice to Scottish Ministers [Secretary of State, NI] that they wish certain information excluded on the grounds of national security; such information will then remain off the register until their decision. The Agency should be advised that a request is being made to Scottish Ministers [Secretary of State, NI] for the exclusion of certain information on grounds of national security.

Where a request is made for information to be excluded on the grounds of commercial confidentiality, the Agency should consider the request within 14 days; if it does not the information automatically receives that classification. If the Agency does not agree that information is commercially confidential, the information should not be placed on the Register for 21 days, thus allowing time for an appeal to Scottish Ministers. The appeal statement should be in writing and should include copies of the material in dispute (see above – Appeals under *Waste Management Licensing Regulations 1994*). A further seven days should elapse following determination or withdrawal of the appeal before information is placed on the Register (*Env. Act 1995*, Sch. 22, para 83(1)). In Northern Ireland the NI Planning Appeals Commission determines appeals in relation to issues of commercial confidentiality.

Information excluded on the grounds of commercial confidentiality loses this classification after four years unless an application is made for classification to continue. Where information has been excluded, a statement to this effect will be included on the register; if the exclusion relates to the terms and conditions of the licence, a statement confirming, or otherwise, compliance with licence conditions will be included.

The registers are open to the public free of charge and copies of information available for a "reasonable charge". Scottish Ministers [DOENI] may prescribe where such Registers are to be available (*Env. Act 1995*, Sch.22, para 82(5); WCLO, Art.34(6)).

Information relating to licences will be retained on the register for the duration of the licence plus twelve months. Rejected applications will be retained on the register for twelve months from the date of rejection. Monitoring data is retained for four years and for four years if superseded by new data.

5.25 ANNUAL REPORTS (NI)

Article 37 of the WCLO requires the DOE(NI) to prepare an annual report covering its regulatory responsibilities for waste management, including income and expenditure; reports should also include information relating to implementation of waste disposal and recycling plans and prosecutions. The report should be published not later than six months after the end of the year to which it relates.

5.26 FIT AND PROPER PERSON (Sc & NI)

Section 36 of the EPA (NI: WCLO, Art.8(4)) states that (subject to compliance with certain other conditions), a WRA cannot reject an application for a waste management licence if it is satisfied that the applicant is a fit and proper person. [Section 36 has been repealed in E & W by *the Environmental Permitting Regulations 2007*.]

A person should not be treated as fit and proper (EPA, s.74, NI: WCLO, Art.3(3)) if they are not technically competent to manage the site concerned, do not have adequate financial resources (to manage the site in compliance with licence conditions) or have been convicted of a relevant offence (see reg.3 of WML Regulations); this includes an offence under various sections of the EPA, COPA, *the Transfrontier Shipment of Waste Regulations 2007*, or *the Controlled Activities Regulations 2005* (Sc); evasion of landfill tax is also a relevant offence, as is an offence under the *Landfill Regulations* (see below 5.29.1 & 3). In such circumstances the regulatory authority has the power to either refuse or revoke a WML or waste carrier's registration.

The *Waste Management Licensing Regulations 1994* (as amended in 1996 – SI 100 in Scotland) prescribe the qualifications and experience necessary for a person to be considered technically competent to manage the activities authorised by the waste management licence; they also detail the transitional arrangements enabling managers to gain the necessary Certificate of Technical Competence (COTC) from the Waste Management Training Board. 1997 Amendment Regulations (SI 2203) specify the type of certificate required to be considered technically competent for different types of treatment plant. 2003 Amendment Regulations, SSI 2003/171 – amend the qualifications for COTCs to take account of recent developments such as the Landfill Directive which classifies sites as hazardous, non-hazardous and inert; the Regulations also clarify that appropriate qualifications from other EU member states are recognised in the UK.

5.27 FEES AND CHARGES (Sc & NI)

Sections 41-43 of the *Environment Act 1995* contain powers for SEPA to make charging schemes for environmental licences; such schemes have to be approved by Scottish Ministers. In Northern Ireland, fees are levied under a charging scheme made under Article 15 of *the Waste and Contaminated Land (Northern Ireland) Order 1997*.

In common with other charging schemes for pollution control, the aim is to enable the regulatory authorities to recover their costs for both assessing applications for licences (or alterations to them) and to fulfil their duty (EPA, s.42) to supervise licensed activities (e.g. inspections and monitoring). Charges are not intended to cover duties such as maintenance of public registers and enforcement activities, but do cover such matters as

- application fee for waste management licence;
- application fee for varying or modifying licence (where the request comes from the licence holder);
- application fee to surrender a licence;
- annual subsistence charge (for landfill sites this is based on whether the site is active, or closed but no certificate of completion issued) – see also final para this section;
- application fee to transfer a licence.

The level of charges and fees is reviewed annually.

In those instances where the regulatory authority has a duty to consult the appropriate planning authority, the fee payable will include an amount to cover their costs. The subsistence charge will also include an amount to cover the costs of the planning authority if it is consulted because of concern that a licensed activity may, or is causing pollution. The fee payable is based on the quantity and type of waste to be dealt with and on the activity to be carried out.

Where the application (or subsequent licence) is for the keeping and treatment of waste, only one charge will be payable and that will be the higher of the two. However, where the activity relates to either the keeping and disposal of waste or the treatment and disposal of waste, separate fees will be payable. An operator requiring both a site licence and a mobile plant licence will also have to pay two fees. However in certain circumstances where more than one site licence is held by the same person only one standard component is payable. Detailed schedules of the amounts which apply in each instance are appended to the Scheme document.

5.28 POWERS OF ENTRY

Sections 108-109 of the *Environment Act 1995* empower "a person who appears suitable to an enforcing authority" to be authorised in writing to enter premises etc where they have reason to believe that an activity for which a licence is required is being, or has been, operated without a licence or where they believe there is a risk of serious pollution. Except in an emergency, a warrant should be obtained if entry has been or is likely to be refused and it is probable that force will be needed to gain entry.

Section 108 is amended by s.55 of *The Anti-Social Behaviour Act 2003* (which inserts s.59A into the EPA) to enable waste collection authorities in England and Wales to investigate incidents of unlawfully deposited waste. (Amendment in force 31.03.04, ASB Act Commencement Order, in Wales No. 1, WSI 2004/999, W.105; in England, No. 2, SI 2004/690).

In Northern Ireland, Article 72 and Schedule 4 of the *Waste and Contaminated Land (Northern Ireland) Order 1997* (effective 1 March 2001) provide similar powers for authorised persons.

5.29 LANDFILL

If not properly managed landfill sites can give rise to a variety of pollution problems, including leachate (a liquid formed when waste is broken down by bacteria which can cause contamination of groundwater). Rodent infestation, production of potentially explosive levels of methane gas, dangerous levels of carbon dioxide, plus trace concentrations of a range of organic gases and vapours, are other problems associated with landfill sites. It should, however, also be noted that landfill gas can be harnessed as a valuable source of renewable energy, resulting in significant energy savings.

The Environment Agency and SEPA have produced a series of guidance documents on various aspects of the management of landfill gas and on monitoring leachate, groundwater and surface water at landfill sites – Appendix 5.1 lists some of the guidance available and see also www.netregs.gov.uk).

5.29.1 Landfill Regulations (Sc & NI)

All landfill sites require a permit under the *Pollution Prevention and Control Regulations* (Scotland, 2000; Northern Ireland, 2003 – see chapter 2, 2.5.1).; Separate *Landfill Regulations* cover those aspects of the Landfill Directive which are outside the PPC Regulations but need to be taken account of in the PPC permit. In Scotland, these are SSI 2003/235, which came into force on 11 April 2003, as amended by SSI 2003/341, effective 1 July 2003; and in Northern Ireland, SR 2003/496, which came into force on 6 January 2004. Conditions attached to the permit will reflect the requirements of both the IPPC and Landfill Directives; the Landfill Regulations also amend the PPC Regulations to cover those landfill sites not covered by the IPPC Directive – i.e. landfills receiving less than 10 tonnes per day, or with a total capacity of less than 25,000 tonnes, or which only take inert waste.

[Similar 2002 Regulations, and their amendments, in England and Wales have been revoked following implementation of *the Environmental Permitting Regulations* on 6 April 2008. Schedule 10 of these Regulations transposes the requirements of the Landfill Directive which are required to be delivered through permits – see chapter 2, 2.4.]

All landfill sites are classified as Part A installations and are thus regulated by SEPA in Scotland or the Northern Ireland Environment Agency (NIEA), as appropriate. Before granting a permit, the regulatory authority must classify the landfill site as either a site for hazardous waste, non-hazardous waste or inert waste.

Landfill sites are defined as a waste disposal site for the deposit of the waste onto or into land; they include any site used for more than a year for the temporary storage of waste and any internal waste disposal site where a producer is carrying out its own waste disposal at the place of production (in Scotland this includes farm dumps). Landfills do not include facilities where waste is unloaded in preparation for its transport elsewhere for recovery, treatment or disposal; sites where waste is stored for less than three years prior to recovery or treatment; or sites where the waste is stored for less than a year prior to disposal. Schedule 2 (NI) and Schedule 3 (Sc) to the Regulations outline the general requirements for landfill sites, including location, geological and other construction requirements to ensure protection for surface and ground water, measures for minimising nuisance, and security requirements.

New landfill sites – i.e. those for which a permit or waste management licence was granted on or after 16 July 2001 and before 15 June 2002 – must comply with the Landfill Directive (and thus with the Regulations) immediately; those also covered by the IPPC Directive and coming into operation after 30 October 1999 also have to comply with that Directive. A PPC permit must be obtained before operating a new landfill site, or making a "substantial change" to an existing site – see chapter 2, 2.5.

Existing landfills are interpreted as those already in operation on 15 June 2002, or had not been brought into operation by that date but the relevant authorisation had already been granted. Existing operational landfills which are not covered by the IPPC Directive will be required to apply for a PPC permit in a phased programme to be completed in Scotland by 31 March 2007 (or 16 July 2009, depending on the size of the landfill) and in NI by 16 July 2009, with those thought to pose the greatest risk being required to apply for a permit first. Non-IPPC landfill sites will not be required to meet the technical requirements of the IPPC Directive and permit applications for such sites will only need advertising in one or more local papers. The waste management licence will cease to have effect for those areas of the site covered by the new permit.

Operators of existing landfill sites which planned to continue accepting waste after 16 July 2002 had to submit a "site conditioning plan" (SCP) to the regulatory authority by that date (NI, 6 February 2004), outlining how they would meet the requirements of the Directive for the type of landfill they intend to operate. On the basis of the SCP, the regulatory authority then decided whether the site could continue in operation or should close and for the former set a date for when application for a new permit should be made, and if the latter ensure that after-care and closure procedures are followed. It should be noted that failure to submit a conditioning plan by the relevant date for a site intending to remain in operation resulted in the relevant authorisation/licence ceasing to have effect; in such instances the site had to stop accepting waste for disposal until an SCP had been submitted and the regulatory authority agreed to consider it.

Where the regulatory authority has decided that a site should close, either because it is of the opinion that it will be unable to comply with the Directive or it has failed to submit an SCP, it should, through the service of a "closure notice" ensure the site is closed as soon as possible and no later than 31 March 2007 (NI, reg. 15 &16; Sc, reg. 18 & sch. 4); this should state the reason for the notice, the steps the operator must take to initiate closure, and the period within which such steps should be taken. The authority may withdraw a closure notice at any time. There is a right of appeal to Scottish Ministers (NI: Planning Appeals Commission) over the serving of a closure notice; in such instances the notice will not take effect until determination of the appeal.

Where it is intended to close a site after 16 July 2002, this should be indicated on the SCP and closure and after-care provisions described. Operators of such sites will not be required to apply for a PPC permit unless the date of closure is after the date on which they would normally be required to obtain a PPC permit.

SCPs were not required for landfill sites which ceased taking waste before 16 July 2001, and such sites remain subject to the surrender provisions of the *Waste Management Licensing Regulations 1994* (NI, 2003). Operators of sites continuing to receive waste after 16 July 2001, but which intend to close before 16 July 2002, remain regulated under the *Waste Management Licensing Regulations* but the closure and after-care requirements of the Directive will apply.

The co-disposal of inert or non-hazardous waste at hazardous waste sites was banned from 16 July 2004; such sites wishing to continue accepting hazardous waste after 16 July 2002 had to apply for interim classification as a hazardous waste site; the site must however have ceased to accept hazardous waste by 15 July 2004 in order to revert to classification as a non-hazardous site or only accept hazardous waste from that date. The SCP may be prepared on the basis of the type of waste to be accepted from 16 July 2004.

Amending Regulations in Northern Ireland (SR 2007/179), which came into force on 1 May 2007, made it an offence if a landfill operator continuing to take agricultural or mines & quarries waste after 31 July 2006 did not comply with the requirements to submit an SCP for the site by 1 June 2007, or other requirements pending determination of the closure or continued operation of the site. Further amending Regulations (SR 2007/258) which came into force on 4 June 2007, ban the landfilling of tyres and other specified wastes from 30 October 2007.

The landfilling of prohibited wastes (including liquid wastes and those which are explosive, corrosive, oxidising, highly flammable, hospital and other clinical wastes) at hazardous waste sites is banned from 16 July 2002. Landfilling of whole used tyres at hazardous waste sites has been banned since 16 July 2003 and of shredded used tyres from 16 July 2006; there are exemptions for whole and shredded tyres used for landfill engineering purposes, bicycle tyres and those with an outside diameter of more than 1400 mm. (New tyres, e.g manufacturing rejects may still be disposed of to landfill as the Regulations apply only to used tyres.) Conditions in the permit will specify that the landfilling of prohibited wastes at existing sites for non-hazardous and inert wastes is banned from 30 October 2007.

All waste must be pre-treated prior to landfilling at new sites from July 2001, at existing hazardous waste sites from July 2004 and at existing sites for inert or non-hazardous waste from 30 October 2007 – checking on this requirement is the responsibility of landfill operators or of waste producers. Exceptions cover inert waste if pre-treatment would be technically infeasible, or that pre-treatment would not reduce the quantity of waste or the hazard the waste poses to the environment or human health. Further guidance available via wwwnetregs.gov.uk.

As from 16 July 2002, the date by which the regulatory agency had by notice served on the operator to have classified hazardous waste sites as such, waste acceptance procedures had to be in place; at other sites such procedures will be required from when the permit is issued. Operators are required to keep a register containing details of the waste accepted for landfill, to include: quantities of waste deposited, its characteristics, its origin, dates of delivery, identity of waste producer (or for municipal waste of the waste collector), and in the case of hazardous waste its exact location on the site. A written receipt must be given to the person delivering the waste; the operator must advise the regulatory authority if he has not accepted a delivery of waste for landfill. Schedule 2 to the Regulations details the waste acceptance criteria to which an operator must have regard; SR 2004/297 in NI, implements from 16 July 2005, the criteria and procedures for acceptance of waste at each class of landfill as specified in the EU Decision on Waste Acceptance Criteria (2003/33/EC) – these define the standards, including limit values for some contaminants, to be met by waste before it can be accepted at one of the three classes of site. As from this date the full waste acceptance criteria will apply to all new landfill sites and existing sites for hazardous waste; the criteria will apply to other existing sites from when they receive their permit. *Guidance on sampling and testing of wastes to meet landfill acceptance criteria* is available on the Environment Agency's website.

An application for a permit must demonstrate that the operator is technically competent to manage the site, has adequate financial security to both set up and operate it and to cover the costs of closure and aftercare at the site for at least 60 years or until the permit is surrendered – these requirements will also be reflected in conditions attached to the permit. Permit conditions must also ensure compliance with the Regulations and will include conditions covering

- Types and total quantities of waste authorised for the site;
- Operational procedures, including ensuring the necessary measures are in place to prevent accidents and to limit their consequences;

- Waste acceptance procedures;
- Closure and after-care procedures.

Conditions will also cover monitoring procedures and operators will be required to put in place a suitable monitoring programme – sch. 4 details minimum monitoring procedures required. Results of monitoring must be reported annually to the regulatory authority, as must data on the types and quantities of waste disposed of. For those landfills which are a Part A installation, a condition will also require that energy be used efficiently.

In Scotland, the following landfill sites are exempt from certain provisions of the Regulations (listed in sch. 1 and relating to conditions to be applied, waste acceptance procedures and criteria, control and monitoring procedures and disposal costs):

- A site which is the only such site on an island and which had remaining capacity as at 21 March 2003, is used exclusively for the disposal of waste generated on the island, is for non-hazardous or inert waste with a total capacity not exceeding 15,000 tonnes or with an annual intake not exceeding 1,000 tonnes; exemption applies only until the site is full.
- A site for non-hazardous or inert waste for the disposal of waste only generated by an isolated settlement (i.e. no more than 5 inhabitants per sq km, and more than 50 km from the nearest urban agglomeration with more than 250 inhabitants per sq km, or with difficult access by road to the nearest such agglomeration.

5.29.2 Landfill Allowances

The Landfill Directive, in defining municipal waste as "waste from households and other waste that, because of its nature or composition is similar to waste from households", requires the amount of biodegradable waste landfilled to be reduced to 75% of the total generated in 1995 by 2006, 50% by 2009 and 35% by 2016 – see also 5.7.

Part 1 of *the Waste and Emissions Trading Act 2003* (which received Royal Assent on 13 November 2003) sets out the framework to require local authorities to progressively reduce the amount of biodegradable municipal waste (BMW) landfilled in line with the requirements of the Landfill Directive – with the definition of municipal waste being that used in the Directive. Much of this Part of the Act was brought into force by commencement orders during 2004. Following consultation, the definition of municipal waste in the Act is to be amended to "encompass all waste which comes into the possession or under the control of waste disposal or waste collection authorities, with the exception of construction and demolition waste".

The Secretary of State is required to set the maximum amount of BMW which may be landfilled by the UK and to have a strategy for England and each of the devolved administrations for reducing the amount of BMW sent to landfill to ensure that the UK meets each of the targets imposed by the landfill Directive – the DOE NI must have a similar strategy for NI. *The Landfill (Scheme Year and Maximum Landfill Amount) Regulations 2004* (SI 2004/1936) came into force on 22 July 2004. The scheme year begins on 1 April each year (from 2005 to 2019), with the target years being 2010, 2013 and 2020. The Regulations specify the maximum amount of municipal waste which may be sent to landfill for the UK, and for each of the four countries the maximum amount in the target years, and also in non-target years. Defra/Welsh Assembly Guidance Clarifying the Definition of Municipal Waste (September 2005) says that municipal waste is only waste collected by a waste collection authority or disposed of by a waste disposal authority in fulfilment of their roles as defined in *the Environmental Protection Act 1990.*

The Secretary of State (Scottish Government / Welsh Assembly / DOENI) will allocate landfill "allowances" to waste disposal authorities in England (Scotland / Wales / NI); in England these have been based on authorities' Municipal Waste Management Survey returns for 2001/02; in England (and in Scotland, from 01.04.08) those authorities landfilling less than their allowance may trade (sell) their unused allowance to another authority (including across the border). Unused annual allowances may be "banked" for use in the following year (not N. Ireland), except in the Directive's target years (2010, 2013 and 2020) when it may only be used in that year, and allowances issued in a previous year would become unusable. Authorities may also apply to the regulatory agency for permission to "borrow" up to 5% of the following year's allowance. Exceedence of a BMW allowance would result in a penalty, as would the absence of appropriate records. The detail of the legislation is set out in regulations:

- *The Landfill Allowances and Trading Scheme (England) Regulations 2004* (SI 2004/3212), which came into force 1 April 2005, as amended by SI 2005/880 (which reduce the exceedence penalty).
- *The Landfill Allowance Scheme (Wales) Regulations 2004* (SI 2004/1490, W155) came into force on 25 June 2004.
- *The Landfill Allowances Scheme (Northern Ireland) Regulations 2004* (SR 2004/416), effective 1 April 2005, and the *Landfill (Maximum Landfill Amount) (Northern Ireland) Regulations 2004* (SI 2004/3027) effective 18 November 2004. Monitoring Guidance for District Councils has been published by the EHS (v.2, February 2007).
- *The Landfill Allowance Scheme (Scotland) Regulations 2005* (SSI 2005/157), effective 1 April 2005, except for reg.8 (transfer of landfill allowances), which comes into force on 1 April 2008. Guidance from Scottish Government, March 2007.

The monitoring authorities are the Environment Agency (E &W), SEPA, and the Northern Ireland Environment Agency, as appropriate. The way in which waste disposal and waste collection authorities must keep records and report on the amounts of waste landfilled are detailed in separate Regulations; these will also require the monitoring authority to establish a landfill allowances register and a "penalties register" (in Wales the latter will be established by the Assembly); the penalty for exceeding a landfill allowance has been set at £150 per tonne in England (Wales £200) and at £1,000 for failure to keep adequate records. In Scotland the penalty for exceeding a landfill allowance has been set at £10 per tonne for a scheme year beginning 1 April 2005, £25 in 2006, £50 in 2007 and £150 per tonne from 1 April 2008, with the penalty for failure to keep adequate records also set at £1,000.

An Electronic Register of Landfill Allowances (available at http://lats.defra.gov.uk) records allowances allocated to each local authority to facilitate banking and borrowing of allowances.

5.29.3 Landfill Tax

As from October 1996 all waste deposited in landfill sites has been subject to a tax. Landfill operators licensed under the EPA or the *Waste Management and Contaminated Land (NI) Order 1997* were required to register their liability for the tax by 31 August 1996. Landfill operators who also use their site for recycling, incineration or sorting waste may apply to have the relevant area designated a tax-free site.

The Landfill Tax Regulations 1996 (SI 1527), as amended (SI 1999/3270, 2002/1, 2004/769, 2005/759, 2006/865, 2007/965, 2008/770) cover registration procedures, credits, accounting and environmental trusts. *The Landfill Tax (Qualifying Material) Order 1996* (SI 1528) defines the categories of waste which are subject to the lower rate of tax. *The Landfill Tax (Contaminated Land) Order 1996* (SI 1529) sets out the provisions for exempting waste generated as a result of cleaning up historically contaminated land; SI 2008/2669 makes provision for phasing out the exemption from landfill tax for waste produced as a result of cleaning up land subject to a Remediation Notice under Part IIA of the EPA.

The tax is based on the weight of the waste to be deposited, thus applying the polluter pays principle; it also aims to promote a more sustainable approach to waste management by providing an incentive to dispose of less waste to landfill and to recover more value from waste, e.g. through recycling etc.

The tax on active waste was increased to £32 per tonne from 1 April 2008, (thereafter increasing by £8 per year). Lower risk wastes have been taxed at £2.50 per tonne from 1 April 2008 (previously £2.00); these include:

- Naturally occurring rocks and soils, including clay, sand, gravel, sandstone, limestone, clean building or demolition stone, topsoil, peat, silt and dredgings.
- Ceramic or cemented materials: glass, ceramics (including bricks, tiles, clay ware, pottery, china, bricks & mortar), concrete. Refractories (bricks lining certain types of furnace).
- Processed or prepared mineral materials which have not been used or contaminated: moulding sands and clays, clay absorbents, manmade mineral fibres, silica, mica and abrasives.
- Furnace slags.
- Certain ash: bottom ash and fly ash from wood or coal combustion, including incineration.
- Low activity inorganic compounds: titanium dioxide, calcium carbonate, magnesium carbonate, magnesium oxide, aluminium hydroxide, magnesium hydroxide, calcium hydroxide and salt (if disposed of in a brine cavity), iron oxide, ferric hydroxide, aluminium oxide, zirconium dioxide.
- Gypsum and calcium sulphate-based plaster, if disposed of in a separate containment cell on a mixed landfill site, and on inactive-only sites.

Certain wastes are exempt from the tax; these include: dredgings arising from the maintenance of inland waterways and ports; naturally occurring minerals arising from mining and quarrying; burial of domestic pets at pet cemeteries; and waste resulting from remediation of historically contaminated land. From October 1999 inert waste used in the restoration of sites has also been exempt (Budget Statement, 17 March 1998). Leachate arising from landfill sites, which is then recirculated – for example being moved to another site for treatment) is also exempt (Customs & Excise ruling, 1999).

Landfill site operators may contribute funds to approved environmental bodies for spending on relevant environmental projects; they may then reclaim 90% of their contributions up to a maximum of 6.0% of their annual landfill tax payment (SI 2008/770). To qualify as "approved", environmental bodies must be non-profit distributing and independent of any local authority or site operator interest and they will need to enrol with ENTRUST (www.entrust.org.uk), the regulatory body set up to oversee distribution of funds on behalf of HM Revenue and Customs. Examples of projects falling within the scheme include the funding of local environmental or amenity improvements and building restoration in the vicinity of a landfill, or the reclamation or remediation of polluted or derelict land. However funding may no longer be made available for waste management or recycling projects – 2003 Amendment Regulations, SI 605. Further amendment Regulations (SI 2003/2313) add projects "based on the conservation or promotion of biological diversity" as defined by the UNEP Convention on Biological Diversity to the list of those which may be financed by landfill credits.

HM Revenue and Customs (www.hmrc.gov.uk) produces *A General Guide to Landfill Tax* (the latest edition, September 2005) covering all aspects of the landfill tax and the way in which it is to be applied, including registration, tax liability and landfill tax credit scheme, calculating the weight of waste, record keeping, and reviews and appeals.

5.30 WASTE INCINERATION

5.30.1 Overview

Incineration is the burning of waste at high temperatures. This reduces the weight of the waste by about two-thirds and its volume by 90%. Uncontrolled burning of waste can give off toxic chemicals such as hydrochloric acid, dioxins and furans and heavy metals. Hydrochloric acid contributes locally to acid rain and is given off by the burning of plastics. If organic matter and plastics are burnt at low temperatures, dioxins and furans may also be emitted. These are carcinogenic substances and it is therefore essential that incinerators operate at high temperatures in order to reduce such emissions to a minimum. Grit, chars and dust must also be controlled by special equipment. Heavy metals – such as cadmium, lead, arsenic and chromium – which originate in the waste will be collected in the dust filter; mercury, also a heavy metal, may however escape as a vapour. This is best controlled by reducing the source of mercury in waste (mainly batteries).

In the UK, around 10% of all waste is incinerated, and currently less than 2% of toxic waste. A growing shortage of suitable landfill sites means that alternative methods of disposal, including incineration, will need to be researched and developed. Incineration can be a valuable source of energy; a number of European cities have successful Combined Heat and Power (CHP) schemes generating heat and electricity using incinerators.

5.30.2 Waste Incineration Regulations (Sc & NI)

The EU's *Waste Incineration Directive* (2000/76/EC) – see 5.8 – has largely been implemented through the permitting regime established in *the Pollution Prevention and Control Regulations 2000* (Sc; NI, 2003); see chapter 2, 2.5). New incineration plant were required to comply with the Directive from 28 December 2002 and existing plant from 28 December 2005. The Directive defines incineration plant as "any stationary or mobile technical unit and equipment dedicated to the thermal treatment of wastes with or without recovery of the combustion heat generated".

The Waste Incineration Regulations 2003 (Sc SSI 2003/170 which came into force on 1 April 2003; and NI SR 2003/390, which came into force on 22 September) bring incineration activities covered by the Directive, within the scope of the PPC Regulations. [In England and Wales SI 2002/2980 was revoked on 6 April 2008 following implementation of *the Environmental Permitting Regulations 2007*. Sch. 13 of these Regulations transposes those requirements of the Waste Incineration Directive which are required to be delivered through permits – see chapter 2, 2.4.]

All hazardous waste incinerators and others operating above 1 tonne/hour are regulated by SEPA or the Northern Ireland Environment Agency, as appropriate. Non-hazardous waste incinerators and those of less than 1 tonne/hour are regulated by local authorities in Northern Ireland and by SEPA in Scotland.

Existing incinerators are defined as those that were in operation and had the required permit before 28 December 2002; or were authorised or registered for incineration or co-incineration before 28.12.02 and put into operation not later than one year from then; or plant for which a full permit application had been submitted before 28.12.02 and were to be put into operation not later than two years from then. Certain co-incineration plant for the generation of energy or manufacture of products were also regarded as existing plant so long as they started co-incinerating waste no later than 28 December 2004.

Operators of existing installations carrying out an incineration activity within section 5.1 (disposal of waste by incineration) of Schedule 1 of the PPC Regulations had to apply for a PPC/WID permit between 1 January and 31 March 2005 (NI: 1-28 February 2005); if an operator already had a PPC permit for an A1 or A2 activity, an application for a variation to come under WID had to be made within the same dates. Operators of IPC authorised processes could choose to apply for a variation to cover WID or for a PPC/WID permit, again between 1 January – 31 March 2005 (NI: 1-28 February 2005); operators of incinerators with capacity less than 50 kg/hour, currently regulated under the WML regulations, also had to apply for a PPC/WID permit between the same dates (note, however, that some of these incinerators now fall within the scope of PPC as a result of the WID).

The Pollution Prevention and Control (Waste Incineration Directive) (Scotland) Direction 2003 (and similar in N. Ireland) requires the regulators to include in permits for waste incineration installations appropriate conditions to ensure compliance with the WID.

Practical guidance (Paper 2003/24) on the scope, regulatory and technical requirements of the Directive and how it should be interpreted and applied was issued by the Scottish Government in September 2003.

5.30.3 Chimney Height

The Clean Air Act 1993 (Clean Air (NI) Order 1981), requires local authorities to take grit, dust, and gas emissions into account when accepting plans for new chimneys serving furnaces of 100 lb/hr or more. Chimney height is broadly calculated in relation to the probable future sulphur content of the refuse and the hourly burning rate. Reference to the *Memorandum on Chimney Heights* provides a general guide, and particular attention should be given to the interrelationship of chimney heights, gas plume height, wind velocity and direction, the nature of the terrain and other industry in the area (see chapter 3, 3.19.2, and also Appendix 3.3).

5.31 END OF LIFE VEHICLES

Those aspects of the End of Life Vehicles Directive (see 5.11) covering restrictions on the use of heavy metals, certificates of destruction, marking of components to aid recycling, free take-back of ELVs put on the market from 1 July 2002 have been implemented through *The End of Life Vehicles Regulations 2003* (SI 2003/2635). Most parts of the Regulations apply throughout the UK and came into force on 3 November 2003 (with the requirements covering certificates of destruction and collection & delivery of ELVs in NI taking effect on 31 December 2003).

In England and Wales the Regulations (as amended by *the Environmental Permitting Regulations 2007*, Sch. 1, para 43) also implement the requirements for authorised treatment facilities (ATFs) to be licensed, with this part also taking effect on 3 November 2003 – existing site licences under the *Waste Management Licensing Regulations 1994* have been modified to ensure compliance with Schedule 5 of the Regulations; other waste operators dealing with scrap cars who were exempt from licensing had to apply for a licence by 1 February 2004 if they wished to continue accepting untreated ELVs. ATFs are also responsible for issuing a vehicle's last owner with a certificate of destruction which confirms that the vehicle has been dismantled in accordance with the Regulations, and "signs off" the owner's financial responsibility for the vehicle. Sch. 11 of *the Environmental Permitting (E & W) Regulations 2007* makes further provision in relation to waste motor vehicles.

Guidance on the Standards for Storage and Treatment of End of Life Vehicles was published by the Environment Agency in November 2003. Scottish Regulations covering the storage and treatment of vehicles (SSI 2003/593) came into force on 7 January 2004.

The Environment (Designation of Relevant Directives) Order (Northern Ireland) 2003 (SR 2003/209), which came into force on 1 May 2003 designates the End of Life Vehicles Directive for the purposes of *the Waste and Contaminated Land Order 1997. The Waste Management Licensing Regulations (Northern Ireland) 2003* (SR 2003/493), which came into force on 19 December 2003, implement those parts of the Directive which relate to conditions for the keeping or treatment of end of life vehicles.

It should be noted that all the above Regulations, unlike *the ELV Producer Responsibility Regulations*, apply to *all* end of life vehicles, including HGVs, buses and coaches. These vehicles, therefore, also require de-polluting safely but are not included in the re-use and recovery/recycling targets. Guidance from Defra, *depollution guidance for end of life vehicles over 3.5 tonnes*

(December 2004), outlines the requirements of the ELV Regulations for depollution, and provides an overview of the facilities and equipment needed.

The End of Life Vehicles (Producer Responsibility) Regulations 2005 (SI 2005/263) which cover the whole UK came into force on 3 March 2005; they largely relate to the Directive's requirements for "free take back" of ELVs of any age from 1 January 2007 and producer obligations for providing accessible places for take-back of ELVs through authorised treatment facilities (ATFs) and collection points, and recovery and recycling targets from 2006. "Producer" is defined as the manufacturer or professional importer of a vehicle (but excluding private individuals) and the vehicles covered are passenger cars and light vans.

Producers were required to apply in writing to the Secretary of State (formerly DTI, now Department of Business, Enterprise & Regulatory Reform) by 30 April 2005 for registration, declaring responsibility for vehicles put on to the market prior to the date of their application, and for those they expect to put on the market after that date. The Secretary of State will ascribe responsibility to a producer (in writing, with his reasons for that decision) for any vehicles for which no responsibility has been declared; the Secretary of State may revoke or amend his decision in the light of representations received from the producer to whom responsibility has been ascribed. By 31 August 2005, each producer had to submit a plan for approval to the SoS demonstrating that they have established an adequate network of collection and ATFs which can deal with take back and treatment of all the vehicles for which they have declared responsibility or been ascribed responsibility, and which were likely to become ELVs during 2006. Plans covering vehicles which were subject to BERR registration after 30 April 2005, must be submitted for approval within 6 months of the vehicles being placed on the market (3 months where a producer has been ascribed responsibility). The Plan should include numbers of vehicles for which the producer is responsible and which it was thought would become ELVs in 2006; details of collection system – number and location of collection points, how it is to be publicised; number and location of ATFs, licence number (issued by the Agency, SEPA, NIEA as appropriate) and capacity of facility.

For the years 2006-2014, producers are required to achieve a reuse and recovery target of 85% and an 80% reuse and recycling target each year of the weight of the vehicles for which they have declared responsibility and for those for which they have been ascribed responsibility; these targets rise to 95% and 85% respectively from 2015 onwards. The targets for vehicles put on the market prior to 1980 have been set at 75% and 70% respectively.

By 1 April each year, beginning 2007, producers (in respect of ELVs for which they have declared responsibility and which are taken back to their network of ATFs), and ATFs (in respect of vehicles accepted for treatment and for which they do not have a producer's contract), must provide evidence that they have met recovery and recycling targets by weight of the vehicles for which they have responsibility and which were accepted into their ATF network in the previous calendar year.

5.32 ELECTRICAL & ELECTRONIC EQUIPMENT

5.32.1 Waste EEE Regulations

The Waste Electrical and Electronic Equipment Regulations 2006 (SI 2006/3289, amended by SI 2007/3454) transposing the EU's waste EEE Directive in the UK finally came into force on 2 January 2007, with the full producer responsibility provisions taking effect on 1 July 2007. The Environment Agency in England and Wales, SEPA in Scotland and the DOE NI are responsible for registration of producer compliance schemes and overseeing compliance with the Regulations. The Regulations have been amended in England and Wales by the *Environmental Permitting Regulations 2007* to refer to permits under those Regulations.

The Department of Business, Enterprise and Regulatory Reform (BERR) has produced non-statutory guidance for producers of WEEE on the Regulations (revised December 2007). The guidance also covers the roles and implications for others, including local authorities, the waste management industry, and the charitable and voluntary sector, as well as the implications of the Regulations for users and consumers of EEE. A Code of Practice for collection of WEEE from DCFs has also been published which deals with the relationship between local authorities and others running DCFs for household WEEE, and the producers arranging clearance of those DCFs.

The Regulations apply to the following product categories specified by the Directive – Sch. 2 of the Regulations lists the products within each of these categories:

- large household appliances
- small household appliances
- IT and telecommunications equipment
- consumer equipment
- lighting equipment
- electrical and electronic tools
- toys, leisure and sports equipment
- medical devices
- monitoring and control instruments
- automatic dispensers

All the following groups have obligations under the Regulations:

- producers (i.e. any business that manufactures, imports or rebrands electrical and electronic products)
- retailers and distributors (any business that sells EEE to end users. It should be noted that these include all retailers however small who sell EEE, including, for example, chemists selling electrical toothbrushes or hairdryers etc.
- local authorities
- waste management industry
- exporters and processors
- businesses and other non-household users of EEE.

Private householders have no legal obligations under the Regulations, but can deposit their unwanted or waste EEE free of charge at a DCF (usually local authority civic amenity site), or arrange for collection (with any associated charge) by their local authority, or charity or voluntary sector organisation which accepts EEE for reuse; some retailers may also offer in-store take-back on a like-for-like basis, but such retailers are not obliged to accept old EEE if no replacement purchase is made.

All producers placing EEE on the UK market during a compliance period had to register initially with an approved compliance scheme by 15 March 2007, providing data on EEE put on the market during 2006, as well as paying the prescribed registration fee – approved compliance schemes are listed on the Agencies' websites. Subsequently they should register on or before 15 October in the year immediately preceding the commencement of the compliance period. A producer who does not put, or form the intention of putting EEE on the market until after 15 October should join a scheme within 28 days of putting or intending to put EEE on the market. The initial compliance period ran from 1 July - 31 December 2007. Subsequent compliance periods will run from 1 January - 31 December.

From 1 April 2007, producers have been required to mark all EEE with a crossed out wheeled bin symbol, a producer identification mark and a date mark.

Producer compliance schemes (PCSs) – which may be either private (e.g. a single producer); limited to a single trading group, regional or general – were required to apply for approval to the appropriate Agency before the end of January 2007 and to have registered all their scheme members by 31 March 2007. The Agency will issue each producer with a producer registration number which must be given to anyone who distributes or sells their products. In subsequent years Schemes will be required to register their members with the appropriate Agency by 31 October, with the Agency confirming to the Scheme by 1 December that its members are registered for the next compliance period.

Producer Compliance Schemes are also responsible for financing the costs of the collection, treatment, recovery and environmentally sound disposal of WEEE from private households that is deposited at designated collection facilities, and obtaining evidence of compliance from scheme members to ensure the Directive's targets for recycling, reuse and treatment of WEEE is met. Compliance will be demonstrated by evidence notes from approved authorised treatment facilities (AATFs) or authorised exporters showing the tonnage of WEEE received for treatment, recovery and recycling by category (listed in Sch.1 to the Regulations, and further classified by whether it is from private households or other users). Evidence of reuse of whole appliances can be counted as part of a scheme's recovery target.

PCS Approval lasts for three compliance periods, unless approval is withdrawn. An application for approval of a proposed scheme should be made between 1 July and 31 August in the year immediately preceding the commencement of the compliance period – Sch. 7 to the Regulations details the documentation to be submitted with the application – together with the application fee.

Distributors – i.e. retailers and distance sellers and manufacturers selling direct to consumers – are required to provide facilities for their customers to return old electrical equipment when they buy a replacement item. They can either join the Distributor Takeback Scheme (DTS) which supports a network of, mainly, local authority collection points (designated collection facilities) for customers or allow customers to bring goods back to the store. BERR has appointed the Vehicle Certification Agency as the enforcement body to ensure distributors meet their obligations.

The WEEE Regulations provide for the application for approval as an authorised treatment facility or exporter to be made to the appropriate agency (EA, SEPA or DOENI), with Sch. 8 to the Regulations specifying the information to be included with the prescribed fee. Approval lasts from 1 January to 31 December. ATFs should have regard to the *Guidance on Best Available Treatment Recovery and Recycling Techniques and Treatment of WEEE* published jointly by Defra, the Welsh Assembly Government and the Scottish Government in November 2006, with similar guidance for NI published by the Department of the Environment in December 2006. AATFs are required to provide the appropriate Agency with reports on or before 1 May, 1 August, 1 November and 1 February. These should cover the previous quarter and detail the amount (by category listed in Sch. 1) of: WEEE received and treated; delivered to another ATF for treatment, or to a reprocessor for recovery or recycling; delivered to an approved exporter for treatment, recovery or recycling. Approved exporters must also provide quarterly reports which should include details of the amount of WEEE exported by category and the total number of evidence notes issued.

The Waste Electrical and Electronic Equipment (Charges) Regulations (Northern Ireland) 2006 (SR 2006/509) prescribe the fees to be paid by producer compliance schemes and their members and those to be paid by authorised treatment facilities

5.32.2 WEEE & Waste Management Licensing (Sc & NI)

The Waste Electrical and Electronic Equipment (Waste Management Licensing) Regulations (Northern Ireland) 2006 (SR 2006/519), which came into force on 5 January 2007, transpose the permit requirements of the WEEE Directive; they also provide for exemptions from permit requirements for treatment of WEEE, and for new exemptions for the treatment and storage of WEEE.

The Waste Management Licensing Amendment (Waste Electrical and Electronic Equipment) (Scotland) Regulations 2007 (SSI 2007/172), which came into force on 30 March 2007, also amend the 1994 Regulations to facilitate the grant or variation of WMLs in respect of WEEE, requiring SEPA to ensure that conditions attached to WMLs will satisfy the requirement of the WEEE Directive.

[Similar 2006 Regulations for England and Wales (SI 2006/3315) were revoked on 6 April 2008 by *the Environmental Permitting Regulations* on 6 April 2008; Sch. 12 of these Regulations makes provision for the requirements of the WEEE Directive which are required to be delivered through permits – see chapter 2, 2.4.]

5.32.3 Hazardous EEE

The EU's 2002 Directive on hazardous substances in EEE (see 5.12.2) has been implemented by *the Restriction of the Use of Certain Hazardous Substances in Electrical and Electronic Equipment Regulations 2008* (SI 2008/37), which came into force on 1 February 2008 – these Regulations replace SI 2006/1463. The new Regulations take account of revisions to the RoHS Directive and also strengthen enforcement powers available for dealing with non-compliant goods and their producers.

The Regulations prohibit producers from placing on the market any new EEE which contains certain substances in excess of concentrations laid down in the Directive (subject to a number of exemptions). Producers must be able to demonstrate compliance with the Directive by submitting, on request, technical documentation to the enforcing authority; technical documentation should be retained for four years after placing the EEE on the market.

Failure to comply with the requirements of the RoHS Regulations – which are enforced by the National Weights and Measures Laboratory (an executive agency of the Department of Business, Enterprise and Regulatory Reform, BERR) – will result in the removal of the manufacturer's products from the market place. BERR has published non-statutory guidance (revised February 2008) on the Regulations, with more information available on www.rohs.gov.uk.

The Commission has clarified that the ban on the use of lead in electric and electronic equipment does not extend to [church] pipe organs which now use a small electric fan to blow air through the sound-making lead pipes, rather than relying on organist foot-power. (DTI – now BERR – press release, 27.06.06)

5.33 PRODUCER RESPONSIBILITY

5.33.1 Packaging Waste Regulations

1994 and 2004 European Union Directives on packaging waste aim to increase the use of reusable and recyclable packaging and set targets to this end – see earlier this chapter 5.14.

Sections 93-95 of the *Environment Act 1995* (which came into force on 21 September 1995) relate to producer responsibility for packaging and packaging waste. The Secretary of State may, after consultation with bodies likely to be affected, make Regulations imposing an obligation on producers to increase the reuse, recycling or recovery of products and materials (at a time when it becomes, or has become, waste). Regulations should only be introduced if they will result in significant economic and environmental benefit, when weighed against the cost of imposing producer responsibility; the Regulations should impose the minimum burden necessary on business to achieve the desired result. The Regulations may specify:

- the classes or descriptions of persons to whom the producer responsibility applies and of products or materials covered;
- targets to be met by weight, volume or other method;
- registration of persons subject to producer responsibility and of exemption schemes and associated requirements; the registers would be the responsibility of the Environment Agency, SEPA and the Northern Ireland Environment Agency (NIEA), as appropriate;
- fees, application for registration and appeals procedure, etc.

The 1994 Directive was originally transposed in England, Scotland and Wales through *the Producer Responsibility Obligations (Packaging Waste) Regulations 1997* (SI 648), as amended. These Regulations, together with all the amendments were revoked on 16 December 2005, when new Regulations came into force – SI 2005/3468, which also implemented the 2004 amendment Directive. These have also been revoked and replaced by SI 2007/871 which came into force on 16 March 2007. Similar Regulations apply in Northern Ireland – SR 2007/198, which revoke and replace SR 2006/356.

The Environment Agency, SEPA and NIEA are responsible, as appropriate, for enforcement of the Regulations.

The 2007 Regulations consolidate the earlier Regulations and their amendments and give the enforcement agencies powers to refuse accreditation to reprocessors and exporters of packaging waste if they have been convicted of fraud or they are not satisfied that they will comply with the regime. They also require companies exporting packaging waste to provide full details of the reprocessor to whom it is being sent. 2008 Amendment Regulations (SI 2008/413), which came into force on 14 March 20078, amend the recovery and recycling targets for 2008, 2009 and 2010 imposed on producers by the 2007 Regulations; similar Amendment Regulations (SR 2008/77) came into force in NI on 6 April 2008. – see Table 5.4 below.

The Regulations apply to all businesses involved in the packaging chain which handle more than 50 tonnes of packaging material and/or packaging in a year, and have an annual turnover of £2 million or more. Such businesses ("obligated businesses") include

- manufacturers of raw materials for packaging
- those that convert raw materials into packaging
- those that put goods or products into packaging
- those that sell packaged goods to the final user
- those that perform a "service provision", e.g. business that lease or hire out packaging, such as wooden or plastic pallets; licensors, including franchisors and pub operating companies*; imported transit packaging
- those that import packaging or packaging material or packaged goods into the UK for any of the above activities.

(*Guidance, published April 2006 is available from Defra for this sector)

Obligated businesses must join a registered compliance scheme or register as an individual business with the Environment Agency, the Scottish Environment Protection Agency or NIEA, as appropriate. (It should be noted that Defra is expected to consult on amending the Regulations to remove the option for individual compliance, to take effect in 2010.)

Compliance schemes are responsible for meeting the aggregated targets of their members (who must provide data to the scheme on the amount of packaging handled) and for providing data demonstrating compliance to the relevant Agency. Businesses joining a compliance scheme pay a reduced fee to the relevant Agency and must provide calculations showing the tonnages of packaging handled. Scheme members are required to provide the scheme with packaging data. Registered compliance schemes are required to let the appropriate Agency have aggregated packaging data and a certificate of compliance with the Regulations by the end of January of the following year to which the data applies; compliance schemes and large producers (more than 500 tonnes of packaging) should let the Agency and Defra have updated operational plans by 31 January – these demonstrate how compliance with the Regulations and with recycling and recovery obligations are to be achieved; details of the information to be provided is available on Defra's website. Records and returns to the Agencies must be kept for at least four years.

A business which chooses to take on the obligation of meeting the recovery and recycling targets on its own must register with the appropriate Agency by 7 April at the latest each year; a registration fee is payable to cover the Agency's costs with regard to monitoring, data analysis and management. At the time of registration details of how much packaging went through the business in the previous year must be provided, together with calculations showing its recovery and recycling obligations for packaging waste for that year. Evidence of compliance must be submitted to the relevant Agency by 31 January of the following year.

Smaller obligated businesses (handling more than 50 tonnes of packaging but with a turnover of between £2 million and £5 million) may choose to have a recycling obligation allocated to them instead of calculating their own obligations – see Table 5.4. If this method is chosen it must be adhered to for three years (providing turnover remains under £5 million), with evidence of compliance provided through PRNs and/or PERNs.

Evidence of compliance must be provided to the appropriate Agency – either by the compliance scheme or individual business – through "Packaging Waste Recovery Notes" (PRNs) and/or "Packaging Waste Export Recovery Notes" (PERNs) which can only be obtained (purchased) from Agency-accredited (i.e. EA, SEPA, NIEA) waste reprocessors and exporters. Thus compliance schemes or individual obligated businesses are responsible for ensuring sufficient packaging waste is delivered to waste reprocessors and exporters to generate the PRNs or PERNs demonstrating compliance with their obligations. The Regulations allow for evidence to be submitted electronically (www.npwd.environment-agency.gov.uk) or by hard copy. The 2007 Regulations for E, S & W and those for NI have been amended to change requirements in relation to the export of packaging and packaging waste; SI 2008/1941 (E, W, S) and SR 2008/373 (NI) give the relevant Environment Agency greater discretion with regard to the type of evidence required to prove that material will be reprocessed under "broadly equivalent" standards to those in the UK. Guidance on exporting metal packaging waste outside the EU is available from the Environment Agency.

Table 5.4: Packaging Recovery & Recycling Business Targets (%)

	2008	2009	2010
Paper/board	67.5	68.5	69.5
Glass	78	80	81
Aluminium	35	38	40
Steel	68	68.5	69
Plastic	26	27	29
Wood	20.5	21	22
Overall recovery	72	73	74

Minimum amount of recovery to be achieved through recycling is 92% each year.

The EU minimum recycling and recovery targets are 55% and 60% respectively.

The percentage obligation for each activity for meeting the targets is currently: manufacturers 6%; convertors 9%; packer/filler 37%; seller 48%; secondary provider 85%; service provider 85%.

Note: The calculation of the recovery and recycling obligation is based on the amount of packaging handled by the business and not packaging waste produced

The Agencies are required to monitor compliance with the Regulations and to maintain a public register containing details of individual businesses and schemes registered under the Regulations, together with details of their recovery and recycling obligations; the register should also contain a statement confirming that a certificate of compliance has been received from each obligated producer, or in relation to a scheme, a statement confirming compliance with its obligations.

A scheme which commits an offence under the Regulations, e.g. by knowingly submitting false data, will be de-registered by the appropriate Agency; individual members of the scheme then become liable for meeting their own recycling and recovery obligations. An individual business which commits an offence under the Regulations, such as knowingly supplying false data to the appropriate Agency, is liable for prosecution; a fine not exceeding the statutory maximum (currently £5,000) may be levied on summary conviction or an unlimited fine on indictment.

Defra and the Devolved Administrations in Scotland and Wales are developing producer responsibility regulations aimed at increasing the collection and recycling of non-packaging plastic waste from farms (Defra press release, 21.03.06) – such waste includes silage wrap, greenhouse and tunnel film, tree guards, mulch film and crop cover.

The Environment Agency, SEPA and NIEA have developed the National Packaging Waste Database website at www.npwd.environment-agency.gov.uk to help companies meet the Regulations; it simplifies the submissions of data and will allow Defra and the UK regulators to monitor progress towards meeting EU targets more closely.

5.33.2 Packaging, Essential Requirements

The Packaging (Essential Requirements) Regulations 2003 (SI 2003/1941), as amended by SI 2004/1188, came into force on 25 August 2003, revoking and replacing SI 1998/1165. The Regulations specify the "essential requirements" to which all packaging must comply, including that

- it is the minimum necessary to meet safety and hygiene criteria for the product concerned;
- it can be re-used, recycled or recovered and will have minimal environmental impact if disposed of;
- it has been made with the minimum use of substances which become noxious or hazardous when incinerated or landfilled;
- limits for concentrations of lead, mercury, cadmium and hexavalent chromium of packaging or its components do not exceed 600 ppm after 30 June 1998, 250 ppm after 30 June 1999 and 100 ppm after 30 June 2001.

Technical documentation showing compliance with the Regulations must be kept for four years following the placing of the packaging on the market.

These Regulations are enforced by trading standards officers who can inspect compliance documentation and issue enforcement notices prohibiting the supply of packaging which it considers does not comply with the Regulations. An offence under the Regulations carries a maximum fine of £5,000, or unlimited on indictment.

Amendment Regulations in 2004 (SI 2004/1188, effective 24 May 2004) and 2006 (SI 2006/1492, effective 1 July 2006) update the Regulations to reflect the 2004 EU Directive (see 5.14). Updated guidance reflecting the 2003 and 2004 amendments was published by the DTI (now Department for Business, Enterprise and Regulatory Reform) in March 2005.

5.34 TRANSPORT OF WASTE

The Control of Pollution (Amendment) Act 1989 deals with the registration of waste carriers and provides powers to control fly-tipping (illegal dumping of waste). The Act empowers the Secretary of State to make Regulations dealing with applications for registration, renewal, maintenance of registers and access to them, and appeal procedures etc. Both the Act and Regulations under it are the responsibility of the Environment Agency in England and Wales and the Scottish Environment Protection Agency in Scotland; *the Anti-Social Behaviour Act 2003*, (s.55) gives waste collection authorities in England and Wales similar powers to the Environment Agency to investigate and prosecute those responsible for fly-tipping. These powers were brought into force on 31 March 2004 (ASB Act, Commencement Order No. 2 SI 690 & in Wales, No. 1, WSI 999, W.105).

In Northern Ireland, Articles 38-43 of *the Waste and Contaminated Land (Northern Ireland) Order 1997* (which came into force on 19 August 1999) cover the registration etc of waste carriers. The legislation is enforced by the Northern Ireland Environment Agency (NIEA). A number of amendments have been made to Articles 38-43 through *the Waste (Amendment) (Northern Ireland) Order 2007*; these have been included in the text below.

The Clean Neighbourhoods and Environment Act 2005 (E & W) also makes various amendments to the Act to strengthen controls for dealing with fly-tipped waste:

- s.35 amends s.1 to remove the defence that waste being transported illegally was done under instruction from an employer (brought into force 7 June 2005); however it is still a defence that the waste was transported in an emergency, or that the carrier had no reason to suspect that the waste was controlled. (Art. 13 of the 2007 NI Order, brought into force on 25.06.07, SR 2007/294, similarly amends Art. 38 of NI WCLO).
- s.36 amends s.2 to enable a charge to be made for providing a certificate of registration, or copies thereof, and also for conditions relating to vehicles to be included on carriers' registrations (brought into force 7 June 2005). (Art. 14 of 2007 Order similarly amends Art. 39 of NI WCLO).
- New sections 5 and 5A (inserted by s.37 of the CN&EA) will replace s.5 of the 1989 Act, enabling an authorised officer of the regulatory authority to stop, search and instantly seize a vehicle that it is reasonably believed is being used unlawfully for the transportation of waste. Draft Regulations (issued for consultation in June 2008 – 5.34.3) will specify how enforcing authorities must deal with any vehicles or contents seized under these powers, time to be allowed for presenting documentation, and what will happen to vehicles when they are seized; statutory guidance is also to be issued to authorities on the use of these powers. A police constable in uniform remains the only person empowered to stop a vehicle on a road. (Art. 15 of the 2007 NI Order substitutes a new Art. 42 of NI WCLO to make similar provision; it also gives an authorised officer of the NIEA, as well as a constable, power to stop vehicles on a road.)
- Section 38 of the Act inserts new sections 5B and 5C into the 1989 Act making failure to produce relevant documentation (e.g. waste carrier registration certificate) subject to a fixed penalty fine (maximum £300). (Similar provision is made in NI in the 2007 Order which inserts Art. 42A & B into the WCLO.) Section 5C enables waste collection authorities to retain receipts from fixed penalty fines; those collected by the Environment Agency must be paid to the Secretary of State (brought into effect in April 2006).

Scotland's *Anti-Social Behaviour Act 2004* increases the fine for certain cases of fly-tipping to £40,000. Local authorities, SEPA and the police have also been given powers to issue fixed penalty notices (£50 up to a maximum of £200) for fly-tipping offences.

Defra has published guidance (2006) on the new powers introduced through the CN&EA.

The Tackling Fly-tipping Guide published (2006) by the National Fly-tipping Prevention Group (which is chaired by the Environment Agency) provides advice for landowners on what to do if they come across fly-tipping, including what can be done to help the authorities catch and prosecute offenders, and tips on how to prevent being a victim of fly-tipping (also see www.nftpg.org).

5.34.1 Registration of Carriers & Seizure of Vehicles

The Controlled Waste (Registration of Carriers and Seizure of Vehicles) Regulations 1991 (SI 1991/1624), as amended by SI 1998/605), apply in England, Scotland and Wales and came into force on 14 October 1991. A Circular published jointly by the Department of Environment (11/91), Welsh Office (34/91) and Scottish Office (18/91) provides guidance on the Regulations. Similar Regulations (SR 362) applied in Northern Ireland from 21 September 1999.

Defra has published draft Regulations for consultation (June 2008) for strengthening the duty of care, and registration and other regulatory requirements for waste carriers (and brokers) in England and Wales with a view to reducing the amount of waste which is illegally disposed; it is hoped that the new Regulations will come into force on 6 April 2009 – see below, 5.34.3.

From April 1992 (NI: 11 March 2000), it became an offence to transport controlled waste in the course of business or for profit unless registered with the regulation authority. Waste carriers – including producers of building and demolition waste (which includes waste arising from improvements, repairs or alterations) – are subject to s.33 (prohibition on unlicensed deposits of waste etc) and s.34 (duty of care) of the EPA – both as amended and added to by *the Clean Neighbourhoods and Environment Act 2005* – see above 5.16 and 5.17 (NI: WCLO, Arts.4 & 5, as amended by 2007 Order).

Waste carriers, including those operating in Great Britain who do not have, or are not proposing to have, a place of business in Great Britain, must apply for registration from the office of the regulatory authority (i.e. Environment Agency or SEPA) appropriate to the location of their principal place of business; Schedule 22, para 37(2), of *the Environment Act 1995* (brought into force on 1 April 1998) adds s.2(3A) to *the Control of Pollution (Amendment) Act 1989*; this empowers the regulatory authority to prescribe the application form and the information to be included (1998 amendment Regulations). An application to register should be made by all (including prospective) partners in the business. Persons wishing to register, or renew registration, as a carrier and as a broker (see below) can make a single application covering both. A charge is made for both registration and annual renewal; the charging scheme is made under s.41 of the *Environment Act 1995* (1998 amendment Regulations).

There are a number of exemptions to registration; these include:

- charities and other voluntary organisations.
- waste regulation, disposal and collection authorities.
- ferry operators in respect of waste on a vehicle being carried on the ferry.
- ship operators carrying waste for disposal at sea under licence.
- a person who transports only waste which comprises animal by-products collected and transported in accordance with certain provisions of the *Animal By-Products Order 2003* (E&W SI 2003/1482; Sc SSI 2003/411) (NI: 1993); (carriers whose sole business is carrying animal waste are not exempt).
- those who transport only mines or quarries waste or waste from premises used for agriculture – (*Waste (Scotland) Regulations 2005* (SSI 2005/22); *Waste Management (England and Wales) Regulations 2006* (SI 2006/937) and in NI SR 2006/280).

The regulatory authority has two months (or longer, as agreed mutually) in which to consider the application. Registrations may only be refused if:

- the application procedure has not been complied with;
- the applicant or another relevant person has been convicted of a "prescribed offence" (e.g. contravening environmental pollution control legislation) and is, therefore, in the opinion of the waste regulation authority not suitable for registration as a waste carrier.

Registrations are valid for three years and apply throughout Great Britain. Transporting waste without a valid registration (which should be carried on the vehicle) makes the offender liable to a maximum fine of £5,000.

The Regulations provide for appeal to the Secretary of State when an application for registration or renewal is refused, or registration revoked; in the latter two cases registrations remain valid until the appeal is determined. (In Northern Ireland appeals will be made to the Planning Appeals Commission, WCLO, Art. 41.) Any appeal must be made within 28 days (NI: 2 months). Section 114 of the *Environment Act 1995* enables the Secretary of State to appoint someone to handle the appeal on his behalf. On 1 April 1996 the Secretary of State transferred determination of registration appeals to the Planning Inspectorate (based in Bristol).

Where authorised officers of the regulatory authority believe that an unregistered carrier is carrying controlled waste, they can stop, search and seize the vehicle for further investigation – only police constables in uniform can stop a vehicle on a road. (NI: WCLO, Art.42; 2007 Order will extend this power to an authorised officer of the NIEA). Failure to produce a registration certificate, or evidence to show that he does not need to be registered, is an offence and is subject to a fixed penalty fine of up to £300 (amendments to s.5 of Act made by CN&EA, s.38 and to Art. 42 of NI WCLO by 2007 Order).

If the regulatory authority, and a waste collection authority in England and Wales, believes a vehicle has been used for the illegal disposal of waste – fly-tipping – the Regulations require that it must first try to obtain the name and address (through the DVLA) of any person who would be able to provide information as to who was using the vehicle at the time the offence was committed – for instance the vehicle might have been hired out to a third party. If this proves unsuccessful, or the regulatory authority (or WCA) has reasonable grounds for believing a particular vehicle has been used for an illegal deposit – and thus committed an offence under s.33 of the EPA – it can apply to a Justice of the Peace (E & W) or Sheriff (Scotland) (*Control of Pollution (Amendment) Act 1989*, s.6) for a warrant to seize the vehicle. In Northern Ireland, a warrant would be obtained from a Justice of the Peace under Art.43 of the WCLO.

The Regulations provide for the return of vehicles and their contents subject to certain conditions, e.g. establishment of ownership and carrying out of such other requirements as may be ordered. The regulatory authority may sell or otherwise dispose of or destroy the vehicle and/or its contents only after efforts have been made to inform the owners and following a notice in a newspaper and after a prescribed period (*Env. Act 1995*, Sch. 22, para 37(4)); however the regulatory authority may arrange immediate disposal where the condition of the vehicle's contents so require. New Regulations for England and Wales will set out the procedure for dealing with seized vehicles under the new powers brought in by the CN&EA.

Details of registrations, amendments, renewals, offences etc, will be filed on a register to be maintained by the regulatory authority; the registers will be open to public inspection free of charge.

5.34.2 Registration of Waste Brokers & Dealers

The European Union Framework Directive on Waste (2006/12/EC – see above 5.6) requires that, in addition to waste carriers, waste brokers and dealers – i.e. those who recover or dispose of waste on behalf of somebody else – should be registered; brokers may also apply to be carriers. This requirement has been implemented in Great Britain through the *Waste Management Licensing Regulations 1994* (Reg.18 & Sch.5), as amended by the *Waste Management Licensing (Amendment) Regulations 1998* (SI 606) which came into force on 1 April 1998 (Sch. 5 is to be revoked in E & W – see below 5.34.3). Brokers of controlled waste must use a form prescribed by the Environment Agency (or SEPA) in applying for registration or for registering exempt activities. The 1998 Regulations also amend the 1994 Regulations to require charges to be paid in accordance with a scheme made under s.41 of the *Environment Act 1995*. Charitable and voluntary organisations, and waste regulation, collection and disposal authorities are exempt from registering.

Waste brokers and dealers include:

- companies which buy and sell scrap metal and other recoverables either operating from a yard, or as middleman finding both buyer and seller and making all the necessary arrangements for the transaction;
- companies which arrange for the disposal of waste to an appropriate facility on behalf of either another company or waste producer;
- waste disposal operators, carriers etc who arrange for the disposal of waste not covered by their own operating licence.

5.34.3 Waste Controls (England & Wales) Regulations [draft]

These Regulations, published in draft in June 2008, are expected to take effect on 6 April 2009. They will give the Environment Agency and local authorities new powers to stop, search and instantly seize vehicles suspected of being involved in flytipping and other waste offences. *The Controlled Waste (Registration of Carriers and Seizure of Vehicles Regulations 1991* (and 1998 amendment regulations), and *the Duty of Care Regulations 1991* (and 2003 amendment regulations) will be revoked in England and Wales.

Waste Brokers and Dealers

The new Regulations make it an offence to act as a broker or dealer in waste unless registered with the Environment Agency; "broker" is defined as any person who arranges for the recovery or disposal of controlled waste on behalf of others; a "dealer" is any person who buys and sells waste as a principal – neither brokers nor dealers need be in physical possession of the waste for an offence to be committed. The registration requirements outlined below for carriers also apply to brokers and dealers.

The Environment Agency is required to make periodic inspections of all registered carriers of, and brokers and dealers in, controlled waste, and must also keep and maintain a register of carriers, brokers and dealers.

Registration

People who do not normally and regularly transport waste are not required to register as carriers. All other businesses and individuals must apply for registration on a form provided by the Environment Agency, together with the appropriate fee. These requirements also apply to brokers and dealers.

Registration for charities and voluntary organisations is indefinite (unless revoked) and carries a one-off charge of £30. The same applies to those carriers who only transport

- waste they produce themselves;
- animal by-products;
- waste from a mine or quarry;
- waste from premises used for agriculture.

For all other waste carriers and brokers, registration must be renewed annually, with new applications subject to a fee of £60, with annual declaration (renewal) carrying a fee of £45.

Registration may only be refused if the Environment Agency is of the opinion that it would be "undesirable" for the applicant to be authorised to transport controlled waste or if the applicant has been convicted of a relevant offence. These include offences under

- s.33 or 34 of the EPA (see 5.16 and 5.17 above);
- s.1 or 7(3) of *the Control of Pollution (Amendment) Act 1999* (5.34 above);
- *the Transfrontier Shipment of Waste Regulations 2007* (5.37 below);
- *the Hazardous Waste Regulations 2005* (5.23.2 above);
- reg.38 of *the Environmental Permitting (E&W) Regulations 2007* (chapter 2, 2.4).

If registration is refused, the Environment Agency should notify the applicant giving the reasons. Otherwise the Agency will issue a Certificate of Registration. Registered carriers should notify any changes or circumstances affecting the details of their registration within 28 days to the Agency.

Registration may be revoked if a registered carrier breaches any conditions of registration, or is convicted of a relevant offence and the Agency is of the opinion that it would be undesirable for the carrier to continue to transport controlled waste. Revocation takes effect 28 days after notification or when the carrier notifies the Agency that he does not intend to either make or continue with an appeal.

Appeals in respect of refusal of registration or revocation should be made to the Secretary of State within 28 days of the refusal or revocation.

It is proposed to make it an offence to provide false and misleading information on an application for registration or to fail to inform the Environment Agency of changes to registration details.

It is also proposed that waste carriers should carry proof of registration on their vehicles (similar to a tax disc, and renewable annually), and that the fine for transporting waste without a valid registration form should be increased from £5,000 to £10,000 – this would apply to all duty of care offences, in that the holder of the waste would be legally responsible for ensuring that waste was transferred only to those authorised to receive it. Both proposals require primary legislation.

Transfer Notes under the EPA

A written description of waste ("Transfer Note") must accompany all transfers of waste to another person. This will include the following details:

- identify the waste using the appropriate codes in *the List of Waste Regulations 2005* (E: SI 2005/895; W: SI 2005/1820, W.148), quantity and whether loose or containerised (and type of container), time and place of transfer.
- name and address of both transferor and transferee, and signed by both.
- state whether the above are the producer or the transporter of the waste; a local authority, a holder of a permit under *the Environmental Permitting Regulations*, and if so give permit number, a person to whom s.33(1) (see above 5.16) of the EPA does not apply by virtue of the EPR; registered as a carrier, or as a broker or dealer of controlled waste (giving registration number).

Waste transfer notes must be retained for two years, and must be produced to an officer of the Environment Agency on demand within seven days.

This procedure does not apply to hazardous waste transferred under *the Hazardous Waste Regulations* and in accordance with a consignment note.

Seizure of Vehicles

Section 5 and 5A of *the Control of Pollution (Amendment) Act 1989* (inserted by s.37 of the CN&EA) and s. 34B of the EPA, enable an authorised officer of the enforcement authority (Environment Agency of local authority) to stop, search and instalntly seize a vehicle if they have reasonable grounds to believe or suspect that it is being used unlawfully for the transportation of waste (only a police officer may stop a vehicle on a road) or for the illegal disposal of waste. The new Regulations enable the enforcement authority to dispose of the vehicle immediately if its condition is such that it should be disposed of without delay. In all other cases, the authority should take all reasonable steps (e.g through the DVLA or other appropriate authority) to trace the registered keeper, and to ascertain whether the vehicle is subject to any hire purchase agreement; the authority should also notify the police force for the area in which the vehicle was seized.

If the registered keeper (or hire purchase company) is found, they should be given details of the vehicle, its contents, how and where it can be claimed, the action that will be taken if the vehicle is not claimed, and contact details for further information. If the vehicle is not claimed within 21 days, the enforcement authority may sell or destroy the vehicle. If claimed within 14 days, the authority may return the vehicle or keep it for a further 14 days pending investigation of any possible criminal offence, or it may apply to a magistrate to retain the vehicle pending enforcement action, notifying the registered keeper that this has been done, and the date and time of the hearing.

If after having taken all reasonable steps the registered keeper cannot be found, the enforcement authority should place a notice either on its website or in a local newspaper stating that the vehicle has been seized and why and that the vehicle will be sold or destroyed if not claimed within seven days of the notice. If the vehicle is claimed within seven days, the procedure in the previous paragraph should be followed.

5.35 BATTERIES AND ACCUMULATORS

European Union Directive *on batteries and accumulators and spent batteries and accumulators* (2006/66/EC), replaces Directive 91/157EEC, which was repealed on 26 September 2008, the date by which Member States had to implement the requirements of the 2006 Directive (see 5.10.1).

The 2006 Directive introduces stricter collection and disposal requirements, setting a target for collection of 25% and 45% of portable batteries to be achieved by 26 September 2012 and 2016 respectively, and all industrial and commercial batteries; it also requires the recovery of all the lead and 65% of the average weight of lead-acid batteries; for nickel-cadmium batteries, all the cadmium should be recovered and at least 75% of the average weight of the batteries; and for all other batteries at least 50% of their average weight – these targets to be achieved by 26 September 2011.

The Batteries and Accumulators (Placing on the Market) Regulations 2008 (SI 2008/2164 came into force on 26 September 2008, replacing and revoking SI 1992/232 and amendments (E, W & S) and SR 1995/122 as amended in NI. They implement those parts of the 2006 Directive covering the composition of new batteries and accumulators, labelling requirements, and requirements for manufacturers to ensure spent batteries can be readily removed from appliances. The Regulations apply to all batteries, whether manufactured in the UK or EU, or imported into the EU from a third country. Non-compliant new batteries and accumulators could not be placed on the EU market on or after 26 September 2008. BERR has published guidance (August 2008) on the application of the Regulations.

Further Regulations will be needed to implement the Directive's recycling and recovery (producer responsibility) requirements – these are expected during 2009; in a consultation document published in May 2008, BERR put forward the following for comment:

- Producers will be required to finance the collection, treatment and recycling of batteries, with their obligations to be discharged via membership of a producer compliance scheme. Schemes would need to provide evidence to demonstrate that collection targets, treatment requirements, recycling efficiencies and other obligations have been met. The environment agencies will be responsible for registering compliance schemes and for receiving evidence of compliance etc.
- Producers will be responsible for ensuring their batteries do not exceed limits for hazardous substances and comply with labelling requirements.
- Compliance schemes will be required to work together on consumer information campaigns to ensure householders know what to do with their spent batteries.
- Interim collection targets for waste portable batteries in the UK are suggested as: 8% by 26 September 2009, 15% by 26.09.10, 20% by 26.09.11, and 25% by 26.09.12.
- The Directive requires distributors (i.e. any shop that sells batteries) to collect waste batteries in store (unless assessment shows that an alternative existing scheme meets the Directive's environmental aims); compliance schemes will need to work with retailers on collection schemes. Battery producer compliance schemes will also be responsible for collection, treatment and recycling of batteries removed from appliances in WEEE; compliance schemes will also need to work with other collection schemes, such as postal schemes, those collected at local authority civic amenity sites, libraries, schools etc.

In England and Wales It is proposed to implement the treatment and recycling requirements for batteries and accumulators through *the Environmental Permitting Regulations 2007* which now regulate waste management activities and licensing – these requirements would need to be transposed separately in Scotland and Northern Ireland.

5.36 DISPOSAL OF PCBs

The Environmental Protection (Disposal of Polychlorinated Biphenyls and other Dangerous Substances) Regulations 2000 (SI 1043) were made under the *European Communities Act 1972*; they cover England and Wales and came into force on 4 May 2000 (SI 1043), with amendment Regulations (SI 3359) taking effect on 1 January 2001. Similar Regulations (SI 95) came into force in Scotland on 8 May and in Northern Ireland

(SR 232) on 7 August. They implement a 1996 EU Directive requiring the phasing out of PCBs. They also apply to polychlorinated terphenyls, monomethyl-dibromo-diphenyl methane, monomethyl-dichloro-diphenyl methane and monomethyl-tetrachloro-diphenyl methane.

Holders of PCB contaminated equipment with a total volume greater than 5 dm^3 (5 litres) or any other PCBs at a concentration greater than 0.005% (50 ppm) weight/weight had to register with the Environment Agency (SEPA or DOE, NI) by 31 July 2000 (NI: 31 October 2000), to be renewed annually. The holding of any substances to which the Regulations apply after these dates is prohibited unless registered.

Registered holders (subject to various exceptions) had to remove from use and dispose of properly all the substances covered by the Regulations by 31 March 2001. E&W amendment Regulations extended the deadline to 31 March 2001 (previously 31.12.00) so long as an application for exemption was received by 1 January 2001. Those permitted to continue holding stocks of PCBs etc after these dates include:

- companies authorised to decontaminate or dispose of PCBs;
- laboratories using PCBs for analytical or research work;
- transformers with PCBs in oil at concentrations of 50-500 ppm (the transformer can continue to be used until the end of its useful life);
- companies specifically permitted to continue holding PCBs by the relevant Agency.

The Regulations required the Agencies to compile an inventory of the PCB contaminated equipment held at each registered location by 30 September 2000, to be updated annually. The inventory is publicly available.

5.37 TRANSFRONTIER SHIPMENT OF WASTE

5.37.1 UK Regulations

In 1996 the UK imported 76,136 tonnes of hazardous waste for treatment and disposal and exported 12,072 tonnes (originating in E, W & NI). (*Digest of Environmental Statistics*, No. 20, 1998).

The Transfrontier Shipment of Waste Regulations 2007 (SI 2007/1711, as amended SI 2008/9), which came into force on 12 July 2007, revoke and replace the 1994 Regulations (SI 1994/1137) to give full effect to the EU's 2006 Regulation on shipments of waste (see 5.9.2 above). They apply throughout the United Kingdom. EU Regulations are binding in their entirety on Member States and thus do not require transposing into national legislation; the UK Regulations are necessary to set out penalties and offences and to designate the competent authorities – the Environment Agency in England & Wales, SEPA in Scotland, and the Department of the Environment in Northern Ireland. The Secretary of State is the competent authority of despatch and destination for movements of waste for the whole offshore marine area for all waste arising from the offshore marine area (it is hoped to delegate his functions to the relevant environment agency).

The Regulations apply to all shipments – those from the UK to another member state; from another member state to the UK; or through the UK between two other member states. It is an offence to ship waste, or undertake the recovery or disposal of waste unless this is done in an environmentally sound manner and without endangering human health; it is also an offence if waste is shipped without the correct documentation. Liability for compliance with the Regulations covers any person involved in the shipment of waste and under UK jurisdiction, including the notifier, any transporter, or any freight-forwarder.

The Regulations amend *the Environment Act 1995* to enable the Environment Agency and SEPA to establish charging schemes to cover their responsibilities under the Regulations for both notifiable and Green List waste; prior to the establishment of this, Sch. 2 to the Regulations details notification fees which apply in England & Wales and in Scotland, and Sch. 3 those which apply in NI. A charge is also made for each shipment made under the notification.

All shipments of waste for disposal or recovery (listed in annexes to the EU Regulation) require written prior notification to the authority responsible for the shipment and the written consent of the despatch and destination authorities All waste shipments will remain under the control of the authorities until final disposal or recovery is effected, including while awaiting treatment or disposal. For Green List wastes (Annex III of the EU Regulation) a movement form must be completed and must accompany each shipment of waste; this should include details of the composition, quantity and destination of the waste, and a declaration that a contract with the consignee has been entered into. The 2008 Amendment Regulations (SI 2008/9), which took effect on 5 February 2008, make it an offence for anyone to export material for recycling to a non-OECD country without their specific agreement and in compliance with Commission Regulation 1418/2007. See also earlier this chapter, 5.9.2.

5.37.2 UK Waste Shipments Plan

The Transfrontier Shipment of Waste Regulations 2007 place an obligation on the Secretary of State to prepare a waste management plan outlining policies on the shipment of waste into or out of the UK, for disposal; in preparing the Plan he should consult all those persons likely to be affected by, or have an interest in, the plan taking account of any representations received in the final plan (as required by the public participation procedures of Directive 2003/35 – see chapter 1, 1.25.2).

The UK Plan for Shipments of Waste, a final version of which was published 12 July 2007, replaces the *1994 UK Management Plan for Exports and Imports of Waste*. It is legally binding and sets out Government Policy on exports out of and imports into the UK for disposal and recovery. It takes account of the UK's international obligations, in particular under the Basel Convention (see 5.9.3 above), and under the EU Regulations on Waste Shipments (1013/2006). The Plan also provides technical guidance to assist the UK competent authorities in taking decisions on proposed shipments notified under the WSR.

As a general rule, shipments of waste to and from the UK for disposal are prohibited, with the following exceptions (which are described in more detail in the Plan):

- emergency situations – i.e. where there is a clear and immediate risk to human health and/or the environment – that may require the shipment of hazardous waste to the UK from any country;

- emergency situations that may require shipment of HW from the UK to other Member States and EFTA countries (provided they are destined for disposal in a facility which applies best available techniques as defined in the IPPC Directive – see chapter 2, 2.1.4);
- trial runs to the UK from any country in order to test specialised disposal technology, except where the technology to be tested exists in the country of despatch;
- trial runs from the UK to other member states or EFTA countries in order to test specialised disposal technology, except where the technology exists in the UK;
- shipments of HW between Northern Ireland and the Republic of Ireland, in either direction, for disposal operations specified in the Plan and where the waste is generated and disposed of within NI or the Republic;
- shipments of HW produced in such a small quantity overall per yer in the UK, another member state or an EFTA country that the provision of new specialised disposal installations in the country of despatch would be uneconomic;
- shipments of waste into the UK from a party to the Basel Convention outside the Community for which a UK competent authority has acceded to a duly reasoned request;
- shipments of waste into the UK from a non-Party to the Basel Convention with which the UK Government has concluded a bilateral agreement.

RADIOACTIVE WASTE

Radioactive materials are used in the generation of electricity, in industry – including the food processing industry – in medicine, research and defence. Their use results in gaseous, liquid and solid wastes which can be divided into three categories:

- *high level (or heat generating) waste (HLW)*: this results from reprocessing of spent nuclear fuels; about 97% is recycled into new fuel;
- *intermediate level waste (ILW)*: this includes scrap metal, sludge and residues from fuel storage ponds, plutonium contaminated materials and fuel cladding;
- *low level waste (LLW)*: this includes paper, plastics, protective clothing, laboratory equipment, building materials and soils. (A further category of Very LLW includes small volume waste from hospitals and universities.)

5.38 INTRODUCTION

5.38.1 Risk Criteria for Disposal

Strict safety standards have been set by the Government with which any disposal scheme for radioactive waste must comply. The National Radiological Protection Board (which is now part of the Health Protection Agency set up in 2005) was responsible for advising the Government on radiological protection criteria to be applied to the disposal of all types of solid radioactive waste; in 1992 NRPB recommended the following objectives for its disposal:

a) future generations should not be subjected to risks which would be considered unacceptable today, i.e. people alive at any time in the future should be given a level of protection at least equivalent to that accorded to members of the public alive now;

b) the radiological risk from one disposal facility to the most exposed group of people should not exceed a risk of serious health effects in the individuals or their descendants of 1 in 100,000 per year;

c) the radiological risks to members of the public should be as low as reasonably achievable (ALARA), economic and social factors being taken into account.

The Health Protection Agency is proposing to update the 1992 guidelines to take account of recent recommendations from the International Commission on Radiological Protection and to ensure compatibility with draft regulatory guidance on requirements for authorisation of disposal facilities published by the environment agencies in May 2008 – see below 5.39.3(b).

5.38.2 UK Strategy on Radioactive Discharges

The UK Strategy on Radioactive Discharges 2001-2020 (Defra, 2002) reiterates the ALARA principle and sets out a framework for radioactive discharges from UK installations for the period which will also meet the UK's obligations under the OSPAR Convention (see chapter 6, 6.21.4). The Strategy adopts the precautionary principle, whereby preventive measures should be taken when there is reasonable grounds for concern that substances or energy directly or indirectly introduced to the marine environment may cause a hazard to human health, harm living ecosystems or the marine environment, damage amenities or interfere with legitimate uses of the sea; and the costs of pollution prevention, control and reduction measures should be borne by the polluter.

In mid-2008, the UK Government and Devolved Administrations published for consultation the draft UK Strategy for Radioactive Discharges 2006-2030. The revised Strategy updates Government policy and describes how the UK will continue to implement agreements reached at OSPAR meetings (see chapter 6, 6.21.4). Specific objectives are

- to implement the UK's obligations rigorously and transparently, in respect of the OSPAR Radioactive Substances Strategy's intermediate objectives for 2020;
- to provide a clear statement of Government policy and a strategic framework for discharge reductions, sector by sector, to inform decision making by industry and regulators.

Expected outcomes of the revised Strategy are:

- progressive and substantial reduction of radioactive discharges and discharge limits, to achieve targets specified in the strategy for each of the main sectors responsible for such discharges;
- progressive reductions in concentrations of radionuclides in the marine environment resulting from radioactive discharges, such that by 2020 they add close to zero to historic levels.
- progressive reductions in human exposures to ionising radiation resulting from radioactive discharges, as a result of the planned reductions in discharges.

Defra has also published (June 2008) draft Statutory Guidance to the Environment Agency on the Regulation of Radioactive Discharges into the Environment; this provides guidance on how the Environment Agency should implement the Discharges Strategy. The Scottish Government issued Statutory Guidance to SEPA in Spring 2008.

5.38.3 Long-Term Disposal of HLW

All high-level wastes must be stored for at least 50 years to allow them to cool. They are currently kept in liquid form in double-walled cooled stainless steel tanks enclosed in thick concrete walls at Sellafield in Cumbria and Dounreay* in Scotland. In 1990 British Nuclear Fuels brought into operation a vitrification plant at Sellafield; HLW is converted into glass blocks and sealed in stainless steel containers and will be stored for at least 90 years before eventual disposal. ILW is also treated and stored in cement based materials within stainless steel drums, pending a decision on its long-term disposal (see below).

(*The facility at Dounreay was closed in 2004 on scientific and economic grounds, and also because of safety concerns. Reprocessing of waste currently stored at the plant was expected to be completed by 2006, with decommissioning and clean-up of the plant expected to take 60 years to complete; however it could be another 300 years before the site returns to a "natural" state – i.e. is completely free of radioactivity – Report from UK Atomic Energy Authority, October 2000.)

An independent body – the Committee on Radioactive Waste Management (CoRWM) – was established in 2003 by the UK Government and Devolved Administrations for Scotland, Wales and N. Ireland, with a remit to review and make recommendations on radioactive waste policy and on how the UK should manage its most highly radioactive waste. The Committee's report, published in July 2006 recommends that, in the long term, radioactive waste be disposed of deep underground – the facility or facilities would be located "several hundred metres underground, making use of the surrounding rock as well as specially engineered structures to protect the environment", and it suggests that about one-third of the land in the UK could be geologically suitable. CoRWM recognises that it could be several decades before such a facility was ready and that thus interim storage facilities must be robust enough to house waste for a significant period of time pending its disposal to the underground repository. CoRWM suggests that host communities should be identified on the basis of a willingness to participate and an equal partnership approach to decision-making. In accepting the Committee's report (October 2006), the Government have stated they prefer that "open and transparent partnerships with potential host communities for disposal facilities" should be established.

CoRWM has now been reconstituted with terms of reference that provide for ongoing independent scrutiny and advice to the UK Government and Devolved Administrations on the long term management, including interim storage and disposal, of radioactive waste.

Planning and Development of geological disposal will be the responsibility of the Nuclear Decommissioning Authority, set up by *the Energy Act 2004* to oversee the decommissioning and clean-up of nuclear facilities. (UK Nirex Ltd, originally set up in 1992 and whose current mission is to develop and advise on safe, environmentally sound and publicly acceptable options for the long-term management of radioactive materials in the UK, is to become part of the NDA.)

In June 2007, the Government, together with the Devolved Administrations in Northern Ireland and Wales published a consultation document seeking views on a framework for identifying an underground repository for the long term disposal of higher activity radioactive waste. As a first step, the Government has invited (June 2008) communities interested in finding out more about hosting an underground disposal facility to open "no commitment" discussions with the Government. Defra is working on the basis that it will be 20-30 years before the procedures for the identification, approval etc of an appropriate site can be completed and waste be deposited. A dedicated website explains how the process will work – www.defra.gov.uk/mrws.

The Scottish Government does not accept that higher activity waste should be buried or that any community should be expected to host such a facility. Instead they support long term "near surface near site" storage facilities so that the waste can be monitored and retrieved, and the need for transporting it over long distances is minimal.

5.38.4 Long-Term Disposal of LLW

Most LLW is disposed of at the national disposal facility at Drigg in Cumbria; this is a "near surface facility" comprising concrete lined trenches which will be capped and earthed over when full. Conditions attached to such authorisations limit the intensity of radiation detectable at the surface of the site to specified limits. There are other conditions, one of which specifies the minimum depth at which the waste must be buried. Very low level solid waste, provided its activity is below authorised limits, is collected and deposited in landfill sites where it is diluted by significant quantities of other non-radioactive waste.

Following consultation in 2006, the UK Government and Devolved Administrations published (March 2007) their policy for managing solid low level radioactive waste responsibly and safely. It seeks to

- allow greater flexibility in managing the wide range of LLW that already exists and will arise in the future;
- maintain a focus on safety, with arrangements supported by the independent regulators (including the HSE and environment agencies);
- seek first to minimise the amount of LLW created before looking at disposal options, through avoiding generation, minimising the amount of radioactive substances used, recycling and reuse;
- create a UK-wide strategy for managing LLW from the nuclear industry, including at what point in the future attention will need to be given to replacing the national disposal facility at Drigg;
- initiating a UK-wide strategy on the management of non-nuclear LLW;
- emphasise the need to involve communities and the wider public in developing and delivering LLW management plans.

The latest UK inventory reports that as at April 2004, the total volume of existing and forecast radioactive waste was 2.3 million cubic metres, of which 2.1m cubic metres is LLW, 220,000 cubic metres is ILW and 1,300 cubic metres HLW.

5.39 REGULATORY CONTROLS

Under *the Radioactive Substances Act 1993*, anybody wishing to keep or use radioactive materials requires to be registered by the Environment Agency in England and Wales, the Scottish Environment Protection Agency in Scotland and the Department of the Environment in Northern Ireland; the accumulation or disposal of radioactive waste also requires an authorisation from the same bodies.

The EU's Basic Safety Standards Directive (96/29/Euratom) which sets legislative dose standards, has been implemented through *the Radioactive Substances (Basic Safety Standards) (England & Wales) Direction 2000* (and similar in Scotland) and by 2003 Regulations in Northern Ireland (SR 2003/208). These require the Agencies to ensure, when exercising their functions and duties under the RSA, that

- all public ionising radiation exposures from radioactive waste disposal are kept ALARA;
- the sum of the doses arising from such exposures does not exceed the individual dose limit of 1 mSv a year;
- the individual dose received from any new discharge source since 13 May 2000 does not exceed 0.3 mSv a year;
- the individual dose received from any single site does not exceed 0.5 mSv a year.

Draft Statutory Guidance to the Environment Agency was published for consultation in June 2008 on the Regulation of Radioactive Waste Discharges to the Environment. In providing guidance on implementing the revised UK Strategy for Radioactive Discharges 2006-2030 (issued in draft in June 2008, see above, 5.38.2), it applies to all authorisations under the RSA93 and to nuclear licensed sites, focusing on the use of Best Available Techniques in the optimisation of doses and the setting of discharge limits. The Scottish Government has also issued Statutory Guidance to SEPA (Spring 2008).

5.39.1 Nuclear Installations

The Nuclear Installations Act 1965, as amended, requires that before a site can be used for a commercial nuclear installation, it must obtain a nuclear site licence. The *Nuclear Installations Regulations 1971* (SI 381) prescribe the types of installations and activities covered by the Act. HM Nuclear Installations Inspectorate (which is part of the Health and Safety Executive) is responsible for licensing (and delicensing) and may attach specific conditions to licences to ensure appropriate standards are maintained. The NII continuously monitors sites for which it is responsible to ensure compliance with conditions and also ensures that emergency procedures are up to date and staff familiar with them. In setting conditions both the appropriate Agency, and the Department for Environment, Food and Rural Affairs will be consulted. This Act does not cover radioactive waste or discharges from nuclear sites, which are licensed separately under the *Radioactive Substances Act 1993.*

In delicensing all or part of a nuclear installation, ss.3(6) and 5(3) of the Act require that the Inspectorate must be satisfied that "there has ceased to be any danger from ionising radiations from anything on the site or, as the case may be, part thereof". HSE's policy statement outlining its criteria for delicensing nuclear sites and its interpretation of the no danger criterion was published in May 2005 (and is also available at www.hse.gov.uk/nuclear/delicensing/.pdf)

The licensing of nuclear installations, published 2007 and which replaces an earlier 1994 HSE document on nuclear site licences, describes how HSE regulates the design, construction, operation and decommissioning of nuclear installations for which a licence is required under the Act. HSE, SEPA and the Environment Agency have also published (February 2007) for consultation and trial use guidance on *the management of radioactive waste on nuclear licensed sites.*

The Energy Bill (introduced in January 2008) will require operators of proposed new nuclear energy facilities to draw up a fully-funded decommissioning programme, which must be approved by the Secretary of State, prior to the start of any construction work.

5.39.2 Ionising Radiation

The Ionising Radiations Regulations 1999 (SI 1999/3232) (made under the *Health and Safety at Work Act 1974*) (NI: SR 2000/375) provide protection for the workforce from ionising radiations, including the handling and storage of radioactive waste. These Regulations came into force on 1 January 2000, replacing 1985 Regulations (see also chapter 3, 3.32.1).

The Justification of Practices Involving Ionising Radiation Regulations 2004 (SI2004/1769), which came into force on 2 August 2004, implement EU Directive 96/29/Euratom. This lays down basic safety standards for the protection of the health of workers and the general public against the dangers arising from ionising radiation, and 97/43/Euratom on health protection of individuals against the dangers of ionising radiation in relation to medical exposure. Justification involves weighing the overall benefits of classes or types of activities which might result in the exposure of people to ionising radiation against the harm likely to be caused by the radiation exposure.

5.39.3 Radioactive Substances Act 1993

The Radioactive Substances Act 1993 (RSA) which consolidates and replaces the 1960 Act came into force on 27 August 1993 and applies to the whole of the UK. It brings together under a single Act the provisions of the 1960 RSA (which was repealed), together with all the various amendments to it made under other Acts. Consolidating Acts, such as this, do not amend legislation but may make minor changes for the purposes of clarification or consistency of approach.

The Act controls the keeping and use of radioactive materials and the accumulation and disposal of radioactive wastes through authorisation and registration systems; it also provides for public access to information held by the enforcing authorities.

This Act applies to Crown premises, except those occupied for navy, army or air force purposes, visiting forces or the Secretary of State for Defence.

The Act is enforced, on behalf of the Secretary of State by the Environment Agency in England and Wales, the Scottish Environment Protection Agency in Scotland, and by the Department of the Environment in Northern Ireland. Sites in England which are also registered under the *Nuclear Installations Act* will be jointly controlled with Defra, and those in Northern Ireland with the Department of Agriculture for N. Ireland.

The *Environment Act 1995*, Schedule 22, paras 200-230 made a number of amendments to the RSA; these are noted in the following text and most came into effect on 1 April 1996.

Further amendments have been made to the RSA by *The High-Activity Sealed Radioactive Sources and Orphan Sources Regulations 2005* (SI 2005/2686) (HASS Regulations) which implement Directive 2003/122/Euratom. These Regulations, which apply throughout the UK (apart from reg.5 in Scotland) and came into force on 20 October 2005, provide for the variation of RSA authorisations and registrations to comply with the Directive and impose requirements relating to site security and record keeping; they also provide for detecting, recovering and dealing with radioactive sources that are not currently under regulatory control ("orphan sources"). *The HASS (England) Directions 2005* (with similar to be made by the Devolved Administrations) require the Environment Agency to satisfy itself that holders of HASS have made adequate provision, by way of financial security or equivalent means, for the safe management of sources when they become disused sources or where the holder becomes insolvent or goes out of business.

(a) Registration for Users of Radioactive Materials

Anybody who keeps or uses radioactive materials, including mobile radioactive apparatus, must apply for registration under the RSA (ss.6-12). The application must include details of the premises (or mobile apparatus), what they are used for, the materials being kept or used and likely quantities, and how the materials are to be used. The application must be accompanied by the appropriate fee (see below). A copy of the application, and the subsequent registration certificate, should be sent to all those local authorities in whose area the premises fall except where the Secretary of State has decided that national security could be compromised (s.25).

In granting a certificate of registration, the appropriate Agency may impose such conditions as it thinks fit; these may include alterations to the premises, requirements as to apparatus or equipment used or to be used, restrictions on selling or supplying radioactive materials from the premises. Sections 23-24 enable the Secretary of State (in the case of nuclear sites, with Defra) to direct the appropriate Agency as to how an application for registration is to be determined – see below. Copies of the certificate of registration should be prominently displayed so that all those with duties on the premises will see it (s.19).

Applications not dealt with within the prescribed determination period (or other period as agreed) should be deemed to have been refused. Section 12 of the Act enables the appropriate Agency to cancel or vary registrations and to notify the operator; the appropriate local authorities should also be advised that such action has been taken.

The appropriate Agency may, by notice, require an operator to retain records of operations for a specified period after ceasing those activities to which registration relates; where registration has been cancelled, authorisation revoked or activities ceased, the operator may by notice be required to give the appropriate Agency copies of relevant records (s.20).

Holders of licences under the *Nuclear Installations Act* do not require registration under s.7 although the appropriate Agency may direct that conditions in accordance with s.7 should be imposed on the licensee; premises which keep or use watches and clocks which are radioactive material do not require registration although their repair or manufacture using luminous materials does (s.8(4) & (5)).

It should be noted that any radioactive materials kept or used on premises for more than three months will be considered to be accumulated for subsequent disposal and thus require authorisation under s.14(4).

(b) Authorisation of Disposal & Accumulation of Radioactive Waste

The disposal of any radioactive waste (including disposal from mobile radioactive apparatus) and accumulation of radioactive waste for subsequent disposal requires a certificate of authorisation (ss.13-18).

Exceptions to the above are the accumulation and disposal of radioactive waste from clocks and watches, although its accumulation and disposal from premises using luminous materials to repair clocks and watches does (s.15). *The Radioactive Substances (Clocks and Watches) (England and Wales) Regulations 2001* (SI 4005), which came into force on 14 January 2002, limit the exemptions: these cover the presence of tritium, promethium or radium so long as the prescribed amount of radionuclides on premises is not exceeded, or the waste is from no more than five watches or clocks.

The disposal of radioactive waste from educational establishments and from hospitals is also exempt from authorisation – conditions and restrictions relating to disposal from such premises are set out, respectively, in the *Radioactive Substances (Schools etc) Exemption Order 1963* (SI 1832) and in the *Radioactive Substances (Hospitals) Exemption Order 1990* (SI 2512, amended 1995, SI 2395).

Applications for authorisations for the accumulation and disposal of radioactive waste will be determined by the appropriate Agency. Applications should be accompanied by the prescribed charge (see 5.39.4a below); this will be refunded in full should the application be withdrawn within eight weeks of the regulator receiving it.

Section 16(3) – determination by the appropriate Minister of authorisations for disposal of radioactive waste from a nuclear site in England, Wales and N. Ireland – has been revoked by the *Environment Act 1995*, Sch. 22, para 205(3); in its place, s.16(4A) (Sch. 22, para 205(5)) requires the appropriate Agency to consult both the relevant Minister, the Health and Safety Executive and the Food Standards Agency prior to making a decision on whether to grant an authorisation and if minded to grant an authorisation on what terms and conditions etc; before granting an authorisation the appropriate Agency should forward a copy to the relevant Minister. Sections 23 and 24 of the Act enable the Secretary of State to give directions to the appropriate Agency as to how a registration or authorisation is to be determined, or conditions to be imposed or to call in an application for his determination.

All applications (and subsequent authorisations), subject to the usual considerations of national security (s.25), should be copied to those local authorities within whose areas it is intended to accumulate or dispose of radioactive waste. In addition, before granting a disposal authorisation in respect of a nuclear site, other local authorities public bodies and water bodies as is thought appropriate should be consulted.

Authorisations will set conditions for the operation of such plant, including discharge limits for radioactive waste and the manner in which it is to be disposed. The authorisation may permit the radioactive waste to be removed from the premises for disposal in a local authority disposal site; in such instances the local authority has a duty to accept the waste and to deal with it as specified in the authorisation. Authorisations will normally take effect not less than 28 days after the relevant local authorities would have received a copy of it.

A copy of the certificate of authorisation must be prominently displayed where all those with duties on the premises will see it (s.19). Failure to do so renders the offender liable to prosecution and a fine (s.33).

Applications (other than those relating to disposal of radioactive waste from a nuclear site – *Env. Act 1995*, Sch. 22, para 205(7)) not dealt with within the prescribed period (or such other period as has been agreed) are deemed to have been refused. Following notice, an authorisation may be varied or revoked and the relevant local authorities advised. However, before varying an authorisation relating to disposal of radioactive waste from a nuclear site, the appropriate Agency must follow the same consultation procedure as for an authorisation – see above: *Env. Act 1995*, Sch. 22, para 206(1). Authorisations and their conditions will normally be reviewed every four years in case updating is needed in the light of new knowledge, experience or technology.

Interim Guidance on the authorisation of discharges of radioactive waste to the environment was published by the Agencies, in collaboration with the NRPB (now part of Health Protection Agency) and Food Standards Agency in December 2002.

The Environment Agency and Northern Ireland Environment Agency (NIEA) have published for consultation (May 2008) guidance on their requirements for granting an authorisation for the disposal of any solid radioactive wastes – i.e. low, intermediate and high level wastes, as well as spent fuel and other waste nuclear materials – in deep geological facilities on land; separate guidance covering requirements for applications for authorisation of disposal of solid radioactive waste in near-surface disposal facilities on land was published for consultation at the same time by the EA, NIEA and SEPA. These documents will replace 1997 guidance covering authorisation for disposal of low and intermediate level wastes in disposal facilities on land.

(c) Enforcement & Prohibition Notices

The Environment Agency and SEPA may issue enforcement, prohibition and revocation notices where conditions of a registration or authorisation are being contravened, or where there is a risk of environmental harm (ss.21-22). The notice will state the problem and action to be taken to remedy the situation and the timescale within which this should be done. If there is considered to be an imminent risk of pollution or harm to human health, the authorisation or registration may be withdrawn in whole or in part. A copy of any notice will be sent to the relevant local authority. Once the appropriate Agency is satisfied that any risk has been dealt with, it can withdraw the notice; the local authority should be advised that this has been done. A similar procedure applies to registrations issued by Defra.

(d) Appeals

An operator may appeal (ss.26-27) to the Secretary of State if an application for authorisation or registration has been refused, conditions are considered unreasonable or in respect of the terms of an enforcement, prohibition, variation or revocation notice. It should be noted that the terms of all notices issued take immediate effect, except for those relating to revocation of authorisation or registration which are dependent on the outcome of any appeal. There is no appeals procedure in respect of

- decisions imposed by direction of the Secretary of State (s.23);
- applications etc determined by the Secretary of State (s.24);
- in N. Ireland, prohibition or other enforcement notices issued under ss.21-22.

The Radioactive Substances (Appeals) Regulations 1990 require that any appeal be made to the Secretary of State within two months of the appropriate Agency's decision (28 days in the case of revocation); the appellant may specify a preference for the appeal to be decided by written representations or by a hearing. Section 27 enables the Secretary of State to refer an appeal (except where it relates to a decision or notice served by SEPA – *Env. Act 1995*, Sch. 22, para 215) to another person appointed by himself. In Scotland, where an appeal has been made in respect of a decision or notice of SEPA, the Secretary of State may appoint someone to handle the appeal on his behalf under s.114 of the *Environment Act 1995*.

Where the matter is to be decided through written representations, the appropriate Agency has 28 days to comment on the appellant's statement, and the appellant a further 17 days to respond to those comments. All public authorities who commented on the original application for a registration or authorisation may also comment on the appeal; both the appropriate Agency and the appellant have 14 days in which to comment on any such statements received.

Where the matter is to be dealt with in a hearing, the Secretary of State must give 28 days' notice of its date, time and place and must also publish details in a local newspaper. Public authorities who commented on the original application must be given 21 days' notice. The person conducting the hearing may decide if it should be held wholly or partly in public depending on whether matters of national security or commercial confidentiality are involved.

(e) Powers of Secretary of State to deal with Accumulation & Disposal of Radioactive Waste

Section 29 empowers the Secretary of State to arrange for adequate facilities to be provided (by the state or another person) for the safe accumulation or disposal of radioactive waste, to consult with relevant local authorities and to charge for the site's use.

Under s.30 the appropriate Agency (*Env. Act 1995*, Sch. 22, para 217) may dispose of radioactive waste in certain circumstances; these include where premises are empty, the owner is insolvent or there is a risk that the waste may be disposed of unlawfully. Costs may be recovered from the owner or occupier of the premises. Section 30A inserted by the 2005 HASS Regulations deals with the Agencies' powers in respect of the recovery and disposal of orphan sources and for sums to be provided for this.

(f) Offences

It is an offence not to comply with the conditions of registration or authorisation, or with the terms of an enforcement or prohibition notice; summary conviction may result in a maximum fine of £20,000 or maximum six months imprisonment or both for each offence or an unlimited fine or maximum five years imprisonment or both on conviction on indictment (ss.32-38).

Failure to display a copy of registration or authorisation is also an offence, punishable by fine; failure to retain or produce records as specified by the appropriate Agency is also punishable by fine and/or imprisonment.

It is also an offence to disclose information relating to a process unless the person giving it has agreed, disclosure is in accordance with a direction from the Secretary of State, or is in connection with legal proceedings. Summary conviction is subject to a fine not exceeding the statutory maximum and/or up to three months in prison; conviction on indictment is subject to an unlimited fine and/or a maximum two year prison sentence.

(g) Public Access to Documents & Records

Section 39 requires the Environment Agency and SEPA to keep, and to make publicly available, copies of all applications, authorisations and registrations under the RSA, and other relevant documentation, including details of enforcement or prohibition notices, and records of any convictions. Information may be withheld on the grounds of commercial confidentiality or national security and a note to this effect put on the file. Local authorities that have been sent copies of registrations and authorisations etc are also required to make these publicly available.

The public should be allowed access to documents at all reasonable times and may be charged for copies of any documents.

5.39.4 Environment Act 1995

(a) Fees & Charging (ss.41-42)

Prior to implementation of these sections of the *Environment Act 1995*, charging schemes for premises regulated by the *Radioactive Substances Act 1993* and the *Nuclear Installations Act 1965* were made under s.43 of the RSA. This has been replaced by ss.41-42 of the *Environment Act* which contain powers for the Environment Agency and SEPA to make charging schemes for environmental licences and for their approval by the Secretary of State. Fees and charges are usually revised annually with new rates applying on 1 April each year.

The *Charging Scheme for Radioactive Substances Act Regulation* enables the enforcing agencies to recover their costs in respect of the following (Scotland and N. Ireland have similar schemes):

- considering applications, issuing authorisations or registration certificates, including related inspection visits;
- ongoing costs related to authorisations and registrations, e.g. inspection, assessment, reviews, monitoring related to discharges (but not general environmental monitoring), enforcement controls and various administrative costs.

From 2009 the Environment Agency is to introduce a charge for applications to transfer radioactive waste.

Operators are categorised into four bands for charging purposes:

- *Band 1:* British Nuclear Fuels Ltd Sellafield plant;
- *Band 2:* All nuclear sites licensed under the *Nuclear Installations Act 1965* (except Sellafield), together with UK Atomic Energy Authority premises.
- *Band 3:* All other premises authorised under ss.13 or 14 of the RSA to dispose of or accumulate radioactive waste (e.g. hospital or university laboratories).
- *Band 4:* factories and other holders of minor radioactive sources, e.g. small hospitals, registered under s.7 of the RSA or s.10 in respect of mobile equipment.

Premises in Bands 1 and 2 are charged individually on the basis of the actual regulatory time and costs incurred.

Premises in Bands 3 and 4 (all regulated by the appropriate Agency) pay an initial application fee and then an annual subsistence charge. A lower charge applies where the registration relates solely to the use of Technetium 99M and the registered holding does not exceed 10 Gbq (Band 3A) or where the registered holding does not exceed 20 MBq (Band 3B). The annual subsistence charge in Band 4 applies only to those operators holding larger closed sources, any one of which exceeds 4 TBq, measuring, testing and investigative mobile equipment, and open sources kept at premises not subject to authorisation (Bands 4C-4F); the majority in Band 4 (A, B & G), with smaller open or closed sources, will not pay an annual charge since these sources do not require routine inspection.

Failure to pay fees and charges will constitute a breach of authorisation or registration and may result in revocation or other enforcement action. A full refund will often be made if an application is withdrawn within eight weeks of receipt.

(b) Powers of Entry (ss.108-110)

Provisions relating to right of entry to premises regulated under the *Radioactive Substances Act 1993* were formerly set out in s.31 of the RSA. This section of the Act was repealed by the *Environment Act 1995* on 1 April 1996.

The Environment Agency or SEPA or other authorised person must be given access to premises to which a registration or authorisation relates in order to carry out their duties under the Act and in emergency (or possible emergency) situations. Photographs and samples may be taken; the occupiers can be required to leave certain parts of the premises undisturbed and to give the Agency (or SEPA or other authorised person) such assistance as is needed. An offence under this section is subject to a fine.

5.39.5 Transfrontier Shipment of Radioactive Waste

Directive 2006/117/Euratom on *the supervision and control of shipments of radioactive waste and spent fuel*, which came into force on 25 December 2006, replaces Directive 92/3/Euratom on the Transfrontier Shipment of Radioactive Waste. This latter was repealed on 25 December 2008, the date by which Member States were required to transpose its requirements into national legislation. The 2006 Directive lays down systems for authorisation, information exchange and monitoring such transfers both between Member States and into and out of the

European Union. The Directive covers all shipments of radioactive waste and spent fuel if both the quantity and concentration exceed thresholds laid down in the Directive. It does not apply to shipments of disused sources to a supplier or manufacturer of radioactive sources or to a recognised installation; nor does it apply to shipments of radioactive materials recovered for further use through reprocessing or to those containing only naturally occurring radioactive materials.

The 2006 Directive is being implemented through *the Transfrontier Shipment of Radioactive Waste and Spent Fuel Regulations 2008*, a draft of which was issued for consultation in February 2008; they were due to come into force on 25 December 2008, and repeal SI 1993/3031 which implemented the 1992 Directive.

The Regulations designate the Environment Agency (in England and Wales), SEPA (in Scotland) and the Chief Inspector, DOENI as the competent authorities.

Under the Regulations, it an offence

- to ship radioactive waste or spent fuel to a destination outside the UK, or into the UK from outside the EU unless authorised by the competent authority;
- to ship radioactive waste or spent fuel from an EU member state into the UK unless it has been authorised by the competent authority for the state concerned;
- to ship radioactive waste or spent fuel into the UK from a third country by way of transit to another member state unless authorised by the destination member state. Similarly it is an offence to receive shipments into the UK from a third country for transit to another third country unless authorised by the competent authority of the first EU point of entry into the EU.

An authorisation may cover more than one shipment so long as it relates to waste with essentially the same physical, chemical and radioactive characteristics, the shipments are from the same holder to the same consignee, and involve the same competent authorities; and where shipments involve transit through third countries are via the same entry and exit points, unless otherwise agreed between the competent authorities involved. An authorisation may not cover a period of more than three years.

Shipments to African, Caribbean or Pacific states party to the Cotonou ACP-EC Agreement may not be authorised, unless it is to return a shipment of radioactive waste to its country of origin following treatment or reprocessing in the UK; shipments are also banned to third countries which it is considered are unable to deal with radioactive waste or spent fuel safely.

The arrival of a consignment of radioactive waste or spent fuel from outside the UK should be notified to the relevant competent authority within 15 days. Notification should include date of arrival, last EU customs post through which the shipment passed, declaration or certification by the consignee stating that the shipment has reached its correct destination the customs point of entry into the country of destination. In the case of transit shipments authorised under the UK Regulations, the person who applied for the authorisation has a duty to notify the competent authority of the date of arrival and last EU customs post through which the shipment passed.

Applications for authorisation should be made in English using the standard document developed by the Commission; applications in other languages must be accompanied by an authenticated English translation – a Schedule to the Regulations details the procedure to be followed. Shipments of radioactive waste or spent fuel may only be authorised if the following conditions are complied with:

- A written assessment of all practicable options for management of the radioactive waste has been made.
- Low level waste shipments may only be made for the purposes of either the recovery of re-usable materials or treatment to make its subsequent storage and disposal more manageable.
- Shipments of intermediate level or high level radioactive waste may only be made for the purposes of recovery of reusable materials or treatment to make its subsequent storage and disposal more manageable in cases where the processes are at a developmental stage or involve quantities which are too small for the process to be practicable in the country of origin.
- Exports may only be made from the UK to another EU member state or a member of the OECD.
- If the recovery or treatment of the radioactive waste would add to the quantity of such wastes needing to be disposed of in the country of destination, the applicant must confirm that they will be returned to the country of origin, to a timescale agreed with the UK competent authorities and those for country of origin or destination.
- Waste may be imported into the UK for treatment and disposal if it is in the form of spent sealed sources that were manufactured in the UK, or it arises from "small users" such as hospitals in another EU member state that produces such small amounts as to make it impracticable to develop specialised facilities, or developing countries that cannot be expected to acquire suitable disposal facilities.

If authorisation is granted, the competent authority must issue a standard document certifying that the authorisation procedure has been complied with, and the date of authorisation. If authorisation is refused, the applicant may appeal the decision within two months to the Secretary of State (England), Welsh or Scottish Ministers, or DOE NI, as appropriate. The Regulations also enable the Secretary of State (E & W), Scottish Ministers or DOENI, as appropriate, to direct the competent authority to refuse or grant a particular authorisation and if granted, the conditions to be applied.

If the competent authority decides that a shipment may not be completed because it is unlawful (i.e. does not comply with the Regulations), it should notify relevant competent authorities in other EU member States, and if there is one, the third country of origin. In the case of unlawful radioactive waste or spent fuel brought into the UK, the authorisation holder is responsible for ensuring its return to the country of origin, to take corrective action, or otherwise dispose of it, as instructed by the competent authority. If a UK competent authority is notified that a shipment of waste sent from the UK is unlawful, it must ensure that the authorisation holder takes back the waste unless other arrangements are made, and returns it to the UK taking any corrective measures specified by the competent authority.

The Regulations amend s.41(1) of *the Environment Act 1995* to enable the Environment Agency and SEPA to charge fees for determining authorisations etc under these Regulations. Regulation 6(3) enables the Chief Inspector to make a similar charge in NI.

The shipment of radioactive *substances* (sealed sources, other sources and radioactive waste) between EU member states is regulated by Euratom Regulation 1493/93 of June 1993; this outlines the strict procedures to be followed, including pre-notifications, documentation, information exchange etc. Commission Decision (2002) 457, adopted on 12 February 2002, lists national authorities authorised to deal with shipments of radioactive waste between member states. The Commission has also issued a proposal for a Directive on the Supervision and Control of Shipments of Radioactive Waste and Spent Fuel (COM(2004) 716, 12.11.2004).

The European Atomic Energy Community was founded in 1957 by the six founder members of the then European Economic Community – see Appendix 7.

LAND QUALITY

Land may become contaminated as a result of a variety of human activities, and the polluted soil present problems for centuries – the toxic effects of spoil from Roman lead and silver mines are still visible in parts of North Wales while the more recent extraction of zinc in the Shipham area of Somerset resulted in high levels of cadmium in soil – and hence in home grown vegetables – in the area. A considerable amount of contamination can be attributed to past industrial practices when little or no account was taken of potential environmental impacts.

The contamination of soil with toxic chemicals may have direct effects on human health if houses and gardens are built on the land in question. Particles of soil handled or ingested by adults or children may carry irritant or poisonous chemicals while the inhalation of such particles, or vapours from the pollutants, provide another absorption route. Vegetable gardens sited on polluted land may produce crops contaminated by the direct uptake of toxins or the deposition of contaminated particles on the growing plants.

The protection of our soils is also essential for the future of our ecosystems, for farming and food production and for reducing carbon dioxide emissions. Poor management of soil structure, or over-development can lead to erosion, which in turn may cause more flooding and water pollution.

CL:AIRE (Contaminated Land: Applications in Real Environments) is an independent organisation established to stimulate the regeneration of contaminated land in the UK by raising awareness of and confidence in practical sustainable remediation technologies. It is part funded by English Partnerships, Defra, Environment Agency, Welsh Assembly Government and the DOENI. As well as helping to develop a strategy for research on contaminated land in the UK, CL:AIRE provides training, online resources as well as developing advice and codes of practice on various aspects of contaminated land – see Appendix 5.1 and Further Reading section of the *Pollution Control Handbook*.

More at http:www.claire.co.uk/

5.40 SOIL PROTECTION

5.40.1 EU Soil Thematic Strategy

The European Commission published its Thematic Strategy on Soil, together with a proposal for a framework Directive in September 2006. This is the last of the Thematic Strategies presented in accordance with the EU's 6th Environmental Action Programme (see also Appendix 7). The aim of the Strategy is to provide a common framework for preserving, protecting and restoring soil. As such it will promote research on soil and raise public awareness and ensure public participation in the preparation and review of the programmes adopted by Member States.

The proposed framework Directive (COM(2006) 232) requires Member States to identify areas where there is a risk of erosion, organic matter decline, compaction, salinisation and landslides. Risk reduction targets must be set for the identified areas and programmes established to achieve them. An inventory of contaminated sites must be established and national remediation strategies aimed at preventing further contamination developed. When a site on which a potentially contaminating activity has or is taking place is to be sold, a soil status report must be provided by the seller or the buyer to the administration and the other party in the transaction. Member States will be required to limit or mitigate the effects of sealing, for instances by rehabilitating brownfield sites. The Commission is expecting it to take up to two years to reach agreement on the Directive; risk areas would have to be identified within five years of transposition with targets and programmes to achieve them adopted within seven years.

The review of the EU's IPPC Directive (see chapter 2, 2.1) will look at strengthening its soil protection and contamination prevention aspects; consideration will be given to the inclusion of a basic obligation to avoid any pollution risk, returning the site of IPPC installations to a "satisfactory state", as well as a requirement to periodically monitor soil on the site.

5.40.2 Draft Soil Strategy (England)

Defra's draft soil strategy for England (published in March 2008) aims "to maximise the benefits that soils can bring to the economic and environmental well-being of today's generation and future generations", and thus reflects the key principles of the UK's Sustainable Development Strategy. It outlines measures to be taken in the following key areas:

- Sustainable soil management in the agriculture and forestry sector
- Halting the decline of soil carbon
- Sustainable soil management in the built environment
- Protection of soil during the recycling of organic matter to land.

In July 2008, Defra published for consultation a draft *Code of Practice for the Sustainable Use of Soils on Construction Sites*. The aim of the Code is to provide a practical guide to assist anyone involved in the construction industry to protect the soil resources with which they work. The draft Code outlines current guidance and legislation concerning the use of soil in construction projects, before offering stage by stage guidance on the use, management and movement of soil on site. Each section explains the need for the advice and makes recommendations for good practice, including method statements, illustrations and a list of Do's and Don'ts. More: http://www.defra.gov.uk

5.40.3 Draft Soil Strategy (Scotland)

The Scottish Government published its draft Soil Framework at the end of June 2008. It aims to protect Scotland's soils from pollution and climate change, promoting "the sustainable management and protection of soils for the economic, social and environmental stability of Scotland". The document promotes a range of activities to manage soils, raise awareness and ensure soils are considered in planning and other decisions. Four main areas are identified:

- Raise awareness of the distinctiveness of Scottish soils
- Integrate soil protection into other policies
- Improve accessibility to information on soils, including the development of a Scottish Soil Monitoring Network
- Improve understanding, highlighting knowledge gaps and research needs.

5.40.4 Soil Guideline Values

These are being developed by the Environment Agency to help local authorities determine whether a site poses a "significant possibility of significant harm". In September 2005 Defra circulated an advice note, CLAN 2/05, on SGVs and their derivation and their use. This includes advice on whether SGVs are a measure of "unacceptable intake"; it states that SGVs mark the concentration at or below which human exposure can be considered to represent a tolerable or minimal level of risk, and that local authorities should decide on a case-by-case basis whether this equated to an "unacceptable intake". CLAN 6/06 (published November 2006)

In November 2006, Defra's Soil Guideline Values Task Force published for consultation *Soil Guideline Values – the Way Forward* (CLAN 6/06); this outlined proposals for achieving a better match of technical guidance with the policy and legal requirements of Part IIA, focusing on how human health risks should be addressed when evaluating sites and in particular, what constituted "significant possibility of significant harm". The Taskforce also set out "emerging conclusions and issues relating to the production of SGVs insofar as these are affected by

- the science on human toxicology and health effects
- exposure pathways
- practical and policy considerations in assessing risks from land contamination
- decisions on individual sites.

As a next step, revisions are planned for technical guidance CLR 9 and CLR 10 and a new version of CLEA software was released by the Environment Agency in August 2008. CLR7-10 guidance is to be withdrawn. The Agency is also to carry out technical reviews of the body weight and height assumptions used within CLEA, and of the physical chemical properties used to derive soil guideline values. Existing SGVs are to be updated and new guideline values may be produced (depending on the availability of good science, agreement on data interpretation, interaction with ongoing work of other authorities, and funding for the programme beyond March 2009).

'Other' supporting technical work includes the development of site-specific vapour intrusion guidance and work on assessing health risk from asbestos in soil. HPA is producing an introduction to contaminated land for public health professionals (2008). LACORS are producing a summary on circumstances where regulators could expect to get expert advice for specific cases (expected end August 2008). More, including guidance documents available at www.defra.gov.uk/environment/land/contaminated/pubs.htm or www.environment-agency.gov.uk

5.41 CONTAMINATED LAND

The total restoration of contaminated land to an "unpolluted" state is rarely achievable and, in many cases, unnecessary. Instead, a remedial approach tailored to the intensity and extent of the contamination found and the intended end-use of the site is adopted. In some cases it may be necessary to zone the land for specified non-residential purposes, endorsing the title deeds to that effect. The use of lightly contaminated land for amenity purposes may be possible if the materials present are largely inorganic and a sealing sward of tolerant grasses can be planted to prevent both human contact with the soil and the windborne transport of polluted dust. A number of documents are available providing guidance on the development of contaminated land and assessing its suitability for a particular purpose – see bibliography.

More heavily contaminated sites may only be usable for purposes such as car parks, warehouses and other industrial developments where a layer of tarmac or concrete would be laid on the surface, thereby sealing off the pollution. In certain instances, such as the occurrence of highly contaminated pockets in otherwise fairly clean sites, or small sites badly contaminated throughout, the polluted soil may be removed in its entirety to a licensed waste disposal site and replaced with cleaner material.

5.41.1 Planning Controls

Prior to introduction of the current contaminated land regime (see below), the main means of ensuring contaminated land was not used for any unsuitable purposes was through the planning system; thus controls or conditions, based on both current use and the proposed new use, were put on the land, or remedial work specified as part of the development permission.

Annex 2 to PPS 23, *Planning Policy Statement: Planning and Pollution Control* (November 2004, England only) provides guidance for the development of land affected by contamination. This clarifies the relationship between planning and Part IIA of the EPA, and gives more advice on land contamination issues; the need to consider possible contamination at all stages of the planning and development process is noted and where the site history would indicate potential for contamination, planning conditions should ensure appropriate investigation, remediation, monitoring and record keeping. Where contamination is suspected, the developer is responsible both for investigating the land to determine what remedial measures are necessary to ensure its safety and suitability for the purpose proposed, and for actual remediation.

In Scotland, Planning Advice Note 33 on the development of contaminated land (Scottish Government, revised October

2000), provides advice for planning authorities on identifying, assessing and developing contaminated land, including planning approach, powers, duties and financial assistance.

5.41.2 Duty of Care

Section 34 (effective 1 April 1992) of *the Environmental Protection Act 1990* places a duty of care on all those involved in dealing with waste from its generation to its disposal (see this chapter, 5.17). The duty of care extends to closed landfill sites where the site operator remains responsible for the site until it has been declared safe and a "certificate of completion" issued.

5.41.3 Contaminated Land Regulations – Overview

A consultation document, *Paying for our Past* (DOE, 1994), reviewing arrangements for dealing with contaminated land proposed a "suitable for use" approach – i.e. it should be "treated to deal with unacceptable actual or perceived threats to health, safety or the environment, taking account of the actual or intended use of the site"; and that where practical such land should be brought back into beneficial use. This would not preclude a developer or owner from carrying out more extensive remedial work, nor a regulatory body from requiring it. These proposals were given effect through s.57 of the *Environment Act 1995*, which added Part IIA (ss.78A-78YC) to the EPA and which contains the legislative framework for identifying and dealing with contaminated land. These sections of the Act, together with Regulations and Statutory Guidance were brought into force in England on 1 April 2000, in Scotland on 14 July 2000 and in Wales on 1 July 2001.

The Regulations, which give effect to Part IIA of the EPA cover the following:

- Land required to be designated as special sites;
- Pollution of controlled waters (i.e. the circumstances in which controlled waters are adversely affected by land, which is thus designated a special site);
- Content of remediation notices, and persons to whom they should be copied;
- Compensation for rights of entry, etc;
- Grounds of appeal against a remediation notice; appeals to the Magistrates Court and to the Secretary of State;
- Registers.

Part III of the *Waste and Contaminated Land (Northern Ireland) Order 1997*, Articles 49-71, makes similar provision for dealing with contaminated land; district councils are the main enforcing authority, except for land designated as a "special site", which is the responsibility of the Northern Ireland Environment Agency (NIEA), an agency of the DOENI. The NIEA is also responsible for preparing Guidance on relevant Articles of the Order in accordance with Article 69. Reference is made throughout the text to the relevant Articles of the Northern Ireland legislation – draft implementing Regulations (mirroring those which apply elsewhere in the UK) were published for consultation in July 2006, but have not yet been finalised.

Statutory Guidance (2000, revised 2006) contains much of the detail of how the regime operates, and includes guidance on the definition of contaminated land, identification of contaminated land and special sites, remediation, exclusion from, and apportionment of, liability for remediation, and the recovery of the costs of remediation.

The Agencies may issue Guidance to a local authority in respect of its functions in relation to a specific piece of contaminated land (EPA: s.78V; NI: WCLO, Art.67); the local authority is required to have regard to any such Guidance.

The Regulations and Guidance are described in more detail in the main text below.

Local authorities (district councils and unitary authorities) are the enforcing authority for contaminated land and the Environment Agency or Scottish Environment Protection Agency for any land designated a special site due to the nature of its contamination. The Agency also carries out technical research, and publishes scientific and technical advice. The Environment Agency, in collaboration with the Local Government Association, Chartered Institute of Public Health and Defra, has produced the *Local Authority Guide to the Application of Part IIA of the EPA 1990*; it can be downloaded free of charge from the websites of the LGA (www.lga.gov.uk) and CIEH (www.cieh.org); it is also available on CD from the CIEH. The guidance details the main features of the legislation and aims to provide local authority officers with the knowledge they need to enforce the legislation consistently and proportionately.

England

The Contaminated Land (England) Regulations 2006 (SI 2006/1380), which came into force on 4 August 2006, consolidate, with amendments, the provisions of the 2000 Regulations (SI 2000/227), and the 2001 amendment Regulations (SI 2001/663); the earlier Regulations are revoked. They also provide for land which is contaminated as a result of radioactive substances to be designated as special sites, and for appeals against remediation notices to be heard by the Secretary of State (instead of in the Magistrates Court).

The Radioactive Contaminated Land (Enabling Powers) (England) Regulations 2005 (SI 2005/3467), which came into force on 20 January 2006, give the Secretary of State powers to make regulations, give directions and to issue statutory guidance in respect of radioactive contaminated land.

The Radioactive Contaminated Land (Modification of Enactments) (England) Regulations 2006 (SI 2006/1379, as amended by SI 2008/520), which came into force on 4 August 2006, modify the wording of Part IIA of the EPA to extend controls on contaminated land to radioactive contaminated land. 2007 Amendment Regulations (SI 2007/3245) amend the 2006 Regulations to apply the modified Part IIA controls to all radioactivity. They insert a definition of "land contaminated by a nuclear occurrence" and for which the Secretary of State is therefore the responsible person for remediation and make other amendments relating to this responsibility.

The Regulations apply only to radioactivity arising from a historical practice or work activity, not naturally occurring (e.g. radon). In summary the Regulations:

- define "harm" in relation to radioactive contamination meaning "lasting exposure to any person resulting from the after-effects of a radiological emergency, past practice or past work activity".
- Limit a local authority's duty to inspect for radioactive contaminated land to "reasonable grounds" for believing particular land to be radioactively contaminated.
- Define "Mixed" sites – i.e. with both radioactively contaminated and non-radioactively contaminated land – as special sites for regulation by the Environment Agency.

- Require the enforcing authority to carry out remediation itself where contaminated land is causing lasting exposure to radiation; the Secretary of State will be able to consider funding for remediation of "orphan" sites.
- Provide that where the HSE has powers to deal with radioactive contaminated land on a site licensed under *the Nuclear Installations Act 1965*, Part IIA controls will not apply.

Revised Statutory Guidance reflects the changes to the Regulations, including how the regime applies to radioactive contaminated land. Defra is also finalising methodology for radioactively contaminated land exposure assessment (RCLEA) – this is intended to apply to historically contaminated sites (and not nuclear licensed sites). Defra Advice Note CLAN 5/06 provides an update on the extension of Part IIA to cover radioactive contaminated land.

Wales

The Contaminated Land (Wales) Regulations 2006 (SI 2006/2989, W.278), came into force on 10 December 2006. They consolidate, with amendments, the provisions of the 2001 Regulations (SI 2001/2197, W.157), which are revoked. They are similar to the Regulations for England, in providing for land which is contaminated as a result of radioactive substances to be designated as special sites, and for appeals against remediation notices to be made to the National Assembly (instead of the Magistrates Court). Similarly, *the Radioactive Contaminated Land (Modification of Enactments) (Wales) Regulations 2006* (SI 2006/2988, W.277), which came into force on 10 December 2006, modify the wording of Part IIA of the EPA to extend controls on contaminated land to radioactive contaminated land – see above. 2007 Amendment Regulations (SI 2007/3250) amend the 2006 Regulations to insert a definition of "land contaminated by a nuclear occurrence" and for which the Secretary of State is therefore the responsible person for remediation and make other amendments relating to this responsibility.

Scotland

Part IIA is implemented in Scotland through SSI 2000/178 and Statutory Guidance which were brought into force on 14 July 2000 – these are similar to the Regulations which apply in England and Wales.

The Contaminated Land (Scotland) Regulations 2005 (SSI 2005/658), which came into force on 1 April 2006 amend Part IIA of the EPA 1990 and the 2000 Regulations to substitute references to "controlled waters" to "the water environment" to ensure consistency with *the Water Environment and Water Services (Scotland) Act 2003* (see chapter 6, 6.7); they amend the definition of contaminated land to introduce a requirement that pollution be "significant" or likely to be "significant" in respect of the water environment, and define "harm" in relation to the water environment; similar amendments are made to s.79(1)B of the EPA (statutory nuisances). The Regulations insert a new sub-section in s.78YB of the EPA to preclude the serving of a remediation notice where enforcement action is to be taken under the *Controlled Activities Regulations* (see chapter 6, 6.11.6).

The Radioactive Contaminated Land (Scotland) Regulations 2007 (SSI 2007/179), came into force on 30 October 2007; they extend Part IIA of the EPA 1990 to radioactive contaminated land, similar to the regime elsewhere in the UK (see above, England). SI 2007/3240, made by the Secretary of State under powers conferred by ss.78(A)9 and 78YC, amend both the 2007 Regulations and the EPA to insert a definition of "land contaminated by a nuclear occurrence" and for which the Secretary of State is therefore the responsible person for remediation and make other amendments relating to this responsibility.

Revised Statutory Guidance published by the Scottish Government in May 2006 reflects the changes made to the regulatory regime by the 2005 Regulations. Planning Advice Note (PAN 33), published in October 2000 by the Scottish Government provides advice on developing contaminated land.

Northern Ireland

As noted above, draft Regulations implementing Part III of *the Waste and Contaminated Land (Northern Ireland) Order 1997* were published for consultation in July 2006, but have not yet been finalised; the draft Regulations are similar to those applying elsewhere in the UK.

The Radioactive Contaminated Land Regulations (Northern Ireland) 2006 (SR 2006/345, as amended by SR 2007/3236), which came into force on 22 September 2006, place a duty on the Chief Inspector to investigate the condition of land where there are reasonable grounds to believe that it is causing lasting exposure to human beings; where such land has been identified and is the result of a radiological emergency, past practice or past works, the Chief Inspector shall require the responsible person (i.e. the polluter or if the polluter cannot be found, the owner or occupier of the land) to monitor the risk and ensure the land or buildings and access to them is demarcated, and to implement any appropriate intervention taking account of the real characteristics of the land giving rise to lasting exposure. Action to be taken to deal with the risk may be set out in an "intervention notice" served on the responsible person who has 21 days in which to appeal against it to a court of summary jurisdiction. Grounds for appeal include that the person named on the intervention notice has been unreasonably identified as the responsible person, or other people also ought to be identified, that the land subject to the notice has been wrongly identified or demarcated or that there is a material defect in the notice; the notice is suspended pending determination or abandonment of the appeal. It is an offence not to comply with an intervention notice within 21 days of its service, and is subject to a fine not exceeding level 5 on the standard scale, together with a daily fine equal to one-tenth of the fine for each day the offence continues following conviction. If no responsible person can be found on whom to serve an intervention notice then the DOE should take responsibility for the necessary action.

The 2007 Amendment Regulations (SI 2007/3236) amend the 2006 Regulations to insert a definition of "land contaminated by a nuclear occurrence" and for which the Secretary of State is therefore the responsible person for remediation, and make other amendments relating to this responsibility.

5.41.4 Contaminated Land Regulations – Detail

(a) Identification of Contaminated Land & Designation of Special Sites

Under ss.78A-78D of the EPA (NI: WCLO, Arts.50-52) local authorities have a duty to inspect their land to identify whether any is contaminated and whether any such land should be designated as a "special site" (see (f) below) because of the nature of the contamination. In identifying contaminated land, local authorities must act in accordance with Guidance from the Secretary of State. This requires them to "take a strategic approach to the identification of land which merits detailed individual inspection". Thus all local authorities are required to draw up a Contaminated Land Strategy which looks at those circumstances – such as evidence of past use of land, evidence that serious harm or serious pollution of controlled waters (Sc: water environment) has or is being caused; records of contamination and remedial action that has already been taken; in developing its Strategy, which should have been formally adopted and published by June 2001, the authority should consult the Environment Agency and other appropriate public bodies, Natural England, English Heritage and Defra. Development of the Strategy will enable local authorities to ensure that resources are used efficiently and effectively and concentrated on identifying and investigating those sites most likely to be most seriously contaminated.

In Scotland local authorities were required to complete their initial strategies for identifying contaminated land by 14 October 2001, consulting as appropriate; in Wales strategies had to be completed by 1 October 2002 – 15 months after the Welsh Regulations came into force.

Section 78A(2) (as amended by SI 2007/3245, E; SI 2007/3250, W) defines contaminated land as

> "land which appears . . . to be in such a condition, by reason of substances in, on or under the land, that -
>
> (a) significant harm is being caused or there is a significant possibility of such harm being caused (SPOSH); or
>
> (b) pollution of controlled waters is being, or is likely to be caused."

Section 86(2)(a) of *the Water Act 2003* will (when brought into force) amend s.78A(2)b of the EPA to read "significant pollution of controlled waters is being caused, or there is a significant possibility of such pollution being caused". (Section 86(f) of *the Water Act 2003* – brought into force 1.10.04 – also clarifies that groundwater for the purposes of the contaminated land regime does not include waters contained in underground strata but above the saturation zone.) In Scotland, *the Contaminated Land (Scotland) Regulations 2005* make a similar change to the definition of contaminated land and replace the term "controlled waters" with "water environment" in line with the Framework Directive on Water (see chapter 6, 6.2); Both the Secretary of State and the Scottish Government are to issue statutory guidance on the criteria to be used in determining what pollution of the water environment is to be considered significant.

Statutory Guidance on Part IIA (issued in 2000, revised 2006) defines what "harm" (and in Scotland, SSI 2005/658, pollution of the water environment) is to be regarded as "significant" to:

- *human beings:* death, disease, serious injury, genetic mutation, birth defects, or the impairment of reproductive functions. Disease is to be taken to mean an unhealthy condition of the body or some part thereof; this might include cancer, liver dysfunction or extensive skin ailments.
- *living organisms or ecological systems:* an irreversible or other substantial adverse change in the functioning of the habitat or site;
- *property (crops (inc. timber), produce grown domestically or on allotments for consumption, livestock, other owned animals, wild animals which are the subject of shooting or fishing rights, crops):* for crops, this means a substantial diminution in yield or loss of value due to death, disease or other physical damage; for domestic pets the benchmark is death, serious disease or serious physical damage; for food, when it is no longer fit for the purpose intended and fails to comply with food safety legislation;
- *property (buildings):* structural failure or substantial damage making them unfit for their intended purpose.

Defra has also published (July 2008) non-statutory *Guidance on the Legal Definition of Contaminated Land* for local authorities in England. This outlines Defra's view on how local authorities should go about deciding what is SPOSH under Part IIA, and how they should proceed in an area "which is unavoidably complex". This is also available on the contaminated land pages of Defra's website.

In relation to radioactive contaminated land "harm" means "lasting exposure to any person resulting from the after-effects of a radiological emergency, past practice or past work activity" (SI 2007/3245, E; SI 2007/3250, W; SI 2007/3240, Sc; SI 2007/3236, NI).

The Contaminated Land Exposure Assessment (CLEA) Model (Environment Agency 2008). Originally published in 2002, the revised package, designed to help in the assessment of potential risks to human health from long-term exposure to soil contamination, includes:

- Human health toxicological assessment of contaminants in soil (the TOX Guidance Report) (replacing CLR9).
- Updated technical background to the CLEA model (the CLEA report) (eplacing CLR10).
- A review of the body weight and height data used within the CLEA model.
- The new CLEA Software and Handbook, version 1.03 beta, which replaces CLEA UK Software for aiding in the estimation of exposure and assessment of risk.

This non-statutory guidance sets out the method for deriving Health Criteria Values (HCVs) and Soil Guideline Values (SGVs) and takes account of developments in scientific knowledge and practice since the original release in 2002. It aims to be more accessible than its predecessor, and should help provide greater clarity in assessing the human health risks posed by contaminated land. The guidance and software can be found at www.environment-agency.gov.uk/clea.

It should be noted that while two or more sites if considered individually would not be considered to be contaminated, but if considered together would, then each site should be treated as being contaminated (s.78X(1)). Also if land adjoining or adjacent to the area of a local authority appears contaminated, it may act as though the land were within its boundary (s.78X(2)). (NI: WCLO, Arts.68(1) and 68(2) respectively.)

Following identification, the local authority should then by notice (s.78B) inform the appropriate agency, the owner of the land, the occupiers of any part of the land and any other appropriate person.

If the local authority considers that any of the land which it has identified as being contaminated warrants designation as a special site, it should seek the appropriate Agency's advice in making its decision (see (f) below). The local authority should also notify the relevant persons that it considers a site should be designated a special site. The Agency may also advise the local authority that it considers certain land should be designated as a special site; only local authorities have powers to designate land as being contaminated.

Following a decision to designate land as a special site, the local authority should notify the appropriate Agency and other relevant persons (as above). The decision then takes effect on the day after the expiration of 21 days of the notification, or the day after notification is received from the appropriate Agency that it agrees with the designation (whichever is the soonest). All the relevant persons should then be notified that the decision has taken effect.

However, if the appropriate Agency disagrees with the local authority over whether land should be designated a special site, it has 21 days from the date of notification of the decision to notify the local authority that it disagrees and to give a statement of its reasons for doing so. The matter is then referred by the local authority to the Secretary of State (NI: Planning Appeals Commission) for decision. The appropriate Agency should send the local authority a copy of its notification and statement. The local authority should also notify all relevant persons that a decision has been referred to the Secretary of State (or Planning Appeals Commission in NI); decisions to designate special sites do not take effect until the referral is determined and then take effect (as confirmed, revised, etc) the day after notice of determination has been sent to the relevant persons and to the local authority.

(b) Remediation Notices

Sections 78E-G, H(1)-(4) (NI: WCLO, Arts.53 & 70) cover remediation notices. Following designation of land as being contaminated land or a special site, the enforcing authority should serve a Remediation Notice on the appropriate person(s) specifying what needs to be done and the period within which the various pieces of work should be done. The *Contaminated Land Regulations 2006* (Sc. 2000; W. 2006) specify the contents of the Remediation Notice. The Notice should include the name and address of all the persons on whom the Notice is being served and the reason for serving it on them – i.e. why the enforcing authority considers them to be an "appropriate person"; it should also include details of the location and extent of the contaminated land in question and the substances by which it is contaminated; particulars of the significant harm, harm or pollution of controlled waters (Sc. water environment) by reason of which the land is contaminated.

The appropriate person so far as responsibility for remediation is concerned will normally be the person(s) who knowingly caused or permitted the presence of a substance in, on or under the land, thereby causing it to become contaminated. However if after inquiry the person who knowingly caused or permitted the contamination cannot be found, then the appropriate person is the person(s) who currently occupy or own the land; however that person will not be responsible for cleaning up adjoining sites if these have been contaminated too. Where more than one person is required to carry out remediation work because of the presence of different substances, a Notice should be served on each person specifying what each needs to do. Where more than one person is to be responsible for cleaning up the same contamination, the Notice should specify the proportion of works and costs to be borne by each in accordance with Guidance from the Secretary of State.

Before serving the remediation notice, however, the enforcing authority should endeavour to consult the appropriate persons about the remediation to be done and, where necessary to consult any other person who might be required to give consent or grant rights (e.g. of access) for certain work to be done; where known, the name and address of all such persons should also be included on the Remediation Notice. Such rights or consent should be granted to the person on whom the remediation notice is served (s.78G). The person granting consent or rights may be entitled to compensation from the person on whom the remediation notice has been served. Regulation 6 and Schedule 2 (E; Sc Sch.3) of the *Contaminated Land Regulations* outline the procedure for making such a claim and the way in which compensation is to be determined.

In specifying what it is reasonable to do by way of remediation, the enforcing authority must take into account the likely costs of the work and the seriousness of the harm, or pollution of controlled waters; the enforcing authorities should have regard to any Guidance from the Secretary of State as to what should be considered as "reasonable", the remediation to be done in any particular case and the standard to which land or waters should be remediated. Guidance on calculating the costs associated with remediating contaminated land was published by English Partnerships in the form of a Best Practice Note on 18 March 2008 (www.englishpartnerships.co.uk)

It should be noted that a Remediation Notice should not normally be served before the expiry of three months following service of the initial notice identifying the contaminated land (see (a) above) (s.78H); exceptions include when the enforcing authority considers that the condition of the land warrants it, there is imminent danger of serious harm, or of serious pollution to controlled waters.

The Remediation Notice should also note that it is an offence (EPA: s.78M; NI: WCLO, Art.59) not to comply with the terms of a Remediation Notice without reasonable excuse and the penalties that non-compliance may incur – i.e. on summary conviction this may result in a fine not exceeding level 5 on the standard scale, plus an amount equal to one-tenth of level 5 for each day the offence continues following conviction. Where the offence relates to industrial, trade or business premises, a fine not exceeding £20,000 may be imposed, plus a further fine of one-tenth of £20,000 for each day the offence continues following conviction.

A Remediation Notice should not be served if contamination has arisen as a result of non-compliance with an IPC authorisation, or enforcement or prohibition notice served under Part I of the EPA (NI: *Industrial Pollution Control Order 1997*) or, as appropriate an installation regulated under the *Pollution Prevention and Control Regulations 2000*. In such circumstances, the appropriate Agency (as IPC/IPPC enforcing authorities) may with the written permission of the Secretary of State and of any other person whose land is affected, take remedial action and recover costs from the offender. Similarly where land becomes contaminated as a result of a breach of a waste management licence issued under Part II of the EPA (NI: Part II of WCLO), the offence should be dealt with under that Part of the Act (EPA: s.78YB (1)-(3); NI: WCLO, Art.70). Contamination due to unlawful waste deposits (e.g. fly-tipping) will normally be dealt with under s.59 of the EPA (see 5.22 above). In Scotland a remediation notice should not be served if contamination is a result of an activity to which the *Controlled Activities Regulations 2005* (see chapter 6, 6.11.6) apply and enforcement action can be taken under those Regulations (*2006 Contaminated Land Regulations*).

A remediation notice should not require any action which could impede or prevent a discharge permitted under a consent granted under Part III of the *Water Resources Act 1991* (EPA: s.78YB (4)). In Northern Ireland Art.70(4) of the WCLO contains the same provisions in relation to the *Water Act (Northern Ireland) 1972* and the *Water and Sewerage Services (Northern Ireland) Order 1973.*

Finally, the Remediation Notice should also state that there is a right of appeal against it, grounds for appeal, and the time within which any appeal should be made, and that such Notices are suspended pending the determination or abandonment of an appeal – see (d) below.

The *Contaminated Land Regulations* require local authorities to send copies of Remediation Notices which they serve to the Environment Agency/SEPA. The Environment Agency/SEPA should copy any Remediation Notices it serves to the local authority in which the contaminated land in question is situated.

(c) Remediation Declarations and Statements

These are covered under s.78H(5)-(9) of the EPA (NI: WCLO, Art.56(5)-(9)). A remediation notice will not be served in the following circumstances (see also above):

(a) where it appears to the enforcing authority that there is nothing by way of remediation that could be specified in the notice taking into account such matters as cost, seriousness of harm or pollution, standard of remediation, what it is reasonable to do, etc;

in this instance the enforcing authority must publish a Remediation Declaration saying what it would have specified in a Remediation Notice if it could have, and its reasons for not doing so.

(b) if the appropriate person is already carrying out the necessary remediation work or it has been done or is planned;

(c) if it appears that the appropriate person on whom to serve the remediation notice is the enforcing authority itself;

(d) if the enforcing authority is to carry out the remediation work itself for various reasons – see below;

in the case of (b) above the person(s) carrying out the remediation should publish a Remediation Statement or in the case of (c) and (d) the enforcing authority. The Statement should say what is, has or is likely to be done and the timescale, and the names and addresses of all those involved in the remediation work.

If, however, the appropriate person in (b) above does not publish a Remediation Statement within a reasonable time after the date on which a Remediation Notice could have been served, then the enforcing authority may publish the Statement, recovering its costs from the appropriate person. The enforcing authority may also then serve a Remediation Notice.

(d) Appeals

A person on whom a Remediation Notice has been served may appeal (EPA, s.78L; WCLO, Art.58) within 21 days to the appropriate authority; In England all appeals should be directed to the Secretary of State (SI 2006/1380) and in Wales to the National Assembly (SI 2006/2989, W.278); In Scotland or N. Ireland, they should be directed to the Sheriff or NI Court of Summary Jurisdiction in the case of a notice served by a local authority, with appeals against notices (relating to special sites) served by SEPA sent to the Secretary of State. (Section 114 of the *Environment Act 1995* enables the Secretary of State to appoint someone to handle the appeal on his behalf.) In Northern Ireland appeals in relation to special sites will be dealt with by the Planning Appeals Commission (Art.58).

*Section 104 of the *Clean Neighbourhoods & Environment Act 2005* amended s.78L in England & Wales to require all appeals against remediation notices to be referred to the Secretary of State or National Assembly of Wales, as appropriate.

The *Contaminated Land Regulations* detail the grounds for appeal against a Remediation Notice which include:

- failure of the enforcing authority to have regard to, or act in accordance with any parts of the Statutory Guidance;
- that action to deal with the contaminated land should be dealt with under other legislation (e.g. breach of an environmental licence);
- a defect in the Remediation Notice (e.g. should have been addressed to another person or persons; where addressed to more than one person, failed to specify what each should do by way of remediation, or to apportion liability or unreasonably apportioned liability);
- that the time allowed for remediation is insufficient to carry out the required works;
- that the enforcing authority in seeking to recover all or part of its costs, failed to take account of the hardship such action would cause or unreasonably determined that it would recover such costs;
- that the enforcing authority has failed to take account of action already being taken to deal with the contamination without service of a Remediation Notice.

In filing a Notice of Appeal at the Magistrates Court (W), the appellant should copy it to the enforcing authority and to any other person mentioned in the Remediation Notice or Notice of Appeal. A copy of the Remediation Notice and names and addresses of all those listed in the Notice of Appeal and Remediation Notice should also be filed and copied as before. Directions for the conduct of the appeal are notified to all concerned by the justices' clerk or the court.

Where an appeal concerns a Remediation Notice in respect of a special site, the appellant should state whether he wishes the appeal to be determined through written representations or a hearing. A copy of the Appeal Notice should be sent to the Environment Agency/SEPA and all those named in the Notice or in the Remediation Notice, together with a statement noting the names and addresses of all those concerned. The Secretary of State decides if the hearing should be held in private or take the form of a local inquiry. A person appointed to conduct a hearing will report his recommendations as to how the appeal should be determined to the Secretary of State unless he has been appointed under s.114 of *the Environment Act 1995*; this enables the Secretary of State to delegate his functions in respect of appeals to another person.

Where it is proposed to modify a Remediation Notice in a manner which would be less favourable to the appellant than the original Notice, a copy of the proposed modification should be sent to all those who received notification of the appeal, including the enforcing authority; all are permitted to make representations regarding the proposed modification if they so wish.

Remediation Notices do not take effect pending determination or abandonment of an appeal. The Secretary of State/NAW (as appropriate) can refuse an appellant's request for an appeal to be abandoned where the appellant has been notified that it is proposed to modify the Remediation Notice.

(e) Powers of Enforcing Authority to carry out Remediation

The enforcing authority may carry out the remediation work itself in the following circumstances (EPA, s.78N & P; WCLO, Arts.60 & 61):

- to prevent occurrence of serious harm, or serious pollution of controlled waters, of which there is imminent danger;
- where the appropriate person has given written agreement that the enforcing authority should carry out the remediation work (at his/her expense);
- failure to comply with remediation notice;
- contamination has resulted from water from an abandoned mine or where the owner or occupier of land has not caused or knowingly permitted the presence of contaminating substances in on or under land which have subsequently contaminated other land;
- where the enforcing authority decides to carry out certain work itself but does not intend to recover all or part of the cost of remediation work;
- where, after reasonable inquiry, the appropriate person cannot be found to do the particular work.

Where the enforcing authority has carried out the remediation work itself, it may recover all or part of its costs, bearing in mind the degree of hardship this may cause to the person concerned and with regard to any Guidance from the Secretary of State. The enforcing authority should serve a Charging Notice on the person concerned detailing the amount of the charge and any interest payable. The charging notice should also be copied to anyone who might have an interest in the premises affected by the charge. The charge becomes a charge on the premises 21 days after the service of the notice or, if there is an appeal, following its determination. (Any appeal should be made within 21 days of its service.) Where the charge is a charge on the premises, the enforcing authority may by order recover the costs over a period of up to 30 years.

(f) Special Sites

The Regulations require contaminated land to be designated as a special site in the following circumstances (EPA, s.78Q; WCLO, Art.62):

- Presence of waste acid tars in, on or under the land;
- Land which has been used for purification (inc. refining of crude petroleum or oil extraction from petroleum, shale or other bituminous substance (except coal), or for manufacture or processing of explosives;
- Land within a nuclear site;
- Land on which an IPC process or IPPC installation designated for control by the Agency (unconnected with a remediation process) has or is being carried out;
- Land owned or occupied by or on behalf of the Secretary of State for Defence or other defence organisation which is being used for naval, military or air force purposes;
- Land on which chemical weapons or biological weapons have been, or are being, manufactured, produced or disposed of, or land which has been designated under the *Atomic Weapons Establishment Act 1991*.

In all the above cases, the adjoining or adjacent land will also be designated a special site if it appears to have been contaminated as a result of an escape from the actual site.

Contaminated land will also be designated as a special site in the following circumstances:

- Land which adversely affects controlled waters (Sc. water environment) which are, or are intended to be used, for the supply of drinking water for human consumption, and which will need treatment before they can be regarded as wholesome under Part III of the *Water Industry Act 1991*;
- Land which adversely affects controlled waters (Sc. water environment) which results in those waters not meeting the relevant quality classifications under s.82 of the *Water Resources Act 1991* (see chapter 6, 6.11.1);
- Land which is or is likely to pollute controlled waters (Sc. water environment) due to the escape of specified dangerous substances and where the waters are contained in specified underground strata – listed in Schedule 1 to the Regulations.

If contaminated land is designated a special site after a remediation notice has been served, the appropriate Agency may adopt the Remediation Notice, notifying the local authority and any appropriate person of its decision to do so. If the local authority has already begun doing some remediation work, it may continue to do so and the appropriate Agency may inspect the land from time to time to review its condition.

Where the Agency considers that contaminated land no longer requires to be designated as a special site, it should notify both the Secretary of State and the local authority. In Northern Ireland, the WCLI will notify the district council. The land may subsequently be redesignated as a special site if its condition requires.

(g) Public Registers

Sections 78R-T (NI: WCLO, Arts.63-65 & Sch.2) detail the requirements for maintaining records of contaminated land. Each enforcing authority is required to maintain a public register containing particulars relating to contaminated land and special sites for which it is the enforcing authority; local authority registers will also contain details relating to special sites in its area. The registers which may be in any form should be available for inspection by the public, with copies of documents available at a reasonable charge. *The Contaminated Land Regulations 2006* (E & W), Sch.3; (Sc 2000, Sch.4) specify that, as well as details of the location and extent of contaminated land, name and address of person(s) responsible for remediation, the registers should contain full details of the following:

- Remediation notices and appeals against such notices;
- Remediation statements and declarations; any other notifications given to the enforcing authority relating to work which has been done, or is claimed to have been done, as a result of a remediation notice;
- Appeals against charging notices;
- Notices relating to the designation of land as a special site, or withdrawal of that designation;
- Convictions for offences under s.78M of the EPA;
- for local authorities – guidance issued by the Environment Agency under s.78V of the EPA; Agency registers should note the date of any such guidance issued;
- Details of land which is contaminated as a result of
  - a breach of an IPC authorisation (Part I of EPA) or IPPC permit;
  - deposit of controlled waste (Part II of EPA);
  - discharge covered by a Consent under the *Water Resources Act 1991,* or *Controlled Activities Regulations 2005*

    and which is therefore dealt with under the relevant legislation. (The enforcing authority is precluded from serving a remediation notice – s.78YB – in these circumstances.)

Information may be excluded from the registers on the grounds of national security or commercial confidentiality. In the former case, the enforcing authority should notify the Secretary of State of any information excluded on these grounds; or the person concerned may notify the Secretary of State direct of the information to be excluded on these grounds and notify the enforcing authority accordingly.

Where a person requests that information be treated as commercially confidential but the enforcing authority disagrees, 21 days should elapse before putting the information on the register. An appeal against the enforcing authority's decision may be made to the Secretary of State (NI Planning Appeals Commission) in which case the information in question will not be put on the register until seven days following determination or withdrawal of the appeal. Where either party requests a hearing, or the Secretary of State (NI Planning Appeals Commission) decides on a hearing, this must be held in private. Information remains commercially confidential for four years, but the person concerned may apply to the enforcing authority for this classification to be extended.

(h) Reports on the State of Contaminated Land

The Environment Agency, SEPA and the NIEA are required to prepare and publish from time to time reports on the state of contaminated land in their respective countries. They may request local authorities to provide information to enable them to prepare their reports. (EPA, s.78U; WCLO, Art.66)

5.41.5 Anti-Pollution Works

Section 161 of the *Water Resources Act 1991* empowers the Agency to serve a "works notice" on any person who has "caused or knowingly permitted" a pollutant to enter controlled waters, including from contaminated land, requiring them to deal with the problem. If urgent action is needed the Agency is empowered to take remedial action to deal with the pollution and to recover the costs from the person responsible for the pollution. The *Control of Pollution Act 1974*, sections 46A-C, give the same powers to SEPA in Scotland, with Articles 16-19 of the *Waste and Contaminated Land Order* providing similar powers for the NIEA in Northern Ireland. See chapter 6.

Guidance from the Agency (Policy and Guidance on the Use of Anti-Pollution Works Notices) suggests that in most cases of actual or potential pollution of controlled waters as a result of contamination, the problem will usually be dealt with under the contaminated land provisions of the EPA.

6 WATER POLLUTION

Historical controls

European Union directives

Bathing water

Drinking water

Consents to discharge

Agricultural pollution

Marine pollution

Pesticides

WATER POLLUTION – GENERAL

The earliest recorded legislation prohibiting the pollution of water dates from 1388 with an Act forbidding the dumping of animal remains, dung and garbage into rivers, ditches and streams because of the "great annoyance damage and peril of the inhabitants" – the penalty for non-compliance being death. *The Bill of Sewers* in 1531 laid a duty on persons responsible for sewers "to cleanse, and purge the trenches, sewers and ditches, in all places necessary" with a view to protecting public health.

6.1 FIRST LEGISLATIVE CONTROLS

6.1.1 1388-1974

It was not until the 19th century when the effects of increasing industrialisation and, as a consequence, the increasing concentration of the population in towns and cities and industrial areas, that more attention was paid to the problems of industrial pollution of water and the need to protect public health through the provision of proper sanitation. A series of Acts in 1847, known as the "Clauses Acts" variously provided a framework for legislative control of both industrial and domestic water pollution. The first *Public Health Act* in 1848 also aimed to improve sanitation by providing for a system of local management of water supply, sewerage etc which would be subject to overall national control; subsequent legislation between 1858 and 1875 amended and extended the provisions of the 1848 Act with the *Public Health Act 1875* remaining the primary piece of legislation in respect of sanitation facilities until the *Public Health Act 1936*.

The first statute specifically aimed at controlling water pollution was the *Salmon Fisheries Act 1861* which made it an offence to discharge sewage into salmon fishing waters; its main objective was however the protection of salmon and fishing interests, rather than protection of public health. This was followed by the *Rivers Pollution Prevention Act 1876* under which it became a criminal offence to pollute water. Part I of the Act dealt with pollution by solid matter, Part II with sewage pollution and Part III with pollution from manufacturing and mining. The Act provided for local authorities to take legal proceedings against offenders after having obtained permission from central government. The court could require the offender to abstain from committing the offence or order him to treat the discharge in a specific manner in future. The Act also included a defence of "best practicable and available means". Thus, if it could be shown that the best practicable and reasonably available means had been used to render harmless sewage or other polluting matter before discharging to water, no offence was committed.

The increasing demand for water for abstraction purposes both for domestic and industrial use in the late 19th century led to the creation of local bodies responsible for the management of the rivers which served the various towns and districts, and for those serving the major urban and industrial areas, and by the 1930s they existed for most of the more important rivers. The *River Boards Act 1948* established river boards for the whole of England except the Thames and London area. They acquired functions relating to land drainage, fisheries and river pollution within defined catchment areas. The boards were superseded by 27 river authorities as a result of the *Water Resources Act 1963*.

Other Acts, such as the *Salmon and Freshwater Fisheries Act 1923*, the *Water Act 1945* and the *River Boards Act 1948* laid down various powers to prevent pollution of water but the procedure outlined in the 1876 Act remained largely intact until the *Rivers (Prevention of Pollution) Act 1951*. This repealed virtually the whole of the 1876 Act and established a procedure whereby anyone wishing to make a new discharge of sewage or trade effluent to a stream was required to seek the consent of the river board (subsequently the regional water authority). *The Public Health Acts* of 1936 and 1961 and the *Public Health (Drainage of Trade Premises) Act 1937* made it an offence to discharge into public sewers any matter which by its nature or its temperature was likely to damage the sewer or the treatment process, with authorities empowered to impose conditions on discharges of trade effluent. *The Clean Rivers (Estuaries and Tidal Waters) Act 1960* extended controls to discharges to estuaries, and *the Rivers (Prevention of Pollution) Act 1961* to pre-1951 discharges for which specific consent had not been required by the earlier Act. *The Water Resources Act 1963* extended the control system to cover discharges to underground strata via wells, pipes or boreholes, and gave river authorities specific powers to take emergency measures in the event of pollution of the water.

The river authorities in England and Wales remained in existence until 1973 when the *Water Act* of that year replaced them with ten regional water authorities. Each Authority was made responsible for one or more of the major river systems (e.g. Thames, Severn Trent), including the management of the entire hydrological cycle within its area and the adjacent coastal waters. The responsibility included water conservation and supply, pollution control, sewerage, sewage treatment, the development and control of aquifers, land drainage, flood prevention, freshwater and sea fisheries, and the use of the water for amenity and recreation.

6.1.2 Control of Pollution Act 1974, Part II

Part II of the *Control of Pollution Act 1974* was the next major statute controlling water pollution, re-enacting and strengthening the earlier controls in the *Rivers (Prevention of Pollution) Acts 1951* and 1961 and ss.72 & 76 of the *Water Resources Act 1963*. The main objectives of COPA were

a) to extend controls to all inland waters, estuaries, tidal rivers, the sea within a three-mile limit (and in certain cases beyond) and specified underground waters;

b) to prohibit, subject to a few exceptions, the entry of poisonous, noxious or other polluting matter through casual or spontaneous acts, (e.g. dumping waste in a river) to water;

c) to permit discharges of trade or sewage effluent providing that consent is obtained and that any conditions attached to that consent observed;

d) to enable water authorities to carry out operations to prevent the pollution of water or, where pollution has already occurred, to remedy its effects and to recover the costs of so doing from those responsible;

e) to expand and open procedures for the control of effluent discharges to public involvement, including provision for public advertisement, third party representations and, exceptionally, call-in by the Secretary of State and public enquiry.

In ***England and Wales***, controls under COPA were replaced by the *Water Act 1989*; This Act paved the way for the sale of the utility functions of the existing water authorities to the private sector, and established the National Rivers Authority (NRA) – a statutory body – to take over the responsibilities of water authorities in relation to water pollution, water resource management, flood defence, fisheries, and in some areas navigation; the Act removed the regulatory functions of water authorities and provided for the appointment of a Director General of Water Services responsible for the provision of water and sewerage services. A new statutory framework for the control of drinking water quality, river quality and other standards, was established and the NRA required to take specific steps, such as controlling polluting discharges in order to attain objectives set by the Secretary of State for maintaining and improving river quality. Most of the *Water Act* has now been repealed by, and re-enacted in, the *Water Industry Act 1991* and the *Water Resources Act 1991*. These Acts took effect on 1 December 1991. The former Act deals with the functions of the water and sewerage undertakers, the Director General of Water Services and local authority responsibilities for water supplies. The latter Act covered the duties of the National Rivers Authority, which from 1 April 1996 became the responsibility of the Environment Agency – see 6.7.2 below. The main purpose of these two Acts, together with three others enacted at the same time, is to consolidate all the legislation covering water, which was previously spread over some 20 statutes.

Until 1 April 2006, Part II of COPA was the main legislation (although substantially added to and amended since 1974) for controlling water pollution in ***Scotland*** (see 6.8 below). Enforcement was the responsibility of the River Purification Authorities until their replacement by the Scottish Environment Protection Agency on 1 April 1996. However, from 1 April 2006, all point source discharges, abstractions and impoundments and engineering work required an authorisation from SEPA under *the Water Environment (Controlled Activities) (Scotland) Regulations 2005* (SSI 2005/348); existing consents under COPA could be transferred to the new regime under transitional arrangements between 1 October 2005 and 31 March 2006 – see below 6.11.6.

THE EU & WATER POLLUTION

The European Commission's 5th Action Programme, *Towards Sustainability*, (see also Appendix 7) outlined a number of objectives aimed at securing sufficient water supplies and maintaining and improving quality, including:

- The prevention of pollution of fresh and marine surface waters and groundwater;
- The restoration of natural ground and surface waters to an ecologically sound condition; and
- ensuring that water demand and water supply are brought into equilibrium on the basis of more rational use and management of water resources.

The 6th Action Programme – *Environment 2010: Our Future, Our Choice* – which covers the period 2001-2010 identifies the sustainable use of natural resources as a priority objective, with, in relation to water, action focusing on the following:

- achieving quality levels of ground and surface water that do not give rise to significant impacts on and risks to human health;
- phasing out the discharge of hazardous substances to water;
- revision of the bathing water Directive;
- ensuring the integration of approach of the water framework Directive and water quality objectives in the Common Agricultural Policy and regional development policy.

The EU's 2000 Framework Directive on Water Policy – see next section – provides the means by which it is hoped to achieve the EU's objectives in this field. EU policies and initiatives, including the *Thematic Strategy on the Protection and Conservation of the Marine Environment*, are summarised later in this chapter of the Pollution Control Handbook at 6.23.

Brief details of EU institutions and legislative processes are given in Appendix 7. Water-related Directives are listed in Appendix 7.6.

6.2 FRAMEWORK DIRECTIVE ON WATER POLICY

Directive 2000/60 of the European Parliament and the Council establishing a *Framework for Community Action in the field of Water Policy* was adopted on 23 October 2000; it entered into force on 22 December 2000 and had to be implemented by Member States by 22 December 2003. The overall purpose of the Directive is to establish a framework for the protection of surface fresh water, estuaries, coastal waters and groundwater in the Community; the objective is to prevent deterioration and protect and enhance the status of aquatic ecosystems; to promote sustainable water consumption; and to contribute to the provision of a supply of water in the qualities and quantities needed for sustainable use of resources. Among the Directive's main features are:

- Member States must take all necessary measures to ensure groundwater quality does not deteriorate and to prevent or limit the input of pollutants to groundwater. A proposal for a Directive (COM(2003) 550 defining measures to achieve good groundwater chemical status within 15 years was published in September 2003 – see 6.4.3 below;
- discharges of hazardous substances must cease or be phased out within 20 years of their identification as a priority hazardous substance (see below);

- by 2010 water pricing policies to reflect the need to encourage efficient use of water resources and attainment of environmental objectives of the Directive; costs should have regard to the polluter pays principle; regional geological conditions, and economic and social conditions may however be taken into account.

The Directive defines guiding principles for the qualitative and quantitative protection of groundwater, adopting a more integrated approach to water policy and applying both the precautionary and polluter pays principles; in addition the economic development of a region must not be allowed to jeopardise the status of surface waters. Environmental quality objectives using Best Available Techniques (BAT) are to be set so as to ensure consistency with emission limits required by other legislation, e.g. the IPPC Directive and that on pollution caused by dangerous substances to water (76/464/EEC) – see 6.4 below.

By 22 December 2006, Member States had to have consulted on, and designated river basin management areas and to have established arrangements for cooperation where use of water within a river basin has transboundary effects. River basin management plans which meet the objectives of the Directive and achieve good ground and surface water status by 22 December 2015 must be in place by 22 December 2010; for groundwater this would be on the basis of chemical and quantitative status and for surface water on the basis of ecological and chemical status – these are defined in an annex to the Directive together with monitoring and other requirements. Member States unable to meet the deadline because improvements within that time are either technically infeasible or disproportionately expensive may apply for a derogation but must not allow water quality to deteriorate further within the period of the derogation.

Other requirements to have been carried out for each river basin district (RBD) by 22 December 2004 included

- an analysis of the characteristics of each river basin;
- a review of the impact of human activity on the status of surface and groundwater for each RBD, to include estimates of point source pollution, diffuse source pollution, and of water abstractions, and an analysis of other anthropogenic influences on the status of water;
- an economic analysis of water use.

A register of those areas within each RBD requiring special protection had to be established by 22 December 2004, as well as monitoring programmes for such areas, and monitoring of surface and ground water status. By 2010, Member States must ensure full cost recovery for all services provided for water use, but may take into account the social, economic and environmental effects of such a policy.

In November 2001, the Council of Environment Ministers formally adopted Decision 2455/2001/EC *establishing the list of priority substances in the field of water policy*, which becomes Annex X of the Framework Directive. The list comprises 33 substances (see Appendix 6.1A) selected on the basis of the risk they pose to the aquatic environment and to human health. The Decision entered into force on 16 December 2001 and had to be implemented by December 2006. The substances have been classified as Priority Hazardous Substances, Priority Substances under Review and Priority Substances. This latter category comprises substances for which "there is no element, in the light of best knowledge available, to prove they are 'toxic, persistent and liable to bio-accumulate' or should be regarded as giving rise to an equivalent level of concern"; production and use should however be progressively reduced and the Commission is required to propose Daughter Directives setting emission controls and EQSs for them – see below 6.4.4.

Releases of those classed as Priority Hazardous Substances must be ceased or phased out by 2020 and the Commission must, within two years of agreement of the list, bring forward proposals for Daughter Directives setting discharge limits and environmental quality standards (EQSs) for them. The status of those classified as Priority Substances under Review was to be considered further in 2004 to enable a second priority list to be drawn up. Under the Framework Directive, if agreement cannot be reached on control measures to be applied to the initial list of priority substances, Member States should by 22 December 2006 establish EQSs for all surface waters affected by discharges of the substances, including controls on principal sources of discharge. For substances added to the list at a later date, the same applies but within five years of its inclusion on the list. The Directive requires the Commission to review the list every four years.

In July 2006, the Commission published a proposal for a *Directive on environmental quality standards in the field of water policy* (COM(2006) 397) which will set limits on concentrations in surface waters for the 33 substances on the priority substances list and will repeal existing daughter Directives under the 1976 dangerous substances Directive – see below 6.4.4 and Appendix 6.1B.

The Commission has published a number of guidance documents relating to implementation of various aspects of the Directive, including: economic analysis; analysis of pressure and impact; identification of water bodies; identification of heavily modified waters; ecological assessment of transitional and coastal waters; selection of networks and processes; monitoring; public participation; and geographical information systems.

The Framework Directive makes provision for the repeal of a number of other Directives, including, seven years after the Directive enters into force that on surface water (75/440/EEC), with the following being repealed 13 years after its entry into force (i.e. 22 December 2007): 78/659/EEC – freshwater fish; 79/923/EEC – shellfish waters; 80/68/EEC – protection of groundwater from dangerous substances; and 76/464/EEC – dangerous substances discharged into the aquatic environment.

In England and Wales, the Directive is being implemented through *The Water Environment (Water Framework Directive) (England and Wales) Regulations 2003* – 6.10.1 & 2, and similar in Northern Ireland (6.10.4); in Scotland it is being implemented through the *Water Environment and Water Services (Scotland) Act 2003*, with the detailed requirements of the Directive being implemented through regulations, codes of practice, guidance etc – see 6.10.3.

6.3 FLOOD RISK MANAGEMENT

In April 2007, the European Parliament and the Council of Ministers agreed, with amendment, the Commission's proposal for a *Directive on the Assessment and Management of Floods* – COM(2006) 15. The primary aim of the Directive is to prevent and limit floods and their damaging effects on human health, the environment, infrastructure and property. Member States will be required to carry out a preliminary flood risk assessment for all river basin districts and coastline stretches to identify all those subject to flood risk by 2011. For those areas for which "no potential significant flood risk exists or is reasonably foreseeable in the future", no further action would be necessary.

For those areas identified as having a significant potential for flooding flood risk maps and flood hazard maps are to be prepared to support the development of flood risk management plans, spatial planning and emergency plans, as well as increasing public awareness of the risk. Flood risk development plans, to be developed at river basin/sub-basin level would include the analysis and assessment of flood risk, the definition and level of protection and identification and implementation of sustainable measures, ensuring that these did not increase the problem up- or downstream, but preferably contributed to reducing those risks. It would be left to Member States to decide on the appropriate level of protection required in each locality.

The Directive identifies the potential link between the purposes and methods of flood risk management and the achievement of water quality objectives under the Water Framework Directive and thus the administrative districts (river basin districts) and competent authorities for both Directives are the same, as are principles of coordination (for shared river basins); the reporting cycles are also synchronised, as are public participation and information requirements. Member States may choose to include flood risk management plans in their river basin management plans. Thus, risk assessments on areas at risk of flooding would need to be carried out by 2011, the areas mapped by 2013, and management plans drawn up by 2015.

6.4 DANGEROUS SUBSTANCES

Discharges to surface and other waters are covered by the Framework Directive on pollution caused by certain dangerous substances discharged into the aquatic environment of the Community (76/464/EEC, to be repealed 22 December 2013), and subsequent "daughter" Directives – the Commission has published a proposal for a Directive (COM(2003) 847) which will codify the 1976 Directive and subsequent amendments into a single instrument. Discharges to groundwater are dealt with by Directive 80/68/EEC. There are also various Directives (76/769/EEC and amendments) which relate to the marketing and use of certain dangerous substances with the primary objective of reducing risks to human health as well as protecting the aquatic environment; these are implemented through Regulations under the *Environmental Protection Act 1990* – the *Environmental Protection (Control of Injurious Substances) Regulations*. See also chapter 5, 5.10, Directive on the disposal of polychlorinated biphenyls and polychlorinated terphenyls.

6.4.1 Black List Substances

The Framework Directive (76/464/EEC) listed 129 substances considered to be so toxic, persistent or bio-accumulative in the environment that priority should be given to eliminating pollution by them. Included are substances such as organohalogens, organophosphorus, cadmium and mercury and their compounds. Control of listed substances is to be achieved largely through the setting of limit values or environmental quality standards in subsequent "Daughter" Directives, thus formally giving the substance List I – or Black List – categorisation. Directive 86/280/EEC (amended by 90/415/EEC) lays down specific limit values for discharges according to the type of industry, and quality objectives for the receiving waters. Daughter Directives, establishing emission limits and quality objectives, and encouraging the use of best available technology in new plant have now been adopted for a number of substances; these include mercury from the chlor-alkali industry (82/176/EEC) and from other industrial sectors (84/156/EEC), cadmium (83/513/EEC), hexachlorocyclohexane (84/491/EEC). Those which have not yet been given formal List I status in a Daughter Directive are in the meantime treated as List II substances for regulatory purposes.

The 1976 Framework Directive is due to be repealed by the Water Framework Directive in 2013. The Daughter Directives are to be repealed by the proposed *Directive setting environmental quality standards in the field of water policy* on 22 December 2012 – see below 6.4.4.

In the UK Sch. 1, para 7 to *the Environmental Permitting (England & Wales) Regulations 2007* and Sch.1, Part 2, para 13 to *the Pollution Prevention and Control Regulations* (Sc, 2000; NI, 2003) list a number of Black List substances whose releases to water are prescribed for Integrated Pollution Prevention and Control (see chapter 2 and Appendix 2.2). The same substances, together with carbon tetrachloride, are also prescribed substances under the *Trade Effluents (Prescribed Processes and Substances) Regulations 1989* (as amended; see later this chapter, 6.19). All installations releasing these substances must use BAT (Best Available Techniques) to control polluting discharges – see chapter 2. Discharge consents will also have to ensure that strict environmental quality standards are met. *The Surface Waters (Dangerous Substances) (Classification) Regulations 1989*, as amended, and similar in Scotland and N. Ireland (see this chapter, 6.11.1a) implement the environmental quality standards laid down in the 1990 Directive.

A Department of Environment Circular (DOE 7/89) gives guidance on the implementation of the various Directives setting specific environmental quality standards and discharge limits. The Circular emphasises that the aim should be "minimisation of inputs of the most dangerous substances to the aquatic environment".

6.4.2 Grey List Substances

List II substances – the "Grey List" – on the 1976 Framework Directive covers those substances considered less harmful when discharged to water. Included here are metals, such as zinc, nickel, chromium, lead, arsenic and copper; various biocides and substances such as cyanide and ammonia. It also covers those substances awaiting formal List I categorisation. The EU Directive requires Member States to establish pollution reduction programmes and to provide environmental quality objectives and standards for List II substances. Department of Environment Circular 7/89 also includes environmental quality standards for all Grey List substances. Concentrations of Grey List substances in freshwater should not be at such a level that freshwater fish cannot be supported.

The *Surface Waters (Dangerous Substances) (Classification) Regulations 1997* (SI 2560) and 1998 (SI 389) set statutory quality objectives for a total of 21 List II substances (see 6.11.1a below).

6.4.3 Groundwater

In 1979, a Directive (80/68/EEC) on the protection of groundwater against pollution caused by dangerous substances was adopted, for compliance in December 1981. This Directive, which is to be repealed on 22 December 2013 (see 6.2 above), requires the prevention of the discharge of List I substances to groundwater, and investigation of List II substances prior to direct or indirect discharge. (The Lists I and II referred to in this Directive are not identical to those contained in the dangerous substances Directive; also, unlike the dangerous substances Directive, List I status is definitive and does not need confirmation in a Daughter Directive.)

The EU has now adopted a new *Directive on the protection of groundwater against pollution and deterioration* (2006/118/EC), which aims to ensure good groundwater quality by 2015, (as required by the Framework Directive on Water, see 6.2 above). This Directive entered into force on 16 January 2007 and must be implemented by Member States by 16 January 2009.

The new Directive sets out specific measures for preventing and controlling groundwater against pollution and deterioration, criteria for assessing good chemical status and for identifying significant and sustained trends in increases of pollutant concentrations (Annexes III & IV). Groundwater is to be considered of good chemical status so long as measured or predicted concentrations of nitrates do not exceed 50 mg/l and of active ingredients in pesticides of 0.1 µg/l (Annex I) – however, the nitrates limit will not apply in designated nitrate vulnerable zones, which will remain subject to the nitrates Directive (see 6.5.5). Annex II sets out the assessment procedure for testing compliance with the requirement for good chemical status of groundwater for those pollutants for which there are no quality standards. The Directive requires Member States to take "all measures necessary to prevent inputs into groundwater of any hazardous substances"; threshold values for all pollutants identified as putting groundwater at risk must be established and notified to the Commission – Annex II lists pollutants for which such threshold values must be established in addition to those identified by individual Member States; these are: ammonium, arsenic, cadmium, chloride, lead, mercury, sulphate, trichloroethylene and tetrachloroethylene. The Directive is to be reviewed six years after its entry into force and thereafter every six years.

UK policy and measures for the protection of groundwater are summarised below at 6.11.3.

6.4.4 Environmental Quality Standards

In July 2006, the Commission published its proposal for a *Directive on environmental quality standards in the field of water policy* – COM(2006) 397. This Directive will implement the Water Framework Directive's requirement that all EU surface waters should achieve good status by 2015.

The proposed Directive sets limits – environmental quality standards – to be met by 2015 on concentrations in surface waters for the pesticides, heavy metals and other dangerous chemical substances that pose a particular risk to animal and plant life in the aquatic environment and to human health – see Appendix 6.1B. Thirty-three of the listed substances are those which have been designated as "priority substances" under EU Decision 2455/2001/EC (Appendix 6.1A), with the remaining eight pollutants covered by existing Directives on dangerous substances in water. Particularly strict limit values have been set for 13 "priority hazardous substances" because they are toxic, persist in the environment without breaking down, and become increasingly concentrated as they move up the food chain – discharges and emissions of these substances to water must cease by 2025, with the EQSs met by 2015.

In June 2008 the Council of Ministers and European Parliament reached agreement at second reading on the proposal. The European Parliament has also called for revision of the list of priority substances and corresponding environmental quality standards, to be proposed within two years of the Directive entering into force; the introduction of transparent criteria for the designation of so-called "mixing zones," or zones in which the standards can be exceeded under certain conditions; they reinforced the objective of ceasing or phasing out emissions of 13 "priority hazardous substances" within 20 years. In particular, the EC will prepare a report on progress towards achieving this objective in 2018. An inventory of emissions, discharges and losses will help the EC in its assessment of progress.

The various Daughter Directives under the 1976 *Framework Directive on Dangerous Substances* are to be repealed on 22 December 2012.

6.5 QUALITY OBJECTIVES

A number of Directives define the acceptable quality of water for particular purposes and make provision for both achieving and monitoring the quality of the water. Directives have been adopted concerning the quality of bathing water, freshwater fish, shellfish and for water intended for human consumption. (The Framework Directive on Water, see. above 6.2 repealed Directive 75/440/EEC on surface water intended for the abstraction of drinking water on 22 December 2007.)

6.5.1 Bathing Water

Directive 76/160/EEC concerning the quality of bathing water was adopted in December 1975, with Member States being given until December 1985 to bring designated bathing waters up to the required quality. It is to be repealed on 31 December 2014 once all the provisions of its 2006 replacement (see below) have been implemented.

In 2007 95% of coastal bathing waters and 89% of bathing sites in rivers and lakes complied with the mandatory standards. 86.1% of coastal waters and 62.7% of sites in rivers and lakes met the more stringent, but non-binding, guide values (European Commission annual report on compliance with 1976 bathing water quality Directive, published 02.06.08).

The 1976 Directive defines bathing waters as fresh or sea water (other than swimming pools) where bathing is "traditionally practised by a large number of bathers". An annex to the Directive sets out sampling frequencies, guideline and mandatory values, as well as analysis and inspection methods for a number of microbiological and physio-chemical parameters (see Appendix 6.2a).

UK implementation of the 1976 Directive is summarised below at 6.11.2.

Directive 2006/7/EC *concerning the management of bathing water quality* was formally adopted on 15 February 2006 and came into force on 24 March 2006. Member States had to transpose the Directive into national legislation by 24 March 2008. As well as updating the 1976 Directive in line with technical and scientific developments, the new Directive also reflects the requirements of the Water Framework Directive (see 6.2 above), laying down provisions for monitoring and classification of bathing water quality, its management, and the provision of information to the public.

The 2006 Directive applies to all surface waters where a large number of people are expected to bathe and where a permanent bathing prohibition has not been imposed or permanent advice against bathing issued. It does not apply to swimming pools, spa pools or other confined waters subject to treatment or used for therapeutic purposes, or artificially created confined waters separated from ground or surface waters. All bathing waters must be formally identified and the length of bathing season defined annually, and initially before the start of the first bathing season after 24 March 2008, and this data communicated to the Commission.

Monitoring is to be carried out at the location where most bathers are expected, or where the greatest risk of pollution is expected according to the bathing water profile. Member States are required to establish a monitoring calendar for each bathing water and to carry out the monitoring within four days of the date specified on the calendar – the first such calendar had to be drawn up prior to the 2008 bathing water season. Member States may commence monitoring in compliance with the standards and methods of the new Directive for the 2006 bathing water season, in which case monitoring of the 1976 parameters may cease.

With the Framework Directive covering the ecological and environmental aspects of water quality, the bathing water Directive focuses on the standards necessary for the protection of human health, using WHO guidelines for two microbiological indicators of faecal contamination – intestinal enterococci and e.coli (see Appendix 6.2). Samples taken during short-term pollution episodes may be disregarded and the monitoring calendar may be suspended in "abnormal situations". The Directive requires one sample to be taken shortly before the start of the bathing season, and at least four samples to be taken and analysed during the season (three if the bathing season is less than eight weeks, or the site is in a region subject to special geographical constraints); the interval between sampling dates should not exceed one month. In the event of short term pollution, an additional sample is to be taken to confirm whether the incident has ended – this should not be used as part of the bathing water quality data, instead a further sample should be taken seven days later to replace a disregarded sample. Bathing waters should also be inspected visually for pollution such as tarry residues, glass, plastic, rubber etc and where found, adequate management measures taken, including, if necessary, information to the public.

Bathing water classifications – poor, sufficient, good, and excellent – will be based on results for three years and must be completed by the end of the 2015 bathing season, with all registered sites required to have achieved at least "sufficient" status. For each site classified as "poor", adequate management measures must be put in place with effect from the bathing season following classification, including

- a bathing prohibition, or advice against bathing;
- identification of causes and reasons for failure to reach "sufficient" status;
- adequate measures to prevent, reduce or eliminate causes of pollution;
- clear and simple warning signs informing public of the causes of pollution and measures taken.

A prohibition, or permanent advice against bathing, should be placed on bathing waters that receive a "poor" classification for five consecutive years.

Bathing water profiles are to be developed by 24 March 2011 – these may cover either a single site or contiguous sites and should describe, map and assess sources of pollution or contamination in the vicinity of the bathing water which may affect bathers' health, and identify the monitoring point. If the assessment has identified a risk of short term pollution, further details should be given, including anticipated nature, frequency and duration of short term pollution, details of remaining causes of pollution and management measures taken and time schedule for elimination, as well as management measures taken during short term pollution and contact details of bodies responsible for taking action. Profiles of "excellent" bathing sites need only be reviewed if classification is downgraded. Those for "good" sites should be reviewed every four years, for "sufficient" sites every three years and for "poor" sites every two years. In the event of significant construction works or other significant changes in the infrastructure in or near the site, the profile should be updated before the start of the next bathing season.

As well as encouraging public participation in the implementation of the Directive, Member States are required to ensure that the following information is in an easily accessible place in the vicinity of each bathing water during the bathing season to enable the public to take an informed decision on whether to bathe or not:

- Description of the bathing water based on the beach profile;
- Current bathing water classification;
- Sign or symbol showing if any prohibition or advice against bathing is in place, and the reason;
- In the case of short-term pollution, notification to this effect, number of days on which bathing was prohibited or advised against during the preceding season, and a warning whenever such pollution is predicted or present;
- Where a bathing site has been declassified – and there is a permanent prohibition or permanent advice against bathing in place – reasons for this.

Member States are to ensure information concerning bathing waters is available via the Internet and other media and when appropriate in several languages. Information is to be disseminated as soon as available, and with effect from the fifth bathing season after 24 March 2008.

6.5.2 Freshwater Fish and Shellfish Waters

Directives adopted in 1978 and 1979 laid down quality requirements for freshwater fish (78/659/EEC) and shellfish waters (79/923/EEC), respectively. The 1978 freshwater fish Directive has been repealed and replaced by Directive 2006/44/EC *on the quality of fresh waters needing protection or improvement in order to support fish life*. This was adopted on

6 September 2006 and is a consolidation of the 1978 Directive and subsequent amendments. Directive 2006/113/EC *on the quality required for shellfish waters*, adopted on 12 December 2006 codifies and replaces the 1979 shellfish waters Directive.

The freshwater fish Directive requires Member States to identify and designate fresh waters needing protection or improvement in order to support fish, requiring compliance with the Directive's standards by 1985; an annex to the Directive lists a number of physical and chemical parameters – some mandatory and some guidelines.

The shellfish waters Directive also requires Member States to designate shellfish waters and lays down water quality objectives aimed at ensuring that such waters "contribute to the high quality of shellfish products directly edible by man"; compliance with the Directive's standards was required by 1987.

Implementation in the UK of both Directives – which are to be repealed by the Framework Directive on Water on 22 December 2013, see above 6.2 – is summarised below at 6.11.1d & e.

6.5.3 Drinking Water

Directive 98/83/EC *concerning the quality of water intended for human consumption* was formally adopted in November 1998, with national legislation implementing it required to be in place by 25 December 2000. This Directive replaces Directive 80/778/EEC which was repealed on 25 December 2003, the date by which most of the new standards had to be met; the exceptions are lead which had a 2008 compliance date and bromate and trihalomethanes, for which the compliance date is 2013.

The 1998 Directive covers all water for domestic use (whether from tap, bottle or container) and water used by the food industry where this affects the final product and, thus consumers' health; it does not apply to water for agricultural purposes, natural mineral water or medicinal waters. An important objective of the Directive is to contribute towards achieving sustainable development by ensuring a sufficient supply of water of adequate quality, with Member States having a duty to ensure that drinking water is wholesome and clean.

The Directive includes 53 parameters considered essential for the protection of water quality and health; these are divided into three groups: microbiological, chemical and indicator parameters and are to be reviewed at least every five years. Chemical parametric values have been based largely on WHO guidelines adopted in 1992, and include the following:

- **Arsenic**: 10 µg/l.
- **Nitrate concentrations**: 0.5 mg/l.
- **Total pesticides**: 0.5 µg/l, and 0.1 µg/l for individual pesticides.
- **PAHs**: 0.1 µg/l (0.01 µg/l for benzo(a)pyrene).
- **Benzene**: 1 µg/l.
- **Chlorinated solvents**: trichloroethene and tetrachloroethene combined -10 µg/l.
- **Lead**: maximum permitted concentration to be reduced from 50 µg/l to 25 µg/l over five years (i.e. by 2008) and to 10 µg/l within 15 years (i.e. by the end of 2013); the lengthy period for compliance takes account of the costs involved in replacing lead pipes and fittings.
- **Bromate**: 25 µg/l to apply from 2003, reducing to 10 µg/l in 2008.
- **Trihalomethanes**: 150 µg/l from 2003, reducing to 100 µg/l in 2008.

Member States are free to set additional parameters in the light of local environmental conditions; enforcement authorities must inform consumers when water quality falls below standard, of any danger to health, and give advice on what to do; consumers must also be advised when action has been taken and water quality restored. Monitoring is required to check on water quality and assess that measures taken to ensure quality are operating correctly; monitoring is also required to verify the efficiency of disinfection treatment of water intended for human consumption and to ensure no contamination from disinfection by-products.

In December 2001, the European Commission adopted a proposal for a Recommendation on *monitoring and reducing radon levels in drinking water*; remedial action is proposed when concentrations exceed 1000 bq/l.

UK implementation of the Directive is summarised below at 6.17.1 and Appendix 6.3.

6.5.4 Urban Waste Water

A Directive (91/271/EEC) adopted in 1991 lays down minimum requirements for the treatment of municipal wastewater and for the disposal of sludge. The Directive also aims to control the discharge of industrial wastewaters which are of a similar nature to municipal wastewater but which do not enter municipal waste water treatment plants before discharge. The release of treated or untreated sludge from treatment plants in fresh or seawater was banned from 31 December 1998.

Receiving waters are classified according to "sensitivity". In most cases secondary biological treatment prior to discharge will be required as a minimum, with tertiary treatment for waters which are, or are likely to become, eutrophic. Primary treatment – such as the settlement of suspended solids – may be considered sufficient for discharges from treatment processes serving less than 150,000 inhabitants into coastal waters less sensitive to pollution. Limit values for various industrial sector discharges to rivers, estuaries and coastal waters should be set in regulations.

Member States were required to complete the designation of sensitive waters, by the end of 1993, both in respect of this Directive and other Directives (e.g. bathing waters and shellfish waters) where tertiary treatment is needed to meet water quality standards; treatment standards had to be in place by the end of 1998.

This Directive has been implemented in England and Wales through *The Urban Waste Water Treatment (England and Wales) Regulations 1994* and by similar Regulations in Scotland and Northern Ireland – see below 6.20.1. Discharges to controlled waters are currently regulated under IPPC (chapter 2) and consents issued under *the Control of Pollution (Applications, Appeals and Registers) Regulations 1996* and similar 2001 regulations in Northern Ireland (see 6.11.5); in Scotland the relevant regulations are *the Controlled Activities Regulations* (6.11.6).

6.5.5 Nitrate Pollution

Directive 91/676/EEC places restrictions on the use of natural and chemical fertilisers to reduce the risk of nitrate pollution of river, coastal and sea waters and ensure a level of no more than 50 mg/l of nitrates in fresh water sources used for abstraction purposes. Catchment areas in which nitrate levels exceed, or are likely to exceed, the requirements of the Directive at any time in the year must be designated vulnerable zones. In such areas, an annual limit of 210 kg/ha of organic manure will be mandatory, reducing to 170 kg/ha four years later. Designation of such zones should have been completed by the end of 1993; action programmes to control nitrate leaching should have been in place by December 1995 to begin no later than December 1998 to ensure annual limits on nitrogen fertilisers were complied with by the implementation date of December 1999; programmes must be reviewed every four years.

The Directive also requires the drawing up of codes of good agricultural practice which must, as a minimum, include

- details of periods when use of fertilisers would be inappropriate;
- guidelines on use of fertilisers on sloping, flooded, frozen or snow covered ground;
- guidelines on use of fertilisers near water courses;
- measures to prevent run-off and seepage into rivers of manure-containing liquids and other effluents;
- conditions for general use of fertilisers.

Finally the Directive requires monitoring programmes to be established and reports on implementation progress to be sent to the Commission every four years.

The Directive has been implemented separately in England and Wales, Scotland and Northern Ireland – see 6.12.1b below.

6.6 INTERNATIONAL MEASURES

6.6.1 World Health Organisation

So far as drinking water is concerned, WHO's aim is that drinking water should not present any significant risk to health over a lifetime of consumption, and to this end it has produced recommended guidelines for drinking water quality; the first and second editions have been extensively used worldwide as the basis for regulations and standard setting, providing health-based guideline values for a large number of chemical hazards. The third edition, published in 2004, takes account of developments in risk assessment and risk management and outlines a "framework for drinking water safety". It also includes a section on applying the guidelines in specific circumstances, such as emergencies and disasters, large buildings, packaged and bottled water, food production and processing, and water safety on ships and in aviation.

For further information see http://www.who.int

6.6.2 Transboundary Watercourses

The UNECE's 1992 Convention on the Protection and Use of Transboundary Watercourses and International Lakes, came into force on 6 October 1996; it requires parties to the Convention to prevent, control and reduce releases of hazardous, acidifying and eutrophying substances into the aquatic environment and promotes transboundary cooperation for monitoring, information exchange and warning systems. The EU ratified the Convention on 14 September 1995.

A Protocol to the Convention on Water and Health, drafted by the WHO Regional Office for Europe and the UN Economic Commission for Europe, was adopted in London in June 1999 and remained open for signature until 18 June 2000. It will enter into force 90 days after the 16th instrument of ratification has been filed. Its objective is "the protection of human health and well-being ...through improving water management, including the protection of water ecosystems, and through preventing, controlling and reducing water-related disease". It applies to surface freshwater, groundwater, estuaries, coastal waters used for recreation or the production of fish and/or harvesting of shellfish, bathing waters, water in the course of abstraction, transport, treatment or supply, and waste water throughout the course of collection, transport, treatment and discharge or reuse. Parties to the Protocol will take all appropriate measures to prevent, control and reduce water related disease within a framework of integrated water management, aimed at sustainable use of water resources. To this end parties to the Protocol will endeavour to provide access to drinking water and sanitation for everyone. Within two years of ratifying the Protocol, Parties to it should establish and publish national and/or local targets in such areas as quality of drinking water, bathing waters or shellfish waters; reductions in outbreaks and incidents of water related disease; identification and remediation of contaminated sites affecting water.

A further Protocol, adopted by the UNECE in February 2003 covers the civil liability of enterprises for pollution of transboundary watercourses, and was signed by the UK in May 2003 in Kiev. Individuals affected by the transboundary impact of industrial accidents on transboundary watercourses will be able to claim compensation; operators of industrial installations covered by the Protocol will need to have adequate financial insurance to cover the cost of potential claims for damages.

6.6.3 Marine Environment

The European Union participates in various international Conventions relating to the protection of the marine environment and the prevention of oil spills; another part of its work is the harmonisation of national programmes regarding the elimination of pollution. With the enlargement of the EU, bringing in countries bordering the Mediterranean, efforts are also being directed towards the protection of that region, where the environment is sensitive and already severely overloaded locally. Further details, including on the EU's *Thematic Strategy for the Conservation and Preservation of the Marine Environment*, are given later in this chapter at 6.23.

REGULATORY FRAMEWORK

The management of water resources and regulations controlling the pollution of water and implementing EU Directives are the responsibility of the Devolved Administrations.

6.7 ENGLAND AND WALES

6.7.1 Overview

Privatisation of the water industry in England and Wales in 1989 resulted in major changes to the legislation dealing with the control of water pollution. *The Water Act 1989*, which received the Royal Assent on 6 July 1989, largely replaced the water section (Part II) of the *Control of Pollution Act 1974* in England and Wales. On 30 November 1991 *the Water Act 1989*, together with a number of other statutes controlling water pollution in England and Wales, was repealed and their provisions consolidated in the following:

* *Water Resources Act 1991* (see 6.11-15)
* *Water Industry Act 1991* (see 6.16-6.19)
* *Land Drainage Act 1991*
* *Statutory Water Companies Act 1991*

A fifth Act – *the Water Consolidation (Consequential Provisions) Act 1991* – provided for consequential amendments and repeals and other transitional matters. These Acts took effect on 1 December 1991.

The Water Act 2003, which received Royal Assent at the end of November 2003, aims to promote the sustainable use of water resources and as such the Consumer Council for Water is required to exercise its powers and duties "in the manner ... best calculated to contribute to the achievement of sustainable development"; the Secretary of State is under a duty to take steps to, where appropriate, encourage water conservation. As well as reforming the water abstraction licensing system (see *the Water Resources (Abstraction and Impounding) Regulations 2006*, SI 2006/641), the Act also introduces more competition and strengthens the voice of consumers through the establishment of a Consumer Council for Water, requires strategic health authorities to consult locally before the fluoridation of drinking water and establishes the Water Services Regulation Authority (in place of the Director General of Water Services).

The Act also inserted new sections 37A-D (brought into force 01.04.07, SI 2007/1021) into *the Water Industry Act* 1991 to require water companies to prepare, maintain and publish water resources management plans; *the Water Resources Management Plan Regulations 2007* (SI 2007/727), also brought into force on 1 April 2007, deal with the process for developing water resources plans, including publishing and consultation on draft plans, dealing with any representations made, and publishing of final plans. In developing their plans and demonstrating how security of supply is to be maintained, water companies will need to take account of the environmental impact of their proposals.

Other regulatory changes made by the Act, which largely applies in England and Wales only, include:

- ***Drinking water:*** the maximum fine for which water companies will be liable for supplying water unfit for human consumption is raised from £5,000 to £20,000; there is a similar increase for failing to supply information to the Drinking Water Inspectorate; the DWI is given powers to prosecute anyone supplying, or causing the supply of, water unfit for human consumption (brought into force 01.04.04, SI 2004/641 Commencement Order).
- ***Contaminated land:*** the definition of when land is contaminated because of the risk it poses to water pollution is amended to "significant" pollution of controlled water is being caused or there is a "significant possibility" of such pollution being caused". The definition of controlled waters is clarified (for the purposes of Part IIA of the EPA) in that groundwaters does not include waters contained in underground strata but above the saturation zone – this latter amendment was brought into force on 01.10.04, SI 2004/2528 in England and WSI 2004/2916, W.255, 11.11.04.
- ***Coal mines:*** the Coal Authority is given powers to take action to prevent and clean up mine water pollution from abandoned coal mines (brought into force 01.04.04, SI 2004/641).
- ***Discharge consents:*** measures to facilitate the process of transferring a discharge consent from an existing consent holder to a new holder (brought into force 01.10.04, SI 2004/2528);
- ***Trade effluents:*** provisions regarding their regulation under the *Water Industry Act 1991* are amended.

Statutory Nuisance

Section 259 of the *Public Health Act 1936* makes the following matters statutory nuisances in England and Wales for the purposes of water pollution:

a) any pond, pool, ditch, gutter or watercourse which is so foul or in such a state as to be prejudicial to health or a nuisance;

b) any part of a watercourse, not being a part ordinarily navigated by vessels employed in the carriage of goods by water, which is so choked or silted up as to obstruct or impede the proper flow of water and thereby to cause a nuisance, or give rise to conditions prejudicial to health.

Complaints about statutory nuisances are dealt with by local authorities who may take action through the Magistrates' Court for abatement of a nuisance (see also chapter 1, 1.29.1).

Section 260 enables parish and community councils to take action to deal with filthy or stagnant water which may be prejudicial to health.

6.7.2 Environment Agency

Overall responsibility for the maintenance and management of water resources in England and Wales rests with the Secretary of State for Environment, Food and Rural Affairs: this includes responsibility for national policy on all matters of conservation and supply, sewerage and sewage disposal; the control of pollution in both inland and coastal waters; the use of inland waters for recreation and navigation; the control of marine pollution by oil; land drainage; the protection of freshwater and marine fisheries, the dumping of waste at sea; and the safe use of agricultural pesticides.

The Environment Agency for England and Wales was established by the *Environment Act 1995* which received the Royal Assent on 19 July 1995. On 1 April 1996 the Agency assumed all the responsibilities and functions of the National Rivers Authority which was abolished. The Agency also has responsibility for Integrated Pollution Control, Integrated Pollution Prevention and Control and waste regulation (see chapters 2 and 5 respectively).

The Environment Agency is charged with preventing deterioration of water quality and seeking its improvement; it has a duty to promote, as it considers desirable, the conservation and enhancement of the water environment (inland and coastal waters). As well as pollution control, the Agency's responsibilities include water resource management, planning and conservation, flood defence, forecasting and warning, abstraction licences, fisheries and in some areas navigation. It also has certain duties in relation to promoting conservation; amenity and recreational facilities. The Agency and SEPA should consult and collaborate on matters of common interest.

As from 1 April 2008, the Environment Agency became responsible for overseeing management of all flood and erosion risk on the English coast. Defra is proposing that this role be extended to cover inland areas. See also 6.7.7 below

6.7.3 Drinking Water Inspectorate

This Inspectorate was established within the Department of Environment (now Department for Environment, Food and Rural Affairs – Defra) in January 1990; its principal task is to ensure that water undertakers in England and Wales are fulfilling their statutory requirements for the supply of wholesome drinking water.

The Inspectorate carries out annual technical audits of each water company; this includes an assessment (based on information supplied by the company) of the quality of water in each supply zone, arrangements for sampling and analysis, and progress made on achieving compliance with UK and EU requirements.

The Inspectorate monitors compliance with the standards set in the *Water Supply (Water Quality) Regulations 2000* – see 6.17 below. The Inspectorate investigates incidents affecting water quality and water companies are required to notify the Inspectorate of any significant pollution incidents – guidance on the notification of events (incidents) is available on the DWI's website (www.dwi.gov.uk). As from 1 April 2004, the DWI has been able to prosecute anyone supplying, or causing the supply of, water unfit for human consumption; the maximum fine is £20,000 (*Water Act 2003*).

The DWI reports annually on compliance with the 2000 Regulations (see 6.17.1 below), with drinking water standards assessed on the basis of compliance within zones (with each zone covering a population of no more than 100,000). In 2007, results for England show a "mean zonal compliance" of 99.96% (the same as for 2006), and in Wales 99.92% (2006: 99.95%). (*Drinking Water in England and Wales 2007*).

A 2005 Memorandum of Understanding between the UK Drinking Water Regulators outlines the roles and responsibilities of each and sets out an agreed framework for co-operation with the aim of minimising duplication of activity. The Memorandum covers such matters as sharing information and expertise, wherever possible consulting each other on draft regulations, guidance etc in advance; co-ordinating consumer consultation and surveys, and public interest information, etc. The Regulators also agree to work together to promote the principles of the World Health Organisation's *Guidelines on Drinking Water Quality* (third edition, 2004) – see 6.6.1 above.

6.7.4 Local Authorities

Under ss.77 & 78 of the *Water Industry Act 1991*, local authorities have a statutory duty to take all such steps as they consider appropriate for keeping themselves informed about the wholesomeness and sufficiency of both public and private supplies of water in their area.

If the public supply is likely to become unwholesome or insufficient, the local authority must advise the water undertaker who must take appropriate remedial action (s.79). In the case of private supplies, the *Water Industry Act* gives local authorities the power to serve a notice requiring improvements (s.80). *The Private Water Supplies Regulations 1991* set out local authorities' duties with regard to safeguarding water quality (see 6.17.2 below).

Local authorities are also responsible for the quality of bottled mineral waters for drinking. Sources have to be approved and standards as set out in the *Natural Mineral Waters Regulations 1985* complied with.

As from April 1992, local authorities also took over responsibility from the former MAFF for enforcing certain provisions of the *Control of Pesticides Regulations 1986*, as amended (see 6.26.4 below).

6.7.5 Water Strategy

Future Water, the Government's water strategy for England was published in February 2008 (replacing that published in 2002). The Strategy sets a vision for 2030, with the aim of improving standards of service and quality, through sustainable water management, whilst achieving a balance between environmental impacts, quality of surface and ground waters, supply and demand, and social and economic impacts. The immediate outcomes of such a strategy are seen as

- helping to deliver the aims of the Water Framework Directive which requires there to be no deterioration in water quality and to achieve good ecological status by 2015. In particular action is to be taken to reduce water pollution by tackling contaminants at source, and to deal with point sources of pollution.
- climate change mitigation and adaptation.
- sustainable use of water resources with no essential supply interruptions during periods of drought. The strategy includes an aim of reducing water usage from 150 litres to 120 litres per person per day by 2030.
- high levels of drinking water quality.

More details at www.defra.gov.uk/environment/water/strategy/index.htm

6.7.6 River Quality

Since 1958 surveys have been periodically carried out to establish the overall state of rivers, estuaries and canals in England and Wales. Waters were classified as good (1A and 1B), suitable for potable supply abstractions, game and high class fisheries, with high amenity value; fair (2) suitable for the same purposes although of not such high quality; poor (3) probably unable to support fish life, and bad (4) "grossly polluted and likely to cause a nuisance".

River and canal water quality is now surveyed using the General Quality Assessments Scheme; this is not a statutory scheme. The Environment Agency is responsible for monitoring in England and Wales. Chemical quality is assessed on the basis of concentrations of dissolved oxygen, biochemical oxygen demand (BOD) and ammonia. Biological quality is assessed on the basis of the number and diversity of tiny animals (macro-invertebrates) which live in or on the beds of rivers. Samples of waters are also analysed for nutrient (nitrates and orthophosphates) concentrations. A new method of measurement is to be developed to meet the requirements of the Water Framework Directive – see above 6.2. Surface waters in England and Wales are surveyed and classified according to the *Surface Waters (River Ecosystem) (Classification) Regulations 1994* – see below 6.11.1b and Appendix 6.6.

In England 71% of river length was of good biological quality in 2006 (2005: 71%); 66% were of good chemical quality (2005: 64%). For Wales the figures are 82% (80%) and 95% (95%) respectively. (Defra Statistical Release, 28.08.07)

6.7.7 Flood Risk Management

The Environment Agency has responsibility for overseeing management of all flood and erosion risk on the English coast, including:

- taking the lead in managing all sea flooding risk in England, funding and overseeing coastal erosion works undertaken by local authorities;
- ensuring proper and sustainable long term "Shoreline Management Plans" are in place for the English coastline.
- working with local authorities to ensure that the resulting flood and coastal erosion works are properly planned, prioritised, procured, delivered and maintained to get maximum value for taxpayers' money;
- ensuring that third party defences are sustainable.

The Agency's role is to be extended to cover management of inland flood risk. In addition Defra is preparing a draft Floods and Water Bill to ensure better protection from surface water flooding.

The role and membership of Regional Flood Defence Committees has also been widened to include coastal erosion, in addition to flood defence.

Guidance published by Defra in November 2006 advises on the need for new flood defences to be fully adaptable to the consequences of climate change and rising sea levels. It says that those responsible for building flood defences should plan for the long term by factoring in increasing rates of sea level rise. It suggests that this is likely to accelerate from the current 2.5mm to 4mm a year to 13mm and 15 mm a year between 2085 and 2115. The Guidance can be found at http://www.defra.gov.uk/environ/fcd/pubs/pagn/default.htm

Planning Policy Statement on development and flood risk (PPS 25), published December 2006, aims to provide clarity on what is required at regional and local levels to ensure that flood risk is properly taken into account in the planning of new developments. It details policies to be adhered to avoid inappropriate development in areas at risk of flooding, and to direct development away from areas at highest risk – so lessening the future impact of climate change on flood risk. In June 2008, the Communities and Local Government Department published a companion practice guide showing how planning policy should be used to manage flood risk and to help create consistency in the way PPS 25 is applied. It can be downloaded from http://www.communities.gov.uk/publications/planningandbuilding/pps25/practiceguide

6.8 SCOTLAND

6.8.1 Overview

In Scotland, legislation dealing with water pollution has, until recently been contained in the *Water (Scotland) Act 1980*, Part II of the *Control of Pollution Act 1974*, both as amended by ss.168 & 169 of the Water Act 1989, the *Environment Act 1995* (Sch. 16 & Sch. 22, para 29) and the *Sewerage (Scotland) Act 1980*. The legislative regime has now been extended and updated by a number of Acts, summarised below.

The Water Industry (Scotland) Act 2002, which received Royal Assent on 1 March 2002,

- establishes, with effect from 1 April 2002, a single water authority – Scottish Water – through the merger of the existing three;
- establishes a Water Industry Commissioner for Scotland with powers to investigate complaints about Scottish Water and to advise Ministers on matters relating to Scottish Water's standards of service;
- establishes Water Customer Consultation Panels to represent the views of, and promote the interests of, Scottish Water's customers;
- establishes a statutory drinking water inspectorate (see below);
- introduces new provisions relating to public registers for trade effluent consents.

The Act also places a general duty on Scottish Water to "act in the way best calculated to contribute to achievement of sustainable development" but only so far as is consistent with its statutory functions. In exercising its functions Scottish Water should also have regard to preserving Scotland's natural heritage and places of beauty.

Part 1 of *The Water Environment and Water Services (Scotland) Act 2003*, as well as implementing the framework Directive on Water (see 6.2 and 6.10.2 below), also sets out the framework for a new regulatory regime for controlling discharges to water, previously regulated under COPA – see below (6.11.6) *the Water Environment (Controlled Activities) (Scotland) Regulations 2005* which came into force on 1 April 2006. The same regulatory regime will also apply to discharges to groundwater. Part 2 of the Act covers water and sewerage services and Part 3 makes general provision for the making of orders and regulations.

The Water Services etc (Scotland) Act 2005 received Royal Assent on 17 March 2005; Part 2 (brought into force 08.01.07, SSI 2006/599) bans the "common carriage" of water and sewage on public networks; only Scottish Water may add treated drinking water to, and draw wastewater from, the public network, and retail water services to households on the public network. Section 25 of the Act (brought into force 10.02.06) provides for Ministers to issue a statutory code of practice on sewerage nuisance, and for monitoring and enforcement of it – nuisance includes odour, insects and any other thing considered to be prejudicial to health or a nuisance (see also chapter 1, 1.37.3). *The Public Health etc (Scotland) Act 2008*, s.115 amends s.26 of the Act to provide local authorities with powers of entry for the purposes of enforcing the sewerage code to establish whether a statutory nuisance exists or for carrying out remedial action. (*The 2008 Public Health Act* also amends s.79 of the 1990 EPA to include as a statutory nuisance any water covering land, or land covered with water, which is in such a state as to be prejudicial to health or a nuisance – see chapter 1, 1.29.)

Part 3 of the 2005 Act includes provisions for dealing with water pollution from coalmines.

The Scottish Government is also proposing a *Marine Bill* aimed at conserving Scotland's marine environment.

Regulations implementing various EU Directives on water quality, similar to those for England and Wales, have been made in Scotland and are referred to throughout this chapter.

6.8.2 Scottish Environment Protection Agency

SEPA was also established by the *Environment Act 1995* and assumed the responsibilities and functions of the river purification authorities on 1 April 1996. Its functions mirror those of the Environment Agency – see 6.7.1 above, with the additional responsibility for regulation of local air pollution under both the EPA 1990 and PPC Regulations 2000.

Sewage disposal is the responsibility of Scottish Water; sewage treatment (regulation of discharges to water) is the responsibility of SEPA, and odour local authorities.

6.8.3 Drinking Water Quality Regulator

The Drinking Water Quality Regulator (DWQR), appointed under the *Water Industry (Scotland) Act 2002*, is responsible for ensuring public water suppliers comply with their drinking water quality duties. Unitary authorities are responsible for the supply of "wholesome" water for domestic and other purposes (*Water (Scotland) Act 1980*).

The DWQR is empowered to obtain information regarding drinking water quality and has powers of entry and inspection to establish whether duties are being complied with. The Regulator has powers to issue enforcement notices where he believes a drinking water quality duty has been, or is being, contravened; such a notice may be served if the Regulator is of the opinion that the contravention is likely to recur and no action has been taken to prevent its recurrence. Enforcement notices will specify the action to be taken and the timescale. Before serving an enforcement notice, the Regulator should give advance notice to the public water supplier and take into account any representations made. The Act provides for copies of enforcement notices to be sent to the Water Industry Commissioner, affected local authorities and Health Board, as well as ensuring any other person affected is aware of the notice. There is a right of appeal against an enforcement notice to the Sheriff within 14 days of its service, during which time the notice is suspended. If the Regulator is of the opinion that a public water supplier has failed to comply with an enforcement notice, the Act gives the Regulator powers to carry out the work himself, recovering the costs. Where there is a risk to public health the Regulator may serve an emergency notice requiring remedial action by a specified date, or carry out the work himself recovering the costs from the supplier. The Regulator is required to maintain a publicly accessible register of enforcement and emergency notices.

In 2007 99.76% (2006: 99.66%) of the tests carried out on over 160,000 samples from customers' taps met the relevant standards. Scottish Water received 26,702 complaints about drinking water quality in 2007, of which 37.4% related to discoloured supplies. (*Drinking Water Quality in Scotland 2007*, The Stationery Office, Edinburgh)

The UK Drinking Water Regulators have signed a Memorandum of Understanding (see 6.7.3 above).

6.8.4 River Quality

In Scotland SEPA is responsible for monitoring river quality; river lengths are classified as: A1 excellent, A2 good, B fair, C poor, and D seriously polluted; coastal waters and estuaries (km^2) are classified as A excellent, B good, C unsatisfactory, and D seriously polluted. Each stretch of river is assigned a monitoring point where chemical, biological nutrient and aesthetic surveys are taken and the results combined into a single classification.

In 2006, 79.4% of monitored river and stream length were classified as either A1 or A2, with 9.5% classified as fair, 2.7% as poor and 0.1% as seriously polluted. The percentages for 2005 are: 78.9%, 9.7%, 2.8% and 0.2%. In 2006, 96.3% of estuarine waters were of excellent or good quality, with 3.6% unsatisfactory and 0.03 seriously polluted. 99.2% of coastal waters were classified as excellent or good, 0.7% as unsatisfactory and 0.09 as seriously polluted. (National Water Quality Classification, SEPA, 2007)

88% (2005: 87%) of monitored river lengths were of good quality. (Defra Statistical Release, 28.08.07)

A review of water quality in Scottish rivers and lochs published by SEPA in March 2006 showed that 46 rivers and 17 lochs are at risk from pollution. The "at risk" waters, which have now been designated "sensitive areas", showed evidence of being damaged by harmful nutrients from sewage effluent, agriculture, forestry, fish farming and urban drainage.

6.8.5 Flood Risk Management

The Flood Prevention (Scotland) Act 1961 is currently the main legislation in Scotland for managing flood risk from rivers and the sea. It is now proposed to update the legislation with the aim of reducing the risk of significant impacts of flooding through sustainable flood management.

The Scottish Government was planning to publish a draft *Flood Risk Management Bill* in September 2008. This follows consultation earlier in 2008 which asked for views on its proposals for the Bill, including identification of a competent authority with overall responsibility for implementation of the EU *Directive on the Assessment and Management of Floods* (see 6.3 above), and of responsible authorities for the purposes of flood risk management planning. The competent authority will be required to

- undertake preliminary flood risk assessments by 22 December 2011 to create a national picture of flood risk in Scotland;
- undertake flood risk and hazard mapping by 22 December 2013 for those areas identified as being at significant risk of flooding;
- produce strategic Area Flood Risk Management Plans that will coordinate flood risk management objectives across catchments, or groups of catchments, and set the framework in which measures are delivered or planned for at a local level by responsible authorities (local authorities and others).

The competent authority will be required to secure the participation of responsible authorities in the preparation of their plans and to consult stakeholders. Plans will be submitted to Scottish Ministers and must be monitored and reviewed in accordance with the EU Directive. Local authorities will be required to draw up detailed Local Flood Risk Management Plans in cooperation with other responsible authorities.

6.9 NORTHERN IRELAND

6.9.1 Overview

The main legislation for controlling water pollution is *the Water (Northern Ireland) Order 1999*, made on 10 March 1999, the water pollution provisions of which were brought into force on 24 August 2001. The 1999 Order updates *the Water Act (Northern Ireland) 1972* to take account of developments both in science and policy and introduces a similar system of discharge consents etc for controlling water pollution similar to that in England and Wales. It also makes provision for anti-pollution works notices, registers, abandoned mines and enforcement, which are covered later in this chapter. Reference is made to the 1999 Order as appropriate as: WNIO: Art.00, and to Regulations made under it.

The Water and Sewerage Services (Northern Ireland) Order 2006 (SI 2006/3336, NI 21) amends the 1999 Order to bring legislation in Northern Ireland partly into line with the similar legislation applying in the rest of Great Britain and to provide an updated environmental regulatory regime for water and sewerage services in Northern Ireland. The 2006 Order also sets out requirements for the maintenance of water quality and enables the DOE to make regulations defining wholesomeness in respect of both public and private water supplies. The Order establishes a Government-owned company, Northern Ireland Water Limited, as a water and sewerage undertaker to deliver water and sewerage services in Northern Ireland; the role of the Northern Ireland Authority for Utility [formerly, Energy] Regulation is expanded to include the economic and consumer regulation of water and sewerage services, with the Department for Regional Development responsible for other regulatory duties; the General Consumer Council for NI becomes the statutory consumer representative for water and sewerage services in Northern Ireland.

The Water and Sewerage Services (Miscellaneous Provisions) (Northern Ireland) Order 2006 (SI 2006/1946, NI15) amends the 1999 Order to enable the DOENI to make regulations relating to the abstraction and impounding of water (see also below 6.10.4).

6.9.2 NI Environment Agency

The NI Environment Agency (NIEA), which replaced the Environment & Heritage Service in July 2008, is responsible for monitoring water quality, preparing water management plans, controlling effluent discharges, and taking action to combat or minimise the effects of water pollution.

From 1 April 2007, water supply and sewerage services have been managed by Northern Ireland Water Limited, a Government-owned company, established under *the Water and Sewerage Services (Northern Ireland) Order 2006* (SI 2006/3336. NI 21).

6.9.3 Drinking Water Inspectorate

The Drinking Water Inspectorate, in having similar responsibilities to its counterparts in the rest of the UK, regulates drinking water quality for both public and private supplies; it assesses drinking water quality against standards, carries out detailed inspections of water sampling and analytical processes, and deals with consumer complaints.

In 2004 98.65% of the 114,819 tests of drinking water samples complied with regulatory standards (9th Annual Report of the Drinking Water Inspectorate, 2005).

The UK Drinking Water Regulators have signed a Memorandum of Understanding (see 6.7.3 above).

6.9.4 River Quality

In Northern Ireland, the Northern Ireland Environment Agency is responsible for monitoring, using a similar scheme to that used for England and Wales – see above 6.7.6. In 2006, 54% (2005: 56%) of monitored river length were of good biological quality and 74% (2005: 63%) of good chemical quality. (Defra Statistical Release, 28.08.07)

6.10 IMPLEMENTATION OF THE FRAMEWORK DIRECTIVE

6.10.1 Transposing Regulations

Separate Regulations transposing the Directive – see above 6.2 – have been made in England & Wales and in Northern Ireland; in Scotland the Directive has been transposed through *the Water Environment and Water Services (Scotland) Act 2003*, and Regulations under it (see 6.10.3). All the legislation provides for the following (with other measures relating to transposition in each country covered in 6.10.2-4 below):

- Identification of RBDs and associated groundwaters and coastal waters (by 2003); maps showing the RBDs and waters allocated to them must be publicly available (e.g. on websites and in government and Agency libraries).
- An analysis of the characteristics of each RBD and review of the impact of human activity on water quality (in accordance with statutory guidance) – to be completed by 22 December 2004, with a first review to be completed by 22 December 2013, and thereafter every six years. Bodies of waters in RBDs which provide more than 10 m^3 of drinking water per day or serve more than 50 persons must be separately identified.

- The Secretary of State (and/or National Assembly for Wales as appropriate, SEPA, DOE NI) is responsible for completion of an economic analysis of water use in each district – to be completed by 22 December 2004, with a first review to be completed by 22 December 2013, and thereafter every six years.
- A register of protected areas within each RBD had to be drawn up by 22 December 2004, and must be regularly reviewed and updated; protected areas include areas designated under EU legislation for the protection of surface or ground water, or of habitats and species, and identified drinking waters.
- Monitoring programmes to be established and maintained by 22 December 2006 in accordance with the Directive (and with statutory guidance) for each RBD; for surface water this should cover volume and level or rate of flow relevant for ecological and chemical status and ecological potential; for groundwater it should cover chemical and quantitative status. The UK Technical Advisory Panel, comprising experts from the UK and Irish environment agencies, is developing the environmental standards and conditions to be used to measure the ecological health of the water environment.
- Environmental objectives for each RBD, together with a programme of measures which aim to achieve the Directive's objectives of good status for surface and ground waters by December 2015 must be drawn up by the EA and SEPA and submitted for approval by the Secretary of State (NAW, Scottish Ministers) – in NI the DOE is responsible for establishing environmental objectives and the associated programme of measures; the programme must be established by 22 December 2009, operational by 22 December 2012, and reviewed and updated as necessary by 22 December 2015, and thereafter every six years. The public should be consulted on proposed environmental objectives and programmes of measures.

All of the requirements listed above will also form part of the River Basin Management Plan (RBMP) which the Agencies (DOENI) are required to prepare for each of their RBDs. Draft plans should be published for consultation in December 2008, and submitted to the Secretary of State (or NAW, Scottish Ministers) for approval and a notice published (in the *London/Belfast Gazette*, as appropriate, and in Scotland in a newspaper with circulation throughout Scotland; a notice must also be published in one or more appropriate local newspapers) saying where the plan can be viewed, with arrangements made for ensuring the requirements of the public participation directive are adhered to (see chapter 1, 1.25.2). The Secretary of State (NAW, Scottish Ministers) may approve, modify or reject plans;

The Environment Agency's *Framework for River Basin Planning* includes a provisional timetable for the main stages of the first river basin management strategy, with similar published by SEPA and the DOE NI.

The first finalised RBMPs are to be published by December 2009 – the Agencies must publicise this fact, stating where a copy of the final plan can be seen. RBMPs must be reviewed and updated every six years or earlier if directed by the Secretary of State (or NAW, Scottish Ministers/DOENI). Thus, a second draft RBMP is scheduled for 2014, for final publication in 2015.

6.10.2 England and Wales

The Water Environment (Water Framework Directive) (England and Wales) Regulations 2003 (SI 2003/3242) came into force on 2 January 2004. The Environment Agency has been designated as the competent authority and is largely responsible for implementation of the Regulations and for submitting detailed river basin management plans to the Secretary of State; together with the Secretary of State and Welsh Assembly, the Agency has a duty to exercise its functions in relation to each river basin district (RBD) so as to achieve the environmental objectives of the Directive, with the Secretary of State and NAW being required to coordinate programmes of measures to achieve the environmental objectives of the Directive for cross-border river basin districts. Nine RBDs have been designated in England and Wales (two of which span the administrative boundary between England and Wales), and two cross-border RBDs with Scotland.

Defra and WAG have published (August 2006) River Basin Planning Guidance which sets out Ministerial expectations in relation to the principles and key steps of the river basin planning process and the content of the documents which must be produced. This is statutory guidance and is issued under reg.20(3) of the Regulations. The Guidance includes a Direction to the Environment Agency to submit, by 22 September 2009, to the appropriate authority the initial River Basin Management Plan for each RBD; by the same date proposals for environmental objectives and a summary of the measures to be applied to achieve the objectives should also be submitted. A second volume of guidance published for consultation in February 2008 outlines the role of EQSs in implementing the Directive and explains the range of new standards which have been developed by the UK environment agencies. The Environment Agency is expected to publish draft RBMPs for consultation before the end of 2008.

In October 2008, Defra and the Welsh Assembly Government published their proposals for developing environmental quality standards and other criteria recommended for use in classification in the first river basin planning cycle. The consultation also includes recommendations on methodologies for the use of those and criteria to classify all WFD water bodies. These will be finalised as Directions from the Secretary of State and WAG to the Agency:

- *River Basin Districts Surface Water Typology and Environmental Standards (WFD) (E & W) Direction* – this will set out the process for assigning water bodies to a type and the standards and other criteria that apply to each type. It also lists threshold values for the classification of groundwater and boundary values for biological quality elements.
- *River Basin District Surface Water and Groundwater Classification (WFD) (E & W) Direction* – this sets out the rules for classification of water bodies.

Controlling diffuse pollution from agriculture will be an important element in achieving the Directive's goal of achieving good ecological status of water by 2015; agricultural activities contribute on average 60% of nitrates entering English river systems, 25% of phosphorus, up to 75% of sediment, and between 25-50% of bacterial pollution. As well as measures promoting catchment-sensitive farming, such as the imposition of water protection zones (consultation document, August 2007), Defra is also looking at the contribution of non-agricultural (e.g. transport, leisure, industrial, forestry and old mine workings) sources to diffuse water pollution; an initial consultation paper was published in February 2007, and a second one proposing mechanisms for dealing with the problem is due. To comply with the WFD protection zones should to be established by 2009. A further consultation paper, also published in February 2007 looks at the range of mechanisms available in E & W for delivering measures to avoid or reduce effects resulting from hydromorphological pressures (i.e. changes to the physical characteristics of water bodies which could affect ecological quality).

The Environment Agency has published on its website a number of Briefing Notes and guidance on various aspects of the water Framework Directive. See bibliography.

Statutory consultees in England and Wales include: the Secretary of State and/or NAW; Director General of Water Services; Joint Nature Conservation Committee; Natural England and/or the Countryside Council for Wales; local authorities, planning authorities, water and sewerage undertakers within the RBD affected by the consultation; and any other persons representative or who might have an interest.

England-Scotland RBDs

The Water Environment (Water Framework Directive) (Northumbria River Basin District) Regulations 2003 (SI 2003/3245) which came into effect on 2 January 2004 adapt and apply the E&W Regulations and *the Water Environment and Water Services (Scotland) Act 2003* to cover the Northumbria RBD which is partly in England and partly in Scotland. Further Regulations (SI 2004/99) came into force on 10 February 2004 covering the cross-border RBD of the Solway Tweed. Statutory Guidance to the Environment Agency and SEPA on the practical implementation of the WFD as regards the Solway Tweed RBD was issued in December 2007. Included in the Statutory Guidance is a Direction from Scottish Ministers and the Secretary of State to the Agencies which requires them to prepare and submit to them by 22 September 2009 a RBMP for the Solway Tweed RBD, including proposing for environmental objectives and a summary of the programme of measures for achieving them.

SEPA and the Environment Agency have published (September 2006) The Solway Tweed River Basin Planning Framework; this establishes the administrative arrangements and working principles to support the protection of the RBD; it sets out how the integration and coordination of the RBMP with other planning systems can be developed, as well as consultation and participation opportunities. In January 2007, SEPA and the EA published for consultation their Plan of Action for engaging stakeholders in the process of river basin planning, timescales, documents to be produced etc.

6.10.3 Scotland

Part 1 of *The Water Environment and Water Services (Scotland) Act 2003* (WEWS), provides the framework for implementing the EU water Framework Directive (see 6.2) and in so doing

- establishes a new system of management and planning for Scotland's water environment, lochs, coastal waters and groundwaters;
- introduces a requirement to control all impacts on the water environment to ensure "good status" of the waters by the dates specified in the Directive.

The WEWS Act enables Scottish Ministers to create one or more river basin districts for Scotland, with special arrangements to be made for those catchment areas that cross the border with England. SEPA has a duty to promote the sustainable use of water and has lead responsibility for drawing up the river basin management plan (RBMP); it also has a duty to monitor the status of the water environment in each RBD and to prepare monitoring programmes. The Act outlines the matters to be included in the RBMP, as well as requirements relating to the setting of environmental objectives and programme of measures in the RBMP. In drawing up the plan, SEPA will be required to consult other relevant authorities and interested bodies, and to have regard to their views. To facilitate this process river basin district advisory groups will be established, with at least one group for each river basin district. The Act also empowers SEPA to draw up "sub river basin plans" to deal with an identified need or problem in a specific catchment area. These sections of the Act were brought into force on 1 June 2005 by Commencement Order (SSI 2005/235), with the various elements of the Directive being implemented as follows:

- *The Water Environment and Water Services (Scotland) Act 2003 (Designation of Scotland River Basin District) Order 2003* (SSI 2003/610), which came into force on 22 December 2003 designate a single RBD for Scotland (there are also two cross-border RBDs with England).
- *The Water Environment (Register of Protected Areas) (Scotland) Regulations 2004* (SSI 2004/516), which came into force on 22 December 2004, require SEPA to establish a register of those areas within each river basin district requiring special protection, as required by the Framework Directive. The register, which can be accessed via SEPA's website, lists waters designated under the shellfish and freshwater fish Directives, the nitrates, habitats and wild birds Directives and the waste water Directive.
- *The Water Environment (Drinking Water Protected Areas) (Scotland) Order 2007* (SSI 2007/529, revoking SSI 2005/88) identifies on maps surface and ground waters used for the abstraction of more than $10m^3$ per day drinking water or which serve more than 50 persons, or are intended for such use; the Order and the maps can be accessed via www.scotland.gov.uk.
- The *Water Framework Directive (Groundwater Quality) Directions 2005* require SEPA by 22 June 2008, to prepare and submit to Scottish Ministers for their approval proposals for threshold values for pollutants, groups of pollutants or indicators of pollution for groundwaters that have been characterised as being at risk of failing to achieve good chemical status by 2015; and by 22 December 2008, to identify any significant and sustained upward trends in pollutant concentrations or indicators (see also 6.11.3 below).
- In carrying out its functions related to both the WFD and the Controlled Activities Regulations, *the Scotland River Basin District (Surface Water Typology and Environmental Standards) (Scotland) Directions 2007* and *the Solway Tweed River Basin District (Surface Water Typology and*

Environmental Standards) (Scotland) Directions 2007 require SEPA to assign a "type" to each of the waters within the Scotland river basin district, and to apply environmental standards according to the assigned type, both as set out in Schedules attached to the Directions. SEPA has issued proposals for further environmental standards and conditions for surface waters and groundwaters, together with proposals for amending the Directions.

- *The Water Environment (Relevant Enactments and Designation of Responsible Authorities and Functions Order 2008* (SSI 2008/263) designates functions held by Scottish Ministers and SEPA under the relevant enactments; it also designates specified public authorities as responsible authorities in respect of their functions relating to the water environment. (SSI2006/554 and SSI 2006/126 are revoked.

SEPA's River Basin Management Planning Strategy was published in December 2005; it outlines administrative arrangements for the production of the RBMP, including opportunities for public participation and consultation, as well as measures for the integration and coordination of the RBMP with other plans and planning. Guidance and policy documents relating to the WFD are available on SEPA's website at www.sepa.org.uk/wfd/. SSI 2007/530 brought into force on 14 December 2007 various sections of the 2003 Act relating to the submission, review and approval RBMPs.

A full river basin management plan is to be published for consultation in 2008, with the final plan completed by 2009.

The Water Environment (Controlled Activities) (Scotland) Regulations 2005, which came into force on 1 April 2006, replaced the system of discharge consents under *the Control of Pollution Act 1974*. The Regulations apply to abstractions, impoundments, as well as point source discharges, and aim to ensure compliance with the pollution control aspects of the water Framework Directive – see 6.11.6 below. *The Water Environment (Relevant Enactments) Order 2006* (SSI 2006/554) specifies the CAR as a relevant enactment for the purposes of s.2 of the WEWS Act to ensure that the requirement for Scottish Ministers and SEPA to "have regard to the social and economic impact of the exercise of their functions" applies to CAR.

In May 2008 SEPA published its proposals for controlling discharges of priority and dangerous substances and specific pollutants in the water environment, in order to meet the WFD's requirement that all EU surface waters should achieve good status by 2015 – see 6.2 and 6.4.4 above. This outlines SEPA's proposed approach to authorisation of point source continuous discharges to surface waters under the CAR, PPC and waste management licensing regimes. As a first step, SEPA will identify whether the discharge is liable to contain a listed substance; once it has been established that a discharge is liable to contain a listed substance, it will then determine if this is above a significant (specified) concentration or load, and if so instigate a monitoring and sampling regime. Licences and conditions will be reviewed to ensure compliance with the WFD and in particular to prevent any increase in discharges and to require an action plan to reduce discharges where necessary.

Statutory consultees in Scotland include: Scottish Natural Heritage; Scottish Water; responsible authorities with functions exercisable in or in relation to the RBD; local authorities, district salmon fishery board, national park authorities within the RBD affected by the consultation; and any other persons representative or who might have an interest.

6.10.4 Northern Ireland

The Water Environment (Water Framework Directive) Regulations (Northern Ireland) 2003 (SR 2003/544) were brought into force on 12 January 2004. They are similar to those applying in England and Wales and establish four river basin districts, three of which are cross-border river basins with Eire. The Department of Environment is the lead authority for drawing up the river basin management plan for the river basin in the North; for the cross-border RBDs it must where possible produce a single plan for each RBD in coordination with the competent authority in the South; if this does not prove possible, the DOE should produce a plan covering that part of the RBD under its jurisdiction. The timetable for implementation of the various elements of the WFD is as set out at 6.10.1 above.

The Water Abstraction and Impoundment (Licensing) Regulations (Northern Ireland) 2006 (SR 2006/482), which came into operation on 1 February 2007, control the abstraction of fresh surface water and groundwater and the impoundment of fresh surface water to ensure compliance with the Directive's environmental objectives for river basin districts. Certain activities which represent a minimal risk to the water environment will generally need to be notified to the Department and to comply with conditions applying to the activity (Sch. 1 to the Regulations); other activities posing a greater risk will require authorisation by licence. A charging scheme for abstraction and impoundment licences was introduced with effect from 1 July 2008.

Statutory consultees in NI include: NI Water Council, Council for Nature Conservation and the Countryside, district councils and harbour authorities in the area of the RBD; Dept. of Agriculture & Rural Development, Dept. of Culture, Arts & Leisure, Dept. for Regional Development; and any other persons representative or who might have an interest.

6.11 CONTROL OF POLLUTION OF WATER RESOURCES

Part III of *the Water Resources Act 1991* (E & W), and regulations made under it, is concerned with maintaining and enhancing the quality of "controlled waters" (i.e. relevant territorial, coastal, inland freshwaters (including lakes and ponds) and groundwaters.

The Controlled Waters (Lakes and Ponds) Order 1989 (SI 1149) which came into force on 1 September 1989 brings within the definition of "controlled waters" reservoirs not discharging into a watercourse, other than those containing water which has been treated prior to entering the supply.

Certain processes discharging to water are subject to Integrated Pollution Prevention and Control as a result of their potential to pollute the air and/or land – see chapter 2, 2.4 and 2.5.

Section 82 of the *Water Resources Act* provides for the establishment, via regulations, of systems of classifying water quality according to various criteria; s.83 provides for the setting of statutory water quality objectives (SWQOs) for individual stretches of water, on the basis of one or more of the classifications, together with a date by which the classification should be achieved. SWQOs will eventually be set for some 40,000 km of rivers and canals, as well as estuaries, coastal waters, lakes and groundwaters.

For Scotland, similar powers are contained in ss.30B, C, D & E of COPA (see Sch. 23, para 4 of *Water Act 1989* as amended by the *Environment Act 1995*, Sch. 22, para 29(1)-(5)). In Northern Ireland, Articles 5 & 6 of the 1999 Order contain the appropriate provisions.

6.11.1 Surface Waters

(a) Dangerous Substances

The Surface Waters (Dangerous Substances) (Classification) Regulations 1989 (SI 2286 apply in England and Wales and came into effect on 1 January 1990. Similar 1990 Regulations (SI 126) apply in Scotland. The Regulations implement various EU Daughter Directives relating to inputs to inland and coastal and territorial waters of specific Black List (List I) substances listed in Directive 76/464 (see 6.4 above). The substances concerned and the standards set are at Appendix 6.4. The E&W Regulations have been amended by 1992 Regulations made under s.83 of the *Water Resources Act 1991*, and similar in Scotland (SI 1992/574) which implement statutory environmental quality standards for four substances covered by the 1990 EU Daughter Directive – ethylene dichloride, trichloroethylene, perchloroethylene and trichlorobenzene (see this chapter, 6.4.1).

The Surface Waters (Dangerous Substances) (Classification) Regulations 1997 (SI 2560) and 1998 (SI 389) for England and Wales set water quality objectives for a total of 20 List II substances in EU Directive 76/464/EEC (see 6.4 above). Schedules to the Regulations specify the limits of dangerous substances which must not be exceeded to meet criteria for classification as inland freshwaters and for coastal waters and relevant territorial waters – see Appendix 6.4; the aim is to reduce pollution by the dangerous substances listed in the Schedules. The 1997 Regulations came into force on 26 November 1997 and the 1998 Regulations on 25 March 1998.

The Environment Agency is responsible for ensuring compliance with the Regulations, including sampling and analysis and monitoring the effect on waters of discharges containing the dangerous substances and of determining whether the requirements of each classification are being met.

In Scotland, *the Surface Waters (Dangerous Substances) (Classification) (Scotland) Regulations 1998* (SI 250), which came into force on 1 April 1998 and further 1998 regulations (SI 1344), which came into force on 1 July 1998, mirror the 1997 and 1998 E&W Regulations. Similar 1998 Regulations (SR 397) took effect in Northern Ireland on 1 February 1999.

(b) River Ecosystems

The Surface Waters (River Ecosystem) (Classification) Regulations 1994 (SI 1057), covering England and Wales, came into force on 10 May 1994. They prescribe a system for classifying inland rivers and watercourses, replacing a previous classification devised by the National Water Council (see also 6.7.4). The classifications are:

RE1 Water of very good quality suitable for all fish species
RE2 Water of good quality suitable for all fish species
RE3 Water of fair quality suitable for high class coarse fish populations
RE4 Water of fair quality suitable for coarse fish populations
RE5 Water of poor quality which is likely to limit coarse fish populations.

Eight parameters are used for assessing quality: dissolved oxygen; BOD; total ammonia; un-ionised ammonia; pH; hardness; dissolved copper; and total zinc – see Appendix 6.5. Frequency, location, methods of sampling and other procedures for compliance were set out in the NRA's *Water Quality Objectives: Procedures used by the National Rivers Authority for the purpose of the Surface Waters (River Ecosystem) (Classification) Regulations 1994* (30 March 1994).

In addition river and canal quality is surveyed using the "General Quality Assessments Scheme" – see 6.4 above. Under s.93 of the WRA, the Secretary of State can designate specific stretches of water as water protection zones to prevent pollution of drinking water sources – see 6.12.3.

(c) Abstraction for Drinking Water

The Surface Waters (Abstraction for Drinking Water) (Classification) Regulations 1996 (SI 3001), made under ss.82 & 102 of the WRA, came into force on 6 January 1997, with similar 1996 (SR 603) legislation in Northern Ireland. The Regulations prescribe the system of classifying the quality of inland freshwaters according to their suitability for abstraction by water undertakers for supply (after treatment) as drinking water (see Appendix 6.6). Sampling must be carried out at points chosen by the Environment Agency with the frequency for each parameter set in Schedule 3 to the Regulations; Schedule 2 details the measurement method to be used for each parameter. Analysis of the samples must be carried out to ascertain compliance with the parameters in Schedule 1 (see Appendix 6.6) – i.e. 95% of the samples should comply with the limit, none exceed it by more than 50%, and where any of the samples exceed the limit, there should be no danger to public health. The Environment Agency can decide to reduce sampling frequency where there is no pollution of waters, no risk of deterioration in quality or it is of a better quality than the parameter required for waters classified as DW1 (see Appendix 6.6).

(d) Fishlife

The Surface Waters (Fishlife) (Classification) Regulations 1997 (SI 1331) (as amended, SI 2003/194) for England and Wales came into force in June 1997, with similar Regulations (SI 1997/2471, as amended SSI 2003/85 & 2007/178) for Scotland coming into force on 18 November 1997, and for Northern Ireland (SR 488). They implement the EU freshwater fish Directive (see above 6.5.2), prescribing a system for classifying the quality of inland waters which need protection or improvement to support fishlife. A Schedule to the Regulations lists 13 parameters and the requirements to be satisfied for salmonid and for cyprinid waters, as well as methods of analysis or inspection and minimum sampling and measurement frequency. The 2003 Scottish Regulations place a duty on SEPA to establish programmes to reduce pollution in waters classified under the 1997 Regulations.

In England and Wales approximately 20,000 km of rivers and canals and more than 100 lakes and reservoirs have been designated under the Directive. In January 2004, proposals were made to designate a further 14,500 km of rivers and canals and 90 lakes and canals. A consultation document published by the DOENI in November 2003 also proposed the designation of a further 3,300 km of rivers, 3 canals and a further 17 reservoirs and lakes; currently 1,200 km of rivers are designated and 3 lakes. In Scotland 205 salmonid and 4 cyprinid waters have been designated (SEPA website).

In 1996, 93% of the total length of designated rivers in England and Wales complied with standards in the freshwater fish Directive (Scotland 98%; Northern Ireland 28%).

(e) Shellfish

The Surface Waters (Shellfish) (Classification) Regulations 1997 (SI 1997/1332) came into force on 12 June 1997 and apply in England and Wales. Similar Regulations (SI 1997/2470) for Scotland came into force on 18 November 1997, and for Northern Ireland (SR 449). They implement the Shellfish Waters Directive (see above 6.5.2) by prescribing a system for classifying the quality of controlled waters which are coastal or brackish waters and which need protection or improvement in order to support shellfish life and growth. A Schedule to the Regulations lists 11 parameters and the mandatory values assigned to them in the Directive; it lists reference methods of measurement, and the minimum frequency required for sampling and analysis, also laid down in the Directive.

In England and Wales 94 shellfish waters have been designated in accordance with the 1979 Directive, 76 of them in 1999 (House of Commons, written answer, 8 July 1999). Northern Ireland has nine designated shellfish waters. *The Surface Waters (Shellfish) (Classification) (Scotland) Direction 2006* designates 108, with eight further proposed in October 2007.

6.11.2 Bathing Waters

(a) 1991 Regulations (NI, 1993)

The Bathing Waters (Classification) Regulations 1991 (SI 1597, as amended, SI 2003/1238) implement EU Directive 76/160/EEC (see above, 6.5.1) and came into force in August 1991 in England and Wales. There are similar 1991 Regulations in Scotland (SI 1609) and 1993 Regulations in Northern Ireland (SR 205). These Regulations are to be revoked in England and Wales on 24 March 2015 following full implementation of *the Bathing Waters Regulations 2008* (E & W only, SI 2008/0197), which implement EU Directive 2006/7/EC *concerning the management of bathing water quality* (see below and 6.5.2). Separate Regulations implementing the 2006 Directive have been made in both Scotland (SSI 2008/170) and Northern Ireland (SR 2008/231).

The 1991 Regulations set out a system for classifying the quality of UK territorial, coastal and inland waters which are designated bathing waters – classified as BW1 waters – in compliance with the Directive. Schedule 1 to the Regulations specifies criteria for BW1 classifications and Schedule 2 sampling requirements. Schedule 3 (see Appendix 6.2b) sets out quality and additional sampling requirements. The 2008 Regulations amend the 1991 Regulations with effect from 24 March 2012 to omit Schedules 1, 2 and 3. The 2008 Regulations also amend the 1991 Regulations with effect from 24 March 2008 to require that the criterion for a bathing water's achieving the classification of BW1 is that at least 95% of samples of waters meet the parametric value of 2,000 per 100 millitres for Escherichia coli. Samples are to be taken and tested in accordance with procedures in the 2008 Regulations.

The current criteria for BW1 classification include:

a) at least 95% of samples taken and tested in accordance with sampling requirements must conform to the parametric values in Schedule 3;

b) no sample which fails when tested for compliance with the phenols parameter by the absorption method or with the transparency parameter should have a value which deviates by more than 50% from that in Schedule 3;

c) consecutive samples taken at suitable intervals should not deviate from the parametric values specified in Schedule 3.

Sampling should be carried out at the same place and as specified in Schedule 3 from 15 May - 30 September (Sc & NI 1 June - 15 September) with a minimum of 20 samples taken at each site; additional samples should be taken if it is suspected that the water quality is deteriorating. Sampling results can be disregarded if they do not meet the requirements due to abnormal weather conditions.

UK Bathing Water Results

The bathing season in E&W is defined as 15 May to 30 September, and in Sc and NI from 1 June to 15 September. Results for both the 2007 and 2008 bathing seasons were affected by the very poor weather.

In England 97.8% (2006: 99.5%) of the 405 designated bathing waters met mandatory water quality standards, with 72.8% (2006: 76.2%) achieving the more stringent guideline standards (which set maximum permitted levels of total and faecal coliforms and faecal streptococci). In Wales 97.5% (2006: 98.8%) of the 80 designated bathing waters met the mandatory standards in 2007 and 86.3% (2006: 88.8%) the more stringent guideline values;

In 2007, 88.1% of Scotland's 59 designated bathing waters met the mandatory standards, with 49.2% meeting the guideline values. Initial results for the 2008 bathing season, indicate that 91% of the 80 designated bathing sites (and 98% of the samples taken) were of good or excellent quality; 9% of sites (and 2% of the samples taken) were classified as poor.

In Northern Ireland 91.3% of the 23 designated bathing waters met the mandatory standards in 2007 (2006: 100%, 16 designated bathing waters), with 47.5% (2006: 87.5%) complying with the guideline values.

Water companies are also required to comply with the Environment Agency's *Policy for Consents for Sewage Effluent Discharges Affecting Bathing Waters* (1997); under this policy treatment plants must use high standards of conventional treatment and discharge from an outfall well away from bathing waters; if an outfall near a bathing water is used, then very high standards of treatment followed by ultraviolet light disinfection is required. The Agency also requires water companies to meet the requirements of the EU's urban waste water Directive (and of the UK regulations) to ensure identified bathing waters meet the standards set in the bathing water Directive – see 6.5.5 and 6.20.

In Scotland, an independent Bathing Water Review Panel considers applications for bathing water designations; applicants (including local authorities, community groups and other interested bodies) must show evidence of peak daily usage of 150 people or more on three separate days during the bathing season, appropriate beach management and the support of the beach owner-operator.

(b) 2008 Regulations

The Bathing Water Regulations 2008 (SI 2008/1097) implement the EU's 2006 bathing waters Directive (see above 6.5.1) in ***England and Wales***. The regulations relating to the identification of bathing waters, length of bathing season, bathing water profiles, monitoring (in part), and the provision of guidance came into force on 14 May 2008. The remainder of the regulation on monitoring, and those on public information and management measures and public information at bathing waters subject to pollution incidents take effect on 24 March 2012, with implementation of the remaining regulations on 24 March 2015 when the 1991 Regulations will be revoked.

There are similar Regulations for ***Scotland*** (SSI 2008/170), with those regulations covering designation of bathing waters, length of bathing season, consultation, bathing water profiles, monitoring, assessment and classification of bathing waters, and management measures for "poor" bathing waters coming into force on 23 May 2008. Further regulations relating to management measures take effect on 24 March 2011, those relating to public information on 24 March 2012, and regulations relating to amendment of other statutory instruments taking effect on 1 January 2015 when the 1991 Regulations will be revoked.

In ***Northern Ireland*** the relevant Regulations are SR 2008/231. Regulations covering the identification of bathing waters, length of bathing season and public participation took effect on 30 June 2008. Regulation 7 relating to bathing water profiles comes into force on 24 March 2011, with regulations covering monitoring and public information on 24 March 2012. Regulations relating to assessment and classification and management of bathing waters are implemented in part on 24 March 2012, with the remainder on 24 March 2015. Regulation 21 revoking the 1993 Regulations takes effect on 30 September 2015.

The Regulations place a general duty on the Secretary of State, Welsh Ministers and the Environment Agency [Scottish Ministers & SEPA / DOENI] to exercise their functions so as to ensure that by the end of the 2015 bathing season (i.e. 30 September 2015; Sc & NI, 15 September 2015) all designated bathing waters are classified as "sufficient" and to take proportionate measures to increase the number classified as "good" or "excellent" (Sch. 4 to the Regulations and see Appendix 6.2). The Regulations also require that the Secretary of State, Welsh Ministers and the Agency [Scottish Ministers / DOE] encourage public participation in the exercising of their functions – the public should have the opportunity to submit suggestions, comments or complaints. The Secretary of State and Welsh Ministers, and the Environment Agency [Scottish Ministers & SEPA / DOENI] must publish lists of identified bathing waters on their respective websites.

Bathing Water Profiles

The Agency [SEPA, DOENI] is required to establish profiles of each bathing water before 24 March 2011. As well as taking account of data collected in compliance with the regulations implementing the Water Framework Directive, the profile should include (Sch.2; Sc Sch.1))

- a description of the physical, geographical and hydrological characteristics of the bathing water and of any other surface water in the catchment area of the bathing water if this could be a source of pollution;
- identification and assessment of pollution that might affect bathing water quality and pose a risk to bathers' heath;
- assess the potential for cyanobacterial proliferation, and of macro-algae or phytoplankton;
- the location of the monitoring point.

The profile of bathing water which has been classified as "poor" should be reviewed and updated as a minimum every two years; the profile of those classified as "sufficient" or "good" should be reviewed and updated as a minimum every three and four years, respectively. If there has been significant construction work or other infrastructure changes around the bathing water, the profile should be reviewed prior to the start of the next bathing season.

Prior to establishing a bathing water profile and prior to any review, SEPA should consult the relevant local authority and health board, Scottish Water; Scottish Natural Heritage if any part of the bathing water falls within a European site designated under the Habitats Directive or an SSSI; where agriculture activities may cause pollution affecting the bathing water quality representatives of agricultural interests.

Monitoring of Bathing Waters

Reg.8 and Sch.3 (Sc. reg. 7 & Sch. 2) set out requirements for monitoring programmes, sampling and analysis. The Agency [SEPA / DOE] must establish a monitoring calendar for every bathing water prior to the start of the bathing season for intestinal enterococci and *escherichia coli*. At least four samples should be taken during the bathing season, with the first taken shortly before the start of the season. Monitoring may be suspended during an "abnormal situation" (an event or combination of events which could impact on bathing water quality and which the Agency would not expect to occur more than once every four years); monitoring should be resumed as soon as possible and additional samples taken to ensure there are sufficient by the end of the season. Where a bathing water has revealed a potential for proliferation of cyanobacteria, macro-algae and marine phytoplankton, the Agency should carry out an investigation with a view to putting appropriate management measures in place. The Agency should also carry out a visual inspection of each bathing water as necessary for waste (including tarry residues, glass, plastic or rubber) to ensure appropriate management measures are put in place. At the end of the bathing season, the Agency should inform the Secretary of State (Welsh Ministers) if monitoring has been suspended and the reasons.

Public Information

From 24 March 2012 each local authority or private controller of a bathing water must ensure that information about the bathing water is made available in the vicinity of the bathing water. This should include

- current classification of the bathing water;
- any advice against bathing which has been introduced, and the reasons;
- description of the bathing water, based on its profile, in non-technical language – the Agency is responsible for preparing a general description for use by local authorities and private controllers;
- information on nature and expected duration of any abnormal situation;
- indication of sources of more complete information provided using a geographic information system.

The local authority or private controller should also include an indication of the number of days during the preceding bathing season that advice against bathing was introduced because of short-term pollution, and a warning whenever short-term pollution is predicted or present.

The Agency [SEPA / DOE] should also publish the above information on its website, together with the bathing water's classification for the last three years, its profile, results of monitoring since the beginning of the current season, and where water has been classified as "poor" the reasons and management measures being taken.

Bathing Water Assessment & Classification

After the end of every bathing season (beginning 2015), the Agency [SEPA / DOE] must prepare bathing water quality data for that season, and carry out an assessment of bathing water quality based on the previous three years' data (or less having consulted the appropriate Minister); on the basis of the assessment the Agency should then classify the water as "poor", "sufficient", "good" or "excellent".

Where a bathing water has been classified as "poor" for five consecutive years, or it is agreed that it would be infeasible or disproportionately expensive for the bathing water to achieve "sufficient" quality, the Agency [SEPA /DOE] must notify the local authority or private controller that a permanent advice against bathing should be introduced. The local authority or private controller must take the necessary steps to bring this to the attention of the public.

Management Measures

From 24 March 2012, the provisions regarding the management of bathing water come into force with the Agency [SEPA / DOE], local authorities, private controllers and sewerage undertakers all having responsibilities to consult each other when they become aware of an incident of pollution (other than short-term pollution), contamination or waste which may pose a risk to bathing water quality or to bathers' health (reg.12). The Agency [SEPA / DOE] must then take the necessary management measures to protect bathers' health, with local authority and private controllers (as relevant) also required to take appropriate management measures to protect bathers' health, and in the case of pollution by waste to remove it. From 2015 additional management measures must be taken at bathing waters classified as "poor", including measures to prevent bathers' exposure to pollution and advice against bathing by means of public information.

Where the Agency [SEPA / DOE] has established procedures for short term pollution (which it has identified the causes of and is not expected to affect bathing water for more than 72 hours), it should ensure that information as to its anticipated nature, frequency and duration, as well as details of any remaining causes, are included in the bathing water profile. Details of the management measures taken, and schedule for elimination for the causes of the short-term pollution should also be given, as well as relevant procedures to be taken during any short-term pollution incident. The Agency [SEPA/ DOE] should notify the local authority or private controller when short-term pollution is predicted and publish details on its website of the conditions likely to lead to short-term pollution; the likelihood of short-term pollution and its duration; causes, and relevant procedures. The Agency [SEPA / DOE] should take one additional sample as soon as possible after the pollution incident tis presumed to have ended to confirm that it has ended. Where there is short-term pollution at a bathing water, the Agency [SEPA / DOE] must notify the local authority or private controller and ensure relevant procedures are put into operation. The local authority or private controller must ensure notification of the short-term pollution is publicly available.

6.11.3 Protection of Groundwater

Section 84 of the *Water Resources Act* gives the Agency a duty to protect the quality of groundwater and conserve its use for water resources.

The Environment Agency's *Groundwater Protection: Policy and Practice* (1998) provides a framework for regulation of groundwater, identifying the Agency's guiding key principle as being "to protect and manage groundwater resources for present and future generations in ways that are appropriate for the risks that we identify". It outlines the Agency's approach for protecting groundwater supplies from various sources of pollution and defines source protection zones as zone I (inner source protection), zone II (outer source protection) and zone III (source catchment); the Agency applies a risk based approach to managing and protecting groundwater, which involves assessing the hazard – e.g. a leak from a landfill development or a major new abstraction – combining this with an assessment of the magnitude of consequence should the hazard occur; this will take account of the proximity of the activity to the groundwater source and the physical, chemical and biological properties of the underlying soil and rocks. In April 2007, the Agency published for consultation an updated version of its Policy, together with a report (*underground, under threat*) on the state of groundwater in England and Wales.

Scotland's strategy for protection of groundwater was published in 1995 and uses the same approach as the Agency in identifying protection zones; the Environment and Heritage Service published a similar document, Policy and Practice for the Protection of Groundwater in Northern Ireland, in July 2001.

The Groundwater Regulations 1998 (SI 2746), covering England and Wales, came into force on 1 January 1999, with similar Regulations applying in Northern Ireland (SR 401) from the same date; the Regulations complete the UK's implementation of the 1980 groundwater Directive (see 6.4.3). (SI 2746 was revoked in Scotland on 1 April 2006 when the *Controlled Activities Regulations 2005* came into force – see 6.11.6.)

The Regulations are designed to protect groundwater from pollution from mainly industrial and farming activities (including disposal of sheepdip or wastes and pesticides to land).

Activities likely to lead to direct or indirect discharges to groundwater of List I or List II substances (defined in the Groundwater Directive) required authorisation from 1 April 1999. Direct discharges of List I substances are prohibited; activities which may result in indirect discharges (from tipping or disposal) may only be authorised if prior investigation shows the groundwater is permanently unsuitable for other uses (e.g. domestic or agricultural). Such authorisations should include conditions to ensure all necessary technical precautions are taken to prevent the indirect discharge of a List I substance. List II discharges will only be authorised, with conditions, if prior investigation shows that groundwater pollution can be prevented. Where a discharge is authorised, the authorisation will specify where the discharge may be made, and the method to be used, precautions to be taken, as well as quantities of any substance and monitoring requirements. The authorisation will be granted for a limited period and must be reviewed every four years.

The Agency may serve a Notice on any person proposing to carry on an activity on or in the ground which might lead to an indirect discharge of a List I substance or pollution of groundwater as a result of an indirect discharge of a List II substance. The Notice may prohibit the activity, or authorise it subject to conditions specified in the Notice. The Regulations also enable statutory codes of practice to be drawn up providing guidance on preventing such discharges and a number are now available covering:

- Sheepdipping (E& W, 2001; Sc*)
- Quarries and other mineral extraction sites (Sc*)
- Petrol stations and other fuel dispensing facilities involving underground storage tanks (E&W, 2002; Sc*)
- Solvent use and storage (E&W, 2004)
- Septic tanks and other non-mains sewerage systems (E&W, draft 2004)

*2003 codes withdrawn in Scotland; new codes are being issued in Scotland under the Controlled Activities Regulations – see below 6,11.6.

The Regulations are enforced by the Environment Agency and in Northern Ireland by the NI Environment Agency; they also include requirements for compliance monitoring, powers to serve variation and enforcement notices and revocation notices and introduce a cost recovery scheme. There is a right of appeal if the Agency refuses an authorisation and against enforcement notices. Details of all authorisations will be kept on the appropriate Agency's public register.

The Environment Agencies (E&W, SEPA & NIEA), together with Defra, the Department of Health and Chemical Industry Association, have established the Joint Agency Groundwater Advisory Group; JAGDAG is responsible for considering whether individual substances within the generic groups of substances in List I and II in the Directive should be List I or List II for purposes of the Regulations; criteria used include toxicity, persistence and bioaccumulation. A full list of JAGDAG's determinations can be found on the Environment Agency's public register.

Guidance on the Regulations was published by Defra in 2001.

In June 2008 Defra issued proposals for amending the 1998 Regulations to implement Article 6 of the Groundwater Directive which requires Member States to prevent inputs of all hazardous substances, to limit the input of non-hazardous substances to avoid pollution, and to take action to address diffuse pollution wherever technically feasible to achieve good chemical status. It was hoped amending Regulations would take effect in November 2008. Defra is then proposing that the groundwater authorisation procedure be brought within the Environmental Permitting regime.

In Scotland, *the Water Framework Directive (Groundwater Quality) Directions 2005*, which came into force on 21 December 2005, required SEPA

- by 22 June 2008, to prepare and submit to Scottish Ministers for their approval proposals for threshold values for pollutants, groups of pollutants or indicators of pollution for groundwaters that have been characterised as being at risk of failing to achieve good chemical status; the proposed methodologies to be applied for deriving threshold values had to be submitted for approval by 22 December 2006.
- by 22 December 2008, to identify any significant and sustained upward trends in pollutant concentrations or indicators, and to determine the starting point for reversal of each identified trend, and to review these every six years.
- to cooperate with the Environment Agency in measures necessary to achieve good groundwater chemical status with regard to the cross-border Solway Tweed river basin district.

6.11.4 Pollution Offences (E & W, NI)

Under s.85 of the WRA it is an offence to cause or knowingly permit

a) any poisonous, noxious or polluting matter or any solid waste matter to enter any controlled waters;

b) any matter, other than trade effluent or sewage effluent, to enter controlled waters by being discharged from a drain or sewer in contravention of a relevant prohibition;

c) any trade effluent or sewage effluent to be discharged -

   into any controlled waters; or

   from land in England and Wales, through a pipe, into the sea outside the seaward limits of controlled waters;

d) any trade effluent or sewage effluent to be discharged, in contravention of any relevant prohibition, from a building or from any fixed plant on to or into any land or into any waters of a lake or pond which are not inland waters;

e) any matter whatever to enter any inland waters so as to tend (either directly or in combination with other matter which he or another person causes or permits to enter those waters) to impede the proper flow of the waters in a manner leading or likely to lead to a substantial aggravation of pollution due to other causes, or of the consequences of such pollution.

Contravention of this section of the Act or of conditions attached to a consent to discharge given under the Act may result in a fine not exceeding £20,000 and/or three months' imprisonment on summary conviction or an unlimited fine and/or two years' imprisonment on conviction on indictment.

Defences include:

- that the discharge or entry of matter into water was caused or permitted in an emergency to safeguard life or health;
- that all such steps as were reasonably practicable were taken to minimise the extent and effect of the discharge or entry;
- that the Agency were given details of the entry or discharge as soon as reasonably practicable.

Under s.88 of the Act, it is a defence to prove that a discharge to controlled waters did not contravene conditions imposed in accordance with other legislation under which it is regulated. Section 89 of the WRA (as amended by s.60 of the *Env. Act 1995*, effective 1 July 1998), excludes from s.85 any person who allows water from a mine or part of a mine abandoned before 31 December 1999 to enter controlled waters; thus, mines abandoned after this date will be subject to control through discharge consents etc.

Northern Ireland

The Water and Sewerage Services (Northern Ireland) Order 2006 (SI 2006/3336, NI 21), which came into force on 1 April 2007, substitutes Article 7 of *the Water (Northern Ireland) Order 1999* with new Articles 7A and 7B which are more in line with the water pollution offences and defences outlined above.

6.11.5 Consents to Discharge (E & W, NI)

The Control of Pollution (Applications, Appeals and Registers) Regulations 1996 (SI 2971) came into force on 31 December 1996 and apply in ***England and Wales***; they were made under ss.90A, 91, 190, 191 and Sch. 10 of the WRA, as amended by the *Environment Act 1995*, Sch. 22, paras 142-144 and 183 (which replaces the original Schedule 10 to the WRA). The Regulations revoke the *Control of Pollution (Consents for Discharges) Regulations 1989, the Control of Pollution (Consents for Discharges by the National Rivers Authority) Regulations 1989* and the *Control of Pollution (Registers) Regulations 1989.*

In ***Northern Ireland Article*** 9(3) and Schedule 1 of the *Water (Northern Ireland) Order 1999* contain similar provisions in respect of applications for consents, their determination, consents without applications, revocations and modifying consents, variations and transfers; Article 12 covers enforcement notices, Article 13 appeals and Articles 30-32 the pollution control registers. These Articles were all brought into force on 24 August 2001, together with Regulations covering advertisements for applications for a consent and public registers – *the Control of Pollution (Applications and Registers) Regulations (Northern Ireland) 2001* (SR 284).

Note: In ***Scotland***, the pollution control regime under COPA was, with effect from 1 April 2006, replaced by new regulations made under the *Water Environment and Water Services Act 2003: the Water Environment (Controlled Activities) Regulations 2005* – see below 6.11.6.

(a) Application for a Consent

An application for a consent to discharge has to be made on a form provided by the Agency, and submitted to it with any other information required together with the appropriate fee (see below, 6.11.5(f)). The Agency may refuse to process the application if all the information requested has not been provided (WRA, s.90A, see *Env. Act.* Sch.22, para 142). In Northern Ireland application is made to the NIEA's Industrial Consents Section.

A notice that an application for, or variation of, a discharge consent has been made must be advertised in one or more local newspapers and in the *London Gazette*; this should normally be done within 42 days, but not before 14 days after its receipt by the Agency. Exceptions to this are:

- where the Agency notifies the applicant that it refuses to proceed with the application or variation until further information is received, the period for advertising begins 14 days after receipt by the Agency of the additional information;
- where it is agreed that the application has implications for national security or is commercially confidential, the period for advertising begins 14 days later; where the applicant requests exclusion from the public register of those parts of the application considered to be commercially confidential, the date for advertising will begin when the Agency has notified the applicant of its decision, the date on which the period allowed for appeal expires, is determined or withdrawn.

The Agency may decide that the application may not be advertised if it is to be excluded from the public register on grounds of national security or that the activities to which the application relate "are unlikely to have an appreciable effect on controlled waters in the locality...".

Where it appears to the Agency that a discharge has been caused in contravention of s.85 or 86 of the WRA and is likely to happen again, it can issue a discharge consent (together with conditions) without having first received an application. Schedule 1 to the Regulations sets out the procedure (advertising, consultation, etc) which is similar to that for normal applications to discharge.

Where the Agency itself wishes to apply for a discharge consent, this should be done in writing to the Secretary of State; Schedule 2 to the Regulations sets out the procedure to be followed; this is similar to that for applicants to the Agency.

(b) Consultation & Determination

A copy of the application should be sent to all affected local authorities and water undertakers in whose area it is proposed to discharge; to relevant ministers if it is proposed to discharge to coastal waters or relevant territorial waters; to any harbour authority or local fisheries committees whose waters might be affected by the discharge. Unless otherwise specified, a period of six weeks will normally be allowed for representations.

Copies of all applications will be put on the public register kept by the Agency. However, an applicant may apply to the Secretary of State for data to be withheld on the grounds of commercial confidentiality or that publishing such data would not be in the public interest.

An application for a consent must be considered within four months of receipt; if it is not and a longer period has not been agreed with the applicant, it is deemed to have been refused. In considering whether to grant a consent, the Agency will take into account whether statutory water quality objectives will be met; it will also need to be assured that the discharge will not result in deterioration of water quality or adversely affect uses of the water downstream.

The Secretary of State may call in the application for determination as a result of objections or representations made; a local inquiry or hearing will then be arranged as a means of enabling a decision to be made on the application.

If consent is granted, conditions will be included to ensure compliance with statutory water quality objectives (including EU legislation), absolute limits set for discharges and other conditions such as

- the place to which the consent to discharge relates;
- the nature, origin, composition, temperature, volume and rate of discharge, and the periods during which the discharge may be made;
- steps to be taken to minimise the polluting effects of the discharge;
- provision of facilities for sampling and monitoring;
- provision, maintenance and testing of meters for measuring or recording discharges;
- keeping of detailed records relating to the discharge and conditions attaching to the consent;
- provision of information to the enforcing authority in respect of the discharge.

The Environment Agency is extending "operator and pollution risk appraisal" (OPRA) and operator self-monitoring to the regulation of discharge consents (see chapter 2, 2.4.9(c)); OPRA is to be used to plan compliance assessment activities, and for public reporting and comparing operators; from April 2009, it is proposing to base charges for consents on OPRA scores.

Consents and their conditions are reviewed from "time to time" and, following review, the Agency may serve a notice revoking or modifying a consent, modifying its conditions or adding some conditions to it; however, apart from a notice revoking a consent, such a notice will not normally be served within four years of the consent being issued without the agreement of the discharger; exceptions to this are to ensure compliance with EU legislation, other international obligations, or for health or environmental protection reasons.

The holder of a consent may apply (on a form provided by the Agency) for it to be varied. A consent may also be transferred to another person: s.87 of *the Water Act 2003* (brought into force 1.10.04) requires both parties to the transfer to jointly notify the Agency of the transfer and the date of the proposed transfer; the Agency will within 21 days of receipt of the notification arrange to amend the consent and notify both parties that this has been done. A person taking responsibility for a consent following the death or bankruptcy of the holder should advise the Agency of the fact within 15 months.

(c) Appeals

Where an application for consent to discharge has been refused, revoked or otherwise varied, a variation has been refused, or an enforcement notice served, the applicant has a right of appeal to the Secretary of State. Any appeal should be submitted in writing (together with supporting documentation) to the Secretary of State within three months of the decision which is being appealed against, with copies also being sent to the Agency at the same time. Appeals in respect of a revocation notice should be submitted before the revocation takes effect; in respect of an enforcement notice or refusal for information to be excluded from the public register, within 21 days.

The Secretary of State may then require the appellant to submit further documentation supporting the appeal within 28 days and will also advise the Agency that an appeal has been lodged. The Agency has 21 days in which to submit its own representations in respect of the appeal.

The Agency should notify all those who made representations concerning the application for discharge consent or variation and statutory consultees within 14 days of receiving notice of the appeal; it must also inform the Secretary of State within a further 14 days to whom it has sent notification of an appeal. Any representations (which will be copied to the Agency, the appellant and put on the register) should be sent to the Secretary of State within 21 days.

The appellant may choose to have the appeal dealt with by written representation or a hearing. In the case of written representation, information supporting an appeal should be submitted within 28 days (or 14 days in respect of an enforcement notice). If there is to be a hearing, the Secretary of State should give both the appellant and the Agency 28 days' notice of its date, time and place and at least 21 days before should publish a notice in a local newspaper and send notice of it to all who have submitted representations. Section 114 and Schedule 20 of the *Environment Act 1995* enable the Secretary of State to appoint someone to carry out all or some aspect of the appeal on his behalf, including arranging a hearing or local inquiry. With effect from 1 January 1997, appeals relating to discharge consents, works and enforcement notices were transferred to the Planning Inspectorate. The Secretary of State will notify the appellant in writing of his decision concerning the appeal, as well as notify the Agency and others who made representations.

It should be noted that the terms of an enforcement notice, must be complied with pending determination of the appeal. Revocations, variations or the addition of conditions to consents will not take effect until determination or withdrawal of the appeal unless the Agency is of the opinion that to delay might result in the entry of any polluting, poisonous or noxious matter into controlled water or harm to human health.

The Planning Inspectorate has published a guide to the appeals procedure (E&W) relating to discharge consents, enforcement and works notices, and determinations that information is not commercially confidential.

(d) Agency Discharges

Applications etc by the Agency itself to discharge into controlled waters are made in writing to the Secretary of State. Schedule 2 of the Regulations outlines the requirements and procedures, which are similar to those for applicants to the Agency.

(e) Pollution Control Registers

The Agency is required to maintain registers, open to public inspection, with details of

- notices of water quality objectives and other Secretary of State notices and directions;
- applications for discharge consents or for their variation and supporting data;
- consents for discharges and any conditions, and information obtained or given in pursuance of conditions, and any variations thereof;
- data relating to samples of water or effluent, including

results of analysis, and steps taken by the Agency or another person as a consequence;
- prohibition and enforcement notices, revocations, appeals and details of relevant convictions;
- results of analysis of samples of effluent and of samples of water taken by the enforcing authority in carrying out their pollution control functions;
- information acquired by the enforcing authority with respect to samples taken by other persons.

Information should normally be placed on the register within 28 days, or seven days in the case of enforcement notices. Information relating to a withdrawn application for a consent or variation should be removed from the register not less than two months and not more than three months after its withdrawal.

The Agency should inform the Secretary of State of any information excluded on the grounds of national security; or a person wishing to have information excluded on these grounds may notify the Secretary of State direct and, pending his decision, the information concerned should not be put on the register (*Env. Act 1995*, Sch. 22, para 170, adds s.191A to the WRA).

Where a person requests that certain information given in connection with an application for a discharge consent be excluded for commercially confidential reasons, the Agency must determine the request within 14 days; if it does not the information automatically becomes commercially confidential. Where the Agency has obtained information in pursuit of its duties which it considers could be commercially confidential, it should notify the appropriate person "giving him reasonable opportunity" to object to its inclusion. If the Agency considers information not to be commercially confidential, it should notify the relevant person, allowing 21 days before putting the information on the register, during which time an appeal may be made to the Secretary of State. Following determination or withdrawal of the appeal, seven days should elapse before the information is placed on the register (*Env. Act 1995*, Sch. 22, para 170 adds new s.191B to the WRA).

Where information is excluded from the register, for instance on the grounds of commercial confidentiality, a notice should appear on the register to that effect.

The Agency is required to keep information on the register for up to four years after it has been superseded; it is also required to keep monitoring information for up to four years.

(f) Fees and Charges

Section 41 of the *Environment Act* empowers the Agencies to make charging schemes in respect of environmental licences (including consents to discharge), variations of such licences etc; such schemes are approved by the Secretary of State under s.42 of the Act. The original Scheme for England and Wales was made under ss.131 & 132 of the *Water Resources Act 1991*.

The purpose of charging for applications and consents for discharges of sewage and trade effluent to controlled waters is to enable the Environment Agency to recover its costs in respect of monitoring, sampling, analysing and other regulatory functions related to discharge consents. The current scheme came into effect on 1 July 1991 and was reviewed in March 1994. Charges are reviewed annually.

Applications for a new or revised discharge consent are subject to a standard charge per effluent. In addition an annual fee, also per effluent, is payable; when treatment or monitoring is carried out together the annual fee will be calculated on the basis of the highest content and aggregated volume of the discharge. The annual fee is based on the volume and content of the discharges as specified in the consent, and on the nature of the receiving waters with the amount payable depending on the complexity of the discharge and the sensitivity of the receiving waters: each of the three categories of the annual charge has been banded into factors reflecting complexity and sensitivity or in the latter case, type of receiving water; the annual charge payable is the number of factors multiplied, multiplied again by the financial factor. From April 2009 the Environment Agency is to link charges to operator and pollution risk appraisal (OPRA) scores (see also chapter 2, 2.4.9(c)).

In Northern Ireland, charges are levied under a scheme made under Article 11 of the *Water (Northern Ireland) Order 1999*.

6.11.6 Controlled Activities Regulations

The Water Environment (Controlled Activities) (Scotland) Regulations 2005 (SSI 2005/348) (CAR) began coming into force on 1 July 2005, taking full effect on 1 April 2006 when it became an offence to carry out controlled activities without an appropriate CAR authorisation. 2007 Amendment Regulations (SSI 2007/219), which came into force on 1 April 2007 replace Schedule 3 relating to General Binding rules, with further modifications made by *the Water Environment (Diffuse Pollution) (Scotland) Regulations 2008* (SSI 2008/54) – see (c) below.

The transitional provisions of the Regulations (reg. 54 & sch.10) were brought into force partly on 1 July 2005 and 1 October 2005 to enable the transfer of existing consents under COPA to the new regime by 31 March 2006. SEPA is responsible for authorising controlled activities and for enforcement of the Regulations. From 1 April 2006 all new (i.e. begun on or after 2 July 2005) point source discharges, abstractions, impoundments (dams, weirs etc) and engineering works required a CAR authorisation, by way of licence, registration or general binding rules, depending on the activity.

The Regulations introduce a new consent system for point source discharges liable to cause pollution to groundwater, wetlands, rivers, lochs, estuaries and coastal waters, and disposal onto land of listed substances and other matter likely to cause pollution of groundwater. The Regulations also control abstraction and impoundment of water and control building, engineering and other works (in or near waterbodies) which can have an adverse impact on the physical quality of aquatic habitats. The Regulations do not apply to any activity for which a licence is required under Part II of *the Food and Environment Protection Act 1985*.

The Water Environment (Relevant Enactments) Order 2006 (SSI 2006/554), which came into force on 14 December 2006, identify CAR as a relevant enactment for the purposes of *the Water Environment and Water Services (Scotland) Act 2003* (WEWS), which implements the *Water Framework Directive*. Thus the requirement in the WEWS for Scottish Ministers and SEPA to 'have regard to the social and economic impact of the exercise of their functions' applies also to CAR; this designation also ensures that they will exercise their functions under CAR to ensure compliance with the *Water Framework Directive*, as well as promoting sustainable flood management and sustainable development in carrying out their functions.

A practical guide to the Regulations as amended is available from SEPA (v.5, June 2008); it is available in hardcopy or can be downloaded from SEPA's website, www.sepa.org.uk/wfd.

CAR Authorisations

As mentioned above, there are three levels of authorisation – by licence, registration or general binding rules, with the latter not requiring to be notified to SEPA (see below). The regime for authorisation by registration or licence is similar to that for other environmental licences and permits; in both cases conditions attached to the authorisation should ensure protection of the water environment; where the activity falls within the provisions of the Groundwater Directive (see 6.4.3), conditions should be attached to ensure compliance with it. The Regulations also enable SEPA to impose an alternative level of authorisation if it is of the opinion this would be more appropriate.

Applications for CAR authorisations will normally be determined by SEPA, with Scottish Ministers having powers to direct SEPA to refer specified classes of, or individual, applications to them for determination either with or without local inquiry.

Application fees, subsistence charges etc for activities covered by the Regulations are specified in *the Water Environment (Controlled Activities) Fees and Charges (Scotland) Scheme*.

Guidance for operators of abstractions and impoundments has been published by SEPA and is available on their website.

(a) Licensed Activities

An application for a licence should be submitted to SEPA with the appropriate fee. As well as full details of the activity and other information required by SEPA, it should identify the "responsible person" (RP) – i.e. the person or corporate body with responsibility for day-to-day operations, and who will be responsible for compliance with the conditions of the licence. Licence applications will normally be determined by SEPA within four months.

If SEPA considers that the controlled activity has or is likely to have a significant adverse impact on the water environment, it will by notice require the applicant to advertise the application (within 28 days), specifying the content of the advertisement and where it is to be placed. The advertisement should state that any representations should be made to SEPA in writing within 28 days, and that these will be taken into account when determining the application. *The Water Environment (Controlled Activities) (Third Party Representations etc) (Scotland) Regulations 2006* (SSI 2006/553, effective 14.12.06) insert a new provision into the 2005 Regulations requiring SEPA to notify third parties who have made representations in respect of applications for authorisations or their variation or surrender of its decision. If a third party objects to SEPA's decision they may notify Scottish Ministers who may request SEPA to forward the application to them for determination. These Regulations also amend the 2005 Regulations to enable an applicant to appeal to Scottish Ministers against a partial refusal of a request for a variation of the conditions imposed in such a variation.

The Water Environment (Controlled Activities) (Scotland) Regulations 2005 (Notices in the Interests of National Security) Order 2006 (SI 2006/661, S.5) empowers the Secretary of State (defence) to issue a non-disclosure order relating to a controlled activity site if he is of the opinion that this would not be in the interests of national security – this would cover both advertising of an application, inclusion on the public register etc.

CAR authorisations will be reviewed periodically, with SEPA also having powers to vary an authorisation (whether it has been reviewed or not) to remove, add or amend any of its conditions. SEPA should give notice of its intention to vary an authorisation, and the date the variation is to take effect (not less than three months from the date of the notice. A responsible person or operator (in the case of registration) may apply to SEPA for a variation of the authorisation – if SEPA agrees it will notify the RP or operator, detailing the variations and the date (not less than three months from the date of the notice) these are to take effect.

Where a Responsible Person proposes to transfer the whole or part of an authorisation to another person, application should be made by both to SEPA who will need to be satisfied that the proposed RP will be able to carry out the activity in compliance with the licence. SEPA will normally determine a transfer application within two months and if agreed send a copy of the amended authorisation to the new RP, specifying the date the transfer takes effect and the RP in respect of the authorisation. For partial transfers, SEPA will issue a new authorisation covering the transferred activities to the new RP, and issue an amended authorisation to the original RP.

An RP may apply to surrender all or part of an authorisation where it is intended to cease an authorised activity or it has ceased. Before agreeing to the surrender, SEPA will need to be satisfied that all necessary steps have been taken to avoid risk of harm to the water environment and to return it to a state that will not cause the failure to meet an environmental objective. An application to surrender all or part of an authorisation will normally be determined within two months; in the case of a partial surrender, SEPA will, if it considers it necessary, issue a varied authorisation.

The Regulations enable SEPA to serve a notice on an RP or operator suspending or revoking an authorisation where it believes an operator is not complying with its terms, or continues not to do so following enforcement action – 28 days notice should be given. SEPA may at any time lift the suspension (or partly lift it) by notice to the RP; the notice should, in the case of partial lifting, specify the extent to which the authorisation remains suspended, the date on which the lifting or partial lifting takes effect, and the reasons for the lifting or partial lifting of the suspension (SSI 2007/553).

SEPA may serve an enforcement notice on an RP or operator if it is of the opinion that a controlled activity is being carried out in contravention of an authorisation, is having, has had or is likely to have a significant adverse impact on the water environment or is causing, has caused or likely to cause a direct or indirect discharge to groundwater. The notice should specify the activity, its adverse impact, and the steps to be taken to prevent, mitigate or remedy the situation, and the time period; the notice may also require cessation of the controlled activity for a specified period. SEPA may revoke its notice once it is satisfied that the required remedial steps have been taken.

SEPA may if necessary carry out the necessary works itself if it considers it necessary to do so "forthwith" or no person on whom to serve the notice can be found. It may recover the costs from the RP or operator.

SEPA has a duty to maintain a public register with details of all applications for, and actual authorisations, any notices and advertisements under the regulations, appeals, directions to SEPA, monitoring information, and convictions for offences under the Regulations, cessation of authorised activity, etc. An application may be made to SEPA for information to be omitted from the register on the grounds of commercial confidentiality. Monitoring information relating to a controlled activity will be retained on the register for six years, as will information relating to a controlled activity which has been superseded.

(b) Registration

Controlled activities that are not considered likely to have a significant adverse impact on the water environment are authorised on the basis of registration, with appropriate conditions. An application for authorisation by registration should be submitted to SEPA, with the appropriate fee; it should describe the activity, its scale and location and will be treated as authorised once these particulars, and any conditions, have been entered on the public register. Such applications will normally be determined within 30 days. Authorisations by Registration will be reviewed by SEPA from time to time to ensure that conditions etc are being complied with and remain appropriate.

If an operator intends to cease an authorised activity, SEPA should be notified of the date as soon as it is known, and where an activity has already ceased, within seven days of cessation.

Other aspects of the control regime, including advertisement of applications, variations, appeals procedure, revocations etc are similar to that for activities controlled by licence and are described above.

(c) General Binding Rules

Certain activities which are considered to present a low risk are controlled through "general binding rules" and do not have to register with SEPA; they do however need to comply with the rules for each activity (see Sch.3 attached to the 2007 amendment Regulations, SSI 2007/219) and with legislation on the protection of wildlife and habitats. GBR authorisations will be reviewed periodically by SEPA to ensure that the level of control is appropriate or they may impose a higher level of authorisation. SEPA is also required to review periodically the activities covered by GBR (Sch. 3) and should make recommendations for any changes to Scottish Ministers. There are currently 17 activities, and associated rules, covered by GBRs – these range from the operation of passive weirs constructed before 1 April 2006 or the placement of one or more boulders in a river or burn, to abstraction from a borehole, and any subsequent discharge of the abstracted water where the total volume of water abstracted is less than 150 m^3 in any period of one year, or laying of pipeline/cable by boring beneath the bed and banks of a river, burn or ditch.

Schedule 3 has been further modified by *the Water Environment (Diffuse Pollution) (Scotland) Regulations 2008* (SSI 2008/54) which came into force on 1 April 2008. These add various agricultural activities to those covered by GBRs that are required to meet specified conditions to control diffuse pollution.

(d) Codes of Practice

The Sheep Dipping Code of Practice (v.1, July 2006) for Scottish farmers, crofters and contractors replaces the 2003 code issued under the 1998 Groundwater Regulations (6.11.3 above). It provides guidance on preventing pollution of the water environment and applies to all operations before, during and after sheep dipping, including purchase, transport and storage of sheep dip; siting and design and use of dippers, use of mobile dippers, and ensuring authorised disposal of waste sheep dip.

SEPA is also consulting on two codes of practice – for owners and operators of underground storage tanks for hydrocarbons, and for owners and operators of quarries and other mineral extraction sites; these will also replace the 2003 codes issued under the Groundwater Regulations which have been repealed in Scotland (see 6.11.3 above). While the codes are largely the same as the earlier ones, they have been updated to take account of the EU's Water Framework Directive. Evidence of compliance with the Codes is not a defence in the event of a pollution incident.

6.11.7 Enforcement Notices

Section 90B of the WRA is added by the *Environment Act 1995*, Sch. 22, para 142. It was brought into force by Commencement Order (No. 8, SI 2909) on 1 January 1997. In Northern Ireland, the relevant Article is Article 12 of the *Water (Northern Ireland) Order 1999*. In Scotland, similar provision for the serving of enforcement notices is made by reg.28 of *the Water Environment (Controlled Activities) (Scotland) Regulations 2005*.

The Agency may serve an enforcement notice if it is of the opinion that the conditions of a consent are being, or may be contravened; the notice will specify the nature of the contravention or why it is considered that a contravention is likely, the steps to be taken to remedy the matter and the period within which this must be done. The Agency will normally give ten days' notice of its intention to serve an enforcement notice. A copy of the enforcement notice should be placed on the public register within seven days of service. There is a right of appeal against an enforcement notice (see 6.11.5c above), which is not suspended during the course of the appeal.

Failure to comply with a notice is an offence subject to a fine not exceeding £20,000 (Sc. £40,000) and/or up to three months' imprisonment on summary conviction or up to two years' imprisonment and/or a fine on conviction on indictment.

6.11.8 Abandoned Mines

Sections 91A and 91B were added to the WRA by s.58 of the *Environment Act 1995* which was brought into force on 1 July 1998. They cover mines (including tourist mines and those for commercial mineral extraction) in England and Wales abandoned on or after 1 January 1999. Article 35 (brought into force on 24 August 2001) of the *Water (Northern Ireland) Order 1999* contains similar legislation for Northern Ireland.

"Abandoned" includes stopping all or some of the mining or related activities at a mine, or stopping or substantially changing operations for the removal of water from a mine.

The Mines (Notice of Abandonment) Regulations 1998 (SI 892) apply in England and Wales and came into force on 1 July 1998, with similar Regulations (SI 1572) taking effect in Scotland on 31 July.

A mine operator is required to give at least six months' notice of intention to abandon a mine to the appropriate Agency. The Notice should include the following details:

- name and address of the mine operator, and of the owner if this is different;
- location and description of the mine, pumping rates, etc;
- consequences of abandonment;
- proposals for monitoring groundwater quality and water quality in the mine;
- proposals for treating, lessening or preventing discharges from the mine.

Details of the proposed closure should also be published in at least one local newspaper.

Failure to give the required notice is an offence except if the mine is abandoned in an emergency to safeguard life or health or in some cases of insolvency. Notice of abandonment should be sent to the appropriate Agency as soon as practicable.

If the appropriate Agency considers that the proposed or actual abandonment is likely to result in surrounding land becoming contaminated as defined by Part IIA of the EPA 1990 (see chapter 5, 5.30.3), the relevant local authority should be advised.

It should be noted that ss.161A-C enable the Environment Agency to serve a works notice requiring certain anti-pollution works to be carried out to prevent or deal with polluting matter entering controlled waters from, *inter alia*, abandoned mines; it applies only to mines abandoned after 31 December 1999 – see below 6.14.

Articles 16-19 of the *Water (Northern Ireland) Order 1999* relating to anti-pollution works and the service of notices has been given effect through *The Anti-Pollution Works Regulations (Northern Ireland) 2003* (SR 2003/7), which came into force on 29 January 2003. Reg.29 of *the Controlled Activities Regulations* in Scotland provide SEPA with powers to carry out works.

6.12 Powers to Prevent & Control Pollution

6.12.1 Agricultural Pollution

(a) Silage, Slurry & Agricultural Fuel Oil

The Control of Pollution (Silage, Slurry and Agricultural Fuel Oil) Regulations 1991 (SI 324) which aim to prevent water pollution from farms, were made under s.110 of the *Water Act 1989* (subsequently replaced by s.92 of the WRA). The Regulations, which apply in England and Wales, came into effect on 1 September 1991. New Regulations are expected to bring them into line with EU requirements.

New silage, slurry and agricultural fuel oil storage facilities have to be designed and built to minimum standards to prevent the pollution of controlled waters. Existing installations – i.e. those in use or constructed before 1 March 1991, or a contract for construction, or actual construction begun, before 1 March 1991 and completed before 1 September 1991 – are largely exempt from the Regulations except where the Agency considers that there is a serious risk of pollution. In such instances the Agency may require (subject to a right of appeal) that all or some of the standards for new installations are met. Substantially extended or reconstructed facilities must meet the new standards.

Silage can only be made in a silo with an impermeable base surrounded by a system to collect any effluent and take it to a storage tank. Silage can also be made in bales in a tower silo or an existing installation. Any other methods of silage making (e.g. field silage making) had to be registered with the Agency under transitional arrangements and could continue until 31 August 1996 unless considered to be a serious pollution risk. Farms producing slurry must have a minimum four months storage capacity unless the Agency is satisfied that less does not represent a serious pollution risk. There are also minimum standards for loading and height limits for slurry lagoons. Tanks for storing more than 1500 litres of fuel oil on a farm must be surrounded by a bund (impervious wall) to prevent leaks reaching watercourses.

1997 Amendment Regulations (SI 547, E & W only) permit field silage making – production and storage is prohibited within a 50 metre radius of the nearest relevant abstraction point of a protected water supply source and within 10 metres of any inland freshwaters or coastal waters. Farmers must give the Environment Agency 14 days' notice of any place where silage is to be made or stored and similar notice of their intention to use a new site for the purpose. The Regulations also enable the Agency to serve a notice (instead of immediate prosecution) on a farmer requiring measures to be taken to prevent pollution of water from all slurry, silage and agricultural fuel oil stores.

The penalty for contravening the Regulations is a fine of up to £20,000 and/or three months in prison on summary conviction or an unlimited fine and/or two years in prison on conviction on indictment.

In ***Scotland*** 2003 Regulations (SSI 2003/531) which came into force on 1 December 2003 (revoking earlier 2001 Regulations) contain similar provisions to those for England and Wales. As well as enabling SEPA to issue works notices to farmers requiring improvements to structures for silage etc to prevent pollution, SEPA can require the farmer to draw up a Farm Waste Management Plan in accordance with the *Code of Good Practice for the Prevention of Pollution from Agricultural Activity*. The period for notifying SEPA before a new or substantially enlarged or reconstructed structure is brought into use is 28 days. The Regulations also take account of the latest British Standard specifying design and construction standards, as well as enabling SEPA to take a site specific view on the type of pollution controls needed. *The Water Environment (Diffuse Pollution) (Scotland) Regulations 2008* (SSI 2008/54), which came into force on 1 April 2008, amend the 2003 Regulations to permit lightly contaminated water from farm yards to be drained to constructed farm wetlands. The 2003 Regulations no longer apply to the storage of agricultural fuel oil – this is now covered by *the Water Environment (Oil Storage) (Scotland) Regulations 2006* (see 6.12.2 below).

Similar Regulations (SR 2003/319) came into force in ***Northern Ireland*** on 21 July 2003; new, substantially enlarged or reconstructed silage, slurry and oil stores had to comply with the tighter standards required by the Regulations from that date. The period for notifying the DOENI before a new or substantially enlarged or reconstructed structure is brought into use is 28 days. Existing stores may be required by notice to upgrade to the new standards if it is thought they pose a pollution risk. There is a right of appeal against such notices to the Water Appeals Commission.

(b) Nitrate Pollution

Directive 91/676/EEC places restrictions on the use of natural and chemical fertilisers to reduce the risk of nitrate pollution of river, coastal and sea waters and ensure a level of no more than 50 mg/l of nitrates in fresh water sources used for abstraction purposes. Catchment areas in which nitrate levels exceed, or are likely to exceed, the requirements of the Directive at any time in the year must be designated vulnerable zones and action programmes to control nitrate leaching implemented. Monitoring programmes must be established and reports on implementation progress sent to the Commission every four years.

The establishment of nitrate sensitive areas (NSAs) in England and Wales is provided for by ss.94-95 and Sch. 12 of the WRA and enables controls to be introduced over agricultural activity in order to reduce the amount of nitrate leaching from agricultural land into water sources. In Scotland, s.31(b) of COPA includes powers to designate nitrate sensitive areas.

England

The Nitrate Pollution Prevention Regulations 2008 (SI 2008/2349) largely came into force on 1 January 2009 (2010 in relation to land within areas not previously designated as an NVZ); the requirements relating to the storage of organic manures (Part 7) and reg.22(1) regarding the equipment to be used for spreading nitrogen fertilisers take effect on 1 January 2012. These Regulations revoke earlier Regulations relating to nitrate sensitive areas and 1998 Regulations on action programmes for nitrate vulnerable zones. The Regulations are enforced by the Environment Agency.

The Regulations increase the amount of land designated as NVZs to around 70% of England – maps identifying the areas can be accessed on http:adasis://nvz.co.uk/mpas/index.html. In such areas a whole farm limit of 170 kg per hectare for nitrogen from livestock manures applies to all land in each calendar year (though Defra plans to seek a derogation to this to enable a higher limit to be set in certain situations). The total amount of nitrogen inorganic manure spread on any given hectare in any twelve month period should not exceed 250 kg.

The Secretary of State is required to keep under review the eutrophic state of fresh surface waters, estuarial and coastal waters; before 1 January 2013 (and then at least every four years) the Secretary of State must monitor the nitrate concentration in freshwaters regularly over a year at sampling stations representative of both surface water and groundwater. If all samples register below 25 mg/l and there have been no change in circumstances likely to increase nitrate concentrations, monitoring need only be carried out every eight years. At the end of each monitoring period the Secretary of State should identify those waters that are or could be affected by pollution, revising the NVZs as appropriate. Every four years the Secretary of State is also required to review the effectiveness of restrictions in NVZs in reducing or preventing water pollution. In carrying out his review the public should be given early and effective opportunities to participate, ensuring adequate arrangements are made for such participation and for consultation.

Before 30 April 2009, occupiers of holdings are required to record the total size of their holding. By the same date they should calculate the amount of manure to be produced by the anticipated number of animals to be kept in a building or on hardstanding during the storage period, and the amount of storage capacity required. Before 30 April each year, an occupier of a holding with livestock must record for the previous storage period the number and category of animals in a building or on hardstanding during the previous storage period. He should also record the sites used for field heaps and the dates of use. Also before 30 April each year, the number and category of animals on the holding during the previous calendar year should be recorded, the days spent on the holding to enable a calculation to be made of the amount of nitrogen in the manure, using the table in Sch. 1 to the Regulations. The Regulations also require records to be kept of the amounts of livestock manure brought on or sent off the holding, of crops sown, when and how much nitrogen fertiliser was spread and the yield achived by an arable crop so spread. Records must be kept for five years.

Before 1 January 2010, an occupier of a holding who spreads organic manure on the land must reduce a map (a "risk map") showing area of each field, identifying surface waters and land within 10 metres of surface waters; bore holes, springs and wells either on the holding or within 50 metres of the boundary and land within 50 metres; land drains; land with an incline of more than 12%; sites suitable for temporary field heaps; and land that has a low run-off risk.

The Regulations require an inspection of land before spreading nitrogen fertiliser to consider the risk of its getting into surface water. Other factors to be taken into account include, slope of land; ground cover, weather conditions, soil type and land drains; nitrogen fertiliser may not be spread if the soil is waterlogged, flooded or snow covered or has been frozen for more than 12 hours in the previous 24 hours.

The spreading of manufactured fertiliser within 2 metres of surface water is prohibited; organic manure may not be spread within 10 metres of surface water. Livestock manure (other than slurry and poultry) may be spread on land managed for breeding wader birds or as a species-rich semi-natural grassland if it has been notified as an SSSI between 1 June and 31 October, so long as it is not spread directly onto surface water and the total amount does not exceed 12.5 tonnes per hectare. Organic manure applied to the surface of bare soils or stubble (other than land which has been sown) should usually be incorporated as soon as possible or within 24 hours.

The Regulations prohibit the spreading on land of "organic manure with high readily available nitrogen" (i.e. organic manure in which more than 30% of the total nitrogen content is available to the crop at the time of spreading) during the following closed periods:

- Sandy or shallow land – grassland: 1 September - 31 December; tillage land: 1 August - 31 December (spreading allowed between 1 August and 15 September for crops sown on or before 15 September).
- All other soils – grassland: 15 October - 15 January; tillage land: 1 October - 15 January.

The amount of slurry that can be spread from the end of the closed period until the end of February at any one time is limited to 50 cubic metres per hectare, with the maximum amount of poultry manure limited to 8 tonnes per hectare; there must be at least three weeks between each spreading. Manufactured nitrogen fertiliser may not be spread on grassland from 15 September to 15 January or on tillage land from 1 September to 15 January (all dates inclusive).

Part 7 of the Regulations relating to the storage of organic manure will come into force on 1 January 2012. Organic manure, or bedding contaminated with organic manure, must be stored in a vessel, in a covered building, on an impermeable surface, or in the case of solid manure that can be stacked in a free standing heap and does not drain liquid, on a temporary field site – this should not be liable to flooding or within 50 meters of a spring borehole or well, or within 10 metres of surface water or a land drain. Occupiers of holdings must have sufficient storage for all pig and poultry slurry to cover the period 1 October to 1 April and for all other slurry for the period 1 October to 1 March. Storage facilities are not needed for slurry or poultry manure sent off the holding or spread on land that has a low run-off risk.

An offence under the Regulations is subject to a fine not exceeding the statutory maximum on summary conviction or a fine on conviction on indictment.

Wales

The Protection of Water Against Agricultural Nitrate Pollution (England and Wales) Regulations 1996 (SI 888), which came into force on 17 April 1996, designate nitrate vulnerable zones (NVZs) – these Regulations have been revoked in England (see above). WSI 2002/2297 amends the 1996 Regulations to bring the total number of NVZs in Wales to ten, covering less than 3% of land in Wales. SI 2006/1289 amends the 1996 Regulations to allow for the public to be consulted over any proposals to review or modify action plans and programmes under the Nitrates Directive (in line with the 2003 public participation Directive, see chapter 1, 1.25.2).

The Action Programme for Nitrate Vulnerable Zones (England and Wales) Regulations 1998 (SI 1998/1202) establish action programmes which occupiers of farms within designated NVZs are required to implement to control nitrate leaching (these Regulations have been revoked in England by SI 2008/2349 – see above).

Required measures include:

- no chemical fertilisers to be applied between 1 September and 7 February, or 15 September and 1 February in the case of grassland;
- no more than 250 kg per hectare of organic manure per year to be applied to grassland; for other agricultural land no more than 210 kg per hectare per year to 19.12.02, and then 170 kg per year from 19.12.06; for individual fields within an NVZ the limit is 250 kg/hectare;
- no applications of organic manure in the form of slurry, poultry manure or liquid digested sewage sludge to sandy or shallow soil between 1 September, or 1 August for other types of soil, and 1 November; sufficient storage capacity to be available when manure and slurry spreading is restricted;
- no fertilisers to be applied if soil is waterlogged, frozen, flooded or snow covered;
- no fertilisers to be applied within 10m of surface water;
- detailed records to be kept of fertiliser and manure use, which should be retained for 5 years

In addition, farmers in NVZs must ensure that they have adequate storage capacity for the closed periods unless other environmentally acceptable means of disposal are available; they must also comply with *the Control of Pollution (Silage, Slurry and Agricultural Fuel Oil) Regulations 1991*, as amended (see above 6.12.1a).

The Environment Agency is required to monitor the concentrations of nitrates in freshwaters at designated sampling stations and to review the eutrophic state of fresh surface waters, estuarial and coastal waters before 19 December 1997 and thereafter on a monthly basis over a period of a year every four years. Where previous monitoring has shown nitrate concentrations to be consistently below 25 mg/l, the dates for monitoring by the Environment Agency are 19 December 2001 and thereafter every eight years. The Regulations designate relevant sections of the *Code of Good Agricultural Practice for the Protection of Water* (see below) as the code of good practice required by the Directive.

Scotland

The Designation of Nitrate Vulnerable Zones (Scotland) Regulations 2002 (SSI 2002/276) and No. 2 Regulations (SSI 2002/546) designate four NVZs, with *the Action Programme for Nitrate Vulnerable Zones (Scotland) Regulations 2008* (SSI 2008/298) establishing a revised action programme. These Regulations came into force on 1 January 2009, revoking SSI 2003/51, as amended by 2003/169.

The 2005 *Code of Good Practice on the Prevention of Environmental Pollution from Agricultural Activity* provides guidance on preventing nitrate pollution, and deals with the content of action programmes for those areas designated NVZs. Those parts of the Code which relate to water pollution have statutory backing through *the Water (Prevention of Pollution) (Code of Practice) (Scotland) Order 2005* (SSI 2005/63). *The Nitrate (Public Participation etc) (Scotland) Regulations 2005* (SSI 2005/305), which came into force on 25 June 2005 require public participation in the drawing up of action plans (as per Directive 2003/35/EC).

Under the 2008 Regulations an occupier of a farm within an NVZ is required to draw up a "fertiliser and manure management plan" before 1 March each year. This should include a risk assessment plan in respect of organic manure, capacity of storage facilities required for livestock manure and a calculation of the quantity of nitrogen fertiliser required for each crop. A risk assessment plan should also be prepared showing each field, its area in hectares, location of surface water, wells and boreholes, areas which slope more than 12%, location of field heaps and any other areas of high risk to the water environment.

There must be sufficient storage facilities for all livestock manure that is likely to have to be stored, as well as capacity to store the total quantity of slurry likely to be produced by housed pigs or housed cattle and manure from housed poultry. The Regulations provide a formula for calculating capacity. Solid manure may only be stored on an impermeable surface which prevents drainage to the water environment; it may be may be temporarily stored on land immediately prior to being spread on land but not in an area identified as being of high risk. Farms without sufficient storage capacity on 1 January 2009 have until 1 January 2012 to comply; written notification must be given to Scottish Ministers by 1 January each year (or no later than 31 March 2009 in respect of 2009).

Nitrogen fertiliser must not be applied to land if there is a significant risk of its getting into surface water, taking account of the slope of land; ground cover and weather conditions, soil type and land drains; nitrogen fertiliser may not be applied if the soil is waterlogged, flooded or snow covered or has been frozen for more than 12 hours in the previous 24 hours. Chemical fertilisers may not be applied to any land where it is likely that it will enter surface water. Organic manure may not be spread within 10 metres of surface water or 50 metres of any well, borehole or similar underground strata for any water supply. Organic manure may not be applied to any field where the application would result in the total nitrogen contained in organic manure applied in any 12 month period exceeding a rate of 250 kg per hectare, excluding that from grazing animals.

The following closed periods apply:

- Chemical fertilisers: Aberdeenshire, Banff, Buchan & Moray NVZ – 15 September to 20 February (grassland); 1 September to 20 February (other land). Other NVZs – 15 September and 15 February (grassland); 1 September and 15 February (other land).
- Organic manure with high available nitrogen content: between 1 September and 31 December (grassland on sandy or shallow soil); 15 October and 15 January (grassland on any other soil); 1 August and 31 December (not grassland, but sandy or shallow soil); 1 October and 15 January (not grassland, any other soil). Such manure may however be spread up to and including 15 September if the land is sown with a cereal crop before that date, and up to and including 30 September if sown with oilseed rape, a catch crop or a cover crop before that date.

Livestock manure (other than slurry and poultry) may be spread on land managed for breeding wader birds or as a species-rich semi-natural grassland if it has been notified as an SSSI between 1 June and 31 October, so long as it is not spread directly onto surface water and the total amount does not exceed 12.5 tonnes per hectare. Organic manure applied to the surface of bare soils or stubble (other than land which has been sown) should usually be incorporated as soon as possible or within 24 hours.

Occupiers of farms are keep detailed records of their farm, including its total area, area of each field and soil type, quantity of each type of nitrogen fertilser applied; number of livestock, species and type and length on farm; quantity of each type of livestock manure produced on farm and nitrogen content; quantity and nitrogen content of organic manures brought onto the farm, date and supplier and quantity and type of chemical fertiliser brought, used and retained on the farm. Records must be kept for five years.

The Regulations also provide for inspections on behalf of the Scottish Government for the purpose of monitoring implementation, for the serving of notices where the requirements of the Regulations have not been met and for appeals against such notices.

An offence under the Regulations is subject to a fine not exceeding the statutory maximum on summary conviction or a fine on conviction on indictment.

The Farm Waste Grant (Nitrate Vulnerable Zones) (Scotland) Scheme 2003 (SSI 2003/52) enables farmers in Scotland within NVZs who need to install new or upgraded storage facilities to apply for a grant towards the costs for expenditure incurred after 14 March 2003 and before 14 March 2010.

The Scottish Government, farming industry and environmentalists have developed a "Four Point Plan" on the management of farm pollution, providing guidance on the minimisation of dirty water in farm steadings, better nutrient use, risk assessment for manure and slurry spreading, managing water margins, and tackling organic waste washing into water courses when it rains. The Plan can be accessed via the Scottish Government's website.

Northern Ireland

The Protection of Water Against Agricultural Nitrate Pollution Regulations (Northern Ireland) 2004 (SR 2004/419), which came into force on 29 October 2004, designate the whole of the Province as an NVZ, and require the establishment and application of an action programme as required by the EU Directive – this has been done through *the Nitrates Action Programme Regulations (Northern Ireland) 2006* (SR 2006/489) which came into force on 1 January 2007, with further measures limiting the application of chemical fertilisers to land contained in *the Phosphorus (Use in Agriculture) Regulations (Northern Ireland) 2006* (SR 2006/488), which also came into force on 01.01.07.

Among the action programme requirements are that no chemical fertiliser should be applied to any grassland between 15 September and 31 January, or to any other land between the same dates unless there is a demonstrable crop requirement; no organic manure (excluding farmyard manure and dirty water) should be applied between 15 October and 31 January. Total nitrogen in organic manure applied to land, including by animals and brought on to the farm, should not exceed 170 kg N/ha/year; organic manure should not be applied to a field which would result in total nitrogen supplied in the manure exceeding 250 kg N/ha in any 12 month period. Chemical fertilisers containing phosphorus may not be applied unless it can be demonstrated that there is a crop requirement; the maximum farm phosphorus surplus limit should be no more than 10 kg P/ha/year by 1 January 2010, reducing to 6 kg by 1 January 2012. Depending on the type and numbers of livestock, farms should have a total livestock manure storage capacity of between 22 and 26 weeks and may be stored on the field on which it is to be applied for no more than 180 days.

Farmers must keep annual records demonstrating compliance with the action programme; from 1 January 2010 those farmers whose records show compliance with the 170 kg N/ha/year limit and do not apply chemical phosphorus fertiliser will no longer be required to keep records demonstrating compliance with the phosphorus limit. Annual records must be retained for five years.

The Code of Good Agricultural Practice for the Prevention of Pollution of Water, Air and Soil is available from the website of the NI Department of Agriculture and Rural Development (www.dardni.gov.uk); a farm waste management grant scheme is also being implemented.

The Farm Nutrient Management Scheme (Northern Ireland) 2006 (SR 2006/537) allows the Department for Agriculture and Rural Development to pay grants for the provision and replacement of facilities to collect and store farm animal waste and farm effluent and for the separation of rainwater from farm waste.

(c) Codes of Practice

Section 97 of *the Water Resources Act* enables the Secretary of State to approve codes of practice which give practical guidance to those engaged in agricultural activities that may affect controlled waters, and to promote desirable practices to control pollution. Breach of a code would not be an offence, but may be taken into account by the Agency in determining whether to serve a notice to prohibit pollution by discharges to water, or to require that precautions are taken against potential pollution.

The Code of Good Agricultural Practice for the Protection of Water, published in 1998, by the former MAFF and the Welsh Office, is a statutory code of practice covering England and Wales. The code provides guidance on avoiding water pollution from various farming practices, such as pesticide and fertiliser use, silage, sheep dip and disposal of animal carcases; it also provides advice on development of a farm waste management plan and outlines the legislation farmers need to be aware of. Breach of the code is not an offence although it can be used as evidence in any prosecution for polluting water. There are also codes covering air pollution from agricultural sources and on soil pollution, revised versions of which were also published in 1998 (see chapter 3, 3.25). Defra has published (August 2007) a draft revised *Code of Good Agricultural Practice* covering England only; this brings together the three separate codes and provides a base line of protection for waters from nitrate and phosphate pollution.

In Scotland the *Code of Practice on the Prevention of Environmental Pollution from Agricultural Activity* covers air, soil and water pollution (latest version published 2005). Those sections covering water pollution have statutory backing under the *Water (Prevention of Pollution) (Code of Practice) (Scotland) Order 2005* (SSI 2005/63). Compliance with the Code is not a defence in any action for water pollution. The water sections of the Code cover similar issues to that for England and Wales, and also recommend good practices to prevent water pollution from nitrates. This is in line with EU Directive 91/676/EEC which places restrictions on the use of certain fertilisers (see this chapter, 6.5.6).

There is a similar Code in Northern Ireland – see above.

(d) Catchment Sensitive Farming

Forty catchments across England have been identified as priority areas for action, to be targeted under a range of measures aimed at improving farm practices and reducing water pollution from farming. Farmers will be encouraged – and given advice on how – to adopt catchment sensitive farming practices, including limiting the use of fertilisers, manures and pesticides; promoting good soil structure and rain infiltration to avoid run-off and erosion; and protecting watercourses from faecal contamination, and from sedimentation and pesticides. More details www.defra.gov.uk/farm/environment/water/csf/index.htm

SEPA is also working closely with farmers and farming representatives to develop solutions to diffuse pollution and is working on a Best Management Practice Manual to help farmers identify what they can do to minimise such pollution.

6.12.2 Oil Storage

The Control of Pollution (Oil Storage) (England) Regulations 2001 (SI 2001/2954) came into force on 1 March 2002. Their objective is to reduce the risk of pollution of water from oil stores in ***England***. They apply to anyone storing more than 200 litres of oil above ground on industrial, commercial and institutional (residential and non-residential) premises (outside only); they do not apply to farms (which are covered by *The Control of Pollution (Silage, Slurry and Agricultural Fuel Oil) Regulations 1991*), to storage containers of up to 3,500 litres at single private dwellings, oil stores at refineries or onward distribution centres and waste oil stores. The storage of oil in underground facilities (e.g. at petrol and diesel stations) is covered by the *Groundwater Regulations 1998*.

The Regulations detail the requirements to be met both for the construction of storage tanks and ancillary equipment, including pipes and valves and also the positioning of tanks; secondary containment measures must be in place sufficient to hold the contents of the oil store. New oil stores had to comply with the Regulations immediately. Existing oil stores in use before 1 September 2001 had until 1 September 2005 to meet the Regulations, with those situated less than 10 metres from any inland freshwaters or coastal waters or less than 50 metres from a well or borehole required to comply by 1 September 2003; however, if the Environment Agency considered there was a "significant risk of pollution of controlled waters", it could serve a notice requiring appropriate works to be carried out to minimise the risk.

Failure to comply with the Regulations carries a fine not exceeding the statutory maximum (currently £5000) on summary conviction or an unlimited fine on indictment.

The Water Environment (Oil Storage) ***(Scotland)*** *Regulations 2006* (SSI 2006/133), which came into force on 1 April 2006, consolidate the control of oil which is not covered by other legislation (e.g. *the Controlled Activities Regulations* – see above 6.11.6); they impose requirements similar to those in England, but also cover the storage of agricultural fuel oil and waste oil, with the Regulations also applying whether the oil is stored inside or outside the building. The storage of oil on premises used wholly or mainly as a single private dwelling with a storage capacity of less than 2500 litres is exempt from the Regulations as is the storage of oil in any container which is underground. Existing oil stores in use prior to 1 April 2006 have until 1 April 2010 to comply with the Regulations, with those situated less than 10 metres from any surface water or wetlands or less than 50 metres from a well or borehole required to comply by 1 April 2008. (New tanks installed after 1 April 2006 had to comply by 1 October 2006.)

Failure to comply with the Regulations is a criminal offence and could result in a fine of up to £40,000. SEPA has issued advice (August 2006) on how industry can comply with the design standards for new and existing above ground oil storage facilities. Further information available at http://www.sepa.org.ukl/regulation/oilstorage2006/index.htm

6.12.3 Water Protection Zones

Section 93 and Sch. 11 of the WRA provide powers for areas to be designated water protection zones, enabling the Agency to have control over the application of pesticides and other potential pollutants in surrounding areas to prevent discharge of polluting matter into controlled waters to protect drinking waters. To date only one stretch of river has been given statutory protection – the River Dee, which flows through North East Wales, Cheshire and Merseyside – under *the Water Protection Zone (River Dee Catchment) Designation Order 1999* (SI 915); this took effect on 21 June 1999 and requires all those carrying on a "controlled activity" – i.e. keeping or using controlled substances – within the zone to apply for a protection zone consent from the Environment Agency. The application and appeals procedures are contained in the *Water Protection Zone (River Dee Catchment) (Procedural and other Provisions) Regulations 1999* (SI 916), which also came into force on 21 June 1999.

An application for a protection zone consent must be made on a form provided by the Agency and must include a site map, substance location map and details of the controlled substances and emergency and safety statement; this latter should describe the potential sources of an incident which might result in a discharge to land or inland waters; how such an incident might occur and the measures taken or proposed to be taken to prevent, control or minimise the consequences of any incident to inland waters; and finally emergency procedures for dealing with an incident at the site. A fee of £250 should accompany the application (£50 if it relates to 10 or fewer controlled substances). The Agency will consult relevant local authorities and water undertakers within the control area, the Health and Safety Executive and, in relation to sites within SSSIs or designated special conservation areas, the Nature Conservancy Council for England or Countryside Council for Wales, as appropriate.

A protection zone consent will include details of the catchment control site, description of the substances to which the consent relates, together with maximum quantities to be kept or used at any one time and storage controls; any conditions attached to the consent should ensure the use of BATNEEC to prevent the direct or indirect release of controlled substances in circumstances which could cause pollution of "inland waters at any point where such waters are abstracted for the purposes of public water supply".

The Agency may, by notice, revoke a consent or alter or add conditions if it considers it expedient to do so. It may also, by notice, revoke a consent if planning permission results in a material change of use of the catchment control site, or in the case of a consent relating to one substance, it has not been kept or used for at least five years at the site to which the consent relates. The Regulations make provision for consents to be varied, for appeals to the Secretary of State and for consents to be kept on a public register – these are similar to those in other Regulations. A protection zone consent ceases to have effect where the person in control of the site changes, unless an application has been made for the continuation of the consent.

Under the River Dee Designation Order it is an offence to carry on a controlled activity within a protection zone without a consent, if the quantity of controlled substances kept or used exceeds that permitted in a consent, or conditions in a consent are not being complied with. Summary conviction may result in imprisonment up to three months and/or a maximum fine of £20,000; conviction on indictment may result in up to two years' imprisonment and/or a fine.

The imposition of further water protection zones as a means of tackling diffuse phosphorus pollution from agriculture has been proposed by Defra (August 2007), to ensure that water quality in such areas can meet the water framework directive's target of good ecological and chemical quality by 2015 – see earlier this chapter, 6.10.2.

6.13 RADIOACTIVE SUBSTANCES

The Control of Pollution (Radioactive Waste) Regulations 1989 (SI 1158) were brought into force on 1 September 1989. They apply in England and Wales and brought radioactive waste within the scope of the following sections of the *Water Act* (now *Water Resources Act*):

- classification of quality of waters (s.82);
- general duties to achieve and maintain objectives etc (s.84);
- offences of polluting controlled waters etc (ss.85-87);
- requirements to take precautions against pollution (s.92);
- consents and application to the Agency (s.99);
- anti-pollution works and operations (s.161);
- registers (s.190);
- information and assistance (s.202);
- exchange of information with respect to pollution incidents etc (s.203);
- local inquiries (s.213).

The radioactive properties of such waste remain subject to control under the *Radioactive Substances Act 1993*.

The corresponding Regulations in Scotland are the *Control of Pollution (Radioactive Waste) (Scotland) Regulations 1991*.

6.14 ANTI-POLLUTION WORKS

Section 161 of the WRA empowers the Environment Agency to take remedial action to deal with actual or potential pollution affecting controlled waters. It could be used to deal with pollution from sources such as contamination from landfill sites, old mine workings, chemical discharges etc.

Section 60(3) of the *Environment Act 1995* (effective 1 July 1997, Commencement Order No. 9) amends this section of the Act to enable the Agency to carry out an investigation to establish the source of the polluting matter and the person responsible. Costs and expenses in respect of the investigation and in dealing with the pollution may be recovered from the person who caused or knowingly permitted the pollution. (It should be noted that the recovery of costs in respect of investigating or dealing with pollution from legally abandoned mines is specifically excluded from the Act s.161(4) & *Env. Act 1995*, s.60(5)).

Sections 161A-D (*Env. Act 1995*, Sch. 22, para 162) – NI: Art.16-19 of the WNIO – enable the Agencies to serve a "works notice" on the person responsible for causing or knowingly permitting poisonous, polluting or noxious matter to enter controlled waters or to be likely to enter controlled waters.

Before serving the notice the Agency should make reasonable efforts to consult the responsible person on the works to be undertaken. If the person responsible for the actual or potential pollution cannot be found, or if the Agency considers the situation should be dealt with "forthwith", it may carry out the works itself and recover the costs and expenses incurred.

A works notice may not be served on an operator in respect of water from mines or part of a mine abandoned before 31 December 1999.

The Environment Agency's *Policy and Guidance on the Use of Anti-Pollution Works Notices* provides guidance on the interface and overlap between the powers contained in s.161 of the WRA and those in Part IIA (contaminated land) of the *Environmental Protection Act 1990*. Where land has been identified as being contaminated and is potentially affecting controlled waters, Part IIA of the EPA requires the service of a remediation notice; the Agency will usually use its powers to serve a works notice to deal with historic pollution.

The Anti-Pollution Works Regulations 1999 (SI 1999/1006) covering England and Wales were brought into force on 29 April 1999), and in Northern Ireland on 29 January 2003 (SR 2003/7). Similar 2003 Regulations were revoked in Scotland on 1 April 2006 with their provisions now covered in reg.28 of *the Water Environment (Controlled Activities) (Scotland) Regulations 2005*.

The Regulations specify the content of the works notice and outline appeal procedures in respect of such notices. As well as the name and address of the person on whom the notice is being served, and identifying the controlled waters to which the notice applies, the notice should specify

- that in the opinion of the Agency, poisonous, noxious or polluting matter or solid waste matter has entered, or is likely to enter, the identified waters;
- the works or operations to be carried out to remedy the situation, and the period within which each action specified in the notice must be carried out;
- that the person on whom the notice is served has a right of appeal to the Secretary of State within 21 days from the date the notice is received;
- the Agency's entitlement to recover costs in carrying out its investigations;
- consequences of non-compliance with a works' notice; i.e. that it is an offence which on summary conviction may result in a fine not exceeding £20,000 and/or three months imprisonment; conviction on indictment may result in up to two years imprisonment and/or a fine. Where a notice has not been complied with the Agency may carry out the work and recover the costs and expenses incurred.

Works Notices are not suspended during determination of an appeal; in sending written notice of appeal to the Secretary of State, the appellant should send relevant documents and should state whether the appeal is to be determined by written representations or a hearing; in the former case the Agency has 14 days to respond to any enquiries from the Secretary of State regarding the matter.

Where a person is required to grant rights of entry to a person on whom a works notice has been served in order that that person can comply with the notice, compensation is payable; the Regulations deal with the way in which compensation payments are to be calculated. The Lands Tribunal determines disputes over compensation payments.

Copies of works notices, appeals and relevant documentation and appeal decisions, as well as convictions under the Regulations are held on the pollution control register maintained by the Environment Agency.

6.15 POWERS OF ENTRY

Sections 169 and 170 and Sch. 20 of the WRA empower an authorised person to enter any premises to determine whether any provisions of the *Water Resources Act* have been or are being contravened. Where entry is required for enforcement purposes, no advance notice need be given though, except in emergencies, entry should be sought at a "reasonable time"; in all other cases seven days' notice should be given (Schedule 20 to the WRA). The inspector may carry out tests, install monitoring equipment or take away samples for further analysis.

Under s.111 of the *Environment Act 1995* any apparatus used for recording or registering information shall be deemed to be accurate unless the contrary is proved. Where an authorised person has been unable to gain entry to monitor observance with a condition, this will be taken as evidence that the condition is not being complied with.

Where entry is required in relation to the Agency's pollution control functions (including s.161-161D) of the WRA, the powers of entry provisions of the *Environment Act 1995* (ss.108-109) apply (*Env. Act*, Sch. 22, para 165); these provisions are set out in chapter 2, 2.7.

6.16 QUALITY & SUFFICIENCY OF SUPPLIES

The Water Industry Act 1991, which applies in England and Wales only, came into force on 1 December 1991. It consolidates various enactments relating to the appointment of water and sewerage undertakers, conditions of appointment, supply of water and the provision of sewerage services. It also sets out the functions of the Director General of Water Services, and the duties of local authorities with regard to the supply of wholesome and sufficient water.

In Scotland, Part 2 of *the Water Environment and Water Services (Scotland) Act 2005* covers sewerage services, with Part 2 of *the Water Services etc (Scotland) Act 2005* banning the "common carriage" of water and sewage on public networks; only Scottish Water may add treated drinking water to, and draw wastewater from, the public network (see also 6.8 above).

6.16.1 General Obligations of Undertakers

Section 68 places a duty on water undertakers to supply only water which is "wholesome at the time of supply" to any premises for domestic purposes. Section 69 empowers the Secretary of State through Regulations to specify minimum requirements for the monitoring and recording of water quality, to forbid or regulate the use of substances, processes and products which might affect the quality of water supplied, and to require provision of certain information to the public about its quality.

Section 70 makes it an offence to supply water which is unfit for human consumption unless it can be shown that there were no reasonable grounds for suspecting that the water would be used for human consumption, or took all reasonable steps to ensure it was fit, or was not used for human consumption.

6.16.2 Waste, Contamination, Misuse of Water

Under ss.71-75, it is an offence for an owner or occupier of premises either through negligence or intentionally to allow water fittings to remain in disrepair so as to cause the contamination, wasting or misuse of water. The Secretary of State is empowered to make Regulations with regard to water fittings, and representatives of water undertakers or local authorities will be able to enter any premises to ensure compliance with those regulations and to check the installation of any water fittings. To prevent damage to persons or property, or contamination or waste of water, water authorities will be entitled either to cut off a supply of water, or to serve a notice on a consumer requiring action to remedy the problem.

6.16.3 Local Authority Functions

Where existing supplies of water are insufficient or unwholesome, and piped water supplies would not be practicable at reasonable cost, s.79 entitles local authorities to require water undertakers to supply water for domestic purposes other than through pipes. Sections 78-82 place a general duty on local authorities to monitor the quality and sufficiency of water supplies in their area and where this is found to be below standard, empowers them to serve a notice on the water undertaker specifying what steps they consider necessary to effect improvements. There is a right of appeal to the Secretary of State against any action by a local authority in relation to private supplies.

Under s.86 the Secretary of State may appoint technical assessors to carry out investigations of drinking water quality matters and advise whether any duty imposed on a water undertaker in relation to such matters has been or might be contravened.

6.17 DRINKING WATER

6.17.1 Water Supply (Water Quality) Regulations

These Regulations consolidate and replace the 1989 *Water Quality Regulations* (E & W; Sc. SI 1990/119, NI 1994 regs) and subsequent amendments and implement the 1998 EU Directive concerning the quality of water intended for human consumption (the drinking water Directive) – see this chapter 6.5.4. The Regulations are similar throughout the UK.

In ***England*** SI 2000/3184 (as amended SI 2001/2885) began coming into force on 1 January 2001. In ***Wales*** SI 2001/3911 (W.323), began coming into force from 1 January 2002; amendment regulations (SI 2007/2734 & WSI 2007/3374), which apply in England and Wales, came into force on 22 December 2007, and are reflected in the text below. An unofficial consolidated version of the Regulations covering England is available on the Drinking Water Inspectorate's website (www.dwi.gov.uk).

In ***Scotland*** SSI 2001/207, as amended (SSI 2001/238), was progressively implemented from 26 June 2001. In ***Northern Ireland***, SR 2007/147), which came into operation on 1 April 2007 replace SR 2002/331, as amended by SR 2003/369) to reflect that water services are now delivered by water undertakers rather than the Department for Regional Development;

The Directive's requirements concerning spring water and bottled drinking water have been implemented through 2003 Regulations in Northern Ireland, SR 2003/182 which came into force on 1 May 2003. In England, these requirements have been implemented through SI 2007/2785, which came into force on 31 October 2007; in Scotland SR 2007/483, which came into force on 20 October 2007 apply – these latter sets of Regulations replace SI 1999/1540 which extended to the whole of Great Britain.

In all Administrations the Regulations are enforced by the appropriate Drinking Water Inspectorate on behalf of the Secretary of State, National Assembly of Wales, Scottish Ministers, DOENI. Guidance on the E & W Regulations is available on the Drinking Water Inspectorate's website (www.dwi.gov.uk).

Part of the Regulations came into force on 1 January 2001 in England (Sc: 26.6.01; W: 1.01.02; NI: 28.10.02) to ensure that water companies undertook the necessary work to meet the standards by the required dates (mostly 25 December 2003 – but see Appendix 6.3); the Regulations affecting monitoring arrangements took effect on 1 January 2004 (Sc: 25.12.03). Schedule 1 to the Regulations lists prescribed concentrations and values (microbiological & chemical parameters); Schedule 2 indicator parameters; Schedule 3 – monitoring, including parameters and circumstances for check monitoring & annual sampling frequencies; Schedule 4 analytical methodology.

Part I of the Regulations details when individual Regulations will come into force and what is meant by the terminology used in the Regulations (how they should be interpreted); Part II of the Regulations (which came into force on 1 June 2003) requires water undertakers, before the beginning of each year, to designate "water supply zones" within its area; these may cover a population of no more than 100,000 and may not be altered during the year.

Part III (which came into force on 25 December 2003) is concerned with the requirements for wholesomeness. Water supplied for domestic purposes and to premises in which food is produced must not contain any micro-organism, parasite or other substance – other than a permitted parameter (listed in schedules to the Regulations) – at a concentration or value which could be a potential danger to human health, or any substances (whether a parameter or not) which together could be a potential danger to human health. SI 2007/2734 (E&W) extends the definition of wholesomeness to include water supplied in bottles or containers by water undertakers or licensed water suppliers in place of a piped supply

Parts IV and V and Schedule 3 of the Regulations came into force on 1 January 2004 (Sc: 25.12.03); SI 2007/2734 & WSI 2007/3374 (E&W) insert Part VA into the Regulations to introduce new requirements for monitoring drinking water at abstraction points to comply with new provisions on disinfection of raw water, other treatment arrangements and risk assessments. The Regulations detail requirements for monitoring of water supplies, sampling points, the way in which samples should be taken and the frequency, as well as analysis of samples. Part VI, which came into force on 25 December 2003, sets out the responsibilities of water undertakers in the event that water fails, or is likely to fail, to meet the required standards. As well as identifying which of the parameters are likely to exceed permitted concentrations, the water undertaker must identify to what the fault can be attributed – i.e. to the domestic distribution system, or to the maintenance of that system, or some other reason. The Secretary of State should be notified of any failures and the action being taken to rectify the faults. Where the failure is attributed to the domestic distribution system or to its maintenance, consumers in the affected area must be notified and given advice on action to be taken to safeguard their health.

If a water undertaker notifies the Secretary of State that he is unable to meet the prescribed standards and that the fault is not due to a failure in the domestic distribution system or its maintenance, the Secretary of State may by notice require the water undertaker to apply for an authorisation permitting a departure from the provisions of Part III of the Regulations for any of its water supply zones. The application should include details of the grounds for the authorisation, for which water supply zones and the number of people affected; the parameters for which the authorisation is sought and results of sampling over the previous 12 months, and the period for which the authorisation is sought. The application should include details of further monitoring of the quality of the water to be supplied during the period of the authorisation. With the application, the water undertaker should submit a programme of works, estimated cost and timetable for bringing the water supply up to standard. A copy of the application and accompanying documents should be sent to the appropriate local authority, the district health authority and to the customer services committee, who all have 30 days in which to make representations on the application to the Secretary of State. (SI 2007/2734, WSI 2007/3374 amend the 2001 Regulations to require notifications to be sent to the Health Protection Agency (E) or National Public Health Service for Wales, to relevant to local authorities and where the water supply zone is partly in England and Wales, the notification should be sent to the HPA.)

A departure will not be authorised unless the Secretary of State is satisfied that there is no potential danger to human health and that there is no other reasonable way in which water can be supplied to the zone. Details of the authorisation should be published in an appropriate local paper as soon as reasonably practicable (and in E & W on the websites of the relevant suppliers, SI 2007/2734) unless the Secretary of State considers the contravention of the standard to be "trivial" and the situation is remedied within 30 days. The Secretary of State should normally give six months' notice of intention to modify or revoke an authorisation to the water undertaker and those to whom it was copied. He may revoke or modify an authorisation without notice in the interests of public health; an authorisation may also be revoked without notice where a water undertaker advises the Secretary of State that the circumstances for which the authorisation was granted no longer exist.

Part VII of the Regulations covers water treatment, including risk assessment and treatment for cryptosporidium and the treatment of water as a result of contamination from copper or lead pipes – not Scotland, see below.

The regulations dealing with cryptosporidium came into force on 1 January 2001 (NI: 28.10.02); those dealing with contamination from lead pipes came into force on 25 December 2003, with the residual regulations in this Part taking effect on 1 January 2004. (Sc: 25.12.03). Water undertakers who had not already carried out a risk assessment for cryptosporidium from treatment works brought into operation after 30 June 1999 and before 1 January 2001 under the 1999 amendment Regulations (SI 1524) were required to do so before 28 February 2001, notifying the Secretary of State of the results. From 1 October 2001 it became an offence to supply water from treatment works unless a risk assessment for cryptosporidium had been carried out and the Secretary of State authorised the supply or any necessary remedial work carried out; the Regulations contain stringent continuous monitoring and analysis requirements to ensure that any water posing a "significant" risk is dealt with quickly; it is an offence to supply treated water with more than one oocyst per ten litres of water.

In England and Wales SI 2007/2734 & WSI 2007/3374, which came into force on 22 December 2007, replace the above provisions in relation to cryptosporidium with a general duty to conduct risk assessments of water treatment works and to take appropriate action to deal with any potential danger to human health which is identified. Supplying water from water treatment works in breach of conditions specified in a notice served under these provisions will be a criminal offence.

*Note: the Scottish Regulations do not cover risk assessment and treatment for cryptosporidium. Revised Directions to Scottish Water – *the Cryptosporidium (Scottish Water) Directions 2003*, which came into force on 31 December 2003 – specify risk assessment methods and monitoring frequencies; continuous monitoring of treated water is required at treatment works supplying over 10 megalitres of water per day where risk assessment identifies a moderate or high cryptosporidium risk with weekly monitoring of raw water; where a lower risk is identified at smaller works, weekly or twice weekly monitoring of treated water is required, with monthly sampling at those showing the lowest risk.

Part VIII, which came into force on 1 January 2004 (Sc: 25.12.03), deals with maintenance of records and provision of information to the public. Water undertakers are required to maintain full records of water supply zones, authorised departures and action being taken to bring supplies up to standard, as well as monitoring and sampling results; most information must be kept for 30 years after first being recorded. Such information should be available for public inspection and copies available free of charge to residents for records relating to their own zone. Each year water undertakers should include with customers' bills a statement of where they may inspect, free of charge, records of water quality. No later than 30 June 2005, and no later than 30 June in subsequent years, water undertakers should send to local authorities full particulars of the water supplied during the previous year to each water supply zone in the authority's area, including number of samples taken, results of analyses, etc (revoked E&W, SI 2007/2734). Water undertakers are also required to notify* the appropriate local authority and district health authority and the customer services committee as soon as possible after an occurrence to the water which it supplies which gives rise, or may give rise, to a significant health risk to people in the area. (*every appropriate local authority, Health Protection Agency, the Council, and the Secretary of State, and where the event concerns Wales, the National Public Health Service for Wales, and Welsh Ministers, SI 2007/2734, WSI 2007/3374.)

Finally, the Regulations required all water undertakers who intend to supply water on or after 25 December 2003 to submit to the Secretary of State, no later than 31 March 2001 (Sc: 25.9.01), a programme of work aimed at compliance with Part III of the Regulations by 25 December 2003 (or for lead by 25 December 2013); if a water undertaker was of the opinion that he would be unable to meet any of the requirements of Part III by 25 December 2003, he had to apply for an authorisation permitting a departure from the standard no later than 25 September 2003.

Regulations requiring Strategic Health Authorities in England to consult the public prior to the fluoridation of water supplies have been made under the *Water Act 2003* – see 6.7 above.

6.17.2 Private Water Supplies

Private supplies are those not provided by a water undertaker (company) – i.e. not connected to a mains pipe (for example, borehole into an underground aquifer, a surface reservoir).

(a) England, Wales, NI

The Private Water Supplies Regulations 1991 (SI 2790) (made under ss.67 & 77 of the *Water Industry Act 1991*) relate to the quality of water from private supplies for drinking, washing, cooking or for food production. They apply in England and Wales only and came into force on 1 January 1992. They contain similar requirements to those of Parts II and III of the *Water Supply (Water Quality) Regulations 1991*. Similar 1994 Regulations apply in Northern Ireland.

Local authorities are required to monitor the quality of private water supplies by reference to two categories: water supplied for domestic purposes; and water supplied for food production, to staff canteens, hospitals, hostels, boarding schools etc and camp sites. The Regulations specify factors to be taken into account in monitoring programmes – sampling frequency, point of sampling, analysis, and charging for sampling and analysis. Where quality is found to be below standard, local authorities must ensure that the necessary remedial action is taken. They are also required to keep a register of all private water supplies in their area.

In August 2008, Defra published proposals for new Regulations in England to ensure that water from private supplies is clean and safe, and implementing EU Directive 98/83/EC. It is proposed that the Regulations include the following provisions:

- Introduce risk assessments, in line with the WHO's 2004 *Guidelines for Drinking Water Quality*.
- Local authorities will be given strengthened and simplified enforcement powers.
- The proposed Regulations and monitoring requirements will apply to domestic private supplies of 10 m^3 per day or more (or serving 50 persons), and to private water supplies of any size that are supplied as part of a commercial or public activity.
- The Regulations will also apply, but with reduced monitoring requirements, to domestic private supplies of less than 10 m^3 per day (or serving less than 50 persons) used for day-to-day domestic needs.
- Local authorities will have limited powers in respect of supplies to single private domestic homes.

(b) Scotland

The Private Water Supplies (Scotland) Regulations 2006 (SSI 2006/209) came into force on 3 July 2006, revoking SI 1992/575. The Regulations place a duty on local authorities to classify and monitor private water supplies to which the drinking water Directive applies; private supplies which provide more than 10 m^3 per day or serve 50 or more persons, or the water is associated with a commercial or public activity are classified "Type A", with all other private water supplies for human consumption classified "Type B". Classifications should be reviewed and updated annually. Responsibility for a private water supply lies with the "relevant person" – that is the person the local authority considers to be responsible for the supply, because they provide it, or occupy the land from which the water originates.

Local authorities were required to take baseline samples of all Type A supplies, with subsequent further monitoring carried out to ensure compliance with the drinking water Directive's standards (see Appendix 6.3, and also Schedules attached to the Regulations), and to carry out a risk assessment (and to assist relevant persons of Type B supplies to do so). Where water is below standard – unwholesome or unfit for human consumption – the local authority must investigate the cause and may authorise a "temporary departure" from the standards, though not if this would compromise public health. The local authority must ensure that all those likely to be affected by an application for a temporary departure have an opportunity to make representations to them, allowing 28 days; there is a right of appeal to the sheriff court where an application for departure has been refused or against conditions to be attached.

In cases where a temporary departure has not been authorised, the local authority must issue an enforcement or improvement notice; failure to comply with an enforcement notice is a criminal offence and failure to comply with an improvement notice a breach of the regulations.

Local authorities are required to maintain detailed registers in respect of private supplies in their area; initial entries to the registers had to be completed within six months of the Regulations coming into force (i.e. by 3 January 2007); included should be details of the supply, its location, premises served by the supply, results of samples taken and any risk assessment, enforcement notices etc. The registers should be reviewed and brought up to date at least once a year, and no later than 31 March; also by 31 March local authorities should provide SEPA with a copy of the register covering the preceding calendar year.

Details of private water supplies must be displayed in relevant public and commercial buildings.

The Private Water Supplies (Grants) (Scotland) Regulations 2006 (SSI 2006/210) set out to whom local authorities may pay grants enabling them to improve their private water supplies and the procedures for determining applications for a grant and its calculations. *The Private Water Supplies (Notices) Scotland) Regulations 2006* (SSI 2006/297) modify *the Water (Scotland) Act 1980* with respect to the serving of notices where supplies do not meet water quality standards requiring those responsible for a private water supply to remedy the failure.

A Technical Manual to assist local authorities in carrying out their responsibilities under the Regulations can be accessed on http://www.privatewatersupplies.gov.uk

6.17.3 Drinking Water in Public Buildings

The Drinking Water Directive also applies to supplies in public buildings where drinking water is supplied for human consumption, such as schools, hospitals, hotels and restaurants. Drinking water standards may only be exceeded if the cause of the exceedence is due to the distribution system within *domestic* premises – e.g. lead piping for which the responsibility for replacing lies with the householder. Compliance with the Directive in England and Wales is to be enforced by the inclusion of affected supplies in water companies' random compliance monitoring programmes (DWI information letter 10/2004).

The Scottish Government issued proposals (March 2002) for implementing the Directive; among its proposals are that local authorities should enforce the new regime; they will need to compile a register of public buildings, and check compliance with a limited number of parameters at specified frequencies; public buildings would be required to display prominently the results of monitoring data and any non-compliance with standards would need to be investigated and remedied; where non-compliance does not pose a threat to human health, the local authority would be able to authorise a departure from the standard set in the Regulations, but for as short a time as possible and not longer than three years. A consultation report (October 2002) confirmed general support for the proposals.

6.18 SEWERAGE SERVICES

Every sewerage undertaker has a general duty to provide and maintain a system of public sewers and sewage disposal, and to have regard to its existing and likely obligations to allow for the discharge of trade effluent into its public sewers, and the disposal of such trade effluent (ss.94-101). The Secretary of State may make Regulations detailing the extent to which breaches of specific sewerage service obligations are to amount to breaches of the general duty. Regulations may also prescribe individual standards of performance in relation to the sewerage service. If these standards are not met, payments are to be made by the undertaker to the persons affected. Sewerage undertakers must comply within six months with any request to provide a public sewer to be used for drainage of premises for domestic purposes, provided that certain financial conditions are met.

6.19 TRADE EFFLUENT

Occupiers of trade or industry premises may not discharge any effluents (liquid wastes) produced wholly or partially in the course of the trade or industrial activity into a public sewer unless authorised by the sewerage undertaker. Such effluents include waste chemicals (including oils), liquid process wastes, detergents, condensate from compressed air installations, cooling water, biodegradable liquids, wash water, liquid wastes or wash waters (other than domestic sewage) discharged using sinks, basins or toilets, and contaminated mine or quarry water.

An application to discharge to foul sewer or to surface or ground waters should include details of the effluent, quantity to be discharged in any one day, and the highest rate at which it is proposed to discharge (ss.118-134); *the Water Act 2003* (s.89) adds to s.119 of the WIA and will require applications to describe the steps to be taken for minimising the polluting effects of the discharge on any controlled waters and for minimising the impact of the discharge on sewerage services. (The sewerage undertaker must refer all applications covering special category effluent – see below – to the Environment Agency who will decide if the discharge should be prohibited or permitted subject to conditions.)

In granting an application, the sewerage undertaker may impose conditions covering the rate, quantity and composition of effluent and the sewer into which it may be discharged, and the time or times of day. Conditions may also relate to provision and maintenance of inspection chambers and meters and of other apparatus for testing the effluent, record keeping and payments to the sewerage undertaker. *The Water Act 2003* (s.89) adds to s.121 of the WIA, requiring conditions to be included on the steps to be taken for minimising the polluting effects of the discharge on any controlled waters and for minimising the impact of the discharge on sewerage services.

There is a right of appeal to the Director of Water Services where an application is refused, has not been considered within two months, or conditions are considered unreasonable. In the case of special category effluent, appeals should be directed to the Secretary of State. A sewerage undertaker may vary conditions attached to a consent; again there is a right of appeal to the Director General of Water Services.

Where an activity regulated under the PPC regime proposes to discharge trade effluent into a sewer, it also requires a consent from the sewerage undertaker.

The Trade Effluents (Prescribed Processes and Substances) Regulations 1989 (SI 1156) (amended 1990, 1992) came into force on 1 September 1989 and were made under ss.74 & 185 of the *Water Act*; they apply to England and Wales only and enable compliance with EU Directives on pollution caused by discharges to the aquatic environment (76/464/EEC) and the prevention and reduction of environmental pollution by asbestos (87/217/EEC).

The Regulations specify two categories of trade effluent which are subject to control over their discharge into public sewers and for which an authorisation to discharge is required from the Environment Agency.

The first category for which authorisation to discharge is required is effluent which contains concentrations of "red list" substances (see Appendix 2.2 – substances controlled for release to water under IPPC) exceeding those that would be present regardless of the activities within the premises from which the effluent is discharged. The second "special" category consists of effluent from the following prescribed processes:

- any process for the production of chlorinated organic chemicals;
- any process for the manufacture of paper pulp;
- any industrial process in which cooling waters or effluents are chlorinated;

- any process for the manufacture of asbestos cement;
- any process for the manufacture of asbestos paper or board.

The 1992 Regulations added effluent containing more than 30 kg/year of trichloroethylene or perchloroethylene to "special category" effluent.

The Regulations also require sewerage undertakers to notify the Environment Agency if they propose to vary existing trade effluent discharge consents so as to permit the discharge of effluent containing "red list" substances at levels in excess of background concentrations. This will enable the Agency to exercise stricter controls over discharges to watercourses via the sewage system, thus enabling the UK to meet its commitment (agreed at the North Sea Conference – see 6.22 below) to reduce such discharges by 50% between 1985 and 1995.

The 1990 amending Regulations tightened up the notification procedure further to require agreements by which sewerage undertakers can accept trade effluents under the *Public Health (Drainage and Trade Premises) Act 1937*, also to be notified to the Agency. Under the Regulations, the Agency must also be notified where it is proposed to vary a consent for a trade effluent discharge containing asbestos or chloroform from any of the processes mentioned above at a level in excess of background concentrations.

In addition, a Department of Environment Circular (7/89) contains advice on implementation of the EU Directive on environmental pollution by asbestos. Water authorities were required to satisfy themselves that producers of asbestos cement, and of paper and board products, had appropriate effluent treatment or recycling technology in operation by mid-1991, or had phased out the use of raw asbestos by then.

In Scotland the 1987 EU Directive on the prevention and reduction of environmental pollution by asbestos has been implemented through the *Trade Effluents (Asbestos) (Scotland) Regulations 1993* (SI 1446). Asbestos discharges should be "reduced at source and prevented ... using BATNEEC, including where appropriate recycling or treatment". Specified limits on asbestos discharges can be included in discharge authorisations, with specific conditions to be included where the discharge relates to "special category" activities (see above).

6.20 URBAN WASTE WATER

6.20.1 Regulations

The Urban Waste Water Treatment (England & Wales) Regulations 1994 (SI 2841), as amended by SI 2003/1788, made under the *European Communities Act 1972*, came into force on 30 November 1994. Similar (1994, SI 2842) Regulations apply in Scotland and in Northern Ireland SR 2007/187 which came into force on 01.01.07 (replacing SR 1995/12, amended by SR 2003/278). They implement the 1991 EU municipal waste water Directive (91/271/EEC) – see above 6.5.4.

Regulation 3 required the Secretary of State to review the identification of "sensitive areas" and "high natural dispersion areas" (i.e. "less sensitive areas") before 31 December 1997 and then at intervals of no more than four years. Schedule 1 to the Regulations lists criteria for identification: sensitive areas include natural freshwater lakes, estuaries and coastal waters and surface freshwaters intended for the abstraction of drinking waters. A high natural dispersion area is a marine body or area where the discharge of waste water will not adversely affect the environment as a result of morphology, hydrology or specific hydraulic conditions in the area; in identifying a high natural dispersion area account should be taken of the discharge being transferred to an area where it could have an adverse effect.

Unless it would have no environmental benefit or would involve excessive cost, collecting systems satisfying Schedule 2 to the Regulations had to be in place (Regulation 4) by

- 31 December 1998 where the discharge is into receiving waters in a sensitive area for agglomeration with a population equivalent (pe) of more than 10,000;
- 31 December 2000 for agglomeration with pe of more than 15,000;
- 31 December 2005 for agglomeration with pe of 2,000-15,000.

Regulation 5 specifies the dates by which suitable treatment plant must be provided, with Schedule 3 detailing the requirements for discharges from treatment plants.

The Regulations also impose requirements relating to discharges of industrial wastewater with sewerage undertakers being empowered to modify trade effluent consents and agreements to ensure compliance. The dumping of sewage sludge at sea had to be phased out by 31 December 1998 and its toxic, persistent and bioaccumulable properties progressively reduced before then.

The appropriate Agency must ensure that monitoring and other studies of discharges are carried out and certain documents made available for public inspection. The 2003 amendment Regulations for E & W and 2007 NI Regulations require the outcome of reviews of sensitive areas and high natural dispersion areas to be publicised in the *London/Belfast Gazette* (as appropriate) and on the websites of the Environment Agency/DOENI, together with up-to-date maps; in Wales a notice should be published in a daily newspaper circulating in the appropriate area.

In England a total of 367 sensitive areas have been identified (132 waters which suffer from eutrophication; 8 drinking water supplies with excess nitrate levels; 180 seaside waters, already identified under the bathing water Directive; and 47 shellfish waters, already identified under the shellfish waters Directive). In Wales 27 sensitive areas (3 eutrophic, 24 bathing waters) have been identified, and in Northern Ireland 16 sensitive areas (eutrophic) have been identified. In Scotland 24 sensitive areas have been identified under the Directive (6 eutrophic, 8 to protect fishlife, 1 to protect shellfish and 9 bathing waters); SEPA has designated a number of additional waters as sensitive to nutrient inputs (eutrophication) under the Water Framework Directive so that measures can be put in place to ensure they meet good ecological status by 2015 (see 6.2 above).

6.20.2 Codes of Practice

*The Water Services etc **(Scotland)** Act 2005* provides powers for the Scottish Government to draw up statutory codes of practice on odour control at sewerage works. *The Sewerage*

Nuisance (Code of Practice) (Scotland) Order 2006 (SSI 2006/155) which came into force on 22 April 2006 includes a statutory code of practice on the assessment and control of odour from waste water treatment works. It applies to existing WWTW in operation on 22 April 2006 and to new ones coming into operation after that date and requires the best practicable means to be applied to control odour emissions from contained and fugitive sources and to ensure that emissions do not create an odour nuisance beyond the boundary. Odour Management Plans must be prepared for all WWTWs which describe the BPM, the complaints administration procedure and procedures for training staff – OMPs, which had to be in place by 1 April 2007, should be regularly reviewed to ensure they remain current and at least once every 12 months. Where there is an odour nuisance, an investigation should be carried out to identify and evaluate the sources and causes; results of the investigation, together with a review of options for solving the problem using the BPM should be documented in an Odour Improvement Plan and submitted to the local authority for approval.

In ***England & Wales*** a voluntary code of practice on odour nuisance from sewage treatment works was published by Defra and the Welsh Assembly Government in April 2006. This provides advice for sewage treatment works operators and regulators on understanding the odour, factors to be taken into account in assessing how much of a nuisance it is to the surrounding area, managing complaints, and ways of abating or limiting the odour. Action against odour from sewage treatment works can be taken under the statutory nuisance legislation – see chapter 1, 1.29.

MARINE POLLUTION

The importance of protecting the marine environment is now recognised as a priority by many governments. Inadequate sewage disposal, the build-up of chlorinated hydrocarbons from pesticides, the dumping of polychlorinated biphenyls (PCBs) and other chemicals, as well as the incineration of hazardous waste at sea all pose a threat to human health and to marine life.

Shipping is also responsible for significant air pollution. A report from the International Council on Clean Transportation, published in 2007, estimates that in 2005 international shipping accounted for up to 3% of global carbon dioxide emissions; it also accounted for 27% of global emissions of nitrogen oxides and 9% of global sulphur dioxide emissions, with 70-80% of emissions occurring within 400 km of land, thus affecting local air quality.

6.21 INTERNATIONAL CONVENTIONS

6.21.1 UN Convention on the Law of the Sea

This Convention defines pollution of the marine environment as

> ". . . the introduction by man, directly or indirectly, of substances or energy into the marine environment, including estuaries, which results or is likely to result in such deleterious effects as harm to living resources and marine life, hazards to human health, hindrance to marine activities, including fishing and other legitimate uses of the sea, impairment of quality for use of sea water and reduction of amenities."

The Convention covers all aspects of the preservation and protection of the marine environment and sets out the duties of states in this respect, as well as "the legal framework within which all activities in the oceans and seas must be carried out". Some of its articles – or aims – are similar to those in other conventions (see below) or have already been incorporated into national legislation.

This Convention came into force in 1994 following ratification by the required number (60) of countries. On 23 March 1998 the Council of Ministers adopted a decision approving the Convention and Agreement of 28 July 1994 on implementation of Part XI of it, and the Convention thus came into force in the European Union on 1 May 1998.

The UN General Assembly discusses the way in which the Convention is working annually; in November 1999 the General Assembly adopted a Resolution providing for an annual consultative meeting open to all UN member states, parties to the Convention and relevant intergovernmental organisations. This meeting will discuss the UN Secretary General's report on oceans and the law with particular "emphasis on identifying areas where coordination and cooperation at intergovernmental and interagency levels should be enhanced".

6.21.2 The London Convention

The London Convention on the Prevention of Marine Pollution by Dumping of Wastes and other Matter was signed in London in 1972; it entered into force on 30 August 1975 and applies worldwide. It aims to ban or restrict the dumping (i.e. deliberate disposal) of various polluting substances according to their potential to harm either human or marine life or otherwise impair the marine environment. As well as the disposal of waste from ships, aircraft and manmade structures at sea, the Convention bans the disposal of ships, aircraft and manmade structures and the residues of incineration at sea.

The London Convention also imposed an immediate ban on the dumping of high-level radioactive waste at sea and in 1983 agreed a 10 year voluntary moratorium on the dumping of low level radioactive waste. A further amendment agreed in November 1993 banned the dumping of low-level radioactive waste at sea, the incineration at sea of industrial wastes and agreed to phase out the dumping of industrial wastes at sea by 31 December 1995.

A 1996 Protocol to the Convention, which when ratified will replace the 1972 Convention, adopts the precautionary principle requiring that "appropriate preventative measures be taken when there is reason to believe that wastes or other matter introduced into the marine environment are likely to cause harm even when there is no conclusive evidence to prove a causal relation between inputs and their effects". As such the Protocol will ban the dumping at sea of all wastes except those listed in Annex I to the Protocol; the list includes dredged material; sewage sludge; fish waste; vessels, platforms and other manmade structures; inert, inorganic geological material; and organic material of natural origin. The Protocol will prohibit the export of waste to another country for dumping or incineration at sea. Finally only those wastes listed in an annex to the Protocol may be dumped at sea.

6.21.3 MARPOL

This Convention – the International Convention for the Prevention of Pollution from Ships – was adopted following the 1973 international conference on marine pollution held under the auspices of the International Maritime Organisation. Included were articles dealing with oil, chemical, sewage and other pollution from ships, and requiring provision of adequate port reception facilities to deal with such wastes. However, it was not ratified because many governments felt it would be too difficult to implement. A further conference in 1978 agreed a Protocol to the Convention which included a number of amendments and improved the annex relating to oil pollution. The Convention and its Protocol (MARPOL 73/78) have now been ratified by a number of governments. In the UK it has been implemented through the 1996 *Merchant Shipping (Prevention of Oil Pollution) Regulations* – see below, 6.24.4.

Further measures aimed at reducing pollution from ships have now been agreed by the IMO and were incorporated in MARPOL for implementation from January 1992. These measures include reducing pollutants in ships' exhaust gases, banning the use of CFCs in refrigeration units and of halons in new fire extinguishers. A new Annex VI to the MARPOL Convention was signed in London in September 1997; this aims to control emissions of air pollutants from shipping – these are summarised in chapter 3, 3.29.1.

The MARPOL Convention, in allowing certain oil discharges from ships, also allows for special areas to be declared in which stricter rules for discharges apply. The IMO has agreed that the seas around the UK and north west European coastline should have special area status; thus from 1 August 1999 almost all oil pollution from ships in this area has been banned. Other special areas include the Gulf of Aden (1989), North Sea (1991), the Antarctic (1992) and the Wider Caribbean (1993).

A further treaty put forward by the IMO was agreed in London at the end of 1990. *The International Convention on Oil Pollution Preparedness Response and Cooperation 1990* lays down the principles of prompt and effective action in the event of an oil spill. Ships, offshore installations, ports and other facilities that handle oil should establish emergency plans and all pollution incidents should be reported. This Convention came into force on 13 May 1995, 12 months after ratification by the required 15 states; as at 13 May 1995, 21 member states had ratified the Convention. In the UK this Convention has been partly implemented through the *Merchant Shipping (Oil Pollution Preparedness, Response and Cooperation Convention) Regulations 1998* (SI 1056); these came into force on 15 May 1998 – see 6.24.4 below.

The IMO's Marine Environment Protection Committee agreed an amendment to Annex 1 to MARPOL at its meeting in July 2005; this was to be formally adopted in March 2006, entering into force on 1 August 2007. This will apply to all ships delivered on or after 1 August 2010 with an aggregate oil fuel capacity of 600 m^3 and above and will limit maximum capacity to 2,500 m^3; strict requirements regarding protected location of fuel tanks or, alternatively, performance standards for accidental oil fuel outflow are included, and Administrations encouraged to consider general safety aspects, including the need for inspection and maintenance of wing and double bottom tanks or spaces, when approving the design and construction of new ships.

A revised Annex IV to MARPOL covering the prevention of pollution of the sea by sewage from ships took effect on 1 August 2005, with existing ships required to comply with its provisions by 27 September 2008. The Annex requires ships to be fitted with either a sewage treatment plant, a sewage comminuting and disinfecting system or a sewage holding tank. The discharge of sewage into the sea is prohibited at a distance of 12 nautical miles, or less, from the nearest land. Ships with an approved sewage treatment plant in operation, or when discharging comminuted and disinfected sewage using an approved system may discharge at a distance of more than three nautical miles from the nearest land. Annex V relating to the disposal of garbage includes a prohibition on disposing into the sea incinerator ashes from plastic products as they may contain toxic or heavy metal residues; and requires the keeping of garbage record books,

The IMO's Marine Environment Protection Committee is also considering a draft text of a new convention providing globally applicable ship recycling regulations for international shipping and recycling activities. It is hoped to finalise this in time for adoption in 2009. The Convention will provide for

- the design, construction, operation and preparation of ships to as to facilitate safe and environmentally sound recycling, without compromising operational safety and efficiency.
- the safe and environmentally sound operation of ship recycling facilities.
- establishment of appropriate enforcement, certification and reporting mechanisms.

6.21.4 The OSPAR Convention

The Convention on the Protection of the North Sea and North East Atlantic, which was signed in Paris in September 1992, entered into force in March 1998. It replaces the 1972 Oslo *Convention for the Prevention of Marine Pollution by Dumping from Ships and Aircraft* and the 1974 Paris *Convention for the Prevention of Marine Pollution from Land-Based Sources*. Both Conventions covered the North Sea and North East Atlantic and were monitored by London based Commissions which also took a leading role in ensuring follow-up of actions agreed at the North Sea Conferences (see below). A Commission – the OSPAR Commission (i.e. drawn from the Oslo and Paris Commissions) – has been set up to implement and monitor the 1992 Convention; the members of OSPAR are Belgium, Denmark, Finland, France, Germany, Iceland, Ireland, Luxembourg, The Netherlands, Norway, Portugal, Spain, Sweden, Switzerland, the UK and the European Community.

The main objective of the 1992 Convention is the protection of the marine environment so as to safeguard human health through the elimination or prevention of pollution. Conservation and repair of the marine ecosystem are equally important objectives and a strategy to combat eutrophication is to be drawn up. Potentially polluting activities covered by the Convention are listed as are the substances for which control programmes must be implemented with a view to continuously reducing marine concentrations to background values and ceasing releases of such substances by 2020; these include pesticides, heavy metals, oil and hydrocarbons, organohalogen compounds and various chemicals. A list of priority substances for control to which the 2020 target will apply has been agreed. Both dumping and incineration at sea are to be banned, as are radioactive discharges thus giving extra weight to the agreements reached at the North Sea Conferences (see below).

At the 1998 meeting of the Commission in Portugal, Ministers agreed a total ban on sea dumping of large steel installations (gas and oil rigs); it was also agreed that

- by the year 2000, the OSPAR Commission would, for the whole maritime area, work towards achieving further substantial reductions or elimination of discharges, emissions and losses of radioactive substances;
- by the year 2020, the OSPAR Commission will ensure that discharges, emissions and losses of radioactive substances are reduced "to levels where the additional concentrations in the marine environment above historic levels resulting from such discharges, emissions and losses, are close to zero". (At the 2003 meeting it was agreed that the baseline by which progress should be measured should be the average of discharges between 1995-2001.)

The UK and France also withdrew their opt-out from a ban on dumping nuclear waste at sea, which is to be phased out by 2020. However both the UK and France did not accept a majority decision (taken at a meeting in Copenhagen in June 2000) that nuclear fuel reprocessing be terminated (and thus nuclear fuel be stored on land). At the 2000 meeting it was also agreed that each country should, by December 2002 (amended to July 2002 at the 2001 meeting), draw up an action plan for reducing radioactive discharges. Parties to the Convention also adopted a Decision (2000/2) on a Harmonised Mandatory Control System for the Use and Reduction of the Discharge of Offshore Chemicals. A further Annex (Annex V) was also added to the Convention dealing with the protection and conservation of maritime ecosystems and biological diversity.

In drawing up action programmes to deal with polluting discharges, signatories must apply the precautionary principle and the polluter pays principle, as well as having regard to both BAT (Best Available Technique) and BEP (Best Environmental Practice); this latter includes information to the public, product labelling and collection and disposal systems. One of the tasks of the Commission implementing and monitoring the Convention will be to draw up documents specifying BAT for disposal, reduction and elimination of a range of substances from various industrial sectors.

The European Community gave its approval to the OSPAR Convention through Council Decision 98/249/EC. The Community formally approved Annex V through a Decision dated 8 May 2000.

The UK Strategy for Radioactive Discharges 2001-2020 (Defra, 2002) outlines how it is intended to implement the OSPAR agreements (see also chapter 5, 5.38.2). *The Offshore Chemicals Regulations 2002* implement the OSPAR Decision for a Harmonised Mandatory Control System for the Use and Reduction of Offshore Chemicals, and OSPAR Recommendations 2000/4 on a harmonised pre-screening scheme for offshore chemicals and 2000/5 on a harmonised off-shore chemical notification format – see 6.24.4 below.

At their 2003 meeting, Ministers adopted 5 recommendations:

- to control the dispersal of mercury emissions from crematoria;
- to promote the use and implementation of environmental management systems by the offshore gas and oil industry;
- to draw up a network of "marine protected areas", together with guidelines for their selection and management;
- to draw up a framework for reporting encounters with marine dumped conventional and chemical munitions;
- to implement a strategy implementing the Joint Assessment and Monitoring Programme.

Ministers also adopted criteria for the identification of habitats and species in need of protection and the selection of 27 species and 10 types of habitat to be protected.

More information: www.ospar.org

6.21.5 Bunkers Convention

The International Convention on Civil Liability for Pollution Damage caused by Bunker Oil (the Bunkers Convention) was adopted by the IMO in London in March 2001. It will come into effect 12 months after ratification by 18 states, including five each with ships whose combined gross tonnage is not less than 1 million. It has so far been ratified by eleven states, including the UK who did so in June 2006.

Under the Convention registered shipowners of vessels with a gross tonnage over 1,000 will be required to maintain insurance covering their liabilities for pollution caused by any spillage of fuel (bunker) oil from their vessels. The Convention allows claimants to take action directly against the insurance provider. *The Merchant Shipping (Oil Pollution) (Bunkers Convention) Regulations 2006* (SI 2006/1244) amend *the Merchant Shipping Act 1995* to implement the Convention.

6.22 NORTH SEA CONFERENCES

In Europe, apart from the Directives which apply throughout the EU (see 6.2-6.5 above and Appendix 7.6), further initiatives have been taken by the countries bordering the North Sea which is the receiving water for a number of effluent-laden rivers which flow into it; these rivers include the Rhine, Elbe, Humber, Forth, Tyne, Tees and the Thames. This growing concern resulted, in 1984, in the Government of the former West Germany convening a conference of countries bordering the North Sea (Belgium, Britain, Denmark, France, The Netherlands, Norway, Sweden and Germany) to draw up measures aimed at reducing pollution of the North Sea.

At a second conference in 1987 in London, among other measures, agreement was reached on ending the dumping of harmful wastes by the end of 1989 and ending incineration by 1994; it was also agreed that nitrogen and phosphorus inputs be cut by 50% by 1995. At the third North Sea Conference (The Hague, 1990), the emissions reductions agreed at the meeting were, for the first time, made legally binding; included were reductions in discharges of hazardous substances via rivers and estuaries and via the air; agreement was made to aim for a substantial reduction in the quantities of pesticides reaching the North Sea, and to phase out and to destroy in an environmentally safe manner all identifiable PCBs as soon as possible with the aim of complete destruction by 1995; and by 1999 at the latest, to ensure that the time between taking out of service and destruction is as short as practicable. It was also agreed that each country should provide its own destruction facilities.

The Urban Waste Water Treatment (England and Wales) Regulations 1994 (and similar in Scotland) required the phasing out of the disposal of sewage sludge to sea by 31 December 1998 (6.20 above).

At the Fourth North Sea Conference in June 1995, there was a further commitment to achieve the reduction targets agreed at the Third Conference with a number of measures agreed for reducing discharges of priority substances (cadmium, mercury etc); there was also a call for on-going reductions in discharges and emissions of hazardous substances leading to their cessation within 25 years.

The Fifth North Sea Conference in Bergen, Norway, in March 2002 adopted a Ministerial Declaration which covered a wide range of issues seen as important for protecting the North Sea, including:

- the use of ecological quality objectives as a tool for setting clear operational environmental objectives and serving as indicators for the ecosystems health;
- by 2010 to designate relevant areas of the North Sea as marine protected areas;
- to take all possible action to ensure that the culture of genetically modified organisms is confined to secure, self-contained land-based facilities in order to prevent their release to the marine environment;
- to reduce the environmental impacts of shipping through the creation of a network of investigators and prosecutors to enforce the internationally agreed rules and standards for the prevention of pollution from ships;
- to increase efforts to meet the target of stopping emissions, discharges and losses of hazardous substances to the marine environment by 2020;
- to achieve full implementation of the EU nitrates, urban waste water and water framework directives to meet the OSPAR target of combating eutrophication – to achieve by 2010 a healthy environment where eutrophication does not occur;
- to progressively reduce discharges from nuclear facilities, thus assisting in effective implementation of OSPAR's strategy for progressive and substantial reductions of discharges, emissions and losses of radioactive substances;
- to increase efforts to stop the dumping of marine litter, through implementation of the EU port reception facilities Directive and information and awareness campaigns.

The Ministerial Declaration agreed at the 2006 North Sea Conference (held in Goteborg, Sweden) includes a 40% reduction in emissions of nitrogen oxides and a further reduction in sulphur in fuels to 1%, as well as measures to strengthen controls on fishing.

6.23 EUROPEAN UNION

6.23.1 Marine Strategy Directive

In October 2005, the Commission published its *Thematic Strategy on the Protection and Conservation of the Marine Environment* (COM(2005) 504) – one of seven developed in the framework of the EU's 6th Action Programme (see Appendix 7). The overall objective of the Strategy is to protect and restore Europe's seas and oceans ensuring that human activities are carried out in a sustainable manner and that seas are safe, clean, healthy and productive.

The Strategy is supported by a *Directive establishing a Framework for Community Action in the field of Marine Environmental Policy* (Marine Strategy Framework Directive), which was adopted on 17 June 2008. The Directive (2008/56/EC) entered into force on 15 July 2008 and must be transposed into national legislation by 15 July 2010. The main objective of the Directive is to protect, conserve and improve the quality of the marine environment, with Member States required to take the necessary measures with the aim of achieving good environmental status by 2020 at the latest; derogations and exemptions would take account of situations where achievement of this objective is either not feasible or practical; nor will Member States be required to do anything where "the costs would be disproportionate taking account of the risks to the marine environment".

The Directive defines four Marine Regions – the Baltic Sea, the North East Atlantic Ocean, the Mediterranean Sea and the Black Sea – and requires Member States to develop a Marine Strategy for their European marine waters in each Marine Region, or sub-region (also identified in the Directive) if that would be more appropriate. There should be close coordination with any Member State or third country sharing the water in the Marine Region. In developing each Marine Strategy, Member States will be required to:

- make an initial assessment of their European marine waters comprising an analysis of the essential characteristics and current environmental status of the water, predominant pressures and impacts, including human activities; and an economic and social analysis of their use and of the cost of degradation of the marine environment – to be completed by 15 July 2012, and reviewed every six years;
- by 15 July 2012, on the basis of the above assessment, to determine a set of characteristics for good environmental status in accordance with generic qualitative descriptors, detailed criteria and standards to be developed by the Commission;
- by 15 July 2012, establish environmental targets and associated indicators;
- by 15 July 2014, establish monitoring programmes to enable regular evaluation of the state of the waters and progress towards achieving good environmental status;
- by 2015, develop a programme of measures designed to achieve good environmental status within the following five years.

Competent authorities for each marine region, or sub-region, should be designated by 15 July 2010, and the Commission provided with details by 15 January 2011.

EU measures on reducing atmospheric emissions from shipping is covered in chapter 3, 3.29.2

6.23.2 Ship Source Pollution

Directive 2005/35 on *ship-source pollution and on the introduction of penalties for infringements*, was adopted on 7 September 2005 with Member States required to implement it by 1 March 2007. The Directive is intended to improve compliance with the MARPOL Convention as well as bring consistency to the way in which it is implemented in EU Member States; it is supplemented by detailed rules on criminal offences and penalties set out in Framework Decision 2005/667/JHA which will enable Member States to impose criminal sanctions on persons found to be responsible for illegal discharges (both accidental and intentional). The definition of illegal discharges goes further than MARPOL in that it includes those resulting from damage to the ship or its equipment. Proposals for amending the 2005 Directive were published in March 2008 – COM(2008) 134 – and will replace the 2005 Framework Decision. The new Directive will clarify that the infringements defined in the 2005 Directive are to be treated as criminal offences with criminal penalties.

The European Safety Maritime Agency (established by EU Regulation 1406/2002) is responsible for preventing and responding to pollution caused by ships.

The 2005 Directive is to be implemented through *the Merchant Shipping (Implementation of Ship-Source Pollution Directive) Regulations*, a draft of which was published for consultation by the Maritime and Coastal Agency in September 2008.

6.23.3 Ship Generated Waste

The *Directive on port reception facilities for ship generated waste and cargo residues* (2000/59/EC) was formally adopted by the European Parliament and Council on 27 November 2000; Member States were required to transpose the Directive into national legislation by 28 December 2002 – see below 6.24.2b. In building on the requirements of MARPOL, which prohibits dumping of waste at sea (see 6.20.3), the Directive aims to reduce the discharge of ship generated waste at sea, by requiring Member States to ensure there are adequate waste reception facilities at ports and marinas. The Directive requires such facilities to be licensed and will also require ports to draw up a waste handling and reception plan. Ships calling at a port will have to deposit ship generated waste in the reception facility unless they have enough storage capacity to do so at a subsequent port. Costs for such facilities will be recovered through fees levied on ships calling at the port.

6.24 UK REGULATORY CONTROLS

In February 2006, thirty-two areas around the UK coast were identified as "Marine Environmental High Risk Areas" – MEHRAs have been identified on the basis of shipping risk, environmental sensitivity and other environmental protection measures already in place and in such areas mariners will be expected to exercise an even greater degree of care when passing through.

The UK Government has also published a *Ship Recycling Strategy* (Defra, 2007) which aims to ensure that government-owned and commercial ships are recycled to acceptable health, safety and environmental standards, particularly in developing countries.

6.24.1 UK Marine Bill

In April 2008, Defra published a draft *Marine Bill* for consultation. The aim of the Bill which it is expected will be included in the Government's legislative programme for 2008-09, is to ensure clean, healthy, safe, productive and biologically diverse oceans and seas, by putting in place better systems for delivering sustainable development of marine and coastal environments. As such the Bill includes provisions for:

- A Marine Management Organisation to deliver the objectives of the Bill.
- A strategic marine planning system – the first stage will involve the drawing up of a UK wide marine policy statement to enable a more integrated approach to marine management and setting of both short and longer term objectives for sustainable use of the marine environment. A June 2008 consultation paper on behalf of the UK Government, Welsh Assembly Government and Northern Ireland Assembly details "high level" objectives as a first stage in the development of the policy statement.
- Designation of Marine Conservation Zones, as part of a Marine Protected Areas network. Draft guidance from Defra published in June 2008 outlines criteria for the selection and designation of MCZs; duties of public authorities in relation to MCZs; and on conservation orders.
- Simplifying licensing of marine developments (e.g. offshore wind farms).
- Improved management of marine and inland fisheries.
- Integrated coastal zone management.

6.24.2 Scottish Marine Bill

The Scottish Government has published (July 2008) proposals for a Scottish Marine Bill, with the aim of providing for

- Improved conservation to ensure a healthy future for Scotland's marine wildlife.
- New marine planning system and a streamlined licensing system to encourage economic investment in areas such as renewable energy.
- Measures to ensure a viable future for Scotland's marine industries.
- Creation of Marine Scotland – this will bring together existing organisations with a view to streamlining governance of Scotland's seas.

6.24.3 Waste and Incineration

(a) Dumping at Sea

Part II of *the Food & Environment Protection Act 1985* repealed the *Dumping at Sea Act 1974* which was the first UK legislation to impose statutory controls on the dumping of waste at sea; it gave effect to various international conventions (London and Oslo – 6.21.2 & 6.21.4 above) imposing obligations relating to the disposal of waste, the dumping of hazardous pollutants and incineration at sea. The Act applies to the whole of the United

Kingdom. (The UK ended incineration at sea in 1990 and the dumping of industrial waste in 1992.) Part II of FEPA, which covers dumping and incineration at sea and/or under the seabed within UK waters, and also the scuttling of vessels, was amended by s.146 of the *Environmental Protection Act 1990*. (The amendment mainly relates to the inclusion of UK controlled waters as well as UK waters.) FEPA applies to dumping etc whether by a UK or foreign ship, aircraft, hovercraft, marine structure or floating container.

Under FEPA the prescribed activities require prior approval by licence from the Department for Environment, Food and Rural Affairs, with the Department of Environment Northern Ireland being the competent authority there. In most instances, the requirement for a licence includes activities prior to the dumping or incineration, such as the loading of ships and aircraft etc within the UK, UK waters or UK controlled waters. Exempted activities are listed in the *Deposits in the Sea (Exemptions) Order 1985* (SI 1699) (NI: SR 1995/234 as amended SR 2003/493 & SR 2008/20); included are activities such as returning to the water matter taken out during fishing and dredging or deposited for coastal protection or harbour works, and the disposal of "victual or domestic waste" from ships and hovercraft etc. As from 31 December 1994 exempted activities had to be registered with Defra – formerly MAFF – (amendment to Exemptions Order by *Waste Management Regulations 1994*).

In considering whether to issue a licence, Defra will take into account the need to protect the marine environment, the living resources which it supports, and human health. The licence may include conditions to meet these requirements, and the Act provides for fees to be charged in respect of administrative expenses, monitoring and sampling. Where a licence has been breached, Defra may revoke it and criminal proceedings may be instituted. Summary conviction carries a fine not exceeding £50,000; conviction on indictment carries a fine and/or imprisonment not exceeding two years. Each licensing authority is required to maintain a public register containing details of licences issued relating to deposits and incineration at sea. This provision has been amended by s.147 of the *Environmental Protection Act 1990* which specifies that the register must contain applications for licences, those issued, variations and revocations as well as details of convictions for breaches of licence conditions. Information may be excluded from the register on grounds of commercial confidentiality or national security. These requirements have been given effect through *The Deposits in the Sea (Public Registers of Information) Regulations 1996* (SI 1427) which came into force on 1 July 1996. The Regulations apply in England and Wales only. The register is held by Defra in London with access available at reasonable times. A charge may be levied for copies of entries on the register.

(b) Ship Generated Waste

The Merchant Shipping and Fishing Vessels (Port Waste Reception Facilities) Regulations 2003 (SI 2003/1809) revoke and replace *The Merchant Shipping (Port Waste Reception Facilities) Regulations 1997*. They came into force on 15 July 2003 and implement the EU Directive on ship-generated waste (see above 6.23.3).

Harbour authorities and terminal operators are required to prepare waste management plans which have to be approved by the Secretary of State. The Regulations also require them to ensure they have adequate facilities (having regard to their waste management plan and to Guidance from the Secretary of State) for receiving prescribed wastes from ships. The Regulations require the master of a ship bound for harbour or terminal to provide the harbour authority or terminal operator with information about the waste (type, quantity, waste storage capacity of vessel) prior to the ship's arrival and require such waste to be delivered to the waste reception facility. Harbour authorities and terminal operators are required to impose charges covering the costs of waste reception facilities for ship-generated waste.

The Merchant Shipping (Prevention of Pollution by Garbage) Regulations 1998 (SI 1377) require all ships above 400 gross tonnage and those certified to carry more than 15 people to have a garbage management plan, and to keep disposal records. Illegal disposal carries a maximum fine of £25,000.

Further Regulations are to be made requiring harbour authorities and terminal operators to provide waste facilities for ship-generated sewage, and for charges to be made for ships using the facilities, and implementing Annexes IV and V of MARPOL. The Regulations will not apply initially to ships built before 27 September 2003. Draft Regulations were issued for consultation by the Maritime and Coastguard Agency (www.mcga.gov.uk) in March 2008.

6.24.4 Oil Pollution

(a) Prevention of Oil Pollution

The Prevention of Oil Pollution Act 1971 makes it an offence (subject to various defences) to discharge oil or any substance containing oil into UK territorial waters, either from a place on land or from sea-based explorations or into harbour waters; The Act introduced requirements for record keeping covering transfers of oil and placed restrictions on when such transfers can take place; it also covers liability for soil pollution and requires compulsory insurance against liability. A 1986 Act extended the 1971 Act to cover discharges from vessels.

Section 148 of the *Environmental Protection Act 1990* amended the 1971 Act to enable action to be taken against foreign owned ships for illegally discharging oil into UK territorial waters or ports and harbours.

Various Regulations have been made implementing the MARPOL Convention (see above); they apply to UK ships worldwide and to others while in UK waters.

The *Merchant Shipping (Prevention of Oil Pollution) (Limits) Regulations 1996* (SI 2128), which came into force on 5 September 1996, establish a UK "pollution zone"; this extends beyond the 12 nautical mile territorial limit up to 200 nautical miles from the coastline. Any foreign ship suspected of committing a pollution offence within the pollution zone which calls at a port in the UK will then be liable to prosecution by the UK courts.

The *Merchant Shipping (Prevention of Oil Pollution) Regulations 1996* (SI 2154), as amended 1997 (SI 1910) came into force on 17 September; they provide for the enforcement of oil pollution offences committed within the pollution zone and consolidate earlier 1983 and 1993 Regulations. These introduced record keeping requirements as well as survey and certification for oil tankers and other UK ships above a certain size. There is a duty to maintain ships and their equipment in compliance with the Regulations; other provisions relate to powers of inspection and penalties. New oil tankers must be fitted with double hulls and existing ones must be fitted with additional protection against grounding or collision once they are 25 years old. Oil pollution emergency plans must be carried and stricter rules regarding discharges of oil complied with. The 1996 Regulations also include measures aimed at improving the provision and use of port waste reception facilities, tighten the regulations on discharges from ships, improve the enforcement of the Regulations and increase the maximum fines for illegal discharges. 2004 Amendment Regulations (SI 2004/303), which came into force on 8 March 2004, implement the EU Regulation 417/2002 regarding the timetable for phasing in double hull oil tankers; they also implement amendments to MARPOL relating to guidelines for the development of shipboard oil pollution emergency plans and provide for such plans to be combined, in certain cases, with those required for noxious liquid substances. 2005 Amendment Regulations (SI 2005/1916) implement various EU Regulations accelerating the timetable for phasing in double hull or equivalent design requirements for single hull tankers.

(b) Salvage and Pollution

The Merchant Shipping (Salvage and Pollution) Act 1994 received Royal Assent in July 1994; from 1 October 1994 the owners of all vessels (except laden oil tankers) became liable for any oil pollution and resultant pollution prevention measures. It also implemented, with effect from 1 January 1995, a number of international marine conventions and declarations, including:

- the 1990 Convention on Oil Pollution Preparedness, Response and Cooperation (see 6.21.3 above);
- a 1992 Declaration from the North Sea Conference calling for implementation of various provisions of the 1982 UN Convention on the Law of the Sea (see above 6.21.1).

(c) Oil Pollution Preparedness

The Merchant Shipping (Oil Pollution Preparedness, Response and Cooperation Convention) Regulations 1998 (SI 1056), which apply throughout the UK, came into force on 15 May 1998; they implement in part the 1990 IMO Convention of the same name (see above 6.21.3). They apply to harbours and oil handling facilities with an annual turnover of more than £1 million, and to those which are able to handle tankers above a certain threshold or are designated by the Secretary of State; they also apply to off-shore installations in UK waters.

The Regulations required harbour authorities and operators of oil handling facilities to prepare and submit an oil spill contingency plan to the Government's Marine Pollution Control Unit by 15 August 1999; off-shore installations had until the same date to submit a similar plan to the Department of Trade and Industry's Oil and Gas Office in Aberdeen; local people and organisations, local authorities and environmental organisations had to be consulted during the plan's preparation. Plans must be reviewed every five years.

Ships' masters and those in charge of harbours, oil handling facilities and off-shore installations must report any oil spills to their nearest Coastguard station.

(d) Merchant Shipping (Pollution) Act 2006

Section 1 of the Act enables the UK to implement revisions which have been made to international arrangements relating to compensation for oil pollution from ships, thus ensuring that victims of oil pollution from tankers are properly and promptly compensated.

Section 2 of the Act amends s.128 of *the Merchant Shipping Act 1995* to make reference to Annex VI of MARPOL which limits atmospheric emissions from ships, thus enabling Regulations to be made (see above 6.21.3, and chapter 3, 3.29).

6.24.5 Merchant Shipping & Maritime Security Act 1997

This Act, which received the Royal Assent in late March 1997, aims to tighten UK controls on shipping which poses a pollution risk, e.g. because of an accident, collision or other internal or external damage. It includes provisions to:

- require the Secretary of State to draw up and implement a national marine pollution contingency plan;
- establish temporary exclusion zones at sea to allow safety zones to be placed around ships or structures on grounds of safety and/or pollution;
- widen the Coastguard's intervention powers to cover any risk of significant pollution;
- permit charging for maritime functions such as emergency pollution response and standard setting activity in line with the polluter pays and user pays principles;
- enable regulations to be made requiring port and harbour authorities to prepare waste management plans and to charge ships for the use of waste reception facilities;
- enable regulations to be made requiring local authorities to draw up contingency plans to deal with oil spills;
- give powers to move ships on if they pose a threat to safety or the environment;
- increase the maximum penalty for marine pollution offences from £50,000 to £250,000.

The European Community has also adopted (December 2000) a Decision establishing a "framework for cooperation in dealing with accidental or deliberate pollution at sea". The Decision aims to strengthen Member States' ability to take action in the event of such pollution to protect human health and the coastline; an information exchange system will enable rapid response in the event of an accident.

The Marine Safety Act 2003, which came into force on 10 September 2003 gives the Secretary of State powers to give a direction to a person in charge of land next to, or accessible from, UK waters or in charge of certain facilities used by ships (e.g. wharves, jetties), to require them to be made available for the purposes of preventing or minimising or dealing with a pollution risk from a ship. The Act provides for the payment of compensation by the Secretary of State to anyone suffering unreasonable loss or damage because of a direction.

6.24.6 Pollution Prevention & Control Act 1999

(a) Offshore Chemicals

The Offshore Chemical Regulations 2002 (SI 1355) were made under the *Pollution Prevention and Control Act 1999*; they implement UK obligations under the OSPAR Convention (see above, 6.20.4) for a harmonised mandatory control system for the use and reduction of the discharge of offshore chemicals in relation to offshore activities (Decision 2000/2). The Regulations took effect on 15 May 2002 and apply throughout the UK.

Operators of offshore oil and gas installations within UK territorial waters are required to apply for a permit from the Secretary of State, BERR, covering their use and discharge of chemicals into the marine environment. The application must be accompanied by a risk assessment of harm to the environment from the use and discharge of the chemicals at the installation; details of the proposed technology and other techniques to be used to prevent or minimise the use or discharge of the chemicals and monitoring proposals must also be included on the application. The Regulations also include provisions relating to consultation, variations, appeals, surrender and revocation of permits. The DTI is to produce guidance on the Regulations.

Permits granted will include conditions relating to:

- quantity, frequency, location and duration of any permitted use or discharge of offshore chemicals;
- a requirement for the operator to seek suitable less hazardous substitute chemicals;
- a requirement to take all appropriate measures to prevent pollution through the use of appropriate technology to limit discharges, emissions and waste;
- the need to take necessary measures to prevent accidents affecting the environment or, where they do occur, to limit their effects on the environment;
- the need to carry out appropriate monitoring on the use and discharge of offshore chemicals (measurement technology and frequency, and evaluation procedures will be specified on the permit);
- the submission of monthly reports to the Secretary of State giving details of the actual quantity, frequency and location of use and discharge of offshore chemicals;
- the minimisation of long-distance or transboundary pollution;
- the need for appropriate measures to deal with abnormal operating conditions (e.g. start up, leaks, malfunctions, temporary stoppages and permanent cessation of operations).

Details of permits granted and monitoring data will be kept on a register.

Operators are required to notify the Secretary of State of any incident or accident involving a chemical which results in a breach of the permit or results, or will result in a significant effect on the environment.

(b) Offshore Combustion Installations

The Offshore Combustion Installations (Prevention & Control of Pollution) Regulations 2001 (SI 1091), cover off-shore combustion installations with over 50 MW capacity and took effect on 19 March 2001 – see chapter 3, 3.29,

(c) Offshore Petroleum Activities

The Offshore Petroleum Activities (Oil Pollution Prevention and Control) Regulations 2005 (SI 2005/2055), which came into force in August 2005, require operators of offshore installations to obtain a permit before discharging oil into UK waters comprising the UK Continental Shelf, except for Scottish controlled waters. BERR is the permitting authority and the system introduced similar to other regimes under PPC. The Regulations also enable the Secretary of State to set an annual discharge allowance for each operator who may then trade surplus allowances to operators who exceed their allowance; operators unable to cover any exceedence of their annual discharge allowance will be liable to a financial penalty.

The Regulations provide for the phasing out of exemptions under the *Prevention of Oil Pollution Act 1971* which permitted certain discharges of oil into the sea.

(d) Offshore Installations – Emergency Controls

The Offshore Installations (Emergency Pollution Control) Regulations 2002 (SI 1861), made under the PPC Act, came into force on 18 July 2002. They enable the Secretary of State's representative to take control of dealing with pollution emergencies as a result of an accident involving an off-shore installation. The Regulations give the Secretary of State's representative wide powers to deal with such emergencies, including the power to order the sinking of an installation to prevent further pollution; failure to comply with a Direction from the SoS's representative is an offence. An offshore operator is entitled to claim compensation from the Government if it can be proved that the action taken to deal with the pollution emergency was in excess of that needed.

PESTICIDES

The term "pesticides" covers a wide range of products for both commercial and domestic use to control insect, rodent and other pest infestations. Pesticides include insecticides, rodenticides, fungicides, herbicides etc as well as substances used as growth regulators, fumigants, insect repellents or attractors and defoliants and desiccants. In addition the European Commission uses the term "biocides" to cover a range of non-agricultural pesticides such as disinfectants, textile and consumer product preservatives and air conditioning biocides; it proposes regulating these in much the same way as the more traditional pesticides. Pesticides can be both chemically derived and derived from natural products (e.g. pyrethrums obtained from chrysanthemums).

Their main use is for agricultural purposes where benefits can be said to include less disease and increased yields. However, their overuse and misuse in some instances has resulted in various pests becoming resistant to certain pesticides; the widespread use of pesticides can also mean that some crops may receive multiple applications; and while workers may risk occupational exposure, the general public may be exposed to spraydrift from aerial applications of pesticides. Pesticide use can also result in environmental damage to rivers, wildlife and vegetation.

In the UK the Pesticides Safety Directorate, which became part of the Health and Safety Executive on 1 April 2008, is responsible for

- providing controls on pesticide used in agriculture, horticulture and domestic use;
- monitoring their use and taking enforcement action against illegal use;
- providing policy advice to Ministers;
- supporting UK and EU reviews of the safety of pesticides on the market;
- working to standardise pesticide regulation within Europe;
- coordinating research and development programmes.

The PSD was formerly an Executive Agency of Defra, and Defra Ministers have retained strategic policy responsibility for pesticides (Defra announcement, March 2008.

More information from www.pesticides.gov.uk

6.25 EU MEASURES

At EU level, there are a number of Directives relating to the discharge of dangerous substances, including pesticides to water (see 6.6 above and Appendix 7.6). In addition there are various Directives covering pesticide residues in products of plant origin (e.g. fruit and vegetables) and in animal feeds and products. A 1979 Directive (79/117/EEC) which has subsequently been amended a number of times prohibits the placing on the market and use of plant protection products (i.e. pesticides etc) containing certain active substances; among the substances so far covered are the "drins" group of pesticides, DDT, chlordane, hexachlorocyclohexane, heptachlor, hexachlorobenzene and various mercury compounds. The marketing and use of various dangerous substances, including pentachlorophenol (PCP), is also controlled under Directive 76/769/EEC; in the UK the marketing and use of PCP and its supply for domestic use (which is banned) is regulated under the *Environmental Protection (Control of Injurious Substances) Regulations 1993*.

At international level, the Third North Sea Conference (1990) (see 6.22 above) also agreed measures to reduce discharges of certain pesticides into the North Sea. In 1985 the United Nations Food and Agriculture Organisation adopted a *Code of Practice on the International Distribution and Use of Pesticides*. This is mainly concerned with ensuring that pesticides restricted or banned in the country of origin are not freely available or dumped on other countries. In September 1998, the International Convention on Trade in Hazardous Chemicals was approved with 95 signatories, including the EU. Amongst other things, the Convention bans the export of any chemical or pesticide which has itself been banned or severely restricted in at least two other countries, unless the importing country gives its specific (prior) consent; the Convention initially applied to a list of 27 hazardous substances with a further 14 being added in 2004.

6.25.1 Thematic Strategy

The European Commission's Fifth Action Programme, Towards Sustainability, included as an objective the "significant reduction of pesticide use . . . and conversion of farmers to methods of integrated pest control". The Sixth Action Programme included as an objective the need to reduce the impact of pesticides on human health and the environment and to achieve a more sustainable use of pesticides, and providing for a thematic strategy on the sustainable use of pesticides to be drawn up.

The Commission's Thematic Strategy published in July 2006 sets out common objectives and requirements aimed at improving the way pesticides are used across the EU. It foresees measures such as national action plans, training for professional users and distributors, certification and control of application equipment, protection of the aquatic environment, and restricting or banning the use of pesticides in specific areas; it proposes banning aerial spraying except in strictly defined circumstances. The Strategy is accompanied by a proposal for a Directive establishing a framework for Community Action to achieve a sustainable use of pesticides (COM(2006) 273). It is also proposed to replace the 1991 Directive on the placing of plant protection products on the market with a Regulation which would aim to streamline approval and authorisation procedures.

The Commission has also published proposals for a *Regulation concerning statistics on plant protection products* (COM(2006) 778. Among the aims of the Regulation are to provide for improved systems for the collection of information on plant protection products across the Community, and to help inform the review of Directive 91/414/EC on authorisation and approval procedures.

6.25.2 Plant Protection Products

In 1991, a Directive (91/414/EEC) concerning the placing of plant protection products on the market was adopted; its main purpose is to regulate agricultural pesticides through the establishment of an EU-wide list of approved active ingredients, harmonise approval procedures and to allow for mutual recognition of individual Member States' authorisation procedures. The list – Annex I – of active ingredients is being drawn up progressively as pesticides currently on sale within the EU are examined on human and environmental health and safety grounds; products containing active ingredients which do not meet the criteria had to be withdrawn from sale in 2003 when the review of all active ingredients should have been completed. The Directive has been amended a number of times to include active substances on Annex I which have been approved following assessment in accordance with the Directive. In December 2001, the Council of Ministers agreed a Common Position on the Commission's proposal that the review deadline be extended to 2008. The Directive has been amended a number of times with Directive 94/43 outlining uniform principles for evaluating and authorising pesticides to ensure uniformity throughout the Community. Following a complaint from the European Parliament, the European Court of Justice annulled Directive 94/43 on the basis that it did not adequately protect groundwater, a requirement of the 1991 Directive. The 1994 Directive has now been replaced by Directive 97/57/EC which was adopted in mid-1997.

These Directives have now been implemented by the *Plant Protection Products Regulations 2005* (SI 2005/1435, as amended SI 2006/1295 and SI 2007/636), which came into force on 1 July 2005, replacing and revoking 2003 Regulations. Similar Regulations apply in Northern Ireland (SR 2005/526), as amended by SR 2008/85. In Scotland the relevant Regulations are SSI 2005331, as amended by SSI 2006/576 and SSI 2007/119.

6.25.3 Biocides

A Directive on the placing of biocidal products on the market – 98/8/EC – was adopted on 16 February 1998 and should have been implemented by Member States by 14 May 2000. This Directive uses a similar approach to the plant products Directive in aiming to standardise the way in which biocides are registered throughout the EU, protecting human health and the environment; the Directive covers chemicals, including disinfectants, "avicides", "piscicides" and dirt inhibitors (but not biocides for agricultural use) not covered by existing legislation. An annex to the Directive sets out the common principles, including risk assessment criteria, to be applied when authorising biocidal products to enable mutual recognition of authorisations throughout Member States. Member States may however apply stricter standards to protect human health or the environment. Only biocidal products containing an active substance approved under the Directive (on the basis of how safe it is for human health and the environment) may be authorised for use. However products already on the market remain authorised for use until the active substance they contain has been reviewed; the Directive gives Member States ten years to review existing active substances of biocides. Member States were required to notify or identify existing substances to the Commission by January 2003 – the environmental and health hazards of those notified to the Commission will then be reviewed by the Commission under its ten year programme.

EU Regulation 2032/2003/EC of 27 November 2003 and which entered into force on 14 December 2003 sets out the EU's ten-year programme for reviewing the safety of active substances in biocides on the EU market notified to them by Member States – these products may remain on the market until they have been reviewed by the Commission. Active substances identified (but not notified) to the Commission may only be used in biocidal products until 1 September 2006; the use of those neither notified nor identified was prohibited from 14 December 2003. The Regulation also details the information which companies manufacturing biocidal products will need to submit as part of the review process. The UK has obtained a derogation (Decision 2007/70/EC of 20.12.06) allowing the continued use of ammonia as a veterinary hygiene biocidal product until the 14 May 2008.

The Commission is reviewing implementation of the Directive and is due to bring forward proposals for its amendment.

6.26 UK MEASURES

The main legislation controlling the use of pesticides is Part III of the *Food and Environment Protection Act 1985* (FEPA), as amended by the Pesticides Act 1998, and Regulations made under it. Also of relevance are the *Control of Substances Hazardous to Health Regulations 2002* which are intended to safeguard employees' health (see this chapter, below, and chapter 2, 2.10). The disposal of pesticides and pesticides containers is covered by Part II of the *Environmental Protection Act 1990* and by the *Water Resources Act 1991* (Scotland – CAR) (see this chapter and chapter 5). *The Water Supply (Water Quality) Regulations 2000* (Sc & W 2001; NI 2007) include maximum admissible concentrations for pesticides and related products – see this chapter 6.17.1c and Appendix 6.3.

Additional controls on the use of pesticides in designated Environmentally Sensitive Areas are contained in the *Agriculture Act 1986*.

In the UK, the pesticide related Directives have been implemented mainly through Part III of the *Food and Environment Protection Act 1985* and the *Control of Pesticide Regulations 1986* and 1997 Amendment Regulations made under the Act, (see below 6.26.1 and 6.26.3) and the *Plant Protection Product Regulations 2003* (E & W; 1995 in Sc & in NI). The pesticide residues Directives are implemented through the *Pesticides (Maximum Residue Levels in Crops, Food and Feeding Stuffs) (E & W) Regulations 1999*, similar 2000 Regulations in Scotland, and separate 2002 Regulations in Northern Ireland (SR 20).

6.26.1 National Pesticides Strategy

The Pesticides Safety Directorate published (2006) a National Pesticides Strategy – *Pesticides and the Environment: a Strategy for the Sustainable Use of Plant Protection Products.* This aims to reach a balance between protecting the environment and sustainable crop production, and reinforces the policy of reducing the risks associated with all uses of plant protection products. It includes five outline action plans (covering water, biodiversity, plant protection products availability, amenity sector and amateur use, and targeted use reduction), together with targets and indicators for each with the aim of contributing to sustainable plant protection use.

Further information on this and all pesticide related issues can be found on www.pesticides.gov.uk

6.26.2 Food & Environment Protection Act 1985, Part III

This Part of the Act aims to protect the health of human beings, creatures and plants, to safeguard the environment and to secure safe, efficient and humane methods of controlling pests. Pesticides are defined as any substance, preparation or organism prepared or used for destroying any pest. The definition therefore includes herbicides, fungicides, insecticides, wood preservatives etc.

The Act provides Ministers with power, by regulation or order, to control the import, sale, supply, storage, use and advertisement of pesticides; to set maximum pesticide residue levels in food, crops and feeding stuffs; to make information supplied in connection with the control of pesticides available to the public. Under the Act an Advisory Committee on Pesticides was established and Codes of Practice published.

The Advisory Committee on Pesticides (ACP) consists of government appointed experts. The ACP considers manufacturers' applications for approval of pesticides and makes recommendations to government; it also reviews existing approvals, reviews safety and can make recommendations regarding revocation or amendments to approvals. A number of expert advisory panels advise the ACP on various issues relating to pesticide approvals.

There is limited public access to information on pesticides data: since 1986 evaluation data on newly approved or reviewed pesticides has been published, with manufacturers' supporting data available on request. In 1992, public access was extended to cover pesticides approved before 1986; fact sheets on older pesticides are also to be published. It should be noted that access to manufacturers' data is limited: information is not available on active ingredients and formulation specification and composition; production methods; names and addresses of laboratories, sites, personnel and individual medical records. People requesting access have to sign an undertaking that they will make no commercial use of the information; access is by "reading room facility" and while notetaking is allowed, photocopying is not; see, however, proposed amendments to the *Control of Pesticides Regulations 1986*, summarised below at 6.26.4.

The monthly Pesticides Register gives information on regulatory and other policy issues and the ACP publishes reports on items discussed at its meetings.

6.26.3 Pesticides Act 1998

This Act, which also extends to Northern Ireland, amends FEPA to enable regulations to be made to give local authorities (as well as Ministers) powers to seize and destroy pesticides, and for information on pesticides to be made available to the public, subject to conditions which Ministers consider appropriate. The *Control of Pesticides Regulations* (see below) are to be amended to provide access to data on pesticides approved before 1986, as well as to provide access to scientific data submitted by manufacturers; the amended Regulations will also clarify local authority seizure powers.

6.26.4 Control of Pesticides Regulations 1986

Prior to the inclusion of an active substance on Annex I of the Plant Protection Products Directive (see 6.25.2) above, approval for pesticide use is given under the *Control of Pesticides Regulations* (SI 1986/1510), which define the types of pesticides subject to control; they apply in England, Scotland and Wales and came into force in October 1986, with similar legislation (SR 1987/414, as amended SR 1991/203 & SR 1997/469) applying in Northern Ireland. Following the inclusion of an existing active substance in Annex I application for approval for the pesticide should be made under *the Plant Protection Products Regulations 2005* (SI 2005/1435) (E & W); these Regulations which replace and revoke SI 2003/3241) implement the procedures and requirements of the 1991 Directive. Products containing new active substances will also be evaluated under PPPR. In Scotland similar 2003 Regulations apply, SSI 2003/579, and in Northern Ireland SR 2005/251, as amended by SR 2007/251.

The COP Regulations implement those sections of FEPA relating to the approval, advertisement, sale, supply, storage and use (including aerial spraying) of pesticides – all these activities are prohibited unless products have been formally approved for use, conditions of approval observed and additional conditions set out in "Consents" are met. Statutory codes of practice (see below) provide guidance on implementation of the Regulations.

The Control of Pesticides (Amendment) Regulations 1997 (SI 188), which came into force on 31 January 1997, amend the 1986 Regulations to:

- clarify that they apply to: any pesticide, or substance, preparation or organism prepared or used as plant or wood protection product; plant growth regulator; to give protection from harmful creatures or to make them harmless; to control organisms with harmful or unwanted effects on water systems or buildings, other structures or manufactured products; to protect animals against ectoparasites;
- require Consents for the advertisement of pesticides, the sale, supply, storage of pesticides, and for their use (including aerial use) to include the conditions which are set out in Schedules 1-4 of the Regulations;
- clarify that in the event of breach of an approval or of a Consent, Ministers may require persons to recover pesticides from the market; empower Ministers to seize and dispose of a pesticide or require the holder of an approval or other person to dispose of it; where a pesticide has been imported into Great Britain or Northern Ireland in contravention of the Regulations (1987 Regulations in NI), to require, in writing, that it is exported again within a specified period;
- amend the rules covering access to information: Ministers may make evaluation documents available to the public and may charge a fee for supplying a copy; a person who has either inspected or paid or received a copy of an evaluation document may then make a request to the Minister to inspect the relevant study report; access to studies commissioned by government (including those produced to meet data requirements set in the course of reviews or approvals are included; the use for commercial purposes or publication of any data is prohibited unless approval in writing has been received from the Minister.

"Consents" are prepared by the Pesticides Safety Directorate, an Executive Agency of the Department for Environment, Food and Rural Affairs; they are regularly updated and published in the Pesticides Register. Defra also publishes The COPR Handbook – a Guide to the Policies, Procedures and Data Requirements relating to their Control under the *Control of Pesticides Regulations 1986*. This also provides detailed information on pesticide legislation and guidance on registration under the Regulations.

Since April 1992 local authorities have been responsible for enforcing those sections of the Regulations which relate to the advertisement, sale, supply, storage and use of pesticides from a variety of commercial premises (e.g. shops, garden centres, sports grounds, hotels) and also domestic premises. Those parts of the Regulations relating to health and safety (e.g. storage and use) are dealt with by environmental health departments; the advertising, sale and supply of pesticides are the responsibility of trading standards departments.

The actual enforcement powers are contained in s.19 of FEPA. Authorised officers of the local authorities have powers of entry and inspection where they suspect infringement of the Regulations. While they may take photocopies of documents and records and photographs, they do not have powers to seize and destroy pesticides – only Ministers have this power, although proposals have now been issued for amending the Regulations to extend this power to local authorities. Where the local authority enforcement officer is of the opinion that an offence has been committed in respect of Regulation 4, e.g. breach of consent, it can serve a notice detailing infringements, and time within which remedial action should be taken.

Both the Pesticides Safety Directorate and the Welsh Assembly (as respects England and Wales) and the Scottish Government have issued proposals for consolidating *the Control of Pesticides Regulations 1986* (as amended) and various plant protection products regulations (February 2008). At the same time it is proposed to include provisions relating to record keeping for all professional users of pesticides, who will be required to keep records for ten years; information to be recorded will include date and time of pesticide application; area of land treated and its location; pesticide used and dose rate; and weather conditions at the time. It is also proposed to give local authorities powers to seize and destroy pesticide products if there is a breach of the law. At the same time it is proposed to repeal *the Farm and Garden Chemicals Act 1967* and related regulations as this has now been overtaken by more recent EU and national legislation.

6.26.5 Biocidal Products

The EU's Biocides Directive has been implemented through *The Biocidal Products Regulations 2001* (SI 880) which came into force on 6 April 2001, as amended by SI 2007/293; they apply in England, Scotland and Wales. Similar Regulations (SR 2001/422, amended by SR 2007/190)) which came into force on 16 January 2002 apply in Northern Ireland. Technical guidance on all aspects of the Regulations is being prepared. A committee of the HSE – the Biocides Consultative Committee – is the UK competent authority, providing expert advice on the evaluation of biocidal products under the Regulations. The HSE has also published three guidance documents – two for importers and suppliers of biocides covering respectively arrangements under the Regulations for active substances already on the market (HSG 198) and substances new to the European market (HSG 208); the third guidance is for users of biocidal products and explains the authorisation process required before biocidal products can be sold for use in the workplace or home (HSG 215). Information on both the EU Directive, UK Regulations and their application is available on www.hse.gov.uk/biocides.

6.26.6 COSHH Regulations

The Control of Substances Hazardous to Health Regulations 2002 (SI 2002/2677) which came into force on 21 November 2002, replacing and consolidating 1999 Regulations, apply in England, Scotland and Wales. They cover a wide range of substances, including agricultural and non-agricultural pesticides; their purpose is to prevent exposure to all toxic or hazardous substances used in the work place which can harm health. A summary of requirements under COSHH is given in chapter 2, 2.11. The Regulations are enforced by the Health and Safety Executive.

Suspected cases of poisoning or death from exposure to pesticides in the course of work should be reported to the HSE and to the Pesticides Incidents Appraisal Panel (PIAP). Incidents within the home or outside should be reported to the local environmental health department and to the PIAP.

6.26.7 Codes of Practice

(a) Using Plant Protection Products

The Code of Practice for Using Plant Protection Products (January 2006) is a statutory code of practice covering England and Wales, with a similar code (January 2007) – for Scotland. They replace the 1998 code of practice on the safe use of pesticides on farms & holdings (the "Green Code"), the forestry sections of the 1991 code on the safe use of pesticides for non-agricultural purposes (the "Blue Code" – c below), as well as a voluntary code on the use of pesticides in amenity and industrial areas (the "Orange Code").

The codes, which can be used by regulatory authorities as evidence for either an enforcement notice or for a court case, or other enforcement action, cover the protection of wildlife and the environment as well as of human health; they set out best practice for the use of pesticides which are controlled under the *Control of Pesticides Regulations* and other plant protection regulations. Both codes provides advice on the laws regarding protection of groundwater and waste management, and on training and certification. They also provide advice on what to do before and during applications of pesticides, including advice on the new legal requirement to keep spray records, and on what should be done if members of the public report being affected by spraying.

The code of practice can be downloaded from the Pesticide Safety Directorate's website at www.pesticides.gov.uk and in Scotland from www.scotland.gov.uk

The Royal Commission on Environmental Pollution has published (September 2005) a special report on crop spraying and the health of residents and bystanders. It makes a number of recommendations aimed at reducing the risks to residents and bystanders and improving their access to information about spraying. In response to the RCEP report the Government is to review the model used to assess resident and bystander exposure as part of the pesticides approval process but does not believe that any further regulatory controls are needed at this time.

(b) Suppliers of Pesticides to Agriculture, Horticulture & Forestry

This Code (the "Yellow Code"), published by MAFF in 1998, covers construction of storage facilities, transportation of pesticides and disposal. Anyone involved in supplying or selling pesticides, or who stores more than 200 kg or litres of pesticides must hold a certificate of competence from the British Agrochemical Standards Inspection Scheme. Before storing pesticides, the relevant water authority, water undertakers, environmental health departments and emergency services must be consulted over safety and pollution control requirements.

While failure to observe both this Code and the one following is not in itself an offence, breach of them may be taken as evidence in any proceedings for breach of the Regulations.

(c) Safe Use of Pesticides for Non-Agricultural Purposes

This Code (the "Blue Code") drawn up by the Health and Safety Executive was published in 1991 and provides advice on compliance with the COSHH Regulations (see chapter 2, 2.11). It is aimed at all those involved in the use of non-agricultural pesticides, including those engaged in amenity horticulture, wood preservation, the application of anti-fouling paints, and public hygiene pest control. Topics covered include undertaking a COSHH assessment, prevention or control of exposure, use of control measures, health surveillance and information, and instruction and training for operators. (The sections relating to commercial forestry have now been updated and included in the new code of practice for using plant protection products – see (a) above.)

APPENDICES

FURTHER READING

INDEX

WEBSITES

APPENDIX 2.1

INTEGRATED POLLUTION PREVENTION AND CONTROL

Annex 1 to the original IPPC Directive (96/61/EC) lists the categories of activities to which the Directive applies and for which a permit is required. This Directive has now been codified into Directive 2008/1/EC to take account of amendments to the Directive since 1996 (see chapter 2, 2.1). It should also be noted that the Commission has published a proposal for a new Directive on industrial emissions (COM(2007) 844 which makes a number of substantive changes to Annex I.

The IPPC Directive has been transposed throughout the UK by separate Regulations, with Schedule 1 in each case listing the activities, installations and mobile plant for which a permit is required, and whether it is to be regulated for all emissions (IPPC) or emissions to air.

The Regulations for ***England and Wales*** are *the Environmental Permitting (England and Wales) Regulations 2007* (SI 2007/3538), which came into force on 6 April 2008, replacing and revoking *the Pollution Prevention and Control (England & Wales) Regulations* 2000 and subsequent amendments – see chapter 2, 2.4.

In ***Scotland***, the IPPC Directive is implemented through *the Pollution Prevention and Control (Scotland) Regulations 2000* (SSI 323) – as amended, SSI 2001/503; SSI 2002/275 & 1702; SSI 2003/146; SSI 2004/110 & 512; SSI 2005/340 &510. *The Solvent Emissions (Scotland) Regulations 2004* (SSI 2004/26) add Section 7 to Schedule 1 on Solvent Emission Directive Activities. All activities (Part A and Part B) are regulated by SEPA.

In ***Northern Ireland*** the relevant Regulations are *the Pollution Prevention and Control Regulations (Northern Ireland) 2003* (SR 2003/46), which came into operation on 31 March 2003, as amended SR 2004/507, SR 2005/285, SR 2007/245. Here activities are categorised as Part A, B or C, with Parts A & B regulated by the Chief Inspector and Part C by district councils. *The Solvent Emissions (Northern Ireland) Regulations 2004* (SR 2004/36) add Section 7 to Schedule 1 on Solvent Emission Directive Activities.

All Acts of Parliament and Statutory Instruments can be accessed via www.opsi.gov.uk

Annex 1 to Directive 2008/1/EC – Categories of Industrial Activities

1. Installations or parts of installations used for research, development and testing of new products and processes are not covered by this Directive.

2. The threshold values given below generally refer to production capacities or outputs. Where one operator carries out several activities falling under the same subheading in the same installation or one the same site, the capacities of such activities are added together.

1. Energy industries
1.1 Combustion installations with a rated thermal input exceeding 50 MW.
1.2 Mineral oil and gas refineries.
1.3 Coke ovens.
1.4 Coal gasification and liquefaction plants.

2. Production and processing of metals
2.1 Metal ore (including sulphide ore) roasting or sintering installations.
2.2 Installations for the production of pig iron or steel (primary or secondary fusion) including continuous casting, with a capacity exceeding 2.5 tonnes per hour.
2.3 Installations for the processing of ferrous metals:
(a) hot rolling mills with a capacity exceeding 20 tonnes of crude per hour;
(b) smitheries with hammers the energy of which exceeds 50 kilojoules per hammer, where the calorific power used exceeds 20 MW;
(c) application of protective fused metal coats with an input exceeding 2 tonnes of crude steel per hour.
2.4 Ferrous metal foundries with a production capacity exceeding 20 tonnes per day.
2.5 Installations:
(a) for the production of non-ferrous crude metals from ore, concentrates or secondary raw materials by metallurgical, chemical or electrolytic processes;
(b) for the smelting, including the alloyage, of non-ferrous metals, including recovered products, (refining, foundry casting, etc) with a melting capacity exceeding 4 tonnes per day for lead and cadmium or 20 tonnes per day for all other metals.
2.6 Installations for surface treatment of metals and plastic materials using an electrolytic or chemical process where the volume of the treatment vats exceeds 30 m^3.

3. Mineral industry
3.1 Installations for the production of cement clinker in rotary kilns with a production capacity exceeding 500 tonnes per day or lime in rotary kilns with a production capacity exceeding 50 tonnes per day or in other furnaces with a production capacity exceeding 50 tonnes per day.
3.2 Installations for the production of asbestos and the manufacture of asbestos-based products.
3.3 Installations for the manufacture of glass including glass fibre with a melting capacity exceeding 20 tonnes per day.
3.4 Installations for melting mineral substances including the production of mineral fibres with a melting capacity exceeding 20 tonnes per day.
3.5 Installations for the manufacture of ceramic products by firing, in particular roofing tiles, bricks, refractory bricks, tiles, stoneware or porcelain, with a production capacity exceeding 75 tonnes per day, and/or with a kiln capacity exceeding 4 m^3 and with a setting density per kiln exceeding 300 kg/m^3.

4. Chemical industry
Production within the meaning of the categories of activities contained in this section means the production on an industrial scale by chemical processing of substances or groups of substances listed in points 4.1 to 4.6.
4.1 Chemical installations for the production of basic organic chemicals, such as:
(a) simple hydrocarbons (linear or cyclic, saturated or unsaturated, aliphatic or aromatic);
(b) oxygen-containing hydrocarbons such as alcohols, aldehydes, ketones, carboxylic acids, esters, acetates, ethers, peroxides, epoxy resins;
(c) sulphuric hydrocarbons;
(d) nitrogenous hydrocarbons such as amines, amides, nitrous compounds, nitro compounds or nitrate compounds, nitriles, cyanates, isocynates;
(e) phosphorus-containing hydrocarbons;
(f) halogenic hydrocarbons;
(g) organometallic compounds;
(h) basic plastic materials (polymers, synthetic fibres and cellulose fibres);
(i) synthetic rubbers;
(j) dyes and pigments;
(k) surface-active agents and surfactants.
4.2 Chemical installations for the production of basic inorganic chemicals, such as:
(a) gases, such as ammonia, chlorine or hydrogen chloride, fluorine or hydrogen fluoride, carbon oxides, sulphur compounds, nitrogen oxides, hydrogen, sulphur dioxide, carbonyl chloride;
(b) acids such as chromic acid, hydrofluoric acid, phosphoric acid, hydrochloric acid, sulphuric acid, oleum, sulphurous acids;
(c) bases, such as ammonium hydroxide, potassium hydroxide, sodium hydroxide;
(d) salts, such as ammonium chloride, potassium carbonate, sodium carbonate, perborate, silver nitrate;
(e) non-metals, metal oxides or other inorganic compounds such as calcium carbide, silicon, silicon carbide.
4.3 Chemical installations for the production of phosphorous-, nitrogen- or potassium-based fertilisers (simple or compound fertilisers).
4.4 Chemical installations for the production of basic plant health products and of biocides.

4.5 Installations using a chemical or biological process for the production of basic pharmaceutical products.
4.6 Chemical installations for the production of explosives.

5. Waste management
Without prejudice to Article 11 of Directive 2006/12/EC or Article 3 of Council Directive 91/689/EEC of 12 December 1991 on hazardous waste:
5.1 Installations for the disposal or recovery of hazardous waste as defined in the list referred to in Article 1(4) of Directive 91/689/EEC, as defined in Annexes IIA and IIB (operations R1, R5, R6, R8, and R9) to Directive 2006/12/EC and in Council Directive 75/439/EEC of 16 June 1975 on the shipment of waste oils, with a capacity exceeding 10 tonnes per day.
5.2 Installations for the incineration of municipal waste (household waste and similar commercial, industrial and institutional wastes) with a capacity exceeding 3 tonnes per hour.
5.3 Installations for the disposal of non-hazardous waste as defined in Annex IIA to Directive 2006/12/EC under headings D8 and D9, with a capacity exceeding 50 tonnes per day.
5.4 Landfills receiving more than 10 tonnes per day or with a total capacity exceeding 25,000 tonnes, excluding landfills of inert waste.

6. Other activities
6.1 Industrial plants for the production of:
(a) pulp from timber or other fibrous materials;
(b) paper and cardboard with a production capacity exceeding 20 tonnes per day.
6.2 Plants for the pre-treatment (operations such as washing, bleaching, mercerisation) or dyeing of fibres or textiles where the treatment capacity exceeds 10 tonnes per day.
6.3 Plants for the tanning of hides and skins where the treatment capacity exceeds 12 tonnes of finished products per day.
6.4 (a) Slaughterhouses with a carcase production capacity greater than 50 tonnes per day.
(b) Treatment and processing intended for the production of food products from:
- animal raw materials (other than milk) with a finished product production capacity greater than 75 tonnes per day,
- vegetable raw materials with a finished product production capacity greater than 300 tonnes per day (average value on a quarterly basis).
(c) Treatment and processing of milk, the quantity of milk received being greater than 200 tonnes per day (average value on an annual basis.
6.5 Installations for the disposal or recycling of animal carcases and animal waste with a treatment capacity exceeding 10 tonnes per day.
6.6 Installations for the intensive rearing of poultry or pigs with more than:
(a) 40,000 places for poultry;
(b) 2,000 places for production pigs (over 30 kg); or
(c) 750 places for sows.
6.7 Installations for the surface treatment of substances, objects or products using organic solvents, in particular for dressing, printing, coating, degreasing, waterproofing, sizing, painting, clearing or impregnating, with a consumption capacity of more than 150 kg per hour or more than 200 tonnes per year.
6.8 Installations for the production of carbon (hard-burnt coal) or electrographite by means of incineration or graphitisation.

APPENDIX 2.2

REGULATED SUBSTANCES

Schedule 1, Part 1, paras 6(3), 7 and 8 of *the Environmental Permitting (England and Wales) Regulations 2007* (SI 2007/3538) list the substances regulated for releases to air, water and land under the Regulations. Schedule 1, Part 2, paras 12, 13 & 14 to *the Pollution Prevention and Control Regulations* for Scotland (SSI 2000/323) and for Northern Ireland (SR 2003/46) list the same substances for regulation under PPC.

Releases into the Air
Oxides of sulphur & other sulphur compounds
Oxides of nitrogen & other nitrogen compounds
Oxides of carbon
Organic compounds & partial oxidation products
Metals, metalloids & their compounds
Asbestos (suspended particulate matter & fibres), glass fibres & mineral fibres
Halogens & their compounds
Phosphorus & its compounds
Particulate matter

Releases into Water

Substance	*Amount greater than the background quantity in any period of 12 months in Grammes)*
Mercury & its compounds	200 (expressed as metal)
Cadmium & its compounds	1000 (expressed as metal)
All isomers of hexachlorocyclohexane	20
All isomers of DDT	5
Pentachlorophenol & its compounds	350 (expressed as PCP)
Hexachlorobenzene	5
Hexachlorobutadiene	20
Aldrin	2
Dieldrin	2
Endrin	1
Polychlorinated biphenyls	1
Dichlorvos	0.2
1,2-Dichloroethane	2000
All isomers of trichlorobenzene	75
Atrazine	350*
Simazine	350*
Tributyltin compounds	4 (expressed as TBT)
Triphenyltin compounds	4 (expressed as TPT)
Trifluralin	20
Fenitrothion	2
Azinphos-methyl	2
Malathion	2
Endosulfan	0.5

*Where both Atrazine and Simazine are released, the figure for both substances in aggregate is 350 grammes.

Releases to Land
Alkali metals and their oxides and alkaline earth metals and their oxides.
Organic solvents
Azides
Halogens & their covalent compounds
Metal carbonyls
Organo-metallic compounds
Oxidising agents
Polychlorinated dibenzofuran & any congener thereof
Polychlorinated dibenzo-p-dioxin & any congener thereof
Polyhalogenated biphenyls, terphenyls & naphthalenes
Phosphorus

Pesticides, i.e, any chemical substance or preparation prepared or used for destroying any pest, including those used for protecting plants or wood or other plant products from harmful organisms; regulating the growth of plants; giving protection against harmful creatures; rendering such creatures harmless; controlling organisms with harmful or unwanted effects on water systems, buildings or other structures, or on manufactured products; or protecting animals against ectoparasites.

APPENDIX 2.3

IPPC Guidance

EU GUIDANCE - BREF NOTES (as at 25 November 2008)

These are drafted by the European IPPC Bureau, which has been established to ensure full exchange of information between Member States on BAT. The following BREFs are available and can be accessed on http://eippcb.jrc.es

(a) Formally adopted

Pulp and paper manufacture (12.01)	*Work started on review 01.07*
Iron and steel production (12.01)	*Working draft revision available 02.08*

Cement and lime production (12.01)	*Working draft revision available 09.07*
Cooling systems (12.01)	
Chlor-Alkali manufacture (12.01)	*Review due 2008*
Ferrous metal processing (12.01)	*Review due 2007*
Non-ferrous metal processes (12.01)	*Working draft revision available 11.08*
Glass manufacture (12.01)	*Working draft revision available 02.08*
Tanning of hides and skins (02.03)	*Work started on review 10.07*
Textile processing (07.03)	
Monitoring systems (07.03)	
Refineries (02.03)	Review due 2008
Large volume organic chemicals (02.03)	*Review due 2008*
Smitheries and foundries (05.05)	
Intensive livestock farming (07.03)	*Review due 2008*
Emissions from storage of bulk or dangerous materials (07.06)	
Common wastewater and waste gas treatment and management systems in the chemical sector (02.03)	*Work started on review 09.08*
Economic and cross media issues under IPPC (07.06)	
Large combustion plant (07.06)	
Large volume inorganic chemicals – ammonia, acids and fertilisers (08.07)	
Large volume inorganic chemicas – solids and others (08.07)	
Slaughterhouses and animal by-products (05.05)	
Food, drink and milk processes (08.06)	
Ceramics (08.07)	
Surface treatment of metals (08.06)	
Surface treatments using solvents (08.07)	
Waste incineration (08.06)	
Waste treatments (08.06)	
Speciality inorganic chemicals (08.07)	
Organic fine chemicals (08.06)	
Polymers (08.07)	

(b) Finalised

Management of tailings and waste-rock in mining activities (07.04)
Energy efficiency (06.08)

ENVIRONMENTAL PERMITTING – ENGLAND AND WALES

Guidance published by Defra, 2008 – www.defra.gov.uk;

Environmental Permitting – General Guidance Manual (A1), Defra.

Guidance on IPPC Directive - Part A1 installations and mobile plant (and includes guidance on the titanium dioxide directive)
Guidance on End of Life Vehicles
Guidance on Large Combustion Plant Directive
Guidance on Solvent Emissions Directive
Guidance on WEEE Directive
Guidance on Framework Directive on Waste
Guidance on Waste Incineration Directive
Guidance on Landfill Directive

Secretary of State's Guidance: General Guidance Manual on Policy and Procedures for A2 and B Installations, together with Guidance on the Directives on Solvents Emissions, Paints, petrol vapour recovery and asbestos.

POLLUTION PREVENTION AND CONTROL – SCOTLAND AND NORTHERN IRELAND

www.sepa.org.uk/regulation/ppcregs.
www.ni-environment.gov.uk

Part A Practical Guide, Issue 2, SEPA
Part B Practical Guide

Low Impact Installations (Criteria for determination under PPC), SEPA, 2004
TG 03 Determining Substantial Change

Integrated Pollution Prevention and Control – A Practical Guide, Department of Environment, NI, 2003

HORIZONTAL GUIDANCE NOTES

Also visit http://www.netregs.gov.uk which provides information and guidance on environmental legislation throughout the UK.
These guidance notes are UK specific.

H1 IPPC: Environmental assessment and appraisal of BAT (July 2003)
H2 IPPC: Energy efficiency (draft, v.3 February 2002)
H3 Noise Guidance: Part I, Regulation & permitting (2002); Part II, Noise assessment and control (2002)
H4 Guidance for Odour (Part A1 activities only) (draft, 2002). Part 1 details tools available to assess the environmental impact of odour, information on human response, and information needed for completing IPPC application. Part 2 details odour impact assessment technologies, collection of samples, control and abatement techniques.
H7 (E & W only) Guidance on the protection of land under the PPC regime: application site report and site protection and monitoring programme (August 2003)
H8 Guidance on the protection of land under the PPC regime: site surrender report (draft, June 2004)

IPPC ACTIVITY SPECIFIC GUIDANCE

These guidance notes are UK specific; for England and Wales, they carry the prefix "EPR" and in Scotland and Northern Ireland "S". They can be accessed, together with further guidance, via the websites of the environment agencies

0.01 IPPC General sector guidance (2001)

1.01 Combustion (v 2.03 July 2005)
1.02 Carbonaceous material, mineral oil, gas, coke & coal (draft, July 2005)

2.01 Guidance for the production of coke, iron & steel (July 2004)
S2.02 Guidance for Non ferrous metals sector (draft 2002)
2.03 Technical guidance for non-ferrous metals and the production of carbon and graphite (July 2004)
2.04 Guidance for the hot rolling of ferrous metals & associated activities sector (July 2004)
2.07 Guidance for the surface treatment of metals and plastics by electrolytic and chemical processes (October 2004)

3.01 Cement and lime sector (May 2001)
3.03 Glass manufacturing sector (draft, October 2001)
S3.04 Mineral Fibres (in progress)

4.01 Large volume organic chemicals sector (April 2003)
4.02 Guidance for the speciality organic chemicals sector (April 2003)
4.03 Inorganic chemicals sector (draft June 2004)

5.01 Guidance for the incineration of waste and fuel manufactured from or including waste (July 2004)
5.02 Landfill (EPR)
5.03 Guidance for the treatment of landfill leachate (Part I & Part II) (February 2007)
5.06 Guidance for the recovery and disposal of hazardous and non-hazardous waste (February 2005)
5.07 Clinical waste (EPR)

6.01 Technical guidance for the pulp and paper sector (November 2000)
6.05 Technical guidance for the textiles sector (July 2002)
6.08 Guidance for the tanning of hides sector (May 2002)
6.09 Intensive farming (EPR)
6.10 Guidance for food and drink sector (modified October 2003)
6.11 Guidance for the poultry processing sector (modified October 2003)
6.12 Guidance for the red meat processing (cattle, pigs & sheep) sector (modified August 2003)
6.13 Guidance for the dairy and milk processing sector (modified October 2003)

IPPC SECTORAL GUIDANCE (A2) (E & W)

Issued under reg. 64 of EPR by the Secretary of State and constitutes statutory guidance on BAT for LA-IPPC

- SG1 Particleboard, oriented strand board and dry process fibreboard sector (September 06)
- SG2 Glass manufacturing activities with melting capacity of more than 20 tonnes per day (revised October 06)
- SG3 A2 ferrous foundries sector (revised January 06)
- SG4 A2 activities in the non-ferrous metals sector (revised January 06)
- SG5 A2 galvanising sector (revised September 06)
- SG6 A2 surface treatment using organic solvents sector (revised March 08)
- SG7 A2 ceramics sector including heavy clay, refractories, calcining clay and whitewear (revised September 07);
- SG8 A2 rendering sector (E) (October 04; revised draft June 08, E & W)
- SG9 A2 roadstone coating, mineral and other processes that burn recovered fuel oil (April 05)
- SG10 A2 animal carcass incineration with capacity of less than 1 tonne per hour (July 2005)

APPENDIX 2.4A

LOCAL AIR POLLUTION PREVENTION AND CONTROL

Secretary of State's Guidance Notes

Issued by the Secretary of State under reg. 64 of the Environmental Permitting Regulations 2007, they constitute statutory guidance on BAT for LA-PPC installations.

PG Notes are published by The Stationery Office for Defra and the Welsh Office.

- PG 1/1(04) Waste oil and recovered oil burners less than 0.4 MW (Oct 04)
- PG 1/2(05) Waste oil or recovered oil burners, 0 4-3 MW net rated thermal input *[status under discussion, see AQ 17(06)]*
- PG 1/3(95) Boilers and furnaces, 20-50 MW net rated thermal input (as amended by AQ 23(04)
- PG 1/4(95) Gas turbines, 20-50 MW net rated thermal input (as amended by AQ 24(04)
- PG 1/5(95) Compression ignition engines, 20-50 MW net rated thermal input *[no update; no processes'; refer queries to LAU]*
- PG 1/12(04) Combustion of fuel manufactured from or comprised of solid waste in appliances (Oct 04)

- PG 1/13(04) Storage, unloading and loading petrol at terminals (Oct 04)
- PG 1/14(06) Unloading of petrol into storage at petrol stations (E, S, W) (Oct 06)
- PG 1/15(04) Odorising natural gas and liquefied petroleum gas (Oct 04)

- PG 2/01(04) Furnaces for the extraction of non-ferrous metal from scrap (Oct 04)
- PG 2/02(04) Hot dip galvanising processes (Oct 04)
- PG 2/03(04) Electrical, crucible and reverberatory furnaces (Oct 04)
- PG 2/04(04) Iron, steel and non-ferrous metal foundry processes (Oct 04)
- PG 2/05(04) Hot and cold blast cupolas & rotary furnaces (Oct 04)
- PG 2/06a(04) Processes for melting & producing aluminium & its alloys (Oct 04)
- PG 2/06b(04) Processes for melting & producing magnesium & its alloys (Oct 04)
- PG 2/07(04) Zinc and zinc alloy processes (Oct 04)
- PG 2/08(04) Copper and copper alloy processes (Oct 04)
- PG 2/09(04) Metal decontamination processes (Oct 04)

- PG 3/01(04) Blending, packing, loading and use of bulk cement (June 04)
- PG 3/02(04) Manufacture of heavy clay goods and refractory goods (June 04)
- PG 3/04(04) Lead glass, glass frit and enamel frit manufacturing processes (June 04)
- PG 3/05(04) Coal, coke, coal product and petroleum coke processes (June 04)
- PG 3/06(04) Polishing or etching of glass or glass products using hydrofluoric acid (June 04)
- PG 3/07(04) Exfoliation of vermiculite and expansion of perlite (June 04)
- PG 3/08(04) Quarry processes (June 04)
- PG 3/12(04) Plaster processes (June 04)
- PG 3/13(95) Asbestos processes; see also AQ 3(96) & 15(04)
- PG 3/14(04) Lime processes (June 04)
- PG 3/15a(04) Roadstone coating processes (June 04)
- PG 3/15b(04) Mineral drying and cooling (June 04, amended by AQ 6(07))
- PG 3/16(04) Mobile crushing and screening processes (June 04)
- PG 3/17(04) China and ball clay processes including the spray drying of ceramics (June 04)

- PG 4/02(05) Fibre re-enforced plastics (July 05)

- PG 5/2(04) Crematoria (Oct 04) – see also statutory guidance AQ 1/05, 12/05
- PG 5/3(04) Animal carcase incineration (Oct 04)

- PG 6/02(04) Manufacture of timber and wood based products (June 04)
- PG 6/03(04) Chemical treatment of timber and wood based products (March 04)
- PG 6/05(05) Maggot breeding processes
- PG 6/07(04) Printing and coating of metal packaging (March 04)
- PG 6/08(04) Textile and fabric coating and finishing (March 04)
- PG 6/09(04) Manufacture of coating powder (Oct 04)
- PG 6/12(05) Sausage casings
- PG 6/13(04) Coil coating processes (March 04)
- PG 6/14(04) Film coating processes (March 04)
- PG 6/15(04) Coating in drum manufacturing & reconditioning (March 04)
- PG 6/16(04) Printing (March 04); see also AQ 21(04)
- PG 6/17(04) Printing of flexible packaging (March 04)
- PG 6/18(04) Paper coating processes (March 04)
- PG 6/19(05) Fish meal processing
- PG 6/20(04) Paint application in vehicle manufacturing (March 04)
- PG 6/21(96) Hide and skin processes *[revised draft 10/04; final due spring/summer 2007]*
- PG 6/22(04) Leather finishing (March 04)
- PG 6/23(04) Coating of metal and plastic (March 04)
- PG 6/24a(05) Wet pet food manufacture
- PG 6/24b Dry pet food manufacturing processes *[draft 10/04]*
- PG 6/25(04) Vegetable oil extraction and fat and oil refining (March 04)
- PG 6/26(05) Animal feed compound
- PG 6/27(05) Vegetable matter drying processes (July 05)
- PG 6/28(04) Rubber (March 04)
- PG 6/29(04) Di-isocyanate processes (June 04)
- PG 6/30(06) Mushroom substrate manufacture (Jan 06)
- PG 6/31(04) Powder coating, including sheradising & vitreous enamelling (Oct 04)
- PG 6/32(04) Adhesive coating including footwear manufacturing (March 04)
- PG 6/33(04) Wood coating (March 04) – amended by AQ25(04)
- PG 6/34a(06) Original coating of road vehicles & trailers (E, S, W) (Oct 06)
- PG 6/34b(06) Respraying of road vehicles (E, S, W) (Oct 06)
- PG 6/36(06) Tobacco processes (Jan 06)
- PG 6/40(04) Coating and recoating of aircraft and aircraft components (March 04)
- PG 6/41(04) Coating and recoating of rail vehicles (March 04)
- PG 6/42(04) Bitumen and tar processes (Oct 04)
- PG 6/43(04) Finishing of pharmaceutical products (March 04)
- PG 6/44(04) Manufacture of coating materials (March 04)
- PG 6/45(04) Surface cleaning (March 04)
- PG 6/46(04) Dry cleaning (August 04)

APPENDIX 2.4B

LOCAL AIR POLLUTION PREVENTION AND CONTROL

Additional Guidance Notes

Additional Guidance Notes are published by the Department for Environment, Food and Rural Affairs and the Welsh Assembly Government. They provide guidance and clarification for local authorities on local industrial pollution control. Some AQ notes have now been incorporated in the updated General Guidance Manual for local authority regulators and thus are not listed here.

AQ Notes can be downloaded from:
http://www.defra.gov.uk/environment/ppc/localauth/pubs/guidance/notes/aqnotes/index.htm

2008

AQ 1(08)	Categories of process regulated by different local authorities (by PG & SG note) [to facilitate contacts between LAs]
AQ 2(08)	Dry cleaners – whether pump or rake out
AQ 3(08)	HAZRED guidance
AQ 4(08)	Defra / TSA drycleaning spreadsheet
AQ 5(08)	Petrol vapour recovery at service stations: explanatory notes on the use of orifice vent devices, pressure vacuum release valves and applications for Stage II
AQ 6(08)	Regulations of biodiesel production

2007

AQ 1(07)	Fuel switching – new advice; see also AQ 5 & 6(06)
AQ 2(07)	Sulphur content of liquid fuels (repeal and replacement of 2000 Regulations by SI 2007/79); references to 2000 Regulations in Defra/WAG publications to be taken as references to the 2007 Regulations
AQ 4(07)	Waste management licensing exemptions: interface with LAPPC & statutory nuisance
AQ 8(07)	Review of waste management licensing exemptions
AQ 10(07)	Crematoria: burden sharing update
AQ 12(07)	Sulphur Content of Liquid Fuels (E&W) Regulations 2007 ("pragmatic" enforcement of new regs where residual stocks of gas oil exceeding 0.1% sulphur) are being used up
AQ 14(07)	Dry cleaning – SLEAT code of practice
AQ 17(07)	roadstone coating paint (regulation of waste oil)
AQ 18(07)	Paints directive and vehicle refinish installations
AQ 19(07)	Cremation standards in the event of mass fatalities
AQ 20(07)	Zinc and ammonia releases from galvanizing installations
AQ 21(07)	Enforcing compliance with the Solvent Emissions Directive

2006

AQ 01(06)	Amendment to PG 6/35(96) – metals and other thermal spraying processes
AQ 04(06)	Roadstone coating plant: use of waste oil (in most cases plant which burn recovered fuel oil and clean fuel oil as waste fuels must comply with waste incineration directive [and with hazardous waste regulations])
AQ 08(06)	The Buncefield explosion
AQ 09(06)	Crematoria – change in calculation for burden sharing
AQ 10(06)	PG 6/46(04) – spillage trays & bunds at dry cleaning installations
AQ 17(06)	Small waste oil burners and the waste incineration Directive (updates AQ 17(05))
AQ 19(06)	Amendment to PG 5/2(04), cremation temperature
AQ 20(06)	Contact details of Defra LA industrial pollution control policy team (amended August 2007)
AQ 21(06)	Chimney height calculations

2005

AQ 1(05)	[statutory guidance on] Control of mercury emissions from crematoria – amends PG 5/2(04)
AQ 6(05)	A2 incineration activities; used cooking oil [if contaminated with animal matter will need a permit]
AQ 7(05)	A2 installations using waste oil (incl. roadstone coating) – permit applications required from operators who wish to continue to burn waste oil after 28 December 2005
AQ 12(05)	PG 5/2(04) well maintained cremators
AQ 16(05)	Calculation of solvent consumption for dry cleaners
AQ 24(05)	Crematoria – burden sharing

2004

AQ 3(04)	Environmental management systems (for LAPC should be in place)
AQ 4(04)	Classification of animal feed manufacturing (as PPC activities)
AQ10(04)	Addendum to SG7 – guidance for the A2 ceremics sector (makes a number of amendments and clarifications to SG7 (E&W)
AQ 15(04)	Amendment to PG 3/13(95): asbestos processes
AQ 16(04)	Performance standards for enquiries
AQ 21(04)	Corrections to PG 6/16(04) (printing processes) (E/W)
AQ 23(04)	Amendments of PG 1/3(95), boilers & furnaces, 20-50 MW net rated thermal input
AQ 24(04)	Amendments of PG1/4(95), gas turbines, 20-50 MW net rated thermal input
AQ 25(04)	Amends PG 6/33(04) date by which certain SED installations should comply with Reduction Scheme to 31.10.05 (not 2007)
AQ 30(04)	SED determination of compliance with reduction scheme
AQ 32(04)	Exemption from LAPC/LAPPC for small petrol stations which unload into stationary storage tanks $100m^3$-$500m^3$ in any 12 month period

2003

AQ 3(03)	Additional guidance on animal rendering processes (E/W)
AQ 7(03)	Link authorities – groupings of LAs for purpose of inter-authority networking on individual LAPC/LAPPC sectors (E/W) www.defra.gov.uk/environment/airquality/lapc/default.htm

For details of earlier AQ notes, please consult previous editions of the Pollution Control Handbook

APPENDIX 3.1

THE RINGELMANN CHART

The method of visual assessment of smoke emission by comparison of the darkness of the smoke with the standard shades of grey on a chart placed in a suitable position was devised by Professor Ringelmann of Paris towards the end of the 19th century. Professor Ringelmann obtained the shades of grey by cross-hatching in black on a white background so that a known percentage of the white was obscured.

The British Standard Ringelmann Chart (BS 2742C) is printed for the BSI so that the shades obtained in use are both consistent and near the average of those to which users were accustomed when using the previously available commercially-printed Ringelmann Charts, including those issued by the US Bureau of Mines.

The Chart consists of a cardboard sheet 581 mm x 127 mm on which are printed five 101 mm squares, four of which are cross-hatched by 20 horizontal and 20 vertical lines so that in use the cross-hatched black lines merge into the white background and produce for each shade, apparently, a uniform grey.

The number of shades – the Ringelmann numbers – range from 0 to 4, each shade increasing by comparison with the previous number by 20% obscuration so that

Ringelmann 1 = 20% obscuration

Ringelmann 2 = 40% obscuration

Ringelmann 3 = 60% obscuration

Ringelmann 4 = 80% obscuration

"Dark smoke" is smoke which is as dark or darker than shade 2 on the Chart; "black smoke" is smoke which is as dark or darker than shade 4 on the Chart.

The Use of the Ringelmann Chart

The use of the Ringelmann Chart is described in BS 2742:1969.

The Ringelmann Chart should be firmly mounted onto a backing holder – aluminium is suitable – similar to that described in BS 2742. Protective coverings must not be applied to the Chart in use and nothing should impair the luminance of its working surface.

It should be used under daylight conditions and held in a vertical plane in line between the observer and the chimney top, at a distance of more than 15 metres to ensure that cross-hatched lines merge into shades of grey.

Where possible the general illumination of the sky should be uniform, but if the sun is shining or the sky bright on one side, the British Standard recommends that the bright source of illumination should be approximately at right angles to the line of vision and not in front of or behind the observer. Under hazy conditions observations should not be taken at extreme distance as there will be a tendency for the readings to be low. The angle of view of the Chart and smoke should be as low as possible; observations at a steep angle should be avoided.

The darkness of the smoke at the chimney terminal should be compared with the Chart; the Ringelmann number which most closely matches the darkness of the smoke, and the time and duration of the emission, should be noted. In favourable conditions it is possible to estimate smoke darkness to the nearest quarter Ringelmann.

Ringelmann Smoke Chart

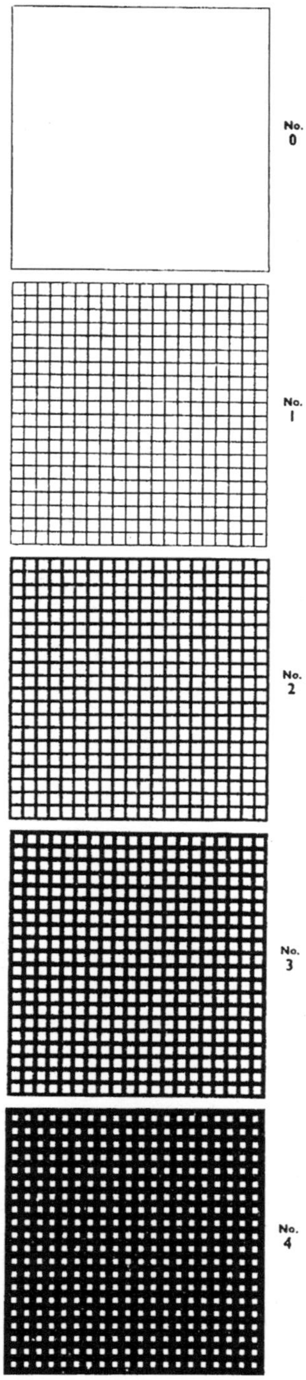

APPENDIX 3.2

THE CLEAN AIR ACT (EMISSION OF GRIT AND DUST FROM FURNACES) REGULATIONS 1971

Schedule 1 – Furnaces Rated by Heat Output

This Schedule applies to boilers and also indirect heating appliances in which the material being heated is a gas or a liquid (e.g. an air heater); where an indirect heating appliance falls also within the provisions of Schedule 2, it should be treated as a Schedule 1 furnace – see page 299.

Schedule 2 – Furnaces Rated by Heat Input

This schedule applies to indirect heating appliances and to furnaces in which combustion gases are in contact with the material being heated, but that material does not in itself contribute to the grit and dust in the combustion gases – see page 299.

The Regulations limit the grit in the sample of grit and dust to the following proportions:

- **Schedule 1 furnaces:** *Grit not to exceed 33%* where maximum continuous rating does not exceed 16,800 lbs/hr of steam or 16,800,000 Btu/hr (4.92 MW);

  Grit not to exceed 20% in any other case.

- **Schedule 2 furnaces:** *Grit not to exceed 33%* where designed heat input of the furnace does not exceed 25,000,000 Btu/hr (7.32 MW);

  Grit not to exceed 20% in any other case.

Notes:

1. Multiflue chimneys are defined and for the purpose of the Regulations each flue is to be taken as a separate chimney for the purpose of calculating acceptable emissions.
2. Where a chimney serves a furnace to which the Regulations apply and also a furnace to which the Regulations do not apply, the emissions from the latter shall be disregarded.
3. The prescribed limits apply to any period of standard operation of the furnace, including periods during which it is at or close to the loading to which it is subject for the greater part of its working time, or at any higher loading to which it is regularly subject for a limited time (whether or not this exceeds the furnace maximum combustion rating or designed heat input).

Schedule 1 Max. continuous rating in lbs of steam/hr (from & at 100°C (212°F) in thousands BTU/hr)		Max. permitted quantities of grit and dust per hour Furnaces burning solid matter		liquid matter	
000 Btu/hr	MW	lbs	kg	lbs	kg
825	0.24	1.10	0.55	0.25	0.11
1,000	0.29	1.33	0.60	0.28	0.12
2,000	0.58	2.67	1.21	0.56	0.25
3,000	0.87	4.00	1.81	0.84	0.38
4,000	1.17	5.33	2.41	1.12	0.50
5,000	1.46	6.67	3.02	1.4	0.63
7,000	2.19	8.50	3.85	2.1	0.95
10,000	2.93	10.00	4.53	2.8	1.27
15,000	4.39	13.33	6.04	4.2	1.90
20,000	5.86	16.67	7.56	5.6	2.54
25,000	7.32	20.0	9.07	7.0	3.17
30,000	8.79	23.4	10.61	8.4	3.81
40,000	11.72	30	13.60	11.2	5.08
50,000	14.65	37	16.78	12.5	5.66
100,000	29.30	66	29.93	18	8.16
150,000	43.96	94	40.97	24	10.88
200,000	58.61	122	55.37	29	13.15
250,000	73.42	149	67.58	36	16.32
300,000	87.92	172	78.01	41	18.59
350,000	102.57	195	88.45	45	20.41
400,000	117.22	217	98.42	50	22.67
450,000	131.88	239	108.40	54.5	24.72
475,000	139.20	250	113.39	57	25.85
1 Watt = 3.41214 Btu/hr					

Schedule 2		Max. permitted quantities of dust emissions per hour Furnaces burning solid matter		liquid matter	
in millions BTU/hr)	MW	lbs	kg	lbs	kg
1.25	0.36	1.1	0.49	0.28	0.12
2.5	0.73	2.1	0.95	0.55	0.24
5.0	1.46	4.3	1.95	1.1	0.49
7.5	2.19	6.8	3.08	1.7	0.77
10	2.93	7.6	3.44	2.2	0.99
15	4.39	9.7	4.39	3.3	1.49
20	5.86	11.9	5.39	4.4	1.99
25	7.32	14.1	6.40	5.5	2.49
30	8.79	16.3	7.39	6.6	2.99
35	10.25	18.4	8.34	7.7	3.49
40	11.72	20.6	9.34	8.8	3.99
45	13.18	22.8	10.34	9.8	4.44
50	14.65	25	11.33	10.9	4.94
100	29.30	45	20.51	16	7.25
200	58.61	90	40.82	26	11.79
300	87.92	132	59.87	35	15.87
400	117.22	175	79.37	44	19.95
500	146.53	218	98.88	54	24.49
575	168.51	250	113.39	57	25.85
Note: typically, a large power station is 2000-4000 MW					

APPENDIX 3.3

CALCULATION OF CHIMNEY HEIGHT

This appendix supplements the details given in chapter 3, 3.19.2 relating to chimney height, ss.14-16 of the *Clean Air Act 1993*, and also has relevance to the requirements for BATNEEC in s.7(2) of the *Environmental Protection Act 1990*.

Chimney heights offer a means of local control of pollutants discharged to the atmosphere from combustion and incineration plant and from industrial processes. They may be used to control the deposition of pollutants to the ground or their ambient concentrations over either long or short time scales, and for a variety of purposes including health effects and nuisance (due to odour for example). There are often additional controls on the scale of discharges from particular processes, both because of the requirements of BATNEEC to minimise discharge where practically possible and also to limit polluting discharges where the accumulation of a multiplicity of discharges may cause problems on larger scales, for example over the background level of pollutants in urban areas or where the accumulation of pollutants on continental or global scales is important. However, even when there are high levels of abatement, there is likely to be a residual discharge which must be effectively dispersed by means of an adequate chimney height.

Chimneys act as a means of control over local pollution levels by raising the polluting source to a sufficient height above the ground for the initial dispersion of the discharge plume, where pollutant concentrations are high, to occur away from the ground where most pollution problems occur. The chimney height is arranged so that the eventual contact of the plume with the ground results in pollutant concentrations at the ground within some prescribed limit value. This process is only effective within distances of about 50-100 chimney heights, much beyond which the effects of discharge height are no longer distinguishable. The highest pollutant concentrations usually occur within this range.

The calculation of effective chimney heights is thus a matter of great practical importance and there are a variety of methods of doing this, either by direct dispersion modelling or by using one of a number of guides for this purpose. For large plant, for example that defined as Part A processes (regulated for Integrated Pollution Control), it is common for individual dispersion modelling studies to be requested as a part of the authorisation process. Direct dispersion modelling may also be used for calculating chimney heights for smaller plant where there are complex problems due to the character of the dispersion or the nature of the discharge. However this is not usually practicable and there are several guides which allow chimney heights to be calculated more quickly and easily. These are discussed below. The majority of direct dispersion modelling uses computer models of varying sophistication. The USEPA models are the most commonly used (in the UK, along with the NRPB models), though they have distinct limitations in dealing with the more complex dispersion problems. Some complex dispersion problems also use small scale wind tunnel modelling to determine chimney heights.

For the great majority of smaller polluting discharges, chimney heights are determined using the published guides for this purpose. For many years chimney heights for conventional combustion plant (mainly heating and steam raising plant running on coal, fuel oil or gas) have been calculated using the 3rd Edition of the *Clean Air Act Memorandum "Chimney Height"* which is still recommended for this purpose. More recently there has been a need to deal with newer types of combustion plant (combined cycle for example) and with the requirements for adequate discharge stack heights for the wide variety of smaller process plant classified as Part B processes under the *Environmental Protection Act*. This has been met with the HMIP Guidance Note D1, *Guidelines for Discharge Stack Heights for Polluting Emissions* (HMSO, 1993), which provides a systematic procedure for dealing with almost any type of polluting discharge. There is also specific advice on minimum chimney heights for discharge stacks in the Process Guidance Notes issued by the Defra Local Authority Unit dealing with the requirements for authorisation of various Part B processes. These also contain maximum pollutant emission limits for the processes. Guidance Note D1 is also used occasionally for setting chimney heights for Part A processes, but this requires specific approval in individual cases.

More information on the *Memorandum of Chimney Heights* and on Guidance Note D1 can be found in the background report *(Background Report to the HMIP Guidelines on Discharge Stack Heights for Polluting Emissions*, D.J. Hall, V. Kukadia, Building Research Establishment Report No. CR200/95); specific advice on calculating chimney heights for odorous discharges can be found in Clean Air, Vol. 24, No. 2, pp74-92, and in AEA Technology Report No. AEA/CS/REMA-038; on dealing with chimneys with low velocity discharges in Clean Air, Vol. 25, No. 3, pp128-339; and on chimney heights for emissions of volatile organic compounds in Clean Air, Vol. 25, No. 4 (NSCA).

(This appendix prepared by D.J. Hall at the Building Research Establishment, 1995)

Addendum – 2002 Edition

Sulphur Dioxide Emissions from Small Boilers – Supplementary Assistance on Stack Height Determination

The Air Quality Strategy (2000) recognises that despite national measures to control SO2 emissions from combustion plant, there may still be exceedances of the 15-minute mean objective in very local areas in the immediate vicinity of small combustion plant less than 20 MW. Currently chimney height calculation cannot be carried out easily for many of these boilers due to a number of limitations in the available methods.

This supplementary assistance has been produced by Stanger Science and Environment at the request of DEFRA and provides a method for estimating the minimum permissible chimney height for small boilers emitting SO_2. This simple screening tool is in the form of an EXCEL spreadsheet, and is also described in Clean Air, vol. 31, no. 3, NSCA, 2001.

APPENDIX 3.4

CLEAN AIR ACT 1993: SMOKE CONTROL

Section 18 of the *Clean Air Act 1993* allows a local authority to make Smoke Control Orders. When operative, it is an offence for an occupier of premises to allow smoke to be emitted from a chimney, unless the smoke is caused by an authorised fuel or the fireplace which the chimney serves is exempt from the Order. The Department for Environment, Food and Rural Affairs (for England) is responsible for authorising fuels and exempted fireplaces, and the Devolved Administrations, as appropriate.

Authorised fuels are listed below at Appendix 3.4a, and exempted fireplaces at Appendix 3.4b.

Full details of the procedure for applying for a fuel to be authorised or for an appliance to be exempted, as well as lists of authorised fuels and exempted fireplaces can be found on www.uksmokecontrolareas.co.uk.

Appendix 3.4a: Authorised Fuels

The Smoke Control (Authorised Fuels) (England) Regulations 2008 (SI 2008/514) came into force on 6 April 2008, consolidating and revoking all the earlier Regulations. There are similar Regulations for Scotland, SSI 2008/295, which came into force on 1 October 2008, revoking earlier Regulations, and in Northern Ireland SR 2008/415, which came into force on 7 November 2008, replacing 2003 Regulations. In Wales the relevant Regulations are SI 2001/3762; W.311, (as amended by SI 2001/3996; W.327 and SI 2002/3160, W295.

While the Schedules for authorised fuels are largely the same in England and the Devolved Administrations, the lists below refer to the Schedules attached to the Regulations for England, and thus readers in Scotland, Wales and NI should check the relevant Regulations.

Anthracite, semi-anthracite, electricity, gas, low volatile steam coals are all authorised fuels, as are those listed below – Schedule 1 to the Regulations.

1. **Aimcor Excel briquettes,** manufactured by Applied Industrial Materials UK Limited at Newfield, County Durham, or manufactured by Oxbow Carbon & Minerals UK Limited, Windsor House, Cornwall Road, Harrogate, North Yorkshire, which –
(a) comprise petroleum coke (as to 60 to 75% of the total weight), low volatile coal and reactive coke (as to 20 to 25% of the total weight) and cold-setting resin binder (as to the remaining weight);
(b) were manufactured from those constituents by a process involving roll-pressing;
(c) are unmarked pillow-shaped briquettes;
(d) have an average weight of 73 grammes per briquette; and
(e) have a sulphur content not exceeding 2% of the total weight.

2. **Aimcor Pureheat briquettes,** manufactured by Applied Industrial Materials UK Limited at Immingham, North East Lincolnshire, or manufactured by Oxbow Carbon & Minerals UK Limited, Windsor House, Cornwall Road, Harrogate, North Yorkshire, which -
(a) comprise anthracite (as to approximately 60% of the total weight), petroleum coke (as to approximately 25% of the total weight) and binder (as to the remaining weight);
(b) were manufactured from those constituents by a process involving roll-pressing and heat treatment at about 250°C;
(c) are pillow-shaped briquettes with a single line indentation on one side and a double line indentation on the reverse side;
(d) have an average weight of 75 grammes per briquette; and
(e) have a sulphur content not exceeding 2% of the total weight.

3. **Ancit briquettes,** manufactured by Coal Products Limited at Immingham Briquetting Works, Immingham, North East Lincolnshire, which -
(a) comprise anthracite (as to approximately 60 to 95% of the total weight), petroleum coke (up to approximately 30% of the total weight), bituminous coal (up to approximately 15% of the total weight) and a molasses and phosphoric acid binder or an organic binder (as to the remaining weight);
(b) were manufactured from those constituents by a process involving roll-pressing and heat treatment at about 300°C;
(c) are unmarked cushion-shaped briquettes;
(d) have an average weight of 48 grammes per briquette; and
(e) have a sulphur content not exceeding 2% of the total weight.

4. **Big K Instant Lighting Fire Logs,** manufactured by Allspan BV at Macroweg 4, 5804 CL Venray, the Netherlands, which
(a) comprise slack wax (as to approximately 58 to 59% of the total weight) and hardwood sawdust (as to approximately 41-42% of the total weight);
(b) were manufactured from those constituents by a process of heat treatment and extrusion;
(c) are firelogs approximately 235 millimetres in length and 80 millimetres in depth with grooves along their faces;
(d) have an average weight of 1.1 kilogrammes per firelog; and
(e) have a sulphur content not exceeding 0.1% of the total weight.

5. **Black Diamond Gem briquettes,** manufactured by Coal Products Limited at Immingham Briquetting Works, Immingham, North East Lincolnshire, which -
(a) comprise anthracite duff (as to 20 to 30% of the total weight), petroleum coke (as to 40 to 45% of the total weight), bituminous coal (as to 12 to 22% of the total weight) and molasses and phosphoric acid binder (as to the remaining weight);
(b) were manufactured from those constituents by a process involving roll-pressing and heat treatment at about 300°C;
(c) are pillow-shaped briquettes marked with two parallel indented lines running latitudinally around the briquette;
(d) have an average weight of 160 grammes per briquette; and
(e) have a sulphur content not exceeding 1.5% of the total weight.

6. **Bord na Móna Firelogs,** manufactured by Bord na Móna Fuels Limited, Newbridge, County Kildare, Republic of Ireland, which -
(a) comprise slack wax (as to approximately 55% of the total weight) and hardwood sawdust (as to approximately 45% of the total weight);
(b) were manufactured from those constituents by a process of heat treatment and extrusion;
(c) are firelogs approximately 255 millimetres in length and 75 millimetres in diameter, with grooves along one longitudinal face;
(d) have an average weight of 1.3 kilogrammes per firelog; and
(e) have a sulphur content not exceeding 0.1% of the total weight.

7. **Bord na Móna Firepak** (also marketed as Arigna Special coal briquettes), manufactured by Bord na Móna Fuels Limited, Newbridge, County Kildare, Republic of Ireland, which –
(a) comprise anthracite (as to approximately 50% of the total weight), petroleum coke (as to approximately 20 to 40% of the total weight), bituminous coal (as to approximately 10 to 30% of the total weight) and starch based binder (as to the remaining weight);
(b) were manufactured from those constituents by a process involving roll-pressing and heat treatment;

(c) are unmarked pillow-shaped briquettes;
(d) have an average weight of 50 grammes per briquette; and
(e) have a sulphur content not exceeding 1.5% of the total weight.

8. Briteflame briquettes, manufactured by Maxibrite Limited at Mwyndy Industrial Estate, Llantrisant, Mid Glamorgan, which –
(a) comprise 10 to 15% bituminous coal, 10 to 15% petroleum coke, 70 to 80% anthracite duff and starch binder (as to the remaining weight);
(b) were manufactured from those constituents by a process involving roll-pressing and heat treatment at about 260°C;
(c) are unmarked pillow-shaped briquettes;
(d) have an average weight of 140 grammes per briquette; and
(e) have a sulphur content not exceeding 1.9 per cent sulphur on a dry basis.

9. Briteheat briquettes, manufactured by Coal Products Limited at Immingham Briquetting Works, North East Lincolnshire, which
(a) comprise anthracite (as to approximately 60 to 85% of the total weight), petroleum coke (up to approximately 30% of the total weight), bituminous coal (up to approximately 15% of the total weight) and a molasses and phosphoric acid binder (as to the remaining weight);
(b) were manufactured from those constituents by a process involving roll-pressing and heat treatment at about 300°C;
(c) are cushion-shaped briquettes with an indented line running longitudinally around the briquette;
(d) have an average weight of 30 grams per briquette, and
(e) have a sulphur content not exceeding 2% of the total weight.

10. Bryant and May Firelogs, manufactured by Swedish Match at Kostenetz, Bulgaria, which
(a) comprise paraffin wax (as to approximately 50% of the total weight), ground poplar wood (as to approximately 25% of the total weight), wheatflour (as to approximately 15% of the total weight), ignitable solids dispersed in gelled paraffin wax (as to approximately 1% of the total weight) and water, swelling agents and preservative (as to the remaining weight);
(b) were manufactured from those constituents by a process involving extrusion;
(c) have a quadrant shaped cross section with a radius of approximately 80 millimetres, a length of approximately 265 millimetres and an ignition strip along one edge;
(d) have an approximate weight of 1.15 kilogrammes per firelog; and
(e) have a sulphur content not exceeding 0.1% of the total weight.

11. Charglow briquettes, manufactured by Polchar Spol/ka z ograniczona odpowiedzialnoscia, Ulica Kuznicka 1, Police, Zachodniepomorskie, Poland, which -
(a) comprise bituminous coal char (as to approximately 45 to 95% of the total weight), anthracite (as to approximately 0 to 20% of the total weight), petroleum coke (as to approximately 0 to 20% of the total weight), bituminous coal (as to approximately 0 to 10% of the total weight) and an organic binder (as to the remaining weight);
(b) were manufactured from those constituents by a process involving roll-pressing and heat treatment at about 110°C;
(c) are unmarked pillow-shaped briquettes;
(d) have an average weight of 100 grammes per briquette; and
(e) have a sulphur content not exceeding 1.5% of the total weight.

12. Coalite manufactured by Coalite Products Limited at Bolsover, near Chesterfield, Derbyshire and at Grimethorpe, South Yorkshire using a low temperature carbonisation process.

13. Coke manufactured by –
(a) Coal Products Limited at Cwm Coking Works, Llantwit Fardre, Pontypridd, Rhondda Cynon Taff, and sold as "**Sunbrite**";
(b) Monckton Coke & Chemical Company Limited at Royston, near Barnsley, South Yorkshire, and sold as "**Sunbrite**" or "**Monckton Boiler Beans**";
(c) Corus UK Limited at Teesside Works, Redcar and sold as "**Redcar Coke Nuts (Doubles)**"; and
(d) Coal Products Limited at Cwm Coking Works, Llantwit Fardre, Pontypridd, Rhondda Cynon Taff and sold as "**Cwm Coke Doubles**".

14. Cosycoke (also marketed as **Lionheart Crusader** or **Sunbrite Plus**), manufactured by Monckton Coke & Chemical Company Limited at Royston, near Barnsley, South Yorkshire, and **Aimcor Supercoke** (also marketed as Supercoke), manufactured by M & G Fuels Limited at Hartlepool Docks, Hartlepool, which in each case -
(a) comprise sized hard coke (as to approximately 45 to 65% of the total weight) and sized petroleum coke (as to the remaining weight);
(b) were manufactured from those constituents by blending;
(c) are unmarked random shapes; and
(d) have a sulphur content not exceeding 2% of the total weight.

15. Dragonglow briquettes, manufactured by Tower Colliery Limited at Aberdare, Mid Glamorgan, South Wales, which
(a) comprise tower duff (as to approximately 95% of the total weight) and a resin-based binder (as to the remaining weight);
(b) were manufactured from those constituents by a process involving cold cure roll pressing;
(c) are unmarked pillow-shaped briquettes;
(d) have an average weight of 100 grammes per briquette; and
(e) have a sulphur content not exceeding 1% of the total weight.

16. Dragonbrite briquettes, manufactured by Tower Colliery Limited at Aberdare, Mid Glamorgan, South Wales, which
(a) comprise tower duff (as to approximately 95% of the total weight) and a resin-based binder (as to the remaining weight);
(b) were manufactured from those constituents by a process involving cold cure roll pressing;
(c) are pillow-shaped briquettes marked with the letter "T" on one side;
(d) have an average weight of 50 grammes per briquette; and
(e) have a sulphur content not exceeding 1% of the total weight.

17. Duraflame firelogs, manufactured by Paramelt BV, Costerstraat 18, PO Box 86, 1700 AB Heerhugowaard, The Netherlands, which
(a) comprise mineral-based petroleum wax (as to approximately 55% of the total weight) and ground hardwood fibre (as to approximately 45% of the total weight);
(b) were manufactured from those constituents by a process of heat treatment and extrusion;
(c) are firelogs approximately 320 millimitres in length 90 millimetres high and 85 millimetres wide;
(d) have an average weight of 1.45 killogrammes per firelog; and
(e) have a sulphur content not exceeding 0.1% of the total weight.

18. Ecoal briquettes, manufactured by Coal Products Limited at Immingham Briquetting Works, Immingham, North East Lincolnshire, which

(a) comprise anthracite fines (as to approximately 40 to 65% of the total weight), petroleum coke (as to approximately 20 to 40% of the total weight), char (as to approximately 0 to 20% of the total weight), biomass (as to approximately 5 to 20% of the total weight) and molasses and phosphoric acid as binder (as to the remaining weight);
(b) were manufactured from those constituents by a process involving roll-pressing;
(c) are unmarked hexagonal briquettes;
(d) have an average weight of 125 grams per briquette; and
(e) have a sulphur content not exceeding 2% of the total weight.

19. Ecobrite briquettes, manufactured by Arigna Fuels Limited at Arigna, Carrick-on-Shannon, County Roscommon, Republic of Ireland, which -
(a) comprise anthracite fines (as to approximately 96% of the total weight) and starch as binder (as to the remaining weight);
(b) were manufactured from those constituents by a process involving roll-pressing and heat treatment at about 250°C;
(c) are unmarked pillow-shaped briquettes in two sizes;
(d) have an average weight per briquette of 37 grammes in the case of the smaller size and 48 grammes in the case of the larger size; and
(e) have a sulphur content not exceeding 1.5% of the total weight.

20. **Extracite briquettes,** manufactured by Sophia-Jacoba Handelsgesellschaft mbH at Hückelhoven, Germany, which
(a) comprise anthracite duff (as to approximately 95.5% of the total weight) and ammonium lignosulphonate lye as binder (as to the remaining weight);
(b) were manufactured from those constituents by a process involving roll-pressing and heat treatment at about 260°C;
(c) are cushion-shaped briquettes with a silvery appearance and are marked with the letters "S" and "J";
(d) have an average weight of 40 grammes per briquette; and
(e) have a sulphur content of approximately 1.2% of the total weight.

21. **Fireglo briquettes,** manufactured by Les Combustibles de Normandie at Caen, France, and by La Société Rouennaise de Defumage at Rouen, France, which -
(a) comprise washed Welsh duffs (as to approximately 92% of the total weight) and coal pitch binder (as to the remaining weight);
(b) were manufactured from those constituents by a process involving roll-pressing and heat treatment at about 330°C;
(c) are ovoids which have three lines on one side and are smooth on the other side;
(d) have an average weight of 30 grammes per briquette; and
(e) have a sulphur content not exceeding 0.8% of the total weight.

22. **Homefire briquettes,** manufactured by Coal Products Limited at Immingham Briquetting Works, Immingham, North East Lincolnshire, which -
(a) comprise anthracite fines (as to approximately 40 to 70% of the total weight), petroleum coke (as to approximately 20 to 45% of the total weight), char (as to approximately 0 to 10% of the total weight), bituminous coal (as to approximately 5 to 30% of the total weight) and an organic binder or a molasses and phosphoric acid binder (as to the remaining weight);
(b) were manufactured from those constituents by a process involving roll-pressing;
(c) have a volatile matter content in the finished briquette of neither less than 9 nor more than 15% of the total weight on a dry basis;
(d) are unmarked hexagonal briquettes;
(e) have an average weight of 140 grammes per briquette; and
(f) have a sulphur content not exceeding 2% of the total weight.

23. **Homefire ovals,** manufactured by Coal Products Limited at Immingham Briquetting Works, Immingham, North East Lincolnshire, which
(a) comprise anthracite duff (as to approximately 57% of the total weight), petroleum coke (as to approximately 17% of the total weight), bituminous coal (as to approximately 13% of the total weight) and molasses and phosphoric acid as binder (as to the remaining weight);
(b) were manufactured from those constituents by a process involving roll-pressing and heat treatment at about 300°C;
(c) are pillow-shaped briquettes with two parallel indented lines running latitudinally around the briquette;
(d) have an average weight of 135 grammes per briquette; and
(e) have a sulphur content not exceeding 2% of the total weight.

24. **Homefire Ovals (R),** manufactured by Coal Products Limited at Immingham Briquetting Works, Immingham, North East Lincolnshire, which -
(a) comprise anthracite fines (as to approximately 50 to 75% of the total weight), petroleum coke (as to approximately 20 to 45% of the total weight), bituminous coal (as to approximately 5 to 17% of the total weight) and an organic binder (as to the remaining weight);
(b) were manufactured from those constituents by process involving roll-pressing;
(c) are pillow-shaped briquettes with two parallel indented lines running latitudinally around the briquette;
(d) have an average weight of 130 grammes per briquette; and
(e) have a sulphur content not exceeding 2% of the total weight.

25. **Island Lump and Island Nuts,** manufactured by Unocal Refinery, California, the United States of America, which
(a) comprise petroleum coke;
(b) were manufactured from the petroleum coke by a process involving heat treatment and steam injection;
(c) are unmarked random shapes;
(d) have an average weight of 80 grammes (per briquette of Island Lump) or 30 grammes (per briquette of Island Nuts); and
(e) have a sulphur content not exceeding 2 per cent of the total weight.

26. **Jewel briquettes,** manufactured by Eldon Colliery Limited at Newfield Works, Bishop Auckland, County Durham, which
(a) comprise anthracite (as to approximately 30 to 50% of the total weight), Long Beach petroleum coke (as to approximately 50 to 70% of the total weight) and a carbohydrate binder (as to the remaining weight);
(b) were manufactured from those constituents by a process involving roll-pressing and heat treatment at about 150°C ;
(c) are unmarked pillow-shaped briquettes;
(d) have an average weight of 33 grammes per briquette; and
(e) have a sulphur content not exceeding 1.5% of the total weight.

27. **La Haciends Easy Logs,** manufactured by Allspan BV at Macroweg 4, 5804 Venray, The Netherlands, which meet the conditions specified for Big K Instant Lighting Fire Logs at paragraph 4.

28. **Long Beach Lump nuts** (otherwise known as LBL nuts), manufactured by Aimcor Carbon Corporation at Long Beach, California, the United States of America, or manufactured by Oxbow Carbon & Minerals LLC, 330 Golden Shore, Suite 210, Long Beach California 90802, the United States of America, which -
(a) comprise petroleum coke (as to approximately 85 to 100% of the total weight), limestone (as to approximately 0 to 10% of the total weight) and coal tar pitch (as to the remaining weight);
(b) were manufactured from those constituents by a process involving heat treatment and steam injection;
(c) are unmarked random shapes; and
(d) have a sulphur content not exceeding 2% of the total weight.

29. **Maxibrite briquettes,** manufactured by Maxibrite Limited at Llantrisant, Rhondda Cynon Taff, which -
(a) comprise anthracite fines (as to approximately 84% of the total weight), petroleum coke (as to approximately 12% of the total weight) and starch as binder (as to the remaining weight);
(b) were manufactured from those constituents by a process involving roll-pressing and heat treatment at 250°C;
(c) are cushion-shaped briquettes marked with the letter "M";
(d) have an average weight of 35 grammes per briquette; and
(e) have a sulphur content not exceeding 2% of the total weight.

30 **Multiheat briquettes,** manufactured by Coal Products Limited at Immingham Briquetting Works, Immingham, North East Lincolnshire, which –
(a) comprise anthracite (as to approximately 60 to 80% of the total weight), petroleum coke (as to approximately 10 to 30% of the total weight) and a molasses and phosphoric acid binder (as to the remaining weight);
(b) were manufactured from those constituents by a process involving roll-pressing and heat treatment at about 300°C;
(c) are unmarked pillow-shaped briquettes;
(d) have average weights per briquette of either 55 or 80 grammes; and
(e) have a sulphur content not exceeding 2% of the total weight.

31. **Newflame briquettes,** manufactured by Maxibrite Limited at Llantrisant, Rhondda Cynon Taff, which -
(a) comprise anthracite fines (as to approximately 84% of the total weight), petroleum coke (as to approximately 12% of the total weight) and starch as binder (as to the remaining weight);
(b) were manufactured from those constituents by a process involving roll-pressing and heat treatment at about 260°C;
(c) are unmarked pillow-shaped briquettes;
(d) have an average weight of 78 grammes per briquette; and
(e) have a sulphur content not exceeding 2% of the total weight.

32. **Optima Fire Logs,** manufactured by Allspan BV at Macroweg 4, 5804 Venray, The Netherlands, which meet the conditions specified for Big K Instant Lighting Fire Logs at paragraph 4.

33. Phurnacite briquettes, manufactured by Coal Products Limited at Immingham Briquetting Works, Immingham, North East Lincolnshire, which -
(a) comprise anthracite duff (as to approximately 65 to 85% of the total weight), petroleum coke (as to approximately 20% of the total weight) and a molasses and phosphoric acid binder (as to the remaining weight);
(b) were manufactured from those constituents by a process involving roll-pressing and heat treatment at about 300°C;
(c) are ovoid-shaped briquettes with two parallel indented lines running longitudinally around the briquette;
(d) have an average weight of 40 grammes per briquette; and
(e) have a sulphur content not exceeding 2% of the total weight.

34. Pyrobloc Fire Logs, manufactured by Allspan BV at Macroweg 4, 5804 CL Venray, the Netherlands, which
(a) comprise slack wax (as to approximately 58 to 59% of the total weight) and hardwood sawdust (as to approximately 41-42% of the total weight);
(b) were manufactured from those constituents by a process of heat treatment and extrusion;
(c) are firelogs approximately 265 millimetres in length and 80 millimetres in depth with grooves along their faces;
(d) have an average weight of 1.1 kilogrammes per firelog; and
(e) have a sulphur content not exceeding 0.1% of the total weight.

35. Safelight Firelogs, manufactured by Advanced Natural Fuels Limited, at Pocklington, East Riding of Yorkshire, which -
(a) comprise woodchip (as to approximately 40 to 55% of the total weight) and Palm Wax binder (as to approximately 45 to 60% of the total weight);
(b) were manufactured from those constituents by a process involving pressing of the mixed ingredients at about 40°C to 50°C;
(c) are rectangular hard finish firelogs with two deep overlapping slots in the top surface and a single continuous slot in the base surface;
(d) have an average weight of 1.8 kilogrammes per firelog; and
(e) have a sulphur content not exceeding 2% of the total weight.

36. Sovereign briquettes, manufactured by Monckton Coke & Chemical Company Limited at Royston, near Barnsley, South Yorkshire, which -
(a) comprise anthracite (as to approximately 75% of the total weight), coal and reactive coke (as to approximately 21% of the total weight) and cold-setting resin binder (as to the remaining weight);
(b) were manufactured from those constituents by a process involving extrusion;
(c) are unmarked hexagonal briquettes;
(d) have an average weight of 130 grammes per briquette; and
(e) have a sulphur content not exceeding 2% of the total weight.

37 Stoveheat Premium briquettes, manufactured by Coal Products Limited at Immingham Briquetting Works, Immingham, North East Lincolnshire, which
(a) comprise anthracite duff (as to approximately 65 to 85% of the total weight), petroleum coke (as to approximately 20% of the total weight) and a molasses and phrophoric acid binder (as to the remaining weight);
(b) were manufactured from those constituents by a process involving roll-pressing and heat treatment at about 300°C;
(c) are unmarked ovoid-shaped briquettes;
(d) have an average weight of 40 grammes per briquette; and
(e) have a sulphur content not exceeding 2% of the total weight.

38. Supabrite Coke Doubles, manufactured by H.J. Banks and Company Limited at Inkerman Road Depot, Tow Law, County Durham, which -
(a) comprise metallurgical coke (as to approximately 40 to 60% of the total weight) and petroleum coke (as to the remaining weight);
(b) were manufactured from those constituents by a process involving blending and screening;
(c) are unmarked random shapes; and
(d) have a sulphur content not exceeding 1.95% of the total weight.

39. Supacite briquettes, manufactured by Maxibrite Limited at Llantrisant, Rhondda Cynon Taff, which -
(a) comprise anthracite fines (as to approximately 84% of the total weight), petroleum coke (as to approximately 12% of the total weight) and starch as binder (as to the remaining weight);
(b) were manufactured from those constituents by a process involving roll-pressing and heat treatment at about 240°C;
(c) are unmarked ovoids;
(d) have an average weight of 45 grammes per briquette; and
(e) have a sulphur content not exceeding 2% of the total weight.

40. Supertherm briquettes, manufactured by Coal Products Limited at Immingham Briquetting Works, Immingham, North East Lincolnshire, which -
(a) comprise a blend (in the proportion of 19:1 by weight) of anthracite and medium volatile coal (as to approximately 93% of the total weight) and cold-setting organic binder or molasses and phosphoric acid binder (as to the remaining weight);
(b) were manufactured from those constituents by a process involving roll-pressing;
(c) are unmarked ovoids;
(d) have an average weight of 160 grammes per briquette; and
(e) have a sulphur content not exceeding 1.5% of the total weight.

41. Supertherm II briquettes, manufactured by Coal Products Limited at Immingham Briquetting Works, Immingham, North East Lincolnshire, which -
(a) comprise anthracite (as to approximately 36 to 51% of the total weight), petroleum coke (as to approximately 40 to 55% of the total weight) and an organic binder or a molasses and phosphoric acid binder (as to the remaining weight);
(b) were manufactured from those constituents by a process involving roll-pressing;
(c) are unmarked ovoids;
(d) have an average weight of 140 grammes per briquette; and
(e) have a sulphur content not exceeding 2% of the total weight.

42. Taybrite briquettes (otherwise known as **Surefire briquettes**), manufactured by Coal Products Limited at Immingham Briquetting Works, Immingham, North East Lincolnshire, which -
(a) comprise anthracite (as to approximately 60 to 80% of the total weight), petroleum coke (as to approximately 10 to 30% of the total weight) and a molasses and phosphoric acid binder (as to the remaining weight);
(b) were manufactured from those constituents by a process involving roll-pressing and heat treatment at about 300°C;
(c) are pillow-shaped briquettes marked with a single indented line running longitudinally along each face, offset from its counterpart by 10 millimetres or unmarked;
(d) have an average weight of 80 grammes per briquette; and
(f) have a sulphur content not exceeding 2% of the total weight.

42A. Therma briquettes, manufactured by Maxbrite Limited at Llantrisant, Rhondda Cynon Taff, which –
(a) comprise anthracite fines (as to approximately 84% of the total weight), petroleum coke (as to approximately 12% of the total weight) and starch as binder (as to the remaining weight);
(b) were manufactured from those constituents by a process involving roll-pressing and heat treatment at 250°C;
(c) are oval/tear shaped briquettes with a line through the centre;
(d) have an average weight of 26 grams per briquette;
(e) have a sulphur content not exceeding 2% of the total weight.

43. Thermac briquettes, manufactured by Coal Products Limited at Shildon, County Durham, which -
(a) comprise anthracite (as to approximately 90% of the total weight) and cold-setting organic binder (as to the remaining weight);
(b) were manufactured from those constituents by a process involving roll-pressing;
(c) are unmarked pillow-shaped briquettes;
(d) have an average weight of 48 grammes per briquette; and
(g) have a sulphur content not exceeding 1.5% of the total weight.

44. Unicite, manufactured by DJ Davies Fuels Ltd at Blaenau Fuel Depot, Ammanford Carmarthenshire, or manufactured by Al Simson at Cabby Latch, Logie by Kirriemuir, Angus, which

(a) is a mixture of the following separate fuels
   (i) anthracite large nuts, and
   (ii) Union briquettes of the description in sub-paragraph (d) manufactured by RWE Power AG at Cologne, Germany (as to no more than approximately 40 to 42% of Unicite's total weight;
(b) has an overall sulphur content not exceeding 0.5% of the total weight;
(c) is declared to be an authorised fuel only in so far as the proportion of Union briquettes combusted in the use of Unicite does not exceed the percentage range specified in sub-paragraph (a)(ii); and
(d) in so far as it is constituted by the Union briquettes referred to in sub-paragraph (a)(ii), must consist of compressed lignite with each briquette measuring approximately 75 millimetres wise, 60 millimetres high and 55 millimetres thick.

45. ZIP Cracklelog firelogs, ZIP Crackle-log firelogs and ZIP Crackling Log firelogs, manufactured by Allspan BV, Macroweg 4, 5804 CL Venray, The Netherlands, which –
(a) comprise slack wax (as to approximately 55% of the total weight) and hardwood sawdust (as to approximately 42% of the total weight) and crackle seeds (as to approximately 3.2% of the total weight);
(b) were manufactured from those constituents by a process of heat tretment and extrusion;
(c) are firelogs approximately 235 millimetres in length and 80 millimetres in diameter, with grooves along the faces;
(d) have an average weight of 1.1 kilogrammes per firelog; and
(e) have a sulphur content not exceeding 0.1% of the total weight.

46. ZIP Firelogs, manufactured by Allspan BV, Macroweg 4, 5804 CL Venray, The Netherlands, which –
(a) comprise slack wax (as to approximately 58 to 59% of the total weight) and hardwood sawdust (as to approximately 41 to 42% of the total weight);
(b) were manufactured from those constituents by a process of heat treatment and extrusion;
(c) are firelogs approximately 265 millimetres in length and 80 millimetres in depth, with grooves along the faces;
(d) have an average weight of 1.3 kilogrammes per firelog; and
(e) have a sulphur content not exceeding 0.1% of the total weight.

47. ZIP Firelogs, manufactured by Woodflame Moerdijk BV, Appolloweg 4, Harbour No. M189A, 4782 SB Moerdijk, The Netherlands, which –
(a) comprise slack wax (as to approximately 55 to 60% of the total weight) and hardwood sawdust (as to approximately 40 to 45% of the total weight);
(b) were manufactured from those constituents by a process of heat treatment and extrusion;
(c) are firelogs approximately 255 millimetres in length and 75 millimetres in diameter, with grooves along one longitudinal face;
(d) have an average weight of 1.3 kilogrammes per firelog; and
(e) have a sulphur content not exceeding 0.1% of the total weight.

48. Zip Firelogs, manufactured by Allspan BV at Macroweg 4, 5804 Venray, The Netherlands which meet the conditions specified for Big K Instant Lighting Fire Logs at paragraph 4.

APPENDIX 3.4B: SMOKE CONTROL (EXEMPTED FIREPLACES) ORDERS

(These Orders cover England and Wales although lattely separate orders have been made in each administration. Similar Orders have been made for Scotland and Northern Ireland From 2000 separate orders are also made for England and Wales)

Class of Fireplace	*Conditions*
1970 Order (SI 1970 No. 615)	
1. Any fireplace specially designed or adapted for combustion of liquid fuel.	No fuel shall be used other than that for which the mechanical stoker was designed.
2. Any fireplace (other than a fireplace fired by pulverised fuel) constructed on or after 31st December 1956 and installed before 1st May 1970 and equipped with mechanical stokers, or adapted between those dates for use with such stokers.	
3. Any fireplace designed to burn coal (other than a fireplace fired by pulverised coal) with a heating capacity exceeding 150,000 Btu/hr constructed and installed on or after 31st December 1956 and equipped with mechanical stokers or adapted on or after that date for use with such stokers.	No fuel shall be used other than that for which the mechanical stoker was designed.
*4. The fireplace known as the Solid Fuel Ductair Unit, manufactured by Radiation Ltd.	
5. The fireplace known as the Fulgora slow combustion Stove, manufactured by Fulgora Stoves Ltd.	No fuel shall be used other than wood waste in clean condition.
*6. The fireplace known as the Housewarmer manufactured for the National Coal Board by Ideal Standard Limited and latterly by Stelrad Group Ltd.	No fuel shall be used other than selected washed coal marketed under the name Housewarm" by agreement with the National Coal Board.
*7. The fireplace known as Wood Chip Fired Air Heater, manufactured by Air Plants (Sales) Ltd.	No fuel shall be used other than clean wood waste of a size within the limits referred to in the manufacturer's instructions.
*8. The fireplace known as the Hounsell Sawdust Burning Stove, manufactured by John Hounsell (Engineers) Ltd.	No fuel shall be used other than wood waste in clean condition.
1970 No. 2 Order (SI 1970 No. 1667)	
The Fireplace known as the Triancomatic T 80 and manufactured by Trianco Ltd.	
****1971 Order (SI 1971 No. 1265)***	
The fireplace known as the Rayburn CB 34 and manufactured by Glynwed Foundries Ltd.	

* no longer in production, but where such fireplaces remain in use they remain exempt.

Class of Fireplace	*Conditions*
1972 Order (SI 1972 No. 438)	
The fireplace known as the Parkray Coalmaster and manufactured by Radiation Parkray Ltd, now T. I. Parkray Ltd.	No fuel shall be used other than selected washed coal marketed under the name "Housewarm" by agreement with the National Coal Board.
1972 No. 2 Order (SI 1972 No. 955)	
The fireplace known as the Trianco TGB 17 and manufactured by Trianco Ltd.	
1973 Order (SI 1973 No. 2166)	
The fireplace known as the Rayburn Prince 101 and the fireplace known as the Rayburn Prince 301, both manufactured by Glynwed Foundries Ltd.	
****1975 No. 2 Order (SI 1975 No. 1001)***	
The fireplace known as the Riley Nihot Woodchip Fired Air Heater type NM011 manufactured by Clarke Chapman Ltd, latterly by NEI International Combustion Ltd, Riley Equipment.	No fuel shall be used other than clean wood waste of a size within the limits of the manufacturer's specifications and containing not more than 5% sander dust.
****1975 No. 3 Order (SI 1975 No. 1111)***	
The fireplace known as the Rayburn Prince 76, manufactured by Glynwed Domestic and Heating Appliances Ltd.	No fuel shall be used other than selected washed coal doubles and trebles.
1978 Order (SI 1978 No. 1609)	
The fireplace known as the Spanex Wood Fired Air Heater (types UL50, UL75 and UL100 only), manufactured by Spanex Sander GmbH KG of Volpriehausen, Solling, in the Federal Republic of Germany	1. The Fireplace shall be installed, maintained and operated so as to minimise the emission of smoke and in accordance with Issue No 1 of the manufacturer's instructions dated May 1978. 2. No fuel shall be used other than clean wood waste contaning not more than 5% sander dust and not more than 1% plastic contamination.
1982 Order (SI 1982 No. 1615)	
The fireplace known as the APE Saffire Boiler (in the sizes, expressed in BTUs per hour, as 1M to 8M, 10M and 12M) manufactured by Air Pollution Engineering Limited, now Air Pollution Equipment Ltd.	1. The fireplace shall be installed, maintained and operated so as to minimise the emission of smoke at all times and in accordance with the manufacturer's instructions dated 27 August 1982 and bearing the reference "SAF 250-3000/HW & /S". 2. No fuel shall be used other than sawdust or shavings, or mixtures of sawdust, shavings and offcuts, being fuel containing (by weight) not more than 1% of plastic material.
1983 Order (SI 1983 No. 277)	
The fireplace known as the Rayburn Coalglo C-30 and manufactured, both as an inset model and as a freestanding model, by Glynwed Domestic & Heating Appliances Ltd, now Glynwed Appliances Ltd.	1. The fireplace shall be installed, maintained and operated so as to minimise the emission of smoke at all times and in accordance with the manufacturer's instructions which – a) in the case of the inset model, bear the date 1 September 1981 and the reference "Code 510" or b) in the case of the freestanding model, bear the date 1 December 1982 and the reference "Code 510 F". 2. No fuel shall be used other than selected washed coal doubles and trebles (also known as nuts and large nuts).
1983 No. 2 Order (SI 1983 No. 426)	
The fireplace known as the Talbott 500 Hot-air Heater (afterburn model) and manufactured by Talbott's Heating Ltd.	1. The fireplace shall be installed, maintained and operated so as to minimise the emission of smoke at all times and in accordance with the manufacturer's instructions dated 1 January 1983 and bearing the reference A1000. 2. No fuel shall be used other than wood off-cuts, woodwaste, pellets, chipboard, plastic covered chip-board (the plastic content of the covering being not more than 1% by weight), cardboard or paper. 3. The afterburn cycle shall last not less than 25 mins. and shall come into operation each time the loading door is opened.

* no longer in production, but where such fireplaces remain in use they remain exempt.

Class of Fireplace	*Conditions*
1983 No. 3 Order (SI 1983 No. 1018)	
The fireplace known as the Trianco Coal King' Boiler and manufactured by Trianco Redfyre Ltd.	1. The fireplace shall be installed, maintained and operated so as to minimise the emission of smoke at all times, and in accordance with the manufacturer's instructions dated May 1983 and which – a) in the case of installation instructions bear the reference '47664' and b) in the case of the user, instructions bear the reference '47665'. 2. No fuel shall be used other than selected washed coal doubles or Union Coal Briketts, manufactured by Rheinbraun AG of Germany and comprising lignite compressed into briquettes of approximately 15 centimetres in length with square ends and with a sulphur content not exceeding 1 per cent.
1984 Order (SI 1984 No. 1649)	
The Parkray Coalmaster II manufactured by T.I. Parkray Limited, both as an inset model and as a free standing model.	a) installation instructions for for the inset model: reference "List No. 1048/June 1983 8402"; b) installation instructions for the free-standing model: reference "List No. 1058/February 1984"; and c) user's instructions for both models: reference "List No. 1049/September 1983".
The Spacewarmer B200 manufactured by Clean Air Systems.	installation operation and maintenance instructions: reference "B200/DEC 1983".
The Spacewarmer B500 manufactured by Clean Air Systems.	installation operation and maintenance instructions: reference "B500/DEC 1983".
1985 Order (SI 1985 No 864)	
The SE60 Coalburner Stove manufactured by Jetmaster Fires Limited.	a) installation instructions: reference "SE60/18/84"; b) operating and maintenance instructions: reference "SE 60/28/84".
The Worcester Centair 40 Solid Fuel Warm Air Heater manufactured by Worcester Engineering Co. Limited	a) installation instructions: reference "BC40.1.2 July 1983"; b) operating and maintenance instructions: reference "BC40.OM2 July 1983".
1986 Order (SI 1986 No. 638)	
The Triancomatic 60 boiler manufactured by Trianco Redfyre Limited.	1. The fireplace shall be installed, maintained and operated so as to minimise the emission of smoke at all times, and in accordance with the manufacturer's instructions dated 24 July 1985 and which – a) in the case of the installation instructions bear the reference "No. 45576"; and b) in the case of the user instructions bear the reference "No. 45574". 2. No fuel shall be used other than the washed coals recommended in the manufacturer's user instructions: reference "No. 45574".
The Worcester Coalstream 17.5 kw underfeed stoker manufactured by Worcester Engineering Company Limited	1. The fireplace shall be installed, maintained and operated so as to minimise the emission of smoke at all times, and in accordance with the manufacturer's instructions dated 1 August 1985 and which a) in the case of the installation instructions bear the reference No. "ZKLIT 140"; and b) in the case of the user instructions bear the reference No." ZKLIT 139". 2. No fuel shall be used other than the washed coals recommended in the manufacturer's user instructions: reference No. "ZK LIT 139".
The Corsair 300HW, 500HW and 750HW boilers manufactured by Erithglen Limited.	The fireplace shall be installed, maintained and operated so as to minimise the emission of smoke at all times, and in accordance with the manufacturer's instructions: reference "001 August 1985".

Class of Fireplace	*Conditions*
1988 Order (SI 1988 No. 2282)	
The Babcock Worsley Fluidised Bed Combuster manufactured by Babcock Robey Limited.	1. The fireplace shall be installed, maintained and operated so as to minimise the emission of smoke at all times, and in accordance with the manufacturer's instructions dated 30 April 1986 and which bear the reference No. "C06/0016". 2. No fuel shall be used other than the waste derived fuel recommended in the manufacturer's instructions. 3. The fireplace shall be operated as a single unit with – a) the Tollemache Drier manufactured by Newell Dunford Limited; and b) The Venturi Scrubber manufactured by Air Pollution Control Limited.
The CBR Flexifuel Heater, models 300, 400 and 600 manufactured by CBR Fabrications Limited	1. The fireplaces shall be installed, maintained and operated so as to minimise the emission of smoke at all times, and in accordance with the manufacturer's instructions, which bear the reference FF/CBR Turbo Heat 88. 2. No fuel shall be used other than hard or soft wood off-cuts, chipboard or plastic coated chipboard. 3. The fireplace shall not be used to burn sawdust in bulk or plastic materials other than plastic coated chipboard.
The Eclipse Junior, Standard, Senior, Jumbo 30 and Jumbo 50 incinerators, manufactured by Northern Incinerators Limited.	1. The fireplaces shall be installed, maintained and operated so as to minimise the emission of smoke at all times, and in accordance with the manufacturer's instructions dated 21 June 1988 and which bear the reference "NORCIN/ TECH/88". 2. No fuel shall be used, other than fuel consisting of paper, cardboard cartons, scrap wood, foliage, combustible floor sweepings and other waste from domestic, commercial and industrial activities containing no more than 20% of restaurant and cafeteria waste, and containing less than 5% by weight of coated papers, plastic or rubber waste.
The Haat LD, MD and HD Incinerators manufactured by Haat Incineration Limited.	1. The fireplaces shall be installed, maintained and operated so as to minimise the emission of smoke at all times, and in accordance with the manufacturer's instructions dated March 1988 and which bear the reference "HAAT/I/CAA". 2. No fuel shall be used, other than fuel consisting of paper, cardboard cartons, scrap wood, foliage, combustible floor sweepings and other waste from domestic, commercial and industrial activities containing no more than 20% of restaurant and cafeteria waste, and containing less than 5% by weight of coated papers, plastic or rubber waste.
The Holden Heat House 29.3kw and 45.4kw underfeed bituminous coal burning boilers manufactured by Holden Heat plc.	1. The fireplaces shall be installed, maintained and operated so as to minimise the emission of smoke at all times, and in accordance with the manufacturer's instructions which – a) in the case of the installation instructions bear the reference "HH861B"; and b) in the case of the user instructions bear the reference "HH861A". 2. No fuel shall be used other than the washed coals recommended in the manufacturer's user instructions.
The RanHeat Boiler Type RHA20, manufactured by RanHeat, Energy A/S.	1. The fireplace shall be installed, maintained and operated so as to minimise the emission of smoke at all times, and in accordance with the manufacturer's instructions dated 30 March 1988 and which bear the reference "RHGBEA 30388". 2. No fuel shall be used other than hard and soft wood shavings.
The Talbott Pirojet P150 Heater manufactured by Talbott's Heating Limited.	1. The fireplace shall be installed, maintained and operated so as to minimise the emission of smoke at all times, and in accordance with the manufacturer's instructions dated February 1987 and which bear the reference No. "P150-2-87". 2. No fuel shall be used other than hard or soft wood off-cuts, chipboard or plastic coated chipboard. 3. When the fireplace is used to burn chipboard and plastic coated chipboard, an afterburner which is supplied with the fireplace and which must be capable of continuously producing 20 megajoules of heat per hour, shall be used. 4. The fireplace shall not be used to burn sawdust in bulk or plastic materials other than plastic coated chipboard.

Class of Fireplace	*Conditions*
The Talbott Pirojet P300 Heater manufactured by Talbott's Heating Limited.	1. The fireplace shall be installed, maintained and operated so as to minimise the emission of smoke at all times, and in accordance with the manufacturer's instructions dated February 1987 and which bear the reference No. "P300-2-87". 2. No fuel shall be used other than hard or soft wood off-cuts, chipboard or plastic coated chipboard. 3. When the fireplace is used to burn chipboard and plastic coated chipboard an afterburner which is supplied with the fireplace and which must be capable of continuously producing 40 megajoules of heat per hour, shall be used. 4. The fireplace shall not be used to burn sawdust in bulk or plastic materials other than plastic coated chipboard.
The Talbott Pirojet P600 Heater manufactured by Talbott's Heating Limited.	1. The Fireplace shall be installed, maintained and operated so as to minimise the emission of smoke at all times, and in accordance with the manufacturer's instructions dated December 1986 and which bear the reference "P600-12-86" 2. No fuel shall be used other than hard or soft wood off-cuts, or chipboard or plastic coated chipboard. 3. When the fireplace is used to burn chipboard, and plastic coated chipboard an afterburner which is supplied with the fireplace and which must be capable of continuously producing 75 megajoules of heat per hour, shall be used. 4. The fireplace shall not be used to burn sawdust in bulk or plastic materials other than plastic coated chipboard.
The Triancomatic 90 boiler manufactured by Trianco Redfyre Limited	1. The fireplace shall be installed, maintained and operated so as to minimise the emission of smoke at all times, and in accordance with the manufacturer's instructions dated July 1987 and which – a) in the case of the installation instructions bear the reference "No. 46576"; and b) in the case of the user instructions bear the reference "No. 46574". 2. No fuel shall be used other than the washed coals recommended in the manufacturer's user instructions.
The Triancomatic 140 boiler manufactured by Trianco Redfyre Limited	1. The fireplace shall be installed, maintained and operated so as to minimise the emission of smoke at all times, and in accordance with the manufacturer's instructions dated November 1986 and which – a) in the case of the installation instructions bear the reference No. 48375"; and b) in the case of the user instructions bear the reference No. 48374". 2. No fuel shall be used other than the washed coals recommended in the manufacturer's user instructions.
1989 Order (SI 1989 No. 1769)	
The Talbott T5-1M Automatic Heater manufactured by Talbott's Heating Ltd.	1. The fireplace shall be installed, maintained and operated so as to minimise the emission of smoke at all times, and in accordance with the manufacturer's instructions bearing the reference Jan. 89 T5-A-1-89. 2. No fuel other than softwood shavings, hardwood shavings or chipboard dust shall be used.
The Nordist Waste Fired System models 10, 11, 12, 13, 14, 15, 16, 17, 18, 19, 20, 21, 22 and 23 manufactured in part by Nordfab A/S and in part by Danstoker A/S.	1. The fireplace shall be installed, maintained and operated so as to minimise the emission of smoke at all times, and in accordance with the manufacturer's instructions bearing the reference P.S.501 9.12.88. 2. No fuel other than softwood shavings, hardwood shavings or sawdust shall be used.
The CBR Flexifuel Heater, models 150 and 200 manufactured by CBR Fabrications Ltd.	1. The fireplace shall be installed, maintained and operated so as to minimise the emission of smoke at all times, and in accordance with the manufacturer's instructions bearing the reference FF/CBR Flexifuel 150/200 1989. 2. No fuel other than hardwood or softwood offcuts, chipboard or plastic coated chipboard shall be used. 3. The fireplace shall not be used to burn sawdust in bulk or plastic materials other than plastic coated chipboard.

Class of Fireplace	*Conditions*
The Silent Glow Incinerator, models CCD 450, 600 and 750 manufactured by Silent Glow Incinerators Ltd.	1. The fireplace shall be maintained and operated so as to minimise the emission of smoke at all times, and in accordance with the manufacturer's instructions dated May 1989. 2. No fuel other than solid waste of the types specified in the manufacturer's instructions dated May 1989 shall be used.
The Hughes Edwards 45 Coalflow Boiler manufactured by Hughes Edwards Heating Ltd.	1. The fireplace shall be installed, maintained and operated so as to minimise the emission of smoke at all times, and in accordance with the manufacturer's instructions which: a) in the case of the installation instructions bear the reference II/02D/07-89, and b) in the case of the user instructions bear the reference UI/02D/07-89. 2. No fuel other than the washed coals recommended in the manufacturer's instructions bearing the reference II/02D/07-89 shall be used.
1990 Order (SI 1990 No. 345)	
The Haat Pioneer ABI Incinerators, models 12, 20, 28 and 36 manufactured by Haat Incineration Limited.	1. The fireplaces shall be installed, maintained and operated so as to minimise the emission of smoke at all times, and in accordance with the manufacturer's instructions bearing the reference "HAAT/2/CAA/10/89". 2. No fuel shall be used other than the following fuels used in the mixtures by volume recommended in Appendix A of the manufacturer's instructions: paper, cardboard, corrugated paper, office or domestic waste, wood, polythene sheet or foam, polypropylene, nylon, polyethylene, rubber foam backed carpet, polystyrene, polyurethane and glass fibre reinforced plastics.
The Triancomatic 45 boiler manufactured by Trianco Redfyre Limited.	1. The fireplace shall be installed, maintained and operated so as to minimise the emission of smoke at all times, and in accordance with the manufacturer's instructions dated September 1989 and which – (a) in the case of the installation instructions bear the reference "No. 44359"; and (b) in the case of the user instructions bear the reference "No. 44358". 2. No fuel shall be used other than the washed coals recommended in the manufacturer's user instructions.
1990 Order (SI No. 2457)	
The Intrepid II woodstove manufactured by Vermont Castings, Inc.	1. The fireplace shall be installed, maintained and operated in accordance with the manufacturer's instructions bearing the reference "AMENDED No. 20917C U.K. MAY 1990". 2. No fuel shall be used other than 16" firewood that has been split, stacked and air-dried.
The Spacewarmer B200 manufactured by Clearair Limited.	1. The fireplace shall be installed, maintained and operated in accordance with the manufacturer's instructions bearing the reference "B200/DEC 83". 2. No fuel shall be used other than hardwood or softwood offcuts, a mixture of offcuts (hardwood or softwood) and sawdust with no more than 50% of the mixture being sawdust, chipboard offcuts or plastic coated chipboard offcuts. 3. The fireplace shall not be used to burn chipboard or plastic coated chipboard unless an afterburner, supplied by the manufacturer of the fireplace and capable of producing 50 megajoules of heat per hour, has been fitted to the fireplace.
The Spacewarmer B500 manufactured by Clearair Limited.	1. The fireplace shall be installed, maintained and operated in accordance with the manufacturer's instructions bearing the reference "B500/DEC 83". 2. No fuel shall be used other than hardwood or softwood offcuts.
The Turbo Heat, models 1000 and 1200 manufactured by CBR Engineers Limited.	1. The fireplace shall be installed, maintained and operated in accordance with the manufacturer's instructions bearing the reference "CBR TURBO' HEAT 1000 TO 1200 1990". 2. No fuel shall be used other than hardwood or softwood offcuts, chipboard offcuts or plastic coated chipboard offcuts.
The Talbott's Energy Saving Wood Waste Burning Automatic Warm Air Heaters, models T1.5/A, T3/A, T5/A and TM/A manufactured by Talbotts Heating Limited.	1. The fireplaces shall be installed, maintained and operated in accordance with the manufacturer's instructions bearing the reference "T/A JANUARY 1990". 2. No fuel shall be used other than hardwood or softwood shaving or sawdust.

Class of Fireplace — *Conditions*

1991 Order (SI No. 2892)

The Hughes Edwards Model 65 hot water boiler manufactured by Hughes Edwards Heating Limited.

1. The fireplace shall be installed, maintained and operated in accordance with the manufacturer's instructions bearing the reference "II/02D/11-90 and UI/02D/11-90".
2. No fuel shall be used other than bituminous Coalflow pearls distributed by the British Coal Corporation.

The Jansen Design stove, using the five channel system, manufactured by Kakkel Ovnsmakeriet as.

1. The fireplace shall be installed, maintained and operated in accordance with the manufacturer's instructions bearing the reference "CSC-JDS-OI dated 21/10/91".
2. No fuel shall be used other than wood.

The Resolute Acclaim woodstove manufactured by Vermont Castings, Inc.

1. The fireplace shall be installed, maintained and operated in accordance with the manufacturer's instructions bearing the reference "AMENDED No. 200 – 0908 U.K. OCTOBER 1991".
2. No fuel shall be used other than 16" firewood that has been split, stacked and air-dried.

The Talbott's Warm Air Heaters Down Firing range models D250, D500 and D700 manufactured by Talbotts Heating Limited.

1. The fireplaces shall be installed, maintained and operated in accordance with the manufacturer's instructions bearing the reference "MAY 1991 Ref No: D257".
2. No fuel shall be used other than hardwood or softwood offcuts, chipboard, plastic coated chipboard or cardboard.

1992 Order (SI No. 2811)

The Cronspisen stove, using the five channel chimney system, manufactured by Cronspisen Produktion AB.

1. The fireplace is installed, maintained and operated in accordance with the manufacturer's instructions bearing the reference "January 1992. Serial No: UM 351B".
2. No fuel is used other than firewood that has been split, stacked and air-dried.

The RanHeat Boiler Types RHA 160, 200 and 250, manufactured by RanHeat Energy GB Limited.

1. The fireplaces are installed, maintained and operated in accordance with the manufacturer's instructions dated 30th March 1988 bearing the reference "RHGBEA 30388".
2. No fuel is used other than hard and soft wood shavings.

The Defiant Encore manufactured by VCW International Limited.

1. The fireplace is installed, maintained and operated in accordance with the manufacturer's instructions dated "November 1991 bearing the reference 200-8632".
2. No fuel is used other than firewood that has been split, stacked and air-dried.

1993 Order (SI No. 2277)

The Apollo incinerator types 10, 25 and 50 manufactured by Apollo Incineration Systems Limited.

1. The incinerators shall be installed, maintained and operated in accordance with the manufacturer's instructions dated 2nd November 1992 and bearing the reference "Apollo 10-25-50".
2. No fuel shall be used other than wood, wood shavings, cardboard and polythene.

The Clearview models 650 and 750, and the Vision 500 model, manufactured by Clearview Stoves.

1. The fireplaces shall be installed, maintained and operated in accordance with the manufacturer's instructions dated 1st January 1993 and, in the case of the Clearview models 650 and 750, bearing the reference "1/42" and, in the case of the Vision 500 model, bearing the reference "V1/42".
2. No fuel shall be used other than firewood which has been split, stacked and air dried.

The Farm 2000 boilers HT60, HT70, HT80 and BB154/2V manufactured by Farm 2000 Limited.

1. The boilers shall be installed, maintained and operated in accordance with the manufacturer's instructions which, in the case of models HT60, HT70 and HT80, are dated 1st July 1993 and bear the reference "HT60/70/80", and, in the case of model BB154/2V, are dated 1st September 1993 and bear the reference "9/93".
2. No fuel shall be used other than cereal straw.

1996 Order (SI No. 1108)

The Dovre woodstove models 500CBW and 700CBW manufactured by Dovre Castings Limited, including those models when referred to by their former reference numbers 500G and 700G respectively.

1. The fireplaces shall be installed, maintained and operated in accordance with the manufacturer's instructions dated October 1993 and bearing the following reference :–
   (a) model 500G: "500G";
   (b) model 700G:"700G".
2. No fuel shall be used other than wood.

Class of Fireplace	*Conditions*
The Farm 2000 boiler models BB154, BB154A and BB154/2 (when used with the Farm 2000 conversion kit BB154/2Vcx), maunufactured by Farm 2000 Limited.	1. The boilers shall be installed in accordance with the manufacturer's instructions dated May 1994 and bearing the reference "BB154/2VC", and maintained and operated in accordance with the manufacturer's instructions dated September 1993 and bearing the reference "BB154/2V". 2. No fuel shall be used other than cereal straw.
The Farm 2000 model HT6 (when used with the Farm 2000 conversion kit HT60cx), the HT6PLUS and HT600 (when each is used with the Farm conversion kit HT70cx), and the HT7 and HT8 (when each is used with the Farm 2000 conversion kit HT80cx), manufactured by Farm 2000 Limited.	1. The boilers shall be installed in accordance with the manufacturer's instructions dated May 1994 and bearing the reference "HT6780" and "HT 6781", and maintained and operated in accordance with the manufacturer's instructions dated 1 July 1993 and bearing the reference "HT60/70/80". 2. No fuel shall be used other than cereal straw.
The Jotul catalyst stove models 8TDIC and 12TDIC manufactured by Jotul Limited.	1. The fireplaces shall be installed, maintained and operated in accordance with the manufacturer's instructions dated November 1995 and bearing the following references :– (a) model 8TDIC: "JT8/001" and "J/S/001"; (b) mode12TDIC: "JT12/001" and "JT/S/001". 2. No fuel shall be used other than wood.
The RanHeat appliances models MSU150, MSU300, MSU500, WA150, WA300 and WA500, manufactured by RanHeat Engineering Limited.	1. The appliances shall be installed, maintained and operated in accordance with the manufacturer's instructions dated July 1995 bearing the following references :– (a) models MSU150, MSU300 and MSU500: "MSU150/5001994"; (b) models WA150, WA300 and WA500: "WA150/5001994". 2. No fuel shall be used other than chipboard, fibreboard, melamine coated chipboard, wood offcuts, softwood or hardwood shavings or dust.
The Talbott's Combustion Units, models C1, C2, C3 and C4, manufactured by Talbott's Heating Limited.	1. The fireplaces shall be installed, maintained and operated in accordance with the manufacturer's instructions dated August 1995 and bearing the reference "Issue: C1000/C/Range". 2. No fuel shall be used other than chipboard, or wood shavings, forestry chips or paper briquettes.
The Talbott's Heating models T75, T150, T300, D250B, D500B and D700B, manufactured by Talbott's Heating Limited.	1. The fireplaces shall be installed, maintained and operated in accordance with the manufacturer's instructions dated 1st January 1995 and bearing the following references :– (a) models T75, T150 and T300: "gen/sttech"; (b) models D250B, D500B and D700B: "Gen/Tech". 2. No fuel shall be used other than the following; (a) models T75, T150 and T300: wood offcuts, woodwaste pallets, chipboard, plastic coated chipboard, cardboard or paper; (b) models D250B, D500B and D700B: hardwood or softwood offcuts, chipboard, plastic coated chipboard or cardboard.
1997 Order (SI No. 3009)	
The Talbott's Energy Saving Wood Waste Burning Automatic Hot Water Heaters, models T1.5/AB, T3/AB, T5/AB, manufactured by Talbott's Heating Limited.	1. The fireplaces shall be installed, maintained and operated in accordance with the manufacturer's instructions dated October 1993 and bearing the reference "Issue: A1001". 2. No fuel shall be used other than hardwood or softwood shavings or sawdust.
The Talbott's Combustion Units, models CM1, CM2, CM3 and CM4, manufactured by Talbott's Heating Limited.	1. The fireplaces shall be installed, maintained and operated in accordance with the manufacturer's instructions dated August 1997 and bearing the reference "Issue: CM897". 2. No fuel shall be used other than wood-waste pallets, medium density fibreboard, melamine faced chipboard, cardboard, paper or untreated and airdried wood offcuts

Class of Fireplace

Conditions

1999 Order (SI No. 1515)

The Dunsley Yorkshire Stove, the Dunsley Yorkshire Multifuel Stove and the Dunsley Yorkshire Multifuel Stove and Boiler manufactured by Dunsley Heat Limited.

1. The fireplaces shall be installed, maintained and operated in accordance with the manufacturer's instructions dated 5th August 1998 and bearing the reference "A/22160" as amended by the manufacturer's instructions dated 4th December 2004 bearing the reference "D13" with regard to the Stove and Multifuel Stove, and by the manufacturer's instructions dated 4th October 2004 bearing the reference GHD/DUN4B/1 with regard to the Multifuel Stove and Boiler.
2. No fuel shall be used other than-
   (a) untreated dry wood;
   (b) peat or peat briquettes with, in either case, less than 25 per cent moisture;
   (c) Union Coal Briketts, manufactured by Rheinbraun AG of Germany and comprising lignite compressed into briquette of approximately 15 centimetres in length with square ends and with a sulphur content not exceeding 1 per cent of the total weight;
   (d) CPL Wildfire, manufactured by Coal Products Limited, Coventry, which comprise bituminous coal with a volatile content of 32 per cent to 36 per cent (as to approximately 96 per cent of the total weight) and a cold cure resin binder (as to the remaining weight); are manufactured from those constituents by a progress involving roll-pressing; have an average weight of between 80 and 90 grammes or between 160 and 170 grammes; and have a sulphur content not exceeding 1.8 per cent of the total weight.

The I.D.L, M and H incinerators, manufactured by the Incinerator Doctor.

1. The fireplaces shall be installed, maintained and operated in accordance with the manufacturer's instructions dated February 1999 and bearing the reference "I.D.E.I".
2. No fuel shall be used other than paper, cardboard, untreated dry wood, foliage, floor sweepings, and other waste containing no more than 20 per cent by weight of restaurant and cafeteria waste and less than 5 per cent by weight of coated paper, plastic or rubber waste.

The Osier wood burning stove, manufactured by Reinhart von Zschock.

1. The fireplaces shall be installed, maintained and operated in accordance with the manufacturer's instructions dated 30th September 1997 and bearing the reference "CSC/OS/001".
2. No fuel shall be used other than untreated dry wood.

The Trash X Incinerator, models TX25, TX35, TX50, TX35L and TX50L manufactured by Roscoe Fabrications Limited.

1. The fireplaces shall be installed, maintained and operated in accordance with the manufacturer's instructions dated 1998 and bearing the reference "TX/1998".
2. No fuel shall be used other than paper, card, untreated dry wood and cotton waste.

The Wood Stone Ovens, models WS-MS-4-W (Mt. Chucknaut), WS-MS-5-W (Mt. Adams), WS-MS-6-W (Mt. Baker) and WS-MS-7-W (Mt. Rainier), manufactured by the Wood Stone Corporation, USA.

1. The fireplaces shall be installed, maintained and operated in accordance with the manufacturer's instructions dated 1998 and bearing the reference "BPWS version 1".
2. No fuel shall be used other than untreated dry wood.

2003 Order (SI No. 2328)

The B9 Energy Wood Chip Gasifier Unit manufactured by B9 Energy Biomass Limited, Unit 22, Northland Industrial Estate, Londonderry, Northern Ireland.

1. The fireplaces shall be installed, maintained and operated in accordance with the manufacturer's instructions dated 2002 and bearing the reference 001/B9EnergyBiomass.
2. No fuel shall be used other than untreated wood chips, shredded wood or small untreated pieces of wood.

The Mawera FU-RIA models 110, 140, 180, 220 and 350 manufactured by Mawera (UK) Limited, 31 Enterprise Industrial Park, Britannia Way, Lichfield Staffordshire WS14 9UY.

1. The fireplaces shall be installed, maintained and operated in accordance with the manufacturer's instructions dated 31 October 2002 and bearing the reference 098234.
2. No fuel shall be used other than wood, wood products, forest residues, forest thinnings, melamine coated or uncoated MDF chipboard, plywood or compound wood products.

2008 No 2 Order (SI 2008/2343, England. SSI 2008/296, Scotland. SR 2008/282, NI)

The Binder Wood Fired Boiler models RRK 22-49 (49 kW output) manufactured by Binder Gesellschaft mit beschrankter Haftung (GMbH), Austria

The manufacturer's and Wood Energy Limited's instructions 'Binder Equipment Installation, Commissioning and Maintenance' dated 17 May 2005, reference WELBinderManualrev0.doc, revision 0 Permitted fuel: wood pellets or wood chips (*see p.318)

Class of Fireplace	*Conditions*
The Binder Wood Fired Boiler models RRK 80-175 (75-149 kW output) with cyclone models ZA 80-175 manufactured by Binder Gesellschaft mit beschrankter Haftung (GMbH), Austria	The manufacturer's and Wood Energy Limited's instructions 'Binder Equipment Installation, Commissioning and Maintenance' dated 17 May 2005, reference WELBinderManualrev0.doc, revision 0 Permitted fuel: wood pellets or wood chips*
The Binder Wood Fired Boiler models RRK 130-250 (185-230 kW output), RRK 200-350 (250-300 kW output), RRK 400-600 (350-500 kW output) and RRK 640-850 (650-840 kW output) with cyclone model type ZA manufactured by Binder Gesellschaft mit beschrankter Haftung (GMbH), Austria	The manufacturer's and Wood Energy Limited's instructions 'Binder Equipment Installation, Commissioning and Maintenance' dated 17 May 2005, reference WELBinderManualrev0.doc, revision 0 Permitted fuel: wood pellets or wood chips*
The Binder Wood Fired Boiler model RRK 1000 (1200 kW output) with multi-cyclone model MZA manufactured by Binder Gesellschaft mit beschrankter Haftung (GMbH), Austria	The manufacturer's and Wood Energy Limited's instructions 'Binder Equipment Installation, Commissioning and Maintenance' dated 17 May 2005, ref WELBinderManualrev0.doc, revision 0 Permitted fuel: wood pellets or wood chips*
Biomatic 220 and 250 fitted with Zyklovent 220-250 cyclone separator manufactured by HERZ Armaturen GmbH, Geschaftsbereich HERZ, Feuerungstechnik, A-8272 Sebersdorf 138, Austria	The manufacturer's instructions dated April 2005, reference Bedienungsanleitung Englisch v2.0 doc Permitted fuel: chipped wood, wood pellets or planed wood shavings*
Biomatic 300, 350, 5400 and 500 fitted with Zyklovent 300-500 cyclone separator manufactured by HERZ Armaturen GmbH, Geschaftsbereich HERZ, Feuerungstechnik, A-8272 Sebersdorf 138, Austria	The manufacturer's instructions dated April 2005, reference Bedienungsanleitung Englisch v2.0 doc Permitted fuel: chipped wood, wood pellets or planed wood shavings*
Scotland & NI only: Broag-Gilles HPK-RA 13 kW, 15 kW, 20 kW, 25 kW, 30 kW, 35 kW, 40 kW, 45 kW, 49 kW, 60 kW, 70 kW, 75 kW, 85 kW, 95 kW, 100 kW, 120 kW and 145 kW, all manufactured by GILLES Energie- und Umwelttechnik GmbH, Koaserbauer Str, 16 A-4810 Gmunden, Austria	The manufacturer's operating instructions dated January 2007, reference Broag-Gilles/0-145-1 Permitted fuel: wood pellets or wood chips*
Scotland & NI only: Broag-Gilles HPK USK 150 kW, 180 kW, 240 kW and 300 kW, all manufactured by GILLES Energie- und Umwelttechnik GmbH, Koaserbauer Str, 16 A-4810 Gmunden, Austria	The manufacturer's operating instructions dated January 2007, reference Broag-Gilles/360/1 Permitted fuel: wood pellets or wood chips*
NI only: Broag-Gilles HPK USK 360 kW manufactured by GILLES Energie- und Umwelttechnik GmbH, Koaserbauer Str, 16 A-4810 Gmunden, Austria	The manufacturer's operating instructions dated January 2008, reference Broag-Gilles/150-300/1 Permitted fuel: wood pellets or wood chips*
CTC Ecoflex 15 & 20 manufactured by Enertech AB, Box 313, S-341 26 Ljungby, Sweden	The manufacturer's instructions dated 20 February 2007, ref. 161 505 27 06/1 Permitted fuel: wood pellets
The Clearview Pioneer 400, Solution 400 and Pioneer Oven wood burning stoves fitted with a tamperproof mechanical stop on the air slide/damper manufactured by Clearview Stoves of More Works, Bishops Castle, Shropshire SY9 5HH	The manufacturer's instructions 'Operating Instructions Clearview Vision/Vision Inset/Pioneer & Solution' dated 1 july 2006, ref V1/42, stamped and signed "smoke control". Permitted fuel: air-dried wood logs*
Dovre 250 (model DV-250) manufactured by Dovre NV, Nijverheidsstraat 18, BE-2381, Weelde, Belgium	The manufacturer's instructions ref PM 250 Issue A. A permanent stop must be in place on the secondary air inlet to prevent closure beyond the 75% open position. Permitted fuel: untreated dry wood
The Dunsley Yorkshire Stove manufactured by Dunsley Heat Ltd	The manufacturer's instructions dated 5 August 1998, reference A/22160 as amended by those dated 4 December 2004, reference D13 Permitted fuels:. 1. Untreated dry wood; 2. Peat or peat briquettes with in either case, less than 25% moisture; 3. Union Coal Briketts, manufactured by Rheinbraum AG of Germany comprising lignite compressed into briquettes of approx. 15 cm in length with square ends and with a sulphur content not exceeding 1% of the total weight; 4. CPL Wildfire, manufactured by Coal Products Ltd, Immingham, which comprise bituminous coal with a volatile content of 32-36% (as to approx. 96% of the total weight) and a cold cure resin binder (as to the remaining weight); are manufactured from those constituents by a process involving roll-pressing; have an average weight of either 80-90 grams or 160-170 grams; and have a sulphur content not exceeding 1.8% of the total weight.

Class of Fireplace	Conditions
The Dunsley Yorkshire Multifuel Stove manufactured by Dunsley Heat Ltd	The manufacturer's instructions dated 5 August 1998, reference A/22160 as amended by those dated 4 December 2004, reference D13 Permitted fuels: 1. Untreated dry wood; 2. Peat or peat briquettes with in either case, less than 25% moisture; 3. Union Coal Briketts, manufactured by Rheinbraum AG of Germany comprising lignite compressed into briquettes of approx. 15 cm in length with square ends and with a sulphur content not exceeding 1% of the total weight; 4. CPL Wildfire, manufactured by Coal Products Ltd, Immingham, which comprise bituminous coal with a volatile content of 32-36% (as to approx. 96% of the total weight) and a cold cure resin binder (as to the remaining weight); are manufactured from those constituents by a process involving roll-pressing; have an average weight of either 80-90 grams or 160-170 grams; and have a sulphur content not exceeding 1.8% of the total weight.
The Dunsley Yorkshire Multifuel Stove and Boiler manufactured by Dunsley Heat Ltd	The manufacturer's instructions dated 5 August 1998, reference A/22160 as amended by those dated 4 October 2004, reference GHD/DUN4B/1 Permitted fuels: 1. Untreated dry wood; 2. Peat or peat briquettes with in either case, less than 25% moisture; 3. Union Coal Briketts, manufactured by Rheinbraum AG of Germany comprising lignite compressed into briquettes of approx. 15 cm in length with square ends and with a sulphur content not exceeding 1% of the total weight; 4. CPL Wildfire, manufactured by Coal Products Ltd, Immingham, which comprise bituminous coal with a volatile content of 32-36% (as to approx. 96% of the total weight) and a cold cure resin binder (as to the remaining weight); are manufactured from those constituents by a process involving roll-pressing; have an average weight of either 80-90 grams or 160-170 grams; and have a sulphur content not exceeding 1.8% of the total weight.
The Dunsley Yorkshire Woodburning Stove manufactured by Dunsley Heat Limited	The manufacturer's instructions dated 4 December 2004, ref D13W. Permitted fuel: untreated dry wood
The Dunsley Yorkshire Woodburning Stove and boiler manufactured by Dunsley Heat Limited	The manufacturer's instructions dated 4 December 2004, reference D13W, together with the manufacturer's instructions dated 4 October 2004, ref GHD/DUN4B/1. Permitted fuel: untreated dry wood
The Extraflame: Babyfiamma; Bella and Bella Lux; Clementina; Comfort Maxi; Contessa; Divina, Divina Plus & Divina Steel; Duchessa & Duchessa Steel; Ecologica; Esmeralda & Esmeralda Crystal; Falo 1CP, Falo 2CP & Falo 1XP; Isabella; Karolina; Preziosa; all of which are manufactured by Extraflame Spa in Montecchio Precalcino, (Vicenza) Italy	The manufacturer's operating instructions dated 5 March 2007, ref 004275101011. Permitted fuel: wood pellets*
The Extraflame Comfort Mini manufactured by Extraflame Spa in Montecchio Precalcino, (Vicenza) Italy	The manufacturer's operating instructions dated 2 October 2006, ref 004275115 REV 005. Permitted fuel: wood pellets*
The Extraflame Idro manufactured by Extraflame Spa in Montecchio Precalcino, (Vicenza) Italy	The manufacturer's operating instructions dated 28 August 2006, ref 004275128 REV 002. Permitted fuel: wood pellets*
The Extraflame Lucrezio Idro and Lucrezia Steel manufactured by Extraflame Spa in Montecchio Precalcino, (Vicenza) Italy	The manufacturer's operating instructions dated 12 July 2006, ref 004275110 REV 011. Permitted fuel: wood pellets*
Firematic 25, 50, 90 and 150 manufactured by HERZ Armaturen GmbH, Geschaftsbereich HERZ, Feuerungstechnik , A-8272 Sebersdorf 138, Austria	The manufacturer's instructions dated June 2005, ref Bedienungsanleitung Firematic V1.1 Englisch doc Permitted fuel: chipped wood, wood pellets or planed wood shavings*
Froling Turbomatic 85, 100 & 110 manufactured by Frolinbg Heizkessel- & Behalterbau GmbH, Industriestr. 12, A-3710, Austria	The manufacturer's instruction manual dated November 2003, ref M 060 01 03; and that dated August 2002 ref 99229a. Permitted Fuel: wood chip less than 35% moisture content & meeting ONORM M 7133: 1998 of either size G30 or G50
Froling Turbomat 150 and 220 manufactured by Froling Heizkessel- und Behalterbau GmbH, Industriestr.12, A-4710 Grieskirchen, Austria	The manufacturer's operating instructions dated August 2003, reference B0310003 Permitted fuel: wood pellets or wood chips*
Froling Turbomat 320 and 500 manufactured by Froling Heizkessel- und Behalterbau GmbH, Industriestr.12, A-4710 Grieskirchen, Austria	The manufacturer's operating instructions dated November 2004, reference B0340004 Permitted fuel: wood pellets or wood chips*

Class of Fireplace	*Conditions*
Gilles HPK-RA 13 kW, 15 kW, 20 kW, 25 kW, 30 kW, 35 kW, 40 kW, 45 kW, 49 kW, 60 kW, 70 kW, 75 kW, 85 kW, 95 kW, 100 kW, 120 kW and 145 kW, manufactured by GILLES Energie- und Umwelttechnik GmbH, Koaserbauer Str, 16 A-4810 Gmunden, Austria	Operating instructions dated January 2007, reference Gilles/0-145-1 Permitted fuel: wood pellets or wood chips*
Gilles HPK USK 150 kW, 180 kW, 240 kW and 300 kW, manufactured by GILLES Energie- und Umwelttechnik GmbH, Koaserbauer Str, 16 A-4810 Gmunden, Austria	Operating instructions dated January 2007, reference Gilles/150-300/1 Permitted fuel: wood pellets or wood chips*
Gilles HPK USK 360 kW manufactured by GILLES Energie- & Umwelttechnik GmbH, Koaserbauer Str 16 A-4810 Gmunden, Austria	Operating instructions dated January 2008, ref Gilles/360/1 Permitted fuel: wood pellets or wood chips
Gilles UTSK 450 kW, 550 kW boilers manufactured by GILLES Energie- & Umwelttechnik GmbH, Koaserbauer Str 16 A-4810 Gmunden, Austria	Operating instructions dated January 2008, ref Gilles/360/1 Permitted fuel: wood pellets or wood chips
Gilles UTSK 700 kW, 900 kW boilers manufactured by GILLES Energie- & Umwelttechnik GmbH, Koaserbauer Str 16 A-4810 Gmunden, Austria	Operating instructions dated January 2008, ref Gilles/360/1 Permitted fuel: wood pellets or wood chips
Gilles UTSK 1200 kW, 1600 kW boilers manufactured by GILLES Energie- & Umwelttechnik GmbH, Koaserbauer Str 16 A-4810 Gmunden, Austria	Operating instructions dated January 2008, ref Gilles/360/1 Permitted fuel: wood pellets or wood chips
The Hoval BioLyt 50 Wood Pellet Boiler (49kW thermal output) manufactured in Slovakia by Hoval of the Principality of Liechtenstein	The manufacturer's technical information and installation instructions dated October 2005, ref 4 204794/00 and operating instructions dated January 2006, ref 4 204 793/01. Permitted fuel: wood pellets*
The Hoval BioLyt 50 and 70 models manufactured by Hoval AG, Austrasse 70, 9490 Vaduz, Liechtenstein	Installation and operation manual dated June 2007, reference 4 204 794/01 and 4 204 793/02 Permitted fuel: 6 mm diameter wood pellets
The Hoval AgroLyt 20, 25, 35, 45, and 50 manufactured by Hoval AG, Austrasse 70, 9490 Vaduz, Liechtenstein	Installation and operation manuals dated January 2006, reference 4 205/ 194/00, and June 2006, reference 205 193/00 Permitted fuel: Wood logs with less than 20% moisture*
The Hoval STU 150, 200, 250, 300, 350, 425 and 500 manufactured by Hoval UK Ltd, Northgate, Neweark-on-Trent, Notts NG24 1JN	Installation and operation manual dated November 2007, reference STUman/01 Permitted fuel: 6 mm diameter wood pellets
The Hoval STU 600, 800 & 1000 manufactured by Hoval UK Ltd, Northgate, Newark-on-Trent, Notts NH24 1JN	Installation & operation manual dated June 2008, ref STU/man/01/1 Permitted fuel: 6 mm diameter wood pellets
Huntingdon clear door (model no. 7057) & Huntingdon 25 tracery door (model no. 7058) manufactured by Stovax Ltd, Flacon Road, Sowerton Industrial Estate, Exeter, Devon EX2 7LF	The manufacturer's instruction manual ref PM 249 issue 1. A permanent stop must be in place on the secondary air inlet to prevent closure beyond the 75% open position. Permitted fuel: untreated dry wood.
The Hwam Beethoven and Hwam Beethoven H (4.5 kW nominal thermal output) fitted with mechanism to prevent closure of the air wash/secondary air control manufactured by HWAM Heat Design AS, Nydamsvej 53, 8362 Horning, Denmark	Euroheat Distributors (HBS) Ltd's operational instructions dated November 2006, ref IN1141 Ed B. Permitted fuel: untreated dry wood*
The Hwam Mozart (4.5 kW nominal thermal output) fitted with mechanism to prevent closure of the air wash/secondary air control manufactured by HWAM Heat Design AS, Nydamsvej 53, 8362 Horning, Denmark	Euroheat Distributors (HBS) Ltd's operational instructions dated November 2006, ref IN1141 Ed B. Permitted fuel: untreated dry wood*
The Hwam Ravel (4.5 kW nominal thermal output) fitted with mechanism to prevent closure of the air wash/secondary air control manufactured by HWAM Heat Design AS, Nydamsvej 53, 8362 Horning, Denmark	Euroheat Distributors (HBS) Ltd's operational instructions dated November 2006, ref IN1141 Ed B. Permitted fuel: untreated dry wood*
The Hwam Vivaldi (4.5 kW nominal thermal output) fitted with mechanism to prevent closure of the air wash/secondary air control manufactured by HWAM Heat Design AS, Nydamsvej 53, 8362 Horning, Denmark	Euroheat Distributors (HBS) Ltd's operational instructions dated November 2006, ref IN1141 Ed B. Permitted fuel: untreated dry wood*
The Hwam 30 (4.5 kW nominal thermal output) fitted with mechanism to prevent closure of the air wash/secondary air control manufactured by HWAM Heat Design AS, Nydamsvej 53, 8362 Horning, Denmark	Euroheat Distributors (HBS) Ltd's operational instructions dated November 2006, ref IN1141 Ed B. Permitted fuel: untreated dry wood*
KWB TDS Powerfire 130, 240 & 300 manufactured by KWB-Draft und Warme aus Biomasse GmbH, Industriestr 235, A-8321 St. Margarethen an der Raab	The manufacturer's instructions dated 14 May 2008, ref BA TDS-05.2008, including specification of recommended fuels. Permitted fuel: wood pellets or wood chips

Class of Fireplace	*Conditions*
KWB TDS Powerfire 150 manufactured by KWB-Kraft und Warme aus Biomasse GmbH of Industriestrasse 235, A-8321 St. Margarethen an der Raab	The manufacturer's instructions dated August 2005, reference BA TDS-0805 Permitted fuel: wood pellets or wood chips*
The Little Thurlow wood burning stove manufactured by Town & Country Fires, Pickering, N. Yorkshire	The manufacturer's instructions dated 30 June 2007, issue no. 3 Permitted fuel: seasoned logs of less than 20% moisture and of a max length of 350 mm*
The Lovenholm manufactured by Hunter Stoves Ltd, Unit 6, Old Mill Industrial Estate, Stoke Canon, Exeter, Devon EX5 4RJ	Lovenholm Multifuel stove installation & operation manual version 07/08 Permitted fuel: untreated wood with a moisture content below 20%
The Morso 1400 series consisting of the 1412 and 1442 manufactured by Morso Jernstoberi A/S DK-7900 Nykobing, Mors, Denmark	Installation and operation manual dated 10 January 2008, reference 72146800 Permitted fuel: wood logs with less than 20% moisture*
The Morso 3100 series consisting of the 3112 and 3142 models manufactured by Morso Jernstoberi A/S DK-7900 Nykobing, Mors, Denmark	Installation and operation manual dated 4 February 2008, reference 72311200 Permitted fuel: wood logs with less than 20% moisture*
The Morso 6140 and 6148 manufactured by Morso Jernstoberi A/S DK-7900 Nykobing, Mors, Denmark	Installation and operation manual dated 4 September 2007, reference 72610400 Permitted fuel: wood logs with less than 20% moisture*
The Morso Owl 3410, 3420 & 3440 for wood burning manufactured by Morso Jernstoberi A/S, Denmark	The manufacturer's instructions dated 1 January 2000, ref 72345600. Permitted Fuel: untreated dry wood which has been split, stacked & air dried; or air dried wood briquettes
The Nestor Martin Harmony Multi-Fuel stove H13 (3.5kW nominal thermal output) and H33 (8kW nominal thermal output) manufactured by Nestor Martin, SA Finnuick Nv, Voie Axiale 5, B-5660 Couvin, Belgium	Euroheat Distributors (HBS) Ltd's operational instructions dated January 2007, ref IN1116 Edition D. Permitted fuel: untreated dry wood
The Nestor Martin Harmony Multi-Fuel stove H23 (6kW nominal thermal output) manufactured by Nestor Martin, SA Finnuick Nv, Voie Axiale 5, B-5660 Couvin, Belgium	Euroheat Distributors (HBS) Ltd's operational instructions dated November 2006, ref IN1141 Ed B. Permitted fuel: untreated dry wood
The Nestor Martin Stanford Multi-Fuel stove S13, S23 & SP23 (6kW nominal thermal output) manufactured by Nestor Martin, SA Finnuick Nv, Voie Axiale 5, B-5660 Couvin, Belgium	Euroheat Distributors (HBS) Ltd's operational instructions dated November 2006, ref IN1141 Ed B. Permitted fuel: untreated dry wood
The Nestor Martin Stanford Multi-Fuel stove S13 (3.5kW nominal thermal output), S33 & SP33 (8kW nominal thermal output) manufactured by Nestor Martin, SA Finnuick Nv, Voie Axiale 5, B-5660 Couvin, Belgium	Euroheat Distributors (HBS) Ltd's operational instructions dated January 2007, ref IN1116 Edition D. Permitted fuel: untreated dry wood
The Nestor Martin R23 stove (6kW nominal thermal output) manufactured by Nestor Martin, SA Finnuick Nv, Voie Axiale 5, B-5660 Couvin, Belgium	Euroheat Distributors (HBS) Ltd's operational instructions dated November 2006, ref IN1141 Ed B. Permitted fuel: untreated dry wood
The Nestor Martin multi-fuel stove IT 13 (3.5kW nominal thermal output) manufactured by Nestor Martin, SA Finnuick Nv, Voie Axiale 5, B-5660 Couvin, Belgium	Euroheat Distributors (HBS) Ltd's operational instructions dated April 2007, ref IN1180 Edition C. Permitted fuel: untreated dry wood
The Nestor Martin multi-fuel stove R33 & D33 (8kW nominal thermal output) manufactured by Nestor Martin, SA Finnuick Nv, Voie Axiale 5, B-5660 Couvin, Belgium	Euroheat Distributors (HBS) Ltd's operational instructions dated January 2007, ref IN1116 Edition D. Permitted fuel: untreated dry wood
OkoFEN Pellematic PE08, PE12, PE16, PEK16, PEK20, PEK25, PEK32, PES12, PES16, PES20, PES25, PES32, PESK12, PESK16, PESK20, PESK25 & PESK32 manufactured by OkoFEN Forschungs- & Entwicklungs GmbH of Gewerbepark 1, 4133, Niederkappel, Austria	OkoFEN operation instructions dated February 2007, ref PE/HB/001.E Permitted fuel: wood pellets of maximum diameter 6mm & maximum length 4cm
OkoFEN Pellematic PE36, PE48, PE56, PESK36, PESK48, PESK56, PEK36, PEK48, PEK56, PES36, PES48 & PES56 manufactured by OkoFEN Forschungs- & Entwicklungs GmbH of Gewerbepark 1, 4133, Niederkappel, Austria	OkoFEN operation instructions, ref PBC 2000 CMP 1.4 (V2.31) & with revision dated March 2008-09-11 Permitted fuel: wood pellets of maximum diameter 6mm & maximum length 4cm
The Orchard Ovens models: FVR Speciale 80, 100, 110, 110x160 & 120; TOP Superiore 100 & 120; GR 100, 120, 140, 120x160, 140x160, 140x180 & 180; OT 100, 120, 140, 120x160, 140x160, 140x180 & 180; the Valoriani Piccolo, all of which are manufactured by Valoriani of Via Caselli alla Fornace, 213, 50066 Reggello, Firenze, Italy	The manufacturer's instructions dated 25 October 2004, references: Instruction OOLinst.1: model FVR 80; OOLinst.2: FVR 100; OOLinst.3: FVR 110; OOLinst.4: FVR 120;OOLinst.5: FVR 110x160; OOLinst.6: TOP 100; OOLinst.7: TOP 120; OOLinst.8: GR 100 & OT 100; OOLinst.9: GR 120 & OT 120; OOLinst.10: GR 140 & OT 140; OOLinst.11: GR 120x160 & OT 120x160; OOLinst.12: GR 140x160 & OT 140x160; OOLinst.13: GR 140x180 & OT 140x180; OOLinst.14: GR 180 & OT 180; OOLinst.15: Valoriani Piccolo. Permitted fuel: untreated dry wood

Class of Fireplace	*Conditions*
The Quadra-Fire 2100 Millennium Freestanding wood burning stove manufactured by Hearth & Home Technologies 1445 North Highway, Colville, WA 99114,	The manufacturer's instructions dated 9 October 2003, ref 250-6931B. Permitted fuel: dry cured untreated wood of max size 38.1 cm x 10.16 cm x 5.08
The Quadra-Fire Yosemite Freestanding wood burning stove manufactured by Hearth & Home Technologies 1445 North Highway, Colville, WA 99114, USA	The manufacturer's instructions dated 4 June 2003, ref 7004-187 Permitted fuel: dry cured untreated wood of max size 38.1 cm x 10.16 cm x 5.08 cm
The Quadra-Fire Cumberland Gap Freestanding wood burning stove manufactured by Hearth & Home Technologies 1445 North Highway, Colville, WA 99114, USA	The manufacturer's instructions dated 8 July 2003, ref 7006-186 Permitted fuel: dry cured untreated wood of max size 38.1 cm x 10.16 cm x 5.08 cm
Scan Andersen 4-5 manufactured by Kog Iversen & Co, A/S DK-5492 Vissenbjerg, Denmark	The manufacturer's instruction manual dated 15 November 2007 entitled "Scan Andersen 4-5", ref Edition 09/05-GB Permitted fuel: untreated dry wood
Stockton 5 Flat top (model no. 7119), Stockton 5 Canopy (model no. 7160) & Stockston 5 Midline (model no. 7133) manufactured by Stovax Ltd, Falcon Road, Sowerton Industrial Estate, Exeter, Devon EX2 7LF.	The manufacturer's instruction manual ref. PM 176 – issue 4. A permanent stop must be in place on the secondary air inlet to prevent closure beyond the 50% open position. Permitted fuel: untreated dry wood
The Type USP wood pellet-fired boilers, model nos. USP-10, 15, 20, 25 & 30, manufactured by KWB – Kraft & Warme aus Biomasse Gesellschaft mit beschrankter Haftung (GmbH) of Industriestrasse 235, A-8321 St Margarethen an der Raab, Austria	The manufacturer's instructions dated 28 July 2003, ref BA-USP 0703. Permitted fuel; wood pellets*
The Type USV wood pellet and woodchipfired boilers, model nos. USV-15, 25, 30, 40, 50, 60, 80 & 100, manufactured by KWB – Kraft & Warme aus Biomasse Gesellschaft mit beschrankter Haftung (GmbH) of Industriestrasse 235, A-8321 St Margarethen an der Raab, Austria	The manufacturer's instructions dated August 2003, ref BA-USV 0803. Permitted fuel; wood pellets or wood chips*
Westfire Uniq 15, 16 and 20 (also known as WF UNIQ 15, 16 and 20) manufactured by Westfire Tomrervey 3 DK-6800 Varde, Denmark	Installation and operating instructions dated 2007, reference WF15-WF16-WF20, together with supplementary instructions dated 2007, reference 01wfuk.2007. A permanent stop must be in place on the secondary air inlet to prevent closure beyond 50 mm Permitted fuel: untreated dry wood*
Winhager BioWin 90 & 120 manufactured by Windhager Zentralheizung GmbH, Anton-Windhager-Str 20, A-5201 Seerkirchen bei Salzburg, Austria.	The manufacturer's instruction manual dated August 2008, ref 092125/00 Permitted fuel: wood pellets
Winhager BioWin 100, 150, 210 & 260 manufactured by Windhager Zentralheizung GmbH, Anton-Windhager-Str 20, A-5201 Seerkirchen bei Salzburg, Austria.	The manufacturer's instruction manual dated May 2008, ref WT10 & WT15 Permitted fuel: hardwood and softwood offcuts
The Wood Waste Technology Heater, models WT10 & WT15, manufactured by Wood Waste Technology	The manufacturer's instructions dated 19 January 2005. ref WT10 & WT15. Permitted fuel: hardwood & softwood offcuts*

* The fuel must not contain halogenated organic compounds or heavy metals as a result of treatment with wood-preservatives or coatings.

APPENDIX 4.1

ENVIRONMENTAL NOISE – IDENTIFICATION OF NOISE SOURCES IN ENGLAND

The Environmental Noise (Identification of Noise Sources) (England) Regulations 2007, SI 2007/415, as amended by SI 2007/2458, identify first round agglomerations, first round major roads, first round major railways and major airports to be mapped in accordance with the Environmental Noise Directive. Maps identifying the agglomerations, roads, railways have been deposited at Defra, Ashdown House, 123 Victoria Street, London SW1e 6DE, and can be inspected during office hours. They are also available at http://www.defra.gov.uk/environment/noise/ambient.htm

The major airports which have been identified in the Regulations are:

Birmingham International
Blackpool Squire's Gate
Bournemouth
Bristol Lulsgate
Leeds Bradford
Liverpool John Lennon
London City
London Gatwick
London Heathrow
London Luton
London Stansted
Manchester International
Newcastle International
Nottingham East Midlands
Southampton Eastleigh

Maps for the above airports, together with maps for Coventry, Southend and Shoreham, were published by Defra in late 2007 and are abvialbe on their website.

Note: In Scotland noise maps have been produced for Aberdeen, Edinburgh, Glasgow and Prestwick Airports.

There are no first round major airports in Wales.

APPENDIX 5.1

GUIDANCE – WASTE MANAGEMENT

Standard Permits for Waste sites (E & W)

The Environment Agency has developed standard rules for a number of low to medium risk activities:

- Waste transfer and treatment operations, including household waste amenity sites and materials recycling facilities
- Biological treatment of waste, including open windrow composting
- Metal recycling, including vehicle depollution and dismantling
- Transfer and treatment of clinical waste
- Mobile remediation plant
- Treatment of WEEE
- Sewage sludge treatment
- Deposit of dredging sludge
- Small scale animal carcass incineration

Consultation on draft standard rules for the following was issued in August 2008:

- Pet cemeteries
- Low-impact installations
- Installations producing biodiesel
- Waste transfer stations for inert and excavation waste, processing up to 250,000 tonnes a year of material
- Sites storing scrap metal for recovery, up to 1 million tonnes a year
- The disposal of inert wastes at mines and quarries.

Operators of waste management facilities covered by the above SRs have the option of applying for either a standard rules permit or a "bespoke permit"; there is no right of appeal against SRs, nor can they be varied; instead if the activities at the site change in a way not covered by SRs, then the operator should apply for a "bespoke permit".

Landfill Guidance

The Environment Agency has produced a number of guidance notes relating to landfill, including those listed below. These are also available on the Environment Agency website, with similar guidance being developed by SEPA. Further sources of guidance are noted in the main text.

Technical Guidance Notes

01 Hydrogeological risk assessments for landfills and derivation of groundwater control and trigger levels (2003)
02 Monitoring landfill leachate, groundwater and surface water (2004)
03 Management of landfill gas (2004)
04 Monitoring trace components in landfill gas (2004)
05 Monitoring enclosed landfill gas flares (2005)
06 Guidance on gas treatment technologies for landfill gas engines (2005)
07 Guidance on monitoring landfill gas surface emissions (2005)
08 Guidance for monitoring landfill gas engine emissions (2005)

Technical guidance on capping and restoration of landfills (draft, 2005)
Technical requirements of the landfill directive and IPPC (draft, 2005)
Technical guidance for the treatment of landfill leachate (draft, 2006)

Guidance on sampling and testing of wastes to meet landfill acceptance criteria (v.4.3a, 2003)

Guidance on the assessment of risks from landfill sites

Regulatory Guidance Notes

01 Classification of sites (2003)
02 Interim waste acceptance criteria and procedures (v4, 2002)
03 Groundwater protection: locational aspects of landfills in planning consultation responses and permitting decisions (v4, 2002)
04 Defining existing landfill sites
05 Habitats regulations and the landfill directive
06 Interpretation of the engineering requirements of sch.2 of the Landfill (E&W) Regulations 2002 (v2, 2004)

07 Requirements for landfills that stop operating: closure & aftercare (v2.0)
11 Disposal in landfills for non-hazardous waste of stable, non-reactive hazardous wastes, asbestos wastes, or wastes with high sulphate or gypsum contents (internal guidance, v2, 2005)
14 Duty of care and European Waste Catalogue
15 Applicability of the landfill directive to the deposit of waste in lagoons
16 Establishing area to be covered by PPC permit for a landfill
17 Ban on the landfilling of whole used tyres and shredded used tyres in accordance with the Landfill (E&W) Regulations 2002 (2003)

General Guidance

Requirements for waste destined for disposal in landfill: a guide for waste producers and waste managers (2004)

A Better Place: guidance for waste destined for disposal in landfills: interpretation of the waste acceptance requirements of the Landfill (E&W) Regulations 2002 (2005)

Guidance on Offences (Nov 05, Defra)

Guidance on the classification and coding of mixed wastes at authorised treatment facilities (Env Agency, 2005)

Classification and Coding of Wastes from treatment facilities (Environment Agency, consultation draft 27.10.06)

Guidance on pet cemeteries and crematoria, Environment Agency , v.1, 25.03.08

Definition of Waste: Development Industry Code of Practice, CL:AIRE Contaminated Land: Applications in Real Environments), 2008

Environment Agency Guidance for Registering an Exempt Activity:
Land Treatment (WMX7)
Waste for Construction (WMX19)
WEEE (WMX49)

Hazardous Waste

WM2 - Interpretation of the definition and classification of hazardous waste (current edition, May 2008) (EA, SEPA, NIEA)

Hazardous Waste Regulations (E & W) 2005 – Guidance on transitional provisions for permitting regimes (Nov 05, Defra)

HWR01 What is a Hazardous Waste, v4 July 2008 (Environment Agency)
HWR02A Do I need to notify my premises
HWR03A Consignment notes: standard procedures
HWR03B Consignment notes: multiple collections
HWR03C Consignment notes: pipelines and ships
HWR03D Consignment notes: cross-border movements
HWR03E Consignment notes: schedule of carriers
HWR03F Consignment notes: rejected loads
HWR07 Mobile services

Treatment and Landfilling of Hazardous Waste (Environment Agency, November 2006)

Regulatory Position Statement: Hazardous Waste Treatment (June 2008)
Guidance on Classifying and Coding Hazardous Waste (June 2008)

Waste Electrical & Electronic Equipment

WEEE Regulations, Government Guidance Notes, (URN 07/781), DTI, February 2007

Code of Practice for collection of WEEE from designated collection facilities, DTI, February 2007

Waste Management Papers

This series of papers provided guidance on all aspects of waste management. Section 35(8) of the *Environmental Protection Act 1990* requires regulatory authorities to take account of guidance (such as WMPs) issued by the Secretary of State in the discharge of their licence functions. WMPs listed here were prepared by the Department of the Environment.

WMP 1 A review of options (second edition 1992)
WMP 2/3 [replaced by DOE Guidance, *Waste Management Planning - Principles and Practice* (1995)
WMP 4 The licensing of waste facilities (1994, amended 2000) (under review, 2003)
WMP 4A Licensing of metal recycling sites (1995, amended 2000)
WMP 5 The relationship between waste disposal authorities and private industry (1976)
WMP 6 Polychlorinated biphenyl (PCB) wastes - a technical memorandum on reclamation, treatment and disposal (1994)
WMP 7 Mineral oil wastes - a technical memorandum on arisings, treatment and disposal (1976)
WMP 8 Heat treatment cyanide wastes - a technical memorandum on arisings, treatment and disposal (second edition 1985)
WMP 9 Halogenated hydrocarbon solvent wastes from cleaning processes - a technical memorandum on reclamation and disposal (1976)
WMP 10 local authority waste disposal statistics 1974/75 (1976)
WMP 11 Metal finishing wastes - a technical memorandum on arisings, treatment and disposal (1976)
WMP 12 Mercury bearing wastes - a technical memorandum on storage, handling, treatment, disposal and recovery (1979)
WMP 13 Tarry and distillation wastes and other chemical based wastes - a technical memorandum on arisings, treatment and disposal (1977)
WMP 14 Solvent wastes (excluding halogenated hydrocarbons) - a technical memorandum on reclamation and disposal (1977)
WMP 15 Halogenated organic wastes - a technical memorandum on arisings, treatment and disposal (1978)
WMP 16 Wood preserving wastes - a technical memorandum on arisings, treatment and disposal (1980)
WMP 17 Wastes from tanning, leather dressing and fellmongering - a technical memorandum on recovery, treatment and disposal (1978)
WMP 18 Asbestos waste - a technical memorandum on arisings and disposal (1979)
WMP 19 Wastes from the manufacture of pharmaceuticals, toiletries and cosmetics - a technical memorandum on arisings, treatment and disposal (1978)
WMP 20 Arsenic bearing wastes - a technical memorandum on recovery, treatment and disposal (1980)
WMP 21 Pesticide wastes - a technical memorandum on arisings and disposal (1980)
WMP 22 Local authority waste disposal statistics 1974/75 to 1977/78 (1978)
WMP 23 Special wastes - a technical memorandum providing guidance on their definition (1981)
Technical guidance note on special wastes (EA, SEPA, DOENI, 1999)
WMP 24 Cadmium bearing wastes - a technical memorandum on arisings, treatment and disposal (1984)
WMP 25 Clinical wastes - a technical memorandum on arisings, treatment and disposal (replaced 1999, HSC Guidance: Safe disposal of clinical waste
WMP 26 Landfilling wastes - a technical memorandum on landfill sites (1986)
WMP 26A Landfill completion - a technical memorandum providing guidance on assessing the completion of licensed landfill sites (1994)
WMP 26B Landfill design, construction and operation (1995)
WMP 26D Landfill monitoring
WMP 26E Landfill restoration and post closure management (1996)
WMP 26F Landfill co-disposal
WMP 27 Landfill gas - a memorandum for local authorities on recycling (1991)
WMP 28 Recycling - a memorandum for local authorities on recycling (1991)

Waste Policy Guidance: Preparing and revising local authority recycling strategies and recycling plans (DETR, 1998)

WMP 401 – Guidance note: Policy on licence surrender (Environment Agency, 1999)

APPENDIX 6.1A

FRAMEWORK DIRECTIVE ON WATER POLICY

Priority List of Water Pollutants

The Framework Directive on water policy (see chapter 6, 6.5) was adopted in October 2000. In 2001 the Council of Environment Ministers formally adopted Decision 2455/2001/EC *establishing the list of priority substances in the field of water policy (Annex X to the Framework Directive).* These substances were selected on the basis of toxicity, persistence and bioaccumulation and thus the risk they pose to the aquatic environment and to human health. Annex II to the proposal for a *Directive on environmental quality standards in the field of water policy* (COM(2006) 397) replaces Annex X with a new list which identifies further priority hazardous substances from those which were under appraisal as required by the 2001 Decision. Annex I sets environmental quality standards for these and eight other pollutants – see Appendix 6.1B.

Priority Hazardous Substances
(Aim is to terminate or phase-out releases within 20 years)
Anthracene
Brominated diphenylethers
Cadmium and its compounds
C_{10-13}-chloralkanes
Endosulfan (alpha-endosulfan)
Hexachlorobenzene
Hexachlorobutadiene
Hexachlorocyclohexane (gamma isomer, lindane)
Mercury and its compounds
Nonylphenols (4-(para)-nonylphenol)
Pentachlorobenzene
Polyaromatic hydrocarbons (Benzo(a)pyrene, Benzo(b)fluoranthene, Benzo(g, h, i)perylene, Benzo(k)fluoranthene, fluoranthene, indeno (1,2,3-cd)pyrene)
Tributyltin compounds (tributyltin-cation)

Priority Substances
Alachlor
Atrazine
Benzene
Chlorfenvinphos
Chlorpyrifos
1,2-Dichloroethane
Dichloromethane
Di(2-ethylhexyl)phthalate (DEHP)
Diuron
Fluoranthene
Isoproturon
Lead and its compounds
Naphthalene
Nickel and its compounds
Octylphenols (para-tert-octylphenol)
Pentachlorophenol
Simazine
Trichlorobenzenes (1,2,4-Trichlorobenzene)
Trichloromethane (Chloroform)
Trifluralin

APPENDIX 6.1B

PROPOSED ENVIRONMENTAL QUALITY STANDARDS FOR PRIORITY SUBSTANCES AND CERTAIN OTHER SUBSTANCES (COM 2006/397)

AA: Annual Average; MAC: Maximum Allowable Concentration
Unit: µg/l

Substance	**Inland Surface Waters**		**Other Surface Waters**	
	AA-EQS	**MAC-EQS**	**AA-EQS**	**MAC-EQS**
1. Alachlor	0.3	0.7	0.3	0.7
2. Anthracene	0.1	0.4	0.1	0.4
3. Atrazine	0.6	2.0	0.6	2.0
4. Benzene	10	50	8	50
5. Pentabromodiphenylether	0.0005	n/a	0.0002	n/a
6. Cadmium & its compounds *(depends on water hardness class)*[1]				
class 1	≤-0.08	≤-0.45	0.2	
class 2	0.08	0.45		
class 3	0.9	0.6		
class 4	0.15	0.9		
class 5	0.25	1.5		
7. C10-13 Chloralkanes	0.4	1.4	0.4	1.4
8. Chlorfenvinphos	0.1	0.3	0.1	0.3
9. Chlorpyrifos	0.03	0.1	0.03	0.1
10. 1,2-Dichloroethane	10	n/a	10	n/a
11. Dichloromethane	20	n/a	20	n.a
12. Di(2-ethylexyl)phthalate (DEHP)	1.3	n/a	1.3	n/a
13. Diuron	0.2	1.8	0.2	1.8
14. Endosulfan	0.005	0.01	0.0005	0.004
15. Fluoranthene	0.1	1	0.1	1
16. Hexachlorobenzene	0.01	0.05	0.01	0.05
17. Hexachlorobutadiene	0.1	0.6	0.1	0.6
18. Hexachlorocyclohexane	0.02	0.04	0.002	0.02
19. Isoproturon	0.3	1.0	0.3	1.0
20. Lead & its compounds	7.2	n/a	7.2	n/a
21. Mercury & its compounds	0.05	0.07	0.05	0.07
22. Naphthalene	2.4	n/a	1.2	n/a
23. Nickel & its compounds	20	n/a	20	n/a
24. Nonylphenol	0.3	2.0	0.3	2.0
25. Octylphenol	0.1	n/a	0.01	n/a
26. Pentachlorobenzene	0.007	n/a	0.0007	n/a
27. Pentachlorophenol	0.4	1	0.4	1
28. Polyaromatic hydrocarbons (PAH)	note 2	note 2	note2	note 2
Benzo(a)pyrene	0.05	0.1	0.05	0.1
Benzo(b)fluoranthene } Benzo(k)fluoroanthene }	Σ=0.03	n/a	Σ=0.03	n/a
Benzo(g.h,I)perylene } Indeno(1,2,3-cd)pyrene }	Σ=0.002	n/a	Σ=0.002	n/a
29. Simazine	1	4	1	4
30. Tributyltin compounds	0.0002	0.0015	0.0002	0.0015
31. Trichlorobenzenes *(all isomers)*	0.4	n/a	0.4	n/a
32. Trichloromethane	2.5	n/a	2.5	n/a
33. Trifluralin	0.03	n/a	0.03	n/a
Environmental Quality Standards for Other Pollutants				
1. DDT total	0.025	n/a	0.025	n/a
Para-para-DDT	0.01	n/a	0.01	n/a
2. Aldrin } 3. Dieldrin } 4. Endrin } 5. Isodrin }	Σ=0.010	n/a	Σ=0.005	n/a
6. Carbontetrachloride	12	n/a	12	n/a
7. Tetrachloroethylene	10	n/a	10	n/a
8. Trichloroethylene	10	n/a	10	n/a

1. Class 1: < 40 mg $CaCO_3$/l; Class 2: 40 to < 50 mg $CaCO_3$/l; Class 3: 50 to < 200 mg $CaCO_3$/l; Class 4: 100 to < 200 mg $CaCO_3$/l; Class 5: ≤/- 200 mg $CaCO_3$/l.

2. For PAHs each individual EQS shall be complied with; i.e. the EQS for Benzo(a)pyrene and the EQS for the sum of Benzo(b)fluoranthene and Benzo(k)fluoroanthene and the EQS for the sum of Benzo(g,h,i)perylene and Indeno(1,2,3-cd)pyrene must be met.

APPENDIX 6.2A

QUALITY REQUIREMENTS FOR BATHING WATER (76/160/EEC)

	Parameters		*G (Guideline)*	*I (Mandatory)*	*Minimum sampling frequency*	*Method of analysis and inspection*
MICROBIOLOGICAL						
1	Total coliforms /100 ml		500	10000	Fortnightly (1)	Fermentation in multiple tubes subculturing of the positive tubes on a confirmation medium.
2	Faecal coliforms /100 ml		100	2000	Fortnightly (1)	Count according to MPN (most probable number) or membrane filtration and culture on an appropriate medium such as Tergitol lactose agar, endo agar. 0.4% Teepol broth, subculturing and identification of the suspect colonies. In the case of 1 and 2, the incubation temperature is variable according to whether total or faecal coliforms are being investigated.
3	Faecal streptococci	/100ml	100	-	(2)	Litsky method. Count according to MPN (most probable number) or filtration on membrane. Culture on an appropriate membrane.
4	Salmonella / 1 litre		-	0	(2)	Concentration by membrane filtration. Inoculation on a standard membrane. Enrichment – subculturing on isolating – agar identification.
5	Entero viruses PFU/10 litres		-	0	(2)	Concentrating by filtration, flocculation or centrifuging and confirmation.
PHYSICO-CHEMICAL:						
6	pH		-	6 to 9 (0)	(2)	Electrometry with calibration at pH 7 and 9.
7	Colour		- -	No abnormal change in colour (0) -	Fortnightly (1) (2)	Visual inspection or photometry with standards on the Pt. Co scale.
8	Mineral oils	mg/litre	- <0.3	No film visible on the surface of the water and no odour -	Fortnightly (1) (2)	Visual and olfactory inspection or extraction using an adequate volume and weighing the dry residue.
9	Surface-active substances reacting with methylene blue	mg/litre (lauryl-sulfate)	- <0.3	No lasting foam	Fortnightly (1) (2)	Visual inspection or absorption spectrophotometry with methylene blue.
10	Phenols (phenolindices)	mg/litre C_6H_5OH	- ≤0.005	No specific odour ≤0.05	Fortnightly (1) (2)	Verification of the absence of specific odour due to phenol or absorption spectrophotometry 4-aminoantipyrine (4 AAP) method.
11	Transparency	m	2	1(0) (1)	Fortnightly	Secchi's disc.
12	Dissolved oxygen	% saturation O_2	80 to 190	-	(2)	Winkler's method or electrometric method (oxygen meter).
13	Tarry residues and floating materials such as wood, plastic articles, bottles, containers of glass, plastic, rubber or any other substance. Waste or splinters.		Absence		Fortnightly (1)	Visual inspection.
14	Ammonia	mg/litre NH4			(3)	Absorption spectrophotometry. Nessler's method, or indophenol blue method.

	Parameters		G (Guideline)	I (Mandatory)	Minimum sampling frequency	Method of analysis and inspection
15	Nitrogen Kjeldahl	mg/litre N			(3)	Kjeldahl method.
16	Pesticides (parathion, HCH, dieldrin)	mg/litre			(2)	Extraction with appropriate solvents and chromatographic determination.
17	Heavy metals such as – arsenic As – cadmium Cd – chrom VI CrVI – lead Pb – mercury Hg		mg/litre			(5) Atomic absorption possible preceded by extraction
18	Cyanides	mg/litre CN			(2)	Absorption spectrophotometry using a specific reagent
19	Nitrates and phosphates	mg/litre NI_1 PO_4			(2)	Absorption spectrophotometry using a specific reagent

G guide; I mandatory; (0) provision exists for exceeding the limits in the event of exceptional geographical or meteorological conditions; (1) When a sampling taken in previous years produced results which are appreciably better than those in this Annex and when no new factor likely to lower the quality of the water has appeared, the competent authorities may reduce the sampling frequency by a factor of 2; (2) Concentration to be checked by the competent authorites when an inspection in the bathing area shows that the substance may be present or that the quality of the water has deteriorated; (3) These parameters must be checked by the competent authorites when there is a tendency towards the eutrophication of the water.

EU DIRECTIVE 2006/7/EC concerning the management of bathing water quality, Annex 1

Member States have until 24 March 2008 to transpose this Directive into national legislation – see chapter 6, 6.5.2. Member States will be required to ensure that bathing water quality is monitored in accordance with the parameters set out in Annex I to the Directive, with analysis of samples carried out in accordance with the reference methods specified.

Parameter	Excellent	Quality Good	Sufficient
Inland Waters			
Intestinal enterococci (cfu/100 ml)[1]	200*	400*	330**
Escherichia coli (cfu/100ml)[2]	500*	1000*	900**
Coastal & transitional waters			
Intestinal enterococci (cfu/100 ml)[1]	100*	200*	185**
Escherichia coli (cfu/100ml)[2]	250*	500*	500**

1. Reference method of analysis: ISO 7899-1 or ISO 7899-2
2. Reference method of analysis: ISO 9308-3 or ISO 9308-1

* Based upon a 95-percentile evaulation
** Based upon a 90-percentile evaluation

Bathing waters are to be classified as "poor" if, in the set of bathing water quality data for the last assessment period, the percentile values for microbiological enumerations are worse than the "sufficient" values.

APPENDIX 6.2B

THE BATHING WATERS (CLASSIFICATION) REGULATIONS 1991

Quality and Additional Sampling Requirements in Compliance with Directive 76/160/EEC

	Parameter	*Parametric value*	*Minimum sampling frequency*	*Method of analysis and inspection*
MICROBIOLOGICAL				
1	Total coliforms	10,000/ 100ml	Fortnightly (1)	Fermentation in multiple tubes. Sub-culturing of the positive tubes on a confirmation medium. Either counting according to MPN (most probable number) or membrane filtration, culturing on an appropriate medium, subculturing and identification of the suspect colonies.
2	Faecal coliforms	2,000/ 100ml	Fortnightly (1)	The incubation temperature is variable according to whether total or faecal coliforms are being investigated.
3	Salmonella	Absent in 1 litre	(2)	Membrane filtration, culturing on an appropriate medium, sub-culturing and identification of the suspect colonies.
4	Entero viruses	No plaque forming units in 10 litres	(2)	Concentration (by filtration, flocculation or centrifuging) and confirmation.
PHYSICO-CHEMICAL:				
5	pH	6 to 9	(2)	Electrometry with calibration at pH 7 and 9.
6	Colour	No abnormal change in colour	Fortnightly (1)	Visual inspection or photometry with standards on the platinum cobalt scale.
7	Mineral oils	No film visible on the surface of the water and no odour	Fortnightly (1)	Visual and olfactory inspection.
8	Surface-active substances reacting with methylene blue	No lasting foam	Fortnightly (1)	Visual inspection.
9	Phenols (phenol indices)	No specific odour	Fortnightly (1)	Olfactory inspection.
		≤0.05 mg/litre (C_6H_5OH)	(2)	Absorption spectrophotometry 4-aminoantipyrine (4 AAP) method.
10	Transparency	1 metre	Fortnightly (1)	Secchi's disc.

Notes

1. Samples may be taken at intervals of four weeks where samples taken in previous years show that the waters are of an appreciably higher standard than that required for the classification in question and the quality of the waters has not subsequently deteriorated and is unlikely to do so.
2. Samples must be taken in relation to this parameter when there are grounds for suspecting that there has been a deterioration in the quality of the waters or the substance is likely to be present in the waters.

THE BATHING WATER REGULATIONS 2008

The Bathing Water Regulations 2008 (SI 2008/1097) apply in England and Wales and began coming into force on 14 May 2008. Schedule 4 to the Regulations details the standards the Environment Agency must use for classification of bathing waters. Designated bathing waters are required to achieve "sufficient" status by the end of the 2015 bathing season. There are similar Regulations for Scotland (SSI 2008/170) and Northern Ireland (SR 2008/231) (Sch. 3 to the Scottish Regulations details the standards for classification.

Standards for inland waters

Parameter	*Excellent*	*Good*	*Sufficient*
Intestinal Enterococci(1)	200(2)	400(2)	330(3)
Escherichia coli(1)	500(2)	1,000(2)	900(3)

Standards for coastal and transitional waters

Parameter	*Excellent*	*Good*	*Sufficient*
Intestinal Enterococci(1)	100(2)	200(2)	185(3)
Escherichia coli(1)	250(2)	500(2)	500(3)

1. Colony forming units per 100 millilitres (cfu/100 ml).
2. Based upon a 95-percentile evaluation.
3. Based upon a 90-percentile evaluation.

APPENDIX 6.3

WATER SUPPLY (WATER QUALITY) REGULATIONS 2000

These Regulations, as amended by SI 2007/2734 began coming into force from 1 January 2001, replacing the 1989 (and amendment) Regulations – see chapter 6, 6.17.1. They implement the 1998 EC Drinking Water Directive (98/83/EC) – see chapter 6, 6.7.5.

Schedules 1 and 2 to the Regulations, shown below, list prescribed concentrations and values, and indicator parameters, respectively. Schedules 3 and 4 cover monitoring and analytical methodology.

Schedule 1: Prescribed concentrations and values

Table A: Microbiological parameters

Item	*Parameters*	*Concentration or value (max)*	*Units of Measurement*
Part I: Directive requirements			
1.	Enterococci	0	number/100 ml
2.	Escherichia coli (E.coli)	0	number/100 ml
[point of compliance: consumers' taps]			
Part II: National requirements			
1.	Coliform bacteria	0	number/100 ml
2.	Escherichia coli (E. coli)	0	number/100 ml
[point of compliance: service reservoirs and water treatment works; 95% of samples to comply from each service reservoir]			

Table B: Chemical parameters

Part I: Directive requirements			
1.	Acrylamide (i)	0.10	μg/l
2.	Antimony	5.0	μgSb/l
3.	Arsenic	10	μgAs/l
4.	Benzene	1.0	μg/l
5.	Benzo(a)pyrene	0.010	μg/l
6.	Boron	1.0	mgB/l
7.	Bromate	10	$\mu gBrO_3/l$
8.	Cadmium	5.0	μgCd/l
9	Chromium	50	μgCr/l
10.	Copper (ii)	2.0	mg Cu/l
11.	Cyanide	50	μgCN/l
12.	1,2 dichloroethane	3.0	μg/l
13.	Epichlorohydrin (i)	0.10	μg/l
14.	Fluoride	1.5	μg F/l
15.	Lead (ii)		
	from 25.12.03-24.12.13	25	μgPb/l
	from 25.12.13	10	
16.	Mercury	1.0	μgHg/l
17.	Nickel (ii)	20	μgNi/l
18.	Nitrate (iii)	50	$mgNO_3/l$
19.	Nitrite (iii)	0.50	$mgNO_2/l$
		0.10	$mgNO_2/l$*
20.	Pesticides (iv, v)		
	aldrin	0.030	μg/l
	dieldrin	0.030	μg/l
	heptachlor	0.030	μg/l
	heptachlor epoxide	0.030	μg/l
	other pesticides	0.10	μg/l
21.	Total pesticides (vi)	0.50	μg/l
22.	Polycyclic aromatic hydrocarbons (vii)	0.10	μg/l
23.	Selenium	10	μgSe/l
24.	Tetrachloroethene & Trichloroethene (viii)	10	μg/l
25.	Trihalomethanes: total (ix)	100	μg/l
26.	Vinyl chloride (i)	0.50	μg/l
[point of compliance: consumers' taps]			

Notes:

i The parametric value refers to the residual monomer concentration in the water as calculated acccording to specifications of the maximum release from the corresponding polymer in contact with the water. This is controlled by product specification.
ii See also Reg. 6(6) [Secretary of State to determine how these parameters to be monitored].
iii See also Reg. 4(2)d [when water is to be regarded as unwholesome].
iv See also definition of "pesticides and related products" in Reg. 2.
v The parametric value applies to each individual pesticide.
vi "Pesticides: total" means the sum of the concentrations of the individual pesticides detected and quantified in the monitoring procedure.
vii The specified compounds are: benzo(b)fluoranthene; benzo(k)fluoranthene; benzo(ghi)perylene; and indeno(1,2,3-cd)pyrene.
The parametric value applies to the sum of the concentrations of the individual compounds detected and quantified in the monitoring process.
viii The parametric value applies to the sum of the concentrations of the individual compounds detected and quantified in the monitoring process.
ix The specified compounds are: chloroform; bromoform; dibromochloromethane; and bromodichloromethane.
The parametric value applies to the sum of the concentrations of the individual compounds detected and quantified in the monitoring process.

Item	*Parameters*	*Concentration or value (max.)*	*Units of Measurement*
Part II: National requirements			
1.	Aluminium	200	µg Al/l
2.	Colour	20	mg/l Pt/Co
3.	omitted: SI 2007/2734		
4.	Iron	200	µgFe/l
5.	Manganese	50	µgMn/l
6.	Odour	***	
7.	Sodium	200	mgNa/l
8.	Taste	***	
9.	Tetrachloromethane	3	µg/l
10.	Turbidity	4	NTU

[point of compliance: consumers' taps]

Schedule 2: Indicator parameters

Item	*Parameters*	*Concentration or Value (max.)*	*Units of Measurement*
1.	Ammonium	0.50	$mgNH_4/l$
2.	Chloride (i)	250	mgCl/l*
3.	Clostridium perfringes (inc. spores)	0	number/100 ml
4.	Coliform bacteria	0	number/100 ml
5.	Colony count	No abnormal change	no./1 ml at 22°C** no./1 ml at 37°C**
6a.	Hydrogen ion	9.5 6.5 min	pH Value pH Value
6.	Conductivity (i)	2500	µS/cm at 20°C*
7.	Sulphate (i)	250	$mgSO_4/l$*
8.	Total indicative dose (for radioactivity) (ii)	0.10	mSv/year*
9.	Total organic carbon (TOC)	No abnormal change	mgC/l*
10.	Tritium (for radioactivity)	100	Bq/l*
11.	Turbidity	1	NTU**

Notes:

(i) The water should not be aggressive.
(ii) Excluding tritium, potassium-40, radon and radon decay products.
1, 4 & 6a Point of monitoring: consumers' taps.
* Point of monitoring: may be monitored from samples of water leaving treatment works or other supply point, as no significant change during distribution.
** Point of monitoring: service reservoirs and treatment works (11, treatment works only).
*** Acceptable to consumers and no abnormal change.

APPENDIX 6.4

THE SURFACES WATERS (DANGEROUS SUBSTANCES) (CLASSIFICATION) REGULATIONS 1989

Environmental quality standards relating to inputs of various EC "Black List" substances to inland, coastal and territorial waters.

	Inland µg/l, annual mean	Coastal and territorial
Aldrin, dieldrin endrin & isodrin*	0.03 for total drins 0.005 for endrin	
Cadmium & compounds	5 (total)	2.5 (dissolved)
Carbon tetrachloride	12	12
Chloroform	12	12
DDT (all isomers)	0.025	0.025
pp-DDT	0.01	0.01
Hexachlorobenzene	0.03	0.03
Hexachlorobutadiene	0.1	0.1
Hexachlorocyclohexane (all isomers)	0.1	0.02
Mercury & compounds	1 (total)	0.3 (dissolved)
Pentachlorophenol & its compounds	2	2

* In July 1997 the Department of the Environment, Transport and the Regions expressed an intention to amend these standards as follows:

- aldrin	10 ng/l
- dieldrin	10 ng/l
- endrin	5 ng/l
- isodrin	5 ng/l

THE SURFACE WATERS (DANGEROUS SUBSTANCES) (CLASSIFICATION) REGULATIONS 1997 AND 1998

Schedules to the Regulations list concentrations of certain dangerous substances which should not be exceeded for meeting criteria for classification of either inland freshwaters or coastal and relevant territorial waters – see chapter 6, 6.15.2.

Schedules 1 and 2 to 1997 Regulations

Dangerous Substance	Inland µg/l, annual mean	Coastal and territorial
Arsenic	50	25
Atrazine & Simazine (for the two substances in total)	2	2
Azinphos-methyl	0.01	0.01
Dichlorvos	0.001	0.04
Endosulphan	0.003	0.003
Fenitrothion	0.01	0.01
Malathion	0.01	0.02
Trifluralin	0.1	0.1
Tributyltin	0.02 µg/l	0.002 µg/l
Triphenyltin & its derivatives	0.02 µg/l	0.008 µg/l
Dichlorvos (used as a treatment For sea-lice infestation)		0.6 µg/l

Schedules 1 and 2 to 1998 Regulations

Dangerous Substance	Inland	Coastal and territorial
	µg/l, annual mean	
4-Chloro-3-methyl-phenol	40	40
2-Chlorophenol	50	50
2,4-Dichlorophenol	20	20
2,4-D (ester)	1	1
(Non-ester)	40	40
1,1,1-Trichloroethane	100	100
1,1,2-Trichloroethane	400	300
Bentazone	500	500
Benzene	30	30
Biphenyl	25	25
Chloronitrotoluenes	10	10
Demeton	0.5	0.5
Dimethoate	1	1
Linuron	2	2
Mecoprop	20	20
Naphthalene	10	5
Omethoate	0.01	-
Toluene	50	40
Triazaphos	0.005	0.005
Xylene	30	30

APPENDIX 6.5

THE SURFACE WATERS (RIVER ECOSYSTEM) (CLASSIFICATION) REGULATIONS 1994

These classifications apply to inland rivers and watercourses in England and Wales.

(1) Class	(2) Dissolved Oxygen % saturation 10 percentile	(3) BOD (ATU) mg/l 90 percentile	(4) Total Ammonia mg N/l 90 percentile	(5) Un-ionised Ammonia mg N/l 95 percentile	(6) pH lower limit as 5 percentile; upper limit as 95 percentile	(7) Hardness mg/l $CaCO_3$	(8) Dissolved Copper µg/l 95 percentile	(9) Total Zinc µg/l 95 percentile
RE1	80	2.5	0.25	0.021	6.0-9.0	≤ 10 > 10 & ≤ 50 > 50 & ≤ 100 > 100	5 22 40 112	30 200 300 500
RE2	70	4.0	0.6	0.021	6.0-9.0	≤ 10 > 10 & ≤ 50 > 50 & ≤ 100 > 100	5 22 40 112	30 200 300 500
RE3	60	6.0	1.3	0.021	6.0-9.0	≤ 10 > 10 & ≤ 50 > 50 & ≤ 100 > 100	5 22 40 112	300 700 1000 2000
RE4	50	8.0	2.5	—	6.0-9.0	≤ 10 > 10 & ≤ 50 > 50 & ≤ 100 > 100	5 22 40 112	300 700 1000 2000
RE5	20	15.0	9.0	—	—	—	—	—

APPENDIX 6.6

THE SURFACE WATERS (ABSTRACTION FOR DRINKING WATER) (CLASSIFICATION) REGULATIONS 1996

Schedule 1: Criteria for Classification of Waters

(The limits set out below are maxima)

Parameters	Units	Limits DW1	DW2	DW3
Coloration (after simple filtration)	mg/1 Pt Scale	20	100	200
Temperature	°C	25	25	25
Nitrates	mg/l NO_3	50	50	50
Fluorides	mg/l F	1.5		
Dissolved iron	mg/l Fe	0.3	2	
Copper	mg/l Cu	0.05		
Zinc	mg/l Zn	3	5	5
Arsenic	mg/l As	0.05	0.05	0.01
Cadmium	mg/l Cd	0.005	0.005	0.005
Total chromium	mg/l Cr	0.05	0.05	0.05
Lead	mg/l Pb	0.05	0.05	0.05
Selenium	mg/l Se	0.01	0.01	0.01
Mercury	mg/l Hg	0.001	0.001	0.001
Barium	mg/l Ba	0.1	1	1
Cyanide	mg/l CN	0.05	0.05	0.05
Sulphates	mg/l SO_4	250	250	250
Phenol (phenol index) paranitraniline 4-aminoantipyrine	mg/l C_6H_5OH	0.001	0.005	0.1
Dissolved or emulsified hydrocarbons	mg/l	0.05	0.2	1
Polycyclic aromatic hydrocarbons	mg/l	0.0002	0.0002	0.001
Total pesticides (parathion, hexachlorocyclohexane, dieldrin	mg/l	0.001	0.0025	0.005
Ammonium	mg/l NH_4		1.5	4

The classifications, DW1, DW2 and DW3, correspond to the 1975 Surface Water Directive's classifications of A1, A2 and A3; the limits specified are the maximum allowable for each classification; an annex to Directive 75/440 specifies the treatment required to bring the surface water up to drinking water standard in each classification:

- DW1: simple physical treatment and disinfection;
- DW2: normal physical treatment, chemical treatment and disinfection;
- DW3: intensive physical and chemical treatment and disinfection

APPENDIX 7

THE EUROPEAN UNION AND THE ENVIRONMENT

The European Economic Community came into existence on 1 January 1958 following the signing of the 1957 *Treaty of Rome* by its six founder members – Belgium, France, West Germany, Italy, Luxembourg and The Netherlands. (These same six countries had, in 1951, formed the European Coal and Steel Community which created a common market in coal and steel.) The European Community was enlarged in 1973 when Denmark, Ireland and the United Kingdom joined; in 1981 Greece became a member, followed in 1986 by Spain and Portugal and by Austria, Finland and Sweden in 1995. The 1992 *Treaty on European Union* (the *Maastricht Treaty*) (see below) created the "European Union", the term by which the 15 member states are more usually called. On 1 May 2004, a further 10 countries joined the EU – Cyprus, the Czech Republic, Estonia, Hungary, Latvia, Lithuania, Malta, Poland, the Slovak Republic and Slovenia, with Bulgaria and Romania joining in 2007.

The presidency of the European Union (EU) rotates every six months between member states (as does chairmanship of the Council of Ministers). From 2007, it was planned that three countries would together hold the presidency of the EU; it is planned that each presidency group should comprise a large and a small member state, and at least one of the countries which joined in 2004.

The 1957 *Treaty of Rome*, which has been amended a number of times, sets out the basic operating principles of the EU with the aim of ensuring a Common Market – that is the free movement of goods, people and services between Member States.

The European Union has become one of the most important international innovators on environmental issues, with more than four-fifths of UK environmental legislation resulting from EU initiatives.

Although some of the EU's earliest environmental protection measures, on for example vehicle emissions, were adopted in 1970, it was not until 1973 that the *First Action Programme on the Environment* was adopted (see below). And it was not until 1986, when the *Single European Act 1986* was adopted, that the *Treaty of Rome* was also amended to make explicit mention of the environment.

Article 174 (formerly 130R), as amended, of the Treaty includes the following points:

a) Action by the Community relating to the environment shall have the following objectives:
   - to preserve, protect and improve the quality of the environment;
   - to contribute towards protecting human health;
   - to ensure a prudent and rational utilisation of natural resources;

A further objective was added by the *Maastricht Treaty* (see below) and that was the need for EC environmental policies to contribute to international environmental measures to deal with regional and worldwide environmental problems:

b) Action by the Community relating to the environment shall be based on the principles that preventive action should be taken, that environmental damage should as a priority be rectified at source, and that the polluter should pay. Environmental protection requirements shall be a component of the Community's other policies.

In preparing its action relating to the environment, the Community shall take account of:
- available scientific and technical data;
- environmental conditions in the various regions of the Community;
- the potential benefits and costs of action or lack of action;
- the economic and social development of the Community as a whole and the balanced development of its regions.

This development, which had both legal and symbolic significance, was given further impetus through the *Treaty on European Union* (the *Maastricht Treaty*) signed in Maastricht in February 1992. Following ratification by all Member States, this came into force on 1 November 1993. The Treaty has as a principle objective "the promotion of sustainable and non-inflationary growth respecting the environment, including among the activities of the Community a policy in the sphere of the environment". The Treaty was further amended in 1997 to require the Union to promote "balanced and sustainable development of economic activities". It stresses that the policy should not only aim at a high level of protection for the environment but that environmental protection should be integrated into other areas of EU policy.

The *Maastricht Treaty* also gives further substance to the principle of "subsidiarity" and refers "to the process of creating an ever closer union among the peoples of Europe in which decisions are taken as closely as possible to the citizen". This principle has been included in the EU's Fifth Action Programme (see below). Finally, the *Maastricht Treaty* created the European Union (i.e. all the countries within the Community) – this is made up of three sections: the European Community, Common Foreign & Security Policy and Home Affairs & Justice Policy. An amendment to the *Treaty of Rome* abolishes the European Economic Community (EEC) which becomes the European Community (EC). The Council of Ministers has become the Council of the European Community and the Commission of the European Communities the European Commission.

The *Amsterdam Treaty*, which came into force on 1 May 1999, strengthens requirements for EU proposals to promote a high level of protection for the environment and improvement in environmental quality, as well as to promote the objectives of sustainable development. The Treaty also enables adoption of legislation through "co-decision" the procedure under which most EU legislation is adopted. Thus most legislation now requires the formal agreement of both the Council of Ministers and the European Parliament. The Treaty also consolidated and renumbered the Articles of the *Treaty of Rome*.

The *Treaty of Nice*, which entered into force on 1 February 2003, focused on the transitional arrangements for EU institutions and procedures as a result of enlargement of the Union to 25 Member States. It allocates the number of seats in the European Parliament to the new Members, votes within the Council; it amends the definition of qualified majority, pending agreement on the threshold in a future Treaty and lists the provisions to which the qualified majority rule will apply. This Treaty remains in force until ratification of the *Constitutional Treaty*.

A new *Constitutional Treaty* was agreed by the 25 EU Heads of State in June 2004, but has to be ratified by individual Member States before it comes into force. The new Treaty will extend the areas subject to qualified voting in the Council, make changes to the size of and way in which the Commission is selected and give more power to the European Parliament.

Polluter Pays Principle

A key principle of the European Union's environment policy is that the cost of preventing pollution or of minimising environmental damage due to pollution should be borne by those responsible for the pollution. The principle originates from the proceedings of the UN Conference on the Human Environment, Stockholm 1972; in the European Community PPP was first mentioned in the First Action Programme on the Environment (1973-77) . As a signatory to the Treaty of Rome, the UK accepts the principle of PPP.

Amongst the elements of the principle as set out by the EU are:

- *payments* in the form of grants or subsidies from the public purse which may be made towards: investment in anti pollution installations and equipment; the introduction of new processes; and the costs of operating and maintaining anti pollution installations.

- *charges* which may be levied directly as taxes on the process which generates the pollution, or alternatively as the purchase price of a licence which entitles the holder to generate specific quantities of pollutants.

Legislative and Institutional Procedures

The *Single European Act 1986*, which came into force in July 1987, had as a principle objective the abolition of all barriers between Member States by 31 December 1992. To this end, new measures have been drafted and legislation amended with the aim of creating a single market in goods, services and employment – i.e. the right, subject to various safeguards, to trade and work in any Member State and the removal of technical, fiscal and other barriers. A fundamental objective to completion of the Single Market is that products and services legally sold or offered in one Member State should not be banned in another.

With regard to legislative proposals relating to the environment, those relating to product standards, competition or the "establishment and functioning of the internal market" (i.e. harmonising measures) are subject to majority voting under Article 95 (formerly 100A). Other Directives relating to action on the environment are proposed under Articles 174 and 175 (formerly 130) and thus require unanimous voting. An amendment agreed in Maastricht enables some environmental measures to be adopted on a qualified majority voting (QMV) – Article 205, and indeed most are. However, those incorporating, for example, fiscal, energy, planning and land use (excluding waste management) still require unanimous approval. Article 176 (formerly 130T) permits Member States to adopt or keep stricter environmental protection legislation under certain conditions.

The European Commission – which can be seen as the "civil service" of the EU – is split into 36 Directorates-General (DG), each responsible for a particular area of the Commission's work, with DG XI responsible for most environmental policy. The Commission is headed by a President, and comprises one representative of each Member State. Commissioners are currently nominated by Member States' governments – once the accession treaty for the 27th Member States is signed, the Council will vote on both the number of Commissioners and the way in which membership should be rotated to ensure the Commission reflects both the demographic and geographic composition of the Union.

Commissioners are expected to act independently of national interests. The Commission is responsible for developing and proposing new legislation and other initiatives. Proposals are published usually with the prefix "COM" – i.e. Communication [from the Commission to the Council and Parliament concerning ...], followed by the year and a reference number, e.g. COM(2009) 00. Following adoption, they are published with a reference number preceded by the year of adoption (2009/00).

EU measures are finalised as either

- **Regulations:** these are binding on all Member States in their entirety and take precedence over national legislation.
- **Directives:** these are binding on Member States as to the result to be achieved, while leaving a degree of flexibility on how the measure will be implemented in national legislation.
- **Decisions:** these are binding in their entirety and may be addressed to government, private enterprise or individual.
- **Recommendations and opinions:** these are not binding.

The Council of the European Community (i.e. Council of Ministers) is responsible for agreeing, amending or referring back to the Commission its draft proposals. The Council formally consists of the Foreign Ministers of the Member States, with the chairmanship changing every six months in line with the Presidency of the Community. Where, for example, environmental matters are to be discussed, then the Council will usually comprise the environmental ministers of the Member States. Initial discussions will have taken place between national governments' officials and also in the Committee of Permanent Representatives (COREPER), who will aim to reach agreement on proposals prior to discussion in the Council of Ministers. COREPER consists of each Member States' Ambassador to the EU.

Members of the European Parliament are directly elected by electors in each Member State. Prior to ratification of the *Maastricht Treaty*, the EP had a rather limited role, being restricted to offering an opinion on Commission proposals; it was unable to amend or reject decisions reached by the Council, but could only ask them to reconsider. The only exception to this was the budget which it could reject.

The European Parliament now has the right to approve the appointment of the European Commission and of its President. On a majority vote it may also ask the Commission to produce a proposal for legislation in furtherance of the objectives of the Treaty. Under the co-decision procedure the Parliament may block a measure, even if approved by the Council of Ministers, if it is of the opinion that the proposal does not meet the purpose for which it is intended (Article 251). If agreement cannot be reached on a proposal it is referred to a conciliation committee consisting of representatives of the EP and of the Council for further negotiation. The cooperation procedure (Article 252) enables the Parliament on a second reading to propose amendments to proposals even though the Council of Ministers has already adopted a common position. The Commission then decides which of the European Parliament's amendments it can accept before the proposal returns to the Council of Ministers. The Council of Ministers can only reject or further amend the European Parliament's amendments which have been accepted by the European Commission by unanimous vote.

Following adoption, the final text of Directives etc are published in the *Official Journal of the European Union*. All legislation includes a date by which Member States must have incorporated it into national legislation. The amount of time allowed largely depends on the impact of the measure on national practice and the cost of implementation. In some instances, individual Member States will be allowed a derogation – i.e. extra time in which to comply.

In the UK EU measures are implemented either under the appropriate primary legislation, or where no suitable legislation exists, under the *European Communities Act 1972*.

The European Commission is also responsible for monitoring the performance of Member States in complying with Directives etc. Where it feels that a Member State is in breach of EU legislation it will first try to resolve the matter through informal discussion, correspondence, etc – there is a procedure for doing this. If however the Commission is still not satisfied, it can initiate legal action in the European Court of Justice. This is situated in Luxembourg with judges drawn from all Member States. Individuals in Member States may also complain to the European Court if they feel EU legislation is being breached. The European Court of Justice has the power to fine or levy a financial penalty on offenders.

European Environment Agency

Although agreement was reached on the establishment of the European Environment Agency (Regulation 1210/90) in 1990, it did not formally begin work until late 1993. It consists of the EU Member States, together with Iceland, Norway, Liechtenstein, Turkey, Bulgaria and Romania.

The EEA's main task is to record, collate and analyse data on the environment, drawing up expert reports. These will enable both the European Commission and Member States to formulate appropriate environmental protection policies. A further aim of the EEA, which is based in Copenhagen, is to develop ways of harmonising measurement and forecasting methods. The Agency's remit initially includes air, water, soil, wildlife, land use, waste management and noise. Following adoption of an amending Regulation (933/1999), the Agency's remit has been extended to include responsibility for ensuring the dissemination of methodologies for assessing environmental impacts, and on the use of environmentally-friendly technologies, particularly telematics; it has also been charged with disseminating results of environmental research in a way which enables the research to assist in policy development.

The Agency's information network – EIONET: European Information and Observation Network – will help to coordinate the work and establish links between policy and research institutes throughout the EC. Information collected by the Agency will be publicly available and a report on the state of the environment will be published every three years.

EC Network of Environmental Enforcement Agencies

The permanent Network of Environmental Enforcement Agencies was founded in 1992 with representatives from all EC Member States. Its aims are to "provide a mechanism for the exchange of information and experience between environmental agencies in the EC in order to address issues of mutual concern and to enhance the quality of the environment". The Network operates mainly through four permanent working groups with specific areas of interest:

- technical aspects of permitting;
- procedural and legal aspects of permitting;
- compliance monitoring, inspection and enforcement;
- managing the enforcement process.

Action Programmes on the Environment

The United Nations Conference on the Human Environment, held in Stockholm in 1972, marked the beginning of global recognition of environmental problems which could only be solved by a sustained effort in all countries.

The European Community responded by drawing up its *First Action Programme on the Environment*; it covered the period 1973-76 and was adopted by the Council of Ministers in November 1973. This *First Action Programme* was a wide-ranging document containing a large number of measures designed to deal with already urgent pollution problems; it also took a longer term view of the future and sought the positive management of human activities in such a way as to prevent further problems arising. The First Action Programme also set out goals and objectives, defined principles, such as the Polluter Pays Principle, and recognised that prevention is better than cure. It adopted a comprehensive approach, tackling both air and water pollution, the management of wastes, noise pollution and the protection of wildlife and habitats. These ideas and principles were given further substance in three subsequent Action Programmes covering 1977-81, 1982-86 and 1987-92.

In 1993 the Commission's *Fifth Action Programme - Towards Sustainability* was approved by the Council of Ministers and the Parliament. In their Resolution adopting the Programme, the EC Environment Ministers asked the Commission to "ensure that all proposals relating to the environment properly reflect the principle of subsidiarity". They also note that further measures will be needed if the desire for sustainable development is to be achieved. The Programme, covering 1993-2000 with a review in late 1995, stressed the need for all sectors of the community – governments, industry and citizens – to become involved in and to take responsibility for the protection of their environment. It says that environmental policies should be integrated into all areas of Community policy if the goal of sustainability is to be achieved. Sustainability in this context is defined as reflecting "a policy and strategy for continued economic and social development without detriment to the environment and the natural resources on the quality of which continued human activity and further development depend".

The 5th Programme also states that "the Community will take action, in accordance with the principle of subsidiarity, only if and insofar as the objectives of the proposed action cannot be sufficiently achieved by the Member States and can therefore, by reason of the scale or effects of the proposed action, be better achieved by the Community". It also recognises that in many areas, objectives will best be achieved through "shared responsibility" using a mix of "actors and instruments at the appropriate levels" – i.e. that action should be taken at Community, national or local level as appropriate.

Five areas requiring major effort because of the stresses each can place on the environment have been targeted: industry, agriculture, energy, transport and tourism. The Commission suggests that these are all areas at which action for improvement can most efficiently and appropriately be taken at Community level. Within each of these target areas the Programme identifies a number of objectives aimed at achieving sustainable development while benefiting the sector concerned. The Programme also identifies a number of environmental issues which because of their transboundary or global nature should be dealt with at Community level. Again objectives and targets are set and a programme to be reviewed and developed is proposed. The environmental issues highlighted are climate change, acidification, air quality, urban environment (noise), waste management, protection of nature and biodiversity, management of water resources and coastal zones. The targets and actions proposed are summarised in the relevant chapters of the *Pollution Handbook*.

The 6th Action Programme – *Environment 2010: Our Future Our Choice* – singles out four priority areas for action:

- ***climate change*** – to ratify the Kyoto protocol; and to make progress towards the long-term requirement established by the Intergovernmental Panel on Climate Change to reduce emissions of greenhouse gases by 70% over 1990 levels;
- ***nature and biodiversity*** – with the aim of protecting and restoring the functioning of natural systems and halting the loss of biodiversity both in the EU and on a global scale;
- ***environment and health*** – to aim at an environment where the levels of manmade contaminants do not give rise to significant impacts on, or unacceptable risks to, human health;
- ***sustainable use of natural resources and management of waste*** – objectives include to ensure that the consumption of renewable and non-renewable resources and their associated impacts do not exceed the carrying capacity of the environment; to achieve a significant reduction in the quantity of waste going to final disposal and in the volumes of hazardous waste produced; better resource efficiency and a shift towards more sustainable consumption patterns.

Thematic Strategies

The 6th Action Programme envisaged the production of seven "thematic strategies", together with supporting legislation. These have now all been published and show how the objectives of the 6th Action Programme are to be delivered:

- Air Pollution – (COM(2005) 446, together with a proposal for an *Ambient Air Quality Directive* (adopted as 2008/50/EC) (COM(2005) 447), September 2005: see chapter 3, 3.9.1
- Prevention and recycling of waste – (COM(2005) 666), together with a proposal for revising the Waste Framework Directive (COM(2005) 667), December 2005: see chapter 5, 5.5.
- Protection and conservation of the marine environment – (COM(2005) 504), together with a proposal for a Directive establishing a framework for Community Action in the field of marine environmental policy (COM(2005) 505): see chapter 6,
- Soil Protection (COM(2006) 231, together with a proposal for a framework Directive (COM(2006)232): September 2006. September 2006: see chapter 5, 5.40.1
- Sustainable use of natural resources, COM(2005) 670: December 2005.
- Urban environment, COM(2005) 718 : January 2006
- Sustainable use of pesticides (COM(2006) 372, together with a proposal for a Directive establishing a framework for Community action to achieve a sustainable use of pesticides (COM(2006) 373: July 2006: see chapter 6, 6.25.1.

On the pages which follow, we list Community measures relating to industrial pollution (7.2) air pollution (7.3), noise (7.4), waste (7.5) and water (7.6). Other measures of relevance are listed at Appendix 7.7.

Further Reading

A number of useful publications providing a more detailed discussion on the development of Community environmental policies and legislation, their implementation and the way in which the various Community institutions operate, are listed in the bibliography (General – European Issues) section of this Handbook. Useful websites are listed in the Bibliography to the Pollution Handbook.

APPENDIX 7.1: ENVIRONMENTAL ACTION PROGRAMMES

Number	Title	Date adopted	Date brought into force	OJ No. and date	Purpose
–	Declaration of the Council of the European Communities and of the Representatives of the Governments of the Member States meeting in the Council on the Action Programme of the European Communities of the Environment. (1973-1976)	22.11.73	22.11.73	C 112 20.12.73	To lay down the objectives and principles of a Community environment policy. To describe measures to be taken to reduce pollution and nuisances and to improve the environment.
–	Resolution of the Council of the European Communities and of the Representatives of the Governments of the Member States meeting within the Council on the continuation and implementation of a European Community policy and Action Programme on the Environment. (1977-1981)	17.05.77	17.05.77	C 139 13.06.77	To restate the policy objectives and principles. To describe measures to be undertaken to reduce pollution and nuisance, protect and manage land, the natural environment and resources and improve the environment.
–	Resolution of the Council of the European Communities and of the Representatives of the Governments of the Member States meeting in the Council on the continuation and implementation of a European Community policy and and Action Programme on the Environment (1982-86)	07.02.83	07.02.83	C 46 17.02.83	To develop an overall strategy. To describe measures to be taken to prevent and reduce pollution and nuisance in the different environments, as well as the protection and rational management of land, the environment and natural resources.
–	Resolution of the Council on the continuation and implementation of a European Community policy and action programme on the environment. (1987-1992)	19.10.87	19.10.87	C328 07.12.87	To identify pollutants; to determine best focus for control measures; to set objectives to combat acid deposition and forest die-back; to reduce ambient pollutant concentrations; to develop appropriate management techniques.
	Resolution of the Council on the fifth Community Policy and Action Programme on the Environment and Sustainable Development. (1993-2000)	18.03.92		C138 17.05.93	Define objectives for tackling a number of issues and for achieving sustainable development. Issues include climate change, acidification, air and water pollution, waste management.
1600/ 2002/ EC	Decision of the EP & Council laying down the Community Environmental Action Programme 2001-2010: Our Future, Our Choice	22.07.02	22.07.02	L242 10.09.02	

APPENDIX 7.2: INDUSTRIAL POLLUTION

Number	Title	Date adopted	Date brought into force	OJ No. and date	Purpose
84/360/ EEC	"Framework" Directive on combating air pollution from industrial plants.	28.06.84	30.06.87	L 188 16.07.84	repealed 30.10.07
96/61/ EC	Council Directive concerning integrated pollution prevention and control	24.09.96	30.10.96	L257 10.10.96	repealed and replaced by 2008/1/EC
2008/1/ EC	EP & Council Directive on integrated pollution prevention and control	15.01.08	18.02.08	L24 29.01.08	Replaces and codifies 96/61/EC which introduced integrated permitting for industrial installations
99/13/EC	Council Directive on limitation of emissions of VOCs due to the use of organic solvents in certain activities & installations	11.03.99	29.03.99	L85 29.03.99	To limit VOCs – part of strategy reducing VOC emissions from other sectors.
2001/80/ EC	Directive on limitation of emissions of certain pollutants into the air from large combustion plant	23.10.01	27.11.01	L309 27.11.01	Updates and repeals 1988 Directive; covers both new and existing LCP
2001/81/ EC	Directive on national emissions ceilings for certain atmospheric pollutants	23.10.01	27.11.01	L309 27.11.01	Sets limits on emissions of sulphur dioxide, nitrogen oxides, VOCs and ammonia
96/82/EC	Directive on the control of major accident hazards – Seveso 2	09.12.96	03.02.97	L10 14.01.97	To establish notification & information system for installation storing & using hazardous substances
2003/105/ EC	EP & Council Directive amending 96/82/EC	16.12.03	31.12.03	L345 16.12.03	Amends thresholds and information requirments
2004/42/ EC	EP & Council Directive on limitation of emissions of VOCs due to use of organic solvents in certain paints & varnishes & vehicle refinishing products	21.04.04	30.04.04	L143 30.04.04	Amends Directive 99/13/EC
2004/35/ EC	EP & Council Directive on environmental liability to the prevention & remedying of environmental damage	21.04.04		L143 30.04.04	

APPENDIX 7.3: AIR POLLUTION

Transport and Fuels

Number	Title	Date adopted	Date brought into force	OJ No. and date	Purpose
715/2007	Regulation on emissions from light passenger and commercial vehicles	20.06.07	02.07.07	L 171/11 29.06.07	Sets emission limits (Euro 5 & 6) to begin coming into force from 01.09.09 & requires access to repair and maintenance information; Directive 70/220/EEC and amendments will be repealed on 02.01.13
629/2008	Regulation implementing and amending 715/2007	18.07.08	31.07.08	L 199 28.07.08	Establishes administrative provisions for EC type approval for light duty vehicles, including provisions for conformity of production and in-service conformity.
70/220/ EEC	Directive relating to measures to be taken against air pollution by gases from positive ignition engines of motor vehicles.	20.03.70	30.06.70	L 76 06.04.70	To set limit values for emissions of CO and HC from petrol-engined vehicles.*
83/351/ EEC	Amendment Directive (4th)	16.06.83	30.11.83	L 197 20.07.83	Reduce limits set in 78/665 of CO by 23%; HC and NOx by 20-30%; also limits pollutants from diesel engines.*
88/76	Amendment Directive (5th): The Luxembourg Agreement".	03.12.87		L36 09.02.88	To set emission limits for CO, HC + NOx and NOx according to engine size.*
89/458/ EEC	Amendment directive (7th) on gaseous emissions from private motor cars with a capacity of less than 1400cc	18.07.89		L226 03.08.89	To set mandatory limits on emissions of 19g/test CO and 5g/test HC + NOx for small sized cars. (Partially supersedes 88/76)*
89/491/ EEC	Amendment Directive (15th)			L 238 15.08.89	Amends various directives with regard to fuel consumption.*
91/441/ EEC	Amendment Directive (8th)	26.06.91	01.04.92	L 242	Extends emission standards set in 89/458*

* to be repealed 02.01.13

Number	Title	Date			Purpose
		adopted	brought into force	OJ No. and date	
94/12/ EC	Amendment Directive (9th)	1994		L 100 19.04.94	Tightens limits for petrol cars & for diesels to be met in two stages.*
98/69/ EC	Amendment Directive (11th)	09.98		L 350 28.12.98	Part of Auto-oil programme emission limits set for 2000 & 2005*
93/59 EEC	Amendment directive. (18th)	28.06.93	01.10.93	L186 28.07.93	To place emission limits on light commercial vehicles, off-road vehicles and heavy cars.*
96/69	Amendment Directive (20th)	8.10.96		L282 01.11.96	Tightens emission limits for light commercial vehicles.*
2001/01/ EC	Amendment Directive	22.01.01	06.02.01	L35 06.02.01	Sets out dates by which new passenger cars & light commercial vehicles should be equipped with OBD system.*
88/436	Amendment Directive (6th) limiting gaseous pollutants from diesel vehicles up to 3.5 tonnes	16.06.88		L 214 06.08.88	To set standards for the emission of particulates of 1.1g/test for new engine types and 1.4g/test for new vehicles.*
2005/55/ EC	Directive relating to emissions from heavy duty vehicles	28.09.05	09.11.05	L275 20.10.05	Consolidates and repeals earlier Directives; covers emission of gaseous and particulate pollutants from compression-ignition engines for use in vehicles, and the emission of gaseous pollutants from positive-ignition engines fuelled with natural gas or liquefied petroleum gas for use in vehicles.
2005/78/ EC	Directive relating to emissions from heavy duty vehicles	14.11.05	19.12.05	L313 29.11.05	Contains and implements technical requirements of 2005/55/EC
2006/51/ EC	Directive amending above HGV Directive	06.06.06	10.06.06	L152 07.06.06	Amends requirements for the emission control monitoring system for use in vehicles and exemptions for gas engines.
2008/74/ EC	Commission Directive amending heavy goods vehicles Directives 2005/55/EC and 2005/78	18.07.08	08.08.08	192/51 19.07.08	Introduces requirements relating to test procedures to enable type approval of heavy duty engines and vehicles with petrol engines, & adds requirements for smoke opacity testing following repeal of 72/306/EEC
97/98	Directive concerning emissions from non-road machinery	16.12.97	19.03.98	L59 27.02.98	To limit emissions of gaseous and particulate pollutants from internal combustion engines to be installed in non-road mobile macinery
2000/25	EP & Council Directive on emissions from agricultural or forestry tractors	22.05.00		L173 12.07.00	Controls emissions of gaseous and particulate pollutants
2002/51/ EC	Directive reducing emissions from 2 & 3 wheel motor cycles			L252 20.09.02	To cut emissions in two stages
2003/44/ EC	EP & Council Directive relating to recreational craft	16.06.03	26.08.03	L214 26.08.03	Tightens noise and exhaust emission limits
2004/26/ EC	EP & Council Directive on emissions of gaseous & particulate pollutants from non-road mobile machinery	21.04.04		L146 30.04.04	Tightens emission limits and extends definition of NRMM to railcars
93/12 EEC	Directive relating to the sulphur content of certain liquid fuels.	23.03.93	1.10.94	L74 27.03.93	Reduces sulphur content of gasoils to 0.2% by weight by 01.10.94 and to 0.05% from 01.10.96.
98/70/ EC	Council Directive relating to quality of petrol & diesel fuels	09.98	28.12.98	L 350 28.12.98	Part of Auto-oil programme; bans marketing of leaded petrol, reduces amount of benzene & other compounds in fuels and amends 93/12
2003/17/ EC	Directive amending 98/70 relating to the quality of petrol & diesel fuels	03.03.03		L76 22.03.03	To require "sulphur free" fuels to be avaliable from 1 Jan 05 and to ban those with more than 10ppm from 1 Jan 09
99/32/ EC	Council Directive relating to sulphur content of certain liquid fuels	26.04.99		L 121 11.05.99	Reduce sulphur content of gas oil & amends 93/12

* to be repealed 02.01.13

Number	Title	Date adopted	Date brought into force	OJ No. and date	Purpose
85/210/ EEC	Directive on the approximation of Member State legislation on lead content of petrol, and the introduction of lead-free petrol.	21.03.85	01.01.86	L 096 03.04.85	Replaces Directive 78/611. Also fixes benzene content of leaded and unleaded petrol at 5% max. by volume.
87/416/ EEC	Amendment of Directive 85/210/EEC.	21.07.87	-	L 225 13.08.87	To permit Member States to ban the sale of regular grade leaded petrol after giving 6 months notice.
92/55 EEC	Directive on vehicle emission testing.	10.06.92		L225 10.08.92	To harmonise in-service testing of vehicle emissions.
93/116/ EEC	Directive relating to the fuel consumption of motor vehicles.			L369 30.12.93	To set down procedure for determining CO2 emissions and require these to be recorded in document kept by vehicle owner.
99/94	Directive relating to consumer information on fuel economy and CO_2 emissions	13.12.99		L12 18.01.00	To require information to be available in respect of marketing new cars.
94/63/ EC	EP and Council Directive on the control of VOCs resulting from the storage of petrol and its distribution from terminals to service stations.	20.12.94	31.12.95	C365 31.12.94	To reduce evaporative fuel losses.
2003/30/ EC	Directive on the promotion of the use of biofuels and other renewable fuels	08.05.03	17.05.03	123/42 17.05.03	To promote use of biofuels and other renewable fuels to replace diesel or petrol and contribute to meeting climate change commitments.
Air Quality					
2008/50 EC	Directive on ambient air quality and cleaner air for Europe	21.05.08	11.06.08	L152 11.06.08	New Framework Directive on assessment and management of air quality
96/62/EC	Directive on air quality assessment and management	24.09.96	21.10.96	L296 01.11.96	Framework Directive outlining system of monitoring and assessing air quality.
99/30/EC	Council Directive relating to limit values for SO_2, NO_2 & oxides of nitrogen particulate matter & lead	22.04.99	19.07.99	L 163 29.06.99	"Daughter" Directive under 96/62
2002/31/ EC	Directive relating to ozone in ambient air	12.02.02	09.03.02	L67 09.03.02	Sets target values for the protection of human health and of vegetation
2000/69/ EC	Directive relating to carbon monoxide and benzine	16.11.00	13.12.00	L313 13.12.00	Sets health-based limits
97/101/EC	Decision (superseding 82/459), establishing a reciprocal exchange of information and data from networks and individual stations measuring ambient air quality within Member States.	27.01.97	01.01.97	L 35 5.02.97	To set up a procedure for exchanging information between the Member States' measuring stations on a number of substances causing air pollution. To nominate a coordinating body. To arrange for the publication of a summary report.
80/779/ EEC	Directive on air quality limit values and guide values for sulphur dioxide and suspended particulates.	15.07.80	17.07.82	L 229 30.08.80	To lay down standards for sulphur dioxide and suspended particulates in the air, as well as the conditions for their application. To lay down the reference methods for sampling and analysis to be used.
–	Resolution relating to transboundary air pollution by sulphur dioxide and suspended particulates.	15.07.80	-	C 222 30.08.80	Calls on Member States to limit, and as far as possible reduce or prevent transboundary pollution in line with Directive
81/462/ EEC	Decision on the conclusion of the Convention on long-range transboundary air pollution. (The Geneva Convention).	11.06.81	-	L 171 27.06.81	To limit and as far as possible reduce and prevent transboundary air pollution.

Number	Title	Date			
		adopted	brought into force	OJ No. and date	Purpose
86/277/ EEC	Council Decision concluding a Protocol to the 1979 Geneva Convention on Long-Distance Transboundary Air pollution concerning the funding of a long-term programme of cooperation for the constant monitoring and evaluation of the long-distance atmospheric transfer of pollutants in Europe (EMEP).	12.06.86	-	L181 04.07.86	See Title.
93/361	Council decision on the accession of the community to the Protocol to the LRTAP Convention on the control of emissions of nitrogen oxides or their transboundary fluxes.	17.05.93		L.149 21.06.93	
2003/507/ EC	Council decision on accession to LRTAP protocol to abate acidification, eutrophication and ground-level ozone	13.06.03		L179 17.07.03	
82/884/ EEC	Directive on a limit value for lead in the air.	03.12.82	10.12.84	L 378 31.12.82	To fix a limit value of 2 micrograms Pb/m^3 expressed as an annual mean concentration. To define characteristics to be complied with for choosing a sampling method and a reference method for analysing the concentration of lead.
85/203/ EEC	Directive on air quality standards for nitrogen dioxide.	07.03.85	01.01.87	L 087 27.03.85	To establish a limit value for NO_2 in air of 200 $\mu g/m^3$,expressed as a 98 percentile of the mean of one-hour measurements made throughout a year; also lower "guide values" for special protection zones.
87/217/ EEC	Directive on prevention and reduction of environmental pollution by asbestos.	28.03.87	31.12.88	L 085 28.03.87	To put controls on the pollution of air, water and land by asbestos from all significant point
sources.					
89/427/ EEC	Directive modifying 80/779 on limit values and guide values of air quality for sulphur dioxide and suspended particulates.	21.06.89			To harmonise measurement methods.
92/72 EEC	Directive on air pollution by ozone.	21.09.92		L297 13.10.92	To establish procedure for monitoring ground-level ozone, give information to the public and issue public alerts when levels are expected to be high. To be repealed 09.09.03
Incinerators					
2000/76/ EC	Directive on waste incinerators.	04.12.00	28.12.00	L 332 28.12.00	Replaces both the MWI Directives.
Ozone Layer/Greenhouse gases					
–	Resolution on chlorofluorocarbons in the environment.	30.05.78	30.05.78	C 133 07.06.78	To limit the production of CFCs in the Community. To encourage the search for substitute
products.					To promote elimination of CFCs.
87/412	Decision on the signing of a Protocol of the Vienna Convention for the protection of the ozone layer, relating to the control of chlorofluorocarbons.	10.09.87	-		See title
88/540/ EEC	Decision implementing the Vienna Convention on protection of the ozone layer and the Montreal Protocol on substances which deplete the ozone layer.	14.10.88	-	L 297 31.10.88	See title
3093/94 EC	Regulation on substances that deplete the ozone layer	15.12.94		L333 22.12.94	Tightens controls on CFCs etc and introduces restrictions on HCFCs and methyl bromide.
2000/22	Commission Decision re. essential uses			L7 12.01.00	Covers allocation of quantities of controlled substances allowed for essential uses under Reg.3093/94.

Number	Title	Date adopted	Date brought into force	OJ No. and date	Purpose
280/2004 /EC	EP & Council decision on a monitoring mechanism for greenhouse gases & for implementing Kyoto Protocol	11.02.04		L49 19.02.04	
2003/87/ /EC	Directive establishing a scheme for greenhouse gas emissions trading in the community			L275 15.10.03	Requires members to report to the Commission how emissions quotas are to be distributed.
2004/101 /EC	Amendment Directive	27.10.05		L338 13.11.04	Amends above in respect of Kyoto Protocol's project mechanism.
2006/40/ EC	Directive relating to emissions from air-conditioning systems in motor vehicles	17.05.06	04.07.06	L161 14.06.06	Will ban the use of HFC134a in new model vehicles from 01.01.11 and all new vehicles from 01.01.17
842/2006	Regulation on certain fluorinated greenhouse gases	17.05.06	04.07.06	L161 14.06.06	To contain, prevent and reduce emissions of HFCs, PFCs and SF_6
1493/2007	Regulation on reports on fluorinated greenhouse gases	17.12.07	07.01.08	L332/7 18.12.07	Specifies format of reports to be submitted to the Commission by producers, importers and exporters.
1497/2007	Regulation establishing standard leakage requirements for fire protection systems containing fluorinated greenhouse gases	18.12.07	08 .01.08	L334/4 19.12.07	Sets out standard leakage requirements for stationary fire protection systems containing more than 3 kg of FGHGs.
Energy Efficiency					
93/76 EEC	Directive to limit carbon dioxide emissions by improving energy efficiency (SAVE).	13.09.93	31.12.94	L237 23.09.93	
2002/91 /EC	Directive on energy performance of buildings	16.12.02		L001 04.01.03	To require buildings to be made more energy efficient
2006/32/ EC	Directive on energy end-use efficiency and energy services	05.04.06	17.05.06	L114 27.04.06	To enhance the cost-effective improvement of energy end-use; sets energy savings target of 9% to be achieved by 2017. Repeals 93/76/EEC (SAVE programme)

APPENDIX 7.4: NOISE POLLUTION

Number	Title	Date adopted	Date brought into force	OJ No. and date	Purpose
2002/49/ EC	Directive relating to the assessment and management of environmental noise	25.06.02	18.07.02	L189 18.07.02	To limit exposure to noise in built-up areas, public parks, quiet areas and in noise sensitive areas.
Vehicles/Construction Plant/Aircraft					
70/157/ EEC	Directive on the approximation of laws in the Member States relating to the permissible sound level and the exhaust system of motor vehicles.	06.02.70	10.08.71	L 42 23.02.70	To fix the permissible limits for the sound level, the equipment, conditions and methods for measuring this level.
73/350/ EEC	Directive adapting 70/157 to technical progress.	07.11.73	01.03.74	L 321 21.11.73	To prescribe measures for exhaust systems.
84/424/ EEC	Council amendment to 70/157	03.09.84	06.09.84	L 238 06.09.84	Amends noise limits set in 77/212; buses and lorries under 3.5 tonnes are categorised together, and those over categorised by horsepower.
92/97 EEC	Amendment to 70/157	10.11.92	01.07.92	L371 19.12.92	Reduces noise limits for cars, buses and HGVs by between 2-5 dB(A).
96/20 EC	Amends 92/97	27.03.96		L92 13.04.96	To make various changes to noise level testing method.

Number	Title	Date adopted	Date brought into force	OJ No. and date	Purpose
78/1015/ EEC	Directive relating to the permissible sound level and exhaust system of motorcycles.	23.11.78	01.10.80	L 349 13.12.78	To fix the permissible limits for the sound level, the equipment, conditions and methods for measuring this level. To prescribe measures for exhaust systems.
87/56/ EEC	Amendment to Directive 78/1015	18.12.86	01.10.88	L 24 27.01.87	To modify the test procedure.
89/235/ EEC	Modification of Directive 78/1015 on sound levels of motorbikes.	13.03.89	01.10.90	L 98 11.04.89	To lay down common technical standards for exhaust systems; to provide a model EC type approval certificate.
74/151/ EEC	Directive relating to certain parts and characteristics of wheeled agricultural or or forestry tractors.	04.03.74	07.04.75	L 84 28.03.74	To fix the permissible limits for the sound level, the equipment, conditions and methods for measuring this level.
77/311/ EEC	Directive related to the driver-perceived noise level of wheeled agricultural or forestry tractors. (Extended by Directive 82/80.)	29.03.77	01.10.79	L 105 28.04.77	To fix the permissible limits for the sound level, the equipment, conditions and methods for measuring this level.
80/51/ EEC	Directive relating to the limitation of noise emissions from subsonic aircraft.	20.12.79	21.06.80	L 18 24.01.80	To make it compulsory for the Member States to apply annex 16 to the Chicago Convention on Subsonic Aircraft to establish mutual recognition of validity certificates. To determine exemptions.
83/206/ EEC	Amendment to 80/51		26.04.84	L117 04.05.83	To ensure that aircraft landing in the Community since 21.04.84 respect standards laid down by ICAO Chicago Convention.
89/629/ EEC	Directive limiting noise emissions from subsonic aircraft.	04.12.89		L 363 13.12.89	To prohibit registraton after 01.11.90 of aircraft unable to meet specific noise limits.
2006/93/ EC	Directive on the regulation of aeroplanes covered by Part II, Chapter 3, vol. 1 of Annex 16 of Chicago Convention	12.12.06	16.01.07	L 374/1 27.12.06	Codifies and repeals Directives 92/14/EEC, 98/20/EC & Regulation 991/2001; aircraft unable to meet Chapter 3 standards from operating into or out of the EC after 01.04.95 and all Chapter 2 aircraft after 01.04.92.
92/14 EEC	Directive limiting aircraft noise.	02.03.92		L 76 23.03.92	To ban all aircraft unable to meet Chapter 3 standards from operating into or out of EC after 01.04.95 and all Chapter 2 aircraft after 01.04.92.
925/99	Council Regulation on registration and use of certain types of jet aeroplanes	24.04.99		L115 4.5.99	Bans use of aeroplanes fitted with hushkits from specific dates.
79/113/ EEC	Directive on the approximation of the laws of the Member States relating to the determination of the noise emission of construction plant and equipment.	19.12.78	21.06.80	L 33 08.02.79	To define the sound level for construction plant and equipment. To define the criteria to use for expressing results, equipment and conditons for carrying out measurements and calculation method.
81/1051/ EEC	Amendment to 79/113	07.12.81	14.06.83	L 376 30.12.81	To determine the noise emitted to the operator's position by all categories of machines.
2000/14	EP & Council Directive relating to noise emissions from equipment for use outdoors	08.05.00	3.7.00	L162 3.7.00	Requires equipment to conform to specific noise level and for manufacturers to guarantee noise level. Equipment covered ranges from concrete breakers to hedge trimmers.

Household Appliances

Number	Title	Date adopted	Date brought into force	OJ No. and date	Purpose
86/594/ EEC	Directive on airborne noise emitted by household appliances.	01.12.86	04.12.89	L 344 06.12.86	To harmonise the methods of measuring the noise, arrangements for checking, general principles for publishing information on the noise emitted by these appliances.

Number	Title	Date adopted	Date brought into force	OJ No. and date	Purpose
Protection of Workers					
86/188/ EEC	Directive on the protection of workers from the risks related to exposure to noise at work.	12.05.86	01.01.90	L 137 24.05.86	To protect workers from risks to hearing by setting limits on noise levels at which preventative action is required.
2003/10 EC	Directive on the minimum health & safety requirements regarding the exposure of workers to the risks arising from physical agents (noise)	06.02.03		L 42 15.02.03	See title

APPENDIX 7.5: WASTE

Waste

Number	Title	Date adopted	Date brought into force	OJ No. and date	Purpose
2006/12 EC	Framework Directive on waste.		17.05.06		Repeals and codifies provision of 75/442 and 91/156. To encourage the prevention and recycling of waste. To determine the arrangements to be made for the harmless disposal of waste. To provide administrative provisions for management and control. To provide a system of authorisations for firms responsible for collection, recycling or disposal. Clarifies definition of waste and lists 16 specific categories.
94/62 EC	EP and Council Directive on packaging waste	20.12.94	31.12.94	C365 31.12.94	To harmonise measures, reduce amount of such waste and its environmental impact; set up collection, recovery and re-use systems.
2004/12/ EC	EP & Council Directive amending 94/62/EC on packaging & packaging waste	11.02.04	18.02.04	L147 18.02.04	
75/439/ EEC	Directive on the disposal of waste oils. amended by 87/01/EEC	16.06.75	18.06.77	L 194 25.07.75	To determine the arrangements to be made for the collection and harmless disposal of waste oils. To require that priority is given to the regeneration of waste oils. To require a system of authorisations for plant either regenerating or burning waste oil. To prevent waste oils being mixed with toxic and dangerous wastes, in particular with PCBs.
99/31/ EC	Council Directive on Landfill of Waste	26.04.99	16.07.99	L 182 16.07.99	To reduce the amount of biodegradable waste sent to landfill
2000/532	Commission Decision establishing a list of hazardous and non-hazardous waste.	03.05.00	–	L226 06.09.00	
2000/76/ EEC	EP & Council Directive on incineration of waste	04.12.00	28.12.00	L 332 28.12.00	To impose tigher emission standards. Will replace the Directives on MWI and incineration of hazardous waste.
2002/96 EC	Directive on waste electrical and electronic equipment	27.01.03	13.02.03	L 37 13.02.03	To prevent waste EEE and promote reuse, recycling and recovery of such wastes.
2003/108 EC	EP & Council Directive amending 2002/96	08/12/03		L345 08.12.03	Limits producer responsibilty for "historic" commercial WEEE

Hazardous Waste

Number	Title	Date adopted	Date brought into force	OJ No. and date	Purpose
91/689/ EEC	Directive on hazardous waste.	12.12.91	11.12.93	L 377 31.12.91	Updates 1978 Directive on hazardous waste management, defines various types of, and constituencies of hazardous wastes.
94/311/	Amendment of hazardous waste directive.				Sets implementation date of 91/689 at 31.3.95.
90/170/ EEC	Decision on acceptance of OECD recommendation on control of trans-frontier shipment of hazardous waste.	02.04.90	-	L 92 07.04.90	See title.

Number	Title	Date adopted	Date brought into force	OJ No. and date	Purpose
91/157/ EEC	Directive on batteries and accumulators containing certain dangerous substances. Adapted by 93/72.	18.03.91	18.09.92	L 78 26.03.91	To reduce the amount of pollution from used batteries containing heavy metals; to encourage recycling and the production of batteries with lower levels of heavy metals.
259/93	Regulation on the supervision and control of shipments of waste into and out of the EC.	01.02.93	01.05.94	L30 06.02.93	Repealed 12.07.07
669/2008	Regulation with regard to Annex IC of transfrontier shipment of waste regulation	15.07.08	19.07.08	L 188/7 16.07.08	Specific instructions for completing notification and movement documents.
2006/117/ EURATOM	Directive on the supervision and control of shipments of radioactive waste and spent fuel	20.11.06	25.12.06	L 337/21 05.12.06	Lays down authorisation procedures for transboundary shipments of radioactive waste.
93/98	Council decision on the conclusion of a convention of transboundary movements of hazardous waste and their disposal (the Basel Convention).	01.02.93		L39 16.02.93	
1420/ 1999	Council Regulation establishing common rules & procedures to apply to waste shipments to certain non-OECD countries	29.04.99		L 166 01.07.99	See title
1013/ 2006	Council Regulation on Shipment of waste.	14.06.06			Clarifies and simplifies proceddures for transfrontier shipments of waste, and implements the Basel Convention.
96/59 EC	Directive on the disposal of PCBs andPCTs	16.09.96	16.09.96	L 243 24.09.96	To identify PCBs and to control their disposal and elimination by 31.12.10
2002/95 EC	Directive on the restriction of use of certain hazardous substances in EEE.	27.01.03	13.02.03	L 37 13.02.03	See title
2008/385/ EC	Commission Decision amending Annex of Directive 2002/95 on hazardous substances in electrical & electronic equipment	24.01.08		L 136/9 24.05.08	Exempts from prohibition under 2002 Directive cadmium alloys, lead in soldering materials, and lead oxide in seal frit.

Recycling

81/972/ EEC	Recommendation concerning the re-use of waste paper and the use of recycled paper.	03.12.81	-	L 355 10.12.81	To define and implement policies to promote the use of recycled paper and board.

Sewage Sludge

86/278/ EEC	Directive on the protection of the environment, and in particular of the soil, when sewage sludge is used in agriculture.	12.06.86	17.06.89	L 181 04.07.86	To provide for a special regime concerning the spreading of sludge in agriculture and to fix sludge and soil analyses.

APPENDIX 7.6: WATER POLLUTION

2000/60/ EC	Framework Directive of the EP & Council on water policy.	23.10.00	22.12.00	L 327 22.12.00	Establishes a framework for community action on water policy
2455/ 2001/EC	Decision of the EP & the Council establishing the list of priority substances in the field of water policy	20.11.01	16.12.01	L331 15.12.01	See appendix 6.1

Surface Water

75/440/ EEC	Directive concerning the quality required of surface water intended for the abstraction of drinking water in Member States.	16.05.75	18.06.77	L 194 25.07.75	To define the quality requirements for surface fresh water used for or intended for abstraction of drinking water. Provision of plans to clean up water. To define the requirements with which the quality measurements must comply.

Number	Title	Date adopted	Date brought into force	OJ No. and date	Purpose
79/869/ EEC	Directive concerning the methods of measurement and frequencies of sampling and analysis of surface water intended for the abstraction of drinking water in the Member States.	09.10.79	09.10.81	L 271 29.10.79	To lay down reference measuring methods for the parameters contained in Directive 75/440/EEC and sampling frequencies. To lay down the requirements with which these measurements must comply.
77/795/ EEC	Decision establishing a common procedure for the exchange of information on the quality of surface fresh water in the Community. Amended by 86/574.	12.12.77	12.12.77	L 334 24.12.77	To designate a coordinating body. To designate the stations taking part in the exchange of information. To define the parameters to be measured. To provide for the publication of a summary report.
Drinking Water					
98/83/ EC	Directive on the quality of water for human consumption.	03.11.98	25.12.98	L330 5.12.98	Tightens standards for drinking water.
Fresh Water					
2006/44/ EC	Directive on the quality of fresh waters needing protection or improvement in order to support fish life	06.09.06	26.06.06		Lays down quality requirements, sampling frequency, measuring methods and requirements with which they must comply.
Shellfish Waters					
2006/113 EC	Directive on the quality required of shellfish waters.	12.12.06			Replaces & codifies 1979 Directive laying down the quality requirements EEC for shellfish waters.
Bathing Water					
76/160/ EEC	Directive concerning the quality of bathing water.	08.12.75	10.12.77	L31 05.02.76	To define the quality requirements for bathing water. To define the sampling frequency, the measuring methods and the requirements with which these measurements must comply. To lay down the conditions in which this quality ` is to achieved.
2006/7/ EC	Directive concerning the management of bathing water quality	15.02.06	24.0306	L64 04.03.06	To lay down provisions for monitoring and classification of bathing water quality, its management and the provision of information to the public. 1976 Directive to be repealed 31.12.14
Groundwater					
80/68/ EEC	Directive on the protection of groundwater against pollution caused by certain dangerous substances.	17.12.79	19.12.81	L 20 26.01.80	To prevent the discharge of List I substances and restrict that of List II substances. To set up a system of authorisations and lay down the conditions for derogations. To prepare an inventory of discharge authorisations granted.
Dangerous Substances					
76/464/ EEC	Framework directive on pollution caused by certain dangerous substances discharged into the aquatic environment of the Community.	04.05.76	04.05.78	L 31 05.02.76	To provide a system of authorisations for the discharge of dangerous substances into water. To provide limit values or quality objectives and monitoring procedures for List I substances. To provide quality objectives for List II substances. To adopt anti-pollution programmes for both types of substances and communicate them to the Commission. To draw up a list of discharges involving List I substances.
82/176/ EEC	Directive on limit values and quality objectives for mercury discharges by the chlor-alkali electrolysis industry.	22.03.82	01.07.83	L 81 27.03.82	To apply the Directive 76/464/EEC to the discharges from this industrial sector.
83/513/ EEC	Directive on limit values and quality objectives for cadmium discharges.	26.09.83	28.09.85	L 291 24.10.83	- as above –

Number	Title	Date adopted	Date brought into force	OJ No. and date	Purpose
84/156/ EEC	Directive on limit values and quality objectives for mercury discharges by sectors other than the chlor-alkali electrolysis industry.	18.03.84	12.03.86	L 74 17.03.84	- as above –
84/491/ EEC	Directive on limit values and quality objectives for discharges of hexachloro-cyclohexane (in particular lindane).	09.10.84	01.04.86	L 274 17.10.84	- as above –
86/280/ EEC	Directive setting out legal provisions applicable to various substances in List I of Annex to 76/464.	12.06.86	01.01.88	L 181 04.07.86	As above. To exend Directive 76/464/ EEC to cover discharges of DDT, carbon tetrachloride and pentachlorophenol.
88/347/ EEC	Directive on limit values and quality objectives for Hexachlorobenzene (HCB), Hexachlorabutadiene (HCBD), chloroform, isodrin, endrin, dieldrin and aldrin, discharges.	16.06.88	01.01.90	L 158 25.06.88	To extend Directive 76/464/EEC to cover discharges of the HCB, HCBD, chloroform and drins.
90/415/ EEC	Directive on limit values and quality objectives for certain List I substances.	31.07.90	01.01.92	L 219 14.08.90	To include 1,2 dichloroethane, trichloroethane, perchloroethane and trichlorobenzene.
Nitrates					
91/676/ EEC	Directive on protection of fresh, coastal and marine waters from nitrate pollution.	12.12.91	19.12.93	L 375 31.12.91	To reduce and prevent nitrate pollution from agriculture, including nitrogen compounds on soils.
Waste Water					
91/271/ EEC	Directive on urban waste water treatment.	21.05.91	30.06.93	L 135 30.05.91	To lay down minimum standards for the treatment of municipal waste water and the disposal of sludge.
98/51	Directive on urban waste water treatment.			L67 07.03.98	
Titanium Dioxide					
78/176/ EEC	Directive on waste from the titanium dioxide industry.	20.02.78	22.02.79	L 54 25.02.78	To promote the prevention and recycling of such waste. To ensure its harmless disposal. To provide a system of authorisations for disposal operations. To lay down provisions governing immersion, discharge, storage and dumping. To enact provisions for monitoring long-established industries and to provide decontamination programmes with possible exceptions for new industries.
83/29/ EEC	Amendment	03.02.83		L 32 03.02.83	To provide a system of authorisation which will include a preliminary impact assessment. To determine the information to be notified to the Commission.
82/883/ EEC	Directive on procedures for the surveil-lance and monitoring of environments concerned by waste from the titanium dioxide industry.	03.12.82	10.12.84	L 378 31.12.78	To determine the parameters to be surveyed, the minimum sampling and analysis frequencies as well as the methods of measurement to be used in controlling the application of the Directive 78/176/EEC.
92/112/ EEC	Directive on procedures for the harmonisation of programmes for the reduction and eventual elimination of pollution caused by waste from the titanium dioxide industry.	15.12.92		L409 31.12.92	See title.
Marine Pollution					
77/585/ EEC	Decision concluding the Convention for the Protection of the Mediterranean Sea against Pollution and the Protocol for the Prevention of the Pollution of the Mediterranean Sea by Dumping from Ships and Aircraft. (Barcelona Convention).	25.07.77	-	L 240 19.09.77	See title.

Number	Title	Date adopted	Date brought into force	OJ No. and date	Purpose
81/420/ EEC	Decision on the conclusion of the Protocol concerning cooperation in combating pollution of the Mediterranean Sea by oil and other harmful substances in cases of emergency.	19.05.81	-	L 162 19.06.81	See title.
83/101/ EEC	Decision concluding the Protocol for the protection of the Mediterranean Sea against pollution from land-based sources.	28.02.83	-	L 67 12.03.83	See title.
87/1125	Decision on mercury and cadmium discharges.	21.09.87	-		To allow the Commission to negotiate on behalf of the EEC, the adoption of measures relating to mercury and cadmium discharges as well as organosilicate components, within the framework of the convention for the protection of the Mediterranean Sea against pollution.
75/437/ EEC	Decision concluding the Convention for the Prevention of Marine pollution from Land-based Sources. (Paris Convention).	03.03.75	-	L 194 25.07.75	Ratification by the Community of the Convention.
75/438/ EEC	Decision concerning Community participation in the Interim Commission established on the basis of Resolution No. III of the Convention for the Prevention of Marine Pollution from Land-based Sources.	03.03.75	-	L 194 25.07.75	To authorise the Commission to represent the Community in the Interim Commission responsible for administering the Paris Convention.
87/57/ EEC	Decision concluding the protocol amending the convention for the prevention of marine pollution from land-based sources.	22.12.86	-	L 24 27.07.87	To extend its scope to include airborne pollution.
98/249	Decision on the conclusion of the Convention for the protection of the marine environment of the north-east Atlantic – OSPAR			L 104 03.04.98	
2000/59/ EC	Directive of EP & Council on ship generated waste	27.11.00		L 332 28.12.00	Requires provision of port reception facilities for ship waste and cargo residues.

APPENDIX 7.7: MISCELLANEOUS

Environmental Assessment/Access to Information

Number	Title	Date adopted	Date brought into force	OJ No. and date	Purpose
75/436/ Euratom ECSC/ EEC	Council Recommendation of cost allocation and action by public authorities on environmental matters.	03.03.75	03.03.75	L 194 25.07.75	To set out in detail the procedures for applying the Polluter Pays Principle, and to provide for some exceptions which may be made to this Principle.
79/3/ EEC	Council Recommendation to the Member States regarding methods of evaluating the cost of pollution control to industry.	09.12.78	09.12.78	L 5 09.01.79	To define the principles and methods to be followed by the Member States in assessing the costs of pollution control to industry.
85/337/ EEC	Directive concerning the assessment of the environmental effects of certain public and private projects.	27.06.85	03.07.88	L 175 05.07.85	To grant planning permission for projects which are likely to have significant effects on the environment only after an appropriate prior assessment of their environmental effects.
2003/35 EC	Directive providing for public participation in respect of the drawing up of certain plans and programmes relating to the environment	26.05.03		L 156 25.06.03	Also amends public participation and access to justice provisions of 85/337/EEC and of IPPC Directive.
97/11	Council directive amending 1985 directive EC			L 73 14.03.97	Amend 1985 Directive
2001/42 EC	Directive of the Council and of the EP on strategic environmental assessment	27.06.01	21.07.01	L 197 21.07.01	To contribute to the integration of environmental considerations into the preparation and adoption of plans and programmes to promote sustainable development.

Number	Title	Date adopted	Date brought into force	OJ No. and date	Purpose
90/313/ EEC	Directive on freedom of access to information on the environment.	07.06.90	31.12.92	L 158 23.06.90	To oblige public authorities to provide access to environmental information, subject to certain restrictions.To be repealed 14.02.05
2003/4/ EC	Directive on public access to environmental information	28.01.03	14.02.03	L 41 14.02.03	To guarantee right of acccess to environmental information held by or for public authorities; to ensure that such information is made available; repeals 1990 Directive.
12/10/90	Regulation on the establishment of the European Environmental Agency and the European Environmental Information and Observation Network.	07.05.90	-	L 120 11.05.90	To set up a body to oversee the information network and ensure collection and objective data on environment.
880/92	Regulation on a Community eco-label award scheme.	23.03.92	23.03.92	L 99 11.04.92	To set up a Community scheme, define criteria and procedures, for awarding eco-labels; to promote the development of products with minimal environmental impact through their whole life-cycle.
1836/93	Regulation allowing voluntary participation by companies in the industry sector in a Community eco-management and audit scheme.			L 168 10.07.93	To establish system of environmental management and auditing at industrial sites and provision of information to the public.
761/2001	Regulation of the EP & Council allowing voluntary participation by organisations in a Community eco-management and audit scheme	19.03.01	27.04.01	L 414 24.04.01	Updates and replaces the 1993 Regulation and allowing participation by a wider range of organisations.
2001/331/ EC	Recommendation providing for minimum criteria for environmental inspections	04.04.01		L118 27.04.01	
Pesticides [see also water pollution: dangerous substances]					
78/631/ EEC	Directive on classification, packaging and labelling of dangerous preparations.			L 206 29.07.78	To specify packaging and labelling labelling requirements for pesticides.
91/414/ EEC	Directive concerning placing of plant protection products on the market.	15.07.91	26.07.93	L 230 19.08.91	To harmonise EC-wide registration of pesticides.
98/8/EC	Directive on placing of biocidal products on the market	16.02.98		L123 24.04.98	To standardise way in which biocides are authorised throughout the EU; common principles, risk assesment criteria, etc.
Radioactivity					
80/836/ Euratom	Directive amending the Directives laying down the basic safety standards for the health protection of the general public and workers against the dangers of ionising radiation.	15.07.80	03.12.83 03.06.84	L 246 17.09.80	To establish a system of prior authorisation. To set doses for controllable expo sures and principles governing operational protection of exposed workers.
87/600/ Euratom	Council Decision on Community arrangments for the early exchange of information in the event of a radiological emergency.	14.12.87	-	L 371 30.12.87	See title.
87/3954/ Euratom	Regulation laying down maximum permitted radioactivity levels for foodstuffs in the event of a nuclear accident.	22.12.87	-	L 371 30.12.87	See title.
89/618/ Euratom	Directive on informing the public in the event of a radiological emergency.	27.11.89	27.01.91	L 357 07.12.89	To require prior information on what to do in an emergency to general public likely to be affected and information about an accident to be communicated without delay to those affected; information on possible health risks to be given to all those giving emergency assistance.
90/143/ Euratom	Recommendation on the protection of the public against indoor exposure to radon.	21.02.90	-	L 80 27.03.90	See title.

Number	Title	Date adopted	Date brought into force	OJ No. and date	Purpose
90/641/ Euratom	Directive concerning protection of outside workers.			L 349 13.12.90	Supplements 80/836 by setting standards for outside workers' exposure to ionising radiation.
96/29	Council Directive laying down basic safety standards relating to ionising radiation	14.05.96	13.05.00	L159 29.06.96	To protect workers and general public from dangers of ionising radiation
92/3/ Euratom	Directive on the supervision and control of shipments of radioactive waste.	13.02.92	01.01.94	L 35 12.02.92	To monitor and control transfrontier shipments of radioactive waste both between Member States and into and out of the Community.
1493/93	Regulation on Euratom shipments of radioactive substances between Member States.	08.06.93		L148 19.06.93	Similar to above.
2003/122 Euratom	Directive on the control of high-activity sealed sources and orphan sources	22.12.03	31.12.03	L346 31.12.03	To prevent exposure of workers and the public to ionising radiation arising from inadequate control of high-activity sealed radioactive sources.
2006/117 Euratom	Supervision and control of shipments of radioactive and spent fuel				Replaces 92/3 which is repealed on 25.12.08.
Genetically Modified Organisms					
90/219/ EEC	Directive on contained use of GMOs.	23.04.90	23.10.91	L 117 08.05.90	To establish notification, consent and emergency system to protect human health and the environment.
98/81/ EC	Directive on contained use of GMOs				Extends controls relating to contained use, including risk assessments requirements.
90/220/ EEC	Directive on deliberate release of GMOs into the environment.	23.04.90	23.10.91	L 117 08.05.90	Repealed 17.10.02
2001/18/ EC	Directive on the deliberate release into the environment of GMOs			L 106 17.04.01	Repeals the 1990 Directive, introducing stricter safeguards.

FURTHER READING

Further sources of guidance are detailed in the appropriate place in the text of individual chapters and the Appendices to the Pollution Control Handbook

Chapter 1 – General

Sustainable Development

Our Common Future - World Commission on Environment and Development, Oxford University Press, 1987.

Biodiversity

Conserving Biodiversity in a Changing Climate: Guidance on Building Capacity to Adapt. Defra, 2007 (available via www.ukbap.org.uk).

England Biodiversity Strategy: Towards Adaptation to Climate Change, Defra, 2007 (available via Defra website).

Monarch: A Synthesis for Biodiversity Conservation, UK Climate Impacts Programme, 2007 (available via www.ukcip.org.uk).

Environmental Management

Focus on Climate Change, J. Vernon. BSI 2007. (Explains how legislative policies and tools, such as taxation, emission targets etc will affect business, and shows what steps can be taken to stay in line with the legislation).

Focus on Waste Management, M. Yoxon, S. Burnley & T. Gladding. BSI 2007. (identifying areas of waste and implementing operational procedures to reduce and control it).

Focus on Environment and Corporate Responsibility Communications, J. Taylor & K. Wastell. BSI, 2007. (Key mechanisms for integrating CR practices into everyday organisational thinking).

Focus on Sustainability and its implications for CSR, Henriques & Engelschmidt. BSI, 2007. (What sustainability means, in particular for business.)

Environmental Risk Assessment

Preparing Safety Reports: Control of Major Accident Hazards Regulations, HSG190, COMAH Competent Authority (1998), HSE Books.

Guidance on the Environmental Risk Assessment Aspects of COMAH Safety Reports, COMAH Competent Authority (1999). Environment Agency at http://www.environment-agency.gov.uk/

Guidelines for Environmental Risk Assessment and Management, DETR, Environment Agency and IEH (2000).

Methodology for the Derivation of Remedial Targets for Soil and Groundwater to Protect Water Resources, Environment Agency (1999). R&D Publication P20.

Introducing Environmental Risk Assessment, Environment Agency (2000).

Risks of Contaminated Land to Buildings, Building Materials and Services, Environment Agency (2001). R&D Technical Report P331.

Development of Appropriate Soil Sampling Strategies for Land Contamination, Environment Agency (2001). R&D Technical Report P5-066/TR.

Reducing Risks, Protecting People, Health and Safety Executive (2001).

Five Steps to Risk Assessment, INDGi63 (rev2), HSE, 2006

Environmental Legislation and Human Health – Guidance for Assessing Risk, SNIFFER, 2007

Planning

England

Planning Together – Local Strategic Partnerships and Spatial Planning: A Practical Guide, Communities & Local Government, 2007.

Development Plans Examination: A Guide to the Process of Assessing the Soundness of Statements of Community Involvement; ODPM, December 2005.

Development Plans Examination: A Guide to the Process of Assessing the Soundness of Development Plan Documents; ODPM, December 2005.

*Planning Policy Guidance Notes** - The Stationery Office. 2: Green Belts; 3: Housing (2000); 4: Industrial & Commercial Development & Small Firms (2001); 5: Simplified Planning Zones; 7: The Countryside: Environmental Quality & Economic and Social Development (2001); 8: Telecommunications (2001); 13: Transport (2001); 14: Development on Unstable Land (1990); 15: Planning & the Historic Environment; 16: Archaelogy & Planning; 17: Sport, Open Space & Recreation (draft 2001); 18: Enforcing Planning Control (1991); 19: Outdoor Advertisement Control; 20: Coastal Planning; 21: Tourism; 24: Planning and Noise (1994); 25: Development & Flood Risk (2001). *PPGs are being replaced by PPSs

Planning Policy Statements - available through The Stationery Office and also on the website of the Department for Communities and Local Government: www.communities.gov.uk
PPS 1: Delivering sustainable development (2005); 3: Housing (draft 2006); 6: Planning for town centres (2005); 7: Sustainable development in rural areas (draft, 2003); 9: Biodiversity & geological conservation (2005); 10: Planning for sustainable waste management (2005); 11: Regional spatial strategies (2004); PPS 12 Local spatial Planning (2008);
22: Renewable energy (2004); 23: Planning for pollution control (inc. Annexes on pollution control, air & water quality, & on contaminated land (2004); 25: Development & flood risk (draft 2005)
Minerals Policy Statements – (England) – MPS 1: Planning & minerals, and good practice guide (2006); 2: Controlling & mitigating the environmental effects of minerals extraction in England (with Annex 1, Dust & 2, Noise) (2005) ODPM

Minerals policy statements can be accessed via www.communities.gov.uk

Scotland – www.scotland.gov.uk/topics/planning/
National Planning Framework for Scotland

SPP - 1: The planning system (web, 02); 2: Economic development (web 02); 3: Housing (revises NPPG 3; draft due summer 07); 4: Planning for minerals (2006); 6: Renewable energy (revises NPPG 6; final due spring 07); 7: Planning & flooding (web 04); 8: Town centers & retailing (2006); 10: Planning for waste management (2007); 16: Opencast coal (web 05); 17: Planning for transport (web 05); 20: Architecture & design Scotland (2005); 21: Green belts (2006); 22: Planning for fish farming (2007); 23 Planning and the Historic Environment (consultation draft Feb 2008; will consolidate and replace NPPGs 5 & 18)

Planning Advice Notes – 33: Development of Contaminated Land, (Revised Oct 2000); 45: Renewable Energy Technologies (2002); 63: Waste Management Planning (Feb 2002); 75: Planning for Transport (2005). 79: Guidance on water infrastructure. 80: Control and management of flyposting (Dec.06).

Wales – www.planningportal.gov.uk
Planning Policy – Wales (2002). Full list of supporting Technical Advice Notes (TAN) available at

Northern Ireland

Planning Policy Statements (PPS) 1: General principles (1998); 2: Planning & Nature Conservation (1997); 3: Access, movement & parking (2005, further clarification 2006); 4: Industrial Development (1997; draft revision Industry, business & distribution, 2003); 5: Retailing & Town Centres (1996; draft revision Retailing, town centres & commercial leisure developments, 2006); 6: Planning, archaeology & the built heritage (1999; addendum Areas of townscape character, 2003); 7: Quality residential environments (2001); 8: Open space, sport & outdoor recreation (2004); 9: The enforcement of planning control (2000); 10: Telecommunications (2002); 11: Planning and Waste Management (2002); 12: Housing in settlements (2005); 13: Transportation and land use (2005); 14: Sustainable development in the Countryside (draft, 2006); 15: Planning and flood risk (2006); 17: Control of outdoor advertisements (2006).

Strategic Environmental Assessment

A Practical Guide to the Strategic Environmental Assessment Directive, ODPM, Scottish Executive, Welsh Assembly Govt & DOE NI, September 2005.

Land Use and Spatial Plans in England, ODPM, 2004. Scotland: Interim Advice, 2003; Wales: Interim good practice guide, 2004.

SEA and Climate Change: Guidance for Practitioners. Environment Agency, English Nature, Countryside Council for Wales, 2004.

SEA Good Practice Guidelines, Environment Agency guidelines 2004: (Environment Agency website).

Strategic Environmental Assessment – core guidance for transport plans and programmes, Department for Transport, 2004.

Environmental Appraisal and Assessment

Good Practice Guide on Scoping Potential Environmental Issues During the Planning of Developments, Environment Agency (July 2002).

Preparation of Environmental Statements for Planning Projects that require Environmental Assessment – A Good Practice Guide, DoE, 1995.

Evaluation of Environmental Information for Planning Projects : A Good Practice Guide, DoE, 1994.

Environmental Impact Assessment. A Guide to Procedures. Thomas Telford Publishing, 2000.

EC DG XI guidance: "EIA – Guidance on Screening" (1996); "EIA – Guidance on Scoping" (1996); "EIA – Review Check List" (1994); "Guidance on the Assessment of indirect and cumulative impacts as well as impact interactions" (undated – c.1999).

Environmental Impact Assessment: A new Planning Practice Standard, RTPI, 2001.

Nuisance

Statutory Nuisance – Law and Practice, Pointing J & Malcolm R. Oxford University Press, 2002.

Clean Neighbourhoods and Environment Act 2005 – Final Guidance on: Abandoned & nuisance vehicles; Litter & refuse; Defacement; Waste; Dog control orders; Noise; Issuing fixed penalty notices to juveniles; statutory nuisance (light & insects); Abandoned trolleys; Fixed penalty notices, Defra 2006.

Local Environmental Enforcement: Guidance on the Use of Fixed Penalty Notices, Defra, April 2007.

Light Pollution and the Law, January 2006; www.skyatnightmagazine.com

Guidance Notes for the Reduction of Obtrusive Light, Institution of Lighting Engineers, 2005. ww.ile.org.uk

Litter

Code of Practice on Litter and Refuse, Defra, 2006.

Odour

Odour Measurement and Control - An Update, Woodfield, M & Hall, D (Eds). AEA Technology, NETCEN, 1994.

Chapter 2 – Industrial Regulation

IPPC (see also Appendix 2.3)

Integrated Pollution Prevention and Control – A Practical Guide (4th edition), Defra, 2005. Similar guidance available in Scotland & N. Ireland.

Guidance on Interpretation and Implementation of the IPPC Directive, first published on website, April 2007: http://ec.europa/environment/ippc/general_guidance.htm

Guidance manual for Local Authorities on Part A2 and Part B Installations, Defra, 2003.

Practical Guide for Part B Activities, SEPA, 2000. (available on website).

Local Air Pollution Control: Management Guide, Chartered Institute of Environmental Health, 2000.

Environmental Permitting *(see also Appendix 2.3)*

Environmental Permitting – General Guidance Manual (A1), Defra.

Guidance on IPPC Directive - Part A1 installations and mobile plant, Defra,
[replaces guidance on IPPC Directive in IPPC: a Practical Guide, 4th ed]

Guidance on Waste Incineration Directive, Defra.

Guidance on End of Life Vehicles, Defra
[complements existing DTI/Defra technical policy guidance on Depollution Guidance for ELVs over 3.5 tonnes; and Depolluting ELV Guidance for Authorised Treatment Facilities]

Guidance on WEEE Directive, Defra
[complements existing Defra and Devolved Administrations' technical policy: Guidance on Best Available Treatment Recovery and Recycling Techniques and Treatment of WEEE]

Guidance on Landfill Directive, Defra
[subsumes current Government interpretation of the Landfill (E&W) Regulations, 2002, and includes EA guidance]

Guidance on Large Combustion Plant Directive, Defra

Guidance on Solvent Emissions Directive, Defra

Guidance on Framework Directive on Waste, Defra
[replaces joint circular 11/94 on WML, and WMPs 4, 4A & 4A Amendment 1]

Guidance on Waste Incineration Directive, Defra
[replaces Guidance on Directive 2000/76/EC on the Incineration of Waste, Defra

UK National Emission Reduction Plan for Implementation of the revised LCP Directive, Defra, 2003.

Secretary of State's Guidance: General Guidance Manual on Policy and Procedures for A2 and B Installations, together with Guidance on the Directives on Solvents Emissions, Paints, petrol vapour recovery and asbestos, Defra.

REACH

Navigating Reach: An Activist's Guide To Using And Improving The New EU Chemicals Legislation, published on behalf of Chemical Reaction, a joint project of the EEB, FoEE and Greanpeace, 2007.

UK Reach Competent Authority (www.hse.gov.uk/reach) – *Creating an inventory for downstream users of chemicals including formulators; The Distributor's role in REACH*

European Chemicals Agency (http://echa.europa.eu) – *Guidance on REACH* (2007); *Guidance for Intermediates; Guidance on Scientific Research & Development and Product & Process Oriented Research Development; Guidance on requirements for substances in articles; Guidance on how to submit data; Guidance for monomers and polymers; Chemical authorizations under REACH; Advice on Fees and Charges* – 2008.

Automotive Industry Guidance on REACH (2008) – available at www.smmt.co.uk/reach

Chapter 3 – Air Pollution

Climate Change

Air Quality and Climate Change, Air Quality Expert Group (AQEG), Defra, 2007 (available as hard copy or as download from Defra's website).

The Economics of Climate Change ("The Stern Review"), Commissioned by The Chancellor of the Exchequer, and published October 2006 (available on HM Treasury website or download from www.sternreview.org.uk).

Intergovernmental Panel on Climate Change (IPCC) Fourth Assessment Report – Climate Change 2007; The Physical Science Basis (Working Group 1); Climate Impacts and Adaptation (Working Group II); Mitigation (Working Group III), Synthesis Report (this provides an integrated view of climate change based on the reports of the three working groups) (www.ipcc.ch)
UK Climate Change Programme, March 2006
www.defra.gov.uk/environment/climatechange/uk/ukccp/index.htm

The Climate of the United Kingdom and Recent Trends – first of five reports to be launched under the umbrella of the UK 21st century Climate Change Scenarios (UKCIP08) – Jenkins, G.J, Perry M.C. and Prior M.J.O, Met Office Hadley Centre, 2007 (key climate variables from 1961-2006; maps & graphs to show long term averages & changes)

Greenhouse Gas Emission Trends and projections in Europe 2006 – European Environment Agency, 2006.

Greenhouse Gas Verification Standards, BS ISO 14064:1-3 – three part international standard to help organisations prepare inventories for GHG emissions and removals. BSI, 2006.

Operator's Guide to the EU Emissions Trading Scheme: the Steps to Compliance, v1.0, Defra, February 2006.

Air Quality

Air Quality Strategy for England, Scotland, Wales and Northern Ireland; Vol 1 (objectives etc), Vol 2 (evidence base), Defra and the Devolved Administrations, 2007.

Air Quality Guidelines for Europe, WHO Regional Office for Europe. WHO Regional Publications, European Series; No. 23, 1987. Revised 2001 & 2006. See also WHO website at www.who.dk

Air Quality Guidelines: Global Update 2005, WHO Regional Office for Europe, 2007.

Air Quality Management

Steps to Better Practice: Guidance for Local Authorities on LAQM Consultation, UWE 2006. http://www.uwe.ac.uk/aqm/esrc

Local Air Quality Management Policy Guidance (LAQM.PG(03), & Addendum (LAQM.PGA(05), Defra & NAW.

Technical Guidance for Review and Assessment (LAQM.TG(03), Defra & Devolved Administrations.

Guidance on Producing Progress Reports (LAQM.PRG(03) – E, Sc. & W) – LAQM.PRGNI(04) – NI.

Air Quality Research

UK Air Pollutants: Key Facts and Monitoring Data, Environment Agency, 2006.
UK Emissions of Air Pollutants 1970-2004, AEA Environment and Energy, December 2006.

Air Quality Expert Group – www.defra.gov.uk/environment/airquality/panels/aqeg
Nitrogen dioxide in the UK (2004); *Particulate Matter in the United Kingdom* (2005); *Air quality and climate change: a UK perspective* (2007); *Trends in primary nitrogen dioxide in the UK* (2007); *Ozone in the United Kingdom* (draft, 2008)
Trends in Primary nitrogen dioxide in the UK, 4th report of Air Quality Experts Group, 2007

Expert Panel on Air Quality Standards (EPAQS) - *Benzene* (1994); *Ozone* (1994); *1,3 butadiene* (1994 & 1992). *Carbon monoxide* (1994); *Sulphur dioxide* (1995); *Particles* (PM_{10} (1995); *Nitrogen dioxide* (1996); *Lead* (1998); *Polycyclic Aromatic Hydrocarbons* (1999); *Airborne particles: what is the most appropriate measurement on which to base a standard?* (2001). *Guidelines for halogen and hydrogen halides in ambient air for protecting human health against acute irritancy effects* (2006; draft addendum 2008). *Guidelines for metals and metalloids* (draft, 2008).

Quality of Urban Air Review Group - 1st report: *Urban Air Quality in the United Kingdom* (1993); 2nd: *Diesel Vehicle Emissions and Urban Air Quality* (1993); 3rd: *Airborne Particulate Matter in the United Kingdom* (1996). HMSO.
UK Air Quality modelling for annual reporting 2004 on ambient air quality assessment under Council Directives 96/62/EC, 1999/30/EC and 2000/69/EC.
link:http://www.airquality.co.uk/archive/reports/reports.php?report_id=416

Ozone

Ground-level ozone in the 21st century: future trends, impacts and policy implications. Science Policy Report 15/08 The Royal Society, 2008.

Dust

The Control Of Dust and Emissions from Construction And Demolition – Best Practice Guidance. Produced in partnership by London Councils and the GLA, & published by GLA, November 2006.

Air Pollution and Health

Environmental Burden of Disease – country by country analysis of the effects of pollution on human health, WHO, 2007.

Department of Health, Advisory Group on the Medical Aspects of Air Pollution Episodes - 1st report: *Ozone* (1991); 2nd: *Sulphur Dioxide, Acid Aerosols and Particulates* (1992); 3rd: *Oxides of Nitrogen* (1993); 4th: *Health Effects of Exposures to Mixtures of Air Pollutants.* HMSO.

Department of Health, Committee on the Medical Effects of Air Pollutants - *Asthma and Outdoor Air Pollution* (1995); *Non-Biological Particles and Health* (1995), HMSO. *Handbook on Air Pollution and Health* (1997); *Quantification of the Effects of Air Pollution on Health in the United Kingdom* (1998), The Stationery Office.

Health Aspects of Air Pollution, WHO, 2004.
Air Pollution and Health
Cardiovascular disease and air pollution, COMEAP, 2006
http://www.advisorybodies.doh.gov.uk/comeap/state.htm#rep)

Health Risks of Heavy Metals from Long Range Transboundary Air Pollution – WHO, 2007; www.euro.who.int/publications

Health, Place and Nature: How Outdoor Environments influence health and wellbeing, Sustainable Development Commission, April 2008; available at: www.sd-commission.org.uk/publications

Assessment of Environmental Legislation and Associated Requirements for the Protection of Human Health – Scotland and Northern Ireland Forum for Environmental Research (SNIFFER), 2007

Transport

Transport Statistics Great Britain - The Stationery Office, annual.

The Eddington Transport Study – report to Government "on the long-term links between transport and the UK's economic productivity, growth and stability, within the context of the Government's commitment to sustainable development". Sir Rod Eddington, 2006 – NEED ref

Air Pollution and Greenhouse Gas Emissions from Ocean-going Ships: Impacts, Mitigation, Options and Opportunities for Managing Growth, International Council on Clean Transportation, 2007.

Mortality from Ship Emissions: a Global Assessment – published online in American Chemical Society journal *Environmental Science and Technology,* 7 November 2007.

Radiation

Radiation Exposure of the UK Population, Hughes, JS & others. HMSO for National Radiological Protection Board, annual.

Chapter 4 – Noise

Guidelines for Community Noise, World Health Organisation, 2000.

WHO Night Noise Guidelines for Europe – paper presented by Martin van den Berg to Inter-Noise 2007.

Noise Control: The Law and its Enforcement, 3rd edition, Penn CN. Shaw & Sons, 2002.

Rating Industrial Noise Affecting Mixed Residential and Industrial Areas, BS 4142. British Standards Institution, 1990.

Framework guide on choosing noise impact assessment methods for projects, BS 9142 (due 2007), British Standards Institution.

The UK National Noise Attitude Survey 1999/2000, BRE for Defra and the Devolved Administrations, 2002.

The UK National Noise Incidence Study 2000/2001, BRE for Defra and the Devolved Administrations, 2002.

Neighbourhood Noise Policies and Practice for Local Authorities – A Management Guide, CIEH, Defra, September 2006.

Delivering the Goods: A Toolkit for Improving Deliveries – Guidance on Quieter Deliveries – www.fta.co.uk

Noise from Arriving Aircraft: An Industry Code of Practice – drawn up by CAA, DfT, BAA, BA, Virgin Airlines, Easyjet, My Travel, 2nd ed, November 2006.

Low Frequency Noise – Some Solutions, Dr. Andy Moorhouse & Dr. D. Bagley. In *Environmental Health Practitioner* vol 115/04, April 2007.

Low Frequency Noise – Technical Research Report for Defra Noise Programme, 2002;
www.defra.gov.uk/environment/noise/casella/pdf/lowfrequencynoise.pdf

The Little Red Book of Acoustics, Hepworth Acoustics, 2007. (Topics covered include basic acoustics, parameters, descriptors, propagation and transmission, as well as UK standards and regulations on environmental and building acoustics).

Traffic noise reduction in Europe: health effects, social costs and technical and policy options to reduce road and rail traffic noise – Transport and Environment, 2008. www.transportenvironment.org

Attitudes to noise from aviation sources in England – final report, DfT, 2007

Noise and Health: Making the Link – The London Health Commission (www.londonhealth.gov.uk),

Chapter 5 – Waste Management

see also appendix 5.1

Waste Electrical & Electronic Equipment Regulations – Government Guidance Notes (revised), BERR, August 2007.

Hazardous Waste Guidance Document - Waste Electrical & Electronic Equipment Regulations – NI EHS, v.2, July 2007.

Guidance on Best Available Treatment, Recovery and Recycling Techniques and Treatment of Waste Electrical and Electronic Equipment. Defra, Scottish Executive & Welsh Assembly Government, November 2006.

Guidance on Storage and Disposal of Waste Oils, Environment Agency, 2007.

Treatment of Non-hazardous Wastes for Landfill, Environment Agency, 2007.

Guidance Relating to Secretary of State Powers to Change the Classification of Hazardous or Non-Hazardous of a Specific Batch Of Waste because it does not Exhibit the Properties Indicated in the EU List of Wastes Decision 2000/532/EC, Environment Agency 2007.

Landfill Allowance Scheme (Northern Ireland) Regulations 2004 – Monitoring Guidance for District Councils, EHS, v2, February 2007.

Landfill Allowance Scheme (Scotland) Regulations 2005, Scottish Government Guidance, March 2007.

Use of Higher Landfill Waste Acceptance Criteria Limit Values for Hazardous Waste – version 2; Update v1, March 2007 (Environment Agency).

Tackling Flytipping Guide, National Flytipping Prevention Group, 2006.

Soil & Contaminated Land

UK Soil and Herbage Pollutant Survey, Environment Agency website (2007).

Contaminated Land Research Reports: *CLR7 – Assessment of Risks to Human Health from Land Contamination: An Overview of the Development of Soil Guidelines Values and Related Research; CLR8 – Priority Contaminants Report; Defra & Environment Agency, 2002.* Available for download at www.defra.gov.uk

Contaminated Land – Managing Legal Liabilities (2nd edition). Ed Martin Warren, Eversheds/Earthscan, 2001.

Guidance for the Safe Development of Housing on Land Affected by Contamination – Environment Agency, CIEH and NHBC, 2008. http://conland.cieh.org/

Guidance on comparing soil contamination data with a critical concentration. Explains the use of statistical tests in the analysis of contaminated land. Developed by CL:AIRE, 2008. Available from www.cieh.org

Communicating Understanding Of Contaminated Land Risks. SNIFFER (1999), Publication SR97(11)F. Available from the Foundation for Water Research (FWR).
Contaminated Land, Statutory Guidance: Edition 2, Paper SE/2006/44, Scottish Government, May 2006.

Reclamation of Contaminated Land, C. Paul Nathanail & R. Paul Bardos, Wiley, 2005 (reprinted 2007). (Part 1 looks at site characterization and the design of site investigations, risk assessment; Part 2 looks at how risks from contaminated land are managed and the role of different remediation approaches.

Contaminated Land Management Handbook, Bardos & Nathanail, Thomas Telford, London, 2004.

Contaminated Land Ready Reference Book, Nathanail, Bardos & Nathanail, EPP & Land Quality Press, 2002.

Radioactive Substances

The Law of Nuclear Installations and Radioactive Substances, Tromans, S & Fitzgerald, J. Sweet & Maxwell, 1997.

Authorising Discharges of Radioactive Waste to the Environment: Principles for the Assessment of Prospective Public Doses, Issue1. Environment Agency, SEPA, DOE NI, December 2002

Chapter 6 – Water

Assessing the Value of Groundwater, Science report SC040016,/SR1, Environment Agency, March 2007.

Groundwater Regulatory Guidance, Note 001: Groundwater Remediation – Groundwater Pump and Treat Involving Water Abstraction/discharge, NI EHS, 2007.

The Water Environment (Controlled Activities) (Scotland) Regulations 2005 – A Practical Guide, SEPA, 2007 (also available on website).

River Basin Planning Guidance, Defra, August 2006
www.environment-agency.gov.uk
www.sepa.org.uk
www.ehsni.gov.uk

Water for Life and Livelihoods: A framework for river basin planning, Environment Agency, December 2005

WFD and Planning: Initial advice to Planning Authorities in England and Wales, Environment Agency, LGA, RTPI & Welsh LGA, February 2006

Shipping

Air Pollution And Greenhouse Gas Emissions From Ocean-Going Ships, International Council on Clean Transportation, 2007 (www.theicct.org)

Pesticides

Pesticides and the Environment: A Strategy for the Sustainable Use of Plant Protection Products, Pesticides Safety Directorate, Scottish Government, Welsh Assembly Government, Defra,

General Reference

The Politics of Clean Air, Ashby, A & Anderson, M. Clarendon Press, Oxford, 1981.

Silent Spring, Carson, R. Hamish Hamilton, 1962.

The Mayor of London's State of Environment Report for London, GLA, June 2007.

Environment in your Pocket, Defra, 2007.

Royal Commission on Environmental Pollution

First Report (1971); 2: Three Issues in Industrial Pollution (1972); 3: Pollution in some British Estuaries and Coastal Waters (1972); 4: Pollution Control: Progress and Problems (1974); 6: Nuclear Power and the Environment (1976); 7: Agriculture and Pollution (1979); 8: Oil Pollution of the Sea (1981); 9: Lead in the Environment (1983); 10: Tackling Pollution - Experience and Prospects (1984); 11: Managing Waste: The Duty of Care (1985); 12: Best Practicable Environmental Option (1988); 13: The Release of Genetically Engineered Organisms to the Environment (1989); 14: GENHAZ - A System for the Critical Appraisal of Proposals to Release GMOs into the Environment (1991); 15: Emissions from Heavy Duty Diesel Vehicles (1991) 16: Freshwater Quality (1992); 17: Incineration of Waste (1993); 18: Transport and the Environment (1994); 19: Sustainable Use of Soil (1996) - all published by HMSO. 20: Transport and the Environment - Developments since 1994 (1997); 21: Setting Environmental Standards (1998); 22: Energy – The Changing Climate (2000); 23: Environmental Planning (2002); special report: The Environmental Effects of Civil Aviation in Flight (2002); 24: Chemicals in Consumer Products (2003); 25: Turning the Tide: Addressing the Impact of Fisheries on the Marine Environment (2004); 26: The Urban Environment (2007) - The Stationery Office.

Europe – General Reference

Access to European Union Law, Economics, Policies, 16th ed, Nicholas Moussis, 2007.

Europe's Environment: The fourth assessment, European Environment Agency, 2007

EU Environment Guide (updated annually), EC Committee of the American Chamber of Commerce, Avenue des Arts 50, Bte 5, B-1040 Brussels.

Manual of Environmental Policy: the EC and Britain, Haigh, N. Longman, 1992. (Updated twice a year).

Wolf and Stanley on Environmental Law, 4th ed, Cavendish Publishing, 2003.

POLLUTION HANDBOOK INDEX

WEBSITES

Government

Department for Environment Food & Rural Affairs: www.defra.gov.uk
Department for Transport: www.dft.gov.uk
National Assembly for Wales: www.wales.gov.uk
Scottish Government: www.scotland.gov.uk
Communities and Local Government – www.communities.gov.uk
Department of Energy & Climate Change: www.decc.gov.uk
Department of the Environment, NI: www.doeni.gov.uk

Regulators

Environment Agency: www.environment-agency.gov.uk
Northern Ireland Environment Agency: www.ni-environment.gov.uk
Scottish Environment Protection Agency: www.sepa.org
Health and Safety Executive: www.hse.gov.uk
Pesticides Safety Directorate: www.pesticides.gov.uk

Europe

European Integrated Pollution Prevention Control Bureau (responsible for IPPC BREF notes): http://eipccb.jrc.es
European Union (press releases, links to draft and adopted official texts and legislation): http://europa.eu.int
http://europa.eu/scadplus/scad_en.htm - European Union website - summaries of EU policies and legislation, details of implementation etc.

Air Quality/ Air Pollution

UK National Air Quality Information Archive (automatic and non-automatic monitoring data and air pollution forecasts; links to UK atmospheric emissions inventories): www.airquality.co.uk
Welsh Air Quality Forum http://www.welshairquality.co.uk

Northern Ireland Air Pollution http://www.airqualityni.co.uk

Scotland: http://www.scottishairquality.co.uk

World Health Organisation (air quality guidelines): www.who.int/peh/air/airqualitygd.htm

Smoke control areas, authorised fuels, exempted fuels, procedures: www.uksmokecontrolareas.co.uk
Information on air pollution – impacts on habitats and species & support tool to model the potential effects of air-borne pollutants on habitats and species – www.apis.ac.uk; supported by all UK regulatory and conservation agencies

Local Authority Air Quality Support Helpdesk – www.laqmsupport.org.uk – provided on behalf of Defra and the Devolved Administrations, contains downloadable tools & guidance; latest updates to Defra's technical guidance, guidance on use of NO2 diffusion tubes, and more

Association for Emissions Control by Catalyst (AECC) – monthly update on legislative and other air quality/clean air developments, worldwide www.aecc.eu

Swedish Secretariat on Acid Rain – regular newsletter on air air pollution in Europe and worldwide – www.acidrain.org

Air Quality Advice
Modelling: modelhelp@stanger.co.uk
Monitoring: aqm.helpline@aeat.co.uk
Emission inventories: aqm.helpline@aeat.co.uk
Pollutant specific guidance: aqm-review@uwe.ac.uk
Review and Assessment Helpdesk: aqm-review@uwe.ac.uk

Reference – General

National Atmospheric Emissions Inventory: www.naei.org.uk

Office of Public Sector Information (incorporates HMSO, The Stationery Office which publishes legislation etc); links to Parliament and Devolved Administrations: www.opsi.gov.uk

UK online (gateway to central and local government): www.ukonline.gov.uk

NetRegs – plain language guidance for businesses on environmental legislation (provided by UK environment agencies) – www.netregs.gov.uk

ENDS – monthly journal on developments in environment policy and pollution control; can also be accessed online via subscription – www.endsreport.com

Planning

England and Wales www.planningportal .gov.uk
Scotland www.scotland.gov.uk/topics/planning
Northern Ireland www.planningni.gov.uk